ALGEBRA'S COMMON GRAPHS

Identity Function
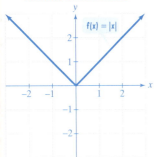

Standard Quadratic Function
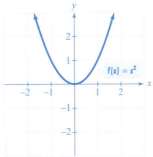

Standard Cubic Function
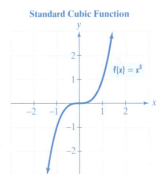

Absolute Value Function

Square Root Function
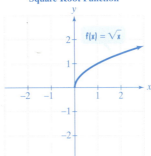

Greatest Integer Function
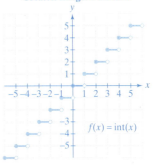

TRANSFORMATIONS

In each case, c represents a positive real number.

Function		Draw the graph of f and:
Vertical translations	$\begin{cases} y = f(x) + c \\ y = f(x) - c \end{cases}$	Shift f upward c units. Shift f downward c units.
Horizontal translations	$\begin{cases} y = f(x - c) \\ y = f(x + c) \end{cases}$	Shift f to the right c units. Shift f to the left c units.
Reflections	$\begin{cases} y = -f(x) \\ y = f(-x) \end{cases}$	Reflect f about the x-axis. Reflect f about the y-axis.
Stretching or Shrinking	$\begin{cases} y = cf(x); c > 1 \\ y = cf(x); 0 < x < 1 \end{cases}$	Stretch f, multiplying each of its y-values by c. Shrink f, multiplying each of its y-values by c.

DISTANCE AND MIDPOINT FORMULAS

1. The distance from (x_1, y_1) to (x_2, y_2) is

$$\sqrt{(x_2 - x_1)^2 + (y_2 - y_1)^2}.$$

2. The midpoint of the line segment with endpoints (x_1, y_1) and (x_2, y_2) is

$$\left(\frac{x_1 + x_2}{2}, \frac{y_1 + y_2}{2} \right).$$

QUADRATIC FORMULA

The solutions to $ax^2 + bx + c = 0$ with $a \neq 0$ are

$$x = \frac{-b \pm \sqrt{b^2 - 4ac}}{2a}.$$

FUNCTIONS

1. Linear Function: $f(x) = mx + b$

Graph is a line with slope m and y-intercept b.

(continued on inside back cover)

College Algebra Essentials

College Algebra
Essentials

Robert Blitzer
Miami-Dade Community College

PEARSON
Prentice Hall

Upper Saddle River, NJ 07458

Library of Congress Cataloging-in-Publication Data

Blitzer, Robert.
   College algebra essentials / Robert Blitzer
      p. cm.
   Includes index.
   ISBN 0-13-109040-2
      1. Algebra.   I. Title.

QA152.3.B645 2004
512.9—dc22   2003062218

Senior Acquisitions Editor: *Eric Frank*
Project Manager: *Dawn Murrin*
Editor-in-Chief : *Sally Yagan*
Vice President/Director of Production and Manufacturing: *David W. Riccardi*
Executive Managing Editor: *Kathleen Schiaparelli*
Senior Managing Editor: *Linda Mihatov Behrens*
Production Management: *Elm Street Publishing Services, Inc./Barbara Mack*
Production Assistant: *Nancy Bauer*
Assistant Managing Editor, Math Media Production: *John Matthews*
Media Production Editor: *Donna Crilly*
Assistant Manufacturing Manager/Buyer: *Michael Bell*
Manufacturing Manager: *Trudy Pisciotti*
Senior Marketing Manager: *Halee Dinsey*
Marketing Assistant: *Rachel Beckman*
Editorial Assistant/Supplements Editor: *Tina Magrabi*
Art Director: *Jon Boylan*
Interior/Cover Designer: *Maureen Eide*
Art Editor: *Thomas Benfatti*
Creative Director: *Carole Anson*
Director of Creative Services: *Paul Belfanti*
Director, Image Resource Center: *Melinda Reo*
Manager, Rights and Permissions: *Zina Arabia*
Interior Image Specialist: *Beth Boyd-Brenzel*
Cover Image Specialist: *Karen Sanatar*
Image Permission Coordinator/Photo Researcher: *Elaine Soares*
Cover Art: *Jalapeño pepper, photographed by John E. Kelly, copyright Getty Images*
Art Studio: *Artworks:*
 Managing Editor, AV Production & Management: *Patty Burns*
 Production Manager: *Ronda Whitson*
 Manager, Production Technologies: *Matt Haas*
 Project Coordinator: *Jessica Einsig*
 Illustrators: *Kathryn Anderson, Mark Landis*
 Quality Assurance: *Pamela Taylor, Timothy Nguyen*
Formatting Manager: *Jim Sullivan*
Assistant Manager of Formatting: *Allyson Graesser*
 Formatters: *Clara Bartunek, Beth Gschwind, Julita Nazario, Nancy Thompson, Judith Wilkens, Vicki L. Croghan*

© 2004 Pearson Education, Inc.
Pearson Prentice Hall
Pearson Education, Inc.
Upper Saddle River, New Jersey 07458

Printed in the United States of America

10 9 8 7 6 5 4 3 2 1

ISBN 0-13-109040-2

Pearson Education LTD., *London*
Pearson Education Australia PTY, Limited, *Sydney*
Pearson Education Singapore, Pte. Ltd
Pearson Education North Asia Ltd, *Hong Kong*
Pearson Education Canada, Ltd., *Toronto*
Pearson Educación de Mexico, S.A. de C.V.
Pearson Education -- Japan, *Tokyo*
Pearson Education Malaysia, Pte. Ltd

Contents

Chapter 2

Chapter 3

Chapter 4

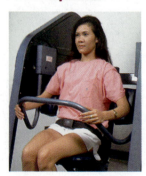

Chapter 5

Systems of Equations and Inequalities   438

Chapter 6

Shaded chapters available in Blitzer, College Algebra, 3rd ed.

Matrices and Determinants   508

Chapter 7

Conic Sections and Analytic Geometry   581

Chapter 8

Sequences, Induction, and Probability   629

Appendix

Where Did That Come From? Selected Proofs   A1

Preface

'**ve written College Algebra Essentials** to help diverse students, with different backgrounds and future goals, to succeed. The book has three fundamental goals:

1. To help students acquire a solid foundation in algebra, preparing them for other courses such as calculus, business calculus, and finite mathematics.
2. To show students how algebra can model and solve authentic real-world problems.
3. To enable students to develop problem-solving skills, while fostering critical thinking, within an interesting setting.

One major obstacle in the way of achieving these goals is the fact that very few students actually read their textbook. This has been a regular source of frustration for my colleagues and me in the classroom. Anecdotal evidence gathered over years highlights two basic reasons that students do not take advantage of their textbook:

- "I'll never use this information."
- "I can't follow the explanations."

As a result, I've written every page of this book with the intent of eliminating these two objections. See the book's Walkthrough, beginning on page xiv, for the ideas and tools I've used to do so.

A Note from the Publisher on Essentials

By publishing this concise version of Robert Blitzer's *College Algebra*, 3rd ed., we provide a lighter, less expensive alternative to the traditional textbook on the subject. It is important to note that *College Algebra Essentials* differs from its predecessor *only in terms of length*, i.e., Chapters 6–8 have been eliminated. The omitted chapters and their sections are highlighted for your reference in the Table of Contents. *College Algebra Essentials* is ideal for courses that do not cover the entire scope of the original textbook. If the complete version of this textbook would better suit your needs, simply contact your Prentice Hall representative for assistance.

A Brief Note on Technology

Technology, and specifically the use of a graphing utility, is covered thoroughly, although its coverage by an instructor is optional. If you require the use of a graphing utility in the course, you will find support for this approach, particularly in the wide selection of clearly designated technology exercises in each exercise set. If you wish to minimize or eliminate the discussion or use of a graphing utility, the book is written to enable you to do so. Regardless of the role technology plays in your course, the technology boxes with TI-83 screens that appear throughout the book should allow your students to understand what graphing utilities can do, enabling them to visualize, verify, or explore what they have already graphed or manipulated by hand. The book's technology coverage is intended to reinforce, but never replace, algebraic solutions.

Supplements

Student Supplements	Instructor Supplements
Student Solutions Manual Fully worked solutions to odd-numbered exercises. 0-13-142312-6	**Instructor's Solutions Manual** Fully worked solutions to all exercises in the text. 0-13-140129-7
CD Lecture Series Four CD-ROMs contain 20 minutes of lectures and tutorials per textbook section; objectives are reviewed and key examples from the textbook are worked out. These are available for students to purchase alone, or in a package with their book. 0-13-140130-0	**Instructor's Edition with Instructor's Resource CD** • Provides answers to all exercises in the back of the text. • Includes Instructor's Resource CD containing TestGen, Instructor's Solutions Manual, Additional Chapter Projects, and Test Item File. ISM is passcode-protected. 0-13-109041-0
VHS Lecture Series Same content as CD Lecture Series in VHS format. Instructors can order the VHS videos and make them available to students in the library or media lab. NOTE: The VHS videos only feature the Right Triangle approach for trigonometry. 0-13-140329-X	**Test Item File** A printed test bank derived from TestGen. 0-13-140337-0
PH Tutor Center Provides students with help when they need it most—while they're doing their homework. PH math tutors (trained college instructors) provide help via a toll-free phone number, fax, and email. 0-13-064604-0	**TestGen** Test-generating software enabling instructors to create tests from the text section objectives. Many questions are algorithmically generated, allowing for unlimited versions of any test. Instructors may also edit problems or create their own. 0-13-140333-8

MathPak 5.0

• Features MathPro 5.0, an online, customizable tutorial and assessment software package, integrated with the text at the learning objective level. The easy-to-use gradebook enables instructors to track and evaluate student performance on tutorial work, quizzes, and tests. An optional Diagnostics module allows students to identify weaknesses in prerequisite material, and to have a customized set of tutorials provided to them for additional practice on those identified weaknesses.

• Includes access to a website containing the Student Solutions Manual, Online Graphing Calculator Help, PowerPoint slides used in the lecture videos, and quizzes and tests allowing students to assess their skills and comprehension of the material.

Student Version: 0-13-140338-9
Instructor Version: 0-13-140795-3

MathPak 4.0

• This interactive tutorial program offers unlimited practice on College Algebra content. Students can watch the author work the problems via videos, view other examples, and see a fully worked-out solution to the problem they are working on.

Student Version: 0-13-140797-X
Instructor Version: 0-13-140800-3

PH Grade Assist

• This online homework and assessment program enables instructors to create customized homework and tests by choosing problems from the text, algorithmic versions of those problems, or creating their own problems.

• PHGA supports multiple question types including free response.

• The built-in parser is sophisticated, grading student responses while recognizing algebraic, numeric, and unit equivalents. The gradebook also allows instructors to easily track student performance.

Student Version: 0-13-140326-5
Instructor Version: 0-13-140332-X

Companion Website

Free website to all text users provides quizzes, chapter tests, PowerPoint slides available for download, and Online Graphing Calculator Help.

URL: www.prenhall.com/blitzer

Acknowledgments

I wish to express my appreciation to all of the reviewers of my precalculus series for their helpful criticisms and suggestions, frequently transmitted with wit, humor, and intelligence. In particular, I would like to thank the following for reviewing **College Algebra**, **Algebra and Trigonometry**, and **Precalculus.**

Reviewers of Current and Previous Editions

Timothy Beaver, *Isothermal Community College*
Bill Burgin, *Gaston College*
Jimmy Chang, *St. Petersburg College*
Donna Densmore, *Bossier Parish Community College*
Disa Enegren, *Rose State College*
Nancy Fisher, *University of Alabama*
Jeremy Haefner, *University of Colorado*
Mary Leesburg, *Manatee Community College*
Joyce Hague, *University of Wisconsin at River Falls*
Christine Heinecke Lehmann, *Purdue University North Central*
Alexander Levichev, *Boston University*
Zongzhu Lin, *Kansas State University*
Benjamin Marlin, *Northwestern Oklahoma State University*
Marilyn Massey, *Collin County Community College*
David Platt, *Front Range Community College*
Janice Rech, *University of Nebraska at Omaha*
Judith Salmon, *Fitchburg State College*
Cynthia Schultz, *Illinois Valley Community College*
Chris Stump, *Bethel College*
Pamela Trim, *Southwest Tennessee Community College*
Chris Turner, *Arkansas State University*
Philip Van Veldhuizen, *University of Nevada at Reno*
Tracy Wienckowski, *Univesity of Buffalo*

Kayoko Yates Barnhill, *Clark College*
Lloyd Best, *Pacific Union College*
Diana Colt, *University of Minnesota-Duluth*
Yvelyne Germain-McCarthy, *University of New Orleans*
Cynthia Glickman, *Community College of Southern Nevada*
Sudhir Kumar Goel, *Valdosta State University*
Donald Gordon, *Manatee Community College*
David L. Gross, *University of Connecticut*
Joel K. Haack, *University of Northern Iowa*
Mike Hall, *Univeristy of Mississippi*
Christopher N. Hay-Jahans, *University of South Dakota*
Celeste Hernandez, *Richland College*
Winfield A. Ihlow, *SUNY College at Oswego*
Nancy Raye Johnson, *Manatee Community College*
James Miller, *West Virginia University*
Debra A. Pharo, *Northwestern Michigan College*
Gloria Phoenix, *North Carolina Agricultural and Technical State University*
Juha Pohjanpelto, *Oregon State University*
Richard E. Van Lommel, *California State University-Sacramento*
Dan Van Peursem, *University of South Dakota*
David White, *The Victoria College*

Additional acknowledgments are extended to Pat Foard, for the Herculean task of preparing the solutions manuals; Teri Lovelace and the team at LaurelTech for preparing the answer section and serving as accuracy checker; Jim Sullivan, Allyson Graesser, and the Prentice Hall formatting team, for the countless hours they spent paging the book; and Cathy Schultz of Elm Street Publishing Services, whose talents as supervisor of production kept every aspect of this complex project moving through its many stages.

I would like to thank my editor at Prentice Hall, Eric Frank, and associate editor, Dawn Murrin, who guided and coordinated the book from manuscript through production. Thanks to the wonderful team of designers, including Jonathan Boylan and Maureen Eide, for the beautiful covers and interior design. Finally, thanks to Halee Dinsey and Patrice Jones, for your innovative marketing efforts, to Sally Yagan for your continuing support, and to the entire Prentice Hall sales force for your confidence and enthusiasm about the book.

To the Student

I've written this book so that you can learn about the power of algebra and how it relates directly to your life outside the classroom. All concepts are carefully explained, important definitions and procedures are set off in boxes, and worked-out examples that present solutions in a step-by-step manner appear in every section. Each example is followed by a similar matched problem, called a Check Point, for you to try so that you can actively participate in the learning process as you read the book. (Answers to all Check Points appear in the back of the book.) Study Tips offer hints and suggestions and often point out common errors to avoid. A great deal of attention has been given to applying algebra to your life to make your learning experience both interesting and relevant.

As you begin your studies, I would like to offer some specific suggestions for using this book and for being successful in this course:

1. **Attend all lectures.** No book is intended to be a substitute for valuable insights and interactions that occur in the classroom. In addition to arriving for lectures on time and being prepared, you will find it useful to read the section before it is covered in the lecture. This will give you a clear idea of the new material that will be discussed.

2. **Read the book.** Read each section with pen (or pencil) in hand. Move through the illustrative examples with great care. These worked-out examples provide a model for doing exercises in the exercise sets. As you proceed through the reading, do not give up if you do not understand every single word. Things will become clearer as you read on and see how various procedures are applied to specific worked-out examples.

3. **Work problems every day and check your answers.** The way to learn mathematics is by doing mathematics, which means working the Check Points and assigned exercises in the exercise sets. The more exercises you work, the better you will understand the material.

4. **Prepare for chapter exams.** After completing a chapter, study the summary, work the exercises in the Chapter Review, and work the exercises in the Chapter Test. Answers to all these exercises are given in the back of the book.

5. **Use the supplements available with this book.** A solutions manual containing worked-out solutions to the book's odd-numbered exercises, all review exercises, and all Check Points; a dynamic web page; and videotapes and CD-ROMs created for every section of the book are among the supplements created to help you tap into the power of mathematics. Ask your instructor or bookstore which supplements are available and where you can find them.

I wrote this book in beautiful and pristine Point Reyes National Seashore, north of San Francisco. It was my hope to convey the beauty of mathematics using nature as a source of inspiration and creativity. Enjoy the pages that follow as you empower yourself with the algebra needed to succeed in college, your career, and your life.

Regards,

Bob

Robert Blitzer

About the Author

Bob Blitzer is a native of Manhattan and received a Bachelor of Arts degree with dual majors in mathematics and psychology (minor: English literature) from the City College of New York. His unusual combination of academic interests led him toward a Master of Arts in mathematics from the University of Miami and a doctorate in behavioral sciences from Nova University. Bob is most energized by teaching mathematics and has taught a variety of mathematics courses at Miami-Dade Community College for nearly 30 years. He has received numerous teaching awards, including Innovator of the Year from the League for Innovations in the Community College, and was among the first group of recipients at Miami-Dade Community College for an endowed chair based on excellence in the classroom. In addition to *College Algebra*, Bob has written *Introductory Algebra for College Students*, *Intermediate Algebra for College Students*, *Introductory and Intermediate Algebra for College Students*, *Algebra for College Students*, *Thinking Mathematically*, *Algebra and Trigonometry*, and *Precalculus*, all published by Prentice Hall.

Finally, Bob loves to spend time with his pal, Harley, pictured to the right. He's so cute (Harley, not Bob) that we couldn't resist including him.

Why Blitzer's College Algebra Essentials?

This text was written to address students' most commonly cited reasons for not using their texts:

- "I'll never use this information ("When will I use this?")"
- "I can't follow the explanations."

"When Will I Use This?"

This text integrates dynamic applications that connect mathematics to the entire spectrum of students' interests.

The interesting and diverse applications . . .

- represent a wide range of disciplines.

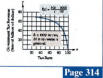

Page 290

Page 314

Page 214

Page 362

- feature unique and interesting data that show students that mathematics can be applied in many settings.

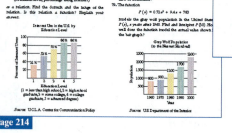

Page 287

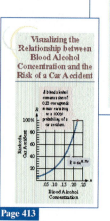

Page 413

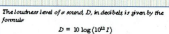

Page 396

- are driven by real and sourced data, illustrating the power of algebra and trigonometry to model contemporary issues and problems.

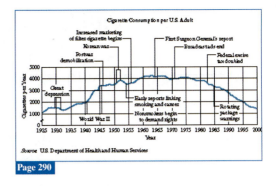

Page 290

Unique Chapter and Section Opening Vignettes

- Each chapter and section begins with a vignette highlighting an everyday scenario, posing a question about it, and exploring how the chapter section subject can be applied to answer the question.
- These are revisited in the course of the chapter or section in an example, discussion, or exercise.

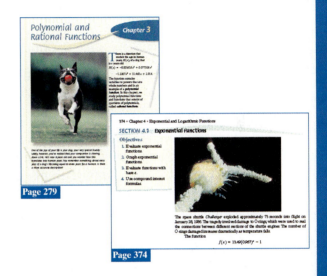

Page 279

Page 374

Enrichment Essay (and other interesting asides)

- Enrichment Essays provide historical, interdisciplinary, and otherwise interesting connections to the math under study,
  showing students that math is an interesting and dynamic discipline.

Page 306

WHEN WILL I USE THIS?

xv

"I Can't Follow the Explanations."

Clear & Friendly Writing Style

- Blitzer's language is clear, direct, and simple. He breaks down concepts in a conversational style, providing analogies and drawing connections to students' experiences whenever possible.

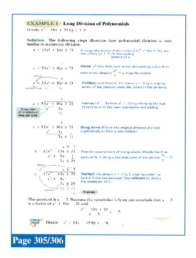

Yogi Berra, catcher and renowned hitter for the New York Yankees (1946-1963), said it best: "Prediction is very hard, especially when it's about the future." At the start of the twenty-first century, we are plagued by questions about the environment. Will we run out of gas? How hot will it get? Will there be neighborhoods where the air is pristine? Can we make garbage disappear? Will there be any wilderness left? Which wild animals will become extinct? These concerns have led to the growth of the environmental industry in the United States.

EXAMPLE 10 The Growth of the Environmental Industry

Page 414

Detailed Illustrations, Examples, and Check Points

Examples:

- are abundant, because students learn by example.
- are thoroughly annotated to the right of the algebraic steps. These annotations are in a conversational style, providing the voice of an instructor in the book, explaining key steps and ideas as the problem is solved.
- offer students the opportunity to stop and test their understanding of the example by working a similar exercise immediately following that example, called a **Check Point.**
- The answers to the **Check Points** are provided in the answer section.

Page 305/306

Exercise Sets that Precisely Parallel Examples (pages 244-247)

- An extensive collection of exercises is included at the end of each section.
- Exercises are organized by level within six category types: **Practice Exercises, Application Exercises, Writing in Mathematics, Technology Exercises, Critical Thinking Exercises,** and **Group Exercises.**
- The order of the practice exercises is exactly the same as the order of the section's illustrative examples. This parallel order enables students to refer to the titled examples and their detailed explanations to achieve success working the practice exercises.

xvi

Explanatory Voice Balloons

Voice balloons are used in a variety of ways to demystify mathematics. They:

- translate mathematical ideas into plain English.
- help clarify problem-solving procedures.
- present alternative ways of understanding concepts.
- connect complex problems to the basic concepts students have already learned.

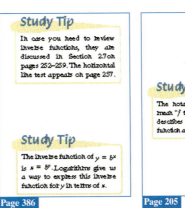

Page 100

Study Tips

Study Tip boxes:

- appear in abundance throughout the book.
- offer suggestions for problem solving.
- point out common mistakes.
- provide informal tips and suggestions.

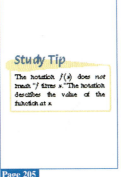

Page 386 **Page 205**

Clearly Stated Section Objectives

Learning objectives:

- are clearly stated at the beginning of each section.
- help students recognize and focus on the most important ideas.
- appear in the margin at their point of use.
- form the foundation for the algorithms in MathPro (tutorial software) and in TestGen (test generator software).

Page 335

Chapter Review Grids

- summarize definitions and concepts for every section of the chapter.
- refer students to the examples that illustrate these key concepts.

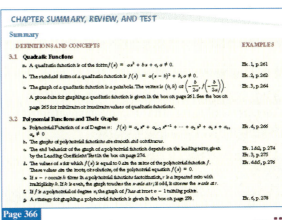

Page 366

xvii

Blitzer M@thP@k

An Integrated Learning Environment

Today's textbooks offer a wide variety of ancillary materials to students, from solutions manuals to tutorial software to text-specific Websites. Making the most of all of these resources can be difficult. Blitzer **M@thP@k** helps students get it together. **M@thP@k** seamlessly integrates the following key products into **an integrated learning environment:**

MathPro 5

MathPro 5 is online, customizable tutorial software integrated with the text at the Learning Objective level. MathPro 5's "watch" feature integrates lecture videos into the algorithmic tutorial environment. The easy-to-use course management system enables instructors to track and assess student performance on tutorial work, quizzes, and tests. A robust reports wizard provides a grade book, individual student reports, and class summaries. The customizable syllabus allows instructors to remove and reorganize chapters, sections, and objectives. MathPro 5's messaging system enhances communication between students and instructors. The combination of MathPro 5's richly integrated tutorial, testing, and robust course management tools provides an unparalleled tutorial experience for students, and new assessment and time-saving tools for instructors.

The Blitzer M@thP@k Website

This robust passcode-protected site features quizzes, homework starters, live animated examples, graphing calculator manuals, and much more. It offers the student many ways to test and reinforce their understanding of the course material.

Student Solutions Manual

The *Student Solutions Manual* offers thorough, accurate solutions that are consistent with the precise mathematics found in the text.

Blitzer M@thP@k.
Helping Students Get it Together.

PHGrade Assist

Problem Solving for Students.
Solving Problems for You.

Students need to practice solving problems—The more they practice, the better problem solvers they become. Professors want relief from the tedium of grading.

That's why we created **PH GradeAssist**. It's...

✓ online—available anytime, anywhere.
✓ text-specific—tied directly to your Prentice Hall text.
✓ algorithmic—contains unlimited questions and assignments for practice and assessment.
✓ customizable—completely unique to your course—edit our questions and add your own.

How does PH GradeAssist work for the instructor?

- You create quizzes or homework assignments from question banks specific to your text. Choose the problems you prefer, edit them, or add your own.
- Your students go online and work the assignments that you have created.
- The problems let students work with real math, not just multiple choice.
- Many problems are algorithmically generated, so each student gets a slightly different problem with a different answer.
- PH GradeAssist scores these assignments for you, using a sophisticated math parser, which recognizes algebraic, numeric, and unit equivalents.
- Results can be easily accessed in a central gradebook.

For a demonstration, contact your local Prentice Hall representative or visit us online at www.prenhall.com/phga

Applications Index

Prerequisites: Fundamental Concepts of Algebra

This chapter reviews fundamental concepts of algebra that are prerequisites for the study of college algebra. Algebra, like all of mathematics, provides the tools to help you recognize, classify, and explore the hidden patterns of your world, revealing its underlying structure. Throughout the new millennium, literacy in algebra will be a prerequisite for functioning in a meaningful way personally, professionally, and as a citizen.

Listening to the radio on the way to work, you hear candidates in the upcoming election discussing the problem of the country's 5.6 trillion dollar deficit. It seems like this is a real problem, but then you realize that you don't really know what that number means. How can you look at this deficit in the proper perspective? If the national debt were evenly divided among all citizens of the country, how much would each citizen have to pay? Does the deficit seem like such a significant problem now?

SECTION P.1 *Real Numbers and Algebraic Expressions*

Objectives

1. Recognize subsets of the real numbers.
2. Use inequality symbols.
3. Evaluate absolute value.
4. Use absolute value to express distance.
5. Evaluate algebraic expressions.
6. Identify properties of the real numbers.
7. Simplify algebraic expressions.

The U.N. Building is designed with three golden rectangles.

The United Nations Building in New York was designed to represent its mission of promoting world harmony. Viewed from the front, the building looks like three rectangles stacked upon each other. In each rectangle, the ratio of the width to height is $\sqrt{5} + 1$ to 2, approximately 1.618 to 1. The ancient Greeks believed that such a rectangle, called a **golden rectangle,** was the most visually pleasing of all rectangles.

The ratio 1.618 to 1 is approximate because $\sqrt{5}$ is an irrational number, a special kind of real number. Irrational? Real? Let's make sense of all this by describing the kinds of numbers you will encounter in this course.

1 Recognize subsets of the real numbers.

The Set of Real Numbers

Before we describe the set of real numbers, let's be sure you are familiar with some basic ideas about sets. A **set** is a collection of objects whose contents can be clearly determined. The objects in a set are called the **elements** of the set. For example, the set of numbers used for counting can be represented by

$$\{1, 2, 3, 4, 5, \ldots\}.$$

The braces, { }, indicate that we are representing a set. This form of representing a set uses commas to separate the elements of the set. The set of numbers used for counting is called the set of **natural numbers.** The three dots after the 5 indicate that there is no final element and that the listing goes on forever.

The sets that make up the real numbers are summarized in Table P.1. We refer to these sets as **subsets** of the real numbers, meaning that all elements in each subset are also elements in the set of real numbers.

Notice the use of the symbol $\approx$ in the examples of irrational numbers. The symbol means "is approximately equal to." Thus,

$$\sqrt{2} \approx 1.414214.$$

We can verify that this is only an approximation by multiplying 1.414214 by itself. The product is very close to, but not exactly, 2:

$$1.414214 \times 1.414214 = 2.0000012378.$$

Technology

A calculator with a square root key gives a decimal approximation for $\sqrt{2}$, not the exact value.

Real numbers ℝ

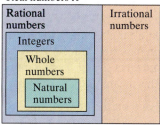

This diagram shows that every real number is rational or irrational.

Table P.1   Important Subsets of the Real Numbers

Name	Description	Examples
Natural numbers ℕ	$\{1, 2, 3, 4, 5, \ldots\}$ These numbers are used for counting.	$2, 3, 5, 17$
Whole numbers 𝕎	$\{0, 1, 2, 3, 4, 5, \ldots\}$ The set of whole numbers is formed by adding 0 to the set of natural numbers.	$0, 2, 3, 5, 17$
Integers ℤ	$\{\ldots, -5, -4, -3, -2, -1, 0, 1, 2, 3, 4, 5, \ldots\}$ The set of integers is formed by adding negatives of the natural numbers to the set of whole numbers.	$-17, -5, -3, -2, 0,$ $2, 3, 5, 17$
Rational numbers ℚ	The set of rational numbers is the set of all numbers which can be expressed in the form $\frac{a}{b}$, where a and b are integers and b is not equal to 0, written $b \neq 0$. Rational numbers can be expressed as terminating or repeating decimals.	$-17 = \frac{-17}{1}, -5 = \frac{-5}{1}, -3, -2,$ $0, 2, 3, 5, 17,$ $\frac{2}{5} = 0.4,$ $\frac{-2}{3} = -0.6666\cdots = -0.\overline{6}$
Irrational numbers 𝕀	This is the set of all numbers whose decimal representations are neither terminating nor repeating. Irrational numbers cannot be expressed as a quotient of integers.	$\sqrt{2} \approx 1.414214$ $-\sqrt{3} \approx -1.73205$ $\pi \approx 3.142$ $-\frac{\pi}{2} \approx -1.571$

Study Tip

Not all square roots are irrational numbers. For example, $\sqrt{25} = 5$ because $5 \times 5 = 25$. Thus, $\sqrt{25}$ is a natural number, a whole number, an integer, and a rational number $\left(\sqrt{25} = \frac{5}{1}\right)$.

The set of **real numbers** is formed by combining the rational numbers and the irrational numbers. Thus, every real number is either rational or irrational.

The Real Number Line

The **real number line** is a graph used to represent the set of real numbers. An arbitrary point, called the **origin**, is labeled 0; units to the right of the origin are **positive** and units to the left of the origin are **negative.** The real number line is shown in Figure P.1.

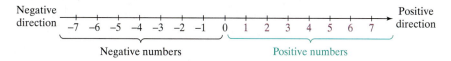

Figure P.1 The real number line

Real numbers are **graphed** on a number line by placing a dot at the correct location for each number. The integers are easiest to locate. In Figure P.2, we've graphed the integers −3, 0, and 4.

Figure P.2 Graphing −3, 0, and 4 on a number line

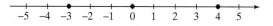

Every real number corresponds to a point on the number line and every point on the number line corresponds to a real number. We say there is a **one-to-one correspondence** between all the real numbers and all points on a real number line. If you draw a point on the real number line corresponding to a real number, you are **plotting** the real number. In Figure P.2, we are plotting the real numbers −3, 0, and 4.

2 Use inequality symbols.

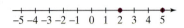

Figure P.3

Study Tip

The symbols < and > always point to the lesser of the two real numbers when the inequality is true.

$2 < 5$ *The symbol points to 2, the lesser number.*

$5 > 2$ *The symbol points to 2, the lesser number.*

Ordering the Real Numbers

On the real number line, the real numbers increase from left to right. The lesser of two real numbers is the one farther to the left on a number line. The greater of two real numbers is the one farther to the right on a number line.

Look at the number line in Figure P.3. The integers 2 and 5 are plotted. Observe that 2 is to the left of 5 on the number line. This means that 2 is less than 5:

$2 < 5$:   2 is less than 5 because 2 is to the *left* of 5 on the number line.

In Figure P.3, we can also observe that 5 is to the right of 2 on the number line. This means that 5 is greater than 2:

$5 > 2$:   5 is greater than 2 because 5 is to the right of 2 on the number line.

The symbols < and > are called **inequality symbols.** They may be combined with an equal sign, as shown in the following table:

Symbols	Meaning	Example	Explanation
$a \leq b$	a is less than or equal to b.	$3 \leq 7$	Because $3 < 7$
		$7 \leq 7$	Because $7 = 7$
$b \geq a$	b is greater than or equal to a.	$7 \geq 3$	Because $7 > 3$
		$-5 \geq -5$	Because $-5 = -5$

3 Evaluate absolute value.

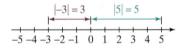

Figure P.4 Absolute value as the distance from 0

Absolute Value

The **absolute value** of a real number a, denoted by $|a|$, is the distance from 0 to a on the number line. This distance is always taken to be nonnegative. For example, the real number line in Figure P.4 shows that

$$|-3| = 3 \quad \text{and} \quad |5| = 5.$$

The absolute value of −3 is 3 because −3 is 3 units from 0 on the number line. The absolute value of 5 is 5 because 5 is 5 units from 0 on the number line. The absolute value of a positive real number or 0 is the number itself. The absolute value of a negative real number, such as −3, is the number without the negative sign.

We can define the absolute value of the real number x without referring to a number line. The algebraic definition of the absolute value of x is given as follows:

Definition of Absolute Value

$$|x| = \begin{cases} x & \text{if } x \geq 0 \\ -x & \text{if } x < 0 \end{cases}$$

If x is nonnegative (that is $x \geq 0$), the absolute value of x is the number itself. For example,

$$|5| = 5 \qquad |\pi| = \pi \qquad \left|\frac{1}{3}\right| = \frac{1}{3} \qquad |0| = 0.$$

Zero is the only number whose absolute value is 0.

If x is a negative number (that is, $x < 0$), the absolute value of x is the opposite of x. This makes the absolute value positive. For example,

$$|-3| = -(-3) = 3 \qquad |-\pi| = -(-\pi) = \pi \qquad \left|-\frac{1}{3}\right| = -\left(-\frac{1}{3}\right) = \frac{1}{3}.$$

This middle step is usually omitted.

EXAMPLE 1 Evaluating Absolute Value

Rewrite each expression without absolute value bars:

    a. $\left|\sqrt{3} - 1\right|$ **b.** $|2 - \pi|$ **c.** $\dfrac{|x|}{x}$ if $x < 0$.

Solution

    a. Because $\sqrt{3} \approx 1.7$, the expression inside the absolute value bars, $\sqrt{3} - 1$, is positive. The absolute value of a positive number is the number itself. Thus,

$$\left|\sqrt{3} - 1\right| = \sqrt{3} - 1.$$

    b. Because $\pi \approx 3.14$, the number inside the absolute value bars, $2 - \pi$, is negative. The absolute value of x when $x < 0$ is $-x$. Thus,

$$|2 - \pi| = -(2 - \pi) = \pi - 2.$$

    c. If $x < 0$, then $|x| = -x$. Thus,

$$\frac{|x|}{x} = \frac{-x}{x} = -1.$$

Study Tip

After working each Check Point, check your answer in the answer section before continuing your reading.

Check Point 1 Rewrite each expression without absolute value bars:

    a. $\left|1 - \sqrt{2}\right|$ **b.** $|\pi - 3|$ **c.** $\dfrac{|x|}{x}$ if $x > 0$.

Listed below are several basic properties of absolute value. Each of these properties can be derived from the definition of absolute value.

Discovery

Verify the triangle inequality if $a = 4$ and $b = 5$. Verify the triangle inequality if $a = 4$ and $b = -5$.

When does equality occur in the triangle inequality and when does inequality occur? Verify your observation with additional number pairs.

4 Use absolute value to express distance.

Properties of Absolute Value

For all real numbers a and b,

    1. $|a| \geq 0$ **2.** $|-a| = |a|$ **3.** $a \leq |a|$

    4. $|ab| = |a||b|$ **5.** $\left|\dfrac{a}{b}\right| = \dfrac{|a|}{|b|}$, $b \neq 0$

    6. $|a + b| \leq |a| + |b|$ (called the triangle inequality)

Distance between Points on a Real Number Line

Absolute value is used to find the distance between two points on a real number line. If a and b are any real numbers, the **distance between a and b** is the absolute value of their difference. For example, the distance between 4 and 10 is 6. Using absolute value, we find this distance in one of two ways:

$$|10 - 4| = |6| = 6 \quad \text{or} \quad |4 - 10| = |-6| = 6.$$

The distance between 4 and 10 on the real number line is 6.

Notice that we obtain the same distance regardless of the order in which we subtract.

> **Distance between Two Points on the Real Number Line**
>
> If a and b are any two points on a real number line, then the distance between a and b is given by
>
> $$|a - b| \text{ or } |b - a|.$$

EXAMPLE 2   Distance between Two Points on a Number Line

Find the distance between −5 and 3 on the real number line.

Solution   Because the distance between a and b is given by $|a - b|$, the distance between −5 and 3 is

$$|-5 - 3| = |-8| = 8.$$

Figure P.5 The distance between −5 and 3 is 8.

Figure P.5 verifies that there are 8 units between −5 and 3 on the real number line. We obtain the same distance if we reverse the order of the subtraction:

$$|3 - (-5)| = |8| = 8.$$

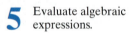   Find the distance between −4 and 5 on the real number line.

Algebraic Expressions

Algebra uses letters, such as x and y, to represent real numbers. Such letters are called **variables.** For example, imagine that you are basking in the sun on the beach. We can let x represent the number of minutes that you can stay in the sun without burning with no sunscreen. With a number 6 sunscreen, exposure time without burning is six times as long, or 6 times x. This can be written $6 \cdot x$, but it is usually expressed as $6x$. Placing a number and a letter next to one another indicates multiplication.

Notice that $6x$ combines the number 6 and the variable x using the operation of multiplication. A combination of variables and numbers using the operations of addition, subtraction, multiplication, or division, as well as powers or roots (which are discussed later in this chapter), is called an **algebraic expression.** Here are some examples of algebraic expressions:

$$x + 6, \quad x - 6, \quad 6x, \quad \frac{x}{6}, \quad 3x + 5, \quad \sqrt{x} + 7.$$

⑤ Evaluate algebraic expressions.

Evaluating Algebraic Expressions

Evaluating an algebraic expression means to find the value of the expression for a given value of the variable. For example, we can evaluate $6x$ (from the sunscreen example) when $x = 15$. We substitute 15 for x. We obtain $6 \cdot 15$, or 90. This means if you can stay in the sun for 15 minutes without burning when you don't put on any lotion, then with a number 6 lotion, you can "cook" for 90 minutes without burning.

Many algebraic expressions involve more than one operation. Evaluating an algebraic expression without a calculator involves carefully applying the following order of operations agreement:

The Order of Operations Agreement

1. Perform operations within the innermost parentheses and work outward. If the algebraic expression involves a fraction, treat the numerator and the denominator as if they were each enclosed in parentheses.
2. Evaluate all exponential expressions.
3. Perform multiplications and divisions as they occur, working from left to right.
4. Perform additions and subtractions as they occur, working from left to right.

EXAMPLE 3   Evaluating an Algebraic Expression

The algebraic expression $2.35x + 179.5$ describes the population of the United States, in millions, x years after 1960. Evaluate the expression for $x = 40$. Describe what the answer means in practical terms.

Solution   We begin by substituting 40 for x. Because $x = 40$, we will be finding the U.S. population 40 years after 1960, in the year 2000.

$$2.35x + 179.5$$

Replace x with 40.

$$= 2.35(40) + 179.5$$
$$= 94 + 179.5 \qquad \text{Perform the multiplication: } 2.35(40) = 94.$$
$$= 273.5 \qquad \text{Perform the addition.}$$

According to the given algebraic expression, in 2000 the population of the United States was 273.5 million.

According to the U.S. Bureau of the Census, in 2000 the population of the United States was 281.4 million. Notice that the algebraic expression in Example 3 provides an approximate, but not an exact, description of the actual population.

 Check Point 3   Evaluate the algebraic expression in Example 3 for $x = 30$. Describe what your answer means in practical terms.

6 Identify properties of the real numbers.

Properties of Real Numbers and Algebraic Expressions

When you use your calculator to add two real numbers, you can enter them in any order. The fact that two real numbers can be added in any order is called the **commutative property of addition.** You probably use this property, as well as other properties of real numbers listed in Table P.2 on the next page, without giving it much thought. The properties of the real numbers are especially useful when working with algebraic expressions. For each property listed in Table P.2, a, b, and c represent real numbers, variables, or algebraic expressions.

The Associative Property and the English Language

In the English language, phrases can take on different meanings depending on the way the words are associated with commas.

Here are two examples.

- *Woman, without her man, is nothing.*
  Woman, without her, man is nothing.
- *What's the latest dope?*
  What's the latest, dope?

Table P.2 Properties of the Real Numbers

Name	Meaning	Examples
Commutative Property of Addition	Two real numbers can be added in any order. $a + b = b + a$	• $13 + 7 = 7 + 13$ • $13x + 7 = 7 + 13x$
Commutative Property of Multiplication	Two real numbers can be multiplied in any order. $ab = ba$	• $\sqrt{2} \cdot \sqrt{5} = \sqrt{5} \cdot \sqrt{2}$ • $x \cdot 6 = 6x$
Associative Property of Addition	If three real numbers are added, it makes no difference which two are added first. $(a + b) + c = a + (b + c)$	• $3 + (8 + x) = (3 + 8) + x$ $= 11 + x$
Associative Property of Multiplication	If three real numbers are multiplied, it makes no difference which two are multiplied first. $(a \cdot b) \cdot c = a \cdot (b \cdot c)$	• $-2(3x) = (-2 \cdot 3)x = -6x$
Distributive Property of Multiplication over Addition	Multiplication distributes over addition. $a \cdot (b + c) = a \cdot b + a \cdot c$	• $7(4 + \sqrt{3}) = 7 \cdot 4 + 7 \cdot \sqrt{3}$ $= 28 + 7\sqrt{3}$ • $5(3x + 7) = 5 \cdot 3x + 5 \cdot 7$ $= 15x + 35$
Identity Property of Addition	Zero can be deleted from a sum. $a + 0 = a$ $0 + a = a$	• $\sqrt{3} + 0 = \sqrt{3}$ • $0 + 6x = 6x$
Identity Property of Multiplication	One can be deleted from a product. $a \cdot 1 = a$ $1 \cdot a = a$	• $1 \cdot \pi = \pi$ • $13x \cdot 1 = 13x$
Inverse Property of Addition	The sum of a real number and its additive inverse gives 0, the additive identity. $a + (-a) = 0$ $(-a) + a = 0$	• $\sqrt{5} + (-\sqrt{5}) = 0$ • $-\pi + \pi = 0$ • $6x + (-6x) = 0$ • $(-4y) + 4y = 0$
Inverse Property of Multiplication	The product of a nonzero real number and its multiplicative inverse gives 1, the multiplicative identity. $a \cdot \dfrac{1}{a} = 1, \quad a \neq 0$ $\dfrac{1}{a} \cdot a = 1, \quad a \neq 0$	• $7 \cdot \dfrac{1}{7} = 1$ • $\left(\dfrac{1}{x - 3}\right)(x - 3) = 1, \ x \neq 3$

Commutative Words and Sentences

The commutative property states that a change in order produces no change in the answer. The words and sentences listed here suggest a characteristic of the commutative property; they read the same from left to right and from right to left!

dad	Draw, o coward!	Revolting is error. Resign it, lover.
repaper	Dennis sinned.	Naomi, did I moan?
never odd or even	Ma is a nun, as I am.	Al lets Della call Ed Stella.

The properties in Table P.2 apply to the operations of addition and multiplication. Subtraction and division are defined in terms of addition and multiplication.

Definitions of Subtraction and Division

Let a and b represent real numbers.

Subtraction: $a - b = a + (-b)$
We call $-b$ the **additive inverse** or **opposite** of b.

Division: $a \div b = a \cdot \frac{1}{b}$, where $b \neq 0$
We call $\frac{1}{b}$ the **multiplicative inverse** or **reciprocal** of b. The quotient of a and b, $a \div b$, can be written in the form $\frac{a}{b}$, where a is the **numerator** and b the **denominator** of the fraction.

Because subtraction is defined in terms of adding an inverse, the distributive property can be applied to subtraction:

$$a(b - c) = ab - ac$$
$$(b - c)a = ba - ca.$$

For example,

$$4(2x - 5) = 4 \cdot 2x - 4 \cdot 5 = 8x - 20.$$

7 Simplify algebraic expressions.

Simplifying Algebraic Expressions

The **terms** of an algebraic expression are those parts that are separated by addition. For example, consider the algebraic expression

$$7x - 9y - 3,$$

which can be expressed as

$$7x + (-9y) + (-3).$$

This expression contains three terms, namely $7x$, $-9y$, and -3.

The numerical part of a term is called its **numerical coefficient.** In the term $7x$, the 7 is the numerical coefficient. In the term $-9y$, the -9 is the numerical coefficient.

A term that consists of just a number is called a **constant term.** The constant term of $7x - 9y - 3$ is -3.

A term indicates a product. The expressions that are multiplied to form the term are called its **factors. Like terms** have the same variable factors with the same exponents on the variables. For example, $7x$ and $3x$ are like terms because they have the same variable factor, x. The distributive property (in reverse) can be used to add these terms:

$$7x + 3x = (7 + 3)x = 10x.$$

Study Tip

To add like terms, add their numerical coefficients. Use this result as the numerical coefficient of the terms' common variable(s).

An algebraic expression is **simplified** when parentheses have been removed and like terms have been combined.

EXAMPLE 4   Simplifying an Algebraic Expression

Simplify: $6(2x - 4y) + 10(4x + 3y)$.

Solution

$$6(2x - 4y) + 10(4x + 3y)$$

$$= 6 \cdot 2x - 6 \cdot 4y + 10 \cdot 4x + 10 \cdot 3y \qquad \text{Use the distributive property to remove the parentheses.}$$

$$= 12x - 24y + 40x + 30y \qquad \text{Multiply.}$$

$$= (12x + 40x) + (30y - 24y) \qquad \text{Group like terms.}$$

$$= 52x + 6y \qquad \text{Combine like terms.}$$

Check Point 4   Simplify: $7(4x - 3y) + 2(5x + y)$.

Properties of Negatives

The distributive property can be extended to cover more than two terms within parentheses. For example,

> This sign represents subtraction.

> This sign tells us that −3 is negative.

$$-3(4x - 2y + 6) = -3 \cdot 4x - (-3) \cdot 2y - 3 \cdot 6$$

$$= -12x - (-6y) - 18$$

$$= -12x + 6y - 18.$$

The voice balloons illustrate that negative signs can appear side by side. They can represent the operation of subtraction or the fact that a real number is negative. Here is a list of properties of negatives and how they are applied to algebraic expressions:

Properties of Negatives

Let a and b represent real numbers, variables, or algebraic expressions.

Property	Examples
1. $(-1)a = -a$	$(-1)4xy = -4xy$
2. $-(-a) = a$	$-(-6y) = 6y$
3. $(-a)b = -ab$	$(-7)4xy = -7 \cdot 4xy = -28xy$
4. $a(-b) = -ab$	$5x(-3y) = -5x \cdot 3y = -15xy$
5. $-(a + b) = -a - b$	$-(7x + 6y) = -7x - 6y$
6. $-(a - b) = -a + b$	$-(3x - 7y) = -3x + 7y$
$\qquad = b - a$	$\qquad = 7y - 3x$

Do you notice that properties 5 and 6 in the box are related? In general, expressions within parentheses that are preceded by a negative can be simplified by dropping the parentheses and changing the sign of every term inside the parentheses. For example,

$$-(3x - 2y + 5z - 6) = -3x + 2y - 5z + 6.$$

EXERCISE SET P.1

Practice Exercises

In Exercises 1–4, list all numbers from the given set that are **a.** *natural numbers,* **b.** *whole numbers,* **c.** *integers,* **d.** *rational numbers,* **e.** *irrational numbers.*

1. $\left\{-9, -\frac{4}{5}, 0, 0.25, \sqrt{3}, 9.2, \sqrt{100}\right\}$

2. $\left\{-7, -0.\overline{6}, 0, \sqrt{49}, \sqrt{50}\right\}$

3. $\left\{-11, -\frac{5}{6}, 0, 0.75, \sqrt{5}, \pi, \sqrt{64}\right\}$

4. $\left\{-5, -0.\overline{3}, 0, \sqrt{2}, \sqrt{4}\right\}$

5. Give an example of a whole number that is not a natural number.

6. Give an example of a rational number that is not an integer.

7. Give an example of a number that is an integer, a whole number, and a natural number.

8. Give an example of a number that is a rational number, an integer, and a real number.

Determine whether each statement in Exercises 9–14 is true or false.

9. $-13 \le -2$
10. $-6 > 2$
11. $4 \ge -7$
12. $-13 < -5$
13. $-\pi \ge -\pi$
14. $-3 > -13$

In Exercises 15–24, rewrite each expression without absolute value bars.

15. $|300|$
16. $|-203|$
17. $|12 - \pi|$
18. $|7 - \pi|$
19. $|\sqrt{2} - 5|$
20. $|\sqrt{5} - 13|$
21. $\dfrac{-3}{|-3|}$
22. $\dfrac{-7}{|-7|}$
23. $\left||-3| - |-7|\right|$
24. $\left||-5| - |-13|\right|$

In Exercises 25–30, evaluate each algebraic expression for $x = 2$ and $y = -5$.

25. $|x + y|$
26. $|x - y|$
27. $|x| + |y|$
28. $|x| - |y|$
29. $\dfrac{y}{|y|}$
30. $\dfrac{|x|}{x} + \dfrac{|y|}{y}$

In Exercises 31–38, express the distance between the given numbers using absolute value. Then find the distance by evaluating the absolute value expression.

31. 2 and 17
32. 4 and 15
33. −2 and 5
34. −6 and 8
35. −19 and −4
36. −26 and −3
37. −3.6 and −1.4
38. −5.4 and −1.2

In Exercises 39–48, evaluate each algebraic expression for the given value of the variable or variables.

39. $5x + 7;\ x = 4$
40. $9x + 6;\ x = 5$
41. $4(x + 3) - 11;\ x = -5$
42. $6(x + 5) - 13;\ x = -7$
43. $\dfrac{5}{9}(F - 32);\ F = 77$
44. $\dfrac{5}{9}(F - 32);\ F = 50$
45. $\dfrac{5(x + 2)}{2x - 14};\ x = 10$
46. $\dfrac{7(x - 3)}{2x - 16};\ x = 9$
47. $\dfrac{2x + 3y}{x + 1};\ x = -2$ and $y = 4$
48. $\dfrac{2x + y}{xy - 2x};\ x = -2$ and $y = 4$

In Exercises 49–58, state the name of the property illustrated.

49. $6 + (-4) = (-4) + 6$
50. $11 \cdot (7 + 4) = 11 \cdot 7 + 11 \cdot 4$
51. $6 + (2 + 7) = (6 + 2) + 7$
52. $6 \cdot (2 \cdot 3) = 6 \cdot (3 \cdot 2)$
53. $(2 + 3) + (4 + 5) = (4 + 5) + (2 + 3)$
54. $7 \cdot (11 \cdot 8) = (11 \cdot 8) \cdot 7$
55. $2(-8 + 6) = -16 + 12$
56. $-8(3 + 11) = -24 + (-88)$
57. $\dfrac{1}{(x + 3)}(x + 3) = 1,\ x \ne -3$
58. $(x + 4) + [-(x + 4)] = 0$

In Exercises 59–68, simplify each algebraic expression.

59. $5(3x + 4) - 4$
60. $2(5x + 4) - 3$
61. $5(3x - 2) + 12x$
62. $2(5x - 1) + 14x$
63. $7(3y - 5) + 2(4y + 3)$
64. $4(2y - 6) + 3(5y + 10)$

65. $5(3y - 2) - (7y + 2)$ **66.** $4(5y - 3) - (6y + 3)$

67. $7 - 4[3 - (4y - 5)]$ **68.** $6 - 5[8 - (2y - 4)]$

In Exercises 69–74, write each algebraic expression without parentheses.

69. $-(-14x)$ **70.** $-(-17y)$

71. $-(2x - 3y - 6)$ **72.** $-(5x - 13y - 1)$

73. $\frac{1}{3}(3x) + [(4y) + (-4y)]$ **74.** $\frac{1}{2}(2y) + [(-7x) + 7x]$

Application Exercises

75. Are first putting on your left shoe and then putting on your right shoe commutative?

76. Are first getting undressed and then taking a shower commutative?

77. Give an example of two things that you do that are not commutative.

78. Give an example of two things that you do that are commutative.

79. The algebraic expression $81 - 0.6x$ approximates the percentage of American adults who smoked cigarettes x years after 1900. Evaluate the expression for $x = 100$. Describe what the answer means in practical terms.

80. The algebraic expression $1527x + 31{,}290$ approximates average yearly earnings for elementary and secondary teachers in the United States x years after 1990. Evaluate the algebraic expression for $x = 10$. Describe what the answer means in practical terms.

81. The optimum heart rate is the rate that a person should achieve during exercise for the exercise to be most beneficial. The algebraic expression

$$0.6(220 - a)$$

describes a person's optimum heart rate, in beats per minute, where a represents the age of the person.

a. Use the distributive property to rewrite the algebraic expression without parentheses.

b. Use each form of the algebraic expression to determine the optimum heart rate for a 20-year-old runner.

Writing in Mathematics

Writing about mathematics will help you learn mathematics. For all writing exercises in this book, use complete sentences to respond to the question. Some writing exercises can be answered in a sentence; others require a paragraph or two. You can decide how much you need to write as long as your writing clearly and directly answers the question in the exercise. Standard references such as a dictionary and a thesaurus should be helpful.

82. How do the whole numbers differ from the natural numbers?

83. Can a real number be both rational and irrational? Explain your answer.

84. If you are given two real numbers, explain how to determine which one is the lesser.

85. How can $\dfrac{|x|}{x}$ be equal to 1 or −1?

86. What is an algebraic expression? Give an example with your explanation.

87. Why is $3(x + 7) - 4x$ not simplified? What must be done to simplify the expression?

88. You can transpose the letters in the word "conversation" to form the phrase "voices rant on." From "total abstainers" we can form "sit not at ale bars." What two algebraic properties do each of these transpositions (called anagrams) remind you of? Explain your answer.

Critical Thinking Exercises

89. Which one of the following statements is true?

a. Every rational number is an integer.

b. Some whole numbers are not integers.

c. Some rational numbers are not positive.

d. Irrational numbers cannot be negative.

90. Which of the following is true?

a. The term x has no numerical coefficient.

b. $5 + 3(x - 4) = 8(x - 4) = 8x - 32$

c. $-x - x = -x + (-x) = 0$

d. $x - 0.02(x + 200) = 0.98x - 4$

In Exercises 91–93, insert either $<$ or $>$ in the box between the numbers to make the statement true.

91. $\sqrt{2} \,\square\, 1.5$ **92.** $-\pi \,\square\, -3.5$

93. $-\dfrac{3.14}{2} \,\square\, -\dfrac{\pi}{2}$

94. A business that manufactures small alarm clocks has a weekly fixed cost of $5000. The average cost per clock for the business to manufacture x clocks is described by

$$\frac{0.5x + 5000}{x}.$$

a. Find the average cost when $x = 100$, 1000, and 10,000.

b. Like all other businesses, the alarm clock manufacturer must make a profit. To do this, each clock must be sold for at least 50¢ more than what it costs to manufacture. Due to competition from a larger company, the clocks can be sold for $1.50 each and no more. Our small manufacturer can only produce 2000 clocks weekly. Does this business have much of a future? Explain.

SECTION P.2 *Exponents and Scientific Notation*

Objectives

1. Understand and use integer exponents.
2. Use properties of exponents.
3. Simplify exponential expressions.
4. Use scientific notation.

1 Understand and use integer exponents.

Powers of Ten

$$10 = 10^1$$
$$100 = 10^2$$
$$1000 = 10^3$$
$$10,000 = 10^4$$
$$100,000 = 10^5$$
$$1,000,000 = 10^6 \quad \text{million}$$
$$10,000,000 = 10^7$$
$$100,000,000 = 10^8$$
$$1,000,000,000 = 10^9 \quad \text{billion}$$

Technology

You can use a calculator to evaluate exponential expressions. For example, to evaluate 5^3, press the following keys:

Many Scientific Calculators

5 $\boxed{y^x}$ 3 $\boxed{=}$

Many Graphing Calculators

5 $\boxed{\wedge}$ 3 $\boxed{\text{ENTER}}$.

Although calculators have special keys to evaluate powers of ten and squaring bases, you can always use one of the sequences shown here.

Although people do a great deal of talking, the total output since the beginning of gabble to the present day, including all baby talk, love songs, and congressional debates, only amounts to about 10 million billion words. This can be expressed as 16 factors of 10, or 10^{16} words.

Exponents such as 2, 3, 4, and so on are used to indicate repeated multiplication. For example,

$$10^2 = 10 \cdot 10 = 100,$$
$$10^3 = 10 \cdot 10 \cdot 10 = 1000, \quad 10^4 = 10 \cdot 10 \cdot 10 \cdot 10 = 10,000.$$

The 10 that is repeated when multiplying is called the **base.** The small numbers above and to the right of the base are called **exponents** or **powers.** The exponent tells the number of times the base is to be used when multiplying. In 10^3, the base is 10 and the exponent is 3.

Any number with an exponent of 1 is the number itself. Thus, $10^1 = 10$.

Multiplications that are expressed in exponential notation are read as follows:

10^1: "ten to the first power"

10^2: "ten to the second power" or "ten squared"

10^3: "ten to the third power" or "ten cubed"

10^4: "ten to the fourth power"

10^5: "ten to the fifth power"

Any real number can be used as the base. Thus,

$$7^2 = 7 \cdot 7 = 49 \quad \text{and} \quad (-3)^4 = (-3)(-3)(-3)(-3) = 81.$$

The bases are 7 and -3, respectively. Do not confuse $(-3)^4$ and -3^4.

$$-3^4 = -3 \cdot 3 \cdot 3 \cdot 3 = -81$$

The negative is not taken to the power because it is not inside parentheses.

Study Tip

-3^4 is read "the opposite of 3 to the fourth power." By contrast, $(-3)^4$ is read "negative 3 to the fourth power."

EXAMPLE 1   Evaluating an Exponential Expression

Evaluate: $(-2)^3 \cdot 3^2$.

Solution

$$(-2)^3 \cdot 3^2 = (-2)(-2)(-2) \cdot 3 \cdot 3 = -8 \cdot 9 = -72$$

This is $(-2)^3$, read "-2 cubed."   This is 3^2, read "3 squared."

Check Point 1   Evaluate: $(-4)^3 \cdot 2^2$.

The formal algebraic definition of a natural number exponent summarizes our discussion:

Definition of a Natural Number Exponent

If b is a real number and n is a natural number,

Exponent

$$b^n = \underbrace{b \cdot b \cdot b \cdots \cdot b}_{\substack{b \text{ appears as a} \\ \text{factor } n \text{ times.}}}$$

Base

b^n is read "the nth power of b" or "b to the nth power." Thus, the nth power of b is defined as the product of n factors of b.
   Furthermore, $b^1 = b$.

Negative Integers as Exponents

A nonzero base can be raised to a negative power using the following definition:

The Negative Exponent Rule

If b is any real number other than 0 and n is a natural number, then

$$b^{-n} = \frac{1}{b^n}.$$

EXAMPLE 2   Evaluating Expressions Containing Negative Exponents

Evaluate: **a.** 5^{-3} **b.** $\dfrac{1}{4^{-2}}$.

Solution

a. $5^{-3} = \dfrac{1}{5^3} = \dfrac{1}{5 \cdot 5 \cdot 5} = \dfrac{1}{125}$

b. $\dfrac{1}{4^{-2}} = \dfrac{1}{\dfrac{1}{4^2}} = 4^2 = 4 \cdot 4 = 16$

Study Tip

When a negative integer appears as an exponent, switch the position of the base (from numerator to denominator or from denominator to numerator) and make the exponent positive.

Check Point 2 Evaluate: **a.** 2^{-3} **b.** $\dfrac{1}{6^{-2}}$.

Zero as an Exponent

A nonzero base can be raised to the 0 power using the following definition:

The Zero Exponent Rule

If b is any real number other than 0,
$$b^0 = 1.$$

Here are three examples involving simplification using the zero exponent rule:

$$7^0 = 1 \qquad (-5)^0 = 1 \qquad -5^0 = -1.$$

Only 5 is raised to the zero power.

2 Use properties of exponents.

The Product Rule

Consider the multiplication of two exponential expressions, such as $2^4 \cdot 2^3$. We are multiplying 4 factors of 2 and 3 factors of 2. We have a total of 7 factors of 2. Thus,

$$2^4 \cdot 2^3 = 2^7.$$

We can quickly find the exponent on the product, 7, by adding 4 and 3, the original exponents. This suggests the following rule:

The Product Rule

$$b^m \cdot b^n = b^{m+n}$$

When multiplying exponential expressions with the same base, add the exponents. Use this sum as the exponent of the common base.

EXAMPLE 3 Using the Product Rule

Use the product rule to simplify each expression:

a. $2^2 \cdot 2^3$ **b.** $4^2 \cdot 4^{-5}$ **c.** $x^{-3} \cdot x^7$.

Solution

a. $2^2 \cdot 2^3 = 2^{2+3} = 2^5 = 32$ **b.** $4^2 \cdot 4^{-5} = 4^{2+(-5)} = 4^{-3} = \dfrac{1}{4^3} = \dfrac{1}{64}$

c. $x^{-3} \cdot x^7 = x^{-3+7} = x^4$

> **Check Point 3** Use the product rule to simplify each expression:
>
> **a.** $3^3 \cdot 3^2$ **b.** $2^4 \cdot 2^{-7}$ **c.** $x^{-5} \cdot x^{11}$.

The Power Rule

The next property of exponents applies when an expression containing a power is itself raised to a power.

The Power Rule (Powers to Powers)

$$(b^m)^n = b^{mn}$$

When an exponential expression is raised to a power, multiply the exponents. Place the product of the exponents on the base and remove the parentheses.

EXAMPLE 4 Using the Power Rule

Use the power rule to simplify each expression:

a. $(2^2)^3$ **b.** $(y^5)^6$ **c.** $(x^{-3})^4$.

Solution

a. $(2^2)^3 = 2^{2 \cdot 3} = 2^6 = 64$ **b.** $(y^5)^6 = y^{5 \cdot 6} = y^{30}$

c. $(x^{-3})^4 = x^{-3 \cdot 4} = x^{-12} = \dfrac{1}{x^{12}}$

> **Check Point 4** Use the power rule to simplify each expression:
>
> **a.** $(3^3)^2$ **b.** $(y^7)^4$ **c.** $(x^{-4})^2$.

The Quotient Rule

The next property of exponents applies when we are dividing exponential expressions with the same base.

The Quotient Rule

$$\frac{b^m}{b^n} = b^{m-n}, \ b \neq 0$$

When dividing exponential expressions with the same nonzero base, subtract the exponent in the denominator from the exponent in the numerator. Use this difference as the exponent of the common base.

EXAMPLE 5 **Using the Quotient Rule**

Use the quotient rule to simplify each expression:

a. $\dfrac{2^8}{2^4}$ **b.** $\dfrac{x^3}{x^7}$ **c.** $\dfrac{y^9}{y^{-5}}$.

Study Tip

$\dfrac{4^3}{4^5}$ and $\dfrac{4^5}{4^3}$ represent different numbers:

$\dfrac{4^3}{4^5} = 4^{3-5} = 4^{-2} = \dfrac{1}{4^2} = \dfrac{1}{16}$

$\dfrac{4^5}{4^3} = 4^{5-3} = 4^2 = 16.$

Solution

a. $\dfrac{2^8}{2^4} = 2^{8-4} = 2^4 = 16$ **b.** $\dfrac{x^3}{x^7} = x^{3-7} = x^{-4} = \dfrac{1}{x^4}$

c. $\dfrac{y^9}{y^{-5}} = y^{9-(-5)} = y^{9+5} = y^{14}$

Check Point 5   Use the quotient rule to simplify each expression:

a. $\dfrac{3^6}{3^4}$ **b.** $\dfrac{x^5}{x^{12}}$ **c.** $\dfrac{y^2}{y^{-7}}$.

Products Raised to Powers

The next property of exponents applies when we are raising a product to a power.

> **Products to Powers**
>
> $$(ab)^n = a^n b^n$$
>
> When a product is raised to a power, raise each factor to that power.

EXAMPLE 6 **Raising a Product to a Power**

Simplify: $(-2y)^4$.

Solution $(-2y)^4 = (-2)^4 y^4 = 16y^4$

Check Point 6   Simplify: $(-4x)^3$.

The rule for products of powers can be extended to cover three or more factors. For example,

$$(-2xy)^3 = (-2)^3 x^3 y^3 = -8x^3 y^3.$$

Quotients Raised to Powers

Our final exponential property applies when we are raising a quotient to a power.

> **Quotients to Powers**
>
> $$\left(\dfrac{a}{b}\right)^n = \dfrac{a^n}{b^n}, \; b \neq 0$$
>
> When a quotient is raised to a power, raise the numerator to that power and divide by the denominator to that power.

EXAMPLE 7 Raising Quotients to Powers

Simplify by raising each quotient to the given power:

a. $\left(\dfrac{2}{5}\right)^4$ **b.** $\left(-\dfrac{3}{x}\right)^3$.

Solution

a. $\left(\dfrac{2}{5}\right)^4 = \dfrac{2^4}{5^4} = \dfrac{16}{625}$ **b.** $\left(-\dfrac{3}{x}\right)^3 = \dfrac{(-3)^3}{x^3} = \dfrac{-27}{x^3}$

Check Point 7 Simplify: **a.** $\left(\dfrac{3}{4}\right)^3$ **b.** $\left(-\dfrac{2}{y}\right)^5$.

3 Simplify exponential expressions.

Simplifying Exponential Expressions

Properties of exponents are used to simplify exponential expressions. Here is a summary of the properties we have discussed.

Properties of Exponents

1. $b^{-n} = \dfrac{1}{b^n}$ **2.** $b^0 = 1$ **3.** $b^m \cdot b^n = b^{m+n}$ **4.** $(b^m)^n = b^{mn}$

5. $\dfrac{b^m}{b^n} = b^{m-n}$ **6.** $(ab)^n = a^n b^n$ **7.** $\left(\dfrac{a}{b}\right)^n = \dfrac{a^n}{b^n}$

An exponential expression is **simplified** when

- No parentheses appear.
- No powers are raised to powers.
- Each base occurs only once.
- No negative exponents appear.

EXAMPLE 8 Simplifying Exponential Expressions

Simplify:

a. $(-3x^4y^5)^3$ **b.** $(-7xy^4)(-2x^5y^6)$ **c.** $\dfrac{-35x^2y^4}{5x^6y^{-8}}$ **d.** $\left(\dfrac{4x^2}{y}\right)^{-3}$.

Solution

a. $(-3x^4y^5)^3 = (-3)^3(x^4)^3(y^5)^3$ Raise each factor inside the parentheses to the third power.

$= (-3)^3 x^{4 \cdot 3} y^{5 \cdot 3}$ Multiply powers to powers.

$= -27x^{12}y^{15}$ $(-3)^3 = (-3)(-3)(-3) = -27$

b. $(-7xy^4)(-2x^5y^6) = (-7)(-2)xx^5y^4y^6$ Group factors with the same base.

$= 14x^{1+5}y^{4+6}$ When multiplying expressions with the same base, add the exponents.

$= 14x^6y^{10}$ Simplify.

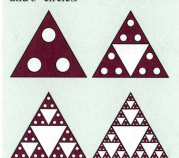

Visualizing Powers of 3

The triangles contain 3, 3^2, 3^3, and 3^4 circles.

c.
$$\frac{-35x^2y^4}{5x^6y^{-8}} = \left(\frac{-35}{5}\right)\left(\frac{x^2}{x^6}\right)\left(\frac{y^4}{y^{-8}}\right)$$

Group factors with the same base.

$$= -7x^{2-6}y^{4-(-8)}$$

When dividing an expression with the same base, subtract the exponents.

$$= -7x^{-4}y^{12}$$

Simplify. Notice that $4 - (-8) = 4 + 8 = 12$.

$$= \frac{-7y^{12}}{x^4}$$

Move x^{-4}, the factor with the negative exponent, from the numerator to the denominator.

d.
$$\left(\frac{4x^2}{y}\right)^{-3} = \frac{4^{-3}(x^2)^{-3}}{y^{-3}}$$

Raise each factor inside the parentheses to the -3 power.

$$= \frac{4^{-3}x^{-6}}{y^{-3}}$$

Multiply powers to powers.

$$= \frac{y^3}{4^3x^6}$$

Move factors with negative exponents from the numerator to the denominator (or vice versa) by changing the sign of the exponent.

$$= \frac{y^3}{64x^6}$$

$4^3 = 4 \cdot 4 \cdot 4 = 64$

 Check Point 8 Simplify:
a. $(2x^3y^6)^4$ **b.** $(-6x^2y^5)(3xy^3)$ **c.** $\frac{100x^{12}y^2}{20x^{16}y^{-4}}$ **d.** $\left(\frac{5x}{y^4}\right)^{-2}$.

Study Tip

Try to avoid the following common errors that can occur when simplifying exponential expressions.

Correct	Incorrect	Description of Error
$b^3 \cdot b^4 = b^7$	$b^3 \cdot b^4 = b^{12}$	The exponents should be added, not multiplied.
$3^2 \cdot 3^4 = 3^6$	$3^2 \cdot 3^4 = 9^6$	The common base should be retained, not multiplied.
$\frac{5^{16}}{5^4} = 5^{12}$	$\frac{5^{16}}{5^4} = 5^4$	The exponents should be subtracted, not divided.
$(4a)^3 = 64a^3$	$(4a)^3 = 4a^3$	Both factors should be cubed.
$b^{-n} = \frac{1}{b^n}$	$b^{-n} = -\frac{1}{b^n}$	Only the exponent should change sign.
$(a + b)^{-1} = \frac{1}{a + b}$	$(a + b)^{-1} = \frac{1}{a} + \frac{1}{b}$	The exponent applies to the entire expression $a + b$.

 Use scientific notation.

Scientific Notation

The national debt of the United States is about \$5.6 trillion. A stack of \$1 bills equaling the national debt would rise to twice the distance from the Earth to the moon. Because a trillion is 10^{12}, the national debt can be expressed as

$$5.6 \times 10^{12}.$$

The number 5.6×10^{12} is written in a form called *scientific notation*. A number in **scientific notation** is expressed as a number greater than or equal to 1 and less than 10 multiplied by some power of 10. It is customary to use the multiplication symbol, $\times$, rather than a dot, to indicate multiplication in scientific notation. Here are two examples of numbers in scientific notation:

• Each day, 2.6×10^7 pounds of dust from the atmosphere settle on Earth.
• The diameter of a hydrogen atom is 1.016×10^{-8} centimeter.

We can use the exponent on the 10 to change a number in scientific notation to decimal notation. If the exponent is *positive*, move the decimal point in the number to the *right* the same number of places as the exponent. If the exponent is *negative*, move the decimal point in the number to the *left* the same number of places as the exponent.

EXAMPLE 9 **Converting from Scientific to Decimal Notation**

Write each number in decimal notation:

a. 2.6×10^7 **b.** 1.016×10^{-8}.

Solution

a. We express 2.6×10^7 in decimal notation by moving the decimal point in 2.6 seven places to the right. We need to add six zeros.

$$2.6 \times 10^7 = 26,000,000$$

b. We express 1.016×10^{-8} in decimal notation by moving the decimal point in 1.016 eight places to the left. We need to add seven zeros to the right of the decimal point.

$$1.016 \times 10^{-8} = 0.00000001016$$

Check Point 9 Write each number in decimal notation:

a. 7.4×10^9 **b.** 3.017×10^{-6}.

To convert from decimal notation to scientific notation, we reverse the procedure of Example 9.

• Move the decimal point in the given number to obtain a number greater than or equal to 1 and less than 10.
• The number of places the decimal point moves gives the exponent on 10; the exponent is positive if the given number is greater than 10 and negative if the given number is between 0 and 1.

EXAMPLE 10 **Converting from Decimal Notation to Scientific Notation**

Write each number in scientific notation:

a. 4,600,000 **b.** 0.00023.

Solution

a. $4,600,000 = 4.6 \times 10^?$ *Decimal point moves 6 places.* $\longrightarrow 4.6 \times 10^6$

b. $0.00023 = 2.3 \times 10^{-?}$ *Decimal point moves 4 places.* $\longrightarrow 2.3 \times 10^{-4}$

Technology

You can use your calculator's [EE] (enter exponent) or [EXP] key to convert from decimal to scientific notation. Here is how it's done for 0.00023:

Many Scientific Calculators

Keystrokes	Display
.00023 [EE] [=]	2.3 − 04

Many Graphing Calculators

Use the mode setting for scientific notation.

Keystrokes	Display
.00023 [ENTER]	2.3E−4

Check Point 10 Write each number in scientific notation:

a. 7,410,000,000   **b.** 0.000000092.

Technology

$(3.4 \times 10^9)(2 \times 10^{-5})$
On a Calculator:

Many Scientific Calculators

3.4 [EE] 9 [×] 2 [EE] 5 [+/−] [=]

Display

      6.8 04

Many Graphing Calculators

3.4 [EE] 9 [×] 2 [EE] [(−)] 5 [ENTER]

Display (in scientific notation mode)

      6.8E 4

Computations with Scientific Notation

The product and quotient rules for exponents can be used to multiply or divide numbers that are expressed in scientific notation. For example, here's how to find the product of 3.4×10^9 and 2×10^{-5}.

$$(3.4 \times 10^9)(2 \times 10^{-5}) = (3.4 \times 2) \times (10^9 \times 10^{-5})$$
$$= 6.8 \times 10^{9+(-5)}$$
$$= 6.8 \times 10^4 \quad \text{or} \quad 68,000$$

In our next example, we use the quotient of two numbers in scientific notation to help put a number into perspective. The number is our national debt. The United States began accumulating large deficits in the 1980s. To finance the deficit, the government had borrowed $5.6 trillion as of the end of 2000. The graph in Figure P.6 shows the national debt increasing over time.

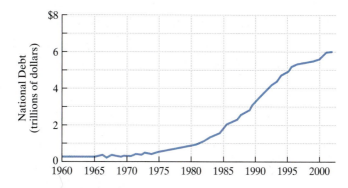

Figure P.6 The national debt

Source: Office of Management and Budget

EXAMPLE 11   The National Debt

As of the end of 2000, the national debt was $5.6 trillion, or 5.6×10^{12} dollars. At that time, the U.S. population was approximately 280,000,000 (280 million), or 2.8×10^8. If the national debt were evenly divided among every individual in the United States, how much would each citizen have to pay?

Solution   The amount each citizen must pay is the total debt, 5.6×10^{12} dollars, divided by the number of citizens, 2.8×10^8.

$$\frac{5.6 \times 10^{12}}{2.8 \times 10^8} = \left(\frac{5.6}{2.8}\right) \times \left(\frac{10^{12}}{10^8}\right)$$
$$= 2 \times 10^{12-8}$$
$$= 2 \times 10^4$$
$$= 20,000$$

Every U.S. citizen would have to pay about $20,000 to the federal government to pay off the national debt. A family of three would owe $60,000.

Technology

Here is the keystroke sequence for solving Example 11 using a calculator:

5.6 [EE] 12 [÷] 2.8 [EE] 8.

The quotient is displayed by pressing [=] on a scientific calculator and [ENTER] on a graphing calculator. The answer can be displayed in scientific or decimal notation. Consult your manual.

Check Point 11 In 2000, Americans spent 3.6×10^9 dollars on full-fat ice cream. At that time, the U.S. population was approximately 280 million, or 2.8×10^8. If ice cream spending is evenly divided, how much did each American spend?

EXERCISE SET P.2

Practice Exercises

Evaluate each exponential expression in Exercises 1–22.

1. $5^2 \cdot 2$ **2.** $6^2 \cdot 2$

3. $(-2)^6$ **4.** $(-2)^4$

5. -2^6 **6.** -2^4

7. $(-3)^0$ **8.** $(-9)^0$

9. -3^0 **10.** -9^0

11. 4^{-3} **12.** 2^{-6}

13. $2^2 \cdot 2^3$ **14.** $3^3 \cdot 3^2$

15. $(2^2)^3$ **16.** $(3^3)^2$

17. $\dfrac{2^8}{2^4}$ **18.** $\dfrac{3^8}{3^4}$

19. $3^{-3} \cdot 3$ **20.** $2^{-3} \cdot 2$

21. $\dfrac{2^3}{2^7}$ **22.** $\dfrac{3^4}{3^7}$

Simplify each exponential expression in Exercises 23–64.

23. $x^{-2}y$ **24.** xy^{-3}

25. x^0y^5 **26.** x^7y^0

27. $x^3 \cdot x^7$ **28.** $x^{11} \cdot x^5$

29. $x^{-5} \cdot x^{10}$ **30.** $x^{-6} \cdot x^{12}$

31. $(x^3)^7$ **32.** $(x^{11})^5$

33. $(x^{-5})^3$ **34.** $(x^{-6})^4$

35. $\dfrac{x^{14}}{x^7}$ **36.** $\dfrac{x^{30}}{x^{10}}$

37. $\dfrac{x^{14}}{x^{-7}}$ **38.** $\dfrac{x^{30}}{x^{-10}}$

39. $(8x^3)^2$ **40.** $(6x^4)^2$

41. $\left(-\dfrac{4}{x}\right)^3$ **42.** $\left(-\dfrac{6}{y}\right)^3$

43. $(-3x^2y^5)^2$ **44.** $(-3x^4y^6)^3$

45. $(3x^4)(2x^7)$ **46.** $(11x^5)(9x^{12})$

47. $(-9x^3y)(-2x^6y^4)$ **48.** $(-5x^4y)(-6x^7y^{11})$

49. $\dfrac{8x^{20}}{2x^4}$ **50.** $\dfrac{20x^{24}}{10x^6}$

51. $\dfrac{25a^{13}b^4}{-5a^2b^3}$ **52.** $\dfrac{35a^{14}b^6}{-7a^7b^3}$

53. $\dfrac{14b^7}{7b^{14}}$ **54.** $\dfrac{20b^{10}}{10b^{20}}$

55. $(4x^3)^{-2}$ **56.** $(10x^2)^{-3}$

57. $\dfrac{24x^3y^5}{32x^7y^{-9}}$ **58.** $\dfrac{10x^4y^9}{30x^{12}y^{-3}}$

59. $\left(\dfrac{5x^3}{y}\right)^{-2}$ **60.** $\left(\dfrac{3x^4}{y}\right)^{-3}$

61. $\left(\dfrac{-15a^4b^2}{5a^{10}b^{-3}}\right)^3$ **62.** $\left(\dfrac{-30a^{14}b^8}{10a^{17}b^{-2}}\right)^3$

63. $\left(\dfrac{3a^{-5}b^2}{12a^3b^{-4}}\right)^0$ **64.** $\left(\dfrac{4a^{-5}b^3}{12a^3b^{-5}}\right)^0$

In Exercises 65–72, write each number in decimal notation.

65. 4.7×10^3 **66.** 9.12×10^5

67. 4×10^6 **68.** 7×10^6

69. 7.86×10^{-4} **70.** 4.63×10^{-5}

71. 3.18×10^{-6} **72.** 5.84×10^{-7}

In Exercises 73–80, write each number in scientific notation.

73. 3600 **74.** 2700

75. 220,000,000 **76.** 370,000,000,000

77. 0.027 **78.** 0.014

79. 0.000763 **80.** 0.000972

In Exercises 81–88, perform the indicated operation and express the answer in decimal notation.

81. $(2 \times 10^3)(3 \times 10^2)$ **82.** $(5 \times 10^2)(4 \times 10^4)$

83. $(4.1 \times 10^2)(3 \times 10^{-4})$ **84.** $(1.2 \times 10^3)(2 \times 10^{-5})$

85. $\dfrac{12 \times 10^6}{4 \times 10^2}$ **86.** $\dfrac{20 \times 10^{26}}{10 \times 10^{15}}$

87. $\dfrac{6.3 \times 10^3}{3 \times 10^5}$ **88.** $\dfrac{9.6 \times 10^2}{3 \times 10^{-3}}$

In Exercises 89–92, write each number in scientific notation and use scientific notation to perform the operation(s). Express the answer in scientific notation.

89. $\dfrac{480,000,000,000}{0.00012}$ **90.** $\dfrac{282,000,000,000}{0.00141}$

91. $\dfrac{0.00072 \times 0.003}{0.00024}$ **92.** $\dfrac{66,000 \times 0.001}{0.003 \times 0.002}$

Application Exercises

Use 10^{12} for one trillion and 2.8×10^8 for the U.S. population in 2000 to solve Exercises 93–95.

93. In 2000, the government collected approximately $1.9 trillion in taxes. What was the per capita tax burden, or the amount that each U.S. citizen paid in taxes? Round to the nearest hundred dollars.

94. In 2000, U.S. personal income was $8 trillion. What was the per capita income, or the income per U.S. citizen? Round to the nearest hundred dollars.

95. In the United States, we spend an average of $4000 per person each year on health care—the highest in the world. What do we spend each year on health care nationwide? Express the answer in scientific notation.

96. Approximately 2×10^4 people run in the New York City Marathon each year. Each runner runs a distance of 26 miles. Write the total distance covered by all the runners (assuming that each person completes the marathon) in scientific notation.

97. The mass of one oxygen molecule is 5.3×10^{-23} gram. Find the mass of 20,000 molecules of oxygen. Express the answer in scientific notation.

98. The mass of one hydrogen atom is 1.67×10^{-24} gram. Find the mass of 80,000 hydrogen atoms. Express the answer in scientific notation.

Writing in Mathematics

99. Describe what it means to raise a number to a power. In your description, include a discussion of the difference between -5^2 and $(-5)^2$.

100. Explain the product rule for exponents. Use $2^3 \cdot 2^5$ in your explanation.

101. Explain the power rule for exponents. Use $(3^2)^4$ in your explanation.

102. Explain the quotient rule for exponents. Use $\dfrac{5^8}{5^2}$ in your explanation.

103. Why is $(-3x^2)(2x^{-5})$ not simplified? What must be done to simplify the expression?

104. How do you know if a number is written in scientific notation?

105. Explain how to convert from scientific to decimal notation and give an example.

106. Explain how to convert from decimal to scientific notation and give an example.

Critical Thinking Exercises

107. Which one of the following is true?
   a. $4^{-2} < 4^{-3}$ **b.** $5^{-2} > 2^{-5}$
   c. $(-2)^4 = 2^{-4}$ **d.** $5^2 \cdot 5^{-2} > 2^5 \cdot 2^{-5}$

108. The mad Dr. Frankenstein has gathered enough bits and pieces (so to speak) for $2^{-1} + 2^{-2}$ of his creature-to-be. Write a fraction that represents the amount of his creature that must still be obtained.

109. If $b^A = MN$, $b^C = M$, and $b^D = N$, what is the relationship among A, C, and D?

Group Exercise

110. Putting Numbers into Perspective. A large number can be put into perspective by comparing it with another number. For example, we put the $5.6 trillion national debt into perspective by comparing it to the number of U.S. citizens. The total distance covered by all the runners in the New York City Marathon (Exercise 96) can be put into perspective by comparing this distance with, say, the distance from New York to San Francisco.

   For this project, each group member should consult an almanac, a newspaper, or the World Wide Web to find a number greater than one million. Explain to other members of the group the context in which the large number is used. Express the number in scientific notation. Then put the number into perspective by comparing it with another number.

SECTION P.3   *Radicals and Rational Exponents*

Objectives

1. Evaluate square roots.
2. Use the product rule to simplify square roots.
3. Use the quotient rule to simplify square roots.
4. Add and subtract square roots.
5. Rationalize denominators.
6. Evaluate and perform operations with higher roots.
7. Understand and use rational exponents.

What is the maximum speed at which a racing cyclist can turn a corner without tipping over? The answer, in miles per hour, is given by the algebraic expression $4\sqrt{x}$, where x is the radius of the corner, in feet. Algebraic expressions containing roots describe phenomena as diverse as a wild animal's territorial area, evaporation on a lake's surface, and Albert Einstein's bizarre concept of how an astronaut moving close to the speed of light would barely age relative to friends watching from Earth. No description of your world can be complete without roots and radicals. In this section, we review the basics of radical expressions and the use of rational exponents to indicate radicals.

1 Evaluate square roots.

Square Roots

The **principal square root** of a nonnegative real number b, written $\sqrt{b}$, is that number whose square equals b. For example,

$$\sqrt{100} = 10 \text{ because } 10^2 = 100 \quad \text{and} \quad \sqrt{0} = 0 \text{ because } 0^2 = 0.$$

Observe that the principal square root of a positive number is positive and the principal square root of 0 is 0.

The symbol $\sqrt{}$ that we use to denote the principal square root is called a **radical sign**. The number under the radical sign is called the **radicand.** Together we refer to the radical sign and its radicand as a **radical.**

The following definition summarizes our discussion:

Definition of the Principal Square Root

If a is a nonnegative real number, the nonnegative number b such that $b^2 = a$, denoted by $b = \sqrt{a}$, is the **principal square root** of a.

In the real number system, negative numbers do not have square roots. For example, $\sqrt{-9}$ is not a real number because there is no real number whose square is -9.

If a number is nonnegative ($a \geq 0$), then $\left(\sqrt{a}\right)^2 = a$. For example,

$$\left(\sqrt{2}\right)^2 = 2, \quad \left(\sqrt{3}\right)^2 = 3, \quad \left(\sqrt{4}\right)^2 = 4, \quad \text{and} \left(\sqrt{5}\right)^2 = 5.$$

A number that is the square of a rational number is called a **perfect square.** For example,

64 is a perfect square because $64 = 8^2$.

$\dfrac{1}{9}$ is a perfect square because $\dfrac{1}{9} = \left(\dfrac{1}{3}\right)^2$.

The following rule can be used to find square roots of perfect squares:

> **Square Roots of Perfect Squares**
> $$\sqrt{a^2} = |a|$$

For example, $\sqrt{6^2} = 6$ and $\sqrt{(-6)^2} = |-6| = 6$.

2 Use the product rule to simplify square roots.

The Product Rule for Square Roots

A square root is **simplified** when its radicand has no factors other than 1 that are perfect squares. For example, $\sqrt{500}$ is not simplified because it can be expressed as $\sqrt{100 \cdot 5}$ and $\sqrt{100}$ is a perfect square. The **product rule for square roots** can be used to simplify $\sqrt{500}$.

> **The Product Rule for Square Roots**
> If a and b represent nonnegative real numbers, then
> $$\sqrt{ab} = \sqrt{a}\,\sqrt{b} \text{ and } \sqrt{a}\,\sqrt{b} = \sqrt{ab}.$$
> The square root of a product is the product of the square roots.

Example 1 shows how the product rule is used to remove from the square root any perfect squares that occur as factors.

EXAMPLE 1   Using the Product Rule to Simplify Square Roots

Simplify: **a.** $\sqrt{500}$ **b.** $\sqrt{6x} \cdot \sqrt{3x}$.

Solution

a. $\sqrt{500} = \sqrt{100 \cdot 5}$ 100 is the largest perfect square factor of 500.

$\phantom{\sqrt{500}} = \sqrt{100}\,\sqrt{5}$ $\sqrt{ab} = \sqrt{a}\,\sqrt{b}$

$\phantom{\sqrt{500}} = 10\sqrt{5}$ $\sqrt{100} = 10$

b. We can simplify $\sqrt{6x} \cdot \sqrt{3x}$ using the power rule only if $6x$ and $3x$ represent nonnegative real numbers. Thus, $x \geq 0$.

$\sqrt{6x} \cdot \sqrt{3x} = \sqrt{6x \cdot 3x}$ $\sqrt{a}\,\sqrt{b} = \sqrt{ab}$

$\phantom{\sqrt{6x} \cdot \sqrt{3x}} = \sqrt{18x^2}$ Multiply.

$\phantom{\sqrt{6x} \cdot \sqrt{3x}} = \sqrt{9x^2 \cdot 2}$ 9 is the largest perfect square factor of 18.

$\phantom{\sqrt{6x} \cdot \sqrt{3x}} = \sqrt{9x^2}\,\sqrt{2}$ $\sqrt{ab} = \sqrt{a}\,\sqrt{b}$

$\phantom{\sqrt{6x} \cdot \sqrt{3x}} = \sqrt{9}\,\sqrt{x^2}\,\sqrt{2}$ Split $\sqrt{9x^2}$ into two square roots.

$\phantom{\sqrt{6x} \cdot \sqrt{3x}} = 3x\sqrt{2}$ $\sqrt{9} = 3$ (because $3^2 = 9$) and $\sqrt{x^2} = x$ because $x \geq 0$.

> **Check Point 1** Simplify:
> **a.** $\sqrt{3^2}$ **b.** $\sqrt{5x} \cdot \sqrt{10x}$.

3 Use the quotient rule to simplify square roots.

The Quotient Rule for Square Roots

Another property for square roots involves division.

> **The Quotient Rule for Square Roots**
>
> If a and b represent nonnegative real numbers and $b \neq 0$, then
> $$\frac{\sqrt{a}}{\sqrt{b}} = \sqrt{\frac{a}{b}} \quad \text{and} \quad \sqrt{\frac{a}{b}} = \frac{\sqrt{a}}{\sqrt{b}}.$$
> The square root of a quotient is the quotient of the square roots.

EXAMPLE 2 **Using the Quotient Rule to Simplify Square Roots**

Simplify: **a.** $\sqrt{\dfrac{100}{9}}$ **b.** $\dfrac{\sqrt{48x^3}}{\sqrt{6x}}$.

Solution

a. $\sqrt{\dfrac{100}{9}} = \dfrac{\sqrt{100}}{\sqrt{9}} = \dfrac{10}{3}$

b. We can simplify the quotient of $\sqrt{48x^3}$ and $\sqrt{6x}$ using the quotient rule only if $48x^3$ and $6x$ represent nonnegative real numbers. Thus, $x \geq 0$.

$$\frac{\sqrt{48x^3}}{\sqrt{6x}} = \sqrt{\frac{48x^3}{6x}} = \sqrt{8x^2} = \sqrt{4x^2}\sqrt{2} = \sqrt{4}\sqrt{x^2}\sqrt{2} = 2x\sqrt{2}$$

$\sqrt{x^2} = x$ because $x \geq 0$.

> **Check Point 2** Simplify: **a.** $\sqrt{\dfrac{25}{16}}$ **b.** $\dfrac{\sqrt{150x^3}}{\sqrt{2x}}$.

4 Add and subtract square roots.

Adding and Subtracting Square Roots

Two or more square roots can be combined provided that they have the same radicand. Such radicals are called **like radicals.** For example,

$$7\sqrt{11} + 6\sqrt{11} = (7 + 6)\sqrt{11} = 13\sqrt{11}.$$

EXAMPLE 3 **Adding and Subtracting Like Radicals**

Add or subtract as indicated:
a. $7\sqrt{2} + 5\sqrt{2}$ **b.** $\sqrt{5x} - 7\sqrt{5x}$.

Solution

a. $7\sqrt{2} + 5\sqrt{2} = (7 + 5)\sqrt{2}$ Apply the distributive property.

$\qquad\qquad\quad = 12\sqrt{2}$ Simplify.

b. $\sqrt{5x} - 7\sqrt{5x} = 1\sqrt{5x} - 7\sqrt{5x}$ Write $\sqrt{5x}$ as $1\sqrt{5x}$.

$\qquad\qquad\quad = (1 - 7)\sqrt{5x}$ Apply the distributive property.

$\qquad\qquad\quad = -6\sqrt{5x}$ Simplify.

Check Point 3   Add or subtract as indicated:

$\qquad$ **a.** $8\sqrt{13} + 9\sqrt{13}$ $\qquad$ **b.** $\sqrt{17x} - 20\sqrt{17x}.$

In some cases, radicals can be combined once they have been simplified. For example, to add $\sqrt{2}$ and $\sqrt{8}$, we can write $\sqrt{8}$ as $\sqrt{4 \cdot 2}$ because 4 is a perfect square factor of 8.

$$\sqrt{2} + \sqrt{8} = \sqrt{2} + \sqrt{4 \cdot 2} = 1\sqrt{2} + 2\sqrt{2} = (1 + 2)\sqrt{2} = 3\sqrt{2}$$

EXAMPLE 4   Combining Radicals That First Require Simplification

Add or subtract as indicated: **a.** $7\sqrt{3} + \sqrt{12}$ $\qquad$ **b.** $4\sqrt{50x} - 6\sqrt{32x}.$

Solution

a. $7\sqrt{3} + \sqrt{12}$

$\qquad = 7\sqrt{3} + \sqrt{4 \cdot 3}$ Split 12 into two factors such that one is a perfect square.

$\qquad = 7\sqrt{3} + 2\sqrt{3}$ $\qquad$ $\sqrt{4 \cdot 3} = \sqrt{4}\sqrt{3} = 2\sqrt{3}$

$\qquad = (7 + 2)\sqrt{3}$ $\qquad$ Apply the distributive property. You will find that this step is usually done mentally.

$\qquad = 9\sqrt{3}$ $\qquad$ Simplify.

b. $4\sqrt{50x} - 6\sqrt{32x}$

$\qquad = 4\sqrt{25 \cdot 2x} - 6\sqrt{16 \cdot 2x}$ 25 is the largest perfect square factor of 50 and 16 is the largest perfect square factor of 32.

$\qquad = 4 \cdot 5\sqrt{2x} - 6 \cdot 4\sqrt{2x}$ $\sqrt{25 \cdot 2} = \sqrt{25}\sqrt{2} = 5\sqrt{2}$ and $\sqrt{16 \cdot 2} = \sqrt{16}\sqrt{2} = 4\sqrt{2}$.

$\qquad = 20\sqrt{2x} - 24\sqrt{2x}$ Multiply.

$\qquad = (20 - 24)\sqrt{2x}$ Apply the distributive property.

$\qquad = -4\sqrt{2x}$ Simplify.

Check Point 4   Add or subtract as indicated:

$\qquad$ **a.** $5\sqrt{27} + \sqrt{12}$ $\qquad$ **b.** $6\sqrt{18x} - 4\sqrt{8x}.$

A Radical Idea: Time Is Relative

What does travel in space have to do with radicals? Imagine that in the future we will be able to travel at velocities approaching the speed of light (approximately 186,000 miles per second). According to Einstein's theory of relativity, time would pass more quickly on Earth than it would in the moving spaceship. The expression

$$R_f\sqrt{1 - \left(\frac{v}{c}\right)^2}$$

gives the aging rate of an astronaut relative to the aging rate of a friend on Earth, R_f. In the expression, v is the astronaut's speed and c is the speed of light. As the astronaut's speed approaches the speed of light, we can substitute c for v:

$$R_f\sqrt{1 - \left(\frac{v}{c}\right)^2}$$ Let $v = c$.

$= R_f\sqrt{1 - 1^2}$

$= R_f\sqrt{0} = 0$

Close to the speed of light, the astronaut's aging rate relative to a friend on Earth is nearly 0. What does this mean? As we age here on Earth, the space traveler would barely get older. The space traveler would return to a futuristic world in which friends and loved ones would be long dead.

5 Rationalize denominators.

Rationalizing Denominators

You can use a calculator to compare the approximate values for $\dfrac{1}{\sqrt{3}}$ and $\dfrac{\sqrt{3}}{3}$.

The two approximations are the same. This is not a coincidence:

$$\frac{1}{\sqrt{3}} = \frac{1}{\sqrt{3}} \cdot \boxed{\frac{\sqrt{3}}{\sqrt{3}}} = \frac{\sqrt{3}}{\sqrt{9}} = \frac{\sqrt{3}}{3}.$$

Any number divided by itself is 1. Multiplication by 1 does not change the value of $\dfrac{1}{\sqrt{3}}$.

This process involves rewriting a radical expression as an equivalent expression in which the denominator no longer contains a radical. The process is called **rationalizing the denominator.** If the denominator contains the square root of a natural number that is not a perfect square, **multiply the numerator and denominator by the smallest number that produces the square root of a perfect square in the denominator.**

EXAMPLE 5   Rationalizing Denominators

Rationalize the denominator: **a.** $\dfrac{15}{\sqrt{6}}$ **b.** $\dfrac{12}{\sqrt{8}}$.

Solution

a. If we multiply numerator and denominator by $\sqrt{6}$, the denominator becomes $\sqrt{6} \cdot \sqrt{6} = \sqrt{36} = 6$. Therefore, we multiply by 1, choosing $\dfrac{\sqrt{6}}{\sqrt{6}}$ for 1.

$$\frac{15}{\sqrt{6}} = \frac{15}{\sqrt{6}} \cdot \frac{\sqrt{6}}{\sqrt{6}} = \frac{15\sqrt{6}}{\sqrt{36}} = \frac{15\sqrt{6}}{6} = \frac{5\sqrt{6}}{2}$$

Multiply by 1.

Simplify: $\dfrac{15}{6} = \dfrac{15 \div 3}{6 \div 3} = \dfrac{5}{2}$.

b. The *smallest* number that will produce a perfect square in the denominator of $\dfrac{12}{\sqrt{8}}$ is $\sqrt{2}$, because $\sqrt{8} \cdot \sqrt{2} = \sqrt{16} = 4$. We multiply by 1, choosing $\dfrac{\sqrt{2}}{\sqrt{2}}$ for 1.

$$\frac{12}{\sqrt{8}} = \frac{12}{\sqrt{8}} \cdot \frac{\sqrt{2}}{\sqrt{2}} = \frac{12\sqrt{2}}{\sqrt{16}} = \frac{12\sqrt{2}}{4} = 3\sqrt{2}$$

Check Point 5   Rationalize the denominator: **a.** $\dfrac{5}{\sqrt{3}}$ **b.** $\dfrac{6}{\sqrt{12}}$.

How can we rationalize a denominator if the denominator contains two terms? In general,

$$(\sqrt{a} + \sqrt{b})(\sqrt{a} - \sqrt{b}) = (\sqrt{a})^2 - (\sqrt{b})^2 = a - b.$$

Notice that the product does not contain a radical. Here are some specific examples.

The Denominator Contains:	Multiply by:	The New Denominator Contains:
$7 + \sqrt{5}$	$7 - \sqrt{5}$	$7^2 - (\sqrt{5})^2 = 49 - 5 = 44$
$\sqrt{3} - 6$	$\sqrt{3} + 6$	$(\sqrt{3})^2 - 6^2 = 3 - 36 = -33$
$\sqrt{7} + \sqrt{3}$	$\sqrt{7} - \sqrt{3}$	$(\sqrt{7})^2 - (\sqrt{3})^2 = 7 - 3 = 4$

EXAMPLE 6 **Rationalizing a Denominator Containing Two Terms**

Rationalize the denominator: $\dfrac{7}{5 + \sqrt{3}}$.

Solution If we multiply the numerator and denominator by $5 - \sqrt{3}$, the denominator will not contain a radical. Therefore, we multiply by 1, choosing $\dfrac{5 - \sqrt{3}}{5 - \sqrt{3}}$ for 1.

$$\frac{7}{5 + \sqrt{3}} = \frac{7}{5 + \sqrt{3}} \cdot \frac{5 - \sqrt{3}}{5 - \sqrt{3}} = \frac{7(5 - \sqrt{3})}{5^2 - (\sqrt{3})^2} = \frac{7(5 - \sqrt{3})}{25 - 3}$$

Multiply by 1.

$$= \frac{7(5 - \sqrt{3})}{22} \quad \text{or} \quad \frac{35 - 7\sqrt{3}}{22}.$$

In either form of the answer, there is no radical in the denominator.

Check Point 6 Rationalize the denominator: $\dfrac{8}{4 + \sqrt{5}}$.

6 Evaluate and perform operations with higher roots.

Other Kinds of Roots

We define the **principal nth root** of a real number a, symbolized by $\sqrt[n]{a}$, as follows:

Definition of the Principal nth Root of a Real Number

$$\sqrt[n]{a} = b \text{ means that } b^n = a.$$

If n, the **index**, is even, then a is nonnegative ($a \geq 0$) and b is also nonnegative ($b \geq 0$). If n is odd, a and b can be any real numbers.

For example,

$$\sqrt[3]{64} = 4 \text{ because } 4^3 = 64 \quad \text{and} \quad \sqrt[5]{-32} = -2 \text{ because } (-2)^5 = -32.$$

The same vocabulary that we learned for square roots applies to nth roots. The symbol $\sqrt[n]{a}$ is called a **radical** and a is called the **radicand.**

A number that is the nth power of a rational number is called a **perfect nth power.** For example, 8 is a perfect third power, or perfect cube, because $8 = 2^3$. In general, one of the following rules can be used to find nth roots of perfect nth powers:

Finding nth Roots of Perfect nth Powers

If n is odd, $\sqrt[n]{a^n} = a$.

If n is even, $\sqrt[n]{a^n} = |a|$.

For example,

$$\sqrt[3]{(-2)^3} = -2 \quad \text{and} \quad \sqrt[4]{(-2)^4} = |-2| = 2.$$

Absolute value is not needed with odd roots, but is necessary with even roots.

The Product and Quotient Rules for Other Roots

The product and quotient rules apply to cube roots, fourth roots, and all higher roots.

The Product and Quotient Rules for nth Roots

For all real numbers, where the indicated roots represent real numbers,

$$\sqrt[n]{a} \cdot \sqrt[n]{b} = \sqrt[n]{ab} \quad \text{and} \quad \frac{\sqrt[n]{a}}{\sqrt[n]{b}} = \sqrt[n]{\frac{a}{b}}, \quad b \neq 0.$$

EXAMPLE 7   Simplifying, Multiplying, and Dividing Higher Roots

Simplify: **a.** $\sqrt[3]{24}$ **b.** $\sqrt[4]{8} \cdot \sqrt[4]{4}$ **c.** $\sqrt[4]{\dfrac{81}{16}}$.

Solution

a. $\sqrt[3]{24} = \sqrt[3]{8 \cdot 3}$ Find the largest *perfect cube* that is a factor of 24. $\sqrt[3]{8} = 2$, so 8 is a perfect cube and is the largest perfect cube factor of 24.

Study Tip

Some higher even and odd roots occur so frequently that you might want to memorize them.

Cube Roots

$\sqrt[3]{1} = 1$	$\sqrt[3]{125} = 5$
$\sqrt[3]{8} = 2$	$\sqrt[3]{216} = 6$
$\sqrt[3]{27} = 3$	$\sqrt[3]{1000} = 10$
$\sqrt[3]{64} = 4$	

Fourth Roots	**Fifth Roots**
$\sqrt[4]{1} = 1$	$\sqrt[5]{1} = 1$
$\sqrt[4]{16} = 2$	$\sqrt[5]{32} = 2$
$\sqrt[4]{81} = 3$	$\sqrt[5]{243} = 3$
$\sqrt[4]{256} = 4$	
$\sqrt[4]{625} = 5$	

$$= \sqrt[3]{8} \cdot \sqrt[3]{3} \qquad \sqrt[n]{ab} = \sqrt[n]{a}\,\sqrt[n]{b}$$
$$= 2\sqrt[3]{3}$$

b. $\sqrt[4]{8} \cdot \sqrt[4]{4} = \sqrt[4]{8 \cdot 4}$ $\qquad \sqrt[n]{a} \cdot \sqrt[n]{b} = \sqrt[n]{ab}$

$$= \sqrt[4]{32}$$ Find the largest *perfect fourth power* that is a factor of 32.

$$= \sqrt[4]{16 \cdot 2}$$ $\sqrt[4]{16} = 2$, so 16 is a perfect fourth power and is the largest perfect fourth power that is a factor of 32.

$$= \sqrt[4]{16} \cdot \sqrt[4]{2} \qquad \sqrt[n]{ab} = \sqrt[n]{a} \cdot \sqrt[n]{b}$$

$$= 2\sqrt[4]{2}$$

c. $\sqrt[4]{\dfrac{81}{16}} = \dfrac{\sqrt[4]{81}}{\sqrt[4]{16}} \qquad \sqrt[n]{\dfrac{a}{b}} = \dfrac{\sqrt[n]{a}}{\sqrt[n]{b}}$

$$= \dfrac{3}{2} \qquad \sqrt[4]{81} = 3 \text{ because } 3^4 = 81 \text{ and } \sqrt[4]{16} = 2 \text{ because } 2^4 = 16.$$

Check Point 7 Simplify: **a.** $\sqrt[3]{40}$ **b.** $\sqrt[5]{8} \cdot \sqrt[5]{8}$ **c.** $\sqrt[3]{\dfrac{125}{27}}$.

We have seen that adding and subtracting square roots often involves simplifying terms. The same idea applies to adding and subtracting nth roots.

EXAMPLE 8 Combining Cube Roots

Subtract: $5\sqrt[3]{16} - 11\sqrt[3]{2}$.

Solution

$$5\sqrt[3]{16} - 11\sqrt[3]{2}$$

$$= 5\sqrt[3]{8 \cdot 2} - 11\sqrt[3]{2}$$ Because $16 = 8 \cdot 2$ and $\sqrt[3]{8} = 2$, 8 is the largest perfect cube that is a factor of 16.

$$= 5 \cdot 2\sqrt[3]{2} - 11\sqrt[3]{2} \qquad \sqrt[3]{8 \cdot 2} = \sqrt[3]{8}\,\sqrt[3]{2} = 2\sqrt[3]{2}$$

$$= 10\sqrt[3]{2} - 11\sqrt[3]{2}$$ Multiply.

$$= (10 - 11)\sqrt[3]{2}$$ Apply the distributive property.

$$= -1\sqrt[3]{2} \text{ or } -\sqrt[3]{2}$$ Simplify.

Check Point 8 Subtract: $3\sqrt[3]{81} - 4\sqrt[3]{3}$.

7 Understand and use rational exponents.

Rational Exponents

Animals in the wild have regions to which they confine their movement, called their territorial area. Territorial area, in square miles, is related to an animal's body weight. If an animal weighs W pounds, its territorial area is

$$W^{141/100}$$

square miles.

W to the *what* power?! How can we interpret the information given by this algebraic expression?

In the last part of this section, we turn our attention to rational exponents such as $\frac{141}{100}$ and their relationship to roots of real numbers.

Definition of Rational Exponents

If $\sqrt[n]{a}$ represents a real number and $n \geq 2$ is an integer, then

$$a^{1/n} = \sqrt[n]{a}.$$

Furthermore,

$$a^{-1/n} = \frac{1}{a^{1/n}} = \frac{1}{\sqrt[n]{a}}, \quad a \neq 0.$$

EXAMPLE 9 **Using the Definition of $a^{1/n}$**

Simplify: **a.** $64^{1/2}$ **b.** $8^{1/3}$ **c.** $64^{-1/3}$.

Solution

a. $64^{1/2} = \sqrt{64} = 8$ **b.** $8^{1/3} = \sqrt[3]{8} = 2$

c. $64^{-1/3} = \dfrac{1}{64^{1/3}} = \dfrac{1}{\sqrt[3]{64}} = \dfrac{1}{4}$

Check Point 9   Simplify: **a.** $81^{1/2}$ **b.** $27^{1/3}$ **c.** $32^{-1/5}$.

Note that every rational exponent in Example 9 has a numerator of 1 or −1. We now define rational exponents with any integer in the numerator.

Definition of Rational Exponents

If $\sqrt[n]{a}$ represents a real number, $\dfrac{m}{n}$ is a rational number reduced to lowest terms, and $n \geq 2$ is an integer, then

$$a^{m/n} = \left(\sqrt[n]{a}\right)^m = \sqrt[n]{a^m}.$$

The exponent m/n consists of two parts: the denominator n is the root and the numerator m is the exponent. Furthermore,

$$a^{-m/n} = \frac{1}{a^{m/n}}.$$

EXAMPLE 10 **Using the Definition of $a^{m/n}$**

Simplify: **a.** $27^{2/3}$ **b.** $9^{3/2}$ **c.** $16^{-3/4}$.

Technology

Here are the calculator keystroke sequences for $27^{2/3}$:

Many Scientific Calculators

$27 \boxed{y^x} \boxed{(} 2 \boxed{\div} 3 \boxed{)} \boxed{=}$

Many Graphing Calculators

$27 \boxed{\wedge} \boxed{(} 2 \boxed{\div} 3 \boxed{)} \boxed{\text{ENTER}}$.

Solution

a. $27^{2/3} = \left(\sqrt[3]{27} \right)^2 = 3^2 = 9$

> The denominator of $\frac{2}{3}$ is the root and the numerator is the exponent.

b. $9^{3/2} = \left(\sqrt{9} \right)^3 = 3^3 = 27$

c. $16^{-3/4} = \dfrac{1}{16^{3/4}} = \dfrac{1}{\left(\sqrt[4]{16} \right)^3} = \dfrac{1}{2^3} = \dfrac{1}{8}$

Check Point 10   Simplify: **a.** $4^{3/2}$ **b.** $32^{-2/5}$.

Properties of exponents can be applied to expressions containing rational exponents.

EXAMPLE 11   Simplifying Expressions with Rational Exponents

Simplify using properties of exponents:

a. $(5x^{1/2})(7x^{3/4})$ **b.** $\dfrac{32x^{5/3}}{16x^{3/4}}$.

Solution

a. $(5x^{1/2})(7x^{3/4}) = 5 \cdot 7x^{1/2} \cdot x^{3/4}$ Group factors with the same base.

$= 35x^{(1/2)+(3/4)}$ When multiplying expressions with the same base, add the exponents.

$= 35x^{5/4}$ $\frac{1}{2} + \frac{3}{4} = \frac{2}{4} + \frac{3}{4} = \frac{5}{4}$

b. $\dfrac{32x^{5/3}}{16x^{3/4}} = \left(\dfrac{32}{16} \right) \left(\dfrac{x^{5/3}}{x^{3/4}} \right)$ Group factors with the same base.

$= 2x^{(5/3)-(3/4)}$ When dividing expressions with the same base, subtract the exponents.

$= 2x^{11/12}$ $\frac{5}{3} - \frac{3}{4} = \frac{20}{12} - \frac{9}{12} = \frac{11}{12}$

Check Point 11   Simplify: **a.** $(2x^{4/3})(5x^{8/3})$ **b.** $\dfrac{20x^4}{5x^{3/2}}$.

Rational exponents are sometimes useful for simplifying radicals by reducing their index.

EXAMPLE 12   Reducing the Index of a Radical

Simplify: $\sqrt[9]{x^3}$.

Solution $\sqrt[9]{x^3} = x^{3/9} = x^{1/3} = \sqrt[3]{x}$

Check Point 12   Simplify: $\sqrt[6]{x^3}$.

EXERCISE SET P.3

 Practice Exercises

Evaluate each expression in Exercises 1–6 or indicate that the root is not a real number.

1. $\sqrt{36}$ **2.** $\sqrt{25}$

3. $\sqrt{-36}$ **4.** $\sqrt{-25}$

5. $\sqrt{(-13)^2}$ **6.** $\sqrt{(-17)^2}$

Use the product rule to simplify the expressions in Exercises 7–16. In Exercises 11–16, assume that variables represent nonnegative real numbers.

7. $\sqrt{50}$ **8.** $\sqrt{27}$

9. $\sqrt{45x^2}$ **10.** $\sqrt{125x^2}$

11. $\sqrt{2x} \cdot \sqrt{6x}$ **12.** $\sqrt{10x} \cdot \sqrt{8x}$

13. $\sqrt{x^3}$ **14.** $\sqrt{y^3}$

15. $\sqrt{2x^2} \cdot \sqrt{6x}$ **16.** $\sqrt{6x} \cdot \sqrt{3x^2}$

Use the quotient rule to simplify the expressions in Exercises 17–26. Assume that $x > 0$.

17. $\sqrt{\dfrac{1}{81}}$ **18.** $\sqrt{\dfrac{1}{49}}$

19. $\sqrt{\dfrac{49}{16}}$ **20.** $\sqrt{\dfrac{121}{9}}$

21. $\dfrac{\sqrt{48x^3}}{\sqrt{3x}}$ **22.** $\dfrac{\sqrt{72x^3}}{\sqrt{8x}}$

23. $\dfrac{\sqrt{150x^4}}{\sqrt{3x}}$ **24.** $\dfrac{\sqrt{24x^4}}{\sqrt{3x}}$

25. $\dfrac{\sqrt{200x^3}}{\sqrt{10x^{-1}}}$ **26.** $\dfrac{\sqrt{500x^3}}{\sqrt{10x^{-1}}}$

In Exercises 27–38, add or subtract terms whenever possible.

27. $7\sqrt{3} + 6\sqrt{3}$ **28.** $8\sqrt{5} + 11\sqrt{5}$

29. $6\sqrt{17x} - 8\sqrt{17x}$ **30.** $4\sqrt{13x} - 6\sqrt{13x}$

31. $\sqrt{8} + 3\sqrt{2}$ **32.** $\sqrt{20} + 6\sqrt{5}$

33. $\sqrt{50x} - \sqrt{8x}$ **34.** $\sqrt{63x} - \sqrt{28x}$

35. $3\sqrt{18} + 5\sqrt{50}$ **36.** $4\sqrt{12} - 2\sqrt{75}$

37. $3\sqrt{8} - \sqrt{32} + 3\sqrt{72} - \sqrt{75}$

38. $3\sqrt{54} - 2\sqrt{24} - \sqrt{96} + 4\sqrt{63}$

In Exercises 39–48, rationalize the denominator.

39. $\dfrac{1}{\sqrt{7}}$ **40.** $\dfrac{2}{\sqrt{10}}$

41. $\dfrac{\sqrt{2}}{\sqrt{5}}$ **42.** $\dfrac{\sqrt{7}}{\sqrt{3}}$

43. $\dfrac{13}{3 + \sqrt{11}}$ **44.** $\dfrac{3}{3 + \sqrt{7}}$

45. $\dfrac{7}{\sqrt{5} - 2}$ **46.** $\dfrac{5}{\sqrt{3} - 1}$

47. $\dfrac{6}{\sqrt{5} + \sqrt{3}}$ **48.** $\dfrac{11}{\sqrt{7} - \sqrt{3}}$

Evaluate each expression in Exercises 49–60, or indicate that the root is not a real number.

49. $\sqrt[3]{125}$ **50.** $\sqrt[3]{8}$

51. $\sqrt[3]{-8}$ **52.** $\sqrt[3]{-125}$

53. $\sqrt[4]{-16}$ **54.** $\sqrt[4]{-81}$

55. $\sqrt[4]{(-3)^4}$ **56.** $\sqrt[4]{(-2)^4}$

57. $\sqrt[5]{(-3)^5}$ **58.** $\sqrt[5]{(-2)^5}$

59. $\sqrt[5]{-\frac{1}{32}}$ **60.** $\sqrt[6]{\frac{1}{64}}$

Simplify the radical expressions in Exercises 61–68.

61. $\sqrt[3]{32}$ **62.** $\sqrt[3]{150}$

63. $\sqrt[3]{x^4}$ **64.** $\sqrt[3]{x^5}$

65. $\sqrt[3]{9} \cdot \sqrt[3]{6}$ **66.** $\sqrt[3]{12} \cdot \sqrt[3]{4}$

67. $\dfrac{\sqrt[5]{64x^6}}{\sqrt[5]{2x}}$ **68.** $\dfrac{\sqrt[4]{162x^5}}{\sqrt[4]{2x}}$

In Exercises 69–76, add or subtract terms whenever possible.

69. $4\sqrt[5]{2} + 3\sqrt[5]{2}$ **70.** $6\sqrt[3]{3} + 2\sqrt[3]{3}$

71. $5\sqrt[3]{16} + \sqrt[3]{54}$ **72.** $3\sqrt[3]{24} + \sqrt[3]{81}$

73. $\sqrt[3]{54xy^3} - y\sqrt[3]{128x}$ **74.** $\sqrt[3]{24xy^3} - y\sqrt[3]{81x}$

75. $\sqrt{2} + \sqrt[3]{8}$ **76.** $\sqrt{3} + \sqrt[3]{15}$

In Exercises 77–84, evaluate each expression without using a calculator.

77. $36^{1/2}$ **78.** $121^{1/2}$

79. $8^{1/3}$ **80.** $27^{1/3}$

81. $125^{2/3}$ **82.** $8^{2/3}$

83. $32^{-4/5}$ **84.** $16^{-5/2}$

In Exercises 85–94, simplify using properties of exponents.

85. $(7x^{1/3})(2x^{1/4})$ **86.** $(3x^{2/3})(4x^{3/4})$

87. $\dfrac{20x^{1/2}}{5x^{1/4}}$ **88.** $\dfrac{72x^{3/4}}{9x^{1/3}}$

89. $(x^{2/3})^3$ **90.** $(x^{4/5})^5$

91. $(25x^4y^6)^{1/2}$ **92.** $(125x^9y^6)^{1/3}$

93. $\dfrac{(3y^{1/4})^3}{y^{1/12}}$ **94.** $\dfrac{(2y^{1/5})^4}{y^{3/10}}$

In Exercises 95–102, simplify by reducing the index of the radical.

95. $\sqrt[4]{5^2}$ **96.** $\sqrt[4]{7^2}$

97. $\sqrt[3]{x^6}$ **98.** $\sqrt[4]{x^{12}}$

99. $\sqrt[6]{x^4}$ **100.** $\sqrt[9]{x^6}$

101. $\sqrt[9]{x^6y^3}$ **102.** $\sqrt[12]{x^4y^8}$

Application Exercises

103. The algebraic expression $2\sqrt{5L}$ is used to estimate the speed of a car prior to an accident, in miles per hour, based on the length of its skid marks, L, in feet. Find the speed of a car that left skid marks 40 feet long, and write the answer in simplified radical form.

104. The time, in seconds, that it takes an object to fall a distance d, in feet, is given by the algebraic expression $\sqrt{\dfrac{d}{16}}$. Find how long it will take a ball dropped from the top of a building 320 feet tall to hit the ground. Write the answer in simplified radical form.

105. The early Greeks believed that the most pleasing of all rectangles were golden rectangles whose ratio of width to height is
$$\frac{w}{h} = \frac{2}{\sqrt{5}-1}.$$
Rationalize the denominator for this ratio and then use a calculator to approximate the answer correct to the nearest hundredth.

106. The amount of evaporation, in inches per day, of a large body of water can be described by the algebraic expression
$$\frac{w}{20\sqrt{a}}$$
where

  a = surface area of the water, in square miles

  w = average wind speed of the air over the water, in miles per hour.

Determine the evaporation on a lake whose surface area is 9 square miles on a day when the wind speed over the water is 10 miles per hour.

107. In the Peanuts cartoon shown below, Woodstock appears to be working steps mentally. Fill in the missing steps that show how to go from $\dfrac{7\sqrt{2\cdot2\cdot3}}{6}$ to $\dfrac{7}{3}\sqrt{3}$.

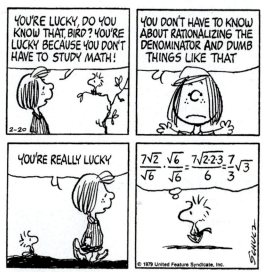

PEANUTS reprinted by permission of United Feature Syndicate, Inc.

108. The algebraic expression $152a^{-1/5}$ describes the percentage of U.S. taxpayers who are a years old who file early. Evaluate the algebraic expression for $a = 32$. Describe what the answer means in practical terms.

109. The algebraic expression $0.07d^{3/2}$ describes the duration of a storm, in hours, whose diameter is d miles. Evaluate the algebraic expression for $d = 9$. Describe what the answer means in practical terms.

Writing in Mathematics

110. Explain how to simplify $\sqrt{10}\cdot\sqrt{5}$.

111. Explain how to add $\sqrt{3}+\sqrt{12}$.

112. Describe what it means to rationalize a denominator. Use both $\dfrac{1}{\sqrt{5}}$ and $\dfrac{1}{5+\sqrt{5}}$ in your explanation.

113. What difference is there in simplifying $\sqrt[3]{(-5)^3}$ and $\sqrt[4]{(-5)^4}$?

114. What does $a^{m/n}$ mean?

115. Describe the kinds of numbers that have rational fifth roots.

116. Why must a and b represent nonnegative numbers when we write $\sqrt{a} \cdot \sqrt{b} = \sqrt{ab}$? Is it necessary to use this restriction in the case of $\sqrt[3]{a} \cdot \sqrt[3]{b} = \sqrt[3]{ab}$? Explain.

 Technology Exercises

117. The algebraic expression

$$\frac{73t^{1/3} - 28t^{2/3}}{t}$$

describes the percentage of people in the United States applying for jobs t years after 1985 who tested positive for illegal drugs. Use a calculator to find the percentage who tested positive from 1986 through 2001. Round answers to the nearest hundredth of a percent. What trend do you observe for the percentage of potential employees testing positive for illegal drugs over time?

118. The territorial area of an animal in the wild is defined to be the area of the region to which the animal confines its movements. The algebraic expression $W^{1.41}$ describes the territorial area, in square miles, of an animal that weighs W pounds. Use a calculator to find the territorial area of animals weighing 25, 50, 150, 200, 250, and 300 pounds. What do the values indicate about the relationship between body weight and territorial area?

 Critical Thinking Exercises

119. Which one of the following is true?

   a. Neither $(-8)^{1/2}$ nor $(-8)^{1/3}$ represent real numbers.

   b. $\sqrt{x^2 + y^2} = x + y$

   c. $8^{-1/3} = -2$

   d. $2^{1/2} \cdot 2^{1/2} = 2$

In Exercises 120–121, fill in each box to make the statement true.

120. $(5 + \sqrt{\square})(5 - \sqrt{\square}) = 22$

121. $\sqrt[\square]{x^{\square}} = 5x^7$

122. Find exact value of $\sqrt{13 + \sqrt{2} + \dfrac{7}{3 + \sqrt{2}}}$ without the use of a calculator.

123. Place the correct symbol, $>$ or $<$, in the box between each of the given numbers. *Do not use a calculator.* Then check your result with a calculator.

   a. $3^{1/2} \,\square\, 3^{1/3}$ **b.** $\sqrt{7} + \sqrt{18} \,\square\, \sqrt{7 + 18}$

SECTION P.4   *Polynomials*

Objectives

1. Understand the vocabulary of polynomials.

2. Add and subtract polynomials.

3. Multiply polynomials.

4. Use FOIL in polynomial multiplication.

5. Use special products in polynomial multiplication.

6. Perform operations with polynomials in several variables.

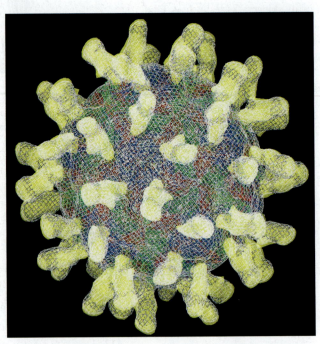

This computer-simulated model of the common cold virus was developed by researchers at Purdue University. Their discovery of how the virus infects human cells could lead to more effective treatment for the illness.

Runny nose? Sneezing? You are probably familiar with the unpleasant onset of a cold. We "catch cold" when the cold virus enters our bodies, where it multiplies. Fortunately, at a certain point the virus begins to die. The algebraic expression $-0.75x^4 + 3x^3 + 5$ describes the billions of viral particles in our bodies after x days of invasion. The expression enables mathematicians to determine the day on which there is a maximum number of viral particles and, consequently, the day we feel sickest.

The algebraic expression $-0.75x^4 + 3x^3 + 5$ is an example of a polynomial. A **polynomial** is a single term or the sum of two or more terms containing variables with whole number exponents. This particular polynomial contains three terms. Equations containing polynomials are used in such diverse areas as science, business, medicine, psychology, and sociology. In this section, we review basic ideas about polynomials and their operations.

1 Understand the vocabulary of polynomials.

The Vocabulary of Polynomials

Consider the polynomial

$$7x^3 - 9x^2 + 13x - 6.$$

We can express this polynomial as

$$7x^3 + (-9x^2) + 13x + (-6).$$

The polynomial contains four terms. It is customary to write the terms in the order of descending powers of the variable. This is the **standard form** of a polynomial.

We begin this section by limiting our discussion to polynomials containing only one variable. Each term of a polynomial in x is of the form ax^n. The **degree** of ax^n is n. For example, the degree of the term $7x^3$ is 3.

> **The Degree of ax^n**
>
> If $a \neq 0$, the degree of ax^n is n. The degree of a nonzero constant is 0. The constant 0 has no defined degree.

Here is an example of a polynomial and the degree of each of its four terms:

$$6x^4 - 3x^3 + 2x - 5.$$

| degree 4 | degree 3 | degree 1 | degree of non-zero constant: 0 |

Notice that the exponent on x for the term $2x$ is understood to be 1: $2x^1$. For this reason, the degree of $2x$ is 1. You can think of -5 as $-5x^0$; thus, its degree is 0.

A polynomial which when simplified has exactly one term is called a **monomial.** A **binomial** is a simplified polynomial that has two terms, each with a different exponent. A **trinomial** is a simplified polynomial with three terms, each with a different exponent. Simplified polynomials with four or more terms have no special names.

The **degree of a polynomial** is the highest degree of all the terms of the polynomial. For example, $4x^2 + 3x$ is a binomial of degree 2 because the degree of the first term is 2, and the degree of the other term is less than 2. Also, $7x^5 - 2x^2 + 4$ is a trinomial of degree 5 because the degree of the first term is 5, and the degrees of the other terms are less than 5.

Up to now, we have used x to represent the variable in a polynomial. However, any letter can be used. For example,

- $7x^5 - 3x^3 + 8$ is a polynomial (in x) of degree 5.
- $6y^3 + 4y^2 - y + 3$ is a polynomial (in y) of degree 3.
- $z^7 + \sqrt{2}$ is a polynomial (in z) of degree 7.

Not every algebraic expression is a polynomial. Algebraic expressions whose variables do not contain whole number exponents such as

$$3x^{-2} + 7 \quad \text{and} \quad 5x^{3/2} + 9x^{1/2} + 2$$

are not polynomials. Furthermore, a quotient of polynomials such as

$$\frac{x^2 + 2x + 5}{x^3 - 7x^2 + 9x - 3}$$

is not a polynomial because the form of a polynomial involves only addition and subtraction of terms, not division.

We can tie together the threads of our discussion with the formal definition of a polynomial in one variable. In this definition, the coefficients of the terms are represented by a_n (read "a sub n"), a_{n-1} (read "a sub n minus 1"), a_{n-2}, and so on. The small letters to the lower right of each a are called **subscripts** and are *not exponents*. Subscripts are used to distinguish one constant from another when a large and undetermined number of such constants are needed.

> ### Definition of a Polynomial in x
>
> A **polynomial in x** is an algebraic expression of the form
> $$a_n x^n + a_{n-1} x^{n-1} + a_{n-2} x^{n-2} + \cdots + a_1 x + a_0,$$
> where $a_n, a_{n-1}, a_{n-2}, \ldots, a_1$, and a_0 are real numbers, $a_n \neq 0$, and n is a nonnegative integer. The polynomial is of **degree n**, a_n is the **leading coefficient,** and a_0 is the **constant term.**

2 Add and subtract polynomials.

Adding and Subtracting Polynomials

Polynomials are added and subtracted by combining like terms. For example, we can combine the monomials $-9x^3$ and $13x^3$ using addition as follows:

$$-9x^3 + 13x^3 = (-9 + 13)x^3 = 4x^3.$$

EXAMPLE 1   Adding and Subtracting Polynomials

Perform the indicated operations and simplify:

     a. $(-9x^3 + 7x^2 - 5x + 3) + (13x^3 + 2x^2 - 8x - 6)$
     b. $(7x^3 - 8x^2 + 9x - 6) - (2x^3 - 6x^2 - 3x + 9).$

Solution

     a. $(-9x^3 + 7x^2 - 5x + 3) + (13x^3 + 2x^2 - 8x - 6)$

$$\begin{aligned}
&= (-9x^3 + 13x^3) + (7x^2 + 2x^2) && \text{Group like terms.}\\
&\quad + (-5x - 8x) + (3 - 6) \\
&= 4x^3 + 9x^2 + (-13x) + (-3) && \text{Combine like terms.}\\
&= 4x^3 + 9x^2 - 13x - 3 && \text{Simplify.}
\end{aligned}$$

Study Tip

You can also arrange like terms in columns and combine vertically:

$$7x^3 - 8x^2 + 9x - 6$$
$$\underline{-2x^3 + 6x^2 + 3x - 9}$$
$$5x^3 - 2x^2 + 12x - 15$$

The like terms can be combined by adding their coefficients and keeping the same variable factor.

b. $(7x^3 - 8x^2 + 9x - 6) - (2x^3 - 6x^2 - 3x + 9)$

$\quad = (7x^3 - 8x^2 + 9x - 6) + (-2x^3 + 6x^2 + 3x - 9)$ Rewrite subtraction as addition of the additive inverse. Be sure to change the sign of each term inside parentheses preceded by the negative sign.

$\quad = (7x^3 - 2x^3) + (-8x^2 + 6x^2)$ Group like terms.
$\qquad + (9x + 3x) + (-6 - 9)$

$\quad = 5x^3 + (-2x^2) + 12x + (-15)$ Combine like terms.

$\quad = 5x^3 - 2x^2 + 12x - 15$ Simplify.

Check Point 1   Perform the indicated operations and simplify:

a. $(-17x^3 + 4x^2 - 11x - 5) + (16x^3 - 3x^2 + 3x - 15)$

b. $(13x^3 - 9x^2 - 7x + 1) - (-7x^3 + 2x^2 - 5x + 9)$.

3   Multiply polynomials.

Multiplying Polynomials

The product of two monomials is obtained by using properties of exponents. For example,

$$(-8x^6)(5x^3) = -8 \cdot 5x^{6+3} = -40x^9.$$

Multiply coefficients and add exponents.

Furthermore, we can use the distributive property to multiply a monomial and a polynomial that is not a monomial. For example,

$$3x^4(2x^3 - 7x + 3) = 3x^4 \cdot 2x^3 - 3x^4 \cdot 7x + 3x^4 \cdot 3 = 6x^7 - 21x^5 + 9x^4.$$

monomial   trinomial

How do we multiply two polynomials if neither is a monomial? For example, consider

$$(2x + 3)(x^2 + 4x + 5).$$

binomial   trinomial

One way to perform this multiplication is to distribute $2x$ throughout the trinomial

$$2x(x^2 + 4x + 5)$$

and 3 throughout the trinomial

$$3(x^2 + 4x + 5).$$

Then combine the like terms that result.

Multiplying Polynomials when Neither is a Monomial

Multiply each term of one polynomial by each term of the other polynomial. Then combine like terms.

EXAMPLE 2 Multiplying a Binomial and a Trinomial

Multiply: $(2x + 3)(x^2 + 4x + 5)$.

Solution

$(2x + 3)(x^2 + 4x + 5)$

$= 2x(x^2 + 4x + 5) + 3(x^2 + 4x + 5)$ Multiply the trinomial by each term of the binomial.

$= 2x \cdot x^2 + 2x \cdot 4x + 2x \cdot 5 + 3 \cdot x^2 + 3 \cdot 4x + 3 \cdot 5$ Use the distributive property.

$= 2x^3 + 8x^2 + 10x + 3x^2 + 12x + 15$ Multiply the monomials: multiply coefficients and add exponents.

$= 2x^3 + 11x^2 + 22x + 15$ Combine like terms: $8x^2 + 3x^2 = 11x^2$ and $10x + 12x = 22x$.

Another method for solving Example 2 is to use a vertical format similar to that used for multiplying whole numbers.

$$
\begin{array}{r}
x^2 + 4x + 5 \\
2x + 3 \\
\hline
3x^2 + 12x + 15 \\
2x^3 + 8x^2 + 10x \\
\hline
2x^3 + 11x^2 + 22x + 15
\end{array}
$$

Write like terms in the same column.

$3(x^2 + 4x + 5)$

$2x(x^2 + 4x + 5)$

Combine like terms.

Check Point 2 Multiply: $(5x - 2)(3x^2 - 5x + 4)$.

4 Use FOIL in polynomial multiplication.

The Product of Two Binomials: FOIL

Frequently we need to find the product of two binomials. We can use a method called FOIL, which is based on the distributive property, to do so. For example, we can find the product of the binomials $3x + 2$ and $4x + 5$ as follows:

$(3x + 2)(4x + 5) = 3x(4x + 5) + 2(4x + 5)$ First, distribute $3x$ over $4x + 5$. Then distribute 2.

$= 3x(4x) + 3x(5) + 2(4x) + 2(5)$

$= 12x^2 + 15x + 8x + 10.$

Two binomials can be quickly multiplied by using the FOIL method, in which F represents the product of the **first** terms in each binomial, O represents the product of the **outside** terms, I represents the product of the two **inside** terms, and L represents the product of the **last,** or second, terms in each binomial.

first   last   F   O   I   L

$(3x + 2)(4x + 5) = 12x^2 + 15x + 8x + 10$

inside

outside

$= 12x^2 + 23x + 10$ Combine like terms.

In general, here is how to use the FOIL method to find the product of $ax + b$ and $cx + d$:

Using the FOIL Method to Multiply Binomials

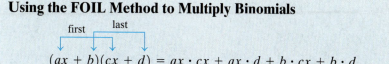

$$(ax + b)(cx + d) = ax \cdot cx + ax \cdot d + b \cdot cx + b \cdot d$$

Product of First terms	Product of Outside terms	Product of Inside terms	Product of Last terms

EXAMPLE 3 Using the FOIL Method

Multiply: $(3x + 4)(5x - 3)$.

Solution

$$(3x + 4)(5x - 3) = 3x \cdot 5x + 3x(-3) + 4 \cdot 5x + 4(-3)$$
$$= 15x^2 - 9x + 20x - 12$$
$$= 15x^2 + 11x - 12 \quad \textit{Combine like terms.}$$

Check Point 3 Multiply: $(7x - 5)(4x - 3)$.

5 Use special products in polynomial multiplication.

Multiplying the Sum and Difference of Two Terms

We can use the FOIL method to multiply $A + B$ and $A - B$ as follows:

$$(A + B)(A - B) = A^2 - AB + AB - B^2 = A^2 - B^2.$$

Notice that the outside and inside products have a sum of 0 and the terms cancel. The FOIL multiplication provides us with a quick rule for multiplying the sum and difference of two terms, referred to as a special-product formula.

The Product of the Sum and Difference of Two Terms

$$(A + B)(A - B) = A^2 - B^2$$

The product of the sum and the difference of the same two terms	is	the square of the first term minus the square of the second term.

EXAMPLE 4 **Finding the Product of the Sum and Difference of Two Terms**

Find each product:

a. $(4y + 3)(4y - 3)$ **b.** $(5a^4 + 6)(5a^4 - 6)$.

Solution Use the special-product formula shown.

$$(A + B)(A - B) = A^2 - B^2$$

First term squared   −   Second term squared   =   Product

a. $(4y + 3)(4y - 3) = (4y)^2 - 3^2 = 16y^2 - 9$

b. $(5a^4 + 6)(5a^4 - 6) = (5a^4)^2 - 6^2 = 25a^8 - 36$

Check Point 4 Find each product:

a. $(7x + 8)(7x - 8)$ **b.** $(2y^3 - 5)(2y^3 + 5)$.

The Square of a Binomial

Let us find $(A + B)^2$, the square of a binomial sum. To do so, we begin with the FOIL method and look for a general rule.

F   O   I   L

$$(A + B)^2 = (A + B)(A + B) = A \cdot A + A \cdot B + A \cdot B + B \cdot B$$
$$= A^2 + 2AB + B^2$$

This result implies the following rule, which is another example of a special-product formula:

Study Tip

Caution! The square of a sum is *not* the sum of the squares.

$(A + B)^2 \neq A^2 + B^2$

The middle term 2AB is missing

$(x + 3)^2 \neq x^2 + 9$

Incorrect

Show that $(x + 3)^2$ and $x^2 + 9$ are not equal by substituting 5 for x in each expression and simplifying.

The Square of a Binomial Sum

$$(A + B)^2 = A^2 + 2AB + B^2$$

The square of a binomial sum   is   first term squared   plus   2 times the product of the terms   plus   last term squared.

EXAMPLE 5 **Finding the Square of a Binomial Sum**

Square each binomial:

a. $(x + 3)^2$ **b.** $(3x + 7)^2$.

Solution Use the special-product formula shown.

$$(A + B)^2 = A^2 + 2AB + B^2$$

Square a Sum	(First Term)^2	+	2 · Product of the Terms	+	(Last Term)^2	= Product
a. $(x + 3)^2 =$	x^2	+	$2 \cdot x \cdot 3$	+	3^2	$= x^2 + 6x + 9$
b. $(3x + 7)^2 =$	$(3x)^2$	+	$2(3x)(7)$	+	7^2	$= 9x^2 + 42x + 49$

Check Point 5 Square each binomial:

    a. $(x + 10)^2$ **b.** $(5x + 4)^2$.

Using the FOIL method on $(A - B)^2$, the square of a binomial difference, we obtain the following rule:

The Square of a Binomial Difference

$$(A - B)^2 = A^2 - 2AB + B^2$$

| The square of a binomial difference | is | first term squared | minus | 2 times the product of the terms | plus | last term squared. |

EXAMPLE 6 Finding the Square of a Binomial Difference

Square each binomial:

    a. $(x - 4)^2$ **b.** $(5y - 6)^2$.

Solution Use the special-product formula shown.

$$(A - B)^2 = A^2 - 2AB + B^2$$

Square a Difference	(First Term)^2	−	2 · Product of the Terms	+	(Last Term)^2	= Product
a. $(x - 4)^2 =$	x^2	−	$2 \cdot x \cdot 4$	+	4^2	$= x^2 - 8x + 16$
b. $(5y - 6)^2 =$	$(5y)^2$	−	$2(5y)(6)$	+	6^2	$= 25y^2 - 60y + 36$

Check Point 6 Square each binomial:

    a. $(x - 9)^2$ **b.** $(7x - 3)^2$.

Special Products

There are several products that occur so frequently that it's convenient to memorize the form, or pattern, of these formulas.

Special Products

Let A and B represent real numbers, variables, or algebraic expressions.

Special Product	Example
Sum and Difference of Two Terms	
$(A + B)(A - B) = A^2 - B^2$	$(2x + 3)(2x - 3) = (2x)^2 - 3^2$
	$= 4x^2 - 9$
Squaring a Binomial	
$(A + B)^2 = A^2 + 2AB + B^2$	$(y + 5)^2 = y^2 + 2 \cdot y \cdot 5 + 5^2$
	$= y^2 + 10y + 25$
$(A - B)^2 = A^2 - 2AB + B^2$	$(3x - 4)^4$
	$= (3x)^2 - 2 \cdot 3x \cdot 4 + 4^2$
	$= 9x^2 - 24x + 16$
Cubing a Binomial	
$(A + B)^3 = A^3 + 3A^2B + 3AB^2 + B^3$	$(x + 4)^3$
	$= x^3 + 3x^2(4) + 3x(4)^2 + 4^3$
	$= x^3 + 12x^2 + 48x + 64$
$(A - B)^3 = A^3 - 3A^2B + 3AB^2 - B^3$	$(x - 2)^3$
	$= x^3 - 3x^2(2) + 3x(2)^2 - 2^3$
	$= x^3 - 6x^2 + 12x - 8$

Study Tip

Although it's convenient to memorize these forms, the FOIL method can be used on all five examples in the box. To cube $x + 4$, you can first square $x + 4$ using FOIL and then multiply this result by $x + 4$. In short, you do not necessarily have to utilize these special formulas. What is the advantage of knowing and using these forms?

6 Perform operations with polynomials in several variables.

Polynomials in Several Variables

The next time you visit the lumber yard and go rummaging through piles of wood, think *polynomials*, although polynomials a bit different from those we have encountered so far. The construction industry uses a polynomial in two variables to determine the number of board feet that can be manufactured from a tree with a diameter of x inches and a length of y feet. This polynomial is

$$\tfrac{1}{4}x^2y - 2xy + 4y.$$

In general, a **polynomial in two variables,** x and y, contains the sum of one or more monomials in the form $ax^n y^m$. The constant, a, is the **coefficient.** The exponents, n and m, represent whole numbers. The **degree** of the monomial $ax^n y^m$ is $n + m$. We'll use the polynomial from the construction industry to illustrate these ideas.

The coefficients are $\frac{1}{4}$, −2, and 4.

$$\tfrac{1}{4}x^2y \qquad -2xy \qquad +4y$$

| Degree of monomial: $2 + 1 = 3$ | Degree of monomial: $1 + 1 = 2$ | Degree of monomial: $0 + 1 = 1$ |

The **degree of a polynomial in two variables** is the highest degree of all its terms. For the preceding polynomial, the degree is 3.

Polynomials containing two or more variables can be added, subtracted, and multiplied just like polynomials that contain only one variable.

EXAMPLE 7 Subtracting Polynomials in Two Variables

Subtract as indicated:

$$(5x^3 - 9x^2y + 3xy^2 - 4) - (3x^3 - 6x^2y - 2xy^2 + 3).$$

Solution

$(5x^3 - 9x^2y + 3xy^2 - 4) - (3x^3 - 6x^2y - 2xy^2 + 3)$

$= (5x^3 - 9x^2y + 3xy^2 - 4) + (-3x^3 + 6x^2y + 2xy^2 - 3)$

Change the sign of each term in the second polynomial and add the two polynomials.

$= (5x^3 - 3x^3) + (-9x^2y + 6x^2y) + (3xy^2 + 2xy^2) + (-4 - 3)$

Group like terms.

$= 2x^3 - 3x^2y + 5xy^2 - 7$ Combine like terms by combining coefficients and keeping the same variable factors.

Check Point 7   Subtract: $(x^3 - 4x^2y + 5xy^2 - y^3) - (x^3 - 6x^2y + y^3).$

EXAMPLE 8 Multiplying Polynomials in Two Variables

Multiply: **a.** $(x + 4y)(3x - 5y)$ **b.** $(5x + 3y)^2.$

Solution   We will perform the multiplication in part (a) using the FOIL method. We will multiply in part (b) using the formula for the square of a binomial sum, $(A + B)^2.$

a. $(x + 4y)(3x - 5y)$ Multiply these binomials using the FOIL method.

     F   O   I   L

$= (x)(3x) + (x)(-5y) + (4y)(3x) + (4y)(-5y)$

$= 3x^2 - 5xy + 12xy - 20y^2$

$= 3x^2 + 7xy - 20y^2$ Combine like terms.

$$(A + B)^2 = A^2 + 2 \cdot A \cdot B + B^2$$

b. $(5x + 3y)^2 = (5x)^2 + 2(5x)(3y) + (3y)^2$

$= 25x^2 + 30xy + 9y^2$

Check Point 8   Multiply:

     a. $(7x - 6y)(3x - y)$ **b.** $(x^2 + 5y)^2.$

EXERCISE SET P.4

Practice Exercises

In Exercises 1–4, is the algebraic expression a polynomial? If it is, write the polynomial in standard form.

1. $2x + 3x^2 - 5$
2. $2x + 3x^{-1} - 5$
3. $\dfrac{2x + 3}{x}$
4. $x^2 - x^3 + x^4 - 5$

In Exercises 5–8, find the degree of the polynomial.

5. $3x^2 - 5x + 4$
6. $-4x^3 + 7x^2 - 11$
7. $x^2 - 4x^3 + 9x - 12x^4 + 63$
8. $x^2 - 8x^3 + 15x^4 + 91$

In Exercises 9–14, perform the indicated operations. Write the resulting polynomial in standard form and indicate its degree.

9. $(-6x^3 + 5x^2 - 8x + 9) + (17x^3 + 2x^2 - 4x - 13)$
10. $(-7x^3 + 6x^2 - 11x + 13) + (19x^3 - 11x^2 + 7x - 17)$
11. $(17x^3 - 5x^2 + 4x - 3) - (5x^3 - 9x^2 - 8x + 11)$
12. $(18x^4 - 2x^3 - 7x + 8) - (9x^4 - 6x^3 - 5x + 7)$
13. $(5x^2 - 7x - 8) + (2x^2 - 3x + 7) - (x^2 - 4x - 3)$
14. $(8x^2 + 7x - 5) - (3x^2 - 4x) - (-6x^3 - 5x^2 + 3)$

In Exercises 15–58, find each product.

15. $(x + 1)(x^2 - x + 1)$
16. $(x + 5)(x^2 - 5x + 25)$
17. $(2x - 3)(x^2 - 3x + 5)$
18. $(2x - 1)(x^2 - 4x + 3)$
19. $(x + 7)(x + 3)$
20. $(x + 8)(x + 5)$
21. $(x - 5)(x + 3)$
22. $(x - 1)(x + 2)$
23. $(3x + 5)(2x + 1)$
24. $(7x + 4)(3x + 1)$
25. $(2x - 3)(5x + 3)$
26. $(2x - 5)(7x + 2)$
27. $(5x^2 - 4)(3x^2 - 7)$
28. $(7x^2 - 2)(3x^2 - 5)$
29. $(8x^3 + 3)(x^2 - 5)$
30. $(7x^3 + 5)(x^2 - 2)$
31. $(x + 3)(x - 3)$
32. $(x + 5)(x - 5)$
33. $(3x + 2)(3x - 2)$
34. $(2x + 5)(2x - 5)$
35. $(5 - 7x)(5 + 7x)$
36. $(4 - 3x)(4 + 3x)$
37. $(4x^2 + 5x)(4x^2 - 5x)$
38. $(3x^2 + 4x)(3x^2 - 4x)$
39. $(1 - y^5)(1 + y^5)$
40. $(2 - y^5)(2 + y^5)$
41. $(x + 2)^2$
42. $(x + 5)^2$
43. $(2x + 3)^2$
44. $(3x + 2)^2$
45. $(x - 3)^2$
46. $(x - 4)^2$
47. $(4x^2 - 1)^2$
48. $(5x^2 - 3)^2$
49. $(7 - 2x)^2$
50. $(9 - 5x)^2$
51. $(x + 1)^3$
52. $(x + 2)^3$
53. $(2x + 3)^3$
54. $(3x + 4)^3$
55. $(x - 3)^3$
56. $(x - 1)^3$
57. $(3x - 4)^3$
58. $(2x - 3)^3$

In Exercises 59–66, perform the indicated operations. Indicate the degree of the resulting polynomial.

59. $(5x^2y - 3xy) + (2x^2y - xy)$
60. $(-2x^2y + xy) + (4x^2y + 7xy)$
61. $(4x^2y + 8xy + 11) + (-2x^2y + 5xy + 2)$
62. $(7x^4y^2 - 5x^2y^2 + 3xy) + (-18x^4y^2 - 6x^2y^2 - xy)$
63. $(x^3 + 7xy - 5y^2) - (6x^3 - xy + 4y^2)$
64. $(x^4 - 7xy - 5y^3) - (6x^4 - 3xy + 4y^3)$
65. $(3x^4y^2 + 5x^3y - 3y) - (2x^4y^2 - 3x^3y - 4y + 6x)$
66. $(5x^4y^2 + 6x^3y - 7y) - (3x^4y^2 - 5x^3y - 6y + 8x)$

In Exercises 67–82, find each product.

67. $(x + 5y)(7x + 3y)$
68. $(x + 9y)(6x + 7y)$
69. $(x - 3y)(2x + 7y)$
70. $(3x - y)(2x + 5y)$
71. $(3xy - 1)(5xy + 2)$
72. $(7x^2y + 1)(2x^2y - 3)$
73. $(7x + 5y)^2$
74. $(9x + 7y)^2$
75. $(x^2y^2 - 3)^2$
76. $(x^2y^2 - 5)^2$
77. $(x - y)(x^2 + xy + y^2)$
78. $(x + y)(x^2 - xy + y^2)$
79. $(3x + 5y)(3x - 5y)$
80. $(7x + 3y)(7x - 3y)$
81. $(7xy^2 - 10y)(7xy^2 + 10y)$
82. $(3xy^2 - 4y)(3xy^2 + 4y)$

Application Exercises

83. The polynomial $0.018x^2 - 0.757x + 9.047$ describes the amount, in thousands of dollars, that a person earning x thousand dollars a year feels underpaid. Evaluate the polynomial for $x = 40$. Describe what the answer means in practical terms.

84. The polynomial $104.5x^2 - 1501.5x + 6016$ describes the death rate per year, per 100,000 men, for men averaging x hours of sleep each night. Evaluate the polynomial for $x = 10$. Describe what the answer means in practical terms.

85. The polynomial $-1.45x^2 + 38.52x + 470.78$ describes the number of violent crimes in the United States, per 100,000 inhabitants, x years after 1975. Evaluate the polynomial for $x = 25$. Describe what the answer means in practical terms. How well does the polynomial describe the crime rate for the appropriate year shown in the bar graph?

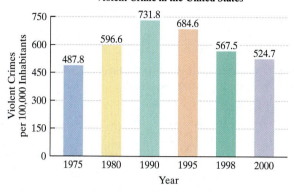

Violent Crime in the United States

Source: F.B.I.

86. The polynomial $-0.02A^2 + 2A + 22$ is used by coaches to get athletes fired up so that they can perform well. The polynomial represents the performance level related to various levels of enthusiasm, from $A = 1$ (almost no enthusiasm) to $A = 100$ (maximum level of enthusiasm). Evaluate the polynomial for $A = 20$, $A = 50$, and $A = 80$. Describe what happens to performance as we get more and more fired up.

87. The number of people who catch a cold t weeks after January 1 is $5t - 3t^2 + t^3$. The number of people who recover t weeks after January 1 is $t - t^2 + \frac{1}{3}t^3$. Write a polynomial in standard form for the number of people who are still ill with a cold t weeks after January 1.

88. The weekly cost, in thousands of dollars, for producing x stereo headphones is $30x + 50$. The weekly revenue, in thousands of dollars, for selling x stereo headphones is $90x^2 - x$. Write a polynomial in standard form for the weekly profit, in thousands of dollars, for producing and selling x stereo headphones.

In Exercises 89–90, write a polynomial in standard form that represents the area of the shaded region of each figure.

89.

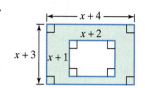

90.

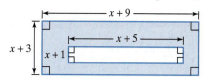

Writing in Mathematics

91. What is a polynomial in x?

92. Explain how to subtract polynomials.

93. Explain how to multiply two binomials using the FOIL method. Give an example with your explanation.

94. Explain how to find the product of the sum and difference of two terms. Give an example with your explanation.

95. Explain how to square a binomial difference. Give an example with your explanation.

96. Explain how to find the degree of a polynomial in two variables.

97. For Exercise 86, explain why performance levels do what they do as we get more and more fired up. If possible, describe an example of a time when you were too enthused and thus did poorly at something you were hoping to do well.

Technology Exercises

98. The common cold is caused by a rhinovirus. The polynomial

$$-0.75x^4 + 3x^3 + 5$$

describes the billions of viral particles in our bodies after x days of invasion. Use a calculator to find the number of viral particles after 0 days (the time of the cold's onset), 1 day, 2 days, 3 days, and 4 days. After how many days is the number of viral particles at a maximum and consequently the day we feel the sickest? By when should we feel completely better?

99. Using data from the National Institute on Drug Abuse, the polynomial

$$0.0032x^3 + 0.0235x^2 - 2.2477x + 61.1998$$

approximately describes the percentage of U.S. high school seniors in the class of x who had ever used marijuana, where x is the number of years after 1980. Use a calculator to find the percentage of high school seniors from the class of 1980 through the class of 2000 who had used marijuana. Round to the nearest tenth of a percent. Describe the trend in the data.

Critical Thinking Exercises

In Exercises 100–103, perform the indicated operations.

100. $(x - y)^2 - (x + y)^2$

101. $[(7x + 5) + 4y][(7x + 5) - 4y]$

102. $[(3x + y) + 1]^2$

103. $(x + y)(x - y)(x^2 + y^2)$

104. Express the area of the plane figure shown as a polynomial in standard form.

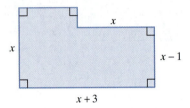

SECTION P.5   *Factoring Polynomials*

Objectives

1. Factor out the greatest common factor of a polynomial.
2. Factor by grouping.
3. Factor trinomials.
4. Factor the difference of squares.
5. Factor perfect square trinomials.
6. Factor the sum and difference of cubes.
7. Use a general strategy for factoring polynomials.
8. Factor algebraic expressions containing fractional and negative exponents.

A two-year-old boy is asked, "Do you have a brother?" He answers, "Yes." "What is your brother's name?" "Tom." Asked if Tom has a brother, the two-year-old replies, "No." The child can go in the direction from self to brother, but he cannot reverse this direction and move from brother back to self.

As our intellects develop, we learn to reverse the direction of our thinking. Reversibility of thought is found throughout algebra. For example, we can multiply polynomials and show that

$$(2x + 1)(3x - 2) = 6x^2 - x - 2.$$

We can also reverse this process and express the resulting polynomial as

$$6x^2 - x - 2 = (2x + 1)(3x - 2).$$

Factoring is the process of writing a polynomial as the product of two or more polynomials. The factors of $6x^2 - x - 2$ are $2x + 1$ and $3x - 2$.

In this section, we will be **factoring over the set of integers,** meaning that the coefficients in the factors are integers. Polynomials that cannot be factored using integer coefficients are called **irreducible over the integers,** or **prime.**

The goal in factoring a polynomial is to use one or more factoring techniques until each of the polynomial's factors is prime or irreducible. In this situation, the polynomial is said to be **factored completely.**

We will now discuss basic techniques for factoring polynomials.

① Factor out the greatest common factor of a polynomial.

Common Factors

In any factoring problem, the first step is to look for the *greatest common factor*. The **greatest common factor,** abbreviated GCF, is an expression of the highest degree that divides each term of the polynomial. The distributive property in the reverse direction

$$ab + ac = a(b + c)$$

can be used to factor out the greatest common factor.

EXAMPLE 1 Factoring out the Greatest Common Factor

Factor: **a.** $18x^3 + 27x^2$ **b.** $x^2(x + 3) + 5(x + 3)$.

Solution

Study Tip

The variable part of the greatest common factor always contains the *smallest* power of a variable or algebraic expression that appears in all terms of the polynomial.

a. We begin by determining the greatest common factor. 9 is the greatest integer that divides 18 and 27. Furthermore, x^2 is the greatest expression that divides x^3 and x^2. Thus, the greatest common factor of the two terms in the polynomial is $9x^2$.

$18x^3 + 27x^2$

$= 9x^2(2x) + 9x^2(3)$ Express each term as the product of the greatest common factor and its other factor.

$= 9x^2(2x + 3)$ Factor out the greatest common factor.

b. In this situation, the greatest common factor is the common binomial factor $(x + 3)$. We factor out this common factor as follows:

$x^2(x + 3) + 5(x + 3) = (x + 3)(x^2 + 5)$.   Factor out the common binomial factor.

Check Point 1   Factor:

a. $10x^3 - 4x^2$ **b.** $2x(x - 7) + 3(x - 7)$.

② Factor by grouping.

Factoring by Grouping

Some polynomials have only a greatest common factor of 1. However, by a suitable rearrangement of the terms, it still may be possible to factor. This process, called **factoring by grouping,** is illustrated in Example 2.

EXAMPLE 2 Factoring by Grouping

Factor: $x^3 + 4x^2 + 3x + 12$.

Solution   Group terms that have a common factor:

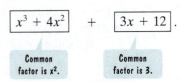

$$\boxed{x^3 + 4x^2} \quad + \quad \boxed{3x + 12}.$$

Common factor is x^2.　　Common factor is 3.

Discovery

In Example 2, group the terms as follows:

$(x^3 + 3x) + (4x^2 + 12)$.

Factor out the greatest common factor from each group and complete the factoring process. Describe what happens. What can you conclude?

We now factor the given polynomial as follows.

$x^3 + 4x^2 + 3x + 12$

$= (x^3 + 4x^2) + (3x + 12)$　　Group terms with common factors.

$= x^2(x + 4) + 3(x + 4)$　　Factor out the greatest common factor from the grouped terms. The remaining two terms have $x + 4$ as a common binomial factor.

$= (x + 4)(x^2 + 3)$　　Factor $(x + 4)$ out of both terms.

Thus, $x^3 + 4x^2 + 3x + 12 = (x + 4)(x^2 + 3)$. Check the factorization by multiplying the right side of the equation using the FOIL method. If the factorization is correct, you will obtain the original polynomial.

Check Point 2　　Factor: $x^3 + 5x^2 - 2x - 10$.

3 Factor trinomials.

Factoring Trinomials

To factor a trinomial of the form $ax^2 + bx + c$, a little trial and error may be necessary.

A Strategy for Factoring $ax^2 + bx + c$

(Assume, for the moment, that there is no greatest common factor.)

1. Find two **First** terms whose product is ax^2:

$$(\square x + \quad)(\square x + \quad) = ax^2 + bx + c.$$

2. Find two **Last** terms whose product is c:

$$(x + \square)(x + \square) = ax^2 + bx + c.$$

3. By trial and error, perform steps 1 and 2 until the sum of the **Outside** product and **Inside** product is bx:

$$(\square x + \square)(\square x + \square) = ax^2 + bx + c.$$

I

O

(sum of O + I)

If no such combinations exist, the polynomial is prime.

EXAMPLE 3   **Factoring Trinomials Whose Leading Coefficients Are 1**

Factor: **a.** $x^2 + 6x + 8$ **b.** $x^2 + 3x - 18$.

Solution

a. The factors of the first term are x and x :

$$x^2 + 6x + 8 = (x \quad)(x \quad).$$

Factors of 8	8, 1	4, 2	−8, −1	−4, −2
Sum of Factors	9	6	−9	−6

This is the desired sum.

To find the second term of each factor, we must find two numbers whose product is 8 and whose sum is 6. From the table in the margin, we see that 4 and 2 are the required integers. Thus,

$$x^2 + 6x + 8 = (x + 4)(x + 2) \text{ or } (x + 2)(x + 4).$$

b. We begin with

$$x^2 + 3x - 18 = (x \quad)(x \quad).$$

Factors of −18	18, −1	−18, 1	9, −2	−9, 2	6, −3	−6, 3
Sum of factors	17	−17	7	−7	3	−3

This is the desired sum.

To find the second term of each factor, we must find two numbers whose product is −18 and whose sum is 3. From the table in the margin, we see that 6 and −3 are the required integers. Thus,

$$x^2 + 3x - 18 = (x + 6)(x - 3)$$
$$\text{or} \quad (x - 3)(x + 6).$$

Check Point 3   Factor:

a. $x^2 + 13x + 40$ **b.** $x^2 - 5x - 14$.

EXAMPLE 4   **Factoring a Trinomial Whose Leading Coefficient Is Not 1**

Factor: $8x^2 - 10x - 3$.

Solution

Step 1   **Find two *First* terms whose product is $8x^2$.**

$$8x^2 - 10x - 3 \stackrel{?}{=} (8x \quad)(x \quad)$$
$$8x^2 - 10x - 3 \stackrel{?}{=} (4x \quad)(2x \quad)$$

Step 2   **Find two *Last* terms whose product is −3.**   The possible factorizations are $1(-3)$ and $-1(3)$.

Step 3   **Try various combinations of these factors.**   The correct factorization of $8x^2 - 10x - 3$ is the one in which the sum of the *O*utside and *I*nside products is equal to $-10x$. Here is a list of the possible factorizations:

Possible Factorizations of $8x^2 - 10x - 3$	Sum of *Outside* and *Inside* Products (Should Equal $-10x$)
$(8x + 1)(x - 3)$	$-24x + x = -23x$
$(8x - 3)(x + 1)$	$8x - 3x = 5x$
$(8x - 1)(x + 3)$	$24x - x = 23x$
$(8x + 3)(x - 1)$	$-8x + 3x = -5x$
$(4x + 1)(2x - 3)$	$-12x + 2x = -10x$
$(4x - 3)(2x + 1)$	$4x - 6x = -2x$
$(4x - 1)(2x + 3)$	$12x - 2x = 10x$
$(4x + 3)(2x - 1)$	$-4x + 6x = 2x$

This is the required middle term.

Thus,
$$8x^2 - 10x - 3 = (4x + 1)(2x - 3) \quad \text{or} \quad (2x - 3)(4x + 1).$$

Show that this factorization is correct by multiplying the factors using the FOIL method. You should obtain the original trinomial.

Check Point 4   Factor: $6x^2 + 19x - 7$.

4   Factor the difference of squares.

Factoring the Difference of Two Squares

A method for factoring the difference of two squares is obtained by reversing the special product for the sum and difference of two terms.

> ### The Difference of Two Squares
> If A and B are real numbers, variables, or algebraic expressions, then
> $$A^2 - B^2 = (A + B)(A - B).$$
> In words: The difference of the squares of two terms factors as the product of a sum and a difference of those terms.

EXAMPLE 5   Factoring the Difference of Two Squares

Factor: **a.** $x^2 - 4$ **b.** $81x^2 - 49$.

Solution   We must express each term as the square of some monomial. Then we use the formula for factoring $A^2 - B^2$.

a. $x^2 - 4 = x^2 - 2^2 = (x + 2)\ (x - 2)$

$$A^2 \ - \ B^2 \ = \ (A \ + \ B)\ (A \ - \ B)$$

b. $81x^2 - 49 = (9x)^2 - 7^2 = (9x + 7)(9x - 7)$

Check Point 5   Factor:

   a. $x^2 - 81$ **b.** $36x^2 - 25$.

We have seen that a polynomial is factored completely when it is written as the product of prime polynomials. To be sure that you have factored completely, check to see whether the factors can be factored.

EXAMPLE 6   A Repeated Factorization

Factor completely: $x^4 - 81$.

Solution

$$
\begin{aligned}
x^4 - 81 &= (x^2)^2 - 9^2 \\
&= (x^2 + 9)(x^2 - 9)
\end{aligned}
$$

Express as the difference of two squares.

The factors are the sum and difference of the squared terms.

Study Tip

Factoring $x^4 - 81$ as
$$(x^2 + 9)(x^2 - 9)$$
is not a complete factorization. The second factor, $x^2 - 9$, is itself a difference of two squares and can be factored.

$$= (x^2 + 9)(x^2 - 3^2) \qquad$$ The factor $x^2 - 9$ is the difference of two squares and can be factored.

$$= (x^2 + 9)(x + 3)(x - 3) \qquad$$ The factors of $x^2 - 9$ are the sum and difference of the squared terms.

Check Point 6 Factor completely: $81x^4 - 16$.

5 Factor perfect square trinomials.

Factoring Perfect Square Trinomials

Our next factoring technique is obtained by reversing the special products for squaring binomials. The trinomials that are factored using this technique are called **perfect square trinomials.**

> ### Factoring Perfect Square Trinomials
> Let A and B be real numbers, variables, or algebraic expressions.
> **1.** $A^2 + 2AB + B^2 = (A + B)^2$
>
> Same sign
>
> **2.** $A^2 - 2AB + B^2 = (A - B)^2$
>
> Same sign

The two items in the box show that perfect square trinomials come in two forms: one in which the middle term is positive and one in which the middle term is negative. Here's how to recognize a perfect square trinomial:

1. The first and last terms are squares of monomials or integers.

2. The middle term is twice the product of the expressions being squared in the first and last terms.

EXAMPLE 7   Factoring Perfect Square Trinomials

Factor: **a.** $x^2 + 6x + 9$ **b.** $25x^2 - 60x + 36$.

Solution

a. $x^2 + 6x + 9 = x^2 + 2 \cdot x \cdot 3 + 3^2 = (x + 3)^2$ The middle term has a positive sign.

$$A^2 + 2AB + B^2 = (A + B)^2$$

b. We suspect that $25x^2 - 60x + 36$ is a perfect square trinomial because $25x^2 = (5x)^2$ and $36 = 6^2$. The middle term can be expressed as twice the product of $5x$ and 6.

$$25x^2 - 60x + 36 = (5x)^2 - 2 \cdot 5x \cdot 6 + 6^2 = (5x - 6)^2$$

$$A^2 - 2AB + B^2 = (A - B)^2$$

Check Point 7   Factor:

  a. $x^2 + 14x + 49$ **b.** $16x^2 - 56x + 49$.

6 Factor the sum and difference of cubes.

Factoring the Sum and Difference of Two Cubes

We can use the following formulas to factor the sum or the difference of two cubes:

> ### Factoring the Sum and Difference of Two Cubes
>
> **1.** Factoring the Sum of Two Cubes
> $$A^3 + B^3 = (A + B)(A^2 - AB + B^2)$$
>
> **2.** Factoring the Difference of Two Cubes
> $$A^3 - B^3 = (A - B)(A^2 + AB + B^2)$$

EXAMPLE 8   Factoring Sums and Differences of Two Cubes

Factor: **a.** $x^3 + 8$ **b.** $64x^3 - 125$.

Solution

  a. $x^3 + 8 = x^3 + 2^3 = (x + 2)(x^2 - x \cdot 2 + 2^2) = (x + 2)(x^2 - 2x + 4)$

$$A^3 + B^3 = (A + B)(A^2 - AB + B^2)$$

  b. $64x^3 - 125 = (4x)^3 - 5^3 = (4x - 5)\left[(4x)^2 + (4x)(5) + 5^2\right]$

$$A^3 - B^3 = (A - B)(A^2 + AB + B^2)$$

$$= (4x - 5)(16x^2 + 20x + 25)$$

Check Point 8   Factor:

  a. $x^3 + 1$ **b.** $125x^3 - 8$.

7 Use a general strategy for factoring polynomials.

A Strategy for Factoring Polynomials

It is important to practice factoring a wide variety of polynomials so that you can quickly select the appropriate technique. The polynomial is factored completely when all its polynomial factors, except possibly for monomial factors, are prime. Because of the commutative property, the order of the factors does not matter.

A Strategy for Factoring a Polynomial

1. If there is a common factor, factor out the GCF.

2. Determine the number of terms in the polynomial and try factoring as follows:

a. If there are two terms, can the binomial be factored by one of the following special forms?

Difference of two squares: $A^2 - B^2 = (A + B)(A - B)$

Sum of two cubes: $A^3 + B^3 = (A + B)(A^2 - AB + B^2)$

Difference of two cubes: $A^3 - B^3 = (A - B)(A^2 + AB + B^2)$

b. If there are three terms, is the trinomial a perfect square trinomial? If so, factor by one of the following special forms:

$$A^2 + 2AB + B^2 = (A + B)^2$$
$$A^2 - 2AB + B^2 = (A - B)^2.$$

If the trinomial is not a perfect square trinomial, try factoring by trial and error.

c. If there are four or more terms, try factoring by grouping.

3. Check to see if any factors with more than one term in the factored polynomial can be factored further. If so, factor completely.

EXAMPLE 9   Factoring a Polynomial

Factor: $2x^3 + 8x^2 + 8x$.

Solution

Step 1   If there is a common factor, factor out the GCF. Because $2x$ is common to all terms, we factor it out.

$$2x^3 + 8x^2 + 8x = 2x(x^2 + 4x + 4) \qquad \textit{Factor out the GCF.}$$

Step 2   Determine the number of terms and factor accordingly. The factor $x^2 + 4x + 4$ has three terms and is a perfect square trinomial. We factor using $A^2 + 2AB + B^2 = (A + B)^2$.

$$2x^3 + 8x^2 + 8x = 2x(x^2 + 4x + 4)$$

$$= 2x(x^2 + 2 \cdot x \cdot 2 + 2^2)$$

$$\underbrace{}$$
$$\text{A}^2 \; + \; 2 \, \text{A} \, \text{B} \; + \; \text{B}^2$$

$$= 2x(x + 2)^2 \qquad \qquad \textit{A}^2 + 2AB + B^2 = (A + B)^2$$

Step 3   Check to see if factors can be factored further. In this problem, they cannot. Thus,

$$2x^3 + 8x^2 + 8x = 2x(x + 2)^2.$$

Check Point 9   Factor: $3x^3 - 30x^2 + 75x$.

EXAMPLE 10 Factoring a Polynomial

Factor: $x^2 - 25a^2 + 8x + 16$.

Solution

Step 1 If there is a common factor, factor out the GCF. Other than 1 or –1, there is no common factor.

Step 2 Determine the number of terms and factor accordingly. There are four terms. We try factoring by grouping. Grouping into two groups of two terms does not result in a common binomial factor. Let's try grouping as a difference of squares.

$$x^2 - 25a^2 + 8x + 16$$
$$= (x^2 + 8x + 16) - 25a^2 \qquad \text{Rearrange terms and group as a perfect square trinomial minus } 25a^2 \text{ to obtain a difference of squares.}$$

$$= (x + 4)^2 - (5a)^2 \qquad \text{Factor the perfect square trinomial.}$$
$$= (x + 4 + 5a)(x + 4 - 5a) \qquad \text{Factor the difference of squares. The factors are the sum and difference of the expressions being squared.}$$

Step 3 Check to see if factors can be factored further. In this case, they cannot, so we have factored completely.

Check Point 10 Factor: $x^2 - 36a^2 + 20x + 100$.

---

8 Factor algebraic expressions containing fractional and negative exponents.

Factoring Algebraic Expressions Containing Fractional and Negative Exponents

Although expressions containing fractional and negative exponents are not polynomials, they can be simplified using factoring techniques.

EXAMPLE 11 Factoring Involving Fractional and Negative Exponents

Factor and simplify: $x(x + 1)^{-3/4} + (x + 1)^{1/4}$.

Solution The greatest common factor is $x + 1$ with the *smallest exponent* in the two terms. Thus, the greatest common factor is $(x + 1)^{-3/4}$.

$$x(x + 1)^{-3/4} + (x + 1)^{1/4}$$

$$= (x + 1)^{-3/4}x + (x + 1)^{-3/4}(x + 1) \qquad \text{Express each term as the product of the greatest common factor and its other factor.}$$

$$= (x + 1)^{-3/4}[x + (x + 1)] \qquad \text{Factor out the greatest common factor.}$$

$$= \frac{2x + 1}{(x + 1)^{3/4}} \qquad\qquad b^{-n} = \frac{1}{b^n}$$

Check Point 11   Factor and simplify: $x(x - 1)^{-1/2} + (x - 1)^{1/2}$.

EXERCISE SET P.5

Practice Exercises

In Exercises 1–10, factor out the greatest common factor.

1. $18x + 27$ **2.** $16x - 24$

3. $3x^2 + 6x$ **4.** $4x^2 - 8x$

5. $9x^4 - 18x^3 + 27x^2$ **6.** $6x^4 - 18x^3 + 12x^2$

7. $x(x + 5) + 3(x + 5)$ **8.** $x(2x + 1) + 4(2x + 1)$

9. $x^2(x - 3) + 12(x - 3)$ **10.** $x^2(2x + 5) + 17(2x + 5)$

In Exercises 11–16, factor by grouping.

11. $x^3 - 2x^2 + 5x - 10$ **12.** $x^3 - 3x^2 + 4x - 12$

13. $x^3 - x^2 + 2x - 2$ **14.** $x^3 + 6x^2 - 2x - 12$

15. $3x^3 - 2x^2 - 6x + 4$ **16.** $x^3 - x^2 - 5x + 5$

In Exercises 17–30, factor each trinomial, or state that the trinomial is prime.

17. $x^2 + 5x + 6$ **18.** $x^2 + 8x + 15$

19. $x^2 - 2x - 15$ **20.** $x^2 - 4x - 5$

21. $x^2 - 8x + 15$ **22.** $x^2 - 14x + 45$

23. $3x^2 - x - 2$ **24.** $2x^2 + 5x - 3$

25. $3x^2 - 25x - 28$ **26.** $3x^2 - 2x - 5$

27. $6x^2 - 11x + 4$ **28.** $6x^2 - 17x + 12$

29. $4x^2 + 16x + 15$ **30.** $8x^2 + 33x + 4$

In Exercises 31–40, factor the difference of two squares.

31. $x^2 - 100$ **32.** $x^2 - 144$

33. $36x^2 - 49$ **34.** $64x^2 - 81$

35. $9x^2 - 25y^2$ **36.** $36x^2 - 49y^2$

37. $x^4 - 16$ **38.** $x^4 - 1$

39. $16x^4 - 81$ **40.** $81x^4 - 1$

In Exercises 41–48, factor any perfect square trinomials, or state that the polynomial is prime.

41. $x^2 + 2x + 1$ **42.** $x^2 + 4x + 4$

43. $x^2 - 14x + 49$ **44.** $x^2 - 10x + 25$

45. $4x^2 + 4x + 1$ **46.** $25x^2 + 10x + 1$

47. $9x^2 - 6x + 1$ **48.** $64x^2 - 16x + 1$

In Exercises 49–56, factor using the formula for the sum or difference of two cubes.

49. $x^3 + 27$ **50.** $x^3 + 64$

51. $x^3 - 64$ **52.** $x^3 - 27$

53. $8x^3 - 1$ **54.** $27x^3 - 1$

55. $64x^3 + 27$ **56.** $8x^3 + 125$

In Exercises 57–84, factor completely, or state that the polynomial is prime.

57. $3x^3 - 3x$ **58.** $5x^3 - 45x$

59. $4x^2 - 4x - 24$ **60.** $6x^2 - 18x - 60$

61. $2x^4 - 162$ **62.** $7x^4 - 7$

63. $x^3 + 2x^2 - 9x - 18$ **64.** $x^3 + 3x^2 - 25x - 75$

65. $2x^2 - 2x - 112$ **66.** $6x^2 - 6x - 12$

67. $x^3 - 4x$ **68.** $9x^3 - 9x$

69. $x^2 + 64$ **70.** $x^2 + 36$

71. $x^3 + 2x^2 - 4x - 8$ **72.** $x^3 + 2x^2 - x - 2$

73. $y^5 - 81y$ **74.** $y^5 - 16y$

75. $20y^4 - 45y^2$ **76.** $48y^4 - 3y^2$

77. $x^2 - 12x + 36 - 49y^2$ **78.** $x^2 - 10x + 25 - 36y^2$

79. $9b^2x - 16y - 16x + 9b^2y$

80. $16a^2x - 25y - 25x + 16a^2y$

81. $x^2y - 16y + 32 - 2x^2$ **82.** $12x^2y - 27y - 4x^2 + 9$

83. $2x^3 - 8a^2x + 24x^2 + 72x$

84. $2x^3 - 98a^2x + 28x^2 + 98x$

In Exercises 85–94, factor and simplify each algebraic expression.

85. $x^{3/2} - x^{1/2}$ **86.** $x^{3/4} - x^{1/4}$

87. $4x^{-2/3} + 8x^{1/3}$ **88.** $12x^{-3/4} + 6x^{1/4}$

89. $(x + 3)^{1/2} - (x + 3)^{3/2}$

90. $(x^2 + 4)^{3/2} + (x^2 + 4)^{7/2}$

91. $(x + 5)^{-1/2} - (x + 5)^{-3/2}$

92. $(x^2 + 3)^{-2/3} + (x^2 + 3)^{-5/3}$

93. $(4x - 1)^{1/2} - \frac{1}{3}(4x - 1)^{3/2}$

94. $-8(4x + 3)^{-2} + 10(5x + 1)(4x + 3)^{-1}$

Application Exercises

95. Your computer store is having an incredible sale. The price on one model is reduced by 40%. Then the sale price is reduced by another 40%. If x is the computer's original price, the sale price can be represented by

$$(x - 0.4x) - 0.4(x - 0.4x).$$

a. Factor out $(x - 0.4x)$ from each term. Then simplify the resulting expression.

b. Use the simplified expression from part (a) to answer these questions: With a 40% reduction followed by a 40% reduction, is the computer selling at 20% of its original price? If not, at what percentage of the original price is it selling?

96. The polynomial $8x^2 + 20x + 2488$ describes the number, in thousands, of high school graduates in the United States x years after 1993.

    a. According to this polynomial, how many students will graduate from U.S. high schools in 2003?

    b. Factor the polynomial.

    c. Use the factored form of the polynomial in part (b) to find the number of high school graduates in 2003. Do you get the same answer as you did in part (a)? If so, does this prove that your factorization is correct? Explain.

97. A rock is dropped from the top of a 256-foot cliff. The height, in feet, of the rock above the water after t seconds is described by the polynomial $256 - 16t^2$. Factor this expression completely.

98. The amount of sheet metal needed to manufacture a cylindrical tin can, that is, its surface area, S, is $S = 2\pi r^2 + 2\pi rh$. Express the surface area, S, in factored form.

In Exercises 99–100, find the formula for the area of the shaded region and express it in factored form.

99. 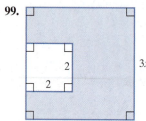   **100.**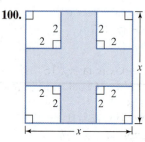

Writing in Mathematics

101. Using an example, explain how to factor out the greatest common factor of a polynomial.

102. Suppose that a polynomial contains four terms. Explain how to use factoring by grouping to factor the polynomial.

103. Explain how to factor $3x^2 + 10x + 8$.

104. Explain how to factor the difference of two squares. Provide an example with your explanation.

105. What is a perfect square trinomial and how is it factored?

106. Explain how to factor $x^3 + 1$.

107. What does it mean to factor completely?

Critical Thinking Exercises

108. Which one of the following is true?

    a. Because $x^2 + 1$ is irreducible over the integers, it follows that $x^3 + 1$ is also irreducible.

    b. One correct factored form for $x^2 - 4x + 3$ is $x(x - 4) + 3$.

    c. $x^3 - 64 = (x - 4)^3$

    d. None of the above is true.

In Exercises 109–112, factor completely.

109. $x^{2n} + 6x^n + 8$ **110.** $-x^2 - 4x + 5$

111. $x^4 - y^4 - 2x^3y + 2xy^3$

112. $(x - 5)^{-1/2}(x + 5)^{-1/2} - (x + 5)^{1/2}(x - 5)^{-3/2}$

In Exercises 113–114, find all integers b so that the trinomial can be factored.

113. $x^2 + bx + 15$ **114.** $x^2 + 4x + b$

Group Exercise

115. Without looking at any factoring problems in the book, create five factoring problems. Make sure that some of your problems require at least two factoring techniques. Next, exchange problems with another person in your group. Work to factor your partner's problems. Evaluate the problems as you work: Are they too easy? Too difficult? Can the polynomials really be factored? Share your response with the person who wrote the problems. Finally, grade each other's work in factoring the polynomials. Each factoring problem is worth 20 points. You may award partial credit. If you take off points, explain why points are deducted and how you decided to take off a particular number of points for the error(s) that you found.

SECTION P.6   *Rational Expressions*

Objectives

1. Specify numbers that must be excluded from the domain of rational expressions.
2. Simplify rational expressions.
3. Multiply rational expressions.
4. Divide rational expressions.
5. Add and subtract rational expressions.
6. Simplify complex rational expressions.

How do we describe the costs of reducing environmental pollution? We often use algebraic expressions involving quotients of polynomials. For example, the algebraic expression

$$\frac{250x}{100 - x}$$

describes the cost, in millions of dollars, to remove x percent of the pollutants that are discharged into a river. Removing a modest percentage of pollutants, say 40%, is far less costly than removing a substantially greater percentage, such as 95%. We see this by evaluating the algebraic expression for $x = 40$ and $x = 95$.

Evaluating $\dfrac{250x}{100 - x}$ for

$x = 40$:	$x = 95$:
Cost is $\dfrac{250(40)}{100 - 40} \approx 167.$	Cost is $\dfrac{250(95)}{100 - 95} = 4750.$

Discovery

What happens if you try substituting 100 for x in

$$\frac{250x}{100 - x}?$$

What does this tell you about the cost of cleaning up all of the river's pollutants?

The cost increases from approximately $167 million to a possibly prohibitive $4750 million, or $4.75 billion. Costs spiral upward as the percentage of removed pollutants increases.

Many algebraic expressions that describe costs of environmental projects are examples of rational expressions. First we will define rational expressions. Then we will review how to perform operations with such expressions.

1 Specify numbers that must be excluded from the domain of rational expressions.

Rational Expressions

A **rational expression** is the quotient of two polynomials. Some examples are

$$\frac{x - 2}{4}, \quad \frac{4}{x - 2}, \quad \frac{x}{x^2 - 1}, \quad \text{and} \quad \frac{x^2 + 1}{x^2 + 2x - 3}.$$

The set of real numbers for which an algebraic expression is defined is the **domain** of the expression. Because rational expressions indicate division and division by zero is undefined, we must exclude numbers from a rational expression's domain that make the denominator zero.

EXAMPLE 1 Excluding Numbers from the Domain

Find all the numbers that must be excluded from the domain of each rational expression:

a. $\dfrac{4}{x-2}$ **b.** $\dfrac{x}{x^2-1}$.

Solution To determine the numbers that must be excluded from each domain, examine the denominators.

a. $\dfrac{4}{x-2}$ **b.** $\dfrac{x}{x^2-1} = \dfrac{x}{(x+1)(x-1)}$

| This denominator would equal zero if x = 2. | This factor would equal zero if x = −1. | This factor would equal zero if x = 1. |

For the rational expression in part (a), we must exclude 2 from the domain. For the rational expression in part (b), we must exclude both −1 and 1 from the domain. These excluded numbers are often written to the right of a rational expression.

$$\frac{4}{x-2}, x \neq 2 \qquad \frac{x}{x^2-1}, x \neq -1, x \neq 1$$

Check Point 1 Find all the numbers that must be excluded from the domain of each rational expression:

a. $\dfrac{7}{x+5}$ **b.** $\dfrac{x}{x^2-36}$.

2 Simplify rational expressions.

Simplifying Rational Expressions

A rational expression is **simplified** if its numerator and denominator have no common factors other than 1 or −1. The following procedure can be used to simplify rational expressions:

> **Simplifying Rational Expressions**
> **1.** Factor the numerator and denominator completely.
> **2.** Divide both the numerator and denominator by the common factors.

EXAMPLE 2 Simplifying Rational Expressions

Simplify: **a.** $\dfrac{x^3+x^2}{x+1}$ **b.** $\dfrac{x^2+6x+5}{x^2-25}$.

Solution

a. $\dfrac{x^3 + x^2}{x + 1} = \dfrac{x^2(x + 1)}{x + 1}$ *Factor the numerator. Because the denominator is x + 1, x ≠ −1.*

$= \dfrac{x^2 \cancel{(x + 1)}^{\,1}}{\cancel{x + 1}_{\,1}}$ *Divide out the common factor, x + 1.*

$= x^2, x \neq -1$ *Denominators of 1 need not be written because $\frac{a}{1} = a$.*

b. $\dfrac{x^2 + 6x + 5}{x^2 - 25} = \dfrac{(x + 5)(x + 1)}{(x + 5)(x - 5)}$ *Factor the numerator and denominator. Because the denominator is $(x + 5)(x - 5)$, x ≠ −5 and x ≠ 5.*

$= \dfrac{\cancel{(x + 5)}^{\,1}(x + 1)}{\cancel{(x + 5)}_{\,1}(x - 5)}$ *Divide out the common factor, x + 5.*

$= \dfrac{x + 1}{x - 5}, \quad x \neq -5, \quad x \neq 5$

Check Point 2   Simplify:

a. $\dfrac{x^3 + 3x^2}{x + 3}$ **b.** $\dfrac{x^2 - 1}{x^2 + 2x + 1}$.

3 Multiply rational expressions.

Multiplying Rational Expressions

The product of two rational expressions is the product of their numerators divided by the product of their denominators. Here is a step-by-step procedure for multiplying rational expressions:

> **Multiplying Rational Expressions**
>
> **1.** Factor all numerators and denominators completely.
> **2.** Divide numerators and denominators by common factors.
> **3.** Multiply the remaining factors in the numerator and multiply the remaining factors in the denominator.

EXAMPLE 3   **Multiplying Rational Expressions**

Multiply and simplify:

$$\frac{x - 7}{x - 1} \cdot \frac{x^2 - 1}{3x - 21}.$$

Solution

$$\frac{x - 7}{x - 1} \cdot \frac{x^2 - 1}{3x - 21}$$

This is the given multiplication problem.

$$= \frac{x - 7}{x - 1} \cdot \frac{(x + 1)(x - 1)}{3(x - 7)}$$

Factor all numerators and denominators. Because the denominator has factors of $x - 1$ and $x - 7$, $x \neq 1$ and $x \neq 7$.

$$= \frac{\overset{1}{\cancel{x - 7}}}{\underset{1}{\cancel{x - 1}}} \cdot \frac{(x + 1)\overset{1}{\cancel{(x - 1)}}}{3\underset{1}{\cancel{(x - 7)}}}$$

Divide numerators and denominators by common factors.

$$= \frac{x + 1}{3}, x \neq 1, x \neq 7$$

Multiply the remaining factors in the numerator and denominator.

> These excluded numbers from the domain must also be excluded from the simplified expression's domain.

Check Point 3 Multiply and simplify:

$$\frac{x + 3}{x^2 - 4} \cdot \frac{x^2 - x - 6}{x^2 + 6x + 9}.$$

4 Divide rational expressions.

Dividing Rational Expressions

We find the quotient of two rational expressions by inverting the divisor and multiplying.

EXAMPLE 4 Dividing Rational Expressions

Divide and simplify:

$$\frac{x^2 - 2x - 8}{x^2 - 9} \div \frac{x - 4}{x + 3}.$$

Solution

$$\frac{x^2 - 2x - 8}{x^2 - 9} \div \frac{x - 4}{x + 3}$$

This is the given division problem.

$$= \frac{x^2 - 2x - 8}{x^2 - 9} \cdot \frac{x + 3}{x - 4}$$

Invert the divisor and multiply.

$$= \frac{(x - 4)(x + 2)}{(x + 3)(x - 3)} \cdot \frac{x + 3}{x - 4}$$

Factor throughout. For nonzero denominators, $x \neq -3$, $x \neq 3$, and $x \neq 4$.

$$= \frac{\overset{1}{\cancel{(x - 4)}}(x + 2)}{\underset{1}{\cancel{(x + 3)}}(x - 3)} \cdot \frac{\overset{1}{\cancel{(x + 3)}}}{\underset{1}{\cancel{(x - 4)}}}$$

Divide numerators and denominators by common factors.

$$= \frac{x + 2}{x - 3}, x \neq -3, x \neq 3, x \neq 4$$

Multiply the remaining factors in the numerator and the denominator.

Check Point 4 Divide and simplify:

$$\frac{x^2 - 2x + 1}{x^3 + x} \div \frac{x^2 + x - 2}{3x^2 + 3}.$$

5 Add and subtract rational expressions.

Adding and Subtracting Rational Expressions with the Same Denominator

We add or subtract rational expressions with the same denominator by (1) adding or subtracting the numerators, (2) placing this result over the common denominator, and (3) simplifying, if possible.

EXAMPLE 5 Subtracting Rational Expressions with the Same Denominator

Subtract: $\dfrac{5x + 1}{x^2 - 9} - \dfrac{4x - 2}{x^2 - 9}$.

Study Tip

Example 5 shows that when a numerator is being subtracted, we must subtract every term in that expression.

Solution

$$\frac{5x + 1}{x^2 - 9} - \frac{4x - 2}{x^2 - 9} = \frac{5x + 1 - (4x - 2)}{x^2 - 9}$$

Subtract numerators and include parentheses to indicate that both terms are subtracted. Place this difference over the common denominator.

$$= \frac{5x + 1 - 4x + 2}{x^2 - 9}$$

Remove parentheses and then change the sign of each term.

$$= \frac{x + 3}{x^2 - 9}$$

Combine like terms.

$$= \frac{\overset{1}{\cancel{x + 3}}}{\underset{1}{\cancel{(x + 3)}}(x - 3)}$$

Factor and simplify ($x \neq -3$ and $x \neq 3$).

$$= \frac{1}{x - 3}, x \neq -3, x \neq 3$$

Check Point 5   Subtract: $\dfrac{x}{x + 1} - \dfrac{3x + 2}{x + 1}$.

Adding and Subtracting Rational Expressions with Different Denominators

Rational expressions that have no common factors in their denominators can be added or subtracted using one of the following properties:

$$\frac{a}{b} + \frac{c}{d} = \frac{ad + bc}{bd} \qquad \frac{a}{b} - \frac{c}{d} = \frac{ad - bc}{bd}, b \neq 0, d \neq 0.$$

The denominator, bd, is the product of the factors in the two denominators. Because we are looking at rational expressions that have no common factors in their denominators, the product bd gives the least common denominator.

EXAMPLE 6 Subtracting Rational Expressions Having No Common Factors in Their Denominators

Subtract: $\dfrac{x + 2}{2x - 3} - \dfrac{4}{x + 3}$.

Solution   We need to find the least common denominator. This is the product of the distinct factors in each denominator, namely $(2x - 3)(x + 3)$. We can therefore use the subtraction property given previously as follows:

$$\frac{a}{b} - \frac{c}{d} = \frac{ad - bc}{bd}$$

$$\frac{x + 2}{2x - 3} - \frac{4}{x + 3} = \frac{(x + 2)(x + 3) - (2x - 3)4}{(2x - 3)(x + 3)}$$

Observe that
$a = x + 2$, $b = 2x - 3$,
$c = 4$, and $d = x + 3$.

$$= \frac{x^2 + 5x + 6 - (8x - 12)}{(2x - 3)(x + 3)}$$

Multiply.

$$= \frac{x^2 + 5x + 6 - 8x + 12}{(2x - 3)(x + 3)}$$

Remove parentheses and then change the sign of each term.

$$= \frac{x^2 - 3x + 18}{(2x - 3)(x + 3)}, x \neq \frac{3}{2}, x \neq -3$$

Combine like terms in the numerator.

Check Point 6   Add: $\dfrac{3}{x + 1} + \dfrac{5}{x - 1}$.

The **least common denominator,** or LCD, of several rational expressions is a polynomial consisting of the product of all prime factors in the denominators, with each factor raised to the greatest power of its occurrence in any denominator. When adding and subtracting rational expressions that have different denominators with one or more common factors in the denominators, it is efficient to find the least common denominator first.

Finding the Least Common Denominator

1. Factor each denominator completely.
2. List the factors of the first denominator.
3. Add to the list in step 2 any factors of the second denominator that do not appear in the list.
4. Form the product of each different factor from the list in step 3. This product is the least common denominator.

EXAMPLE 7   Finding the Least Common Denominator

Find the least common denominator of

$$\frac{7}{5x^2 + 15x} \quad \text{and} \quad \frac{9}{x^2 + 6x + 9}.$$

Solution

Step 1   Factor each denominator completely.

$$5x^2 + 15x = 5x(x + 3)$$

$$x^2 + 6x + 9 = (x + 3)^2$$

Step 2   List the factors of the first denominator.

$$5, x, (x + 3)$$

Step 3 **Add any unlisted factors from the second denominator.** The second denominator is $(x + 3)^2$ or $(x + 3)(x + 3)$. One factor of $x + 3$ is already in our list, but the other factor is not. We add $x + 3$ to the list. We have

$$5, x, (x + 3), (x + 3).$$

Step 4 **The least common denominator is the product of all factors in the final list.** Thus,

$$5x(x + 3)(x + 3), \quad \text{or} \quad 5x(x + 3)^2$$

is the least common denominator.

Check
Point
7

Find the least common denominator of

$$\frac{3}{x^2 - 6x + 9} \quad \text{and} \quad \frac{7}{x^2 - 9}.$$

Finding the least common denominator for two (or more) rational expressions is the first step needed to add or subtract the expressions.

Adding and Subtracting Rational Expressions That Have Different Denominators with Shared Factors

1. Find the least common denominator.

2. Write all rational expressions in terms of the least common denominator. To do so, multiply both the numerator and the denominator of each rational expression by any factor(s) needed to convert the denominator into the least common denominator.

3. Add or subtract the numerators, placing the resulting expression over the least common denominator.

4. If necessary, simplify the resulting rational expression.

EXAMPLE 8 **Adding Rational Expressions with Different Denominators**

Add: $\dfrac{x + 3}{x^2 + x - 2} + \dfrac{2}{x^2 - 1}.$

Solution

Step 1 **Find the least common denominator.** Start by factoring the denominators.

$$x^2 + x - 2 = (x + 2)(x - 1)$$
$$x^2 - 1 = (x + 1)(x - 1)$$

The factors of the first denominator are $x + 2$ and $x - 1$. The only factor from the second denominator that is not listed is $x + 1$. Thus, the least common denominator is

$$(x + 2)(x - 1)(x + 1).$$

Step 2   Write all rational expressions in terms of the least common denominator. We do so by multiplying both the numerator and the denominator by any factor(s) needed to convert the denominator into the least common denominator.

$$\frac{x + 3}{x^2 + x - 2} + \frac{2}{x^2 - 1}$$

$$= \frac{x + 3}{(x + 2)(x - 1)} + \frac{2}{(x + 1)(x - 1)}$$

The least common denominator is $(x + 2)(x - 1)(x + 1)$.

$$= \frac{(x + 3)(x + 1)}{(x + 2)(x - 1)(x + 1)} + \frac{2(x + 2)}{(x + 2)(x - 1)(x + 1)}$$

Multiply each numerator and denominator by the extra factor required to form $(x + 2)(x - 1)(x + 1)$, the least common denominator.

Step 3   Add numerators, putting this sum over the least common denominator.

$$= \frac{(x + 3)(x + 1) + 2(x + 2)}{(x + 2)(x - 1)(x + 1)}$$

$$= \frac{x^2 + 4x + 3 + 2x + 4}{(x + 2)(x - 1)(x + 1)}$$

Perform the multiplications in the numerator.

$$= \frac{x^2 + 6x + 7}{(x + 2)(x - 1)(x + 1)}, x \neq -2, x \neq 1, x \neq -1$$

Combine like terms in the numerator.

Step 4   If necessary, simplify. Because the numerator is prime, no further simplification is possible.

Check Point 8   Subtract: $\dfrac{x}{x^2 - 10x + 25} - \dfrac{x - 4}{2x - 10}$.

6 Simplify complex rational expressions.

Complex Rational Expressions

Complex rational expressions, also called **complex fractions,** have numerators or denominators containing one or more rational expressions. Here are two examples of such expressions:

$$\frac{1 + \dfrac{1}{x}}{1 - \dfrac{1}{x}}$$

Separate rational expressions occur in the numerator and denominator.

$$\frac{\dfrac{1}{x + h} - \dfrac{1}{x}}{h}$$

Separate rational expressions occur in the numerator.

One method for simplifying a complex rational expression is to combine its numerator into a single expression and combine its denominator into a single expression. Then perform the division by inverting the denominator and multiplying.

EXAMPLE 9   **Simplifying a Complex Rational Expression**

Simplify: $\dfrac{1 + \dfrac{1}{x}}{1 - \dfrac{1}{x}}$.

Solution

$$\dfrac{1 + \dfrac{1}{x}}{1 - \dfrac{1}{x}} = \dfrac{\dfrac{x}{x} + \dfrac{1}{x}}{\dfrac{x}{x} - \dfrac{1}{x}}, \; x \neq 0$$

The terms in the numerator and in the denominator are each combined by performing the addition and subtraction. The least common denominator is x.

$$= \dfrac{\dfrac{x + 1}{x}}{\dfrac{x - 1}{x}}$$

Perform the addition in the numerator and the subtraction in the denominator.

$$= \dfrac{x + 1}{x} \div \dfrac{x - 1}{x}$$

Rewrite the main fraction bar as ÷.

$$= \dfrac{x + 1}{x} \cdot \dfrac{x}{x - 1}$$

Invert the divisor and multiply (x ≠ 0 and x ≠ 1).

$$= \dfrac{x + 1}{\cancel{x}_{1}} \cdot \dfrac{\cancel{x}^{1}}{x - 1}$$

Divide a numerator and denominator by the common factor, x.

$$= \dfrac{x + 1}{x - 1}, \; x \neq 0, \; x \neq 1$$

Multiply the remaining factors in the numerator and in the denominator.

Check Point 9   Simplify: $\dfrac{\dfrac{1}{x} - \dfrac{3}{2}}{\dfrac{1}{x} + \dfrac{3}{4}}$.

A second method for simplifying a complex rational expression is to find the least common denominator of all the rational expressions in its numerator and denominator. Then multiply each term in its numerator and denominator by this least common denominator. Here we use this method to simplify the complex rational expression in Example 9.

$$\dfrac{1 + \dfrac{1}{x}}{1 - \dfrac{1}{x}} = \dfrac{\left(1 + \dfrac{1}{x}\right)}{\left(1 - \dfrac{1}{x}\right)} \cdot \dfrac{x}{x}$$

The least common denominator of all the rational expressions is x. Multiply the numerator and denominator by x. Because $\dfrac{x}{x} = 1$, we are not changing the complex fraction (x ≠ 0).

$$= \dfrac{1 \cdot x + \dfrac{1}{x} \cdot x}{1 \cdot x - \dfrac{1}{x} \cdot x}$$

Use the distributive property. Be sure to distribute x to every term.

$$= \dfrac{x + 1}{x - 1}, \; x \neq 0, \; x \neq 1$$

Multiply. The complex rational expression is now simplified.

EXERCISE SET P.6

Practice Exercises

In Exercises 1–6, find all numbers that must be excluded from the domain of each rational expression.

1. $\dfrac{7}{x-3}$

2. $\dfrac{13}{x+9}$

3. $\dfrac{x+5}{x^2-25}$

4. $\dfrac{x+7}{x^2-49}$

5. $\dfrac{x-1}{x^2+11x+10}$

6. $\dfrac{x-3}{x^2+4x-45}$

In Exercises 7–14, simplify each rational expression. Find all numbers that must be excluded from the domain of the simplified rational expression.

7. $\dfrac{3x-9}{x^2-6x+9}$

8. $\dfrac{4x-8}{x^2-4x+4}$

9. $\dfrac{x^2-12x+36}{4x-24}$

10. $\dfrac{x^2-8x+16}{3x-12}$

11. $\dfrac{y^2+7y-18}{y^2-3y+2}$

12. $\dfrac{y^2-4y-5}{y^2+5y+4}$

13. $\dfrac{x^2+12x+36}{x^2-36}$

14. $\dfrac{x^2-14x+49}{x^2-49}$

In Exercises 15–32, multiply or divide as indicated.

15. $\dfrac{x-2}{3x+9}\cdot\dfrac{2x+6}{2x-4}$

16. $\dfrac{6x+9}{3x-15}\cdot\dfrac{x-5}{4x+6}$

17. $\dfrac{x^2-9}{x^2}\cdot\dfrac{x^2-3x}{x^2+x-12}$

18. $\dfrac{x^2-4}{x^2-4x+4}\cdot\dfrac{2x-4}{x+2}$

19. $\dfrac{x^2-5x+6}{x^2-2x-3}\cdot\dfrac{x^2-1}{x^2-4}$

20. $\dfrac{x^2+5x+6}{x^2+x-6}\cdot\dfrac{x^2-9}{x^2-x-6}$

21. $\dfrac{x^3-8}{x^2-4}\cdot\dfrac{x+2}{3x}$

22. $\dfrac{x^2+6x+9}{x^3+27}\cdot\dfrac{1}{x+3}$

23. $\dfrac{x+1}{3}\div\dfrac{3x+3}{7}$

24. $\dfrac{x+5}{7}\div\dfrac{4x+20}{9}$

25. $\dfrac{x^2-4}{x}\div\dfrac{x+2}{x-2}$

26. $\dfrac{x^2-4}{x-2}\div\dfrac{x+2}{4x-8}$

27. $\dfrac{4x^2+10}{x-3}\div\dfrac{6x^2+15}{x^2-9}$

28. $\dfrac{x^2+x}{x^2-4}\div\dfrac{x^2-1}{x^2+5x+6}$

29. $\dfrac{x^2-25}{2x-2}\div\dfrac{x^2+10x+25}{x^2+4x-5}$

30. $\dfrac{x^2-4}{x^2+3x-10}\div\dfrac{x^2+5x+6}{x^2+8x+15}$

31. $\dfrac{x^2+x-12}{x^2+x-30}\cdot\dfrac{x^2+5x+6}{x^2-2x-3}\div\dfrac{x+3}{x^2+7x+6}$

32. $\dfrac{x^3-25x}{4x^2}\cdot\dfrac{2x^2-2}{x^2-6x+5}\div\dfrac{x^2+5x}{7x+7}$

In Exercises 33–54, add or subtract as indicated.

33. $\dfrac{4x+1}{6x+5}+\dfrac{8x+9}{6x+5}$

34. $\dfrac{3x+2}{3x+4}+\dfrac{3x+6}{3x+4}$

35. $\dfrac{x^2-2x}{x^2+3x}+\dfrac{x^2+x}{x^2+3x}$

36. $\dfrac{x^2-4x}{x^2-x-6}+\dfrac{4x-4}{x^2-x-6}$

37. $\dfrac{4x-10}{x-2}-\dfrac{x-4}{x-2}$

38. $\dfrac{2x+3}{3x-6}-\dfrac{3-x}{3x-6}$

39. $\dfrac{x^2+3x}{x^2+x-12}-\dfrac{x^2-12}{x^2+x-12}$

40. $\dfrac{x^2-4x}{x^2-x-6}-\dfrac{x-6}{x^2-x-6}$

41. $\dfrac{3}{x+4}+\dfrac{6}{x+5}$

42. $\dfrac{8}{x-2}+\dfrac{2}{x-3}$

43. $\dfrac{3}{x+1}-\dfrac{3}{x}$

44. $\dfrac{4}{x}-\dfrac{3}{x+3}$

45. $\dfrac{2x}{x+2}+\dfrac{x+2}{x-2}$

46. $\dfrac{3x}{x-3}-\dfrac{x+4}{x+2}$

47. $\dfrac{x+5}{x-5}+\dfrac{x-5}{x+5}$

48. $\dfrac{x+3}{x-3}+\dfrac{x-3}{x+3}$

49. $\dfrac{4}{x^2+6x+9}+\dfrac{4}{x+3}$

50. $\dfrac{3}{5x+2}+\dfrac{5x}{25x^2-4}$

51. $\dfrac{3x}{x^2+3x-10}-\dfrac{2x}{x^2+x-6}$

52. $\dfrac{x}{x^2-2x-24}-\dfrac{x}{x^2-7x+6}$

53. $\dfrac{4x^2+x-6}{x^2+3x+2}-\dfrac{3x}{x+1}+\dfrac{5}{x+2}$

54. $\dfrac{6x^2+17x-40}{x^2+x-20}+\dfrac{3}{x-4}-\dfrac{5x}{x+5}$

In Exercise 55–64, simplify each complex rational expression.

55. $\dfrac{\dfrac{x}{3} - 1}{x - 3}$

56. $\dfrac{\dfrac{x}{4} - 1}{x - 4}$

57. $\dfrac{1 + \dfrac{1}{x}}{3 - \dfrac{1}{x}}$

58. $\dfrac{8 + \dfrac{1}{x}}{4 - \dfrac{1}{x}}$

59. $\dfrac{\dfrac{1}{x} + \dfrac{1}{y}}{x + y}$

60. $\dfrac{1 - \dfrac{1}{x}}{xy}$

61. $\dfrac{x - \dfrac{x}{x + 3}}{x + 2}$

62. $\dfrac{x - 3}{x - \dfrac{3}{x - 2}}$

63. $\dfrac{\dfrac{3}{x - 2} - \dfrac{4}{x + 2}}{\dfrac{7}{x^2 - 4}}$

64. $\dfrac{\dfrac{x}{x - 2} + 1}{\dfrac{3}{x^2 - 4} + 1}$

Application Exercises

65. The rational expression

$$\frac{130x}{100 - x}$$

describes the cost, in millions of dollars, to inoculate x percent of the population against a particular strain of flu.

  a. Evaluate the expression for $x = 40$, $x = 80$, and $x = 90$. Describe the meaning of each evaluation in terms of percentage inoculated and cost.

  b. For what value of x is the expression undefined?

  c. What happens to the cost as x approaches 100%? How can you interpret this observation?

66. Doctors use the rational expression

$$\frac{DA}{A + 12}$$

to determine the dosage of a drug prescribed for children. In this expression, A = child's age, and D = adult dosage. What is the difference in the child's dosage for a 7-year-old child and a 3-year-old child? Express the answer as a single rational expression in terms of D. Then describe what your answer means in terms of the variables in the rational expression.

67. Anthropologists and forensic scientists classify skulls using

$$\frac{L + 60W}{L} - \frac{L - 40W}{L}$$

where L is the skull's length and W is its width.

  a. Express the classification as a single rational expression.

  b. If the value of the rational expression in part (a) is less than 75, a skull is classified as long. A medium skull has a value between 75 and 80, and a round skull has a value over 80. Use your rational expression from part (a) to classify a skull that is 5 inches wide and 6 inches long.

68. The polynomial

$$6t^4 - 207t^3 + 2128t^2 - 6622t + 15{,}220$$

describes the annual number of drug convictions in the United States t years after 1984. The polynomial

$$28t^4 - 711t^3 + 5963t^2 - 1695t + 27{,}424$$

describes the annual number of drug arrests in the United States t years after 1984. Write a rational expression that describes the conviction rate for drug arrests in the United States t years after 1984.

69. The average speed on a round-trip commute having a one-way distance d is given by the complex rational expression

$$\frac{2d}{\dfrac{d}{r_1} + \dfrac{d}{r_2}}$$

in which r_1 and r_2 are the speeds on the outgoing and return trips, respectively. Simplify the expression. Then find the average speed for a person who drives from home to work at 30 miles per hour and returns on the same route averaging 20 miles per hour. Explain why the answer is not 25 miles per hour.

Writing in Mathematics

70. What is a rational expression?

71. Explain how to determine which numbers must be excluded from the domain of a rational expression.

72. Explain how to simplify a rational expression.

73. Explain how to multiply rational expressions.

74. Explain how to divide rational expressions.

75. Explain how to add or subtract rational expressions with the same denominators.

76. Explain how to add rational expressions having no common factors in their denominators. Use $\dfrac{3}{x+5} + \dfrac{7}{x+2}$ in your explanation.

77. Explain how to find the least common denominator for denominators of $x^2 - 100$ and $x^2 - 20x + 100$.

78. Describe two ways to simplify $\dfrac{\dfrac{3}{x} + \dfrac{2}{x^2}}{\dfrac{1}{x^2} + \dfrac{2}{x}}$.

Explain the error in Exercises 79–81. Then rewrite the right side of the equation to correct the error that now exists.

79. $\dfrac{1}{a} + \dfrac{1}{b} = \dfrac{1}{a+b}$

80. $\dfrac{1}{x} + 7 = \dfrac{1}{x+7}$

81. $\dfrac{a}{x} + \dfrac{a}{b} = \dfrac{a}{x+b}$

82. A politician claims that each year the conviction rate for drug arrests in the United States is increasing. Explain how to use the polynomials in Exercise 68 to verify this claim.

Technology Exercises

83. How much are your monthly payments on a loan? If P is the principal, or amount borrowed, i is the monthly interest rate, and n is the number of monthly payments, then the amount, A, of each monthly payment is

$$A = \dfrac{Pi}{1 - \dfrac{1}{(1+i)^n}}.$$

a. Simplify the complex rational expression for the amount of each payment.

b. You purchase a $20,000 automobile at 1% monthly interest to be paid over 48 months. How much do you pay each month? Use the simplified rational expression from part (a) and a calculator. Round to the nearest dollar.

Critical Thinking Exercises

84. Which one of the following is true?

a. $\dfrac{x^2 - 25}{x - 5} = x - 5$

b. $\dfrac{x}{y} \div \dfrac{y}{x} = 1$, if $x \neq 0$ and $y \neq 0$.

c. The least common denominator needed to find $\dfrac{1}{x} + \dfrac{1}{x+3}$ is $x + 3$.

d. The rational expression

$$\dfrac{x^2 - 16}{x - 4}$$

is not defined for $x = 4$. However, as x gets closer and closer to 4, the value of the expression approaches 8.

In Exercises 85–86, find the missing expression.

85. $\dfrac{3x}{x-5} + \dfrac{\boxed{}}{5-x} = \dfrac{7x+1}{x-5}$

86. $\dfrac{4}{x-2} - \dfrac{\boxed{}}{} = \dfrac{2x+8}{(x-2)(x+1)}$

87. In one short sentence, five words or less, explain what

$$\dfrac{\dfrac{1}{x} + \dfrac{1}{x^2} + \dfrac{1}{x^3}}{\dfrac{1}{x^4} + \dfrac{1}{x^5} + \dfrac{1}{x^6}}$$

does to each number x.

CHAPTER SUMMARY, REVIEW, AND TEST

Summary: Basic Formulas

Definition of Absolute Value

$$|x| = \begin{cases} x & \text{if } x \geq 0 \\ -x & \text{if } x < 0 \end{cases}$$

Distance between Points a and b on a Number Line

$$|a - b| \quad \text{or} \quad |b - a|$$

Properties of Algebra

Commutative	$a + b = b + a, \quad ab = ba$
Associative	$(a + b) + c = a + (b + c)$ $(ab)c = a(bc)$
Distributive	$a(b + c) = ab + ac$
Identity	$a + 0 = a \quad a \cdot 1 = a$
Inverse	$a + (-a) = 0 \quad a \cdot \dfrac{1}{a} = 1, a \neq 0$

Properties of Exponents

$$b^{-n} = \frac{1}{b^n}, \quad b^0 = 1, \quad b^m \cdot b^n = b^{m+n},$$

$$(b^m)^n = b^{mn}, \quad \frac{b^m}{b^n} = b^{m-n}, \quad (ab)^n = a^n b^n, \quad \left(\frac{a}{b}\right)^n = \frac{a^n}{b^n}$$

Product and Quotient Rules for nth Roots

$$\sqrt[n]{a} \cdot \sqrt[n]{b} = \sqrt[n]{ab} \qquad \frac{\sqrt[n]{a}}{\sqrt[n]{b}} = \sqrt[n]{\frac{a}{b}}$$

Rational Exponents

$$a^{1/n} = \sqrt[n]{a}, \quad a^{-1/n} = \frac{1}{a^{1/n}} = \frac{1}{\sqrt[n]{a}},$$

$$a^{m/n} = \left(\sqrt[n]{a}\right)^m = \sqrt[n]{a^m}, \quad a^{-m/n} = \frac{1}{a^{m/n}}$$

Special Products

$$(A + B)(A - B) = A^2 - B^2$$
$$(A + B)^2 = A^2 + 2AB + B^2$$
$$(A - B)^2 = A^2 - 2AB + B^2$$
$$(A + B)^3 = A^3 + 3A^2B + 3AB^2 + B^3$$
$$(A - B)^3 = A^3 - 3A^2B + 3AB^2 - B^3$$

Factoring Formulas

$$A^2 - B^2 = (A + B)(A - B)$$
$$A^2 + 2AB + B^2 = (A + B)^2$$
$$A^2 - 2AB + B^2 = (A - B)^2$$
$$A^3 + B^3 = (A + B)(A^2 - AB + B^2)$$
$$A^3 - B^3 = (A - B)(A^2 + AB + B^2)$$

Review Exercises

You can use these review exercises, like the review exercises at the end of each chapter, to test your understanding of the chapter's topics. However, you can also use these exercises as a prerequisite test to check your mastery of the fundamental algebra skills needed in this book.

P.1

1. Consider the set:

$$\left\{-17, -\tfrac{9}{13}, 0, 0.75, \sqrt{2}, \pi, \sqrt{81}\right\}.$$

List all numbers from the set that are **a.** natural numbers, **b.** whole numbers, **c.** integers, **d.** rational numbers, **e.** irrational numbers.

In Exercises 2–4, rewrite each expression without absolute value bars.

2. $|-103|$

3. $|\sqrt{2} - 1|$

4. $|3 - \sqrt{17}|$

5. Express the distance between the numbers -17 and 4 using absolute value. Then evaluate the absolute value.

In Exercises 6–7, evaluate each algebraic expression for the given value of the variable.

6. $\frac{5}{9}(F - 32)$; $F = 68$

7. $\frac{8(x + 5)}{3x + 8}$, $x = 2$

In Exercises 8–13, state the name of the property illustrated.

8. $3 + 17 = 17 + 3$

9. $(6 \cdot 3) \cdot 9 = 6 \cdot (3 \cdot 9)$

10. $\sqrt{3}(\sqrt{5} + \sqrt{3}) = \sqrt{15} + 3$

11. $(6 \cdot 9) \cdot 2 = 2 \cdot (6 \cdot 9)$

12. $\sqrt{3}(\sqrt{5} + \sqrt{3}) = (\sqrt{5} + \sqrt{3})\sqrt{3}$

13. $(3 \cdot 7) + (4 \cdot 7) = (4 \cdot 7) + (3 \cdot 7)$

In Exercises 14–15, simplify each algebraic expression.

14. $3(7x - 5y) - 2(4y - x + 1)$

15. $\frac{1}{5}(5x) + \left[(3y) + (-3y)\right] - (-x)$

P.2

Evaluate each exponential expression in Exercises 16–19.

16. $(-3)^3(-2)^2$

17. $2^{-4} + 4^{-1}$

18. $5^{-3} \cdot 5$

19. $\frac{3^3}{3^6}$

Simplify each exponential expression in Exercises 20–23.

20. $(-2x^4y^3)^3$

21. $(-5x^3y^2)(-2x^{-11}y^{-2})$

22. $(2x^3)^{-4}$

23. $\frac{7x^5y^6}{28x^{15}y^{-2}}$

In Exercises 24–25, write each number in decimal notation.

24. 3.74×10^4

25. 7.45×10^{-5}

In Exercises 26–27, write each number in scientific notation.

26. 3,590,000

27. 0.00725

In Exercises 28–29, perform the indicated operation and write the answer in decimal notation.

28. $(3 \times 10^3)(1.3 \times 10^2)$

29. $\frac{6.9 \times 10^3}{3 \times 10^5}$

30. If you earned $1 million per year ($10^6$), how long would it take to accumulate $1 billion ($10^9$)?

31. If the population of the United States is 2.8×10^8 and each person spends about $150 per year going to the movies (or renting movies), express the total annual spending on movies in scientific notation.

P.3

Use the product rule to simplify the expressions in Exercises 32–35. In Exercises 34–35, assume that variables represent nonnegative real numbers.

32. $\sqrt{300}$

33. $\sqrt{12x^2}$

34. $\sqrt{10x} \cdot \sqrt{2x}$

35. $\sqrt{r^3}$

Use the quotient rule to simplify the expressions in Exercises 36–37.

36. $\sqrt{\frac{121}{4}}$

37. $\frac{\sqrt{96x^3}}{\sqrt{2x}}$ (Assume that $x > 0$.)

In Exercises 38–40, add or subtract terms whenever possible.

38. $7\sqrt{5} + 13\sqrt{5}$

39. $2\sqrt{50} + 3\sqrt{8}$

40. $4\sqrt{72} - 2\sqrt{48}$

In Exercises 41–44, rationalize the denominator.

41. $\frac{30}{\sqrt{5}}$

42. $\frac{\sqrt{2}}{\sqrt{3}}$

43. $\frac{5}{6 + \sqrt{3}}$

44. $\frac{14}{\sqrt{7} - \sqrt{5}}$

Evaluate each expression in Exercises 45–48 or indicate that the root is not a real number.

45. $\sqrt[3]{125}$

46. $\sqrt[5]{-32}$

47. $\sqrt[4]{-125}$

48. $\sqrt[4]{(-5)^4}$

Simplify the radical expressions in Exercises 49–53.

49. $\sqrt[3]{81}$

50. $\sqrt[3]{y^5}$

51. $\sqrt[4]{8} \cdot \sqrt[4]{10}$

52. $4\sqrt[3]{16} + 5\sqrt[3]{2}$

53. $\frac{\sqrt[4]{32x^5}}{\sqrt[4]{16x}}$ (Assume that $x > 0$.)

In Exercises 54–59, evaluate each expression.

54. $16^{1/2}$

55. $25^{-1/2}$

56. $125^{1/3}$

57. $27^{-1/3}$

58. $64^{2/3}$

59. $27^{-4/3}$

In Exercises 60–62, simplify using properties of exponents.

60. $(5x^{2/3})(4x^{1/4})$

61. $\frac{15x^{3/4}}{5x^{1/2}}$

62. $(125x^6)^{2/3}$

63. Simplify by reducing the index of the radical: $\sqrt[6]{y^3}$.

P.4

In Exercises 64–65, perform the indicated operations. Write the resulting polynomial in standard form and indicate its degree.

64. $(-6x^3 + 7x^2 - 9x + 3) + (14x^3 + 3x^2 - 11x - 7)$

65. $(13x^4 - 8x^3 + 2x^2) - (5x^4 - 3x^3 + 2x^2 - 6)$

In Exercises 66–72, find each product.

66. $(3x - 2)(4x^2 + 3x - 5)$ **67.** $(3x - 5)(2x + 1)$

68. $(4x + 5)(4x - 5)$ **69.** $(2x + 5)^2$

70. $(3x - 4)^2$ **71.** $(2x + 1)^3$

72. $(5x - 2)^3$

In Exercises 73–74, perform the indicated operations. Indicate the degree of the resulting polynomial.

73. $(7x^2 - 8xy + y^2) + (-8x^2 - 9xy - 4y^2)$

74. $(13x^3y^2 - 5x^2y - 9x^2) - (-11x^3y^2 - 6x^2y + 3x^2 - 4)$

In Exercises 75–79, find each product.

75. $(x + 7y)(3x - 5y)$ **76.** $(3x - 5y)^2$

77. $(3x^2 + 2y)^2$ **78.** $(7x + 4y)(7x - 4y)$

79. $(a - b)(a^2 + ab + b^2)$

P.5

In Exercises 80–96, factor completely, or state that the polynomial is prime.

80. $15x^3 + 3x^2$ **81.** $x^2 - 11x + 28$

82. $15x^2 - x - 2$ **83.** $64 - x^2$

84. $x^2 + 16$ **85.** $3x^4 - 9x^3 - 30x^2$

86. $20x^7 - 36x^3$ **87.** $x^3 - 3x^2 - 9x + 27$

88. $16x^2 - 40x + 25$ **89.** $x^4 - 16$

90. $y^3 - 8$ **91.** $x^3 + 64$

92. $3x^4 - 12x^2$ **93.** $27x^3 - 125$

94. $x^5 - x$ **95.** $x^3 + 5x^2 - 2x - 10$

96. $x^2 + 18x + 81 - y^2$

In Exercises 97–99, factor and simplify each algebraic expression.

97. $16x^{-3/4} + 32x^{1/4}$

98. $(x^2 - 4)(x^2 + 3)^{1/2} - (x^2 - 4)^2(x^2 + 3)^{3/2}$

99. $12x^{-1/2} + 6x^{-3/2}$

P.6

In Exercises 100–102, simplify each rational expression. Also, list all numbers that must be excluded from the domain.

100. $\dfrac{x^3 + 2x^2}{x + 2}$ **101.** $\dfrac{x^2 + 3x - 18}{x^2 - 36}$

102. $\dfrac{x^2 + 2x}{x^2 + 4x + 4}$

In Exercises 103–105, multiply or divide as indicated.

103. $\dfrac{x^2 + 6x + 9}{x^2 - 4} \cdot \dfrac{x + 3}{x - 2}$ **104.** $\dfrac{6x + 2}{x^2 - 1} \div \dfrac{3x^2 + x}{x - 1}$

105. $\dfrac{x^2 - 5x - 24}{x^2 - x - 12} \div \dfrac{x^2 - 10x + 16}{x^2 + x - 6}$

In Exercises 106–109, add or subtract as indicated.

106. $\dfrac{2x - 7}{x^2 - 9} - \dfrac{x - 10}{x^2 - 9}$ **107.** $\dfrac{3x}{x + 2} + \dfrac{x}{x - 2}$

108. $\dfrac{x}{x^2 - 9} + \dfrac{x - 1}{x^2 - 5x + 6}$

109. $\dfrac{4x - 1}{2x^2 + 5x - 3} - \dfrac{x + 3}{6x^2 + x - 2}$

In Exercises 110–112, simplify each complex rational expression.

110. $\dfrac{\dfrac{1}{x} - \dfrac{1}{2}}{\dfrac{1}{3} - \dfrac{x}{6}}$ **111.** $\dfrac{3 + \dfrac{12}{x}}{1 - \dfrac{16}{x^2}}$ **112.** $\dfrac{3 - \dfrac{1}{x + 3}}{3 + \dfrac{1}{x + 3}}$

Chapter P Test

1. List all the rational numbers in this set:

$$\left\{-7, -\tfrac{4}{5}, 0, 0.25, \sqrt{3}, \sqrt{4}, \tfrac{22}{7}, \pi\right\}.$$

In Exercises 2–3, state the name of the property illustrated.

2. $3(2 + 5) = 3(5 + 2)$ **3.** $6(7 + 4) = 6 \cdot 7 + 6 \cdot 4$

4. Express in scientific notation: 0.00076.

Simplify each expression in Exercises 5–11.

5. $9(10x - 2y) - 5(x - 4y + 3)$

6. $\dfrac{30x^3y^4}{6x^9y^{-4}}$

7. $\sqrt{6r}\,\sqrt{3r}$ (Assume that $r \geq 0$.)

8. $4\sqrt{50} - 3\sqrt{18}$ **9.** $\dfrac{3}{5 + \sqrt{2}}$

10. $\sqrt[3]{16x^4}$ **11.** $\dfrac{x^2 + 2x - 3}{x^2 - 3x + 2}$

12. Evaluate: $27^{-5/3}$.

In Exercises 13–14, find each product.

13. $(2x - 5)(x^2 - 4x + 3)$ **14.** $(5x + 3y)^2$

In Exercises 15–20, factor completely, or state that the polynomial is prime.

15. $x^2 - 9x + 18$

16. $x^3 + 2x^2 + 3x + 6$

17. $25x^2 - 9$

18. $36x^2 - 84x + 49$

19. $y^3 - 125$

20. $x^2 + 10x + 25 - 9y^2$

21. Factor and simplify:

$$x(x + 3)^{-3/5} + (x + 3)^{2/5}.$$

In Exercises 22–25, perform the operations and simplify, if possible.

22. $\dfrac{2x + 8}{x - 3} \div \dfrac{x^2 + 5x + 4}{x^2 - 9}$

23. $\dfrac{x}{x + 3} + \dfrac{5}{x - 3}$

24. $\dfrac{2x + 3}{x^2 - 7x + 12} - \dfrac{2}{x - 3}$

25. $\dfrac{\dfrac{1}{x} - \dfrac{1}{3}}{\dfrac{1}{x}}$

Equations, Inequalities, and Mathematical Models

Formulas like those that describe the height a child will attain as an adult are frequently obtained from actual data. Formulas can be used to explain what is happening in the present and to make predictions about what might occur in the future. Knowing how to create and use formulas will help you recognize patterns, logic, and order in a world that can appear chaotic to the untrained eye. In many ways, algebra will provide you with a new way of looking at your world.

Sitting in the biology department office, you overhear two of the professors discussing the possible adult heights of their respective children. Looking at the blackboard that they've been writing on, you see that there are formulas that can estimate the height a child will attain as an adult. If the child is x years old and h inches tall, that child's adult height, H, in inches, is approximated by one of the following formulas:

Girls: $$H = \frac{h}{0.00028x^3 - 0.0071x^2 + 0.0926x + 0.3524}$$

Boys: $$H = \frac{h}{0.00011x^3 - 0.0032x^2 + 0.0604x + 0.3796}.$$

SECTION 1.1 *Graphs and Graphing Utilities*

Objectives

1. Plot points in the rectangular coordinate system.
2. Graph equations in the rectangular coordinate system.
3. Interpret information about a graphing utility's viewing rectangle.
4. Use a graph to determine intercepts.
5. Interpret information given by graphs.

The beginning of the seventeenth century was a time of innovative ideas and enormous intellectual progress in Europe. English theatergoers enjoyed a succession of exciting new plays by Shakespeare. William Harvey proposed the radical notion that the heart was a pump for blood rather than the center of emotion. Galileo, with his new-fangled invention called the telescope, supported the theory of Polish astronomer Copernicus that the sun, not the Earth, was the center of the solar system. Monteverdi was writing the world's first grand operas. French mathematicians Pascal and Fermat invented a new field of mathematics called probability theory.

Into this arena of intellectual electricity stepped French aristocrat René Descartes (1596–1650). Descartes, propelled by the creativity surrounding him, developed a new branch of mathematics that brought together algebra and geometry in a unified way—a way that visualized numbers as points on a graph, equations as geometric figures, and geometric figures as equations. This new branch of mathematics, called *analytic geometry*, established Descartes as one of the founders of modern thought and among the most original mathematicians and philosophers of any age. We begin this section by looking at Descartes's deceptively simple idea, called the **rectangular coordinate system** or (in his honor) the **Cartesian coordinate system.**

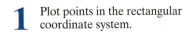

Plot points in the rectangular coordinate system.

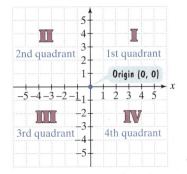

Figure 1.1 The rectangular coordinate system

Points and Ordered Pairs

Descartes used two number lines that intersect at right angles at their zero points, as shown in Figure 1.1. The horizontal number line is the **x-axis.** The vertical number line is the **y-axis.** The point of intersection of these axes is their zero points, called the **origin.** Positive numbers are shown to the right and above the origin. Negative numbers are shown to the left and below the origin. The axes divide the plane into four quarters, called **quadrants.** The points located on the axes are not in any quadrant.

Each point in the rectangular coordinate system corresponds to an **ordered pair** of real numbers, (x, y). Examples of such pairs are $(4, 2)$ and $(-5, -3)$. The first number in each pair, called the **x-coordinate,** denotes the distance and direction from the origin along the x-axis. The second number, called the **y-coordinate,** denotes vertical distance and direction along a line parallel to the y-axis or along the y-axis itself.

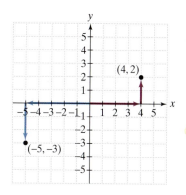

Figure 1.2 Plotting $(4, 2)$ and $(-5, -3)$

Figure 1.2 shows how we **plot,** or locate, the points corresponding to the ordered pairs $(4, 2)$ and $(-5, -3)$. We plot $(4, 2)$ by going 4 units from 0 to the right along the x-axis. Then we go 2 units up parallel to the y-axis. We plot $(-5, -3)$ by going 5 units from 0 to the left along the x-axis and 3 units down parallel to the y-axis. The phrase "the point corresponding to the ordered pair $(-5, -3)$" is often abbreviated as "the point $(-5, -3)$."

EXAMPLE 1   Plotting Points in the Rectangular Coordinate System

Plot the points: $A(-3, 5)$, $B(2, -4)$, $C(5, 0)$, $D(-5, -3)$, $E(0, 4)$, and $F(0, 0)$.

Solution   See Figure 1.3. We move from the origin and plot the points in the following way:

$A(-3, 5)$:	3 units left, 5 units up
$B(2, -4)$:	2 units right, 4 units down
$C(5, 0)$:	5 units right, 0 units up or down
$D(-5, -3)$:	5 units left, 3 units down
$E(0, 4)$:	0 units right or left, 4 units up
$F(0, 0)$:	0 units right or left, 0 units up or down

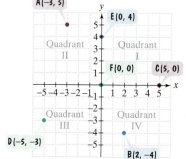

Figure 1.3 Plotting points

The phrase *ordered pair* is used because **order is important.** For example, the points $(2, 5)$ and $(5, 2)$ are not the same. To plot $(2, 5)$, move 2 units right and 5 units up. To plot $(5, 2)$, move 5 units right and 2 units up. The points $(2, 5)$ and $(5, 2)$ are in different locations. **The order in which coordinates appear makes a difference in a point's location.**

Check Point 1   Plot the points:

$$A(-2, 4),\ B(4, -2),\ C(-3, 0),\ \text{and}\ D(0, -3).$$

Graphs of Equations

A relationship between two quantities can be expressed as an **equation in two variables,** such as

$$y = x^2 - 4.$$

A **solution** of this equation is an ordered pair of real numbers with the following property: When the x-coordinate is substituted for x and the y-coordinate is substituted for y in the equation, we obtain a true statement. For example, if we let $x = 3$, then $y = 3^2 - 4 = 9 - 4 = 5$. The ordered pair $(3, 5)$ is a solution of the equation $y = x^2 - 4$. We also say that $(3, 5)$ **satisfies** the equation.

We can generate as many ordered-pair solutions as desired of $y = x^2 - 4$ by substituting numbers for x and then finding the values for y. The **graph of the equation** is the set of all points whose coordinates satisfy the equation.

One method for graphing an equation such as $y = x^2 - 4$ is the **point-plotting method.** First, we find several ordered pairs that are solutions of the equation. Next, we plot these ordered pairs as points in the rectangular coordinate system. Finally, we connect the points with a smooth curve or line. This often gives us a picture of all ordered pairs that satisfy the equation.

Study Tip

Reminder: Answers to all Check Point exercises are given in the answer section. Check your answer before continuing your reading to verify that you understand the concept.

2   Graph equations in the rectangular coordinate system.

EXAMPLE 2 Graphing an Equation Using the Point-Plotting Method

Graph $y = x^2 - 4$. Select integers for x, starting with -3 and ending with 3.

Solution For each value of x we find the corresponding value for y.

x	$y = x^2 - 4$	(x, y)
-3	$y = (-3)^2 - 4 = 9 - 4 = 5$	$(-3, 5)$
-2	$y = (-2)^2 - 4 = 4 - 4 = 0$	$(-2, 0)$
-1	$y = (-1)^2 - 4 = 1 - 4 = -3$	$(-1, -3)$
0	$y = 0^2 - 4 = 0 - 4 = -4$	$(0, -4)$
1	$y = 1^2 - 4 = 1 - 4 = -3$	$(1, -3)$
2	$y = 2^2 - 4 = 4 - 4 = 0$	$(2, 0)$
3	$y = 3^2 - 4 = 9 - 4 = 5$	$(3, 5)$

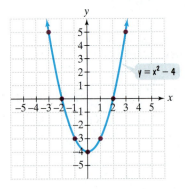

Figure 1.4 The graph of $y = x^2 - 4$

Now we plot the ordered pairs that are solutions of the equation and join the points with a smooth curve, as shown in Figure 1.4. The graph of $y = x^2 - 4$ is a curve where the part of the graph to the right of the y-axis is a reflection of the part to the left of it, and vice versa. The arrows on the left and the right of the curve indicate that it extends indefinitely in both directions.

Check Point 2 Graph $y = 2x - 4$. Select integers for x, starting with -1 and ending with 3.

Do you see a difference between the equations in Example 2 and Check Point 2? The equation in Example 2, $y = x^2 - 4$, involves a polynomial of degree 2. All such equations have graphs that are shaped like cups, such as the graph in Figure 1.4. These U-shaped "cups" can open upward, like the one in Figure 1.4, or downward. By contrast, the equation in Check Point 2, $y = 2x - 4$, involves a polynomial of degree 1. All such equations have graphs that are straight lines.

Study Tip

In Chapters 2 and 3, we will be studying graphs of equations in two variables in which

$$y = \text{a polynomial in } x.$$

Do not be concerned that we have not yet learned techniques, other than plotting points, for graphing such equations. As you solve some of the equations in this chapter, we will display graphs simply to enhance your visual understanding of your work. For now, think of graphs of first-degree polynomials as lines and graphs of second-degree polynomials as symmetric U-shaped cups.

3 Interpret information about a graphing utility's viewing rectangle.

Graphing Equations Using a Graphing Utility

Graphing calculators or graphing software packages for computers are referred to as **graphing utilities** or graphers. A graphing utility is a powerful tool that quickly generates the graph of an equation in two variables. Figure 1.5 shows two such graphs for the equations in Example 2 and Check Point 2.

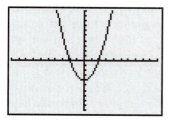

Figure 1.5(a)
The graph of $y = x^2 - 4$

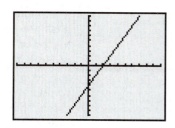

Figure 1.5(b)
The graph of $y = 2x - 4$

What differences do you notice between these graphs and graphs that we draw by hand? They do seem a bit "jittery." Arrows do not appear on the left and right ends of the graphs. Furthermore, numbers are not given along the axes. For both graphs in Figure 1.5, the x-axis extends from -10 to 10 and the y-axis also extends from -10 to 10. The distance represented by each consecutive tick mark is one unit. We say that the **viewing rectangle** is $[-10, 10, 1]$ by $[-10, 10, 1]$.

$$[-10, \qquad 10, \qquad 1] \quad \text{by} \quad [-10, \qquad 10, \qquad 1]$$

| The minimum x-value along the x-axis is -10. | The maximum x-value along the x-axis is 10. | Distance between consecutive tick marks on the x-axis is one unit. | The minimum y-value along the y-axis is -10. | The maximum y-value along the y-axis is 10. | Distance between consecutive tick marks on the y-axis is one unit. |

To graph an equation in x and y using a graphing utility, enter the equation and specify the size of the viewing rectangle. The size of the viewing rectangle sets minimum and maximum values for both the x- and y-axes. Enter these values, as well as the values between consecutive tick marks on the respective axes. The $[-10, 10, 1]$ by $[-10, 10, 1]$ viewing rectangle used in Figure 1.5 is called the **standard viewing rectangle.**

EXAMPLE 3   Understanding the Viewing Rectangle

What is the meaning of a $[-2, 3, 0.5]$ by $[-10, 20, 5]$ viewing rectangle?

Solution   We begin with $[-2, 3, 0.5]$, which describes the x-axis. The minimum x-value is -2 and the maximum x-value is 3. The distance between consecutive tick marks is 0.5.

Next, consider $[-10, 20, 5]$, which describes the y-axis. The minimum y-value is -10 and the maximum y-value is 20. The distance between consecutive tick marks is 5.

Figure 1.6 illustrates a $[-2, 3, 0.5]$ by $[-10, 20, 5]$ viewing rectangle. To make things clearer, we've placed numbers by each tick mark. These numbers do not appear on the axes when you use a graphing utility to graph an equation.

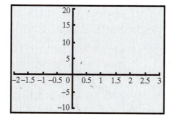

Figure 1.6 A $[-2, 3, 0.5]$ by $[-10, 20, 5]$ viewing rectangle

Check Point 3   What is the meaning of a $[-100, 100, 50]$ by $[-80, 80, 10]$ viewing rectangle? Create a figure like the one in Figure 1.6 that illustrates this viewing rectangle.

On most graphing utilities, the display screen is about two-thirds as high as it is wide. By using a square setting, you can make the distance of one unit along the x-axis the same as the distance of one unit along the y-axis. (This does not occur in the standard viewing rectangle.) Graphing utilities can also *zoom in* and *zoom out*. When you zoom in, you see a smaller portion of the graph, but you see it in greater detail. When you zoom out, you see a larger portion of the graph. Thus, zooming out may help you to develop a better understanding of the overall character of the graph. With practice, you will become more comfortable with graphing equations in two variables using your graphing utility. You will also develop a better sense of the size of the viewing rectangle that will reveal needed information about a particular graph.

Intercepts

An **x-intercept** of a graph is the x-coordinate of a point where the graph intersects the x-axis. For example, look at the graph of $y = x^2 - 4$ in Figure 1.7. The graph crosses the x-axis at $(-2, 0)$ and $(2, 0)$. Thus, the x-intercepts are -2 and 2. **The y-coordinate corresponding to a graph's x-intercept is always zero.**

A **y-intercept** of a graph is the y-coordinate of a point where the graph intersects the y-axis. The graph of $y = x^2 - 4$ in Figure 1.7 shows that the graph crosses the y-axis at $(0, -4)$. Thus, the y-intercept is -4. **The x-coordinate corresponding to a graph's y-intercept is always zero.**

Figure 1.8 illustrates that a graph may have no intercepts or several intercepts.

4 Use a graph to determine intercepts.

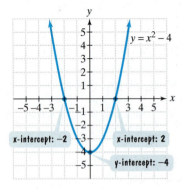

Figure 1.7 Intercepts of $y = x^2 - 4$

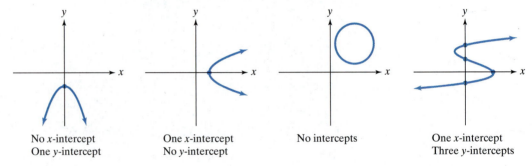

| No x-intercept One y-intercept | One x-intercept No y-intercept | No intercepts | One x-intercept Three y-intercepts |

Figure 1.8

5 Interpret information given by graphs.

Interpreting Information Given by Graphs

Magazines and newspapers often display information using **line graphs** like the one in Figure 1.9. The graph shows the average age at which women in the United States married for the first time over a 110-year period. The years are listed on the horizontal axis and the ages are listed on the vertical axis.

Women's Average Age of First Marriage

Figure 1.9 Average age at which U.S. women married for the first time

Source: U.S. Census Bureau

Like the graph in Figure 1.9, line graphs are often used to illustrate trends over time. Some measure of time, such as months or years, frequently appears on the horizontal axis. Amounts are generally listed on the vertical axis.

A line graph displays information in the first quadrant of a rectangular coordinate system. By identifying points on line graphs and their coordinates, you can interpret specific information given by the graph.

For example, Figure 1.10 shows how to find the average age at which women married for the first time in 1930. (Only the part of the graph that reveals what occurred through about 1940 is shown in the margin because we are interested in 1930.)

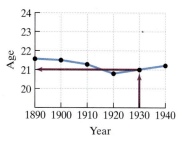

Figure 1.10 In 1930, women were 21 on average when they married for the first time.

Step 1 Locate 1930 on the horizontal axis.

Step 2 Locate the point above 1930.

Step 3 Read across to the corresponding age on the vertical axis.

The age is 21. The coordinates (1930, 21) tell us that in 1930, women in the United States married for the first time at an average age of 21.

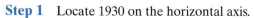

EXAMPLE 4 Applying Estimation Techniques to a Line Graph

Use Figure 1.9 to estimate the maximum average age at which U.S. women married for the first time. When did this occur?

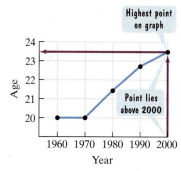

Figure 1.9 Shown again to show only 1960–2000

Solution The maximum average age at which U.S. women married for the first time can be found by locating the highest point on the graph. This point lies above 2000 on the horizontal axis. Read across to the corresponding age on the vertical axis. The age falls approximately midway between 23 and 24, at $23\frac{1}{2}$. The coordinates of the point are approximately $(2000, 23\frac{1}{2})$. Thus, according to the graph, the maximum average age at which U.S. women married for the first time is about $23\frac{1}{2}$. This occurred in 2000. Take another look at the complete line graph in Figure 1.9 at the bottom of page 80 that includes the years 1890 through 2000. Can you see that $23\frac{1}{2}$ is the oldest average age of first marriage over the 110-year period?

Check Point 4 Use the complete line graph in Figure 1.9 to estimate the maximum average age, for the period from 1900 through 1950, at which U.S. women married for the first time. When did this occur?

EXERCISE SET 1.1

Practice Exercises

In Exercises 1–12, plot the given point in a rectangular coordinate system.

1. $(1, 4)$ **2.** $(2, 5)$

3. $(-2, 3)$ **4.** $(-1, 4)$

5. $(-3, -5)$ **6.** $(-4, -2)$

7. $(4, -1)$ **8.** $(3, -2)$

9. $(-4, 0)$ **10.** $(0, -3)$

11. $(\frac{7}{2}, -\frac{3}{2})$ **12.** $(-\frac{5}{2}, \frac{3}{2})$

Graph each equation in Exercises 13–28. Let x = −3, −2, −1, 0, 1, 2, and 3.

13. $y = x^2 - 2$ **14.** $y = x^2 + 2$

15. $y = x - 2$ **16.** $y = x + 2$

17. $y = 2x + 1$ **18.** $y = 2x - 4$

19. $y = -\frac{1}{2}x$ **20.** $y = -\frac{1}{2}x + 2$

21. $y = |x|$ **22.** $y = 2|x|$

23. $y = |x| + 1$ **24.** $y = |x| - 1$

25. $y = 4 - x^2$ **26.** $y = 9 - x^2$

27. $y = x^3$ **28.** $y = x^3 - 1$

In Exercises 29–32, match the viewing rectangle with the correct figure. Then label the tick marks in the figure to illustrate this viewing rectangle.

29. $[-5, 5, 1]$ by $[-5, 5, 1]$

30. $[-10, 10, 2]$ by $[-4, 4, 2]$

31. $[-20, 80, 10]$ by $[-30, 70, 10]$

32. $[-40, 40, 20]$ by $[-1000, 1000, 100]$

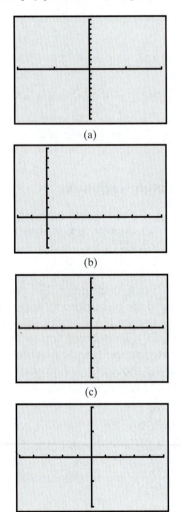

(a)

(b)

(c)

(d)

In Exercises 33–38, use the graph and **a.** *determine the x-intercepts, if any;* **b.** *determine the y-intercepts, if any. For each graph, tick marks along the axes represent one unit each.*

33.

34.

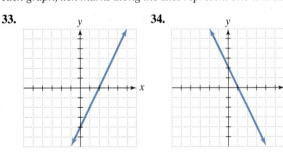

35.

36.

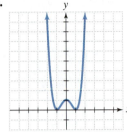

37.

38.

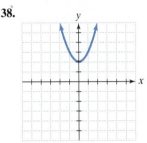

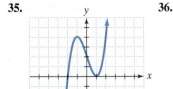

Application Exercises

A football is thrown by a quarterback to a receiver. The points in the figure show the height of the football, in feet, above the ground in terms of its distance, in yards, from the quarterback. Use this information to solve Exercises 39–44.

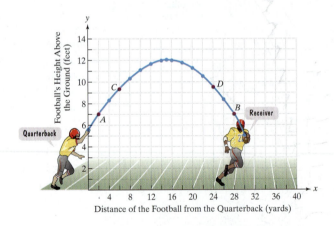

39. Find the coordinates of point A. Then interpret the coordinates in terms of the information given.

40. Find the coordinates of point B. Then interpret the coordinates in terms of the information given.

41. Estimate the coordinates of point C.

42. Estimate the coordinates of point D.

43. What is the football's maximum height? What is its distance from the quarterback when it reaches its maximum height?

44. What is the football's height when it is caught by the receiver? What is the receiver's distance from the quarterback when he catches the football?

The graph shows the percent distribution of divorces in the United States by number of years of marriage. Use the graph to solve Exercises 45–48.

Percent Distribution of Divorces by Number of Years of Marriage

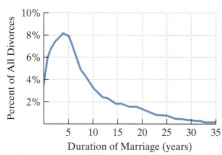

Duration of Marriage (years)

Source: Divorce Center

45. During which years of marriage is the chance of divorce increasing?

46. During which years of marriage is the chance of divorce decreasing?

47. During which year of marriage is the chance of divorce the highest? Estimate, to the nearest percent, the percentage of divorces that occur during this year.

48. During which year of marriage is the chance of divorce the lowest? Estimate, to the nearest percent, the percentage of divorces that occur during this year.

Writing in Mathematics

49. What is the rectangular coordinate system?

50. Explain how to plot a point in the rectangular coordinate system. Give an example with your explanation.

51. Explain why $(5, -2)$ and $(-2, 5)$ do not represent the same point.

52. Explain how to graph an equation in the rectangular coordinate system.

53. What does a $[-20, 2, 1]$ by $[-4, 5, 0.5]$ viewing rectangle mean?

54. Describe the trend shown in the graph for Exercises 45–48. What explanations can you offer for this trend?

Technology Exercises

55. Use a graphing utility to verify each of your hand-drawn graphs in Exercises 13–28. Experiment with the size of the viewing rectangle to make the graph displayed by the graphing utility resemble your hand-drawn graph as much as possible.

56. The stated intent of the 1994 "don't ask, don't tell" policy was to reduce the number of discharges of gay men and lesbians from the military. The equation

$$y = 45.48x^2 - 334.35x + 1237.9$$

describes the number of gay service members, y, discharged from the military for homosexuality x years after 1990. Graph the equation in a $[0, 10, 1]$ by $[0, 2200, 200]$ viewing rectangle. Then describe something about the relationship between x and y that is revealed by looking at the graph that is not obvious from the equation. What does the graph reveal about the success or lack of success of "don't ask, don't tell"?

A graph of an equation is a complete graph *if it shows all of the important features of the graph. Use a graphing utility to graph the equations in Exercises 57–59 in each of the given viewing rectangles. Then choose which viewing rectangle gives a complete graph.*

57. $y = x^2 + 10$
   a. $[-5, 5, 1]$ by $[-5, 5, 1]$
   b. $[-10, 10, 1]$ by $[-10, 10, 1]$
   c. $[-10, 10, 1]$ by $[-50, 50, 1]$

58. $y = 0.1x^4 - x^3 + 2x^2$
   a. $[-5, 5, 1]$ by $[-8, 2, 1]$
   b. $[-10, 10, 1]$ by $[-10, 10, 1]$
   c. $[-8, 16, 1]$ by $[-16, 8, 1]$

59. $y = x^3 - 30x + 20$
   a. $[-10, 10, 1]$ by $[-10, 10, 1]$
   b. $[-10, 10, 1]$ by $[-50, 50, 10]$
   c. $[-10, 10, 1]$ by $[-50, 100, 10]$

Critical Thinking Exercises

60. Which one of the following is true?
   a. If the coordinates of a point satisfy the inequality $xy > 0$, then (x, y) must be in quadrant I.
   b. The ordered pair $(2, 5)$ satisfies $3y - 2x = -4$.
   c. If a point is on the x-axis, it is neither up nor down, so $x = 0$.
   d. None of the above is true.

In Exercises 61–64, match the story with the correct figure. The figures are labeled (a), (b), (c), and (d).

61. As the blizzard got worse, the snow fell harder and harder.

62. The snow fell more and more softly.

63. It snowed hard, but then it stopped. After a short time, the snow started falling softly.

64. It snowed softly, and then it stopped. After a short time, the snow started falling hard.

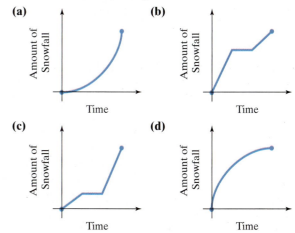

SECTION 1.2   *Linear Equations*

Objectives

1. Solve linear equations in one variable.
2. Solve equations with constants in denominators.
3. Solve equations with variables in denominators.
4. Recognize identities, conditional equations, and inconsistent equations.

Unfortunately, many of us have been fined for driving over the speed limit. The amount of the fine depends on how fast we are speeding. Suppose that a highway has a speed limit of 60 miles per hour. The amount that speeders are fined, F, is described by the statement of equality

$$F = 10x - 600$$

where x is the speed, in miles per hour. We can use this statement to determine the fine, F, for a speeder traveling at, say, 70 miles per hour. We substitute 70 for x in the given statement and then find the corresponding value for F.

$$F = 10(70) - 600 = 700 - 600 = 100$$

Thus, a person caught driving 70 miles per hour gets a $100 fine.

A friend, whom we shall call Leadfoot, borrows your car and returns a few hours later with a $400 speeding fine. Leadfoot is furious, protesting that the car was barely driven over the speed limit. Should you believe Leadfoot?

In order to decide if Leadfoot is telling the truth, use $F = 10x - 600$. Leadfoot was fined $400, so substitute 400 for F:

$$400 = 10x - 600.$$

In Example 1, we will find the value for x. This variable represents Leadfoot's speed, which resulted in the $400 fine.

An **equation** consists of two algebraic expressions joined by an equal sign. Thus, $400 = 10x - 600$ is an example of an equation. The equal sign divides the equation into two parts, the left side and the right side:

$$\boxed{400} \quad = \quad \boxed{10x - 600}.$$

Left side   Right side

The two sides of an equation can be reversed. So, we can also express this equation as

$$10x - 600 = 400.$$

Notice that the highest exponent on the variable is 1. Such an equation is called a *linear equation in one variable*. In this section, we will study how to solve linear equations.

1 Solve linear equations in one variable.

Solving Linear Equations in One Variable

We begin with a general definition of a linear equation in one variable.

Definition of a Linear Equation

A **linear equation in one variable** x is an equation that can be written in the form

$$ax + b = 0$$

where a and b are real numbers, and $a \neq 0$.

An example of a linear equation in one variable is $4x + 12 = 0$. **Solving an equation** in x involves determining all values of x that result in a true statement when substituted into the equation. Such values are **solutions,** or **roots,** of the equation. For example, substitute -3 into $4x + 12 = 0$. We obtain $4(-3) + 12 = 0$, or $-12 + 12 = 0$. This simplifies to the true statement $0 = 0$. Thus, -3 is a solution of the equation $4x + 12 = 0$. We also say that -3 **satisfies** the equation $4x + 12 = 0$, because when we substitute -3 for x, a true statement results. The set of all such solutions is called the equation's **solution set.** For example, the solution set of the equation $4x + 12 = 0$ is $\{-3\}$.

Equations that have the same solution set are called **equivalent equations.** For example, the equations $4x + 12 = 0$, $4x = -12$, and $x = -3$ are equivalent equations because the solution set for each is $\{-3\}$. To solve a linear equation in x, we transform the equation into an equivalent equation one or more times. Our final equivalent equation should be in the form $x = d$, where d is a real number. By inspection, we can see that the solution set of this equation is $\{d\}$.

To generate equivalent equations, we will use the following principles:

Study Tip

We can solve equations such as $3(x - 6) = 5x$ for a variable. However, we cannot solve for a variable in an algebraic expression such as $3(x - 6)$. We *simplify* algebraic expressions.

Correct

Simplify: $3(x - 6)$.

$3(x - 6) = 3x - 18$

Incorrect

~~Simplify: $3(x - 6)$.~~
~~$3(x - 6) = 0$~~
~~$3x - 18 = 0$~~
~~$3x - 18 + 18 = 0 + 18$~~
~~$3x = 18$~~
~~$\dfrac{3x}{3} = \dfrac{18}{3}$~~
~~$x = 6$~~

Generating Equivalent Equations

An equation can be transformed into an equivalent equation by one or more of the following operations:

Example

1. Simplify an expression by removing grouping symbols and combining like terms.

$$3(x - 6) = 6x - x$$
$$3x - 18 = 5x$$

2. Add (or subtract) the same real number or variable expression on *both* sides of the equation.

$$3x - 18 = 5x$$

Subtract 3x from both sides of the equation.

$$3x - 18 - 3x = 5x - 3x$$
$$-18 = 2x$$

3. Multiply (or divide) on *both* sides of the equation by the same *nonzero* quantity.

$$-18 = 2x$$

Divide both sides of the equation by 2.

$$\frac{-18}{2} = \frac{2x}{2}$$
$$-9 = x$$

4. Interchange the two sides of the equation.

$$-9 = x$$
$$x = -9$$

If you look closely at the equations in the box, you will notice that we have solved the equation $3(x - 6) = 6x - x$. The final equation, $x = -9$, with x isolated by itself on the left side, shows that $\{-9\}$ is the solution set. The idea in solving a linear equation is to get the variable by itself on one side of the equal sign and a number by itself on the other side.

EXAMPLE 1   Solving a Linear Equation (Is Leadfoot Telling the Truth?)

Solve the equation: $10x - 600 = 400$.

Solution   Remember that x represents Leadfoot's speed that resulted in the $400 fine. Our goal is to get x by itself on the left side. We do this by adding 600 to both sides to get $10x$ by itself. Then we isolate x from $10x$ by dividing both sides of the equation by 10.

$$10x - 600 = 400 \qquad \text{This is the given equation.}$$
$$10x - 600 + 600 = 400 + 600 \qquad \text{Add 600 to both sides.}$$
$$10x = 1000 \qquad \text{Combine like terms.}$$
$$\frac{10x}{10} = \frac{1000}{10} \qquad \text{Divide both sides by 10.}$$
$$x = 100 \qquad \text{Simplify.}$$

Can this possibly be correct? Was Leadfoot doing 100 miles per hour in the car he borrowed from you? To find out, check the proposed solution, 100, in the original equation. In other words, evaluate for $x = 100$.

Check 100:

$$10x - 600 = 400 \qquad \text{This is the original equation.}$$
$$10(100) - 600 \overset{?}{=} 400 \qquad \text{Substitute 100 for x. The question mark indicates that we do not yet know if the two sides are equal.}$$
$$1000 - 600 \overset{?}{=} 400 \qquad \text{Multiply: } 10(100) = 1000.$$
$$400 = 400 \qquad \text{Subtract: } 1000 - 600 = 400.$$

This statement is true.

The true statement $400 = 400$ indicates that 100 is the solution. This verifies that the solution set is $\{100\}$. Leadfoot was doing an outrageous 100 miles per hour, and lied by claiming that your car was barely driven over the speed limit.

Check Point 1   Solve and check: $5x - 8 = 72$.

We now present a step-by-step procedure for solving a linear equation in one variable. Not all of these steps are necessary to solve every equation.

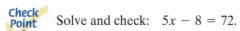

Solving a Linear Equation

1. Simplify the algebraic expression on each side.
2. Collect the variable terms on one side and the constant terms on the other side.
3. Isolate the variable and solve.
4. Check the proposed solution in the original equation.

Study Tip

If your proposed solution is incorrect, you will get a false statement when you check your answer. For example, 65 is not a solution of $10x - 600 = 400$. Look what happens when we substitute 65 for x:

$$10x - 600 = 400$$
$$10(65) - 600 \overset{?}{=} 400$$
$$650 - 600 \overset{?}{=} 400$$
$$50 = 400 \qquad \text{False.}$$

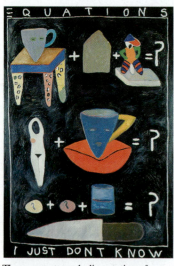

The compact, symbolic notation of algebra enables us to use a clear step-by-step method for solving equations, designed to avoid the confusion shown in the painting.

EXAMPLE 2   Solving a Linear Equation

Solve the equation: $2(x - 3) - 17 = 13 - 3(x + 2)$.

Solution

Step 1   Simplify the algebraic expression on each side.

$$2(x - 3) - 17 = 13 - 3(x + 2) \qquad \text{This is given equation.}$$
$$2x - 6 - 17 = 13 - 3x - 6 \qquad \text{Use the distributive property.}$$
$$2x - 23 = -3x + 7 \qquad \text{Combine like terms.}$$

Discovery

Solve the equation in Example 2 by collecting terms with the variable on the right and numerical terms on the left. What do you observe?

Step 2   Collect variable terms on one side and constant terms on the other side.   We will collect variable terms on the left by adding $3x$ to both sides. We will collect the numbers on the right by adding 23 to both sides.

$$2x - 23 + 3x = -3x + 7 + 3x \qquad \text{Add 3x to both sides.}$$
$$5x - 23 = 7 \qquad \text{Simplify.}$$
$$5x - 23 + 23 = 7 + 23 \qquad \text{Add 23 to both sides.}$$
$$5x = 30 \qquad \text{Simplify.}$$

Step 3   Isolate the variable and solve.   We isolate x by dividing both sides by 5.

$$\frac{5x}{5} = \frac{30}{5} \qquad \text{Divide both sides by 5.}$$

$$x = 6 \qquad \text{Simplify.}$$

Step 4   Check the proposed solution in the original equation.   Substitute 6 for x in the original equation.

$$2(x - 3) - 17 = 13 - 3(x + 2) \qquad \text{This is the original equation.}$$
$$2(6 - 3) - 17 \overset{?}{=} 13 - 3(6 + 2) \qquad \text{Substitute 6 for x.}$$
$$2(3) - 17 \overset{?}{=} 13 - 3(8) \qquad \text{Simplify inside parentheses.}$$
$$6 - 17 \overset{?}{=} 13 - 24 \qquad \text{Multiply.}$$
$$-11 = -11 \qquad \text{Subtract.}$$

The true statement $-11 = -11$ verifies that the solution set is $\{6\}$.

Technology

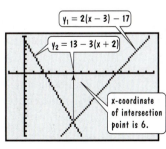

$y_1 = 2(x - 3) - 17$

$y_2 = 13 - 3(x + 2)$

x-coordinate of intersection point is 6.

$[-2, 16, 1]$ by $[-14, 8, 1]$

You can use a graphing utility to check the solution to a linear equation in one variable. **Graph the left side and graph the right side. The solution is the x-coordinate of the point where the graphs intersect.** For example, to verify that 6 is the solution of

$$2(x - 3) - 17 = 13 - 3(x + 2),$$

graph these two equations in the same viewing rectangle:

$$y_1 = 2(x - 3) - 17$$
$$\text{and} \quad y_2 = 13 - 3(x + 2).$$

Choose a large enough viewing rectangle so that you can see where the graphs intersect. The viewing rectangle on the left shows that the x-coordinate of the intersection point is 6, verifying that $\{6\}$ is the solution set for the equation in Example 2.

Check Point 2 Solve and check: $4(2x + 1) - 29 = 3(2x - 5)$.

2 Solve equations with constants in denominators.

Linear Equations with Fractions

Equations are easier to solve when they do not contain fractions. How do we solve equations involving fractions? We begin by multiplying both sides of the equation by the least common denominator of all fractions in the equation. The least common denominator is the smallest number that all the denominators will divide into. Multiplying every term on both sides of the equation by the least common denominator will eliminate the fractions in the equation. Example 3 shows how we "clear an equation of fractions."

EXAMPLE 3   Solving a Linear Equation Involving Fractions

Solve the equation: $\dfrac{3x}{2} = \dfrac{x}{5} - \dfrac{39}{5}$.

Solution   The denominators are 2, 5, and 5. The smallest number that is divisible by 2, 5, and 5 is 10. We begin by multiplying both sides of the equation by 10, the least common denominator.

$$\frac{3x}{2} = \frac{x}{5} - \frac{39}{5}$$ This is the given equation.

$$10 \cdot \frac{3x}{2} = 10\left(\frac{x}{5} - \frac{39}{5}\right)$$ Multiply both sides by 10.

$$10 \cdot \frac{3x}{2} = 10 \cdot \frac{x}{5} - 10 \cdot \frac{39}{5}$$ Use the distributive property and multiply each term by 10.

$$\overset{5}{10} \cdot \frac{3x}{\underset{1}{2}} = \overset{2}{10} \cdot \frac{x}{\underset{1}{5}} - \overset{2}{10} \cdot \frac{39}{\underset{1}{5}}$$ Divide out common factors in each multiplication.

$$15x = 2x - 78$$ Complete the multiplications. The fractions are now cleared.

At this point, we have an equation similar to those we previously solved. Collect the variable terms on one side and the constant terms on the other side.

$$15x - 2x = 2x - 2x - 78$$ Subtract 2x to get the variable terms on the left.

$$13x = -78$$ Simplify.

Isolate x by dividing both sides by 13.

$$\frac{13x}{13} = \frac{-78}{13}$$ Divide both sides by 13.

$$x = -6$$ Simplify.

Check the proposed solution. Substitute -6 for x in the original equation. You should obtain $-9 = -9$. This true statement verifies that the solution set is $\{-6\}$.

Check Point 3 Solve and check: $\dfrac{x}{4} = \dfrac{2x}{3} + \dfrac{5}{6}$.

3 Solve equations with variables in denominators.

Equations Involving Rational Expressions

In Example 3 we solved a linear equation with constants in denominators. Now, let's consider an equation such as

$$\frac{1}{x} = \frac{1}{5} + \frac{3}{2x}.$$

Can you see how this equation differs from the fractional equation that we solved earlier? The variable, x, appears in two of the denominators. The procedure for solving this equation still involves multiplying each side by the least common denominator. However, we must avoid any values of the variable that make a denominator zero. For example, examine the denominators in the equation

$$\frac{1}{x} = \frac{1}{5} + \frac{3}{2x}.$$

This denominator would equal zero if x = 0.

This denominator would equal zero if x = 0.

We see that x cannot equal zero. With this in mind, let's solve the equation.

EXAMPLE 4   Solving an Equation Involving Rational Expressions

Solve: $\dfrac{1}{x} = \dfrac{1}{5} + \dfrac{3}{2x}.$

Solution   The denominators are $x, 5$, and $2x$. The least common denominator is $10x$. We begin by multiplying both sides of the equation by $10x$. We will also write the restriction that x cannot equal zero to the right of the equation.

$$\frac{1}{x} = \frac{1}{5} + \frac{3}{2x}, \quad x \neq 0 \qquad \text{This is the given equation.}$$

$$10x \cdot \frac{1}{x} = 10x\left(\frac{1}{5} + \frac{3}{2x}\right) \qquad \text{Multiply both sides by 10x.}$$

$$10x \cdot \frac{1}{x} = 10x \cdot \frac{1}{5} + 10x \cdot \frac{3}{2x} \qquad \text{Use the distributive property and multiply each term by 10x.}$$

$$10\overset{}{x} \cdot \frac{1}{\cancel{x}} = \overset{2}{\cancel{10}}x \cdot \frac{1}{\underset{1}{\cancel{5}}} + \overset{5}{\cancel{10}}x \cdot \frac{3}{\underset{1}{2\cancel{x}}} \qquad \text{Divide out common factors in each multiplication.}$$

$$10 = 2x + 15 \qquad \text{Complete the multiplications.}$$

Observe that the resulting equation,

$$10 = 2x + 15,$$

is now cleared of fractions. With the variable term, $2x$, already on the right, we will collect constant terms on the left by subtracting 15 from both sides.

$$10 - 15 = 2x + 15 - 15 \qquad \text{Subtract 15 from both sides.}$$

$$-5 = 2x \qquad \text{Simplify.}$$

Finally, we isolate the variable, x, in $-5 = 2x$ by dividing both sides by 2.

$$\frac{-5}{2} = \frac{2x}{2} \qquad \text{Divide both sides by 2.}$$

$$-\frac{5}{2} = x \qquad \text{Simplify.}$$

We check our solution by substituting $-\frac{5}{2}$ into the original equation or by using a calculator. With a calculator, evaluate each side of the equation for $x = -\frac{5}{2}$, or for $x = -2.5$. Note that the original restriction that $x \neq 0$ is met. The solution set is $\{-\frac{5}{2}\}$.

Check Point 4   Solve: $\dfrac{5}{2x} = \dfrac{17}{18} - \dfrac{1}{3x}$.

EXAMPLE 5   Solving an Equation Involving Rational Expressions

Solve: $\dfrac{x}{x - 3} = \dfrac{3}{x - 3} + 9$.

Solution   We must avoid any values of the variable x that make a denominator zero.

$$\frac{x}{x - 3} = \frac{3}{x - 3} + 9$$

> These denominators are zero if $x - 3 = 0$, or equivalently, if $x = 3$.

We see that x cannot equal 3. With denominators of $x - 3$, $x - 3$, and 1, the least common denominator is $x - 3$. We multiply both sides of the equation by $x - 3$. We also write the restriction that x cannot equal 3 to the right of the equation.

$$\frac{x}{x - 3} = \frac{3}{x - 3} + 9, \quad x \neq 3 \qquad \text{This is the given equation.}$$

$$(x - 3) \cdot \frac{x}{x - 3} = (x - 3)\left[\frac{3}{x - 3} + 9\right] \qquad \text{Multiply both sides by } x - 3.$$

$$(x - 3) \cdot \frac{x}{x - 3} = (x - 3) \cdot \frac{3}{x - 3} + (x - 3) \cdot 9 \qquad \text{Use the distributive property.}$$

$$\cancel{(x - 3)} \cdot \frac{x}{\cancel{x - 3}} = \cancel{(x - 3)} \cdot \frac{3}{\cancel{x - 3}} + (x - 3) \cdot 9 \qquad \text{Divide out common factors in each multiplication.}$$

$$x = 3 + (x - 3) \cdot 9 \qquad \text{Simplify.}$$

The resulting equation, which can be expressed as

$$x = 3 + 9(x - 3),$$

is cleared of fractions. We now solve for x.

$$x = 3 + 9x - 27 \qquad \text{Use the distributive property.}$$

$$x = 9x - 24 \qquad \text{Combine numerical terms.}$$

$$x - 9x = 9x - 24 - 9x \qquad \text{Subtract 9x from both sides.}$$
$$-8x = -24 \qquad \text{Simplify.}$$
$$\frac{-8x}{-8} = \frac{-24}{-8} \qquad \text{Solve for x, dividing both sides by } -8.$$
$$x = 3 \qquad \text{Simplify.}$$

Study Tip

Reject any proposed solution that causes any denominator in an equation to equal 0.

The proposed solution, 3, is *not* a solution because of the restriction that $x \neq 3$. There is *no solution to this equation*. The solution set for this equation contains no elements and is called the empty set, written $\varnothing$.

Check Point 5 Solve: $\dfrac{x}{x - 2} = \dfrac{2}{x - 2} - \dfrac{2}{3}.$

4 Recognize identities, conditional equations, and inconsistent equations.

Types of Equations

We tend to place things in categories, allowing us to order and structure the world. For example, you can categorize yourself by your age group, your ethnicity, your academic major, or your gender. Equations can be placed into categories that depend on their solution sets.

An equation that is true for all real numbers for which both sides are defined is called an **identity.** An example of an identity is

$$x + 3 = x + 2 + 1.$$

Every number plus 3 is equal to that number plus 2 plus 1. Therefore, the solution set to this equation is the set of all real numbers. Another example of an identity is

$$\frac{2x}{x} = 2.$$

Because division by 0 is undefined, this equation is true for all real number values of x except 0. The solution set is the set of nonzero real numbers.

An equation that is not an identity, but that is true for at least one real number, is called a **conditional equation.** The equation $10x - 600 = 400$ is an example of a conditional equation. The equation is not an identity and is true only if x is 100.

An **inconsistent equation** is an equation that is not true for even one real number. An example of an inconsistent equation is

$$x = x + 7.$$

There is no number that is equal to itself plus 7. Some inconsistent equations are less obvious than this. Consider the equation in Example 5,

$$\frac{x}{x - 3} = \frac{3}{x - 3} + 9.$$

This equation is not true for any real number and has no solution. Thus, it is inconsistent.

EXAMPLE 6 Categorizing an Equation

Determine whether the equation

$$2(x + 1) = 2x + 3$$

is an identity, a conditional equation, or an inconsistent equation.

Technology

The graphs of $y_1 = 2(x + 1)$ and $y_2 = 2x + 3$ are parallel lines with no intersection point. This shows that the equation

$$2(x + 1) = 2x + 3$$

has no solution and is inconsistent.

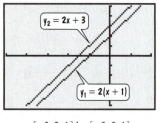

$[-5, 2, 1]$ by $[-5, 5, 1]$

Solution Let's see what happens if we try solving $2(x + 1) = 2x + 3$. Applying the distributive property on the left side, we obtain

$$2x + 2 = 2x + 3.$$

Does something look strange? Can doubling a number and increasing the product by 2 give the same result as doubling the same number and increasing the product by 3? No. Let's continue solving the equation by subtracting $2x$ from both sides.

$$2x + 2 - 2x = 2x + 3 - 2x$$
$$2 = 3$$

The false statement $2 = 3$ verifies that the given equation is inconsistent.

Check Point 6 Determine whether the equation

$$2(x + 1) = 2x + 2$$

is an identity, a conditional equation, or an inconsistent equation.

EXERCISE SET 1.2

Practice Exercises

In Exercises 1–16, solve and check each linear equation.

1. $7x - 5 = 72$ **2.** $6x - 3 = 63$

3. $11x - (6x - 5) = 40$ **4.** $5x - (2x - 10) = 35$

5. $2x - 7 = 6 + x$ **6.** $3x + 5 = 2x + 13$

7. $7x + 4 = x + 16$ **8.** $13x + 14 = 12x - 5$

9. $3(x - 2) + 7 = 2(x + 5)$

10. $2(x - 1) + 3 = x - 3(x + 1)$

11. $3(x - 4) - 4(x - 3) = x + 3 - (x - 2)$

12. $2 - (7x + 5) = 13 - 3x$

13. $16 = 3(x - 1) - (x - 7)$

14. $5x - (2x + 2) = x + (3x - 5)$

15. $25 - [2 + 5y - 3(y + 2)] =$
$$-3(2y - 5) - [5(y - 1) - 3y + 3]$$

16. $45 - [4 - 2y - 4(y + 7)] =$
$$-4(1 + 3y) - [4 - 3(y + 2) - 2(2y - 5)]$$

Exercises 17–30 contain equations with constants in denominators. Solve each equation.

17. $\dfrac{x}{3} = \dfrac{x}{2} - 2$ **18.** $\dfrac{x}{5} = \dfrac{x}{6} + 1$

19. $20 - \dfrac{x}{3} = \dfrac{x}{2}$ **20.** $\dfrac{x}{5} - \dfrac{1}{2} = \dfrac{x}{6}$

21. $\dfrac{3x}{5} = \dfrac{2x}{3} + 1$ **22.** $\dfrac{x}{2} = \dfrac{3x}{4} + 5$

23. $\dfrac{3x}{5} - x = \dfrac{x}{10} - \dfrac{5}{2}$ **24.** $2x - \dfrac{2x}{7} = \dfrac{x}{2} + \dfrac{17}{2}$

25. $\dfrac{x + 3}{6} = \dfrac{3}{8} + \dfrac{x - 5}{4}$ **26.** $\dfrac{x + 1}{4} = \dfrac{1}{6} + \dfrac{2 - x}{3}$

27. $\dfrac{x}{4} = 2 + \dfrac{x - 3}{3}$ **28.** $5 + \dfrac{x - 2}{3} = \dfrac{x + 3}{8}$

29. $\dfrac{x + 1}{3} = 5 - \dfrac{x + 2}{7}$ **30.** $\dfrac{3x}{5} - \dfrac{x - 3}{2} = \dfrac{x + 2}{3}$

*Exercises 31–50 contain equations with variables in denominators. For each equation, **a.** Write the value or values of the variable that make a denominator zero. These are the restrictions on the variable. **b.** Keeping the restrictions in mind, solve the equation.*

31. $\dfrac{4}{x} = \dfrac{5}{2x} + 3$ **32.** $\dfrac{5}{x} = \dfrac{10}{3x} + 4$

33. $\dfrac{2}{x} + 3 = \dfrac{5}{2x} + \dfrac{13}{4}$ **34.** $\dfrac{7}{2x} - \dfrac{5}{3x} = \dfrac{22}{3}$

35. $\dfrac{2}{3x} + \dfrac{1}{4} = \dfrac{11}{6x} - \dfrac{1}{3}$ **36.** $\dfrac{5}{2x} - \dfrac{8}{9} = \dfrac{1}{18} - \dfrac{1}{3x}$

37. $\dfrac{x - 2}{2x} + 1 = \dfrac{x + 1}{x}$ **38.** $\dfrac{4}{x} - \dfrac{9}{5} = \dfrac{7x - 4}{5x}$

39. $\dfrac{1}{x - 1} + 5 = \dfrac{11}{x - 1}$ **40.** $\dfrac{3}{x + 4} - 7 = \dfrac{-4}{x + 4}$

41. $\dfrac{8x}{x + 1} = 4 - \dfrac{8}{x + 1}$ **42.** $\dfrac{2}{x - 2} = \dfrac{x}{x - 2} - 2$

43. $\dfrac{3}{2x - 2} + \dfrac{1}{2} = \dfrac{2}{x - 1}$

44. $\dfrac{3}{x + 3} = \dfrac{5}{2x + 6} + \dfrac{1}{x - 2}$

45. $\dfrac{3}{x+2} + \dfrac{2}{x-2} = \dfrac{8}{(x+2)(x-2)}$

46. $\dfrac{5}{x+2} + \dfrac{3}{x-2} = \dfrac{12}{(x+2)(x-2)}$

47. $\dfrac{2}{x+1} - \dfrac{1}{x-1} = \dfrac{2x}{x^2-1}$

48. $\dfrac{4}{x+5} + \dfrac{2}{x-5} = \dfrac{32}{x^2-25}$

49. $\dfrac{1}{x-4} - \dfrac{5}{x+2} = \dfrac{6}{x^2-2x-8}$

50. $\dfrac{6}{x+3} - \dfrac{5}{x-2} = \dfrac{-20}{x^2+x-6}$

In Exercises 51–58, determine whether each equation is an identity, a conditional equation, or an inconsistent equation.

51. $4(x-7) = 4x - 28$ **52.** $4(x-7) = 4x + 28$

53. $2x + 3 = 2x - 3$ **54.** $\dfrac{7x}{x} = 7$

55. $4x + 5x = 8x$ **56.** $8x + 2x = 9x$

57. $\dfrac{2x}{x-3} = \dfrac{6}{x-3} + 4$ **58.** $\dfrac{3}{x-3} = \dfrac{x}{x-3} + 3$

The equations in Exercises 59–70 combine the types of equations we have discussed in this section. Solve each equation or state that it is true for all real numbers or no real numbers.

59. $\dfrac{x+5}{2} - 4 = \dfrac{2x-1}{3}$ **60.** $\dfrac{x+2}{7} = 5 - \dfrac{x+1}{3}$

61. $\dfrac{2}{x-2} = 3 + \dfrac{x}{x-2}$ **62.** $\dfrac{6}{x+3} + 2 = \dfrac{-2x}{x+3}$

63. $8x - (3x + 2) + 10 = 3x$

64. $2(x+2) + 2x = 4(x+1)$

65. $\dfrac{2}{x} + \dfrac{1}{2} = \dfrac{3}{4}$ **66.** $\dfrac{3}{x} - \dfrac{1}{6} = \dfrac{1}{3}$

67. $\dfrac{4}{x-2} + \dfrac{3}{x+5} = \dfrac{7}{(x+5)(x-2)}$

68. $\dfrac{1}{x-1} = \dfrac{1}{(2x+3)(x-1)} + \dfrac{4}{2x+3}$

69. $\dfrac{4x}{x+3} - \dfrac{12}{x-3} = \dfrac{4x^2+36}{x^2-9}$

70. $\dfrac{4}{x^2+3x-10} - \dfrac{1}{x^2+x-6} = \dfrac{3}{x^2-x-12}$

 ## Application Exercises

71. The equation $d = 5000c - 525{,}000$ describes the relationship between the annual number of deaths, d, in the United States from heart disease and the average cholesterol level, c, of blood. (Cholesterol level, c, is expressed in milligrams per deciliter of blood.)

a. In 2000, 725,000 Americans died from heart disease. Substitute 725,000 for d in the given equation and then solve for c to determine the average cholesterol level in 2000.

b. Suppose that the average cholesterol level for people in the United States could be reduced to 180. Substitute 180 for c in the given equation and then compute the value for d to determine the number of annual deaths from heart disease with this reduced cholesterol level. Compared to the number of deaths in 2000, how many lives would be saved by this cholesterol reduction?

72. There is a relationship between the vocabulary of a child and the child's age. The equation $60A - V = 900$ describes this relationship, where A is the age of the child, in months, and V is the number of words that the child uses. Suppose that a child uses 1500 words. Determine the child's age, in months.

73. The equation

$$p = 15 + \dfrac{15d}{33}$$

describes the pressure of sea water, p, in pounds per square foot, at a depth of d feet below the surface. The record depth for breath-held diving, by Francisco Ferreras (Cuba) off Grand-Bahama Island, on November 14, 1993, involved pressure of 201 pounds per square foot. To what depth did Ferreras descend on this ill-advised venture? (He was underwater for 2 minutes and 9 seconds!)

74. The equation $P = -0.5d + 100$ describes the percentage, P, of lost hikers found in search and rescue missions when members of the search team walk parallel to one another separated by a distance of d yards. If a search and rescue team finds 70% of lost hikers, find the parallel distance of separation between members of the search party.

Writing in Mathematics

75. What is a linear equation in one variable? Give an example of this type of equation.

76. What does it mean to solve an equation?

77. What is the solution set of an equation?

78. What are equivalent equations? Give an example.

79. What is the difference between solving an equation such as $2(x-4) + 5x = 34$ and simplifying an algebraic expression such as $2(x-4) + 5x$? If there is a difference, which topic should be taught first? Why?

80. Suppose that you solve $\dfrac{x}{5} - \dfrac{x}{2} = 1$ by multiplying both sides by 20, rather than the least common denominator of 5 and 2 (namely, 10). Describe what happens. If you get the correct solution, why do you think we clear the equation of fractions by multiplying by the *least* common denominator?

81. Suppose you are an algebra teacher grading the following solution on an examination:

$$-3(x - 6) = 2 - x$$
$$-3x - 18 = 2 - x$$
$$-2x - 18 = 2$$
$$-2x = -16$$
$$x = 8.$$

You should note that 8 checks, and the solution set is $\{8\}$. The student who worked the problem therefore wants full credit. Can you find any errors in the solution? If full credit is 10 points, how many points should you give the student? Justify your position.

82. Explain how to determine the restrictions on the variable for the equation

$$\frac{3}{x + 5} + \frac{4}{x - 2} = \frac{7}{(x + 5)(x - 2)}.$$

83. What is an identity? Give an example.

84. What is a conditional equation? Give an example.

85. What is an inconsistent equation? Give an example.

Technology Exercises

For Exercises 86–89, use your graphing utility to graph each side of the equations in the same viewing rectangle. Based on the resulting graph, label each equation as conditional, inconsistent, or an identity. If the equation is conditional, use the x-coordinate of the intersection point to find the solution set. Verify this value by direct substitution into the equation.

86. $2(x - 6) + 3x = x + 6$

87. $9x + 3 - 3x = 2(3x + 1)$

88. $2(x + \frac{1}{2}) = 5x + 1 - 3x$

89. $\dfrac{2x - 1}{3} - \dfrac{x - 5}{6} = \dfrac{x - 3}{4}$

Critical Thinking Exercises

90. Which one of the following is true?
  a. The equation $-7x = x$ has no solution.
  b. The equations $\dfrac{x}{x - 4} = \dfrac{4}{x - 4}$ and $x = 4$ are equivalent.
  c. The equations $3y - 1 = 11$ and $3y - 7 = 5$ are equivalent.
  d. If a and b are any real numbers, then $ax + b = 0$ always has one number in its solution set.

91. Solve for x: $ax + b = c$.

92. Write three equations that are equivalent to $x = 5$.

93. If x represents a number, write an English sentence about the number that results in an inconsistent equation.

94. Find b such that $\dfrac{7x + 4}{b} + 13 = x$ will have a solution set given by $\{-6\}$.

95. Find b such that $\dfrac{4x - b}{x - 5} = 3$ will have a solution set given by $\varnothing$.

Group Exercise

96. In your group, describe the best procedure for solving the following equation:

$$0.47x + \frac{19}{4} = -0.2 + \frac{2}{5}x.$$

Use this procedure to actually solve the equation. Then compare procedures with other groups working on this problem. Which group devised the most streamlined method?

SECTION 1.3 *Formulas and Applications*

Objectives

1. Solve problems using formulas.
2. Use linear equations to solve problems.
3. Solve for a variable in a formula.

Could you live to be 125? The number of Americans ages 100 or older could approach 850,000 by 2050. Some scientists predict that by 2100, our descendants could live to be 200 years of age. In this section, we will see how equations can be used to make these kinds of predictions as we turn to applications of linear equations.

1 Solve problems using formulas.

Formulas and Modeling Data

The graph in Figure 1.11 shows life expectancy in the United States by year of birth. For example, we can use the graph to find life expectancy for women born in 1980. Find the two bars for 1980 and then look at the bar on the right, representing females. The number printed on this bar is 77.4. Thus, the life expectancy for women born in 1980 is 77.4 years.

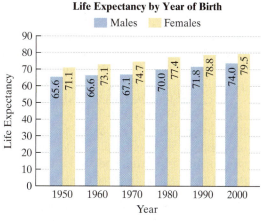

Life Expectancy by Year of Birth

Source: U.S. Bureau of the Census

Figure 1.11 Life expectancy by year of birth

The data for U.S. women in Figure 1.11 can be approximated by the equation

$$E = 0.177t + 71.35$$

where the variable E represents life expectancy for women born t years after 1950. This equation is an example of a *formula*. A **formula** is an equation that uses letters to express a relationship between two or more variables. The given formula expresses the relationship between the number of years born after 1950, t, and life expectancy for U.S. women, E.

EXAMPLE 1 Using a Formula

Use the formula

$$E = 0.177t + 71.35$$

to determine the year of birth for which U.S. women can expect to live 82 years.

Solution We are given that the life expectancy for women is 82 years, so substitute 82 for E in the formula and solve for t.

$E = 0.177t + 71.35$	This is the given formula.
$82 = 0.177t + 71.35$	Replace E with 82 and solve for t.
$82 - 71.35 = 0.177t + 71.35 - 71.35$	Isolate the term containing t by subtracting 71.35 from both sides.
$10.65 = 0.177t$	Simplify.
$\dfrac{10.65}{0.177} = \dfrac{0.177t}{0.177}$	Divide both sides by 0.177.
$60 \approx t$	Simplify. Round to the nearest whole number.

The formula indicates that U.S. women born approximately 60 years after 1950, or in 2010, can expect to live 82 years.

The process of finding equations and formulas to describe real-world phenomena is called **mathematical modeling.** Such equations and formulas, together with the meaning assigned to the variables, are called **mathematical models.** One method of creating a mathematical model is to use available data and construct an equation that describes the behavior of the data. For example, consider the formula

$$E = 0.177t + 71.35$$

in which E is the life expectancy of the U.S. women born t years after 1950. This formula, or mathematical model, can be obtained from the data for women's life expectancy given in the bar graph in Figure 1.11 on the previous page. In Chapter 2, you will learn a modeling technique that will enable you to obtain the formula.

In creating mathematical models from data, we strive for both accuracy and simplicity. The formula $E = 0.177t + 71.35$ is relatively simple to use, but as we can see from Table 1.1, it is not an entirely accurate description of the data. Sometimes a mathematical model gives an estimate that is not a good approximation or is extended too far into the future, resulting in a prediction that does not make sense. In these cases, we say that **model breakdown** has occurred.

Table 1.1 Life Expectancy for U.S. Women

Birth Year	Actual Value	Value Predicted by $E = 0.177t + 71.35$
1950	71.1	71.35
1960	73.1	73.12
1970	74.7	74.89
1980	77.4	76.66
1990	78.8	78.43
2000	79.5	80.2

 Check Point 1 The formula $W = 0.3x + 46.6$ models the average number of hours per week, W, that Americans worked x years after 1980. When will we average 55 hours of work per week?

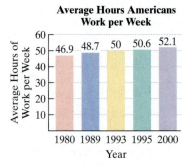

Average Hours Americans Work per Week

46.9   48.7   50   50.6   52.1

1980 1989 1993 1995 2000
Year

Source: U.S.A. Today

2 Use linear equations to solve problems.

Problem Solving with Linear Equations

Americans love their pets. The number of cats in the United States exceeds the number of dogs by 7.5 million. The number of cats and dogs combined is 114.7 million. So, how many dogs and cats are there in the United States?

Before answering the question, let's see if we can write a critical sentence that describes, or *models*, the problems conditions. The **verbal model** is

The number of dogs in the U.S.	plus	The number of cats in the U.S.	equals	114.7 million.
?	+	?	=	114.7 (million).

The question marks under the voice balloons indicate that we need algebraic expressions for these unknowns. Once we obtain these expressions, we will have an equation that models the verbal conditions. Because we are finding equations to describe real-world phenomena, we are engaged in mathematical modeling. The resulting equation, or mathematical model, is formed from a verbal model. Earlier, we mentioned that a mathematical model can be formed using actual data.

Here is a step-by-step strategy for solving problems using mathematical models that are created from verbal models:

Strategy for Problem Solving

Step 1 Read the problem carefully. Attempt to state the problem in your own words and state what the problem is looking for. Let x (or any variable) represent one of the quantities in the problem.

Step 2 If necessary, write expressions for any other unknown quantities in the problem in terms of x.

Step 3 Form a verbal model of the problems conditions and then write an equation in x that translates the verbal model.

Step 4 Solve the equation and answer the question in the problem.

Step 5 Check the proposed solution in the *original wording* of the problem, not in the equation obtained from the words.

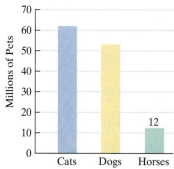

U.S. Pet Population

Source: American Veterinary Medical Association

Americans spend more than $21 billion a year on their pets. 31.4% of households have cats and 34.3% have dogs.

EXAMPLE 2 Pet Population

The number of cats in the United States exceeds the number of dogs by 7.5 million. The number of cats and dogs combined is 114.7 million. Determine the number of dogs and cats in the United States.

Solution

Step 1 Let x represent one of the quantities. We know something about the number of cats; the cat population exceeds the dog population by 7.5 million. This means that there are 7.5 million more cats than dogs. We will let

x = the number, in millions, of dogs in the United States.

Step 2 Represent other quantities in terms of x. The other unknown quantity is the number of cats. Because there are 7.5 million more cats than dogs, let

$x + 7.5$ = the number, in millions, of cats in the United States.

Step 3 Write an equation in x that describes the conditions. The number of cats and dogs combined is 114.7 million.

The number (in millions) of dogs in the U.S.	plus	the number (in millions) of cats in the U.S.	equals	114.7 million.
x	$+$	$x + 7.5$	$=$	114.7

Step 4   Solve the equation and answer the question.

$$x + x + 7.5 = 114.7$$ This is the equation that models the verbal conditions.

$$2x + 7.5 = 114.7$$ Combine like terms on the left side.

$$2x + 7.5 - 7.5 = 114.7 - 7.5$$ Subtract 7.5 from both sides.

$$2x = 107.2$$ Simplify.

$$\frac{2x}{2} = \frac{107.2}{2}$$ Divide both sides by 2.

$$x = 53.6$$ Simplify.

Because x represents the number, in millions, of dogs, there are 53.6 million dogs in the United States. Because $x + 7.5$ represents the number, in millions, of cats, there are $53.6 + 7.5$, or 61.1 million cats in the United States.

Step 5   Check the proposed solution in the original wording of the problem.
The problem states that the number of cats and dogs combined is 114.7 million. By adding 53.6 million, the dog population, and 61.1 million, the cat population, we do, indeed, obtain a sum of 114.7 million.

 Check Point 2 Two of the top-selling music albums of all time are *Jagged Little Pill* (Alanis Morissette) and *Saturday Night Fever* (Bee Gees). The Morissette album sold 5 million more copies than that of the Bee Gees. Combined, the two albums sold 27 million copies. Determine the number of sales for each of the albums.

EXAMPLE 3   Selecting a Long-Distance Carrier

You are choosing between two long-distance telephone plans. Plan A has a monthly fee of $20 with a charge of $0.05 per minute for all long-distance calls. Plan B has a monthly fee of $5 with a charge of $0.10 per minute for all long-distance calls. For how many minutes of long-distance calls will the costs for the two plans be the same?

Solution

Step 1   Let x represent one of the quantities. Let

$$x = \text{the number of minutes of long-distance calls}$$
$$\text{for the two plans to cost the same.}$$

Step 2   Represent other quantities in terms of x. There are no other unknown quantities, so we can skip this step.

Step 3   Write an equation in x that describes the conditions. The monthly cost for plan A is the monthly fee, $20, plus the per minute charge, $0.05, times the number of minutes of long-distance calls, x. The monthly cost for plan B is the monthly fee, $5, plus the per-minute charge, $0.10, times the number of minutes of long-distance calls, x.

The monthly cost for plan A | must equal | the monthly cost for plan B.

$$20 + 0.05x \quad = \quad 5 + 0.10x$$

Step 4 Solve the equation and answer the question.

$20 + 0.05x = 5 + 0.10x$	This is the equation that models the verbal conditions.
$20 + 0.05x - 0.05x = 5 + 0.10x - 0.05x$	Subtract 0.05x from both sides.
$20 = 5 + 0.05x$	Simplify.
$20 - 5 = 5 + 0.05x - 5$	Subtract 5 from both sides.
$15 = 0.05x$	Simplify.
$\dfrac{15}{0.05} = \dfrac{0.05x}{0.05}$	Divide both sides by 0.05.
$300 = x$	Simplify.

Because x represents the number of minutes of long-distance calls for the two plans to cost the same, the costs will be the same with 300 minutes of long-distance calls.

Step 5 Check the proposed solution in the original wording of the problem.
The problem states that the costs for the two plans should be the same. Let's see if they are with 300 minutes of long-distance calls:

$$\text{Cost for plan A} = \$20 + \$0.05(300) = \$20 + \$15 = \$35$$

Monthly fee | Per-minute charge

$$\text{Cost for plan B} = \$5 + \$0.10(300) = \$5 + \$30 = \$35$$

With 300 minutes, or 5 hours, of long-distance chatting, both plans cost $35 for the month. Thus, the proposed solution, 300 minutes, satisfies the problems conditions.

Check Point 3 You are choosing between two long-distance telephone plans. Plan A has a monthly fee of $15 with a charge of $0.08 per minute for all long-distance calls. Plan B has a monthly fee of $3 with a charge of $0.12 per minute for all long-distance calls. For how many minutes of long-distance calls will the costs for the two plans be the same?

Our next example involves simple interest. The annual simple interest that an investment earns is given by the formula

$$I = Pr$$

where I is the simple interest, P is the principal, and r is the simple interest rate, expressed in decimal form. Suppose, for example, that you deposit $2000 ($P = 2000$) in a savings account that has a simple interest rate of 6% ($r = 0.06$). The annual simple interest is computed as follows:

$$I = Pr = (2000)(0.06) = 120.$$

The annual interest is $120.

EXAMPLE 4 Solving a Simple Interest Problem

You inherit $16,000 with the stipulation that for the first year the money must be placed in two investments paying 6% and 8% annual interest, respectively. How much should be invested at each rate if the total interest earned for the year is to be $1180?

Solution

Step 1 **Let *x* represent one of the quantities.**

Let x = the amount invested at 6%.

Step 2 **Represent other quantities in terms of *x*.** The other quantity that we seek is the amount to be invested at 8%. Because the total amount to be invested is $16,000, and we already used up x,

$$16,000 - x = \text{the amount invested at } 8\%.$$

Step 3 **Write an equation in *x* that describes the conditions.** The interest for the two investments combined must be $1180. Interest is Pr or rP for each investment.

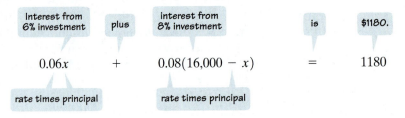

Step 4 **Solve the equation and answer the question.**

$0.06x + 0.08(16,000 - x) = 1180$	This is the equation that models the verbal conditions.
$0.06x + 1280 - 0.08x = 1180$	Use the distributive property.
$-0.02x + 1280 = 1180$	Combine like terms.
$-0.02x + 1280 - 1280 = 1180 - 1280$	Subtract 1280 from both sides.
$-0.02x = -100$	Simplify.
$\dfrac{-0.02x}{-0.02} = \dfrac{-100}{-0.02}$	Divide both sides by -0.02.
$x = 5000$	Simplify.

Because x represents the amount invested at 6%, $5000 should be invested at 6%. Because $16,000 - x$ represents the amount invested at 8%, $16,000 - $5000, or $11,000, should be invested at 8%.

Step 5 **Check the proposed solution in the original wording of the problem.** The problem states that the total interest should be $1180. The interest earned on $5000 at 6% is ($5000)(0.06), or $300. The interest earned on $11,000 at 8% is ($11,000)(0.08), or $880. The total interest is $300 + $880, or $1180, exactly as it should be.

Check Point 4 Suppose that you invest $25,000, part at 9% simple interest and the remainder at 12%. If the total yearly interest from these investments was $2550, find the amount invested at each rate.

Solving geometry problems usually requires a knowledge of basic geometric ideas and formulas. Formulas for area, perimeter, and volume are given in Table 1.2.

Table 1.2   Common Formulas for Area, Perimeter, and Volume

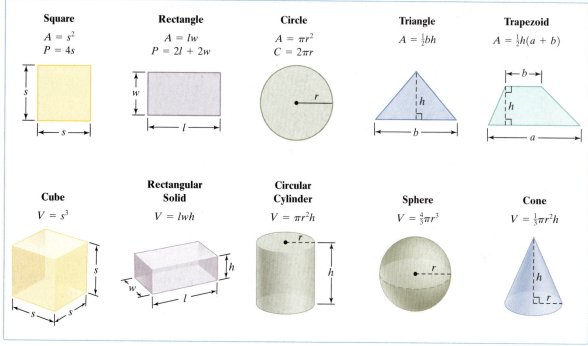

We will be using the formula for the perimeter of a rectangle, $P = 2l + 2w$, in our next example. A helpful verbal model for this formula is 2 times length plus 2 times width is a rectangles perimeter.

EXAMPLE 5   Finding the Dimensions of an American Football Field

The length of an American football field is 200 feet more than the width. If the perimeter of the field is 1040 feet, what are its dimensions?

Solution

Step 1   Let x represent one of the quantities.   We know something about the length; the length is 200 feet more than the width. We will let

$$x = \text{the width.}$$

Step 2   Represent other quantities in terms of x.   Because the length is 200 feet more than the width, let

$$x + 200 = \text{the length.}$$

Figure 1.12 illustrates an American football field and its dimensions.

Step 3   Write an equation in x that describes the conditions.   Because the perimeter of the field is 1040 feet,

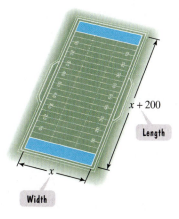

Figure 1.12 An American football field

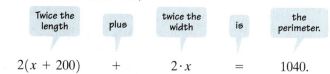

Twice the length	plus	twice the width	is	the perimeter.
$2(x + 200)$	$+$	$2 \cdot x$	$=$	$1040.$

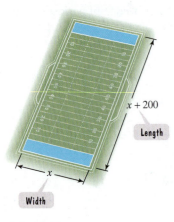

Figure 1.12, repeated

Step 4   **Solve the equation and answer the question.**

$$2(x + 200) + 2x = 1040$$ This is the equation that models the verbal conditions.

$$2x + 400 + 2x = 1040$$ Apply the distributive property.

$$4x + 400 = 1040$$ Combine like terms: $2x + 2x = 4x$.

$$4x + 400 - 400 = 1040 - 400$$ Subtract 400 from both sides.

$$4x = 640$$ Simplify.

$$\frac{4x}{4} = \frac{640}{4}$$ Divide both sides by 4.

$$x = 160$$ Simplify.

Thus,

$$\text{width} = x = 160.$$

$$\text{length} = x + 200 = 160 + 200 = 360.$$

The dimensions of an American football field are 360 feet by 160 feet. (The 360-foot length is usually described as 120 yards.)

Step 5   **Check the proposed solution in the original wording of the problem.**
The perimeter of the football field using the dimensions that we found is

$$2(360 \text{ feet}) + 2(160 \text{ feet}) = 720 \text{ feet} + 320 \text{ feet} = 1040 \text{ feet}.$$

Because the problems wording tells us that the perimeter is 1040 feet, our dimensions are correct.

Check Point 5   The length of a rectangular basketball court is 44 feet more than the width. If the perimeter of the basketball court is 288 feet, what are its dimensions?

3   Solve for a variable in a formula.

Solving for a Variable in a Formula

When working with formulas, such as the geometric formulas shown in Table 1.2 on the previous page, it is often necessary to solve for a specified variable. This is done by isolating the specified variable on one side of the equation. Begin by isolating all terms with the specified variable on one side of the equation and all terms without the specified variable on the other side. The next example shows how to do this.

EXAMPLE 6   Solving for a Variable in a Formula

Solve the formula $2l + 2w = P$ for w.

Solution   First, isolate $2w$ on the left by subtracting $2l$ from both sides. Then solve for w by dividing both sides by 2.

We need to isolate w.

$$2l + 2w = P$$ This is the given formula.

$$2l - 2l + 2w = P - 2l \qquad \text{Isolate } 2w \text{ by subtracting } 2l \text{ from both sides.}$$

$$2w = P - 2l \qquad \text{Simplify.}$$

$$\frac{2w}{2} = \frac{P - 2l}{2} \qquad \text{Isolate } w \text{ by dividing both sides by 2.}$$

> You can divide both P and 2l by 2, expressing the answer as $w = \frac{P}{2} - l$.

$$w = \frac{P - 2l}{2} \qquad \text{Simplify.}$$

Check Point 6   Solve $y = mx + b$ for m.

EXAMPLE 7   **Solving for a Variable That Occurs Twice in a Formula**

Solve the formula $A = P + Prt$ for P.

Solution   Notice that all terms with P already occur on the right side of the equation. Factor P from the two terms on the right to isolate P.

$$A = P + Prt \qquad \text{This is the given formula.}$$

$$A = P(1 + rt) \qquad \text{Factor P on the right side of the equation.}$$

$$\frac{A}{1 + rt} = \frac{P(1 + rt)}{1 + rt} \qquad \text{Divide both sides by 1 + rt.}$$

$$\frac{A}{1 + rt} = P \qquad \text{Simplify: } \frac{P\cancel{(1 + rt)}}{\cancel{(1 + rt)}} = \frac{P}{1} = P.$$

Check Point 7   Solve the formula $P = C + MC$ for C.

Study Tip

You cannot solve $A = P + Prt$ for P by subtracting Prt from both sides and writing

$$A - Prt = P.$$

When a formula is solved for a specified variable, that variable must be isolated on one side. The variable P occurs on both sides of

$$A - Prt = P.$$

EXERCISE SET 1.3

Practice Exercises

In Exercises 1–14, let x represent the number. Write each English phrase as an algebraic expression.

1. The sum of a number and 9
2. A number increased by 13
3. A number subtracted from 20
4. 13 less than a number
5. 8 decreased by 5 times a number
6. 14 less than the product of 6 and a number
7. The quotient of 15 and a number
8. The quotient of a number and 15
9. The sum of twice a number and 20
10. Twice the sum of a number and 20
11. 30 subtracted from 7 times a number
12. The quotient of 12 and a number, decreased by 3 times the number

13. Four times the sum of a number and 12
14. Five times the difference of a number and 6

In Exercises 15–20, let x represent the number. Use the given conditions to write an equation. Solve the equation and find the number.

15. A number increased by 40 is equal to 450. Find the number.
16. The sum of a number and 29 is 54. Find the number.
17. Seven subtracted from five times a number is 123. Find the number.
18. Eight subtracted from six times a number is 184. Find the number.
19. Nine times a number is 30 more than three times that number. Find the number.
20. Five more than four times a number is that number increased by 35. Find the number.

Application Exercises

Medical researchers have found that the desirable heart rate, R, in beats per minute, for beneficial exercise is modeled by the formulas

$$R = 143 - 0.65A \quad \text{for women}$$

$$R = 165 - 0.75A \quad \text{for men}$$

where A is the person's age. Use these formulas to solve Exercises 21–22.

21. If the desirable heart rate for a woman is 117 beats per minute, how old is she? How is the solution shown on the accompanying line graph?

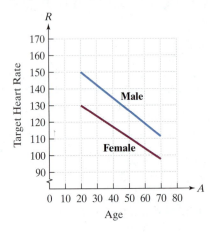

22. If the desirable heart rate for a man is 147 beats per minute, how old is he? How is the solution shown on the line graph?

Growth in human populations and economic activity threatens the continued existence of salmon in the Pacific Northwest. The bar graph shows the Pacific salmon population for various years. The data can be modeled by the formula

$$P = -0.22t + 9.6$$

in which P is the salmon population, in millions, t years after 1960. Use the formula to solve Exercises 23–24. Round to the nearest year.

Pacific Salmon Population

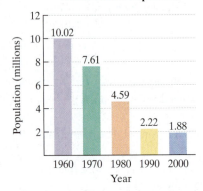

Source: U.S. Department of the Interior

23. When will the salmon population be reduced to 0.5 million?

24. When will there be no Pacific salmon?

25. The formula

$$\frac{W}{2} - 3H = 53$$

models the recommended weight W, in pounds, for a male, where H represents the man's height, in inches, over 5 feet. What is the recommended weight for a man who is 6 feet, 3 inches tall?

26. The International Panel on Climate Change is a U.N.-sponsored body made up of more than 1500 leading experts from 60 nations. According to their recent findings, increased levels of atmospheric carbon dioxide are affecting our climate. Global warming is under way and the effects could be catastrophic. The formula $C = 1.44t + 280$ models carbon dioxide concentration, C, in parts per million, t years after 1939. The preindustrial carbon dioxide concentration of 280 parts per million remained fairly constant until World War II, increasing after that due primarily to the burning of fossil fuels related to energy consumption. When will the concentration be double the preindustrial level? Round to the nearest year.

In Exercises 27–56, use the five-step strategy given in the box on page 97 to solve each problem.

27. Two of the most expensive movies ever made were *Titanic* and *Waterworld*. The cost to make *Titanic* exceeded the cost to make *Waterworld* by $40 million. The combined cost to make the two movies was $360 million. Find the cost of making each of these movies.

28. In 2001, the most populous countries in the world were China and India. In that year, China's population exceeded India's by 260 million. Combined, the two countries had a population of 2310 million. Determine the 2001 population for China and India.

29. Each year, Americans in 68 urban areas waste almost 7 billion gallons of fuel sitting in traffic. The bar graph shows the number of hours in traffic per year for the average motorist in ten cities. The average motorist in Los Angeles spends 32 hours less than twice that of the average motorist in Miami stuck in traffic each year. Together, the average motorist in Miami and the average motorist in Los Angeles spend 139 hours per year in traffic. How many hours are wasted in traffic by the average motorist in Los Angeles and Miami?

Hours in Traffic per Year for the Average Motorist

Source: Texas Transportation Institute

30. The graph shows the five costliest natural disasters in U.S. history. The cost of the Northridge, California, earthquake exceeded Hurricane Hugo by $5.5 billion and the cost of Hurricane Andrew exceeded twice that of Hugo by $6 billion. The combined cost of the three natural disasters was $39.5 billion. Determine the cost of each.

Costliest Natural Disasters in U.S. History

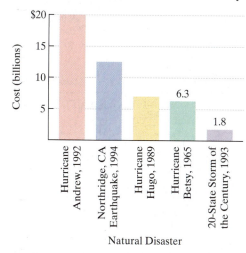

Source: Federal Emergency Management Agency

31. A car rental agency charges $200 per week plus $0.15 per mile to rent a car. How many miles can you travel in one week for $320?

32. A car rental agency charges $180 per week plus $0.25 per mile to rent a car. How many miles can you travel in one week for $395?

According to the National Center for Health Statistics, in 1990, 28% of babies in the United States were born to parents who were not married. Throughout the 1990s, this increased by approximately 0.6% per year. Use this information to solve Exercises 33–34.

33. If this trend continues, in which year will 37% of babies be born out of wedlock?

34. If this trend continues, in which year will 40% of babies be born out of wedlock?

35. The bus fare in a city is $1.25. People who use the bus have the option of purchasing a monthly coupon book for $21.00. With the coupon book, the fare is reduced to $0.50.
  a. Let x represent the number of times in a month the bus is used. Write algebraic expressions for the total monthly costs of using the bus x times both with and without the coupon book.
  b. Determine the number of times in a month the bus must be used so that the total monthly cost without the coupon book is the same as the total monthly cost with the coupon book.

36. A coupon book for a bridge costs $21 per month. The toll for the bridge is normally $2.50, but it is reduced to $1 for people who have purchased the coupon book.

a. Let x represent the number of times in a month the bridge is used. Write algebraic expressions for the total monthly costs of using the bridge x times both with and without the coupon book.
b. Determine the number of times in a month the bridge must be crossed so that the total monthly cost without the coupon book is the same as the total monthly cost with the coupon book.

37. You are choosing between two plans at a discount warehouse. Plan A offers an annual membership fee of $100 and you pay 80% of the manufacturer's recommended list price. Plan B offers an annual membership fee of $40 and you pay 90% of the manufacturer's recommended list price. How many dollars of merchandise would you have to purchase in a year to pay the same amount under both plans? What will be the cost for each plan?

38. You are choosing between two plans at a discount warehouse. Plan A offers an annual membership fee of $300 and you pay 70% of the manufacturer's recommended list price. Plan B offers an annual membership fee of $40 and you pay 90% of the manufacturer's recommended list price. How many dollars of merchandise would you have to purchase in a year to pay the same amount under both plans? What will be the cost for each plan?

39. Your grandmother needs your help. She has $50,000 to invest. Part of this money is to be invested in noninsured bonds paying 15% annual interest. The rest of this money is to be invested in a government-insured certificate of deposit paying 7% annual interest. She told you that she requires $6000 per year in extra income from both of these investments. How much money should be placed in each investment?

40. You inherit $18,750 with the stipulation that for the first year the money must be placed in two investments paying 10% and 12% annual interest, respectively. How much should be invested at each rate if the total interest earned for the year is to be $2117?

41. Things did not go quite as planned. You invested $8000, part of it in stock that paid 12% annual interest. However, the rest of the money suffered a 5% loss. If the total annual income from both investments was $620, how much was invested at each rate?

42. Things did not go quite as planned. You invested $12,000, part of it in stock that paid 14% annual interest. However, the rest of the money suffered a 6% loss. If the total annual income from both investments was $680, how much was invested at each rate?

43. The length of the rectangular tennis court at Wimbledon is 6 feet longer than twice the width. If the court's perimeter is 228 feet, what are the court's dimensions?

44. A rectangular soccer field is twice as long as it is wide. If the perimeter of the soccer field is 300 yards, what are its dimensions?

45. The height of the bookcase in the figure is 3 feet longer than the length of a shelf. If 18 feet of lumber is available for the entire unit, find the length and height of the unit.

height

shelf length

46. A bookcase is to be constructed as shown in the figure. The length is to be 3 times the height. If 60 feet of lumber is available for the entire unit, find the length and height of the bookcase.

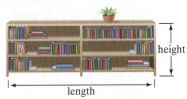

height

length

47. An automobile repair shop charged a customer $448, listing $63 for parts and the remainder for labor. If the cost of labor is $35 per hour, how many hours of labor did it take to repair the car?

48. A repair bill on a yacht came to $1603, including $532 for parts and the remainder for labor. If the cost of labor is $63 per hour, how many hours of labor did it take to repair the yacht?

The graph shows median, or average, income by level of education. Exercises 49–50 use the information in the bar graph.

Income by Level of Education

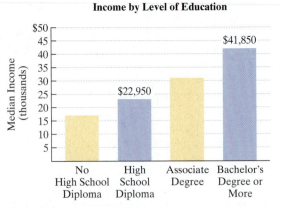

$41,850

$22,950

Median Income (thousands)

No High School Diploma   High School Diploma   Associate Degree   Bachelor's Degree or More

Source: U.S. Department of Commerce

49. The annual salary for people with a bachelor's degree or more is an increase of 35% of the annual salary for people with an associate degree. What is the average annual salary with an associate degree?

50. The annual salary for high school graduates is an increase of 35% of the annual salary for people without a high school diploma. What is the average annual salary without a high school diploma?

51. Answer the question in the following *Peanuts* cartoon strip. (*Note:* You may not use the answer given in the cartoon!)

PEANUTS reprinted by permission of United Features Syndicate, Inc.

52. After a graphing calculator's price is reduced by $\frac{1}{3}$ of its original price, you purchase it for $64. What was the graphing calculator's price before the reduction?

53. After a 12% price reduction, a car sold for $17,600. What was the car's price before the reduction?

54. Including 8% sales tax, an inn charges $162 per night. Find the inn's nightly cost before the tax is added.

55. An HMO pamphlet contains the following recommended weight for women: "Give yourself 100 pounds for the first 5 feet plus 5 pounds for every inch over 5 feet tall." Using this description, what height corresponds to a recommended weight of 135 pounds?

56. A job pays an annual salary of $33,150, which includes a holiday bonus of $750. If paychecks are issued twice a month, what is the gross amount for each paycheck?

In Exercises 57–76, solve each formula for the specified variable. Do you recognize the formula? If so, what does it describe?

57. $A = lw$ for w

58. $D = RT$ for R

59. $A = \frac{1}{2}bh$ for b

60. $V = \frac{1}{3}Bh$ for B

61. $I = Prt$ for P

62. $C = 2\pi r$ for r

63. $E = mc^2$ for m

64. $V = \pi r^2 h$ for h

65. $T = D + pm$ for p

66. $P = C + MC$ for M

67. $A = \frac{1}{2}h(a + b)$ for a

68. $A = \frac{1}{2}h(a + b)$ for b

69. $S = P + Prt$ for r

70. $S = P + Prt$ for t

71. $B = \dfrac{F}{S - V}$ for S

72. $S = \dfrac{C}{1 - r}$ for r

73. $IR + Ir = E$ for I

74. $A = 2lw + 2lh + 2wh$ for h

75. $\dfrac{1}{p} + \dfrac{1}{q} = \dfrac{1}{f}$ for f

76. $\dfrac{1}{R} = \dfrac{1}{R_1} + \dfrac{1}{R_2}$ for R_1

Writing in Mathematics

77. What is a formula?

78. We discussed formulas in this section after we considered procedures for solving linear equations. Doesn't working with a formula simply mean substituting given numbers into the formula and using the order of operations? Is it necessary to know how to solve equations to work with formulas? Explain.

79. In your own words, describe a step-by-step approach for solving algebraic word problems.

80. Did you have some difficulties solving some of the problems that were assigned in this exercise set? Discuss what you did if this happened to you. Did your course of action enhance your ability to solve algebraic word problems?

Technology Exercises

81. The formula $y = 28 + 0.6x$ models the percentage, y, of U.S. babies born out of wedlock x years after 1990. Graph the formula in a $[0, 20, 5]$ by $[0, 50, 10]$ viewing rectangle. Then use the TRACE or ZOOM feature to verify your answer in Exercise 33 or 34.

82. A tennis club offers two payment options. Members can pay a monthly fee of $30 plus $5 per hour for court rental time. The second option has no monthly fee, but court time costs $7.50 per hour.

   a. Write a mathematical model representing total monthly costs for each option for x hours of court rental time.

   b. Use a graphing utility to graph the two models in a $[0, 15, 1]$ by $[0, 120, 20]$ viewing rectangle.

   c. Use your utility's trace or intersection feature to determine where the two graphs intersect. Describe what the coordinates of this intersection point represent in practical terms.

   d. Verify part (c) using an algebraic approach by setting the two models equal to one another and determining how many hours one has to rent the court so that the two plans result in identical monthly costs.

Critical Thinking Exercises

83. At the north campus of a performing arts school, 10% of the students are music majors. At the south campus, 90% of the students are music majors. The campuses are merged into one east campus. If 42% of the 1000 students at the east campus are music majors, how many students did the north and south campuses have before the merger?

84. The price of a dress is reduced by 40%. When the dress still does not sell, it is reduced by 40% of the reduced price. If the price of the dress after both reductions is $72, what was the original price?

85. In a film, the actor Charles Coburn plays an elderly "uncle" character criticized for marrying a woman when he is 3 times her age. He wittily replies, "Ah, but in 20 years time I shall only be twice her age." How old is the "uncle" and the woman?

86. Suppose that we agree to pay you 8¢ for every problem in this chapter that you solve correctly and fine you 5¢ for every problem done incorrectly. If at the end of 26 problems we do not owe each other any money, how many problems did you solve correctly?

87. It was wartime when the Ricardos found out Mrs. Ricardo was pregnant. Ricky Ricardo was drafted and made out a will, deciding that $14,000 in a savings account was to be divided between his wife and his child-to-be. Rather strangely, and certainly with gender bias, Ricky stipulated that if the child were a boy, he would get twice the amount of the mother's portion. If it were a girl, the mother would get twice the amount the girl was to receive. We'll never know what Ricky was thinking of, for (as fate would have it) he did not return from war. Mrs. Ricardo gave birth to twins—a boy and a girl. How was the money divided?

88. Solve for C: $V = C - \dfrac{C - S}{L} N$.

Group Exercise

89. One of the best ways to learn how to *solve* a word problem in algebra is to *design* word problems of your own. Creating a word problem makes you very aware of precisely how much information is needed to solve the problem. You must also focus on the best way to present information to a reader and on how much information to give. As you write your problem, you gain skills that will help you solve problems created by others.

   The group should design five different word problems that can be solved using linear equations. All of the problems should be on different topics. For example, the group should not have more than one problem on simple interest. The group should turn in both the problems and their algebraic solutions.

SECTION 1.4 *Complex Numbers*

Objectives

1. Add and subtract complex numbers.
2. Multiply complex numbers.
3. Divide complex numbers.
4. Perform operations with square roots of negative numbers.

THE KID WHO LEARNED ABOUT MATH ON THE STREET

Who is this kid warning us about our eyeballs turning black if we attempt to find the square root of −9? Don't believe what you hear on the street. Although square roots of negative numbers are not real numbers, they do play a significant role in algebra. In this section, we move beyond the real numbers and discuss square roots with negative radicands.

The Imaginary Unit *i*

In the next section, we'll be studying equations whose solutions involve the square roots of negative numbers. Because the square of a real number is never negative, there is no real number x such that $x^2 = -1$. To provide a setting in which such equations have solutions, mathematicians invented an expanded system of numbers, the complex numbers. The *imaginary number i*, defined to be a solution of the equation $x^2 = -1$, is the basis of this new set.

The Imaginary Unit *i*

The **imaginary unit** *i* is defined as

$$i = \sqrt{-1}, \quad \text{where} \quad i^2 = -1.$$

Using the imaginary unit *i*, we can express the square root of any negative number as a real multiple of *i*. For example,

$$\sqrt{-25} = i\sqrt{25} = 5i.$$

We can check this result by squaring $5i$ and obtaining -25.

$$(5i)^2 = 5^2 i^2 = 25(-1) = -25$$

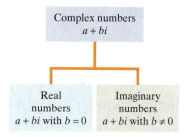

Complex numbers
$a + bi$

Real numbers	Imaginary numbers
$a + bi$ with $b = 0$	$a + bi$ with $b \neq 0$

Figure 1.13 The complex number system

A new system of numbers, called *complex numbers*, is based on adding multiples of i, such as $5i$, to the real numbers.

Complex Numbers

The set of all numbers in the form

$$a + bi$$

with real numbers a and b, and i, the imaginary unit, is called the set of **complex numbers.** The real number a is called the **real part,** and the real number b is called the **imaginary part,** of the complex number $a + bi$. If $b \neq 0$, then the complex number is called an **imaginary number** (Figure 1.13). An imaginary number in the form bi is called a **pure imaginary number.**

Here are some examples of complex numbers. Each number can be written in the form $a + bi$.

$$-4 + 6i \qquad\qquad 2i = 0 + 2i \qquad\qquad 3 = 3 + 0i$$

a, the real part, is −4.	b, the imaginary part, is 6.	a, the real part, is 0.	b, the imaginary part, is 2.	a, the real part, is 3.	b, the imaginary part, is 0.

Can you see that b, the imaginary part, is not zero in the first two complex numbers? Because $b \neq 0$, these complex numbers are imaginary numbers. Furthermore, the imaginary number $2i$ is a pure imaginary number. By contrast, the imaginary part of the complex number on the right is zero. This complex number is not an imaginary number. The number 3, or $3 + 0i$, is a real number.

A complex number is said to be **simplified** if it is expressed in the **standard form** $a + bi$. If b is a radical, we usually write i before b. For example, we write $7 + i\sqrt{5}$ rather than $7 + \sqrt{5}i$, which could easily be confused with $7 + \sqrt{5i}$.

Expressed in standard form, two complex numbers are equal if and only if their real parts are equal and their imaginary parts are equal.

Equality of Complex Numbers

$a + bi = c + di$ if and only if $a = c$ and $b = d$.

1 Add and subtract complex numbers.

Operations with Complex Numbers

The form of a complex number $a + bi$ is like the binomial $a + bx$. Consequently, we can add, subtract, and multiply complex numbers using the same methods we used for binomials, remembering that $i^2 = -1$.

Adding and Subtracting Complex Numbers

1. $(a + bi) + (c + di) = (a + c) + (b + d)i$
In words, this says that you add complex numbers by adding their real parts, adding their imaginary parts, and expressing the sum as a complex number.

2. $(a + bi) - (c + di) = (a - c) + (b - d)i$
In words, this says that you subtract complex numbers by subtracting their real parts, subtracting their imaginary parts, and expressing the difference as a complex number.

EXAMPLE 1 Adding and Subtracting Complex Numbers

Perform the indicated operations, writing the result in standard form:

a. $(5 - 11i) + (7 + 4i)$ **b.** $(-5 + 7i) - (-11 - 6i)$.

Study Tip

The following examples, using the same integers as in Example 1, show how operations with complex numbers are just like operations with polynomials.

a. $(5 - 11x) + (7 + 4x)$
$= 12 - 7x$

b. $(-5 + 7x) - (-11 - 6x)$
$= -5 + 7x + 11 + 6x$
$= 6 + 13x$

Solution

a. $(5 - 11i) + (7 + 4i)$

$\quad = 5 - 11i + 7 + 4i$ Remove the parentheses.

$\quad = 5 + 7 - 11i + 4i$ Group real and imaginary terms.

$\quad = (5 + 7) + (-11 + 4)i$ Add real parts and add imaginary parts.

$\quad = 12 - 7i$ Simplify.

b. $(-5 + 7i) - (-11 - 6i)$

$\quad = -5 + 7i + 11 + 6i$ Remove the parentheses.

$\quad = -5 + 11 + 7i + 6i$ Group real and imaginary terms.

$\quad = (-5 + 11) + (7 + 6)i$ Add real parts and add imaginary parts.

$\quad = 6 + 13i$ Simplify.

Check Point 1   Add or subtract as indicated:

a. $(5 - 2i) + (3 + 3i)$ **b.** $(2 + 6i) - (12 - 4i)$.

2   Multiply complex numbers.

Multiplication of complex numbers is performed the same way as multiplication of polynomials, using the distributive property and the FOIL method. After completing the multiplication, we replace i^2 with -1. This idea is illustrated in the next example.

EXAMPLE 2 Multiplying Complex Numbers

Find the products:

a. $4i(3 - 5i)$ **b.** $(7 - 3i)(-2 - 5i)$.

Solution

a. $4i(3 - 5i) = 4i(3) - 4i(5i)$ Distribute 4i throughout the parentheses.

$\quad\quad\quad\quad = 12i - 20i^2$ Multiply.

$\quad\quad\quad\quad = 12i - 20(-1)$ Replace i^2 with -1.

$\quad\quad\quad\quad = 20 + 12i$ Simplify to 12i + 20 and write in standard form.

b. $(7 - 3i)(-2 - 5i)$

    F   O   I   L

$\quad = -14 - 35i + 6i + 15i^2$ Use the FOIL method.

$\quad = -14 - 35i + 6i + 15(-1)$ $i^2 = -1$

$\quad = -14 - 15 - 35i + 6i$ Group real and imaginary terms.

$\quad = -29 - 29i$ Combine real and imaginary terms.

Check Point 2   Find the products:

a. $7i(2 - 9i)$ **b.** $(5 + 4i)(6 - 7i)$.

3 Divide complex numbers.

Complex Conjugates and Division

It is possible to multiply complex numbers and obtain a real number. This occurs when we multiply $a + bi$ and $a - bi$.

$$(a + bi)(a - bi) = a^2 - abi + abi - b^2i^2 \quad \text{Use the FOIL method.}$$
$$= a^2 - b^2(-1) \quad\quad\quad\quad i^2 = -1$$
$$= a^2 + b^2 \quad\quad\quad\quad\quad \text{Notice that this product eliminates } i.$$

For the complex number $a + bi$, we define its *complex conjugate* to be $a - bi$. The multiplication of complex conjugates results in a real number.

> **Conjugate of a Complex Number**
>
> The **complex conjugate** of the number $a + bi$ is $a - bi$, and the complex conjugate of $a - bi$ is $a + bi$. The multiplication of complex conjugates gives a real number.
>
> $$(a + bi)(a - bi) = a^2 + b^2$$
> $$(a - bi)(a + bi) = a^2 + b^2$$

Complex conjugates are used when dividing complex numbers. By multiplying the numerator and the denominator of the division by the complex conjugate of the denominator, you will obtain a real number in the denominator.

EXAMPLE 3 Using Complex Conjugates to Divide Complex Numbers

Divide and express the result in standard form: $\dfrac{7 + 4i}{2 - 5i}$.

Solution The complex conjugate of the denominator, $2 - 5i$, is $2 + 5i$. Multiplication of both the numerator and the denominator by $2 + 5i$ will eliminate i from the denominator.

$$\frac{7 + 4i}{2 - 5i} = \frac{(7 + 4i)}{(2 - 5i)} \cdot \frac{(2 + 5i)}{(2 + 5i)} \quad \text{Multiply the numerator and the denominator by the complex conjugate of the denominator.}$$

$$= \frac{14 + 35i + 8i + 20i^2}{2^2 + 5^2} \quad \text{Use the FOIL method in the numerator and } (a - bi)(a + bi) = a^2 + b^2 \text{ in the denominator.}$$

$$= \frac{14 + 43i + 20(-1)}{29} \quad \text{Combine imaginary terms and replace } i^2 \text{ with } -1.$$

$$= \frac{-6 + 43i}{29} \quad \text{Combine real terms in the numerator.}$$

$$= -\frac{6}{29} + \frac{43}{29}i \quad \text{Express the answer in standard form.}$$

Observe that the quotient is expressed in the standard form $a + bi$, with $a = -\frac{6}{29}$ and $b = \frac{43}{29}$.

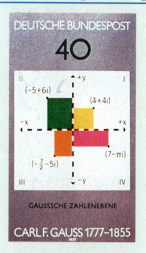

Complex Numbers on a Postage Stamp

DEUTSCHE BUNDESPOST

40

$(-5+6i)$
$(4+4i)$
$(7-\pi i)$
$\left(-\frac{7}{2}-5i\right)$

GAUSSSCHE ZAHLENEBENE

CARL F. GAUSS 1777–1855

This stamp honors the work done by the German mathematician Carl Friedrich Gauss (1777–1855) with complex numbers. Gauss represented complex numbers as points in the plane.

Check Point 3 Divide and express the result in standard form: $\dfrac{5 + 4i}{4 - 2i}$.

4 Perform operations with square roots of negative numbers.

Roots of Negative Numbers

The square of $4i$ and the square of $-4i$ both result in -16.

$$(4i)^2 = 16i^2 = 16(-1) = -16 \qquad (-4i)^2 = 16i^2 = 16(-1) = -16$$

Consequently, in the complex number system -16 has two square roots, namely, $4i$ and $-4i$. We call $4i$ the **principal square root** of -16.

Principal Square Root of a Negative Number

For any positive number real number b, the **principal square root** of the negative number $-b$ is defined by

$$\sqrt{-b} = i\sqrt{b}.$$

EXAMPLE 4 Operations Involving Square Roots of Negative Numbers

Perform the indicated operations and write the result in standard form:

a. $\sqrt{-18} - \sqrt{-8}$ **b.** $\left(-1 + \sqrt{-5}\right)^2$ **c.** $\dfrac{-25 + \sqrt{-50}}{15}$.

Solution Begin by expressing all square roots of negative numbers in terms of i.

a. $\sqrt{-18} - \sqrt{-8} = i\sqrt{18} - i\sqrt{8} = i\sqrt{9 \cdot 2} - i\sqrt{4 \cdot 2}$

$$= 3i\sqrt{2} - 2i\sqrt{2} = i\sqrt{2}$$

$(A + B)^2 = A^2 + 2\,A\,B + B^2$

b. $\left(-1 + \sqrt{-5}\right)^2 = \left(-1 + i\sqrt{5}\right)^2 = (-1)^2 + 2(-1)\left(i\sqrt{5}\right) + \left(i\sqrt{5}\right)^2$

$$= 1 - 2i\sqrt{5} + 5i^2$$
$$= 1 - 2i\sqrt{5} + 5(-1)$$
$$= -4 - 2i\sqrt{5}$$

c. $\dfrac{-25 + \sqrt{-50}}{15}$

$= \dfrac{-25 + i\sqrt{50}}{15}$ $\sqrt{-b} = i\sqrt{b}$

$= \dfrac{-25 + 5i\sqrt{2}}{15}$ $\sqrt{50} = \sqrt{25 \cdot 2} = 5\sqrt{2}$

$= \dfrac{-25}{15} + \dfrac{5i\sqrt{2}}{15}$ Write the complex number in standard form.

$= -\dfrac{5}{3} + i\dfrac{\sqrt{2}}{3}$ Simplify.

Study Tip

Do not apply the properties

$$\sqrt{b}\,\sqrt{c} = \sqrt{bc}$$

and

$$\frac{\sqrt{b}}{\sqrt{c}} = \sqrt{\frac{b}{c}}$$

to the pure imaginary numbers because these properties can only be used when b and c are positive.

Correct:

$\sqrt{-25}\,\sqrt{-4} = i\sqrt{25}\,i\sqrt{4}$
$= (5i)(2i)$
$= 10i^2$
$= 10(-1)$
$= -10$

Incorrect:

$\sqrt{-25}\,\sqrt{-4} = \sqrt{(-25)(-4)}$
$= \sqrt{100}$
$= 10$

One way to avoid confusion is to represent square roots of negative numbers in terms of i before performing any operations.

Check Point 4 Perform the indicated operations and write the result in standard form:

a. $\sqrt{-27} + \sqrt{-48}$ **b.** $\left(-2 + \sqrt{-3}\right)^2$ **c.** $\dfrac{-14 + \sqrt{-12}}{2}$.

EXERCISE SET 1.4

Practice Exercises

In Exercises 1–8, add or subtract as indicated and write the result in standard form.

1. $(7 + 2i) + (1 - 4i)$ **2.** $(-2 + 6i) + (4 - i)$

3. $(3 + 2i) - (5 - 7i)$ **4.** $(-7 + 5i) - (-9 - 11i)$

5. $6 - (-5 + 4i) - (-13 - 11i)$

6. $7 - (-9 + 2i) - (-17 - 6i)$

7. $8i - (14 - 9i)$ **8.** $15i - (12 - 11i)$

In Exercises 9–20, find each product and write the result in standard form.

9. $-3i(7i - 5)$ **10.** $-8i(2i - 7)$

11. $(-5 + 4i)(3 + 7i)$ **12.** $(-4 - 8i)(3 + 9i)$

13. $(7 - 5i)(-2 - 3i)$ **14.** $(8 - 4i)(-3 + 9i)$

15. $(3 + 5i)(3 - 5i)$ **16.** $(2 + 7i)(2 - 7i)$

17. $(-5 + 3i)(-5 - 3i)$ **18.** $(-7 - 4i)(-7 + 4i)$

19. $(2 + 3i)^2$ **20.** $(5 - 2i)^2$

In Exercises 21–28, divide and express the result in standard form.

21. $\dfrac{2}{3 - i}$ **22.** $\dfrac{3}{4 + i}$

23. $\dfrac{2i}{1 + i}$ **24.** $\dfrac{5i}{2 - i}$

25. $\dfrac{8i}{4 - 3i}$ **26.** $\dfrac{-6i}{3 + 2i}$

27. $\dfrac{2 + 3i}{2 + i}$ **28.** $\dfrac{3 - 4i}{4 + 3i}$

In Exercises 29–44, perform the indicated operations and write the result in standard form.

29. $\sqrt{-64} - \sqrt{-25}$ **30.** $\sqrt{-81} - \sqrt{-144}$

31. $5\sqrt{-16} + 3\sqrt{-81}$ **32.** $5\sqrt{-8} + 3\sqrt{-18}$

33. $(-2 + \sqrt{-4})^2$ **34.** $(-5 - \sqrt{-9})^2$

35. $(-3 - \sqrt{-7})^2$ **36.** $(-2 + \sqrt{-11})^2$

37. $\dfrac{-8 + \sqrt{-32}}{24}$ **38.** $\dfrac{-12 + \sqrt{-28}}{32}$

39. $\dfrac{-6 - \sqrt{-12}}{48}$ **40.** $\dfrac{-15 - \sqrt{-18}}{33}$

41. $\sqrt{-8}(\sqrt{-3} - \sqrt{5})$ **42.** $\sqrt{-12}(\sqrt{-4} - \sqrt{2})$

43. $(3\sqrt{-5})(-4\sqrt{-12})$ **44.** $(3\sqrt{-7})(2\sqrt{-8})$

Writing in Mathematics

45. What is i?

46. Explain how to add complex numbers. Provide an example with your explanation.

47. Explain how to multiply complex numbers and give an example.

48. What is the complex conjugate of $2 + 3i$? What happens when you multiply this complex number by its complex conjugate?

49. Explain how to divide complex numbers. Provide an example with your explanation.

50. A stand-up comedian uses algebra in some jokes, including one about a telephone recording that announces "You have just reached an imaginary number. Please multiply by i and dial again." Explain the joke.

Explain the error in Exercises 51–52.

51. $\sqrt{-9} + \sqrt{-16} = \sqrt{-25} = i\sqrt{25} = 5i$

52. $(\sqrt{-9})^2 = \sqrt{-9} \cdot \sqrt{-9} = \sqrt{81} = 9$

Critical Thinking Exercises

53. Which one of the following is true?

   a. Some irrational numbers are not complex numbers.

   b. $(3 + 7i)(3 - 7i)$ is an imaginary number.

   c. $\dfrac{7 + 3i}{5 + 3i} = \dfrac{7}{5}$

   d. In the complex number system, $x^2 + y^2$ (the sum of two squares) can be factored as $(x + yi)(x - yi)$.

In Exercises 54–56, perform the indicated operations and write the result in standard form.

54. $(8 + 9i)(2 - i) - (1 - i)(1 + i)$

55. $\dfrac{4}{(2 + i)(3 - i)}$ **56.** $\dfrac{1 + i}{1 + 2i} + \dfrac{1 - i}{1 - 2i}$

57. Evaluate $x^2 - 2x + 2$ for $x = 1 + i$.

SECTION 1.5   *Quadratic Equations*

Objectives

1. Solve quadratic equations by factoring.
2. Solve quadratic equations by the square root method.
3. Solve quadratic equations by completing the square.
4. Solve quadratic equations using the quadratic formula.
5. Use the discriminant to determine the number and type of solutions.
6. Determine the most efficient method to use when solving a quadratic equation.
7. Solve problems modeled by quadratic equations.

Serpico, 1973, starring Al Pacino, is a movie about police corruption.

In 2000, a police scandal shocked Los Angeles. A police officer who had been convicted of stealing cocaine held as evidence described how members of his unit behaved in ways that resembled the gangs they were targeting, assaulting and framing innocent people.

Is police corruption on the rise? The graph in Figure 1.14 shows the number of convictions of police officers throughout the United States for seven years.

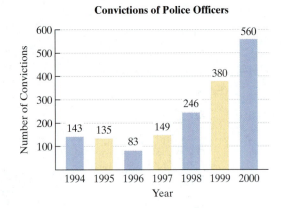

Convictions of Police Officers

Figure 1.14

Source: F.B.I.

The data can be modeled by the formula

$$N = 23.4x^2 - 259.1x + 815.8$$

where N is the number of police officers convicted of felonies x years after 1990. If present trends continue, in which year will 1000 police officers be convicted of felonies? To answer the question, it is necessary to substitute 1000 for N in the formula and solve for x, the number of years after 1990:

$$1000 = 23.4x^2 - 259.1x + 815.8.$$

Do you see how this equation differs from a linear equation? The exponent on x is 2. Solving such an equation involves finding the set of numbers that make the equation a true statement. In this section, we study a number of methods for solving equations in the form $ax^2 + bx + c = 0$. We also look at applications of these equations.

The General Form of a Quadratic Equation

We begin by defining a quadratic equation.

> **Definition of a Quadratic Equation**
>
> A **quadratic equation** in x is an equation that can be written in the **general form**
>
> $$ax^2 + bx + c = 0$$
>
> where a, b, and c are real numbers, with $a \neq 0$. A quadratic equation in x is also called a **second-degree polynomial equation** in x.

An example of a quadratic equation in general form is $x^2 - 7x + 10 = 0$. The coefficient of x^2 is $1\,(a = 1)$, the coefficient of x is $-7\,(b = -7)$, and the constant term is $10\,(c = 10)$.

1 Solve quadratic equations by factoring.

Solving Quadratic Equations by Factoring

We can factor the left side of the quadratic equation $x^2 - 7x + 10 = 0$. We obtain $(x - 5)(x - 2) = 0$. If a quadratic equation has zero on one side and a factored expression on the other side, it can be solved using the **zero-product principle.**

> **The Zero-Product Principle**
>
> If the product of two algebraic expressions is zero, then at least one of the factors is equal to zero.
>
> $$\text{If } AB = 0, \quad \text{then} \quad A = 0 \text{ or } B = 0.$$

For example, consider the equation $(x - 5)(x - 2) = 0$. According to the zero-product principle, this product can be zero only if at least one of the factors is zero. We set each individual factor equal to zero and solve each resulting equation for x.

$$(x - 5)(x - 2) = 0$$
$$x - 5 = 0 \quad \text{or} \quad x - 2 = 0$$
$$x = 5 \qquad\qquad x = 2$$

We can check each of these proposed solutions in the original quadratic equation, $x^2 - 7x + 10 = 0$.

Check 5:

$$5^2 - 7 \cdot 5 + 10 \stackrel{?}{=} 0$$
$$25 - 35 + 10 \stackrel{?}{=} 0$$
$$0 = 0 \checkmark$$

Check 2:

$$2^2 - 7 \cdot 2 + 10 \stackrel{?}{=} 0$$
$$4 - 14 + 10 \stackrel{?}{=} 0$$
$$0 = 0 \checkmark$$

The resulting true statements, indicated by the checks, show that the solutions are 5 and 2. The solution set is $\{5, 2\}$. Note that with a quadratic equation, we can have two solutions, compared to the conditional linear equation that had one.

> **Solving a Quadratic Equation by Factoring**
>
> **1.** If necessary, rewrite the equation in the form $ax^2 + bx + c = 0$, moving all terms to one side, thereby obtaining zero on the other side.
>
> **2.** Factor.   (continues on the next page)

Solving a Quadratic Equation by Factoring (continued)

3. Apply the zero-product principle, setting each factor equal to zero.
4. Solve the equations in step 3.
5. Check the solutions in the original equation.

EXAMPLE 1 Solving Quadratic Equations by Factoring

Solve by factoring:

a. $4x^2 - 2x = 0$ **b.** $2x^2 + 7x = 4$.

Solution

a. We begin with $4x^2 - 2x = 0$.

Step 1 Move all terms to one side and obtain zero on the other side. All terms are already on the left and zero is on the other side, so we can skip this step.

Step 2 Factor. We factor out $2x$ from the two terms on the left side.

$$4x^2 - 2x = 0 \quad \text{This is the given equation.}$$

$$2x(2x - 1) = 0 \quad \text{Factor.}$$

Steps 3 and 4 Set each factor equal to zero and solve the resulting equations.

$$2x = 0 \quad \text{or} \quad 2x - 1 = 0$$

$$x = 0 \qquad\qquad 2x = 1$$

$$x = \tfrac{1}{2}$$

Step 5 Check the solutions in the original equation.

Check 0:

$$4x^2 - 2x = 0$$
$$4 \cdot 0^2 - 2 \cdot 0 \stackrel{?}{=} 0$$
$$0 - 0 \stackrel{?}{=} 0$$
$$0 = 0 \checkmark$$

Check $\tfrac{1}{2}$:

$$4x^2 - 2x = 0$$
$$4\left(\tfrac{1}{2}\right)^2 - 2\left(\tfrac{1}{2}\right) \stackrel{?}{=} 0$$
$$4\left(\tfrac{1}{4}\right) - 2\left(\tfrac{1}{2}\right) \stackrel{?}{=} 0$$
$$1 - 1 \stackrel{?}{=} 0$$
$$0 = 0 \checkmark$$

The solution set is $\left\{0, \tfrac{1}{2}\right\}$.

b. Next, we solve $2x^2 + 7x = 4$.

Step 1 Move all terms to one side and obtain zero on the other side. Subtract 4 from both sides and write the equation in general form.

$$2x^2 + 7x = 4 \qquad \text{This is the given equation.}$$
$$2x^2 + 7x - 4 = 4 - 4 \qquad \text{Subtract 4 from both sides.}$$
$$2x^2 + 7x - 4 = 0 \qquad \text{Simplify.}$$

Step 2 Factor.

$$2x^2 + 7x - 4 = 0$$
$$(2x - 1)(x + 4) = 0$$

Steps 3 and 4 Set each factor equal to zero and solve each resulting equation.

$$2x - 1 = 0 \quad \text{or} \quad x + 4 = 0$$
$$2x = 1 \qquad\qquad x = -4$$
$$x = \tfrac{1}{2}$$

Step 5 Check the solutions in the original equation.

Check $\tfrac{1}{2}$:

$$2x^2 + 7x = 4$$
$$2\left(\tfrac{1}{2}\right)^2 + 7\left(\tfrac{1}{2}\right) \overset{?}{=} 4$$
$$\tfrac{1}{2} + \tfrac{7}{2} \overset{?}{=} 4$$
$$4 = 4 \ \checkmark$$

Check -4:

$$2x^2 + 7x = 4$$
$$2(-4)^2 + 7(-4) \overset{?}{=} 4$$
$$32 + (-28) \overset{?}{=} 4$$
$$4 = 4 \ \checkmark$$

The solution set is $\left\{-4, \tfrac{1}{2}\right\}$.

> **Check Point 1**
>
> Solve by factoring:
> **a.** $3x^2 - 9x = 0$ **b.** $2x^2 + x = 1.$

Technology

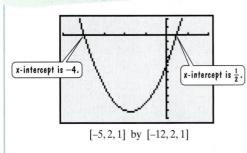

x-intercept is -4.

x-intercept is $\tfrac{1}{2}$.

$[-5, 2, 1]$ by $[-12, 2, 1]$

You can use a graphing utility to check the real solutions of a quadratic equation. **The solutions of $ax^2 + bx + c = 0$ correspond to the x-intercepts of the graph of $y = ax^2 + bx + c$.** For example, to check the solutions of $2x^2 + 7x = 4$, or $2x^2 + 7x - 4 = 0$, graph $y = 2x^2 + 7x - 4$. The cuplike U-shaped graph is shown on the left. Note that it is important to have all nonzero terms on one side of the quadratic equation before entering it into the graphing utility. The x-intercepts are -4 and $\tfrac{1}{2}$, and the graph of $y = 2x^2 + 7x - 4$ passes through $(-4, 0)$ and $(\tfrac{1}{2}, 0)$. This verifies that $\left\{-4, \tfrac{1}{2}\right\}$ is the solution set of $2x^2 + 7x - 4 = 0$.

2 Solve quadratic equations by the square root method.

Solving Quadratic Equations by the Square Root Method

Quadratic equations of the form $u^2 = d$, where $d > 0$ and u is an algebraic expression, can be solved by the **square root method.** First, isolate the squared expression u^2 on one side of the equation and the number d on the other side. Then take the square root of both sides. Remember, there are two numbers whose square is d. One number is positive and one is negative.

We can use factoring to verify that $u^2 = d$ has two solutions.

$$u^2 = d \qquad \text{\color{teal}{This is the given equation.}}$$
$$u^2 - d = 0 \qquad \text{\color{teal}{Move all terms to one side and obtain zero on the other side.}}$$
$$\left(u + \sqrt{d}\right)\left(u - \sqrt{d}\right) = 0 \qquad \text{\color{teal}{Factor.}}$$
$$u + \sqrt{d} = 0 \quad \text{or} \quad u - \sqrt{d} = 0 \qquad \text{\color{teal}{Set each factor equal to zero.}}$$
$$u = -\sqrt{d} \qquad\qquad u = \sqrt{d} \qquad \text{\color{teal}{Solve the resulting equations.}}$$

Because the solutions differ only in sign, we can write them in abbreviated notation as $u = \pm\sqrt{d}$. We read this as "u equals positive or negative the square root of d" or "u equals plus or minus the square root of d."

Now that we have verified these solutions, we can solve $u^2 = d$ directly by taking square roots. This process is called **the square root method.**

The Square Root Method

If u is an algebraic expression and d is a positive real number, then $u^2 = d$ has exactly two solutions:

$$\text{If } u^2 = d, \text{ then } u = \sqrt{d} \text{ or } u = -\sqrt{d}.$$

Equivalently,

$$\text{If } u^2 = d, \text{ then } u = \pm\sqrt{d}.$$

EXAMPLE 2 Solving Quadratic Equations by the Square Root Method

Solve by the square root method:

a. $4x^2 = 20$ **b.** $(x - 2)^2 = 6$.

Solution

a. In order to apply the square root method, we need a squared expression by itself on one side of the equation.

$$4x^2 = 20$$

> We want x^2 by itself.

We can get x^2 by itself if we divide both sides by 4.

$$\frac{4x^2}{4} = \frac{20}{4}$$

$$x^2 = 5$$

Now, we can apply the square root method.

$$x = \pm\sqrt{5}$$

By checking both values in the original equation, we can confirm that the solution set is $\{-\sqrt{5}, \sqrt{5}\}$.

b. $(x - 2)^2 = 6$

> The squared expression is by itself.

With the squared expression by itself, we can apply the square root method.

$$x - 2 = \pm\sqrt{6}$$

We solve for x by adding 2 to both sides.

$$x = 2 \pm \sqrt{6}$$

By checking both values in the original equation, we can confirm that the solution set is $\{2 + \sqrt{6}, 2 - \sqrt{6}\}$ or $\{2 \pm \sqrt{6}\}$.

Check Point 2 Solve by the square root method:

 a. $3x^2 = 21$ **b.** $(x + 5)^2 = 11$.

3 Solve quadratic equations by completing the square.

Completing the Square

How do we solve an equation in the form $ax^2 + bx + c = 0$ if the trinomial $ax^2 + bx + c$ cannot be factored? We cannot use the zero-product principle in such a case. However, we can convert the equation into an equivalent equation that can be solved using the square root method. This is accomplished by **completing the square.**

Completing the Square

If $x^2 + bx$ is a binomial, then by adding $\left(\dfrac{b}{2}\right)^2$, which is the square of half the coefficient of x, a perfect square trinomial will result. That is,

$$x^2 + bx + \left(\frac{b}{2}\right)^2 = \left(x + \frac{b}{2}\right)^2.$$

EXAMPLE 3   Completing the Square

What term should be added to the binomial $x^2 + 8x$ so that it becomes a perfect square trinomial? Then write and factor the trinomial.

Solution   The term that should be added is the square of half the coefficient of x. The coefficient of x is 8. Thus, we will add $\left(\frac{8}{2}\right)^2 = 4^2$. A perfect square trinomial is the result.

$$x^2 + 8x + 4^2 = x^2 + 8x + 16 = (x + 4)^2$$

$$\text{(half)}^2$$

Check Point 3 What term should be added to the binomial $x^2 - 14x$ so that it becomes a perfect square trinomial? Then write and factor the trinomial.

We can solve any quadratic equation by completing the square. If the coefficient of the x^2-term is one, we add the square of half the coefficient of x to both sides of the equation. **When you add a constant term to one side of the equation to complete the square, be certain to add the same constant to the other side of the equation.** These ideas are illustrated in Example 4.

EXAMPLE 4   Solving a Quadratic Equation by Completing the Square

Solve by completing the square: $x^2 - 6x + 2 = 0$.

Solution   We begin the procedure of solving $x^2 - 6x + 2 = 0$ by isolating the binomial, $x^2 - 6x$, so that we can complete the square. Thus, we subtract 2 from both sides of the equation.

$$x^2 - 6x + 2 = 0 \qquad \text{This is the given equation.}$$

$$x^2 - 6x + 2 - 2 = 0 - 2 \qquad \text{Subtract 2 from both sides.}$$

$$x^2 - 6x = -2 \qquad \text{Simplify.}$$

> We need to add a constant to this binomial that will make it a perfect square trinomial.

What constant should we add? Add the square of half the coefficient of x.

$$x^2 - 6x = -2$$

> −6 is the coefficient of x.

$$\left(\frac{-6}{2}\right)^2 = (-3)^2 = 9$$

Study Tip

When factoring perfect square trinomials, the constant in the factorization is always half the coefficient of x.

$$x^2 - 6x + 9 = (x - 3)^2$$

> Half the coefficient of x, −6, is −3.

Thus, we need to add 9 to $x^2 - 6x$. In order to obtain an equivalent equation, we must add 9 to both sides.

$$x^2 - 6x = -2 \qquad \text{This is the quadratic equation with the binomial isolated.}$$

$$x^2 - 6x + 9 = -2 + 9 \qquad \text{Add 9 to both sides to complete the square.}$$

$$(x - 3)^2 = 7 \qquad \text{Factor the perfect square trinomial.}$$

> In this step we have converted our equation into one that can be solved by the square root method.

$$x - 3 = \pm\sqrt{7} \qquad \text{Apply the square root method.}$$

$$x = 3 \pm \sqrt{7} \qquad \text{Add 3 to both sides.}$$

The solution set is $\{3 + \sqrt{7}, 3 - \sqrt{7}\}$ or $\{3 \pm \sqrt{7}\}$.

Check Point 4   Solve by completing the square: $x^2 - 2x - 2 = 0$.

If the coefficient of the x^2-term in a quadratic equation is not one, you must divide each side of the equation by this coefficient before completing the square. For example, to solve $3x^2 - 2x - 4 = 0$ by completing the square, first divide every term by 3:

$$\frac{3x^2}{3} - \frac{2x}{3} - \frac{4}{3} = \frac{0}{3}$$

$$x^2 - \frac{2}{3}x - \frac{4}{3} = 0.$$

Now that the coefficient of x^2 is one, we can solve by completing the square using the method of Example 4.

4 Solve quadratic equations using the quadratic formula.

Solving Quadratic Equations Using the Quadratic Formula

We can use the method of completing the square to derive a formula that can be used to solve all quadratic equations. The derivation given here also shows a particular quadratic equation, $3x^2 - 2x - 4 = 0$, to specifically illustrate each of the steps.

Deriving the Quadratic Formula

General Form of a Quadratic Equation	Comment	A Specific Example
$ax^2 + bx + c = 0, \quad a > 0$	This is the given equation.	$3x^2 - 2x - 4 = 0$
$x^2 + \dfrac{b}{a}x + \dfrac{c}{a} = 0$	Divide both sides by the coefficient of x^2.	$x^2 - \dfrac{2}{3}x - \dfrac{4}{3} = 0$
$x^2 + \dfrac{b}{a}x = -\dfrac{c}{a}$	Isolate the binomial by adding $-\dfrac{c}{a}$ on both sides.	$x^2 - \dfrac{2}{3}x = \dfrac{4}{3}$
$x^2 + \dfrac{b}{a}x + \left(\dfrac{b}{2a}\right)^2 = -\dfrac{c}{a} + \left(\dfrac{b}{2a}\right)^2$ (half)^2	Complete the square. Add the square of half the coefficient of x to both sides.	$x^2 - \dfrac{2}{3}x + \left(-\dfrac{1}{3}\right)^2 = \dfrac{4}{3} + \left(-\dfrac{1}{3}\right)^2$ (half)^2
$x^2 + \dfrac{b}{a}x + \dfrac{b^2}{4a^2} = -\dfrac{c}{a} + \dfrac{b^2}{4a^2}$		$x^2 - \dfrac{2}{3}x + \dfrac{1}{9} = \dfrac{4}{3} + \dfrac{1}{9}$
$\left(x + \dfrac{b}{2a}\right)^2 = -\dfrac{c}{a}\cdot\dfrac{4a}{4a} + \dfrac{b^2}{4a^2}$	Factor on the left side and obtain a common denominator on the right side.	$\left(x - \dfrac{1}{3}\right)^2 = \dfrac{4}{3}\cdot\dfrac{3}{3} + \dfrac{1}{9}$
$\left(x + \dfrac{b}{2a}\right)^2 = \dfrac{-4ac + b^2}{4a^2}$	Add fractions on the right side.	$\left(x - \dfrac{1}{3}\right)^2 = \dfrac{12 + 1}{9}$
$\left(x + \dfrac{b}{2a}\right)^2 = \dfrac{b^2 - 4ac}{4a^2}$		$\left(x - \dfrac{1}{3}\right)^2 = \dfrac{13}{9}$
$x + \dfrac{b}{2a} = \pm\sqrt{\dfrac{b^2 - 4ac}{4a^2}}$	Apply the square root method.	$x - \dfrac{1}{3} = \pm\sqrt{\dfrac{13}{9}}$
$x + \dfrac{b}{2a} = \pm\dfrac{\sqrt{b^2 - 4ac}}{2a}$	Take the square root of the quotient, simplifying the denominator.	$x - \dfrac{1}{3} = \pm\dfrac{\sqrt{13}}{3}$
$x = \dfrac{-b}{2a} \pm \dfrac{\sqrt{b^2 - 4ac}}{2a}$	Solve for x by subtracting $\dfrac{b}{2a}$ from both sides.	$x = \dfrac{1}{3} \pm \dfrac{\sqrt{13}}{3}$
$x = \dfrac{-b \pm \sqrt{b^2 - 4ac}}{2a}$	Combine fractions on the right.	$x = \dfrac{1 \pm \sqrt{13}}{3}$

The formula shown at the bottom of the left column is called the *quadratic formula*. A similar proof shows that the same formula can be used to solve quadratic equations if a, the coefficient of the x^2-term, is negative.

To Die at Twenty

Can the equations

$$7x^5 + 12x^3 - 9x + 4 = 0$$

and

$$8x^6 - 7x^5 + 4x^3 - 19 = 0$$

be solved using a formula similar to the quadratic formula? The first equation has five solutions and the second has six solutions, but they cannot be found using a formula. How do we know? In 1832, a 20-year-old Frenchman, Evariste Galois, wrote down a proof showing that there is no general formula to solve equations when the exponent on the variable is 5 or greater. Galois was jailed as a political activist several times while still a teenager. The day after his brilliant proof he fought a duel over a woman. The duel was a political setup. As he lay dying, Galois told his brother, Alfred, of the manuscript that contained his proof: "Mathematical manuscripts are in my room. On the table. Take care of my work. Make it known. Important. Don't cry, Alfred. I need all my courage—to die at twenty." (Our source is Leopold Infeld's biography of Galois, *Whom the Gods Love*. Some historians, however, dispute the story of Galois's ironic death the very day after his algebraic proof. Mathematical truths seem more reliable than historical ones!)

The Quadratic Formula

The solutions of a quadratic equation in standard form $ax^2 + bx + c = 0$, with $a \neq 0$, are given by the **quadratic formula**

$$x = \frac{-b \pm \sqrt{b^2 - 4ac}}{2a}.$$

x equals negative b, plus or minus the square root of b² − 4ac, all divided by 2a.

To use the quadratic formula, write the quadratic equation in general form if necessary. Then determine the numerical values for a (the coefficient of the squared term), b (the coefficient of the x-term), and c (the constant term). Substitute the values of a, b, and c in the quadratic formula and evaluate the expression. The $\pm$ sign indicates that there are two solutions of the equation.

EXAMPLE 5   Solving a Quadratic Equation Using the Quadratic Formula

Solve using the quadratic formula: $2x^2 - 6x + 1 = 0$.

Solution   The given equation is in general form. Begin by identifying the values for a, b, and c.

$$2x^2 - 6x + 1 = 0$$

$a = 2$ $b = -6$ $c = 1$

$$x = \frac{-b \pm \sqrt{b^2 - 4ac}}{2a}$$ Use the quadratic formula.

$$= \frac{-(-6) \pm \sqrt{(-6)^2 - 4(2)(1)}}{2 \cdot 2}$$ Substitute the values for a, b, and c: a = 2, b = −6, and c = 1.

$$= \frac{6 \pm \sqrt{36 - 8}}{4}$$ −(−6) = 6 and (−6)² = (−6)(−6) = 36.

$$= \frac{6 \pm \sqrt{28}}{2}$$ Complete the subtraction under the radical.

$$= \frac{6 \pm 2\sqrt{7}}{4}$$ $\sqrt{28} = \sqrt{4 \cdot 7} = \sqrt{4}\sqrt{7} = 2\sqrt{7}$

$$= \frac{2(3 \pm \sqrt{7})}{4}$$ Factor out 2 from the numerator.

$$= \frac{3 \pm \sqrt{7}}{2}$$ Divide the numerator and denominator by 2.

The solution set is $\left\{ \dfrac{3 + \sqrt{7}}{2}, \dfrac{3 - \sqrt{7}}{2} \right\}$ or $\left\{ \dfrac{3 \pm \sqrt{7}}{2} \right\}$.

Check Point 5   Solve using the quadratic formula:

$$2x^2 + 2x - 1 = 0.$$

We have seen that a graphing utility can be used to check the solutions of the quadratic equation $ax^2 + bx + c = 0$. The x-intercepts of the graph of

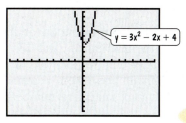

Figure 1.15 This graph has no x-intercepts.

$y = ax^2 + bx + c$ are the solutions. However, take a look at the graph of $y = 3x^2 - 2x + 4$, shown in Figure 1.15. Notice that the graph has no x-intercepts. Can you guess what this means about the solutions of the quadratic equation $3x^2 - 2x + 4 = 0$? If you're not sure, we'll answer this question in the next example.

EXAMPLE 6 **Solving a Quadratic Equation Using the Quadratic Formula**

Solve using the quadratic formula: $3x^2 - 2x + 4 = 0$.

Solution   The given equation is in general form. Begin by identifying the values for a, b, and c.

$$3x^2 - 2x + 4 = 0$$

$$a = 3 \qquad b = -2 \qquad c = 4$$

$$x = \frac{-b \pm \sqrt{b^2 - 4ac}}{2a}$$

Use the quadratic formula.

$$= \frac{-(-2) \pm \sqrt{(-2)^2 - 4(3)(4)}}{2(3)}$$

Substitute the values for a, b, and c: $a = 3$, $b = -2$, and $c = 4$.

$$= \frac{2 \pm \sqrt{4 - 48}}{6}$$

$-(-2) = 2$ and $(-2)^2 = (-2)(-2) = 4$.

$$= \frac{2 \pm \sqrt{-44}}{6}$$

Subtract under the radical.
Because the number under the radical sign is negative, the solutions will not be real numbers.

$$= \frac{2 \pm 2i\sqrt{11}}{6}$$

$\sqrt{-44} = \sqrt{4(11)(-1)}$
$\qquad = 2i\sqrt{11}$

$$= \frac{2(1 \pm i\sqrt{11})}{6}$$

Factor 2 from the numerator.

$$= \frac{1 \pm i\sqrt{11}}{3}$$

Divide numerator and denominator by 2.

$$= \frac{1}{3} \pm i\frac{\sqrt{11}}{3}$$

Write the complex numbers in standard form.

The solutions are complex conjugates, and the solution set is $\left\{ \frac{1}{3} + i\frac{\sqrt{11}}{3}, \frac{1}{3} - i\frac{\sqrt{11}}{3} \right\}$ or $\left\{ \frac{1}{3} \pm i\frac{\sqrt{11}}{3} \right\}$.

If $ax^2 + bx + c = 0$ has complex imaginary solutions, the graph of $y = ax^2 + bx + c$ will not have x-intercepts. This is illustrated by the imaginary solutions of $3x^2 - 2x + 4 = 0$ in Example 6 and the graph in Figure 1.15.

Study Tip

Checking irrational and complex imaginary solutions can be time-consuming. The solutions given by the quadratic formula are always correct, unless you have made a careless error. Checking for computational errors or errors in simplification is sufficient.

Check Point 6   Solve using the quadratic formula:

$$x^2 - 2x + 2 = 0.$$

5 Use the discriminant to determine the number and type of solutions.

The Discriminant

The quantity $b^2 - 4ac$, which appears under the radical sign in the quadratic formula, is called the **discriminant.** In Example 5 the discriminant was 28, a positive number that is not a perfect square. The equation had two solutions that were irrational numbers. In Example 6, the discriminant was -44, a negative number. The equation had solutions that were complex imaginary numbers. These observations are generalized in Table 1.3.

Table 1.3   The Discriminant and the Kinds of Solutions to $ax^2 + bx + c = 0$

Discriminant $b^2 - 4ac$	Kinds of Solutions to $ax^2 + bx + c = 0$	Graph of $y = ax^2 + bx + c$
$b^2 - 4ac > 0$	**Two unequal real solutions;** If a, b, and c are rational numbers and the discriminant is a perfect square, the solutions are rational. If the discriminant is not a perfect square, the solutions are irrational.	Two x-intercepts
$b^2 - 4ac = 0$	**One solution(a repeated solution) that is a real number;** If a, b, and c are rational numbers, the repeated solution is also a rational number.	One x-intercepts
$b^2 - 4ac < 0$	**No real solution; two complex imaginary solutions;** The solutions are complex conjugates.	No x-intercepts

EXAMPLE 7   Using the Discriminant

Compute the discriminant of $4x^2 - 8x + 1 = 0$. What does the discriminant indicate about the number and type of solutions?

Solution   Begin by identifying the values for a, b, and c.

$$4x^2 - 8x + 1 = 0$$

$a = 4$ $b = -8$ $c = 1$

Substitute and compute the discriminant:

$$b^2 - 4ac = (-8)^2 - 4 \cdot 4 \cdot 1 = 64 - 16 = 48.$$

The discriminant is 48. Because the discriminant is positive, the equation $4x^2 - 8x + 1 = 0$ has two unequal real solutions.

Check Point 7 Compute the discriminant of $3x^2 - 2x + 5 = 0$. What does the discriminant indicate about the number and type of solutions?

⑥ Determine the most efficient method to use when solving a quadratic equation.

Determining Which Method to Use

All quadratic equations can be solved by the quadratic formula. However, if an equation is in the form $u^2 = d$, such as $x^2 = 5$ or $(2x + 3)^2 = 8$, it is faster to use the square root method, taking the square root of both sides. If the equation is not in the form $u^2 = d$, write the quadratic equation in general form ($ax^2 + bx + c = 0$). Try to solve the equation by the factoring method. If $ax^2 + bx + c$ cannot be factored, then solve the quadratic equation by the quadratic formula.

Because we used the method of completing the square to derive the quadratic formula, we no longer need it for solving quadratic equations. However, we will use completing the square later in the book to help graph certain kinds of equations.

Table 1.4 summarizes our observations about which technique to use when solving a quadratic equation.

Table 1.4   Determining the Most Efficient Technique to Use When Solving a Quadratic Equation

Description and Form of the Quadratic Equation	Most Efficient Solution Method	Example
$ax^2 + bx + c = 0$ and $ax^2 + bx + c$ can be factored easily.	Factor and use the zero-product principle.	$3x^2 + 5x - 2 = 0$ $(3x - 1)(x + 2) = 0$ $3x - 1 = 0$ or $x + 2 = 0$ $x = \dfrac{1}{3} \qquad x = -2$
$ax^2 + c = 0$ The quadratic equation has no x-term. ($b = 0$)	Solve for x^2 and apply the square root method.	$4x^2 - 7 = 0$ $4x^2 = 7$ $x^2 = \dfrac{7}{4}$ $x = \pm \dfrac{\sqrt{7}}{2}$
$(ax + c)^2 = d$; $ax + c$ is a first-degree polynomial.	Use the square root method.	$(x + 4)^2 = 5$ $x + 4 = \pm\sqrt{5}$ $x = -4 \pm \sqrt{5}$
$ax^2 + bx + c = 0$ and $ax^2 + bx + c$ cannot be factored or the factoring is too difficult.	Use the quadratic formula: $$x = \dfrac{-b \pm \sqrt{b^2 - 4ac}}{2a}.$$	$x^2 - 2x - 6 = 0$ $a = 1 \quad b = -2 \quad c = -6$ $x = \dfrac{-(-2) \pm \sqrt{(-2)^2 - 4(1)(-6)}}{2(1)}$ $= \dfrac{2 \pm \sqrt{4 - 4(1)(-6)}}{2(1)}$ $= \dfrac{2 \pm \sqrt{28}}{2} = \dfrac{2 \pm \sqrt{4}\sqrt{7}}{2}$ $= \dfrac{2 \pm 2\sqrt{7}}{2} = \dfrac{2(1 \pm \sqrt{7})}{2}$ $= 1 \pm \sqrt{7}$

7 Solve problems modeled by quadratic equations.

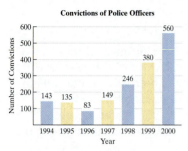

Convictions of Police Officers

Figure 1.14, repeated

Source: F.B.I.

Applications

We opened this section with a graph (Figure 1.14, repeated in the margin) showing the number of convictions of police officers throughout the United States from 1994 through 2000. The data can be modeled by the formula

$$N = 23.4x^2 - 259.1x + 815.8$$

where N is the number of police officers convicted of felonies x years after 1990. Notice that this formula contains an expression in the form $ax^2 + bx + c$ on the right side. If a formula contains such an expression, we can write and solve a quadratic equation to answer questions about the variable x. Our next example shows how this is done.

EXAMPLE 8   Convictions of Police Officers

Use the formula $N = 23.4x^2 - 259.1x + 815.8$ to answer this question: In which year will 1000 police officers be convicted of felonies?

Solution   Because we are interested in 1000 convictions, we substitute 1000 for N in the given formula. Then we solve for x, the number of years after 1990.

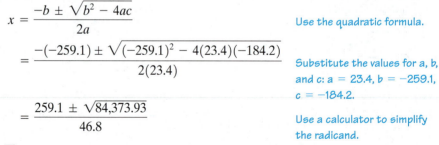

$$N = 23.4x^2 - 259.1x + 815.8 \qquad \text{This is the given formula.}$$

$$1000 = 23.4x^2 - 259.1x + 815.8 \qquad \text{Substitute 1000 for N.}$$

$$0 = 23.4x^2 - 259.1x - 184.2 \qquad \text{Subtract 1000 from both sides and write the quadratic equation in general form.}$$

$$\boxed{a = 23.4} \quad \boxed{b = -259.1} \quad \boxed{c = -184.2}$$

Because the trinomial on the right side of the equation is prime, we solve using the quadratic formula.

$$x = \frac{-b \pm \sqrt{b^2 - 4ac}}{2a} \qquad \text{Use the quadratic formula.}$$

$$= \frac{-(-259.1) \pm \sqrt{(-259.1)^2 - 4(23.4)(-184.2)}}{2(23.4)} \qquad \text{Substitute the values for } a, b, \text{ and } c\text{: } a = 23.4, b = -259.1, c = -184.2.$$

$$= \frac{259.1 \pm \sqrt{84{,}373.93}}{46.8} \qquad \text{Use a calculator to simplify the radicand.}$$

Thus,

$$x = \frac{259.1 + \sqrt{84{,}373.93}}{46.8} \quad \text{or} \quad x = \frac{259.1 - \sqrt{84{,}373.93}}{46.8}$$

$$x \approx 12 \qquad\qquad\qquad x \approx -1 \qquad \text{Use a calculator and round to the nearest integer.}$$

The model describes the number of convictions x years *after* 1990. Thus, we are interested only in the positive solution, 12. This means that approximately 12 years after 1990, in 2002, 1000 police officers will be convicted of felonies.

Check Point 8   Use the formula in Example 8 to answer this question: In which year after 1993 were 250 police officers convicted of felonies? How well does the formula model the actual number of convictions for that year shown in Figure 1.14?

Technology

On most calculators, here is how to approximate

$$\frac{259.1 + \sqrt{84{,}373.93}}{46.8}:$$

Many Scientific Calculators

| (| 259.1 | + | 84373.93 | √ |

|) | ÷ | 46.8 | = |

Many Graphing Calculators

| (| 259.1 | + | √ | 84373.93 |

|) | ÷ | 46.8 | ENTER |.

Similar keystrokes can be used to approximate the other irrational solution in Example 8.

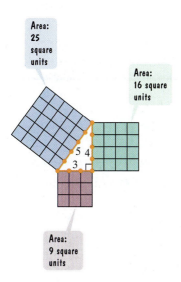

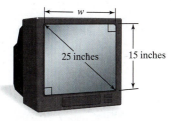

Figure 1.16 The area of the large square equals the sum of the areas of the smaller squares.

In our next example, we will be using the *Pythagorean Theorem* to obtain a verbal model. The ancient Greek philosopher and mathematician Pythagoras (approximately 582–500 B.C.) founded a school whose motto was "All is number." Pythagoras is best remembered for his work with the **right triangle,** a triangle with one angle measuring 90°. The side opposite the 90° angle is called the **hypotenuse.** The other sides are called **legs.** Pythagoras found that if he constructed squares on each of the legs, as well as a larger square on the hypotenuse, the sum of the areas of the smaller squares is equal to the area of the larger square. This is illustrated in Figure 1.16.

This relationship is usually stated in terms of the lengths of the three sides of a right triangle and is called the **Pythagorean Theorem.**

The Pythagorean Theorem

The sum of the squares of the lengths of the legs of a right triangle equals the square of the length of the hypotenuse.
   If the legs have lengths a and b, and the hypotenuse has length c, then

$$a^2 + b^2 = c^2.$$

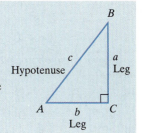

EXAMPLE 9   Using the Pythagorean Theorem

In a 25-inch television set, the length of the screen's diagonal is 25 inches. If the screen's height is 15 inches, what is its width?

Solution   Figure 1.17 shows a right triangle that is formed by the height, width, and diagonal. We can find w, the screen's width, using the Pythagorean Theorem.

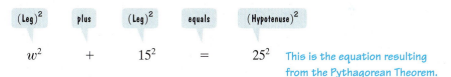

$$w^2 \quad + \quad 15^2 \quad = \quad 25^2 \qquad \text{This is the equation resulting from the Pythagorean Theorem.}$$

Figure 1.17 A right triangle is formed by the television's height, width, and diagonal.

The equation $w^2 + 15^2 = 25^2$ can be solved most efficiently by the square root method.

$$w^2 + 15^2 = 25^2 \qquad \text{This is the equation that models the verbal conditions.}$$

$$w^2 + 225 = 625 \qquad \text{Square 15 and 25.}$$

$$w^2 + 225 - 225 = 625 - 225 \qquad \text{Isolate } w^2 \text{ by subtracting 225 from both sides.}$$

$$w^2 = 400 \qquad \text{Simplify.}$$

$$w = \pm\sqrt{400} \qquad \text{Apply the square root method.}$$

$$w = \pm 20 \qquad \text{Simplify.}$$

Because w represents the width of the television's screen, this dimension must be positive. We reject -20. Thus, the width of the television is 20 inches.

Check Point 9   What is the width in a 15-inch television set whose height is 9 inches?

EXERCISE SET 1.5

Practice Exercises

Solve each equation in Exercises 1–14 by factoring.

1. $x^2 - 3x - 10 = 0$ **2.** $x^2 - 13x + 36 = 0$

3. $x^2 = 8x - 15$ **4.** $x^2 = -11x - 10$

5. $6x^2 + 11x - 10 = 0$ **6.** $9x^2 + 9x + 2 = 0$

7. $3x^2 - 2x = 8$ **8.** $4x^2 - 13x = -3$

9. $3x^2 + 12x = 0$ **10.** $5x^2 - 20x = 0$

11. $2x(x - 3) = 5x^2 - 7x$ **12.** $16x(x - 2) = 8x - 25$

13. $7 - 7x = (3x + 2)(x - 1)$

14. $10x - 1 = (2x + 1)^2$

Solve each equation in Exercises 15–26 by the square root method.

15. $3x^2 = 27$ **16.** $5x^2 = 45$

17. $5x^2 + 1 = 51$ **18.** $3x^2 - 1 = 47$

19. $(x + 2)^2 = 25$ **20.** $(x - 3)^2 = 36$

21. $(3x + 2)^2 = 9$ **22.** $(4x - 1)^2 = 16$

23. $(5x - 1)^2 = 7$ **24.** $(8x - 3)^2 = 5$

25. $(3x - 4)^2 = 8$ **26.** $(2x + 8)^2 = 27$

In Exercises 27–38, determine the constant that should be added to the binomial so that it becomes a perfect square trinomial. Then write and factor the trinomial.

27. $x^2 + 12x$ **28.** $x^2 + 16x$

29. $x^2 - 10x$ **30.** $x^2 - 14x$

31. $x^2 + 3x$ **32.** $x^2 + 5x$

33. $x^2 - 7x$ **34.** $x^2 - 9x$

35. $x^2 - \dfrac{2}{3}x$ **36.** $x^2 + \dfrac{4}{5}x$

37. $x^2 - \dfrac{1}{3}x$ **38.** $x^2 - \dfrac{1}{4}x$

Solve each equation in Exercises 39–54 by completing the square.

39. $x^2 + 6x = 7$ **40.** $x^2 + 6x = -8$

41. $x^2 - 2x = 2$ **42.** $x^2 + 4x = 12$

43. $x^2 - 6x - 11 = 0$ **44.** $x^2 - 2x - 5 = 0$

45. $x^2 + 4x + 1 = 0$ **46.** $x^2 + 6x - 5 = 0$

47. $x^2 + 3x - 1 = 0$ **48.** $x^2 - 3x - 5 = 0$

49. $2x^2 - 7x + 3 = 0$ **50.** $2x^2 + 5x - 3 = 0$

51. $4x^2 - 4x - 1 = 0$ **52.** $2x^2 - 4x - 1 = 0$

53. $3x^2 - 2x - 2 = 0$ **54.** $3x^2 - 5x - 10 = 0$

Solve each equation in Exercises 55–64 using the quadratic formula.

55. $x^2 + 8x + 15 = 0$ **56.** $x^2 + 8x + 12 = 0$

57. $x^2 + 5x + 3 = 0$ **58.** $x^2 + 5x + 2 = 0$

59. $3x^2 - 3x - 4 = 0$ **60.** $5x^2 + x - 2 = 0$

61. $4x^2 = 2x + 7$ **62.** $3x^2 = 6x - 1$

63. $x^2 - 6x + 10 = 0$ **64.** $x^2 - 2x + 17 = 0$

Compute the discriminant of each equation in Exercises 65–72. What does the discriminant indicate about the number and type of solutions?

65. $x^2 - 4x - 5 = 0$ **66.** $4x^2 - 2x + 3 = 0$

67. $2x^2 - 11x + 3 = 0$ **68.** $2x^2 + 11x - 6 = 0$

69. $x^2 - 2x + 1 = 0$ **70.** $3x^2 = 2x - 1$

71. $x^2 - 3x - 7 = 0$ **72.** $3x^2 + 4x - 2 = 0$

Solve each equation in Exercises 73–98 by the method of your choice.

73. $2x^2 - x = 1$ **74.** $3x^2 - 4x = 4$

75. $5x^2 + 2 = 11x$ **76.** $5x^2 = 6 - 13x$

77. $3x^2 = 60$ **78.** $2x^2 = 250$

79. $x^2 - 2x = 1$ **80.** $2x^2 + 3x = 1$

81. $(2x + 3)(x + 4) = 1$ **82.** $(2x - 5)(x + 1) = 2$

83. $(3x - 4)^2 = 16$ **84.** $(2x + 7)^2 = 25$

85. $3x^2 - 12x + 12 = 0$ **86.** $9 - 6x + x^2 = 0$

87. $4x^2 - 16 = 0$ **88.** $3x^2 - 27 = 0$

89. $x^2 - 6x + 13 = 0$ **90.** $x^2 - 4x + 29 = 0$

91. $x^2 = 4x - 7$ **92.** $5x^2 = 2x - 3$

93. $2x^2 - 7x = 0$ **94.** $2x^2 + 5x = 3$

95. $\dfrac{1}{x} + \dfrac{1}{x + 2} = \dfrac{1}{3}$ **96.** $\dfrac{1}{x} + \dfrac{1}{x + 3} = \dfrac{1}{4}$

97. $\dfrac{2x}{x - 3} + \dfrac{6}{x + 3} = -\dfrac{28}{x^2 - 9}$

98. $\dfrac{3}{x - 3} + \dfrac{5}{x - 4} = \dfrac{x^2 - 20}{x^2 - 7x + 12}$

Application Exercises

A driver's age has something to do with his or her chance of getting into a fatal car crash. The bar graph shows the number of fatal vehicle crashes per 100 million miles driven for drivers of various age groups. For example, 25-year-old drivers are involved in 4.1 fatal crashes per 100 million miles driven. Thus, when a group of 25-year-old Americans have driven a total of 100 million miles, approximately 4 have been in accidents in which someone died.

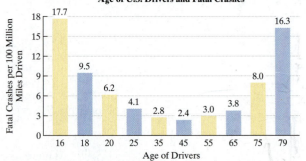

Age of U.S. Drivers and Fatal Crashes

Source: Insurance Institute for Highway Safety

The number of fatal vehicle crashes per 100 million miles, N, for drivers of age x can be modeled by the formula

$$N = 0.013x^2 - 1.19x + 28.24.$$

Use the formula to solve Exercises 99–100.

99. What age groups are expected to be involved in 10 fatal crashes per 100 million miles driven? How well does the formula model the trend in the actual data shown in the bar graph?

100. What age groups are expected to be involved in 3 fatal crashes per 100 million miles driven? How well does the formula model the trend in the actual data shown in the bar graph?

The Food Stamp Program is America's first line of defense against hunger for millions of families. Over half of all participants are children; one out of six is a low-income older adult. Exercises 101–104 involve the number of participants in the program from 1990 through 2000.
The formula

$$y = -\frac{1}{2}x^2 + 4x + 19$$

models the number of people, y, in millions, receiving food stamps x years after 1990. Use the formula to solve Exercises 101–102.

101. In which year did 27 million people receive food stamps?

102. In which years did 19 million people receive food stamps?

The graph of the formula in Exercises 101–102 is shown. Use the graph to solve Exercises 103–104.

Number of People Receiving Food Stamps

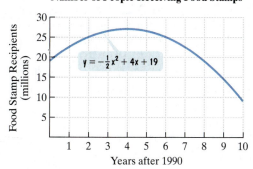

Source: New York Times

103. Identify your solution in Exercise 101 as a point on the graph. Describe what is significant about this point.

104. Identify your solution in Exercise 102 as one or more points on the graph. Then describe the trend shown by the graph.

The formula

$$N = 2x^2 + 22x + 320$$

models the number of inmates, N, in thousands, in U.S. state and federal prisons x years after 1980. The graph of the formula is shown in a [0, 20, 1] by [0, 1600, 100] viewing rectangle at the top of the next column. Use the formula to solve Exercises 105–106.

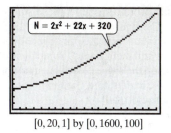

[0, 20, 1] by [0, 1600, 100]

105. In which year were there 740 thousand inmates in U.S. state and federal prisons? Identify the solution as a point on the graph shown.

106. In which year were there 1100 thousand inmates in U.S. state and federal prisons? Identify the solution as a point on the graph shown?

107. A baseball diamond is actually a square with 90-foot sides. What is the distance from home plate to second base?

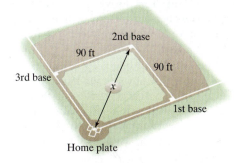

108. A 20-foot ladder is 15 feet from the house. How far up the house does the ladder reach?

109. An 8-foot tree is supported by two wires that extend from the top of the tree to a point on the ground located 15 feet from the base of the tree. Find the total length of the two support wires.

110. A vertical pole is supported by three wires. Each wire is 13 yards long and is anchored 5 yards from the base of the pole. How far up the pole will the wires be attached?

111. The length of a rectangular garden is 5 feet greater than the width. The area of the garden is 300 square feet. Find the length and the width.

112. A rectangular parking lot has a length that is 3 yards greater than the width. The area of the rectangular lot is 180 square yards. Find the length and the width.

113. A machine produces open boxes using square sheets of metal. The figure illustrates that the machine cuts equal-sized squares measuring 2 inches on a side from the corners and then shapes the metal into an open box by turning up the sides. If each box must have a volume of 200 cubic inches, find the length of the side of the open square-bottom box.

114. A machine produces open boxes using square sheets of metal. The machine cuts equal-sized squares measuring 3 inches on a side from the corners and then shapes the metal into an open box by turning up the sides. If each box must have a volume of 75 cubic inches, find the length of the side of the open square-bottom box.

115. A rain gutter is made from sheets of aluminum that are 20 inches wide. As shown in the figure, the edges are turned up to form right angles. Determine the depth of the gutter that will allow a cross-sectional area of 13 square inches. Show that there are two different solutions to the problem. Round to the nearest tenth of an inch.

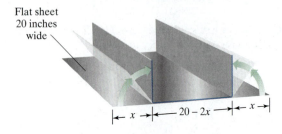

116. A piece of wire is 8 inches long. The wire is cut into two pieces and then each piece is bent into a square. Find the length of each piece if the sum of the areas of these squares is to be 2 square inches.

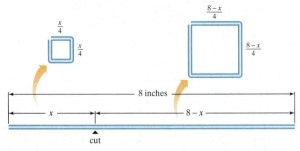

117. A painting measuring 10 inches by 16 inches is surrounded by a frame of uniform width. If the combined area of the painting and the frame is 280 square inches, determine the width of the frame.

Writing in Mathematics

118. What is a quadratic equation?

119. Explain how to solve $x^2 + 6x + 8 = 0$ using factoring and the zero-product principle.

120. Explain how to solve $x^2 + 6x + 8 = 0$ by completing the square.

121. Explain how to solve $x^2 + 6x + 8 = 0$ using the quadratic formula.

122. How is the quadratic formula derived?

123. What is the discriminant and what information does it provide about a quadratic equation?

124. If you are given a quadratic equation, how do you determine which method to use to solve it?

125. If $(x + 2)(x - 4) = 0$ indicates that $x + 2 = 0$ or $x - 4 = 0$, explain why $(x + 2)(x - 4) = 6$ does not mean $x + 2 = 6$ or $x - 4 = 6$. Could we solve the equation using $x + 2 = 3$ and $x - 4 = 2$ because $3 \cdot 2 = 6$?

126. Describe the trend shown by the data for the convictions of police officers in the graph in Figure 1.14 on page 114. Do you believe that this trend is likely to continue or might something occur that would make it impossible to extend the model into the future? Explain your answer.

Technology Exercises

127. If you have access to a calculator that solves quadratic equations, consult the owner's manual to determine how to use this feature. Then use your calculator to solve any five of the equations in Exercises 55–64.

128. Use a graphing utility and x-intercepts to verify any of the real solutions that you obtained for three of the quadratic equations in Exercises 55–64.

Critical Thinking Exercises

129. Which one of the following is true?

  a. The equation $(2x - 3)^2 = 25$ is equivalent to $2x - 3 = 5$.

  b. Every quadratic equation has two distinct numbers in its solution set.

  c. A quadratic equation whose coefficients are real numbers can never have a solution set containing one real number and one complex nonreal number.

  d. The equation $ax^2 + c = 0$ cannot be solved by the quadratic formula.

130. Solve the equation: $x^2 + 2\sqrt{3}x - 9 = 0$.

131. Write a quadratic equation in general form whose solution set is $\{-3, 5\}$.

132. A person throws a rock upward from the edge of an 80-foot cliff. The height, h, in feet, of the rock above the water at the bottom of the cliff after t seconds is described by the formula

$$h = -16t^2 + 64t + 80.$$

How long will it take for the rock to reach the water?

133. A rectangular swimming pool is 12 meters long and 8 meters wide. A tile border of uniform width is to be built around the pool using 120 square meters of tile. The tile is from a discontinued stock (so no additional materials are available), and all 120 square meters are to be used. How wide should the border be? Round to the nearest tenth of a meter. If zoning laws require at least a 2-meter-wide border around the pool, can this be done with the available tile?

 **Group Exercise**

134. Each group member should find an "intriguing" algebraic formula that contains an expression in the form $ax^2 + bx + c$ on one side. Consult college algebra books or liberal arts mathematics books to do so. Group members should select four of the formulas. For each formula selected, write and solve a problem similar to Exercises 99–102 in this exercise set.

SECTION 1.6 *Other Types of Equations*

Objectives

1. Solve polynomial equations by factoring.
2. Solve radical equations.
3. Solve equations with rational exponents.
4. Solve equations that are quadratic in form.
5. Solve equations involving absolute value.

Marine iguanas of the Galápagos Islands

The Galápagos Islands are a volcanic chain of islands lying 600 miles west of Ecuador. They are famed for their extraordinary wildlife, which includes a rare flightless cormorant, marine iguanas, and giant tortoises weighing more than 600 pounds. It was here that naturalist Charles Darwin began to formulate his theory of evolution. Darwin made an enormous collection of the islands' plant species. The formula

$$S = 28.5\sqrt[3]{x}$$

describes the number of plant species, S, on the various islands of the Galápagos chain in terms of the area, x, in square miles, of a particular island.

How can we find the area of a Galápagos island with 57 species of plants? Substitute 57 for S in the formula and solve for x:

$$57 = 28.5\sqrt[3]{x}.$$

The resulting equation contains a variable in the radicand and is called a *radical equation*. In this section, in addition to radical equations, we will show you how to solve certain kinds of polynomial equations, equations involving rational exponents, and equations involving absolute value.

1 Solve polynomial equations by factoring.

Polynomial Equations

The linear and quadratic equations that we studied in the first part of this chapter can be thought of as polynomial equations of degrees 1 and 2, respectively. By contrast, consider the following polynomial equations of degree greater than 2:

$$3x^4 = 27x^2 \qquad\qquad x^3 + x^2 = 4x + 4$$

> This equation is of degree 4 because 4 is the largest exponent.

> This equation is of degree 3 because 3 is the largest exponent.

We can solve these equations by moving all terms to one side, thereby obtaining zero on the other side. We then use factoring and the zero-product principle.

EXAMPLE 1 Solving a Polynomial Equation by Factoring

Solve by factoring: $3x^4 = 27x^2$.

Solution

Step 1 Move all terms to one side and obtain zero on the other side. Subtract $27x^2$ from both sides.

$$
\begin{aligned}
3x^4 &= 27x^2 & &\text{This is the given equation.}\\
3x^4 - 27x^2 &= 27x^2 - 27x^2 & &\text{Subtract } 27x^2 \text{ from both sides.}\\
3x^4 - 27x^2 &= 0 & &\text{Simplify.}
\end{aligned}
$$

Step 2 Factor. We can factor $3x^2$ from each term.

$$3x^4 - 27x^2 = 0$$
$$3x^2(x^2 - 9) = 0$$

Steps 3 and 4 Set each factor equal to zero and solve the resulting equations.

$$
\begin{array}{lcl}
3x^2 = 0 & \text{or} & x^2 - 9 = 0\\
x^2 = 0 & & x^2 = 9\\
x = \pm\sqrt{0} & & x = \pm\sqrt{9}\\
x = 0 & & x = \pm 3
\end{array}
$$

Step 5 Check the solutions in the original equation. Check the three solutions, $0, -3,$ and 3, by substituting them into the original equation. Can you verify that the solution set is $\{-3, 0, 3\}$?

Study Tip

In solving $3x^4 = 27x^2$, be careful not to divide both sides by x^2. If you do, you'll lose 0 as a solution. In general, do not divide both sides of an equation by a variable because that variable might take on the value 0 and you cannot divide by 0.

Check Point 1 Solve by factoring: $4x^4 = 12x^2$.

EXAMPLE 2   Solving a Polynomial Equation by Factoring

Solve by factoring: $x^3 + x^2 = 4x + 4$.

Solution

Step 1   Move all terms to one side and obtain zero on the other side.   Subtract $4x$ and subtract 4 from both sides.

$$x^3 + x^2 = 4x + 4 \qquad \text{This is the given equation.}$$

$$x^3 + x^2 - 4x - 4 = 4x + 4 - 4x - 4 \qquad \text{Subtract 4x and 4 from both sides.}$$

$$x^3 + x^2 - 4x - 4 = 0 \qquad \text{Simplify.}$$

Step 2   Factor.   Because there are four terms, we use factoring by grouping. Group terms that have a common factor.

$$\boxed{x^3 + x^2} + \boxed{-4x - 4} = 0$$

Common factor is x^2.

Common factor is -4.

$$x^2(x + 1) - 4(x + 1) = 0 \qquad \text{Factor } x^2 \text{ from the first two terms and } -4 \text{ from the last two terms.}$$

$$(x + 1)(x^2 - 4) = 0 \qquad \text{Factor out the common binomial, } x + 1, \text{ from each term.}$$

Steps 3 and 4   Set each factor equal to zero and solve the resulting equations.

$$x + 1 = 0 \qquad \text{or} \qquad x^2 - 4 = 0$$
$$x = -1 \qquad\qquad\qquad x^2 = 4$$
$$\qquad\qquad\qquad\qquad x = \pm\sqrt{4} = \pm 2$$

Step 5   Check the solutions in the original equation.   Check the three solutions, $-1, -2,$ and 2, by substituting them into the original equation. Can you verify that the solution set is $\{-2, -1, 2\}$?

Technology

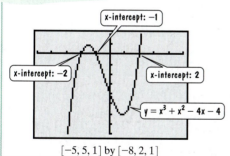

$[-5, 5, 1]$ by $[-8, 2, 1]$

You can use a graphing utility to check the solutions of $x^3 + x^2 - 4x - 4 = 0$. Graph $y = x^3 + x^2 - 4x - 4$, as shown on the left. The x-intercepts are $-2, -1,$ and 2, corresponding to the equations solutions.

Check Point 2   Solve by factoring: $2x^3 + 3x^2 = 8x + 12$.

2   Solve radical equations.

Equations Involving Radicals

A **radical equation** is an equation in which the variable occurs in a square root, cube root, or any higher root. An example of a radical equation is

$$28.5\sqrt[3]{x} = 57.$$

The variable occurs in a cube root.

The equation $28.5\sqrt[3]{x} = 57$ can be used to find the area, x, in square miles, of a Galápagos island with 57 species of plants. First, we isolate the radical by dividing both sides of the equation by 28.5.

$$\frac{28.5\sqrt[3]{x}}{28.5} = \frac{57}{28.5}$$

$$\sqrt[3]{x} = 2$$

Next, we eliminate the radical by raising each side of the equation to a power equal to the index of the radical. Because the index is 3, we cube both sides of the equation.

$$\left(\sqrt[3]{x}\right)^3 = 2^3$$

$$x = 8$$

Thus, a Galápagos island with 57 species of plants has an area of 8 square miles.

The Galápagos equation shows that solving equations involving radicals involves raising both sides of the equation to a power equal to the radicals index. All solutions of the original equation are also solutions of the resulting equation. However, the resulting equation may have some extra solutions that do not satisfy the original equation. Because the resulting equation may not be equivalent to the original equation, we must check each proposed solution by substituting it into the original equation. Let's see exactly how this works.

Study Tip

Be sure to square *both sides* of an equation. Do *not* square each term.

Correct:

$$\left(\sqrt{26 - 11}\right)^2 = (4 - x)^2$$

Incorrect:

$$\left(\sqrt{26 - 11}\right)^2 = 4^2 - x^2$$

Technology

The graph of

$$y = x + \sqrt{26 - 11x} - 4$$

is shown in a $[-10, 3, 1]$ by $[-4, 3, 1]$ viewing rectangle. The x-intercepts are -5 and 2, verifying $\{-5, 2\}$ as the solution set of

$$x + \sqrt{26 - 11x} = 4.$$

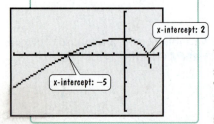

x-intercept: 2

x-intercept: −5

EXAMPLE 3 Solving an Equation Involving a Radical

Solve: $x + \sqrt{26 - 11x} = 4$.

Solution To solve this equation, we isolate the radical expression $\sqrt{26 - 11x}$ on one side of the equation. By squaring both sides of the equation, we can then eliminate the square root.

$x + \sqrt{26 - 11x} = 4$	This is the given equation.
$x + \sqrt{26 - 11x} - x = 4 - x$	Isolate the radical by subtracting x from both sides.
$\sqrt{26 - 11x} = 4 - x$	Simplify.
$\left(\sqrt{26 - 11x}\right)^2 = (4 - x)^2$	Square both sides.
$26 - 11x = 16 - 8x + x^2$	Use $(A - B)^2 = A^2 - 2AB + B^2$ to square $4 - x$.

Next, we need to write this quadratic equation in general form. We can obtain zero on the left side by subtracting 26 and adding $11x$ on both sides.

$$26 - 26 - 11x + 11x = 16 - 26 - 8x + 11x + x^2$$

$0 = x^2 + 3x - 10$		Simplify.
$0 = (x + 5)(x - 2)$		Factor.
$x + 5 = 0$ or $x - 2 = 0$		Set each factor equal to zero.
$x = -5$ $x = 2$		Solve for x.

We have not completed the solution process. Although -5 and 2 satisfy the squared equation, there is no guarantee that they satisfy the original equation. Thus, we must check the proposed solutions. We can do this using a graphing utility (see the technology box in the margin) or by substituting both proposed solutions into the given equation.

CHECK -5:

$$x + \sqrt{26 - 11x} = 4$$
$$-5 + \sqrt{26 - 11(-5)} \stackrel{?}{=} 4$$
$$-5 + \sqrt{81} \stackrel{?}{=} 4$$
$$-5 + 9 \stackrel{?}{=} 4$$
$$4 = 4 \checkmark$$

CHECK 2:

$$x + \sqrt{26 - 11x} = 4$$
$$2 + \sqrt{26 - 11 \cdot 2} \stackrel{?}{=} 4$$
$$2 + \sqrt{4} \stackrel{?}{=} 4$$
$$2 + 2 \stackrel{?}{=} 4$$
$$4 = 4 \checkmark$$

The solution set is $\{-5, 2\}$.

Check Point 3   Solve and check: $\sqrt{6x + 7} - x = 2$.

When solving a radical equation, extra solutions may be introduced when you raise both sides of the equation to an even power. Such solutions, which are not solutions of the given equation, are called **extraneous solutions.**

The solution of radical equations with two or more square root expressions involves isolating a radical, squaring both sides, and then repeating this process. Let's consider an equation containing two square root expressions.

EXAMPLE 4   Solving an Equation Involving Two Radicals

Solve: $\sqrt{3x + 1} - \sqrt{x + 4} = 1.$

Solution

$$\sqrt{3x + 1} - \sqrt{x + 4} = 1 \qquad \text{This is the given equation.}$$

$$\sqrt{3x + 1} = \sqrt{x + 4} + 1 \qquad \text{Isolate one of the radicals by adding } \sqrt{x+4} \text{ to both sides.}$$

$$\left(\sqrt{3x + 1}\right)^2 = \left(\sqrt{x + 4} + 1\right)^2 \qquad \text{Square both sides.}$$

Squaring the expression on the right side of the equation can be a bit tricky. We need to use the formula

$$(A + B)^2 = A^2 + 2AB + B^2.$$

Focusing on just the right side, here is how the squaring is done:

$$(A + B)^2 \quad = \quad A^2 \quad + \quad 2 \quad \cdot \quad A \quad \cdot \quad B + B^2$$

$$\left(\sqrt{x + 4} + 1\right)^2 = \left(\sqrt{x + 4}\right)^2 + 2 \cdot \sqrt{x + 4} \cdot 1 + 1^2.$$

This simplifies to $x + 4 + 2\sqrt{x + 4} + 1$. Thus, our equation $\left(\sqrt{3x + 1}\right)^2 = \left(\sqrt{x + 4} + 1\right)^2$ can be written as follows:

$$3x + 1 = x + 4 + 2\sqrt{x + 4} + 1.$$

$$3x + 1 = x + 5 + 2\sqrt{x + 4} \qquad \text{Combine numerical terms on the right.}$$

$$2x - 4 = 2\sqrt{x + 4} \qquad \text{Isolate } 2\sqrt{x + 4}, \text{ the radical term, by subtracting } x + 5 \text{ from both sides.}$$

$$(2x - 4)^2 = \left(2\sqrt{x + 4}\right)^2 \qquad \text{Square both sides.}$$

$$4x^2 - 16x + 16 = 4(x + 4)$$ Use $(A - B)^2 = A^2 - 2AB + B^2$ to square the left side. Use $(AB)^2 = A^2 B^2$ to square the right side.

$$4x^2 - 16x + 16 = 4x + 16$$ Use the distributive property.

$$4x^2 - 20x = 0$$ Write the quadratic equation in general form by subtracting $4x + 16$ from both sides.

$$4x(x - 5) = 0$$ Factor.

$$4x = 0 \quad \text{or} \quad x - 5 = 0$$ Set each factor equal to zero.

$$x = 0 \qquad\qquad x = 5$$ Solve for x.

Technology

The graph of
$$y = \sqrt{3x + 1} - \sqrt{x + 4} - 1$$
has only one x-intercept at 5. This verifies that the solution set of $\sqrt{3x + 1} - \sqrt{x + 4} = 1$ is $\{5\}$.

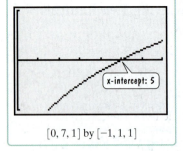

$[0, 7, 1]$ by $[-1, 1, 1]$

Complete the solution process by checking both proposed solutions. We can do this using a graphing utility (see the technology box in the margin) or by substituting both proposed solutions in the given equation.

Check 0:
$$\sqrt{3x + 1} - \sqrt{x + 4} = 1$$
$$\sqrt{3 \cdot 0 + 1} - \sqrt{0 + 4} \stackrel{?}{=} 1$$
$$\sqrt{1} - \sqrt{4} \stackrel{?}{=} 1$$
$$1 - 2 \stackrel{?}{=} 1$$
$$-1 = 1 \quad \text{False}$$

Check 5:
$$\sqrt{3x + 1} - \sqrt{x + 4} = 1$$
$$\sqrt{3 \cdot 5 + 1} - \sqrt{5 + 4} \stackrel{?}{=} 1$$
$$\sqrt{16} - \sqrt{9} \stackrel{?}{=} 1$$
$$4 - 3 \stackrel{?}{=} 1$$
$$1 = 1 \quad \checkmark$$

The false statement $-1 = 1$ indicates that 0 is not a solution. It is an extraneous solution brought about by squaring each side of the equation. The only solution is 5, and the solution set is $\{5\}$.

Check Point 4   Solve and check: $\sqrt{x + 5} - \sqrt{x - 3} = 2$.

Radicals and Windchill

The way that we perceive the temperature on a cold day depends on both air temperature and wind speed. The windchill temperature is what the air temperature would have to be with no wind to achieve the same chilling effect on the skin. The formula that describes windchill temperature, W, in terms of the velocity of the wind, v, in miles per hour, and the actual air temperature, t, in degrees Fahrenheit, is

$$W = 91.4 - \frac{(10.5 + 6.7\sqrt{v} - 0.45v)(457 - 5t)}{110}.$$

Use your calculator to describe how cold the air temperature feels (that is, the windchill temperature) when the temperature is 15° Fahrenheit and the wind is 5 miles per hour. Contrast this with a temperature of 40° Fahrenheit and a wind blowing at 50 miles per hour.

3 Solve equations with rational exponents.

Because $\sqrt[n]{b}$ can be expressed as $b^{1/n}$, radical equations can be written using rational exponents. For example, the Galápagos equation

$$28.5\sqrt[3]{x} = 57$$

can be written

$$28.5x^{1/3} = 57.$$

We solve this equation exactly as we did when it was expressed in radical form. First, isolate $x^{1/3}$.

$$\frac{28.5x^{1/3}}{28.5} = \frac{57}{28.5}$$

$$x^{1/3} = 2$$

Complete the solution process by raising both sides to the third power.

$$(x^{1/3})^3 = 2^3$$

$$x = 8$$

Solving Radical Equations of the Form $x^{m/n} = k$

Assume that m and n are positive integers, $\frac{m}{n}$ is in lowest terms, and k is a real number.

1. Isolate the expression with the rational exponent.
2. Raise both sides of the equation to the $\frac{n}{m}$ power.

If m is even:	**If m is odd:**
$x^{m/n} = k$	$x^{m/n} = k$
$(x^{m/n})^{n/m} = \pm k^{n/m}$	$(x^{m/n})^{n/m} = k^{n/m}$
$x = \pm k^{n/m}$	$x = k^{n/m}$

It is incorrect to insert the $\pm$ symbol when the numerator of the exponent is odd. An odd index has only one root.

3. Check all proposed solutions in the original equation to find out if they are actual solutions or extraneous solutions.

EXAMPLE 5 **Solving Equations Involving Rational Exponents**

Solve:

a. $3x^{3/4} - 6 = 0$ **b.** $x^{2/3} - \frac{3}{4} = -\frac{1}{2}$.

Solution

a. Our goal is to isolate $x^{3/4}$. Then we can raise both sides of the equation to the $\frac{4}{3}$ power because $\frac{4}{3}$ is the reciprocal of $\frac{3}{4}$.

$3x^{3/4} - 6 = 0$ This is the given equation; we will isolate $x^{3/4}$.

$3x^{3/4} = 6$ Add 6 to both sides.

$\dfrac{3x^{3/4}}{3} = \dfrac{6}{3}$ Divide both sides by 3.

$x^{3/4} = 2$ Simplify.

$(x^{3/4})^{4/3} = 2^{4/3}$ Raise both sides to the $\frac{4}{3}$ power. Because $\frac{m}{n} = \frac{3}{4}$ and m is odd, we do not use the $\pm$ symbol.

$x = 2^{4/3}$ Simplify the left side: $(x^{3/4})^{4/3} = x^{\frac{3 \cdot 4}{4 \cdot 3}} = x^{\frac{12}{12}} = x^1 = x.$

The proposed solution is $2^{4/3}$. Complete the solution process by checking this value in the given equation.

$$3x^{3/4} - 6 = 0 \qquad \text{This is the original equation.}$$

$$3\left(2^{4/3}\right)^{3/4} - 6 \stackrel{?}{=} 0 \qquad \text{Substitute the proposed solution.}$$

$$3 \cdot 2 - 6 \stackrel{?}{=} 0 \qquad \left(2^{4/3}\right)^{3/4} = 2^{\frac{4 \cdot 3}{3 \cdot 4}} = 2^{\frac{12}{12}} = 2^1 = 2.$$

$$0 = 0 \checkmark \qquad \text{The true statement shows that } 2^{4/3} \text{ is a solution.}$$

The solution is $2^{4/3} = \sqrt[3]{2^4} \approx 2.52$. The solution set is $\left\{2^{4/3}\right\}$.

b. To solve $x^{2/3} - \frac{3}{4} = -\frac{1}{2}$, our goal is to isolate $x^{2/3}$. Then we can raise both sides of the equation to the $\frac{3}{2}$ power because $\frac{3}{2}$ is the reciprocal of $\frac{2}{3}$.

$$x^{2/3} - \frac{3}{4} = -\frac{1}{2} \qquad \text{This is the given equation.}$$

$$x^{2/3} = \frac{1}{4} \qquad \text{Add } \frac{3}{4} \text{ to both sides.}$$

$$\left(x^{2/3}\right)^{3/2} = \pm\left(\tfrac{1}{4}\right)^{3/2} \qquad \text{Raise both sides to the } \frac{3}{2} \text{ power. Because } \frac{m}{n} = \frac{2}{3}$$
$$\text{and } m \text{ is even, the } \pm \text{ symbol is necessary.}$$

$$x = \pm\frac{1}{8} \qquad \left(\tfrac{1}{4}\right)^{3/2} = \left(\sqrt{\tfrac{1}{4}}\right)^3 = \left(\tfrac{1}{2}\right)^3 = \tfrac{1}{8}$$

Take a moment to verify that the solution set is $\left\{-\frac{1}{8}, \frac{1}{8}\right\}$.

Check Point 5   Solve and check:

a. $5x^{3/2} - 25 = 0$ **b.** $x^{2/3} - 8 = -4$.

4   Solve equations that are quadratic in form.

Equations That Are Quadratic in Form

Some equations that are not quadratic can be written as quadratic equations using an appropriate substitution. Here are some examples:

Given Equation	Substitution	New Equation
$x^4 - 8x^2 - 9 = 0$ or $\left(x^2\right)^2 - 8x^2 - 9 = 0$	$t = x^2$	$t^2 - 8t - 9 = 0$
$5x^{2/3} + 11x^{1/3} + 2 = 0$ or $5\left(x^{1/3}\right)^2 + 11x^{1/3} + 2 = 0$	$t = x^{1/3}$	$5t^2 + 11t + 2 = 0$

An equation that is **quadratic in form** is one that can be expressed as a quadratic equation using an appropriate substitution. Both of the preceding given equations are quadratic in form.

    Equations that are quadratic in form contain an expression to a power, the same expression to that power squared, and a constant term. By letting t equal the expression to the power, a quadratic equation in t will result. Now it's easy. Solve this quadratic equation for t. Finally, use your substitution to find the values for the variable in the given equation. Example 6 shows how this is done.

EXAMPLE 6 **Solving an Equation Quadratic in Form**

Solve: $x^4 - 8x^2 - 9 = 0$.

Technology

The graph of

$$y = x^4 - 8x^2 - 9$$

has x-intercepts at -3 and 3. This verifies that the real solutions of

$$x^4 - 8x^2 - 9 = 0$$

are -3 and 3. The imaginary solutions, $-i$ and i, are not shown as intercepts.

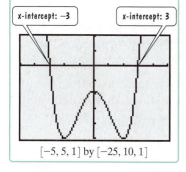

x-intercept: −3   x-intercept: 3

$[-5, 5, 1]$ by $[-25, 10, 1]$

Solution Notice that the equation contains an expression to a power, x^2, the same expression to that power squared, x^4 or $(x^2)^2$, and a constant term, -9. We let t equal the expression to the power. Thus,

let $t = x^2$.

Now we write the given equation as a quadratic equation in t and solve for t.

$x^4 - 8x^2 - 9 = 0$	This is the given equation.
$(x^2)^2 - 8x^2 - 9 = 0$	The given equation contains x^2 and x^2 squared.
$t^2 - 8t - 9 = 0$	Replace x^2 with t.
$(t - 9)(t + 1) = 0$	Factor.
$t - 9 = 0$ or $t + 1 = 0$	Apply the zero-product principle.
$t = 9$ $t = -1$	Solve for t.

We're not done! Why not? We were asked to solve for x and we have values for t. We use the original substitution, $t = x^2$, to solve for x. Replace t with x^2 in each equation shown, namely $t = 9$ and $t = -1$.

$$x^2 = 9 \qquad\qquad x^2 = -1$$
$$x = \pm\sqrt{9} \qquad\qquad x = \pm\sqrt{-1}$$
$$x = \pm 3 \qquad\qquad x = \pm i$$

The solution set is $\{-3, 3, -i, i\}$.

Check Point 6 Solve: $x^4 - 5x^2 + 6 = 0$.

EXAMPLE 7 **Solving an Equation Quadratic in Form**

Solve: $5x^{2/3} + 11x^{1/3} + 2 = 0$.

Solution Notice that the equation contains an expression to a power, $x^{1/3}$, the same expression to that power squared, $x^{2/3}$ or $(x^{1/3})^2$, and a constant term, 2. We let t equal the expression to the power. Thus,

let $t = x^{1/3}$.

Now we write the given equation as a quadratic equation in t and solve for t.

$5x^{2/3} + 11x^{1/3} + 2 = 0$	This is the given equation.
$5(x^{1/3})^2 + 11x^{1/3} + 2 = 0$	The given equation contains $x^{1/3}$ and $x^{1/3}$ squared.
$5t^2 + 11t + 2 = 0$	Replace $x^{1/3}$ with t.
$(5t + 1)(t + 2) = 0$	Factor.
$5t + 1 = 0$ or $t + 2 = 0$	Set each factor equal to 0.
$5t = -1$ $t = -2$	Solve for t.
$t = -\dfrac{1}{5}$	

Use the original substitution, $t = x^{1/3}$, to solve for x. Replace t with $x^{1/3}$ in each of the preceding equations, namely $t = -\frac{1}{5}$ and $t = -2$.

$$x^{1/3} = -\frac{1}{5} \qquad\qquad x^{1/3} = -2 \qquad \text{\color{magenta}Replace } t \text{ \color{magenta}with } x^{1/3}.$$

$$(x^{1/3})^3 = \left(-\frac{1}{5}\right)^3 \qquad (x^{1/3})^3 = (-2)^3 \qquad \text{\color{magenta}Solve for } x \text{ \color{magenta}by cubing both sides of each equation.}$$

$$x = -\frac{1}{125} \qquad\qquad x = -8$$

Check these values to verify that the solution set is $\{-\frac{1}{125}, -8\}$.

Check Point 7 Solve: $3x^{2/3} - 11x^{1/3} - 4 = 0$.

Equations Involving Absolute Value

5 Solve equations involving absolute value.

We have seen that the absolute value of x, $|x|$, describes the distance of x from zero on a number line. Now consider **absolute value equations,** such as

$$|x| = 2.$$

This means that we must determine real numbers whose distance from the origin on the number line is 2. Figure 1.18 shows that there are two numbers such that $|x| = 2$, namely, 2 or -2. We write $x = 2$ or $x = -2$. This observation can be generalized as follows:

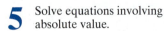
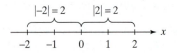

Figure 1.18 If $|x| = 2$, then $x = 2$ or $x = -2$.

Rewriting an Absolute Value Equation without Absolute Value Bars

If c is a positive real number and X represents any algebraic expression, then $|X| = c$ is equivalent to $X = c$ or $X = -c$.

Technology

You can use a graphing utility to verify the solution of an absolute value equation. Consider, for example,

$$|2x - 3| = 11$$

Graph $y_1 = |2x - 3|$ and $y_2 = 11$. The graphs are shown in a $[-10, 10, 1]$ by $[-1, 15, 1]$ viewing rectangle. The x-coordinates of the intersection points are -4 and 7, verifying that $\{-4, 7\}$ is the solution set.

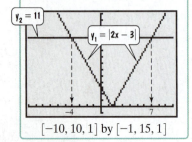

$[-10, 10, 1]$ by $[-1, 15, 1]$

EXAMPLE 8 Solving an Equation Involving Absolute Value

Solve: $|2x - 3| = 11$.

Solution

$$|2x - 3| = 11 \qquad \text{\color{blue}This is the given equation.}$$

$$2x - 3 = 11 \quad \text{or} \quad 2x - 3 = -11 \qquad \text{\color{blue}Rewrite the equation without absolute value bars.}$$

$$2x = 14 \qquad\qquad 2x = -8 \qquad \text{\color{blue}Add 3 to both sides of each equation.}$$

$$x = 7 \qquad\qquad x = -4 \qquad \text{\color{blue}Divide both sides of each equation by 2.}$$

Check 7:	**Check -4:**	
$\|2x - 3\| = 11$	$\|2x - 3\| = 11$	This is the original equation.
$\|2(7) - 3\| \overset{?}{=} 11$	$\|2(-4) - 3\| \overset{?}{=} 11$	Substitute the proposed solutions.
$\|14 - 3\| \overset{?}{=} 11$	$\|-8 - 3\| \overset{?}{=} 11$	Perform operations inside the absolute value bars.
$\|11\| \overset{?}{=} 11$	$\|-11\| \overset{?}{=} 11$	
$11 = 11$ ✓	$11 = 11$ ✓	These true statements indicate that 7 and -4 are solutions.

The solution set is $\{-4, 7\}$.

Check Point 8   Solve: $|2x - 1| = 5$.

The absolute value of a number is never negative. Thus, if X is an algebraic expression and c is a negative number, then $|X| = c$ has no solution. For example, the equation $|3x - 6| = -2$ has no solution because $|3x - 6|$ cannot be negative. The solution set is $\varnothing$, the empty set.

The absolute value of 0 is 0. Thus, if X is an algebraic expression and $|X| = 0$, the solution is found by solving $X = 0$. For example, the solution of $|x - 2| = 0$ is obtained by solving $x - 2 = 0$. The solution is 2 and the solution set is $\{2\}$.

To solve some absolute value equations, it is necessary to first isolate the expression containing the absolute value symbols. For example, consider the equation

$$3|2x - 3| - 8 = 25.$$

> We need to isolate $|2x - 3|$.

How can we isolate $|2x - 3|$? Add 8 to both sides of the equation and then divide both sides by 3.

$$3|2x - 3| - 8 = 25 \qquad \text{This is the given equation.}$$
$$3|2x - 3| = 33 \qquad \text{Add 8 to both sides.}$$
$$|2x - 3| = 11 \qquad \text{Divide both sides by 3.}$$

This results in the equation we solved in Example 8.

EXERCISE SET 1.6

Practice Exercises

Solve each polynomial equation in Exercises 1–10 by factoring and then using the zero-product principle.

1. $3x^4 - 48x^2 = 0$
2. $5x^4 - 20x^2 = 0$
3. $3x^3 + 2x^2 = 12x + 8$
4. $4x^3 - 12x^2 = 9x - 27$
5. $2x - 3 = 8x^3 - 12x^2$
6. $x + 1 = 9x^3 + 9x^2$
7. $4y^3 - 2 = y - 8y^2$
8. $9y^3 + 8 = 4y + 18y^2$
9. $2x^4 = 16x$
10. $3x^4 = 81x$

Solve each radical equation in Exercises 11–28. Check all proposed solutions.

11. $\sqrt{3x + 18} = x$
12. $\sqrt{20 - 8x} = x$
13. $\sqrt{x + 3} = x - 3$
14. $\sqrt{x + 10} = x - 2$
15. $\sqrt{2x + 13} = x + 7$
16. $\sqrt{6x + 1} = x - 1$
17. $x - \sqrt{2x + 5} = 5$
18. $x - \sqrt{x + 11} = 1$
19. $\sqrt{3x + 10} = x + 4$
20. $\sqrt{x - 3} = x - 9$
21. $\sqrt{x + 8} - \sqrt{x - 4} = 2$
22. $\sqrt{x + 5} - \sqrt{x - 3} = 2$
23. $\sqrt{x - 5} - \sqrt{x - 8} = 3$
24. $\sqrt{2x - 3} - \sqrt{x - 2} = 1$
25. $\sqrt{2x + 3} + \sqrt{x - 2} = 2$
26. $\sqrt{x + 2} + \sqrt{3x + 7} = 1$
27. $\sqrt{3\sqrt{x + 1}} = \sqrt{3x - 5}$
28. $\sqrt{1 + 4\sqrt{x}} = 1 + \sqrt{x}$

Solve and check each equation with rational exponents in Exercises 29–38.

29. $x^{3/2} = 8$
30. $x^{3/2} = 27$
31. $(x - 4)^{3/2} = 27$
32. $(x + 5)^{3/2} = 8$
33. $6x^{5/2} - 12 = 0$
34. $8x^{5/3} - 24 = 0$
35. $(x - 4)^{2/3} = 16$
36. $(x + 5)^{2/3} = 4$
37. $(x^2 - x - 4)^{3/4} - 2 = 6$
38. $(x^2 - 3x + 3)^{3/2} - 1 = 0$

Solve each equation in Exercises 39–58 by making an appropriate substitution.

39. $x^4 - 5x^2 + 4 = 0$

40. $x^4 - 13x^2 + 36 = 0$

41. $9x^4 = 25x^2 - 16$

42. $4x^4 = 13x^2 - 9$

43. $x - 13\sqrt{x} + 40 = 0$

44. $2x - 7\sqrt{x} - 30 = 0$

45. $x^{-2} - x^{-1} - 20 = 0$

46. $x^{-2} - x^{-1} - 6 = 0$

47. $x^{2/3} - x^{1/3} - 6 = 0$

48. $2x^{2/3} + 7x^{1/3} - 15 = 0$

49. $x^{3/2} - 2x^{3/4} + 1 = 0$

50. $x^{2/5} + x^{1/5} - 6 = 0$

51. $2x - 3x^{1/2} + 1 = 0$

52. $x + 3x^{1/2} - 4 = 0$

53. $(x - 5)^2 - 4(x - 5) - 21 = 0$

54. $(x + 3)^2 + 7(x + 3) - 18 = 0$

55. $(x^2 - x)^2 - 14(x^2 - x) + 24 = 0$

56. $(x^2 - 2x)^2 - 11(x^2 - 2x) + 24 = 0$

57. $\left(y - \dfrac{8}{y}\right)^2 + 5\left(y - \dfrac{8}{y}\right) - 14 = 0$

58. $\left(y - \dfrac{10}{y}\right)^2 + 6\left(y - \dfrac{10}{y}\right) - 27 = 0$

In Exercises 59–74, solve each absolute value equation or indicate the equation has no solution.

59. $|x| = 8$

60. $|x| = 6$

61. $|x - 2| = 7$

62. $|x + 1| = 5$

63. $|2x - 1| = 5$

64. $|2x - 3| = 11$

65. $2|3x - 2| = 14$

66. $3|2x - 1| = 21$

67. $7|5x| + 2 = 16$

68. $7|3x| + 2 = 16$

69. $|x + 1| + 5 = 3$

70. $|x + 1| + 6 = 2$

71. $|2x - 1| + 3 = 3$

72. $|3x - 2| + 4 = 4$

Hint for Exercises 73–74: Absolute value expressions are equal when the expressions inside the absolute value bars are equal to or opposites of each other.

73. $|3x - 1| = |x + 5|$

74. $|2x - 7| = |x + 3|$

Solve each equation in Exercises 75–84 by the method of your choice.

75. $x + 2\sqrt{x} - 3 = 0$

76. $x^3 + 3x^2 - 4x - 12 = 0$

77. $(x + 4)^{3/2} = 8$

78. $(x^2 - 1)^2 - 2(x^2 - 1) = 3$

79. $\sqrt{4x + 15} - 2x = 0$

80. $x^{2/5} - 1 = 0$

81. $|x^2 + 2x - 36| = 12$

82. $\sqrt{3x + 1} - \sqrt{x - 1} = 2$

83. $x^3 - 2x^2 = x - 2$

84. $|x^2 + 6x + 1| = 8$

Application Exercises

First the good news: The graph shows that U.S. seniors' scores in standard testing in science have improved since 1982. Now the bad news: The highest possible score is 500, and in 1970, the average test score was 304.

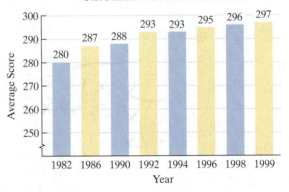

U.S. Seniors' Test Scores in Science

Source: National Assessment of Educational Progress

The formula

$$S = 4\sqrt{x} + 280$$

models the average science test score, S, x years after 1982. Use the formula to solve Exercises 85–86.

85. When will the average science score return to the 1970 average of 304?

86. When will the average science test score be 300?

Out of a group of 50,000 births, the number of people, y, surviving to age x is modeled by the formula

$$y = 5000\sqrt{100 - x}.$$

The graph of the formula is shown. Use the formula to solve Exercises 87–88.

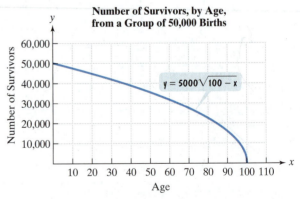

Number of Survivors, by Age, from a Group of 50,000 Births

$$y = 5000\sqrt{100 - x}$$

87. To what age will 40,000 people in the group survive? Identify the solution as a point on the graph of the formula.

88. To what age will 35,000 people in the group survive? Identify the solution as a point on the graph of the formula.

For each planet in our solar system, its year is the time it takes the planet to revolve once around the sun. The formula

$$E = 0.2x^{3/2}$$

models the number of Earth days in a planet's year, E, where x is the average distance of the planet from the sun, in millions of kilometers. Use the formula to solve Exercises 89–90.

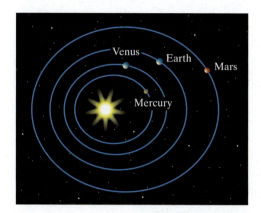

89. We, of course, have 365 Earth days in our year. What is the average distance of Earth from the sun? Use a calculator and round to the nearest million kilometers.

90. There are approximately 88 Earth days in the year of the planet Mercury. What is the average distance of Mercury from the sun? Use a calculator and round to the nearest million kilometers.

Use the Pythagorean Theorem to solve Exercises 91–92.

91. Two vertical poles of lengths 6 feet and 8 feet stand 10 feet apart (see the figure). A cable reaches from the top of one pole to some point on the ground between the poles and then to the top of the other pole. Where should this point be located to use 18 feet of cable?

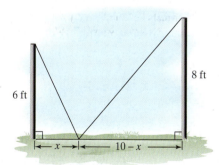

92. Towns A and B are located 6 miles and 3 miles, respectively, from a major expressway. The point on the expressway closest to town A is 12 miles from the point on the expressway closest to town B. Two new roads are to be built from A to the expressway and then to B. (See the figure at the top of the next column.)

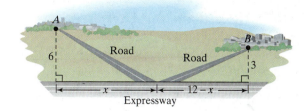

a. Find x if the length of the new roads is 15 miles.

b. Write a verbal description for the road crew telling them where to position the new roads based on your answer to part (a).

Writing in Mathematics

93. Without actually solving the equation, give a general description of how to solve $x^3 - 5x^2 - x + 5 = 0$.

94. In solving $\sqrt{3x + 4} - \sqrt{2x + 4} = 2$, why is it a good idea to isolate a radical term? What if we don't do this and simply square each side? Describe what happens.

95. What is an extraneous solution to a radical equation?

96. Explain how to recognize an equation that is quadratic in form. Provide two original examples with your explanation.

97. Describe two methods for solving this equation: $x - 5\sqrt{x} + 4 = 0$.

98. Explain how to solve an equation involving absolute value.

99. Explain why the procedure that you explained in Exercise 98 does not apply to the equation $|x - 2| = -3$. What is the solution set for this equation?

100. Describe the trend shown by the graph in Exercises 87–88. When is the rate of decrease most rapid? What does this mean about survival rate by age?

Technology Exercises

In Exercises 101–103, use a graphing utility and the graph's x-intercepts to solve each equation. Check by direct substitution. A viewing rectangle is given.

101. $x^3 + 3x^2 - x - 3 = 0$
    $[-6, 6, 1]$ by $[-6, 6, 1]$

102. $-x^4 + 4x^3 - 4x^2 = 0$
    $[-6, 6, 1]$ by $[-9, 2, 1]$

103. $\sqrt{2x + 13} - x - 5 = 0$
    $[-5, 5, 1]$ by $[-5, 5, 1]$

104. Use a graphing utility to obtain the graph of the formula in Exercises 87–88. Then use the $\boxed{\text{TRACE}}$ feature to trace along the curve until you reach the point that visually shows the solution to Exercise 87 or 88.

Critical Thinking Exercises

105. Which one of the following is true?

a. Squaring both sides of $\sqrt{y + 4} + \sqrt{y - 1} = 5$ leads to $y + 4 + y - 1 = 25$, an equation with no radicals.

b. The equation $(x^2 - 2x)^9 - 5(x^2 - 2x)^3 + 6 = 0$ is quadratic in form and should be solved by letting

$$t = (x^2 - 2x)^3.$$

c. If a radical equation has two proposed solutions and one of these values is not a solution, the other value is also not a solution.

d. None of these statements is true.

106. Solve: $\sqrt{6x - 2} = \sqrt{2x + 3} - \sqrt{4x - 1}$.

107. Solve *without* squaring both sides:

$$5 - \frac{2}{x} = \sqrt{5 - \frac{2}{x}}.$$

108. Solve for x: $\sqrt[3]{x\sqrt{x}} = 9$.

109. Solve for x: $x^{5/6} + x^{2/3} - 2x^{1/2} = 0$.

SECTION 1.7 *Linear Inequalities*

Objectives

1. Graph an inequality's solution set.

2. Use set-builder and interval notations.

3. Use properties of inequalities to solve inequalities.

4. Solve compound inequalities.

5. Solve inequalities involving absolute value.

Rent-a-Heap, a car rental company, charges $125 per week plus $0.20 per mile to rent one of their cars. Suppose you are limited by how much money you can spend for the week: You can spend at most $335. If we let x represent the number of miles you drive the heap in a week, we can write an inequality that models the given conditions.

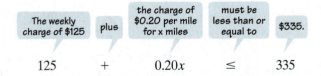

The weekly charge of $125	plus	the charge of $0.20 per mile for x miles	must be less than or equal to	$335.
125	+	0.20x	≤	335

Using the commutative property of addition, we can express this inequality as $0.20x + 125 \leq 335$. The form of this inequality is $ax + b \leq c$, with $a = 0.20$, $b = 125$, and $c = 335$. Any inequality in this form is called a **linear inequality in one variable.** The greatest exponent on the variable in such an inequality is 1. The symbol between $ax + b$ and c can be $\leq$ (is less than or equal to), $<$ (is less than), $\geq$ (is greater than or equal to), or $>$ (is greater than).

In this section, we will study how to solve linear inequalities such as $0.20x + 125 \leq 335$. **Solving an inequality** is the process of finding the set of numbers that make the inequality a true statement. These numbers are called the **solutions** of the inequality, and we say that they **satisfy** the inequality. The set of all solutions is called the **solution set** of the inequality. We begin by discussing how to graph and how to represent these solution sets.

1 Graph an inequality's solution set.

Graphs of Inequalities; Interval Notation

There are infinitely many solutions to the inequality $x > -4$, namely all real numbers that are greater than -4. Although we cannot list all the solutions, we can make a drawing on a number line that represents these solutions. Such a drawing is called the **graph of the inequality.**

Graphs of solutions to linear inequalities are shown on a number line by shading all points representing numbers that are solutions. Parentheses indicate endpoints that are not solutions. Square brackets indicate endpoints that are solutions.

EXAMPLE 1   Graphing Inequalities

Graph the solutions of:

a. $x < 3$ **b.** $x \geq -1$ **c.** $-1 < x \leq 3$.

Study Tip

An inequality symbol points to the smaller number. Thus, another way to express $x < 3$ (x is less than 3) is $3 > x$ (3 is greater than x).

Solution

a. The solutions of $x < 3$ are all real numbers that are less than 3. They are graphed on a number line by shading all points to the left of 3. The parenthesis at 3 indicates that 3 is not a solution, but numbers such as 2.9999 and 2.6 are. The arrow shows that the graph extends indefinitely to the left.

b. The solutions of $x \geq -1$ are all real numbers that are greater than or equal to -1. We shade all points to the right of -1 and the point for -1 itself. The bracket at -1 shows that -1 is a solution of the given inequality. The arrow shows that the graph extends indefinitely to the right.

c. The inequality $-1 < x \leq 3$ is read "-1 is less than x *and x is less than or equal to 3*," or "x is greater than -1 *and* less than or equal to 3." The solutions of $-1 < x \leq 3$ are all real numbers between -1 and 3, not including -1 but including 3. The parenthesis at -1 indicates that -1 is not a solution. By contrast, the bracket at 3 shows that 3 is a solution. Shading indicates the other solutions.

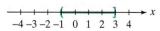

Check Point 1 Graph the solutions of:

a. $x \le 2$ **b.** $x > -4$ **c.** $2 \le x < 6$.

2 Use set-builder and interval notations.

Now that we know how to graph the solution set of an inequality such as $x > -4$, let's see how to represent the solution set. One method is with **set-builder notation.** Using this method, the solution set of $x > -4$ can be expressed as

$$\{x \,|\, x > -4\}.$$

The set of all x such that

We read this as "the set of all real numbers x such that x is greater than -4."

Another method used to represent solution sets of inequalities is **interval notation.** Using this notation, the solution set of $x > -4$ is expressed as $(-4, \infty)$. The parenthesis at -4 indicates that -4 is not included in the interval. The infinity symbol, ∞, does not represent a real number. It indicates that the interval extends indefinitely to the right.

Table 1.5 lists nine possible types of intervals used to describe subsets of real numbers.

Table 1.5   Intervals on the Real Number Line

Let a and b be real numbers such that $a < b$.

Interval Notation	Set-Builder Notation	Graph
(a,b)	$\{x \,|\, a < x < b\}$	
$[a,b]$	$\{x \,|\, a \le x \le b\}$	
$[a,b)$	$\{x \,|\, a \le x < b\}$	
$(a,b]$	$\{x \,|\, a < x \le b\}$	
(a,∞)	$\{x \,|\, x > a\}$	
$[a,\infty)$	$\{x \,|\, x \ge a\}$	
$(-\infty,b)$	$\{x \,|\, x < b\}$	
$(-\infty,b]$	$\{x \,|\, x \le b\}$	
$(-\infty,\infty)$	$\mathbb{R}$ (set of all real numbers)	

EXAMPLE 2   Intervals and Inequalities

Express the intervals in terms of inequalities and graph:

a. $(-1,4]$ **b.** $[2.5,4]$ **c.** $(-4,\infty)$.

Solution

a. $(-1,4] = \{x \,|\, -1 < x \le 4\}$

b. $[2.5,4] = \{x \,|\, 2.5 \le x \le 4\}$

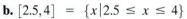

c. $(-4,\infty) = \{x \,|\, x > -4\}$

Check Point 2 Express the intervals in terms of inequalities and graph:

    a. $[-2, 5)$ **b.** $[1, 3.5]$ **c.** $(-\infty, -1)$.

3 Use properties of inequalities to solve inequalities.

Solving Linear Inequalities

Back to our question: How many miles can you drive your Rent-a-Heap car if you can spend at most $335 per week? We answer the question by solving

$$0.20x + 125 \le 335$$

for x. The solution procedure is nearly identical to that for solving

$$0.20x + 125 = 335.$$

Our goal is to get x by itself on the left side. We do this by subtracting 125 from both sides to isolate $0.20x$:

$$0.20x + 125 \le 335$$

$$0.20x + 125 - 125 \le 335 - 125$$

$$0.20x \le 210.$$

Finally, we isolate x from $0.20x$ by dividing both sides of the inequality by 0.20:

$$\frac{0.20x}{0.20} \le \frac{210}{0.20}$$

$$x \le 1050.$$

With at most $335 per week to spend, you can travel at most 1050 miles.

We started with the inequality $0.20x + 125 \le 335$ and obtained the inequality $x \le 1050$ in the final step. Both of these inequalities have the same solution set, namely $\{x \mid x \le 1050\}$. Inequalities such as these, with the same solution set, are said to be **equivalent.**

We isolated x from $0.20x$ by dividing both sides of $0.20x \le 210$ by 0.20, a positive number. Let's see what happens if we divide both sides of an inequality by a negative number. Consider the inequality $10 < 14$. Divide 10 and 14 by -2:

$$\frac{10}{-2} = -5 \quad \text{and} \quad \frac{14}{-2} = -7.$$

Because -5 lies to the right of -7 on the number line, -5 is greater than -7:

$$-5 > -7.$$

Notice that the direction of the inequality symbol is reversed:

$$10 < 14$$

$$-5 > -7.$$

In general, **when we multiply or divide both sides of an inequality by a negative number, the direction of the inequality symbol is reversed.** When we reverse the direction of the inequality symbol, we say that we change the *sense* of the inequality.

We can isolate a variable in a linear inequality the same way we can isolate a variable in a linear equation. The following properties are used to create equivalent inequalities:

Study Tip

English phrases such as "at least" and "at most" can be modeled by inequalities.

English Sentence	Inequality
x is at least 5.	$x \ge 5$
x is at most 5.	$x \le 5$
x is between 5 and 7.	$5 < x < 7$
x is no more than 5.	$x \le 5$
x is no less than 5.	$x \ge 5$

Properties of Inequalities

Property	The Property in Words	Example
Addition and Subtraction Properties If $a < b$, then $a + c < b + c$. If $a < b$, then $a - c < b - c$.	If the same quantity is added to or subtracted from both sides of an inequality, the resulting inequality is equivalent to the original one.	$2x + 3 < 7$ Subtract 3: $2x + 3 - 3 < 7 - 3$. Simplify: $2x < 4$.
Positive Multiplication and Division Properties If $a < b$ and c is positive, then $ac < bc$. If $a < b$ and c is positive, then $\dfrac{a}{c} < \dfrac{b}{c}$.	If we multiply or divide both sides of an inequality by the same positive quantity, the resulting inequality is equivalent to the original one.	$2x < 4$ Divide by 2: $\dfrac{2x}{2} < \dfrac{4}{2}$. Simplify: $x < 2$.
Negative Multiplication and Division Properties If $a < b$ and c is negative, then $ac > bc$. If $a < b$ and c is negative, then $\dfrac{a}{c} > \dfrac{b}{c}$.	If we multiply or divide both sides of an inequality by the same negative quantity and reverse the direction of the inequality symbol, the result is an equivalent inequality.	$-4x < 20$ Divide by -4 and reverse the sense of the inequality: $\dfrac{-4x}{-4} > \dfrac{20}{-4}$. Simplify: $x > -5$.

EXAMPLE 3 Solving a Linear Inequality

Solve and graph the solution set on a number line:

$$3 - 2x < 11.$$

Solution

$3 - 2x < 11$	This is the given inequality.
$3 - 2x - 3 < 11 - 3$	Subtract 3 from both sides.
$-2x < 8$	Simplify.
$\dfrac{-2x}{-2} > \dfrac{8}{-2}$	Divide both sides by -2 and reverse the sense of the inequality.
$x > -4$	Simplify.

The solution set consists of all real numbers that are greater than -4, expressed as $\{x \mid x > -4\}$ in set-builder notation. The interval notation for this solution set is $(-4, \infty)$. The graph of the solution set is shown as follows:

Discovery

As a partial check, select one number from the solution set for the inequality in Example 3. Substitute that number into the original inequality. Perform the resulting computations. You should obtain a true statement.

Is it possible to perform a partial check using a number that is not in the solution set? What should happen in this case? Try doing this.

Check Point 3 Solve and graph the solution set on a number line:

$$2 - 3x \leq 5.$$

EXAMPLE 4   Solving a Linear Inequality

Solve and graph the solution set: $7x + 15 \geq 13x + 51$.

Solution   We will collect variable terms on the left and constant terms on the right.

$7x + 15$	$\geq$	$13x + 51$	This is the given inequality.
$7x + 15 - 13x$	$\geq$	$13x + 51 - 13x$	Subtract $13x$ from both sides.
$-6x + 15$	$\geq$	51	Simplify.
$-6x + 15 - 15$	$\geq$	$51 - 15$	Subtract 15 from both sides.
$-6x$	$\geq$	36	Simplify.
$\dfrac{-6x}{-6}$	$\leq$	$\dfrac{36}{-6}$	Divide both sides by -6 and reverse the sense of the inequality.
x	$\leq$	-6	Simplify.

The solution set consists of all real numbers that are less than or equal to -6, expressed as $\{x \mid x \leq -6\}$. The interval notation for this solution set is $(-\infty, -6]$. The graph of the solution set is shown as follows:

$$\xleftarrow{\hspace{1cm}} \overset{\quad -8 \quad -7 \quad -6 \quad -5 \quad -4 \quad -3 \quad -2 \quad -1 \quad 0 \quad 1 \quad 2}{\rule{8cm}{0.4pt}} \xrightarrow{\hspace{0.3cm}} x$$

Study Tip

You can solve

$$7x + 15 \geq 13x + 51$$

by isolating x on the right side. Subtract $7x$ from both sides.

$$7x + 15 - 7x$$
$$\geq 13x + 51 - 7x$$
$$15 \geq 6x + 51$$

Now subtract 51 from both sides.

$$15 - 51 \geq 6x + 51 - 51$$
$$-36 \geq 6x$$

Finally, divide both sides by 6.

$$\frac{-36}{6} \geq \frac{6x}{6}$$
$$-6 \geq x$$

This last inequality means the same thing as

$$x \leq -6.$$

Check Point 4   Solve and graph the solution set: $6 - 3x \leq 5x - 2$.

Technology

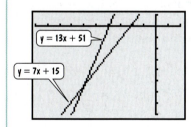

$y = 13x + 51$

$y = 7x + 15$

You can use a graphing utility to verify that $(-\infty, -6]$ is the solution set of

$$7x + 15 \qquad \geq \qquad 13x + 51.$$

For what values of x does the graph of $y = 7x + 15$ lie above or on the graph of $y = 13x + 51$?

The graphs are shown on the left in a $[-10, 2, 1]$ by $[-40, 5, 5]$ viewing rectangle. Look closely at the graphs. Can you see that the graph of $y = 7x + 15$ lies above or on the graph of $y = 13x + 51$ when $x \leq -6$, or on the interval $(-\infty, -6]$?

4 Solve compound inequalities.

Solving Compound Inequalities

We now consider two inequalities such as

$$-3 < 2x + 1 \text{ and } 2x + 1 \leq 3$$

expressed as a **compound inequality**

$$-3 < 2x + 1 \leq 3.$$

The word "and" does not appear when the inequality is written in the shorter form, although it is implied. The shorter form enables us to solve both inequalities at once. By performing the same operation on all three parts of the inequality, our goal is to **isolate x in the middle.**

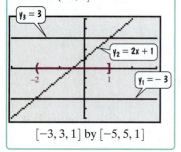
EXAMPLE 5 Solving a Compound Inequality

Solve and graph the solution set:

$$-3 < 2x + 1 \le 3.$$

Solution We would like to isolate x in the middle. We can do this by first subtracting 1 from all three parts of the compound inequality. Then we isolate x from $2x$ by dividing all three parts of the inequality by 2.

$-3 < 2x + 1 \le 3$	This is the given inequality.
$-3 - 1 < 2x + 1 - 1 \le 3 - 1$	Subtract 1 from all three parts.
$-4 < 2x \le 2$	Simplify.
$\dfrac{-4}{2} < \dfrac{2x}{2} \le \dfrac{2}{2}$	Divide each part by 2.
$-2 < x \le 1$	Simplify.

The solution set consists of all real numbers greater than -2 and less than or equal to 1, represented by $\{x \mid -2 < x \le 1\}$ in set-builder notation and $(-2, 1]$ in interval notation. The graph is shown as follows:

Check Point 5 Solve and graph the solution set: $1 \le 2x + 3 < 11$.

---

5 Solve inequalities involving absolute value.

Solving Inequalities with Absolute Value

We know that $|x|$ describes the distance of x from zero on a real number line. We can use this geometric interpretation to solve an inequality such as

$$|x| < 2.$$

Figure 1.19 $|x| < 2$, so $-2 < x < 2$.

This means that the distance of x from 0 is *less than* 2, as shown in Figure 1.19. The interval shows values of x that lie less than 2 units from 0. Thus, x can lie between -2 and 2. That is, x is greater than -2 and less than 2. We write $(-2, 2)$ or $\{x \mid -2 < x < 2\}$.

Some absolute value inequalities use the "greater than" symbol. For example, $|x| > 2$ means that the distance of x from 0 is *greater than* 2, as shown in Figure 1.20. Thus, x can be less than -2 *or* greater than 2. We write $x < -2$ or $x > 2$.

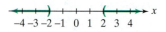

Figure 1.20 $|x| > 2$, so $x < -2$ or $x > 2$.

These observations suggest the following principles for solving inequalities with absolute value:

> **Solving an Absolute Value Inequality**
>
> If X is an algebraic expression and c is a positive number,
>
> 1. The solutions of $|X| < c$ are the numbers that satisfy $-c < X < c$.
> 2. The solutions of $|X| > c$ are the numbers that satisfy $X < -c$ or $X > c$.
>
> These rules are valid if $<$ is replaced by $\le$ and $>$ is replaced by $\ge$.

EXAMPLE 6   **Solving an Absolute Value Inequality with** $<$

Solve and graph the solution set: $|x - 4| < 3$.

Solution

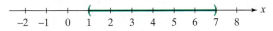

$|X| < c$ means $-c < X < c$.

$|x - 4| < 3$ means $-3 < x - 4 < 3$.

We solve the compound inequality by adding 4 to all three parts.

$$-3 < x - 4 < 3$$
$$-3 + 4 < x - 4 + 4 < 3 + 4$$
$$1 < x < 7$$

The solution set is all real numbers greater than 1 and less than 7, denoted by $\{x \,|\, 1 < x < 7\}$ or $(1, 7)$. The graph of the solution set is shown as follows:

Check Point 6   Solve and graph the solution set: $|x - 2| < 5$.

EXAMPLE 7   **Solving an Absolute Value Inequality with** $\geq$

Solve and graph the solution set: $|2x + 3| \geq 5$.

Solution

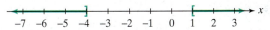

$|X| \geq c$ means $X \leq -c$ or $X \geq c$.

$|2x + 3| \geq 5$ means $2x + 3 \leq -5$ or $2x + 3 \geq 5$.

We solve each of these inequalities separately.

$2x + 3 \leq -5$	or	$2x + 3 \geq 5$	These are the inequalities without absolute value bars.
$2x + 3 - 3 \leq -5 - 3$	or	$2x + 3 - 3 \geq 5 - 3$	Subtract 3 from both sides.
$2x \leq -8$	or	$2x \geq 2$	Simplify.
$\dfrac{2x}{2} \leq \dfrac{-8}{2}$	or	$\dfrac{2x}{2} \geq \dfrac{2}{2}$	Divide both sides by 2.
$x \leq -4$	or	$x \geq 1$	Simplify.

The solution set is $\{x \,|\, x \leq -4 \text{ or } x \geq 1\}$, that is, all x in $(-\infty, -4]$ or $[1, \infty)$. The graph of the solution set is shown as follows:

Study Tip

The graph of the solution set for $|X| > c$ will be divided into two intervals. The graph of the solution set for $|X| < c$ will be a single interval.

Check Point 7 Solve and graph the solution set: $|2x - 5| \geq 3$.

Applications

Our next example shows how to use an inequality to select the better deal between two pricing options. We will use our five-step strategy for solving problems using mathematical models.

EXAMPLE 8 **Creating and Comparing Mathematical Models**

Acme Car rental agency charges $4 a day plus $0.15 a mile, whereas Interstate rental agency charges $20 a day and $0.05 a mile. Under what conditions is the daily cost of an Acme rental a better deal than an Interstate rental?

Solution

Step 1 Let x represent one of the quantities. We are looking for the number of miles driven in a day to make Acme the better deal. Thus,

$$\text{let } x = \text{the number of miles driven in a day.}$$

Step 2 Represent other quantities in terms of x. We are not asked to find another quantity, so we can skip this step.

Step 3 Write an inequality in x that describes the conditions.

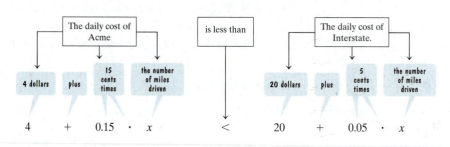

Technology

The graphs of the daily cost models for the car rental agencies

$$y_1 = 4 + 0.15x$$

$$\text{and } y_2 = 20 + 0.05x$$

are shown in a $[0, 300, 10]$ by $[0, 40, 4]$ viewing rectangle. The graphs intersect at $(160, 28)$. To the left of $x = 160$, the graph of Acme's daily cost lies below that of Interstate's daily cost. This shows that for fewer than 160 miles per day, Acme offers the better deal.

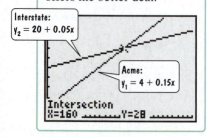

Step 4 Solve the inequality and answer the question.

$$4 + 0.15x < 20 + 0.05x \qquad \text{This is the inequality that models the verbal conditions.}$$

$$4 + 0.15x - 0.05x < 20 + 0.05x - 0.05x \qquad \text{Subtract 0.05x from both sides.}$$

$$4 + 0.1x < 20 \qquad \text{Simplify.}$$

$$4 + 0.1x - 4 < 20 - 4 \qquad \text{Subtract 4 from both sides.}$$

$$0.1x < 16 \qquad \text{Simplify.}$$

$$\frac{0.1x}{0.1} < \frac{16}{0.1} \qquad \text{Divide both sides by 0.1.}$$

$$x < 160 \qquad \text{Simplify.}$$

Thus, driving fewer than 160 miles per day makes Acme the better deal.

Step 5 **Check the proposed solution in the original wording of the problem.**
One way to do this is to take a mileage less than 160 miles per day to see if Acme is the better deal. Suppose that 150 miles are driven in a day.

$$\text{Cost for Acme} = 4 + 0.15(150) = 26.50$$

$$\text{Cost for Interstate} = 20 + 0.05(150) = 27.50$$

Acme has a lower daily cost, making it the better deal.

Check Point 8 A car can be rented from Basic Rental for $260 per week with no extra charge for mileage. Continental charges $80 per week plus 25 cents for each mile driven to rent the same car. Under what conditions is the rental cost for Basic Rental a better deal than Continental's?

EXERCISE SET 1.7

Practice Exercises

In Exercises 1–12, graph the solutions of each inequality on a number line.

1. $x > 6$ **2.** $x > -2$

3. $x < -4$ **4.** $x < 0$

5. $x \geq -3$ **6.** $x \geq -5$

7. $x \leq 4$ **8.** $x \leq 7$

9. $-2 < x \leq 5$ **10.** $-3 \leq x < 7$

11. $-1 < x < 4$ **12.** $-7 \leq x \leq 0$

In Exercises 13–26, express each interval in terms of an inequality and graph the interval on a number line.

13. $(1, 6]$ **14.** $(-2, 4]$

15. $[-5, 2)$ **16.** $[-4, 3)$

17. $[-3, 1]$ **18.** $[-2, 5]$

19. $(2, \infty)$ **20.** $(3, \infty)$

21. $[-3, \infty)$ **22.** $[-5, \infty)$

23. $(-\infty, 3)$ **24.** $(-\infty, 2)$

25. $(-\infty, 5.5)$ **26.** $(-\infty, 3.5]$

Solve each linear inequality in Exercises 27–48 and graph the solution set on a number line. Express the solution set using interval notation.

27. $5x + 11 < 26$ **28.** $2x + 5 < 17$

29. $3x - 7 \geq 13$ **30.** $8x - 2 \geq 14$

31. $-9x \geq 36$ **32.** $-5x \leq 30$

33. $8x - 11 \leq 3x - 13$ **34.** $18x + 45 \leq 12x - 8$

35. $4(x + 1) + 2 \geq 3x + 6$

36. $8x + 3 > 3(2x + 1) + x + 5$

37. $2x - 11 < -3(x + 2)$ **38.** $-4(x + 2) > 3x + 20$

39. $1 - (x + 3) \geq 4 - 2x$ **40.** $5(3 - x) \leq 3x - 1$

41. $\dfrac{x}{4} - \dfrac{3}{5} \leq \dfrac{x}{2} + 1$ **42.** $\dfrac{3x}{10} + 1 \geq \dfrac{1}{5} - \dfrac{x}{10}$

43. $1 - \dfrac{x}{2} > 4$ **44.** $7 - \dfrac{4}{5}x < \dfrac{3}{5}$

45. $\dfrac{x - 4}{6} \geq \dfrac{x - 2}{9} + \dfrac{5}{18}$ **46.** $\dfrac{4x - 3}{6} + 2 \geq \dfrac{2x - 1}{12}$

47. $4(3x - 2) - 3x < 3(1 + 3x) - 7$

48. $3(x - 8) - 2(10 - x) > 5(x - 1)$

Solve each inequality in Exercises 49–56 and graph the solution set on a number line. Express the solution set using interval notation.

49. $6 < x + 3 < 8$ **50.** $7 < x + 5 < 11$

51. $-3 \leq x - 2 < 1$ **52.** $-6 < x - 4 \leq 1$

53. $-11 < 2x - 1 \leq -5$ **54.** $3 \leq 4x - 3 < 19$

55. $-3 \leq \dfrac{2}{3}x - 5 < -1$ **56.** $-6 \leq \dfrac{1}{2}x - 4 < -3$

Solve each inequality in Exercises 57–84 by first rewriting each one as an equivalent inequality without absolute value bars. Graph the solution set on a number line. Express the solution set using interval notation.

57. $|x| < 3$ **58.** $|x| < 5$

59. $|x - 1| \leq 2$ **60.** $|x + 3| \leq 4$

61. $|2x - 6| < 8$ **62.** $|3x + 5| < 17$

63. $|2(x - 1) + 4| \leq 8$ **64.** $|3(x - 1) + 2| \leq 20$

65. $\left|\dfrac{2y + 6}{3}\right| < 2$ **66.** $\left|\dfrac{3(x - 1)}{4}\right| < 6$

67. $|x| > 3$

68. $|x| > 5$

69. $|x - 1| \geq 2$

70. $|x + 3| \geq 4$

71. $|3x - 8| > 7$

72. $|5x - 2| > 13$

73. $\left|\dfrac{2x + 2}{4}\right| \geq 2$

74. $\left|\dfrac{3x - 3}{9}\right| \geq 1$

75. $\left|3 - \dfrac{2}{3}x\right| > 5$

76. $\left|3 - \dfrac{3}{4}x\right| > 9$

77. $3|x - 1| + 2 \geq 8$

78. $-2|4 - x| \geq -4$

79. $3 < |2x - 1|$

80. $5 \geq |4 - x|$

81. $12 < \left|-2x + \dfrac{6}{7}\right| + \dfrac{3}{7}$

82. $1 < \left|x - \dfrac{11}{3}\right| + \dfrac{7}{3}$

83. $4 + \left|3 - \dfrac{x}{3}\right| \geq 9$

84. $\left|2 - \dfrac{x}{2}\right| - 1 \leq 1$

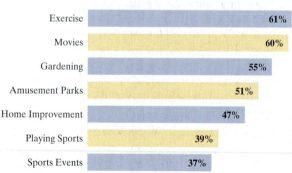

Application Exercises

The bar graph shows how we spend our leisure time. Let x represent the percentage of the population regularly participating in an activity. In Exercises 85–92, write the name or names of the activity described by the given inequality or interval.

Percentage of U.S. Population Participating in Each Activity on a Regular Basis

Activity	Percentage
Exercise	61%
Movies	60%
Gardening	55%
Amusement Parks	51%
Home Improvement	47%
Playing Sports	39%
Sports Events	37%

Source: U.S. Census Bureau

85. $x < 40\%$

86. $x < 50\%$

87. $[51\%, 61\%]$

88. $[47\%, 60\%]$

89. $(51\%, 61\%)$

90. $(47\%, 60\%)$

91. $(39\%, 55\%]$

92. $(37\%, 47\%]$

The line graph at the top of the next column shows the declining consumption of cigarettes in the United States. The data shown by the graph can be modeled by

$$N = 550 - 9x$$

where N is the number of cigarettes consumed, in billions, x

years after 1988. Use this formula to solve Exercises 93–94.

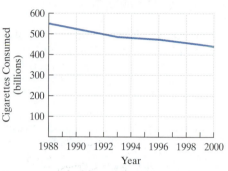

Consumption of Cigarettes in the U.S.

Source: Economic Research Service, USDA

93. How many years after 1988 will cigarette consumption be less than 370 billion cigarettes each year? Which years does this describe?

94. Describe how many years after 1988 cigarette consumption will be less than 325 billion cigarettes each year. Which years are included in your description?

95. The formula for converting Fahrenheit temperature, F, to Celsius temperature, C, is

$$C = \dfrac{5}{9}(F - 32).$$

If Celsius temperature ranges from $15°$ to $35°$, inclusive, what is the range for the Fahrenheit temperature? Use interval notation to express this range.

96. The formula

$$T = 0.01x + 56.7$$

models the global mean temperature, T, in degrees Fahrenheit, of Earth x years after 1905. For which range of years was the global mean temperature at least 56.7°F and at most 57.2°F?

The three television programs viewed by the greatest percentage of U.S. households in the twentieth century are shown in the table. The data are from a random survey of 4000 TV households by Nielsen Media Research. In Exercises 97–98, let x represent the actual viewing percentage in the U.S. population.

TV Programs with the Greatest U.S. Audience Viewing Percentage of the Twentieth Century

Program	Viewing Percentage in Survey
1. "M*A*S*H" Feb. 28, 1983	60.2%
2. "Dallas" Nov. 21, 1980	53.3%
3. "Roots" Part 8 Jan. 30, 1977	51.1%

Source: Nielsen Media Research

97. The inequality $|x - 60.2| \le 1.6$ describes the actual viewing percentage for "M*A*S*H" in the U.S. population. Solve the inequality and interpret the solution. Explain why the survey's *margin of error* is ±1.6%.

98. The inequality $|x - 51.1| \le 1.6$ describes the actual viewing percentage for "Roots" Part 8 in the U.S. population. Solve the inequality and interpret the solution. Explain why the survey's *margin of error* is ±1.6%.

99. If a coin is tossed 100 times, we would expect approximately 50 of the outcomes to be heads. It can be demonstrated that a coin is unfair if h, the number of outcomes that result in heads, satisfies $\left|\dfrac{h - 50}{5}\right| \ge 1.645$. Describe the number of outcomes that determine an unfair coin that is tossed 100 times.

100. The inequality $|T - 57| \le 7$ describes the range of monthly average temperature, T, in degrees Fahrenheit, for San Francisco, California. The inequality $|T - 50| \le 22$ describes the range of monthly average temperature, T, in degrees Fahrenheit, for Albany, New York. Solve each inequality and interpret the solution. Then describe at least three differences between the monthly average temperatures for the two cities.

In Exercises 101–110, use the five-step strategy for solving word problems. Give a linear inequality that models the verbal conditions and then solve the problem.

101. A truck can be rented from Basic Rental for $50 a day plus $0.20 per mile. Continental charges $20 per day plus $0.50 per mile to rent the same truck. How many miles must be driven in a day to make the rental cost for Basic Rental a better deal then Continental's?

102. You are choosing between two long-distance telephone plans. Plan A has a monthly fee of $15 with a charge of $0.08 per minute for all long-distance calls. Plan B has a monthly fee of $3 with a charge of $0.12 per minute for all long-distance calls. How many minutes of long-distance calls in a month make plan A the better deal?

103. A city commission has proposed two tax bills. The first bill requires that a homeowner pay $1800 plus 3% of the assessed home value in taxes. The second bill requires taxes of $200 plus 8% of the assessed home value. What price range of home assessment would make the first bill a better deal?

104. A local bank charges $8 per month plus 5¢ per check. The credit union charges $2 per month plus 8¢ per check. How many check should be written each month to make the credit union a better deal?

105. A company manufactures and sells blank audiocassette tapes. The weekly fixed cost is $10,000 and it cost $0.40 to produce each tape. The selling price is $2.00 per tape. How many tapes must be produced and sold each week for the company to have a profit gain?

106. A company manufactures and sells personalized stationery. The weekly fixed cost is $3000 and it cost $3.00 to produce each package of stationery. The selling price is $5.50 per package. How many packages of stationery must be produced and sold each week for the company to have a profit gain?

107. An elevator at a construction site has a maximum capacity of 2800 pounds. If the elevator operator weighs 265 pounds and each cement bag weighs 65 pounds, how many bags of cement can be safely lifted on the elevator in one trip?

108. An elevator at a construction site has a maximum capacity of 3000 pounds. If the elevator operator weighs 245 pounds and each cement bag weighs 95 pounds, how many bags of cement can be safely lifted on the elevator in one trip?

109. On two examinations, you have grades of 86 and 88. There is an optional final examination, which counts as one grade. You decide to take the final in order to get a course grade of A, meaning a final average of at least 90.
  a. What must you get on the final to earn an A in the course?
  b. By taking the final, if you do poorly, you might risk the B that you have in the course based on the first two exam grades. If your final average is less than 80, you will lose your B in the course. Describe the grades on the final that will cause this happen.

110. Parts for an automobile repair cost $175. The mechanic charges $34 per hour. If you receive an estimate for at least $226 and at most $294 for fixing the car, what is the time interval that the mechanic will be working on the job?

Writing in Mathematics

111. When graphing the solutions of an inequality, what does a parenthesis signify? What does a bracket signify?

112. When solving an inequality, when is it necessary to change the sense of the inequality? Give an example.

113. Describe ways in which solving a linear inequality is similar to solving a linear equation.

114. Describe ways in which solving a linear inequality is different than solving a linear equation.

115. What is a compound inequality and how is it solved?

116. Describe how to solve an absolute value inequality involving the symbol $<$. Give an example.

117. Describe how to solve an absolute value inequality involving the symbol $>$. Give an example.

118. Explain why $|x| < -4$ has no solution.

119. Describe the solution set of $|x| > -4$.

120. The formula
$$V = 3.5x + 120$$
models Super Bowl viewers, V, in millions, x years after 1990. Use the formula to write a word problem that can be solved using a linear inequality. Then solve the problem.

Technology Exercises

In Exercises 121–122, solve each inequality using a graphing utility. Graph each side separately. Then determine the values of x for which the graph on the left side lies above the graph on the right side.

121. $-3(x - 6) > 2x - 2$ **122.** $-2(x + 4) > 6x + 16$

Use the same technique employed in Exercises 121–122 to solve each inequality in Exercises 123–124. In each case, what conclusion can you draw? What happens if you try solving the inequalities algebraically?

123. $12x - 10 > 2(x - 4) + 10x$

124. $2x + 3 > 3(2x - 4) - 4x$

125. A bank offers two checking account plans. Plan A has a base service charge of \$4.00 per month plus 10¢ per check. Plan B charges a base service charge of \$2.00 per month plus 15¢ per check.

   a. Write models for the total monthly costs for each plan if x checks are written.

   b. Use a graphing utility to graph the models in the same $[0, 50, 10]$ by $[0, 10, 1]$ viewing rectangle.

   c. Use the graphs (and the TRACE or intersection feature) to determine for what number of checks per month plan A will be better than plan B.

   d. Verify the result of part (c) algebraically by solving an inequality.

Critical Thinking Exercises

126. Which one of the following is true?

   a. The first step in solving $|2x - 3| > -7$ is to rewrite the inequality as $2x - 3 > -7$ or $2x - 3 < 7$.

   b. The smallest real number in the solution set of $2x > 6$ is 4.

   c. All irrational numbers satisfy $|x - 4| > 0$.

   d. None of these statements is true.

127. What's wrong with this argument? Suppose x and y represent two real numbers, where $x > y$.

$2 > 1$	This is a true statement.
$2(y - x) > 1(y - x)$	Multiply both sides by $y - x$.
$2y - 2x > y - x$	Use the distributive property.
$y - 2x > -x$	Subtract y from both sides.
$y > x$	Add 2x to both sides.

The final inequality, $y > x$, is impossible because we were initially given $x > y$.

128. The graphs of $y = 6$, $y = 3(-x - 5) - 9$, and $y = 0$ are shown in the figure. The graphs were obtained using a graphing utility and a $[-12, 1, 1]$ by $[-2, 8, 1]$ viewing rectangle. Use the graphs to write the solution set for the compound inequality. Express the solution set using interval notation.

$$0 < 3(-x - 5) - 9 < 6.$$

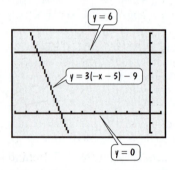

129. The percentage, p, of defective products manufactured by a company is given by $|p - 0.3\%| \leq 0.2\%$. If 100,000 products are manufactured and the company offers a \$5 refund for each defective product, describe the company's cost for refunds.

Group Exercise

130. Each group member should research one situation that provides two different pricing options. These can involve areas such as public transportation options (with or without coupon books) or long-distance telephone plans or anything of interest. Be sure to bring in all the details for each option. At a second group meeting, select the two pricing situations that are most interesting and relevant. Using each situation, write a word problem about selecting the better of the two options. The word problem should be one that can be solved using a linear inequality. The group should turn in the two problems and their solutions.

SECTION 1.8   *Quadratic and Rational Inequalities*

Objectives

1. Solve quadratic inequalities.
2. Solve rational inequalities.
3. Solve problems modeled by nonlinear inequalities.

Not afraid of heights and cutting-edge excitement? How about sky diving? Behind your exhilarating experience is the world of algebra. After you jump from the airplane, your height above the ground at every instant of your fall can be described by a formula involving a variable that is squared. At some point, you'll need to open your parachute. How can you determine when you must do so? Let x represent the number of seconds you are falling. You can compute when to open the parachute by solving an inequality that takes on the form $ax^2 + bx + c < 0$. Such an inequality is called a **quadratic inequality.**

Definition of a Quadratic Inequality

A **quadratic inequality** is any inequality that can be put in one of the forms

$$ax^2 + bx + c < 0 \qquad ax^2 + bx + c > 0$$
$$ax^2 + bx + c \leq 0 \qquad ax^2 + bx + c \geq 0$$

where a, b, and c are real numbers and $a \neq 0$.

In this section we establish the basic techniques for solving quadratic inequalities. We will use these techniques to solve inequalities containing quotients, called **rational inequalities.** Finally, we will consider a formula that models the position of any free-falling object. As a sky diver, you could be that free-falling object!

1 Solve quadratic inequalities.

Solving Quadratic Inequalities

Graphs can help us to visualize the solutions of quadratic inequalities. The cuplike graph of $y = x^2 - 7x + 10$ is shown in Figure 1.21. The x-intercepts, 2 and 5, are **boundary points** between where the graph lies above the x-axis, shown in blue, and where the graph lies below the x-axis, shown in red. These boundary points play a critical role in solving quadratic inequalities.

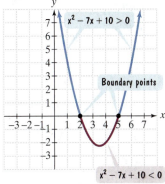

Figure 1.21

Procedure for Solving Quadratic Inequalities

1. Express the inequality in the general form

   $$ax^2 + bx + c > 0 \quad \text{or} \quad ax^2 + bx + c < 0.$$

2. Solve the equation $ax^2 + bx + c = 0$. The real solutions are the **boundary points.**

3. Locate these boundary points on a number line, thereby dividing the number line into **test intervals.**

4. Choose one representative number within each test interval. If substituting that value into the original inequality produces a true statement, then all real numbers in the test interval belong to the solution set. If substituting that value into the original inequality produces a false statement, then no real numbers in the test interval belong to the solution set.

5. Write the solution set, selecting the interval(s) that produced a true statement. The graph of the solution set on a number line usually appears as

   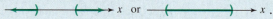

This procedure is valid if $<$ is replaced by $\leq$ and $>$ is replaced by $\geq$.

Study Tip

The five-step procedure for solving quadratic inequalities does not require graphing equations of the form $y = ax^2 + bx + c$. As we have done throughout the chapter, we'll show you these cuplike U-shaped graphs to enhance your visual understanding of solution sets.

EXAMPLE 1   **Solving a Quadratic Inequality**

Solve and graph the solution set on a real number line: $x^2 - 7x + 10 < 0$.

Solution

Step 1   Write the inequality in general form.   The inequality is given in this form, so this step has been done for us.

Step 2   Solve the related quadratic equation.   This equation is obtained by replacing the inequality sign by an equal sign. Thus, we will solve $x^2 - 7x + 10 = 0$.

$$x^2 - 7x + 10 = 0 \quad \text{\color{blue}This is the related quadratic equation.}$$

$$(x - 2)(x - 5) = 0 \quad \text{\color{blue}Factor.}$$

$$x - 2 = 0 \quad \text{or} \quad x - 5 = 0 \quad \text{\color{blue}Set each factor equal to 0.}$$

$$x = 2 \quad \text{or} \quad x = 5 \quad \text{\color{blue}Solve for x.}$$

The boundary points are 2 and 5.

Step 3   Locate the boundary points on a number line.   The number line with the boundary points is shown as follows:

The boundary points divide the number line into three test intervals, namely $(-\infty, 2)$, $(2, 5)$, and $(5, \infty)$.

Step 4 **Take one representative number within each test interval and substitute that number into the original inequality.**

Test Interval	Representative Number	Substitute into $x^2 - 7x + 10 < 0$	Conclusion
$(-\infty, 2)$	0	$0^2 - 7 \cdot 0 + 10 \overset{?}{<} 0$ $10 < 0,$ False	$(-\infty, 2)$ does not belong to the solution set.
$(2, 5)$	3	$3^2 - 7 \cdot 3 + 10 \overset{?}{<} 0$ $9 - 21 + 10 \overset{?}{<} 0$ $-2 < 0,$ True	$(2, 5)$ belongs to the solution set.
$(5, \infty)$	6	$6^2 - 7 \cdot 6 + 10 \overset{?}{<} 0$ $36 - 42 + 10 \overset{?}{<} 0$ $4 < 0,$ False	$(5, \infty)$ does not belong to the solution set.

Step 5 **The solution set consists of the intervals that produce a true statement.** Our analysis shows that the solution set is the interval $(2, 5)$. The graph in Figure 1.22 confirms that $x^2 - 7x + 10 < 0$ (lies below the x-axis) in this interval. The graph of the solution set on a number line is shown as follows:

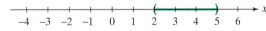

The graph lies below the x-axis between the boundary points 2 and 5, in the interval $(2, 5)$.

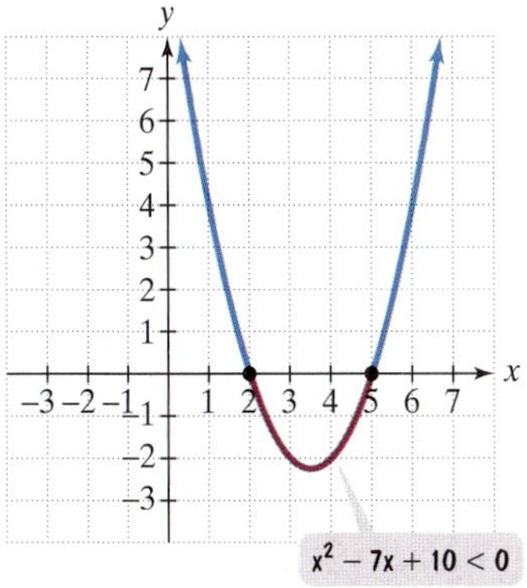

Figure 1.22 The graph lies below the x-axis between the boundary points 2 and 5, in the interval $(2, 5)$.

Check Point 1 Solve and graph the solution set on a real number line:

$$x^2 + 2x - 3 < 0.$$

EXAMPLE 2 Solving a Quadratic Inequality

Solve and graph the solution set: $2x^2 + x \geq 15$.

Solution

Step 1 **Write the inequality in general form.** We can write $2x^2 + x \geq 15$ in standard form by subtracting 15 from both sides. This will give us zero on the right.

$$2x^2 + x - 15 \geq 15 - 15$$

$$2x^2 + x - 15 \geq 0$$

Step 2 **Solve the related quadratic equation.** This equation is obtained by replacing the inequality sign by an equal sign. Thus, we will solve $2x^2 + x - 15 = 0$.

$2x^2 + x - 15 = 0$ This is the related quadratic equation.

$(2x - 5)(x + 3) = 0$ Factor.

$2x - 5 = 0$ or $x + 3 = 0$ Set each factor equal to 0.

$x = \frac{5}{2}$ or $x = -3$ Solve for x.

The boundary points are -3 and $\frac{5}{2}$.

Step 3   Locate the boundary points on a number line. The number line with the boundary points is shown as follows:

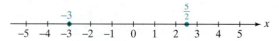

The boundary points divide the number line into three test intervals. Including the boundary points (because of the given greater than or *equal to* sign), the intervals are $(-\infty, -3]$, $[-3, \frac{5}{2}]$, and $[\frac{5}{2}, \infty)$.

Step 4   Take one representative number within each test interval and substitute that number into the original inequality.

Test Interval	Representative Number	Substitute into $2x^2 + x \geq 15$	Conclusion
$(-\infty, -3]$	-4	$2(-4)^2 + (-4) \overset{?}{\geq} 15$ $28 \geq 15$, True	$(-\infty, -3]$ belongs to the solution set.
$\left[-3, \dfrac{5}{2}\right]$	0	$2 \cdot 0^2 + 0 \overset{?}{\geq} 15$ $0 \geq 15$, False	$\left[-3, \dfrac{5}{2}\right]$ does not belong to the solution set.
$\left[\dfrac{5}{2}, \infty\right)$	3	$2 \cdot 3^2 + 3 \overset{?}{\geq} 15$ $21 \geq 15$, True	$\left[\dfrac{5}{2}, \infty\right)$ belongs to

Step 5   The solution set consists of the intervals that produce a true statement. Our analysis shows that the solution set is

$$(-\infty, -3] \text{ or } \left[\frac{5}{2}, \infty\right).$$

The graph of the solution set on a number line is shown as follows:

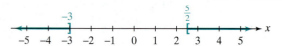

Check Point 2   Solve and graph the solution set: $x^2 - x \geq 20$.

Technology

The solution set for

$$2x^2 + x \geq 15$$

or, equivalently,

$$2x^2 + x - 15 \geq 0$$

can be verified with a graphing utility. The graph of $y = 2x^2 + x - 15$ was obtained using a $[-10, 10, 1]$ by $[-16, 6, 1]$ viewing rectangle. The graph lies above or on the x-axis, representing $\geq$, for all x in $(-\infty, -3]$ or $[\frac{5}{2}, \infty)$.

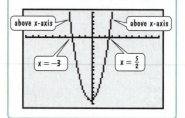

2 Solve rational inequalities.

Solving Rational Inequalities

Inequalities that involve quotients can be solved in the same manner as quadratic inequalities. For example, the inequalities

$$(x + 3)(x - 7) > 0 \quad \text{and} \quad \frac{x + 3}{x - 7} > 0$$

are similar in that both are positive under the same conditions. To be positive, each of these inequalities must have two positive linear expressions

$$x + 3 > 0 \quad \text{and} \quad x - 7 > 0$$

or two negative linear expressions

$$x + 3 < 0 \quad \text{and} \quad x - 7 < 0.$$

Consequently, we solve $\dfrac{x + 3}{x - 7} > 0$ using boundary points to divide the number line into test intervals. Then we select one representative number in each interval to determine whether that interval belongs to the solution set. Example 3 illustrates how this is done.

EXAMPLE 3 Using Test Numbers to Solve a Rational Inequality

Solve and graph the solution set: $\dfrac{x + 3}{x - 7} > 0$.

Solution We begin by finding values of x that make the numerator and denominator 0.

$$x + 3 = 0 \qquad x - 7 = 0 \qquad \textit{Set the numerator and denominator equal to 0.}$$

$$x = -3 \qquad\quad x = 7 \qquad \textit{Solve.}$$

The boundary points are -3 and 7. We locate these numbers on a number line as follows:

These boundary points divide the number line into three test intervals, namely $(-\infty, -3)$, $(-3, 7)$, and $(7, \infty)$. Now, we take one representative number from each test interval and substitute that number into the original inequality.

Test Interval	Representative Number	Substitute into $\dfrac{x + 3}{x - 7} > 0$	Conclusion
$(-\infty, -3)$	-4	$\dfrac{-4 + 3}{-4 - 7} \overset{?}{>} 0$ $\dfrac{-1}{-11} \overset{?}{>} 0$ $\dfrac{1}{11} > 0, \text{True}$	$(-\infty, -3)$ belongs to the solution set.
$(-3, 7)$	0	$\dfrac{0 + 3}{0 - 7} \overset{?}{>} 0$ $-\dfrac{3}{7} > 0, \text{False}$	$(-3, 7)$ does not belong to the solution set.
$(7, \infty)$	8	$\dfrac{8 + 3}{8 - 7} \overset{?}{>} 0$ $11 > 0, \text{True}$	$(7, \infty)$ belongs to the solution set.

Our analysis shows that the solution set is

$$(-\infty, -3) \quad \text{or} \quad (7, \infty).$$

The graph of the solution set on a number line is shown as follows:

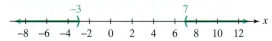

Check Point 3 Solve and graph the solution set: $\dfrac{x - 5}{x + 2} > 0$.

The first step in solving a rational inequality is to bring all terms to one side, obtaining zero on the other side. Then express the nonzero side as a single quotient. At this point, we follow the same procedure as in Example 3, finding values of the variable that make the numerator and denominator 0. These values serve as boundary points that separate the number line into intervals.

EXAMPLE 4 Solving a Rational Inequality

Solve and graph the solution set: $\dfrac{x + 1}{x + 3} \leq 2$.

Solution

Step 1 Express the inequality so that one side is zero and the other side is a single quotient. We subtract 2 from both sides to obtain zero on the right.

$$\frac{x + 1}{x + 3} \leq 2 \qquad \text{\textcolor{blue}{This is the given inequality.}}$$

$$\frac{x + 1}{x + 3} - 2 \leq 0 \qquad \text{\textcolor{blue}{Subtract 2 from both sides, obtaining 0 on the right.}}$$

$$\frac{x + 1}{x + 3} - \frac{2(x + 3)}{x + 3} \leq 0 \qquad \text{\textcolor{blue}{The least common denominator is x + 3. Express 2 in terms of this denominator.}}$$

$$\frac{x + 1 - 2(x + 3)}{x + 3} \leq 0 \qquad \text{\textcolor{blue}{Subtract rational expressions.}}$$

$$\frac{x + 1 - 2x - 6}{x + 3} \leq 0 \qquad \text{\textcolor{blue}{Apply the distributive property.}}$$

$$\frac{-x - 5}{x + 3} \leq 0 \qquad \text{\textcolor{blue}{Simplify.}}$$

Step 2 Find boundary points by setting the numerator and the denominator equal to zero.

$$-x - 5 = 0 \qquad x + 3 = 0 \qquad \text{\textcolor{blue}{Set the numerator and denominator equal to 0. These are the values that make the previous quotient zero or undefined.}}$$

$$x = -5 \qquad\quad x = -3 \qquad \text{\textcolor{blue}{Solve for x.}}$$

The boundary points are -5 and -3. Because equality is included in the given less-than-or-equal-to symbol, we include the value of x that causes the quotient $\dfrac{-x - 5}{x + 3}$ to be zero. Thus, -5 is included in the solution set. By contrast, we do not include -3 in the solution set because -3 makes the denominator zero.

Study Tip

Do not begin solving

$$\frac{x + 1}{x + 3} \leq 2$$

by multiplying both sides by $x + 3$. We do not know if $x + 3$ is positive or negative. Thus, we do not know whether or not to reverse the sense of the inequality.

Study Tip

Any values obtained by setting a denominator equal to zero should never be included in the solution set of a rational inequality. Division by zero is undefined.

Step 3   Locate boundary points on a number line.   The number line, with the boundary points, is shown as follows:

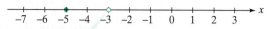

The open dot at −3 indicates −3 is not to be included in the solution set. We can't divide by zero.

The boundary points divide the number line into three test intervals, namely $(-\infty, -5]$, $[-5, -3)$, and $(-3, \infty)$.

Step 4   Take one representative number within each test interval and substitute that number into the original inequality.

Test Interval	Representative Number	Substitute into $\dfrac{x+1}{x+3} \le 2$	Conclusion
$(-\infty, -5]$	-6	$\dfrac{-6+1}{-6+3} \overset{?}{\le} 2$ $\dfrac{5}{3} \le 2$, True	$(-\infty, -5]$ belongs to the solution set.
$[-5, -3)$	-4	$\dfrac{-4+1}{-4+3} \overset{?}{\le} 2$ $3 \le 2$, False	$[-5, -3)$ does not belong to the solution set.
$(-3, \infty)$	0	$\dfrac{0+1}{0+3} \overset{?}{\le} 2$ $\dfrac{1}{3} \le 2$, True	$(-3, \infty)$ belongs to the solution set.

Discovery

Because $(x+3)^2$ is positive, it is possible so solve

$$\frac{x+1}{x+3} \le 2$$

by first multiplying both sides by $(x+3)^2$ (where $x \ne -3$). This will not reverse the sense of the inequality and will clear the fraction. Try using this solution method and compare it to the solution on pages 162–163.

Step 5   The solution set consists of the intervals that produce a true statement. Our analysis shows that the solution set is

$$(-\infty, -5] \quad \text{or} \quad (-3, \infty).$$

The graph of the solution set on a number line is shown as follows:

Check Point 4   Solve and graph the solution set: $\dfrac{2x}{x+1} \le 1$.

 3 Solve problems modeled by nonlinear inequalities.

Applications

We are surrounded by evidence that the world is profoundly mathematical. For example, did you know that every time you throw an object vertically upward, its changing height above the ground can be described by a mathematical formula? The same formula can be used to describe objects that are falling, such as the sky divers shown in the opening to this section.

> ### The Position Formula for a Free-Falling Object Near Earth's Surface
>
> An object that is falling or vertically projected into the air has its height above the ground, s, in feet, given by
>
> $$s = -16t^2 + v_0 t + s_0$$
>
> where v_0 is the original velocity (initial velocity) of the object, in feet per second, t is the time that the object is in motion, in seconds, and s_0 is the original height (initial height) of the object, in feet.

In Example 5, we solve a quadratic inequality in a problem about the position of a free-falling object.

$t = 0$
$s_0 = 176$
$v_0 = 96$

176 ft

EXAMPLE 5   Using the Position Model

A ball is thrown vertically upward from the top of the Leaning Tower of Pisa (176 feet high) with an initial velocity of 96 feet per second (Figure 1.23). During which time period will the ball's height exceed that of the tower?

Figure 1.23 Throwing a ball from 176 feet with a velocity of 96 feet per second

Solution

$s = -16t^2 + v_0 t + s_0$ This is the position formula for a free-falling object.

$s = -16t^2 + 96t + 176$ Because v_0 (initial velocity) $= 96$ and s_0 (initial position) $= 176$, substitute these values into the formula.

| When will the ball's height | exceed that | of the tower? |

$-16t^2 + 96t + 176 \quad > \quad 176$

$-16t^2 + 96t + 176 > 176$ This is the inequality implied by the problem's question. We must find t.

$-16t^2 + 96t > 0$ Subtract 176 from both sides.

$-16t^2 + 96t = 0$ Solve the related quadratic equation.

$-16t(t - 6) = 0$ Factor.

$-16t = 0 \quad \text{or} \quad t - 6 = 0$ Set each factor equal to 0.

$t = 0 \qquad\qquad t = 6$ Solve for t. The boundary points are 0 and 6.

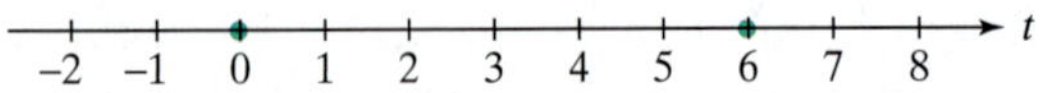

Locate these values on a number line, with $t \geq 0$.

The intervals are $(-\infty, 0)$, $(0, 6)$ and $(6, \infty)$. For our purposes, the mathematical model is useful only from $t = 0$ until the ball hits the ground. (By setting $-16t^2 + 96t + 176$ equal to zero, we find $t \approx 7.47$; the ball hits the ground after approximately 7.47 seconds.) Thus, we use $(0, 6)$ and $(6, 7.47)$ for our test intervals.

Technology

The graphs of
$$y_1 = -16x^2 + 96x + 176$$
and
$$y_2 = 176$$
are shown in a
$$[0, 8, 1] \text{ by } [0, 320, 32]$$

seconds in motion   height, in feet

viewing rectangle. The graphs show that the ball's height exceeds that of the tower between 0 and 6 seconds.

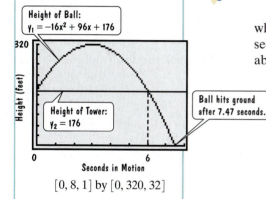

Height of Ball:
$y_1 = -16x^2 + 96x + 176$

320

Height (feet)

Height of Tower:
$y_2 = 176$

Ball hits ground after 7.47 seconds.

0   6
Seconds in Motion
$[0, 8, 1] \text{ by } [0, 320, 32]$

Test Interval	Representative Number	Substitute into $-16t^2 + 96t + 176 > 176$	Conclusion
$(0, 6)$	1	$-16 \cdot 1^2 + 96 \cdot 1 + 176 \overset{?}{>} 176$ $256 > 176$, True	$(0, 6)$ belongs to the solution set.
$(6, 7.47)$	7	$-16 \cdot 7^2 + 96 \cdot 7 + 176 \overset{?}{>} 176$ $64 > 176$, False	$(6, 7.47)$ does not belong to the solution set.

The ball's height exceeds that of the tower between 0 and 6 seconds, excluding $t = 0$ and $t = 6$.

Check Point 5
An object is propelled straight up from ground level with an initial velocity of 80 feet per second. Its height at time t is described by
$$s = -16t^2 + 80t$$
where the height, s, is measured in feet and the time, t, is measured in seconds. In which time interval will the object be more than 64 feet above the ground?

EXERCISE SET 1.8

Practice Exercises

Solve each quadratic inequality in Exercises 1–28, and graph the solution set on a real number line. Express each solution set in interval notation.

1. $(x - 4)(x + 2) > 0$
2. $(x + 3)(x - 5) > 0$
3. $(x - 7)(x + 3) \leq 0$
4. $(x + 1)(x - 7) \leq 0$
5. $x^2 - 5x + 4 > 0$
6. $x^2 - 4x + 3 < 0$
7. $x^2 + 5x + 4 > 0$
8. $x^2 + x - 6 > 0$
9. $x^2 - 6x + 9 < 0$
10. $x^2 - 2x + 1 > 0$
11. $x^2 - 6x + 8 \leq 0$
12. $x^2 - 2x - 3 \geq 0$
13. $3x^2 + 10x - 8 \leq 0$
14. $9x^2 + 3x - 2 \geq 0$
15. $2x^2 + x < 15$
16. $6x^2 + x > 1$
17. $4x^2 + 7x < -3$
18. $3x^2 + 16x < -5$
19. $5x \leq 2 - 3x^2$
20. $4x^2 + 1 \geq 4x$
21. $x^2 - 4x \geq 0$
22. $x^2 + 2x < 0$
23. $2x^2 + 3x > 0$
24. $3x^2 - 5x \leq 0$
25. $-x^2 + x \geq 0$
26. $-x^2 + 2x \geq 0$
27. $|x^2 + 2x - 36| > 12$
28. $|x^2 + 6x + 1| > 8$

Solve each rational inequality in Exercises 29–48, and graph the solution set on a real number line. Express each solution set in interval notation.

29. $\dfrac{x - 4}{x + 3} > 0$
30. $\dfrac{x + 5}{x - 2} > 0$
31. $\dfrac{x + 3}{x + 4} < 0$
32. $\dfrac{x + 5}{x + 2} < 0$

33. $\dfrac{-x + 2}{x - 4} \geq 0$

34. $\dfrac{-x - 3}{x + 2} \leq 0$

35. $\dfrac{4 - 2x}{3x + 4} \leq 0$

36. $\dfrac{3x + 5}{6 - 2x} \geq 0$

37. $\dfrac{x}{x - 3} > 0$

38. $\dfrac{x + 4}{x} > 0$

39. $\dfrac{x + 1}{x + 3} < 2$

40. $\dfrac{x}{x - 1} > 2$

41. $\dfrac{x + 4}{2x - 1} \leq 3$

42. $\dfrac{1}{x - 3} < 1$

43. $\dfrac{x - 2}{x + 2} \leq 2$

44. $\dfrac{x}{x + 2} \geq 2$

45. $\dfrac{3}{x + 3} > \dfrac{3}{x - 2}$

46. $\dfrac{1}{x + 1} > \dfrac{2}{x - 1}$

47. $\dfrac{x^2 - x - 2}{x^2 - 4x + 3} > 0$

48. $\dfrac{x^2 - 3x + 2}{x^2 - 2x - 3} > 0$

Application Exercises

Use the position formula

$$s = -16t^2 + v_0 t + s_0$$

$(v_0 = \text{initial velocity}, s_0 = \text{initial position}, t = \text{time})$

to answer Exercises 49–52. If necessary, round answers to the nearest hundredth of a second.

49. A projectile is fired straight upward from ground level with an initial velocity of 80 feet per second. During which interval of time will the projectile's height exceed 96 feet?

50. A projectile is fired straight upward from ground level with an initial velocity of 128 feet per second. During which interval of time will the projectile's height exceed 128 feet?

51. A ball is thrown vertically upward with a velocity of 64 feet per second from the top edge of a building 80 feet high. For how long is the ball higher than 96 feet?

52. A diver leaps into the air at 20 feet per second from a diving board that is 10 feet above the water. For how many seconds is the diver at least 12 feet above the water?

53. The formula

$$H = \frac{15}{8}x^2 - 30x + 200$$

models heart rate, H, in beats per minute, x minutes after a strenuous workout.

a. What is the heart rate immediately following the workout?

b. According to the model, during which intervals of time after the strenuous workout does the heart rate exceed 110 beats per minute? For which of these intervals has model breakdown occurred? Which interval provides a more realistic answer? How did you determine this?

The bar graph at the top of the next column shows the cost of Medicare, in billions of dollars, projected through 2005. The data can be modeled by

a linear model, $y = 27x + 163$;

a quadratic model, $y = 1.2x^2 + 15.2x + 181.4$.

In each formula, x represent the number of years after 1995 and y represents Medicare spending, in billions of dollars. Use these formulas to solve Exercises 54–56.

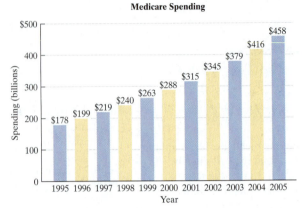

Medicare Spending

Source: Congressional Budget Office

54. The graph indicates that Medicare spending will reach $458 billion in 2005. Find the amount predicted by each of the formulas for that year. How well do the formulas model the value in the graph? Which formula serves as a better model for that year?

55. For which years does the quadratic model indicate that Medicare spending will exceed $536.6 billion?

56. For which years does the quadratic model indicate that Medicare spending will exceed $629.4 billion?

A company manufactures wheelchairs. The average cost, y, of producing x wheelchairs per month is given by

$$y = \frac{500,000 + 400x}{x}.$$

The graph of the formula is shown. Use the formula to solve Exercises 57–58.

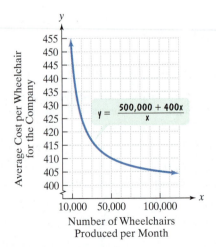

57. Describe the company's production level so that the average cost of producing each wheelchair does not exceed $425. Use a rational inequality to solve the problem. Then explain how your solution is shown on the graph.

58. Describe the company's production level so that the average cost of producing each wheelchair does not exceed $410. Use a rational inequality to solve the problem. Then explain how your solution is shown on the graph.

Writing in Mathematics

59. What is a quadratic inequality?

60. What is a rational inequality?

61. Describe similarities and differences between the solutions of

$$(x - 2)(x + 5) \geq 0 \quad \text{and} \quad \frac{x - 2}{x + 5} \geq 0.$$

Technology Exercises

Solve each inequality in Exercises 62–65 using a graphing utility.

62. $x^2 + 3x - 10 > 0$ **63.** $2x^2 + 5x - 3 \leq 0$

64. $x^3 + x^2 - 4x - 4 > 0$ **65.** $\dfrac{x - 4}{x - 1} \leq 0$

Critical Thinking Exercises

66. Which one of the following is true?

   a. The solution set of $x^2 > 25$ is $(5, \infty)$.

   b. The inequality $\dfrac{x - 2}{x + 3} < 2$ can be solved by multiplying both sides by $x + 3$, resulting in the equivalent inequality $x - 2 < 2(x + 3)$.

   c. $(x + 3)(x - 1) \geq 0$ and $\dfrac{x + 3}{x - 1} \geq 0$ have the same solution set.

   d. None of these statements is true.

67. Write a quadratic inequality whose solution set is $[-3, 5]$.

68. Write a rational inequality whose solution set is $(-\infty, -4)$ or $[3, \infty)$.

In Exercises 69–72, use inspection to describe each inequality's solution set. Do not solve any of the inequalities.

69. $(x - 2)^2 > 0$ **70.** $(x - 2)^2 \leq 0$

71. $(x - 2)^2 < -1$ **72.** $\dfrac{1}{(x - 2)^2} > 0$

In Exercises 73–74, use the method for solving quadratic inequalities to solve each higher-order polynomial inequality.

73. $x^3 + x^2 - 4x - 4 > 0$

74. $x^3 + 2x^2 - x - 2 \geq 0$

75. The graphing utility screen shows the graph of $y = 4x^2 - 8x + 7$.

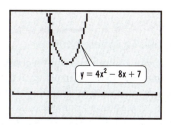

   a. Use the graph to describe the solution set of $4x^2 - 8x + 7 > 0$.

   b. Use the graph to describe the solution set of $4x^2 - 8x + 7 < 0$.

   c. Use an algebraic approach to verify each of your descriptions in parts (a) and (b).

76. The graphing utility screen shows the graph of $y = \sqrt{27 - 3x^2}$. Write and solve a quadratic inequality that explains why the graph only appears for $-3 \leq x \leq 3$.

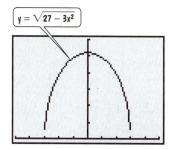

Group Exercise

77. This exercise is intended as a group learning experience and is appropriate for groups of three to five people. Before working on the various parts of the problem, reread the description of the position formula on page 164.

   a. Drop a ball from a height of 3 feet, 6 feet, and 12 feet. Record the number of seconds it takes for the ball to hit the ground.

   b. For each of the three initial positions, use the position formula to determine the time required for the ball to hit the ground.

   c. What factors might result in differences between the times that you recorded and the times indicated by the formula?

   d. What appears to be happening to the time required for a free-falling object to hit the ground as its initial height is doubled? Verify this observation algebraically and with a graphing utility.

   e. Repeat part (a) using a sheet of paper rather than a ball. What differences do you observe? What factor seems to be ignored in the position formula?

   f. What is meant by the acceleration of gravity and how does this number appear in the position formula for a free-falling object?

CHAPTER SUMMARY, REVIEW, AND TEST

Summary

DEFINITIONS AND CONCEPTS	EXAMPLES

1.1 Graphs and Graphing Utilities

a. The rectangular coordinate system consists of a horizontal number line, the x-axis, and a vertical number line, the y-axis, intersecting at their zero points, the origin. Each point in the system corresponds to an ordered pair of real numbers (x, y). The first number in the pair is the x-coordinate; the second number is the y-coordinate. See Figure 1.1 on page 76.

Ex. 1, p. 77

b. An ordered pair is a solution of an equation in two variables if replacing the variables by the corresponding coordinates results in a true statement. The ordered pair is said to satisfy the equation. The graph of the equation is the set of all points whose coordinates satisfy the equation. One method for graphing an equation is to plot ordered-pair solutions and connect them with a smooth curve or line.

Ex. 2, p. 78

c. An x-intercept of a graph is the x-coordinate of a point where the graph intersects the x-axis. The y-coordinate corresponding to a graph's x-intercept is always zero.

d. A y-intercept of a graph is the y-coordinate of a point where the graph intersects the y-axis. The x-coordinate corresponding to a graph's y-intercept is always zero.

1.2 Linear Equations

a. A linear equation in one variable x can be written in the form $ax + b = 0, a \neq 0$.

b. The procedure for solving a linear equation is given in the box on page 86.

Ex. 1, p. 86;
Ex. 2, p. 87

c. If an equation contains fractions, begin by multiplying both sides by the least common denominator, thereby clearing fractions.

Ex. 3, p. 88

d. If an equation contains rational expressions with variable denominators, avoid in the solution set any values of the variable that make a denominator zero.

Ex. 4, p. 89;
Ex. 5, p. 90

e. An identity is an equation that is true for all real numbers for which both sides are defined. A conditional equation is not an identity and is true for at least one real number. An inconsistent equation is an equation that is not true for even one real number.

Ex. 6, p. 91

1.3 Formulas and Applications

a. A formula is an equation that uses letters to express a relationship between two or more variables.

Ex. 1, p. 96

b. Mathematical modeling is the process of finding equations and formulas to describe real-world phenomena. Such equations and formulas, together with the meaning assigned to the variables, are called mathematical models. Mathematical models can be formed from verbal models or from actual data.

c. A five-step procedure for solving problems using mathematical models is given in the box on page 97.

Ex. 2, p. 97;
Ex. 3, p. 98;
Ex. 4, p. 100;
Ex. 5, p. 101

1.4 Complex Numbers

a. The imaginary unit i is defined as

$$i = \sqrt{-1}, \text{ where } i^2 = -1.$$

The set of numbers in the form $a + bi$ is called the set of complex numbers; a is the real part and b is the imaginary part. If $b = 0$, the complex number is a real number. If $b \neq 0$, the complex number is an imaginary number. Complex numbers in the form bi are called pure imaginary numbers.

b. Rules for adding and subtracting complex numbers are given in the box on page 109.

Ex. 1, p. 110

c. To multiply complex numbers, multiply as if they are polynomials. After completing the multiplication, replace i^2 with -1.

Ex. 2, p. 110

DEFINITIONS AND CONCEPTS	EXAMPLES

d. The complex conjugate of $a + bi$ is $a - bi$ and vice versa. The multiplication of complex conjugates gives a real number:

$$(a + bi)(a - bi) = a^2 + b^2.$$

e. To divide complex numbers, multiply the numerator and the denominator by the complex conjugate of the denominator.

Ex. 3, p. 111

f. When performing operations with square roots of negative numbers, begin by expressing all square roots in terms of i. The principal square root of $-b$ is defined by

$$\sqrt{-b} = i\sqrt{b}.$$

Ex. 4, p. 112

1.5 Quadratic Equations

a. A quadratic equation in x can be written in the general form $ax^2 + bx + c = 0, a \neq 0$.

b. The procedure for solving a quadratic equation by factoring and the zero-product principle is given in the box on pages 115–116.

Ex. 1, p. 116

c. The procedure for solving a quadratic equation by the square root method is given in the box on page 118.

Ex. 2, p. 118

d. All quadratic equations can be solved by completing the square. Isolate the binomial with the two variable terms on one side of the equation. If the coefficient of the x^2-term is not one, divide each side of the equation by this coefficient. Then add the square of half the coefficient of x to both sides.

Ex. 4, p. 119

e. All quadratic equations can be solved by the quadratic formula

$$x = \frac{-b \pm \sqrt{b^2 - 4ac}}{2a}.$$

Ex. 5, p. 122;
Ex. 6, p. 123

The formula is derived by completing the square of the equation $ax^2 + bx + c = 0$.

f. The discriminant, $b^2 - 4ac$, indicates the number and type of solutions to the quadratic equation $ax^2 + bx + c = 0$, shown in Table 1.3 on page 124.

Ex. 7, p. 124

g. Table 1.4 on page 125 shows the most efficient technique to use when solving a quadratic equation.

Ex. 8, p. 126;
Ex. 9, p. 127

1.6 Other Types of Equations

a. Some polynomial equations of degree 3 or greater can be solved by moving all terms to one side, obtaining zero on the other side, factoring, and using the zero-product principle. Factoring by grouping is often used.

Ex. 1, p. 132;
Ex. 2, p. 133

b. A radical equation is an equation in which the variable occurs in a square root, cube root, and so on. A radical equation can be solved by isolating the radical and raising both sides of the equation to a power equal to the radicals index. When raising both sides to an even power, check all proposed solutions in the original equation. Eliminate extraneous solutions from the solution set.

Ex. 3, p. 134;
Ex. 4, p. 135

c. A radical equation with rational exponents can be solved by isolating the expression with the rational exponent and raising both sides of the equation to a power that is the reciprocal of the rational exponent. See the details in the box on page 137.

Ex. 5, p. 137

d. An equation is quadratic in form if it can be written in the form $at^2 + bt + c = 0$, where t is an algebraic expression and $a \neq 0$. Solve for t and use the substitution that resulted in this equation to find the values for the variable in the given equation.

Ex. 6, p. 139;
Ex. 7, p. 139

e. Absolute value equations in the form $|X| = c, c > 0$, can be solved by rewriting the equation without absolute value bars: $X = c$ or $X = -c$.

Ex. 8, p. 140

DEFINITIONS AND CONCEPTS	EXAMPLES

1.7 Linear Inequalities

a. A linear inequality in one variable x can be expressed as
$ax + b \leq c$, $ax + b < c$, $ax + b \geq c$, or $ax + b > c$, $a \neq 0$.

Ex. 1, p. 144

b. Graphs of solutions to inequalities are shown on a number line by shading all points representing numbers that are solutions. Parentheses exclude endpoints and square brackets include endpoints.

Ex. 2, p. 146

c. Solution sets of inequalities can be expressed in set-builder or interval notation. Table 1.5 on page 146 compares the notations.

Ex. 3, p. 148;
Ex. 4, p. 149

d. A linear inequality is solved using a procedure similar to solving a linear equation. However, when multiplying or dividing by a negative number, reverse the sense of the inequality.

Ex. 5, p. 150

e. A compound inequality with three parts can be solved by isolating x in the middle.

f. Inequalities involving absolute value can be solved by rewriting the inequalities without absolute value bars. The ways to do this are shown in the box on page 150.

Ex. 6, p. 151;
Ex. 7, p. 151

1.8 Quadratic and Rational Inequalities

a. A quadratic inequality can be expressed as
$ax^2 + bx + c < 0$, $ax^2 + bx + c > 0$,
$ax^2 + bx + c \leq 0$, or $ax^2 + bx + c \geq 0$, $a \neq 0$.

Ex. 1, p. 158;

b. A procedure for solving quadratic inequalities is given in the box on page 158.

Ex. 2, p. 159

c. Inequalities involving quotients are called rational inequalities. The procedure for solving such inequalities begins with expressing them so that one side is zero and the other side is a single quotient. Find boundary points by setting the numerator and denominator equal to zero. Then follow a procedure similar to that for solving quadratic inequalities.

Ex. 3, p. 161;
Ex. 4, p. 162

Review Exercises

1.1

Graph each equation in Exercises 1–4.
Let $x = -3, -2, -1, 0, 1, 2,$ and 3.

1. $y = 2x - 2$ **2.** $y = x^2 - 3$
3. $y = x$ **4.** $y = |x| - 2$

5. What does a $[-20, 40, 10]$ by $[-5, 5, 1]$ viewing rectangle mean? Draw axes with tick marks and label the tick marks to illustrate this viewing rectangle.

In Exercises 6–8, use the graph and determine the x-intercepts, if any, and the y-intercepts, if any. For each graph, tick marks along the axes represent one unit each.

6.
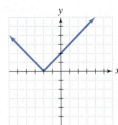

7.
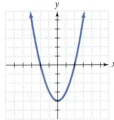

8.
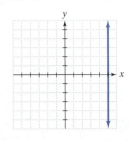

The caseload of Alzheimer's disease in the United States is expected to explode as baby boomers head into their later years. The graph shows the percentage of Americans with the disease, by age. Use the graph to solve Exercises 9–11.

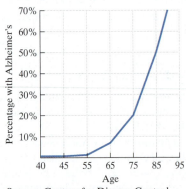

Alzheimer's Prevalence in the U.S., by Age

Source: Centers for Disease Control

9. What percentage of Americans who are 75 have Alzheimer's disease?

10. What age represents 50% prevalence of Alzheimer's disease?

11. Describe the trend shown by the graph.

1.2

In Exercises 12–17, solve and check each linear equation.

12. $2x - 5 = 7$ **13.** $5x + 20 = 3x$
14. $7(x - 4) = x + 2$ **15.** $1 - 2(6 - x) = 3x + 2$

16. $2(x - 4) + 3(x + 5) = 2x - 2$

17. $2x - 4(5x + 1) = 3x + 17$

Exercises 18–22 contain equations with constants in denominators. Solve each equation and check by the method of your choice.

18. $\dfrac{2x}{3} = \dfrac{x}{6} + 1$

19. $\dfrac{x}{2} - \dfrac{1}{10} = \dfrac{x}{5} + \dfrac{1}{2}$

20. $\dfrac{2x}{3} = 6 - \dfrac{x}{4}$

21. $\dfrac{x}{4} = 2 + \dfrac{x - 3}{3}$

22. $\dfrac{3x + 1}{3} - \dfrac{13}{2} = \dfrac{1 - x}{4}$

*Exercises 23–26 contain equations with variables in denominators. **a.** List the value or values representing restriction(s) on the variable. **b.** Solve the equation.*

23. $\dfrac{9}{4} - \dfrac{1}{2x} = \dfrac{4}{x}$

24. $\dfrac{7}{x - 5} + 2 = \dfrac{x + 2}{x - 5}$

25. $\dfrac{1}{x - 1} - \dfrac{1}{x + 1} = \dfrac{2}{x^2 - 1}$

26. $\dfrac{4}{x + 2} + \dfrac{2}{x - 4} = \dfrac{30}{x^2 - 2x - 8}$

In Exercises 27–29, determine whether each equation is an identity, a conditional equation, or an inconsistent equation.

27. $\dfrac{1}{x + 5} = 0$

28. $7x + 13 = 4x - 10 + 3x + 23$

29. $7x + 13 = 3x - 10 + 2x + 23$

1.3

30. The percentage, P, of U.S. adults who read the daily newspaper can be modeled by the formula

$$P = -0.7x + 80$$

where x is the number of years after 1965. In which year will 52% of U.S. adults read the daily newspaper?

31. Suppose you were to list in order, from least to most, the family income for every U.S. family. The median income is the income in the middle of this list of ranked data. This income can be modeled by the formula

$$I = 1321.7(x - 1980) + 21{,}153.$$

In this formula, I represents median family income in the United States and x is the actual year, beginning in 1980. When was the median income $47,587?

In Exercises 32–39, use the five-step strategy given in the box on page 97 to solve each problem.

32. The cost of raising a child through the age of 17 varies by income group. The cost in middle-income families exceeds that of low-income families by $63 thousand, and the cost of high-income families is $3 thousand less than twice that of low-income families. Three children, one in a low-income family, one in a middle-income family, and one in a high-income family, will cost a total of $756 thousand to raise through the age of 17. Find the cost of raising a child in

each of the three income groups. (*Source: The World Almanac;* low annual income is less than $36,800, middle is $36,800–$61,900, and high exceeds $61,900.)

33. In 2000, the average weekly salary for workers in the United States was $567. If this amount is increasing by $15 yearly, in how many years after 2000 will the average salary reach $702. In which year will that be?

34. You are choosing between two long-distance telephone plans. One plan has a monthly fee of $15 with a charge of $0.05 per minute. The other plan has a monthly fee of $5 with a charge of $0.07 per minute. For how many minutes of long-distance calls will the costs for the two plans be the same?

35. You inherit $10,000 with the stipulation that for the first year the money must be placed in two investments paying 8% and 12% annual interest, respectively. How much should be invested at each rate if the total interest earned for the year is to be $950?

36. The length of a rectangular football field is 14 meters more than twice the width. If the perimeter is 346 meters, find the field's dimensions.

37. The bus fare in a city is $1.50. People who use the bus have the option of purchasing a monthly coupon book for $25.00. With the coupon book, the fare is reduced to $0.25. Determine the number of times in a month the bus must be used so that the total monthly cost without the coupon book is the same as the total monthly cost with the coupon book.

38. A salesperson earns $300 per week plus 5% commission of sales. How much must be sold to earn $800 in a week?

39. A study entitled *Performing Arts—The Economic Dilemma* documents the relationship between the number of concerts given by a major orchestra and the attendance per concert. For each additional concert given per year, attendance per concert drops by approximately eight people. If 50 concerts are given, attendance per concert is 2987 people. How many concerts should be given to ensure an audience of 2627 people at each concert?

In Exercises 40–42, solve each formula for the specified variable.

40. $V = \dfrac{1}{3} Bh$ for h

41. $F = f(1 - M)$ for M

42. $T = gr + gvt$ for g

1.4

In Exercises 43–52, perform the indicated operations and write the result in standard form.

43. $(8 - 3i) - (17 - 7i)$

44. $4i(3i - 2)$

45. $(7 - 5i)(2 + 3i)$

46. $(3 - 4i)^2$

47. $(7 + 8i)(7 - 8i)$

48. $\dfrac{6}{5 + i}$

49. $\dfrac{3 + 4i}{4 - 2i}$

50. $\sqrt{-32} - \sqrt{-18}$

51. $(-2 + \sqrt{-100})^2$

52. $\dfrac{4 + \sqrt{-8}}{2}$

1.5

Solve each equation in Exercises 53–54 by factoring.

53. $2x^2 + 15x = 8$ **54.** $5x^2 + 20x = 0$

Solve each equation in Exercises 55–56 by the square root method.

55. $2x^2 - 3 = 125$ **56.** $(3x - 4)^2 = 18$

In Exercises 57–58, determine the constant that should be added to the binomial so that it becomes a perfect square trinomial. Then write and factor the trinomial.

57. $x^2 + 20x$ **58.** $x^2 - 3x$

Solve each equation in Exercises 59–60 by completing the square.

59. $x^2 - 12x + 27 = 0$ **60.** $3x^2 - 12x + 11 = 0$

Solve each equation in Exercises 61–63 using the quadratic formula.

61. $x^2 = 2x + 4$ **62.** $x^2 - 2x + 19 = 0$
63. $2x^2 = 3 - 4x$

Compute the discriminant of each equation in Exercises 64–65. What does the discriminant indicate about the number and type of solutions?

64. $x^2 - 4x + 13 = 0$ **65.** $9x^2 = 2 - 3x$

Solve each equation in Exercises 66–71 by the method of your choice.

66. $2x^2 - 11x + 5 = 0$ **67.** $(3x + 5)(x - 3) = 5$
68. $3x^2 - 7x + 1 = 0$ **69.** $x^2 - 9 = 0$
70. $(x - 3)^2 - 25 = 0$ **71.** $3x^2 - x + 2 = 0$

72. The weight of a human fetus is modeled by the formula $W = 3t^2$, where W is the weight, in grams, and t is the time, in weeks, $0 \le t \le 39$. After how many weeks does the fetus weigh 1200 grams?

73. The alligator, an endangered species, is the subject of a protection program. The formula
$$P = -10x^2 + 475x + 3500$$
models the alligator population, P, after x years of the protection program, where $0 \le x \le 12$. After how many years is the population up to 7250?

74. The graph of the alligator population described in Exercise 73 is shown over time. Identify your solution in Exercise 73 as a point on the graph.

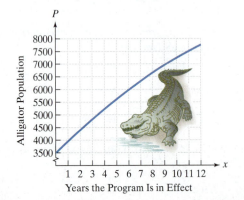

Years the Program Is in Effect

75. An architect is allowed 15 square yards of floor space to add a small bedroom to a house. Because of the room's design in relationship to the existing structure, the width of the rectangular floor must be 7 yards less than two times the length. Find the length and width of the rectangular floor that the architect is permitted.

76. A building casts a shadow that is double the length of its height. If the distance from the end of the shadow to the top of the building is 300 meters, how high is the building? Round to the nearest meter.

1.6

Solve each polynomial equation in Exercises 77–78.

77. $2x^4 = 50x^2$ **78.** $2x^3 - x^2 - 18x + 9 = 0$

Solve each radical equation in Exercises 79–80.

79. $\sqrt{2x - 3} + x = 3$ **80.** $\sqrt{x - 4} + \sqrt{x + 1} = 5$

Solve the equations with rational exponents in Exercises 81–82.

81. $3x^{3/4} - 24 = 0$ **82.** $(x - 7)^{2/3} = 25$

Solve each equation in Exercises 83–84 by making an appropriate substitution.

83. $x^4 - 5x^2 + 4 = 0$ **84.** $x^{1/2} + 3x^{1/4} - 10 = 0$

Solve the equations containing absolute value in Exercises 85–86.

85. $|2x + 1| = 7$ **86.** $2|x - 3| - 6 = 10$

Solve each equation in Exercises 87–90 by the method of your choice.

87. $3x^{4/3} - 5x^{2/3} + 2 = 0$ **88.** $2\sqrt{x - 1} = x$
89. $|2x - 5| - 3 = 0$ **90.** $x^3 + 2x^2 = 9x + 18$

91. The distance to the horizon that you can see, D, in miles, from the top of a mountain H feet high is modeled by the formula $D = \sqrt{2H}$. You've hiked to the top of a mountain with views extending 50 miles to the horizon. How high is the mountain?

1.7

In Exercises 92–94, graph the solutions of each inequality on a number line.

92. $x > 5$ **93.** $x \le 1$ **94.** $-3 \le x < 0$

In Exercises 95–97, express each interval in terms of an inequality, and graph the interval on a number line.

95. $(-2, 3]$ **96.** $[-1.5, 2]$ **97.** $(-1, \infty)$

Solve each linear inequality in Exercises 98–103 and graph the solution set on a number line. Express each solution set in interval notation.

98. $-6x + 3 \le 15$ **99.** $6x - 9 \ge -4x - 3$
100. $\dfrac{x}{3} - \dfrac{3}{4} - 1 > \dfrac{x}{2}$ **101.** $6x + 5 > -2(x - 3) - 25$
102. $3(2x - 1) - 2(x - 4) \ge 7 + 2(3 + 4x)$
103. $7 < 2x + 3 \le 9$

Solve each inequality in Exercises 104–106 by first rewriting each one as an equivalent inequality without absolute value bars. Graph the solution set on a number line. Express each solution set in interval notation.

104. $|2x + 3| \le 15$

105. $\left|\dfrac{2x + 6}{3}\right| > 2$

106. $|2x + 5| - 7 \ge -6$

107. Approximately 90% of the population sleeps h hours daily, where h is modeled by the inequality $|h - 6.5| \le 1$. Write a sentence describing the range for the number of hours that most people sleep. Do *not* use the phrase "absolute value" in your description.

108. The formula for converting Fahrenheit temperature, F, to Celsius temperature, C, is $C = \frac{5}{9}(F - 32)$. If Celsius temperature ranges from $10°$ to $25°$, inclusive, what is the range for the Fahrenheit temperature?

109. A person can choose between two charges on a checking account. The first method involves a fixed cost of $11 per month plus 6¢ for each check written. The second method involves a fixed cost of $4 per month plus 20¢ for each check written. How many checks should be written to make the first method a better deal?

110. A student has grades on three examinations of 75, 80, and 72. What must the student earn on a fourth examination in order to have an average of at least 80?

1.8

Solve each quadratic inequality in Exercises 111–112, and graph the solution set on a real number line. Express each solution set in interval notation.

111. $2x^2 + 7x \le 4$

112. $2x^2 > 6x - 3$

Solve each rational inequality in Exercises 113–114, and graph the solution set on a real number line. Express each solution set in interval notation.

113. $\dfrac{x - 6}{x + 2} > 0$

114. $\dfrac{x + 3}{x - 4} \le 5$

115. Use the position formula

$$s = -16t^2 + v_0 t + s_0$$

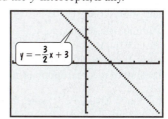

initial velocity   initial height

to solve this problem. A projectile is fired vertically upward from ground level with an initial velocity of 48 feet per second. During which time period will the projectile's height exceed 32 feet?

Chapter 1 Test

1. Graph $y = x^2 - 4$ by letting x equal integers from -3 through 3.

2. The graph of $y = -\frac{3}{2}x + 3$ is shown in a $[-6, 6, 1]$ by $[-6, 6, 1]$ viewing rectangle. Determine the x-intercepts, if any, and the y-intercepts, if any.

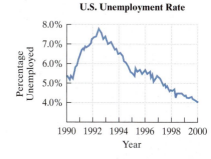

$y = -\frac{3}{2}x + 3$

3. The graph shows the unemployment rate in the United States from 1990 through 2000. For the period shown, during which year did the unemployment rate reach a maximum? Estimate the percentage of the work force unemployed, to the nearest tenth of a percent, at that time.

U.S. Unemployment Rate

Percentage Unemployed

8.0%
7.0%
6.0%
5.0%
4.0%

1990 1992 1994 1996 1998 2000
Year

Source: Bureau of Labor Statistics

Find the solution set for each equation in Exercises 4–16.

4. $7(x - 2) = 4(x + 1) - 21$

5. $\dfrac{2x - 3}{4} = \dfrac{x - 4}{2} - \dfrac{x + 1}{4}$

6. $\dfrac{2}{x - 3} - \dfrac{4}{x + 3} = \dfrac{8}{x^2 - 9}$

7. $2x^2 - 3x - 2 = 0$

8. $(3x - 1)^2 = 75$

9. $x(x - 2) = 4$

10. $4x^2 = 8x - 5$

11. $x^3 - 4x^2 - x + 4 = 0$

12. $\sqrt{x - 3} + 5 = x$

13. $\sqrt{x + 4} + \sqrt{x - 1} = 5$

14. $5x^{3/2} - 10 = 0$

15. $x^{2/3} - 9x^{1/3} + 8 = 0$

16. $\left|\dfrac{2}{3}x - 6\right| = 2$

Solve each inequality in Exercises 17–22. Express the answer in interval notation and graph the solution set on a number line.

17. $3(x + 4) \ge 5x - 12$

18. $\dfrac{x}{6} + \dfrac{1}{8} \le \dfrac{x}{2} - \dfrac{3}{4}$

19. $-3 \le \dfrac{2x + 5}{3} < 6$

20. $|3x + 2| \ge 3$

21. $x^2 < x + 12$

22. $\dfrac{2x + 1}{x - 3} > 3$

In Exercises 23–25, perform the indicated operations and write the result in standard form.

23. $(6 - 7i)(2 + 5i)$

24. $\dfrac{5}{2 - i}$

25. $2\sqrt{-49} + 3\sqrt{-64}$

In Exercises 26–27, solve each formula for the specified variable.

26. $V = \dfrac{1}{3}lwh$ for h

27. $y - y_1 = m(x - x_1)$ for x

The male minority? The graphs show enrollment in U.S. colleges, with projections from 2000 to 2009. The trend indicated by the graphs is among the hottest topics of debate among college-admission officers. Some private liberal arts colleges have quietly begun special efforts to recruit men— including admissions preferences for them.

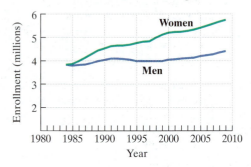

Enrollment in U.S. Colleges

Source: U.S. Department of Education

Exercises 28–29 are based on the data shown by the graphs.

28. The data for the men can be modeled by the formula

$$N = 0.01x + 3.9$$

where N represents enrollment, in millions, x years after 1984. According to the formula, when will the projected enrollment for men be 4.1 million? How well does the formula describe enrollment for that year shown by the line graph?

29. The data for the women can be modeled from the following verbal description:

> In 1984, 4.1 million women were enrolled. Female enrollment has increased by 0.07 million per year since then.

According to the verbal model, when will the projected enrollment for women be 5.71 million? How well does the verbal model describe enrollment for that year shown by the line graph?

30. On average, the number of unhealthy air days per year in Los Angeles exceeds three times that of New York City by 48 days. If Los Angeles and New York City combined have 268 unhealthy air days per year, determine the number of unhealthy days for the two cities. (*Source:* U.S. Environmental Protection Agency)

31. The costs for two different kinds of heating systems for a three-bedroom home are given in the following table. After how many years will total costs for solar heating and electric heating be the same? What will be the cost at that time?

System	Cost to Install	Operating Cost/Year
Solar	$29,700	$150
Electric	$5000	$1100

32. You placed $10,000 in two investments paying 8% and 10% annual interest, respectively. At the end of the year, the total interest from these investments was $940. How much was invested at each rate?

33. The length of a rectangular carpet is 4 feet greater than twice its width. If the area is 48 square feet, find the carpet's length and width.

34. A vertical pole is to be supported by a wire that is 26 feet long and anchored 24 feet from the base of the pole. How far up the pole should the wire be attached?

35. You take a summer job selling medical supplies. You are paid $600 per month plus 4% of the sales price of all the supplies you sell. If you want to earn more than $2500 per month, what value of medical supplies must you sell?

Functions and Graphs

Chapter 2

The cost of mailing a package depends on its weight. The probability that you and another person in a room share the same birthday depends on the number of people in the room. In both these situations, the relationship between variables can be described by a *function*. Understanding this concept will give you a new perspective on many ordinary situations.

'Tis the season and you've waited until the last minute to mail your holiday gifts. Your only option is overnight express mail. You realize that the cost of mailing a gift depends on its weight, but the mailing costs seem somewhat odd. Your packages that weigh 1.1 pounds, 1.5 pounds, and 2 pounds cost $15.75 each to send overnight. Packages that weigh 2.01 pounds and 3 pounds cost you $18.50 each. Finally, your heaviest gift is barely over 3 pounds and its mailing cost is $21.25. What sort of system is this in which costs increase by $2.75, stepping from $15.75 to $18.50 and from $18.50 to $21.25?

SECTION 2.1 Lines and Slope

Objectives

1. Compute a line's slope.
2. Write the point-slope equation of a line.
3. Write and graph the slope-intercept equation of a line.
4. Recognize equations of horizontal and vertical lines.
5. Recognize and use the general form of a line's equation.
6. Find slopes and equations of parallel and perpendicular lines.
7. Model data with linear equations.

Is there a relationship between literacy and child mortality? As the percentage of adult females who are literate increases, does the mortality of children under five decrease? Figure 2.1, based on data from the United Nations, indicates that this is, indeed, the case. Each point in the figure represents one country.

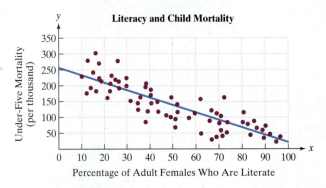

Figure 2.1

Source: United Nations

Data presented in a visual form as a set of points is called a **scatter plot.** Also shown in Figure 2.1 is a line that passes through or near the points. A line that best fits the data points in a scatter plot is called a **regression line.** By writing the equation of this line, we can obtain a model of the data and make predictions about child mortality based on the percentage of adult females in a country who are literate.

Data often fall on or near a line. In this section we will use equations to model such data and make predictions. We begin with a discussion of a line's steepness.

 Calculate a line's slope.

The Slope of a Line

Mathematicians have developed a useful measure of the steepness of a line, called the *slope* of the line. Slope compares the vertical change (the **rise**) to the horizontal change (the **run**) when moving from one fixed point to another along the line. To calculate the slope of a line, we use a ratio that compares the change in y (the rise) to the corresponding change in x (the run).

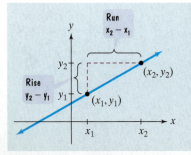

Definition of Slope

The **slope** of the line through the distinct points (x_1, y_1) and (x_2, y_2) is

$$\frac{\text{Change in } y}{\text{Change in } x} = \frac{\text{Rise}}{\text{Run}}$$

$$= \frac{y_2 - y_1}{x_2 - x_1}$$

where $x_2 - x_1 \neq 0$.

Slope and the Streets of San Francisco

San Francisco's Filbert Street has a slope of 0.613, meaning that for every horizontal distance of 100 feet, the street ascends 61.3 feet vertically. With its 31.5° angle of inclination, the street is too steep to pave and is only accessible by wooden stairs.

It is common notation to let the letter m represent the slope of a line. The letter m is used because it is the first letter of the French verb *monter*, meaning to rise, or to ascend.

EXAMPLE 1 Using the Definition of Slope

Find the slope of the line passing through each pair of points:

a. $(-3, -1)$ and $(-2, 4)$ **b.** $(-3, 4)$ and $(2, -2)$.

Solution

a. Let $(x_1, y_1) = (-3, -1)$ and $(x_2, y_2) = (-2, 4)$. We obtain a slope of

$$m = \frac{\text{Change in } y}{\text{Change in } x} = \frac{y_2 - y_1}{x_2 - x_1} = \frac{4 - (-1)}{-2 - (-3)} = \frac{5}{1} = 5.$$

The situation is illustrated in Figure 2.2(a). The slope of the line is 5, indicating that there is a vertical change, a rise, of 5 units for each horizontal change, a run, of 1 unit. The slope is positive, and the line rises from left to right.

Study Tip

When computing slope, it makes no difference which point you call (x_1, y_1) and which point you call (x_2, y_2). If we let $(x_1, y_1) = (-2, 4)$ and $(x_2, y_2) = (-3, -1)$, the slope is still 5:

$$m = \frac{y_2 - y_1}{x_2 - x_1} = \frac{-1 - 4}{-3 - (-2)} = \frac{-5}{-1} = 5.$$

However, you should not subtract in one order in the numerator $(y_2 - y_1)$ and then in a different order in the denominator $(x_1 - x_2)$. The slope is *not*

$$\frac{-1 - 4}{-2 - (-3)} = \frac{-5}{1} = -5. \quad \text{Incorrect}$$

b. We can let $(x_1, y_1) = (-3, 4)$ and $(x_2, y_2) = (2, -2)$. The slope of the line shown in Figure 2.2(b) is computed as follows:

$$m = \frac{-2 - 4}{2 - (-3)} = \frac{-6}{5} = -\frac{6}{5}.$$

The slope of the line is $-\frac{6}{5}$. For every vertical change of -6 units (6 units down), there is a corresponding horizontal change of 5 units. The slope is negative and the line falls from left to right.

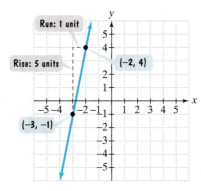

(a)

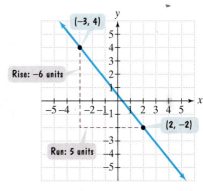

(b)

Figure 2.2 Visualizing slope

Check Point 1 Find the slope of the line passing through each pair of points:

      a. $(-3, 4)$ and $(-4, -2)$ **b.** $(4, -2)$ and $(-1, 5)$.

Example 1 illustrates that a line with a positive slope is rising from left to right and a line with a negative slope is falling from left to right. By contrast, a horizontal line neither rises nor falls and has a slope of zero. A vertical line has no horizontal change, so $x_2 - x_1 = 0$ in the formula for slope. Because we cannot divide by zero, the slope of a vertical line is undefined. This discussion is summarized in Table 2.1.

Table 2.1 Possibilities for a Line's Slope

Positive Slope	Negative Slope	Zero Slope	Undefined Slope
$m > 0$	$m < 0$	$m = 0$	m is undefined.
Line rises from left to right.	Line falls from left to right.	Line is horizontal.	Line is vertical.

2 Write the point-slope equation of a line.

The Point-Slope Form of the Equation of a Line

We can use the slope of a line to obtain various forms of the line's equation. For example, consider a nonvertical line with slope m that contains the point (x_1, y_1). Now, let (x, y) represent any other point on the line, shown in Figure 2.3. Keep in mind that the point (x, y) is arbitrary and is not in one fixed position. By contrast, the point (x_1, y_1) is fixed. Regardless of where the point (x, y) is located, the shape of the triangle in Figure 2.3 remains the same. Thus, the ratio for slope stays a constant m. This means that for all points along the line,

$$m = \frac{y - y_1}{x - x_1}, \quad x \neq x_1.$$

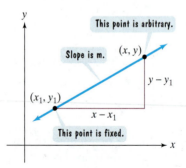

Figure 2.3 A line passing through (x_1, y_1) with slope m

We can clear the fraction by multiplying both sides by $x - x_1$.

$$m(x - x_1) = \frac{y - y_1}{x - x_1} \cdot x - x_1$$

$$m(x - x_1) = y - y_1 \qquad \text{Simplify.}$$

Now, if we reverse the two sides, we obtain the *point-slope form* of the equation of a line.

> **Point-Slope Form of the Equation of a Line**
>
> The **point-slope equation** of a nonvertical line with slope m that passes through the point (x_1, y_1) is
>
> $$y - y_1 = m(x - x_1).$$

For example, an equation of the line passing through $(1, 5)$ with slope 2 $(m = 2)$ is

$$y - 5 = 2(x - 1).$$

After we obtain the point-slope form of a line, it is customary to express the equation with y isolated on one side of the equal sign. Example 2 illustrates how this is done.

EXAMPLE 2 Writing the Point-Slope Equation of a Line

Write the point-slope form of the equation of the line passing through $(-1, 3)$ with slope 4. Then solve the equation for y.

Solution We use the point-slope equation of a line with $m = 4$, $x_1 = -1$, and $y_1 = 3$.

$$y - y_1 = m(x - x_1) \qquad \text{This is the point-slope form of the equation.}$$

$$y - 3 = 4[x - (-1)] \qquad \text{Substitute the given values.}$$

$$y - 3 = 4(x + 1) \qquad \text{We now have the point-slope form of the equation for the given line.}$$

We can solve this equation for y by applying the distributive property on the right side.

$$y - 3 = 4x + 4$$

Finally, we add 3 to both sides.

$$y = 4x + 7$$

Check Point 2 Write the point-slope form of the equation of the line passing through $(2, -5)$ with slope 6. Then solve the equation for y.

EXAMPLE 3 Writing the Point-Slope Equation of a Line

Write the point-slope form of the equation of the line passing through the points $(4, -3)$ and $(-2, 6)$. (See Figure 2.4.) Then solve the equation for y.

Solution To use the point-slope form, we need to find the slope. The slope is the change in the y-coordinates divided by the corresponding change in the x-coordinates.

$$m = \frac{6 - (-3)}{-2 - 4} = \frac{9}{-6} = -\frac{3}{2} \qquad \text{This is the definition of slope using } (4, -3) \text{ and } (-2, 6).$$

We can take either point on the line to be (x_1, y_1). Let's use $(x_1, y_1) = (4, -3)$. Now, we are ready to write the point-slope equation.

$$y - y_1 = m(x - x_1) \qquad \text{This is the point-slope form of the equation.}$$

$$y - (-3) = -\tfrac{3}{2}(x - 4) \qquad \text{Substitute: } (x_1, y_1) = (4, -3) \text{ and } m = -\tfrac{3}{2}.$$

$$y + 3 = -\tfrac{3}{2}(x - 4) \qquad \text{Simplify.}$$

We now have the point-slope form of the equation of the line shown in Figure 2.4. Now, we solve this equation for y.

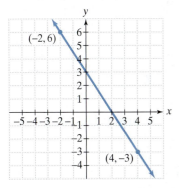

Figure 2.4 Write the point-slope equation of this line.

Discovery

You can use either point for (x_1, y_1) when you write a line's point-slope equation. Rework Example 3 using $(-2, 6)$ for (x_1, y_1). Once you solve for y, you should still obtain

$$y = -\tfrac{3}{2}x + 3.$$

$y + 3 = -\tfrac{3}{2}(x - 4)$	This is the point-slope form of the equation.
$y + 3 = -\tfrac{3}{2}x + 6$	Use the distributive property.
$y = -\tfrac{3}{2}x + 3$	Subtract 3 from both sides.

Check Point 3 Write the point-slope form of the equation of the line passing through the points $(-2, -1)$ and $(-1, -6)$. Then solve the equation for y.

3 Write and graph the slope-intercept equation of a line.

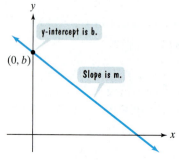

Figure 2.5 A line with slope m and y-intercept b

The Slope-Intercept Form of the Equation of a Line

Let's write the point-slope form of the equation of a nonvertical line with slope m and y-intercept b. The line is shown in Figure 2.5. Because the y-intercept is b, the line passes through $(0, b)$. We use the point-slope form with $x_1 = 0$ and $y_1 = b$.

$$y - y_1 = m(x - x_1)$$

Let $y_1 = b$.   Let $x_1 = 0$.

We obtain

$$y - b = m(x - 0).$$

Simplifying on the right side gives us

$$y - b = mx.$$

Finally, we solve for y by adding b to both sides.

$$y = mx + b$$

Thus, if a line's equation is written with y isolated on one side, the x-coefficient is the line's slope and the constant term is the y-intercept. This form of a line's equation is called the *slope-intercept form* of a line.

Slope-Intercept Form of the Equation of a Line

The **slope-intercept equation** of a nonvertical line with slope m and y-intercept b is

$$y = mx + b.$$

EXAMPLE 4   **Graphing by Using the Slope and y-Intercept**

Graph the line whose equation is $y = \tfrac{2}{3}x + 2$.

Solution   The equation of the line is in the form $y = mx + b$. We can find the slope, m, by identifying the coefficient of x. We can find the y-intercept, b, by identifying the constant term.

$$y = \frac{2}{3}x + 2$$

The slope is $\tfrac{2}{3}$.   The y-intercept is 2.

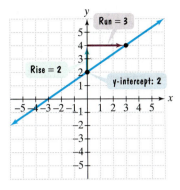

Figure 2.6 The graph of $y = \frac{2}{3}x + 2$

We need two points in order to graph the line. We can use the y-intercept, 2, to obtain the first point $(0, 2)$. Plot this point on the y-axis, shown in Figure 2.6.

We know the slope and one point on the line. We can use the slope, $\frac{2}{3}$, to determine a second point on the line. By definition,

$$m = \frac{2}{3} = \frac{\text{Rise}}{\text{Run}}.$$

We plot the second point on the line by starting at $(0, 2)$, the first point. Based on the slope, we move 2 units *up* (the rise) and 3 units to the *right* (the run). This puts us at a second point on the line, $(3, 4)$, shown in Figure 2.6.

We use a straightedge to draw a line through the two points. The graph of $y = \frac{2}{3}x + 2$ is shown in Figure 2.6.

Graphing $y = mx + b$ by Using the Slope and y-Intercept

1. Plot the y-intercept on the y-axis. This is the point $(0, b)$.
2. Obtain a second point using the slope, m. Write m as a fraction, and use rise over run, starting at the point containing the y-intercept, to plot this point.
3. Use a straightedge to draw a line through the two points. Draw arrowheads at the ends of the line to show that the line continues indefinitely in both directions.

Check Point 4 Graph the line whose equation is $y = \frac{3}{5}x + 1$.

4 Recognize equations of horizontal and vertical lines.

Equations of Horizontal and Vertical Lines

Some things change very little. For example, Figure 2.7 shows that the percentage of people in the United States satisfied with their lives remains relatively constant for all age groups. Shown in the figure is a horizontal line that passes near most tops of the six bars.

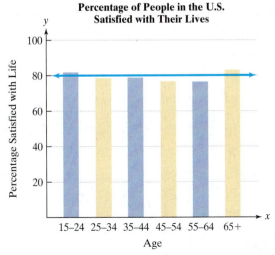

Percentage of People in the U.S. Satisfied with Their Lives

Figure 2.7
Source: *Culture Shift in Advanced Industrial Society,* Princeton University Press

We can use $y = mx + b$, the slope-intercept form of a line's equation, to write the equation of the horizontal line in Figure 2.7. We need the line's slope, m, and its y-intercept, b. Because the line is horizontal, $m = 0$. The line intersects the y-axis at $(0, 80)$, so its y-intercept is 80: $b = 80$.

Thus, an equation in the form $y = mx + b$ that models the percentage, y, of people at age x satisfied with their lives is

$$y = 0x + 80, \quad \text{or} \quad y = 80.$$

The percentage of people satisfied with their lives remains relatively constant in the United States for all age groups, at approximately 80%.

In general, if a line is horizontal, its slope is zero: $m = 0$. Thus, the equation $y = mx + b$ becomes $y = b$, where b is the y-intercept. All horizontal lines have equations of the form $y = b$.

EXAMPLE 5 Graphing a Horizontal Line

Graph $y = -4$ in the rectangular coordinate system.

Solution   All points on the graph of $y = -4$ have a value of y that is always -4. No matter what the x-coordinate is, the y-coordinate for every point on the line is -4. Let us select three of the possible values for x: $-2, 0,$ and 3. So, three of the points on the graph $y = -4$ are $(-2, -4)$, $(0, -4)$, and $(3, -4)$. Plot each of these points. Drawing a line that passes through the three points gives the horizontal line shown in Figure 2.8.

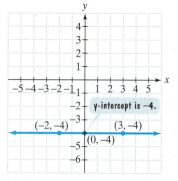

Figure 2.8 The graph of $y = -4$

Check Point 5   Graph $y = 3$ in the rectangular coordinate system.

Next, let's see what we can discover about the graph of an equation of the form $x = a$ by looking at an example.

EXAMPLE 6 Graphing a Vertical Line

Graph $x = 5$ in the rectangular coordinate system.

Solution   All points on the graph of $x = 5$ have a value of x that is always 5. No matter what the y-coordinate is, the corresponding x-coordinate for every point on the line is 5. Let us select three of the possible values of y: $-2, 0,$ and 3. So, three of the points on the graph of $x = 5$ are $(5, -2)$, $(5, 0)$, and $(5, 3)$. Plot each of these points. Drawing a line that passes through the three points gives the vertical line shown in Figure 2.9.

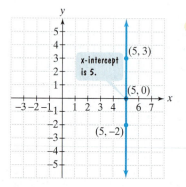

Figure 2.9 The graph $x = 5$

Horizontal and Vertical Lines

The graph of $y = b$ is a horizontal line. The y-intercept is b.

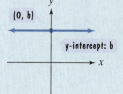

The graph of $x = a$ is a vertical line. The x-intercept is a.

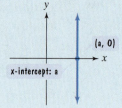

Check Point 6   Graph $x = -1$ in the rectangular coordinate system.

5 Recognize and use the general form of a line's equation.

The General Form of the Equation of a Line

The vertical line whose equation is $x = 5$ cannot be written in slope-intercept form, $y = mx + b$, because its slope is undefined. However, every line has an equation that can be expressed in the form $Ax + By + C = 0$. For example, $x = 5$ can be expressed as $1x + 0y - 5 = 0$, or $x - 5 = 0$. The equation $Ax + By + C = 0$ is called the *general form* of the equation of a line.

General Form of the Equation of a Line

Every line has an equation that can be written in the **general form**

$$Ax + By + C = 0$$

where A, B, and C are real numbers, and A and B are not both zero.

If the equation of a line is given in general form, it is possible to find the slope, m, and the y-intercept, b, for the line. We solve the equation for y, transforming it into the slope-intercept form $y = mx + b$. In this form, the coefficient of x is the slope of the line, and the constant term is its y-intercept.

EXAMPLE 7   Finding the Slope and the y-Intercept

Find the slope and the y-intercept of the line whose equation is $2x - 3y + 6 = 0$.

Solution   The equation is given in general form. We begin by rewriting it in the form $y = mx + b$. We need to solve for y.

$2x - 3y + 6 = 0$	This is the given equation.
$2x + 6 = 3y$	To isolate the y-term, add 3y to both sides.
$3y = 2x + 6$	Reverse the two sides. (This step is optional.)
$y = \dfrac{2}{3}x + 2$	Divide both sides by 3.

The coefficient of x, $\frac{2}{3}$, is the slope and the constant term, 2, is the y-intercept. This is the form of the equation that we graphed in Figure 2.6 on page 181.

Check Point 7   Find the slope and the y-intercept of the line whose equation is $3x + 6y - 12 = 0$. Then use the y-intercept and the slope to graph the equation.

We've covered a lot of territory. Let's take a moment to summarize the various forms for equations of lines.

Equations of Lines

1. Point-slope form:	$y - y_1 = m(x - x_1)$
2. Slope-intercept form:	$y = mx + b$
3. Horizontal line:	$y = b$
4. Vertical line:	$x = a$
5. General form:	$Ax + By + C = 0$

6 Find slopes and equations of parallel and perpendicular lines.

Parallel and Perpendicular Lines

Two nonintersecting lines that lie in the same plane are **parallel.** If two lines do not intersect, the ratio of the vertical change to the horizontal change is the same for each line. Because two parallel lines have the same "steepness," they must have the same slope.

> **Slope and Parallel Lines**
>
> 1. If two nonvertical lines are parallel, then they have the same slope.
> 2. If two distinct nonvertical lines have the same slope, then they are parallel.
> 3. Two distinct vertical lines, both with undefined slopes, are parallel.

EXAMPLE 8 **Writing Equations of a Line Parallel to a Given Line**

Write an equation of the line passing through $(-3, 2)$ and parallel to the line whose equation is $y = 2x + 1$. Express the equation in point-slope form and slope-intercept form.

Solution The situation is illustrated in Figure 2.10. We are looking for the equation of the line shown on the left. How do we obtain this equation? Notice that the line passes through the point $(-3, 2)$. Using the point-slope form of the line's equation, we have $x_1 = -3$ and $y_1 = 2$.

$$y - y_1 = m(x - x_1)$$

$y_1 = 2$ $x_1 = -3$

The equation of this line is given: $y = 2x + 1$.

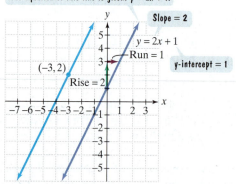

Slope = 2

$y = 2x + 1$

Run = 1

y-intercept = 1

$(-3, 2)$

Rise = 2

We must write the equation of this line.

Figure 2.10 Writing equations of a line parallel to a given line

Now, the only thing missing from the equation is m, the slope of the line on the left. Do we know anything about the slope of either line in Figure 2.10? The answer is yes; we know the slope of the line on the right, whose equation is given.

$$y = 2x + 1$$

The slope of the line on the right in Figure 2.10 is 2.

Parallel lines have the same slope. Because the slope of the line with the given equation is 2, $m = 2$ for the line whose equation we must write.

$$y - y_1 = m(x - x_1)$$

$y_1 = 2$ $m = 2$ $x_1 = -3$

The point-slope form of the line's equation is

$$y - 2 = 2[x - (-3)] \text{ or}$$

$$y - 2 = 2(x + 3).$$

Solving for y, we obtain the slope-intercept form of the equation.

$$y - 2 = 2x + 6 \quad \text{Apply the distributive property.}$$

$$y = 2x + 8 \quad \text{Add 2 to both sides. This is the slope-intercept form,}$$
$$y = mx + b, \text{ of the equation.}$$

Check Point 8 Write an equation of the line passing through $(-2, 5)$ and parallel to the line whose equation is $y = 3x + 1$. Express the equation in point-slope form and slope-intercept form.

Two lines that intersect at a right angle (90°) are said to be **perpendicular,** shown in Figure 2.11. There is a relationship between the slopes of perpendicular lines.

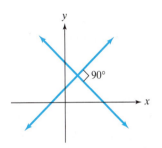

Figure 2.11 Perpendicular lines

Slope and Perpendicular Lines

1. If two nonvertical lines are perpendicular, then the product of their slopes is −1.
2. If the product of the slopes of two lines is − 1, then the lines are perpendicular.
3. A horizontal line having zero slope is perpendicular to a vertical line having undefined slope.

An equivalent way of stating this relationship is to say that one line is perpendicular to another line if its slope is the *negative reciprocal* of the slope of the other. For example, if a line has slope 5, any line having slope $-\frac{1}{5}$ is perpendicular to it. Similarly, if a line has slope $-\frac{3}{4}$, any line having slope $\frac{4}{3}$ is perpendicular to it.

EXAMPLE 9 **Finding the Slope of a Line Perpendicular to a Given Line**

Find the slope of any line that is perpendicular to the line whose equation is $x + 4y - 8 = 0$.

Solution   We begin by writing the equation of the given line, $x + 4y - 8 = 0$, in slope-intercept form. Solve for y.

$$x + 4y - 8 = 0 \qquad \text{This is the given equation.}$$
$$4y = -x + 8 \qquad \text{To isolate the y-term, subtract x and add 8 on both sides.}$$
$$y = -\tfrac{1}{4}x + 2 \qquad \text{Divide both sides by 4.}$$

> Slope is
> $-\dfrac{1}{4}$.

The given line has slope $-\frac{1}{4}$. Any line perpendicular to this line has a slope that is the negative reciprocal of $-\frac{1}{4}$. Thus, the slope of any perpendicular line is 4.

Check Point 9   Find the slope of any line that is perpendicular to the line whose equation is $x + 3y - 12 = 0$.

7 Model data with linear equations.

Applications

Slope is defined as the ratio of a change in y to a corresponding change in x. Our next example shows how slope can be interpreted as a **rate of change** in an applied situation.

EXAMPLE 10   Slope as a Rate of Change

A best guess at the look of our nation in the next decade indicates that the number of men and women living alone will increase each year. Figure 2.12 shows line graphs for the number of U.S. men and women living alone, projected through 2010. Find the slope of the line segment for the women. Describe what the slope represents.

Number of People in the U.S Living Alone

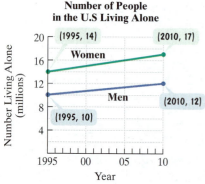

Figure 2.12 *Source:* Forrester Research

Solution   We let x represent a year and y the number of women living alone in that year. The two points shown on the line segment for women have the following coordinates:

$$(1995, 14) \qquad \text{and} \qquad (2010, 17).$$

> In 1995, 14 million U.S. women lived alone.

> In 2010, 17 million U.S. women are projected to live alone.

Now we compute the slope:

$$m = \frac{\text{Change in } y}{\text{Change in } x} = \frac{17 - 14}{2010 - 1995}$$

> The unit in the numerator is million people.

> The unit in the denominator is year.

$$= \frac{3}{15} = \frac{1}{5} = \frac{0.2 \text{ million people}}{\text{year}}.$$

The slope indicates that the number of U.S. women living alone is projected to increase by 0.2 million each year. The rate of change is 0.2 million women per year.

Check Point 10   Use the graph in Example 10 to find the slope of the line segment for the men. Express the slope correct to two decimal places and describe what it represents.

If an equation in slope-intercept form models relationships between variables, then the slope and y-intercept have physical interpretations. For the equation $y = mx + b$, the y-intercept, b, tells us what is happening to y when x is 0. If x represents time, the y-intercept describes the value of y at the beginning, or when time equals 0. The slope represents the rate of change in y per unit change in x.

Using these ideas, we can develop a model for the data for women living alone, shown in Figure 2.12 on the previous page. We let x = the number of years after 1995. At the beginning of our data, or 0 years after 1995, 14 million women lived alone. Thus, $b = 14$. In Example 10, we found that $m = 0.2$ (rate of change is 0.2 million women per year). An equation of the form $y = mx + b$ that models the data is

$$y = 0.2x + 14,$$

where y is the number, in millions, of U.S. women living alone x years after 1995.

Linear equations are useful for modeling data in scatter plots that fall on or near a line. For example, Table 2.2 gives the population of the United States, in millions, in the indicated year. The data are displayed in a scatter plot as a set of six points in Figure 2.13.

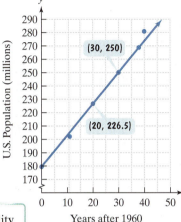

Table 2.2

Year	x (Years after 1960)	y (U.S. Population) (in millions)
1960	0	179.3
1970	10	203.3
1980	20	226.5
1990	30	250.0
1998	38	268.9
2000	40	281.4

Figure 2.13

Also shown in Figure 2.13 is a line that passes through or near the six points. By writing the equation of this line, we can obtain a model of the data and make predictions about the population of the United States in the future.

Technology

You can use a graphing utility to obtain a model for a scatter plot in which the data points fall on or near a straight line. The line that best fits the data is called the **regression line.** After entering the data in Table 2.2, a graphing utility displays a scatter plot of the data and the regression line.

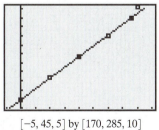

$[-5, 45, 5]$ by $[170, 285, 10]$

Also displayed is the regression line's equation.

```
LinReg
y=ax+b
a=2.45748031496
b=178.377952756
```

EXAMPLE 11   Modeling U.S. Population

Write the slope-intercept equation of the line shown in Figure 2.13. Use the equation to predict U.S. population in 2010.

Solution   The line in Figure 2.13 passes through $(20, 226.5)$ and $(30, 250)$. We start by finding the slope.

$$m = \frac{\text{Change in } y}{\text{Change in } x} = \frac{250 - 226.5}{30 - 20} = \frac{23.5}{10} = 2.35$$

The slope indicates that the rate of change in the U.S. population is 2.35 million people per year. Now we write the line's slope-intercept equation.

$$y - y_1 = m(x - x_1) \qquad \text{Begin with the point-slope form.}$$

$$y - 250 = 2.35(x - 30) \qquad \text{Either ordered pair can be } (x_1, y_1).$$
$$\text{Let } (x_1, y_1) = (30, 250). \text{ From above, } m = 2.35.$$

$$y - 250 = 2.35x - 70.5 \qquad \text{Apply the distributive property on the right.}$$

$$y = 2.35x + 179.5 \qquad \text{Add 250 to both sides and solve for y.}$$

A linear equation that models U.S. population, y, in millions, x years after 1960 is

$$y = 2.35x + 179.5.$$

Now, let's use this equation to predict U.S. population in 2010. Because 2010 is 50 years after 1960, substitute 50 for x and compute y.

$$y = 2.35(50) + 179.5 = 297$$

Our equation predicts that the population of the United States in the year 2010 will be 297 million. (The projected figure from the U.S. Census Bureau is 297.716 million.)

Check Point 11 Use the data points $(10, 203.3)$ and $(20, 226.5)$ from Table 2.2 to write an equation that models U.S. population x years after 1960. Use the equation to predict U.S. population in 2020.

Cigarettes and Lung Cancer

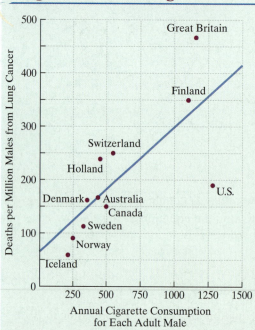

This scatter plot shows a relationship between cigarette consumption among males and deaths due to lung cancer per million males. The data are from 11 countries and date back to a 1964 report by the U.S. Surgeon General. The scatter plot can be modeled by a line whose slope indicates an increasing death rate from lung cancer with increased cigarette consumption. At that time, the tobacco industry argued that in spite of this regression line, tobacco use is not the cause of cancer. Recent data do, indeed, show a causal effect between tobacco use and numerous diseases.

Source: Smoking and Health, Washington, D.C., 1964

EXERCISE SET 2.1

Practice Exercises

In Exercises 1–10, find the slope of the line passing through each pair of points or state that the slope is undefined. Then indicate whether the line through the points rises, falls, is horizontal, or is vertical.

1. $(4, 7)$ and $(8, 10)$
2. $(2, 1)$ and $(3, 4)$
3. $(-2, 1)$ and $(2, 2)$
4. $(-1, 3)$ and $(2, 4)$
5. $(4, -2)$ and $(3, -2)$
6. $(4, -1)$ and $(3, -1)$
7. $(-2, 4)$ and $(-1, -1)$
8. $(6, -4)$ and $(4, -2)$
9. $(5, 3)$ and $(5, -2)$
10. $(3, -4)$ and $(3, 5)$

In Exercises 11–38, use the given conditions to write an equation for each line in point-slope form and slope-intercept form. only do 1 way

11. Slope = 2, passing through $(3, 5)$
12. Slope = 4, passing through $(1, 3)$
13. Slope = 6, passing through $(-2, 5)$
14. Slope = 8, passing through $(4, -1)$
15. Slope = -3, passing through $(-2, -3)$

16. Slope $= -5$, passing through $(-4, -2)$
17. Slope $= -4$, passing through $(-4, 0)$
18. Slope $= -2$, passing through $(0, -3)$
19. Slope $= -1$, passing through $(-\frac{1}{2}, -2)$
20. Slope $= -1$, passing through $(-4, -\frac{1}{4})$
21. Slope $= \frac{1}{2}$, passing through the origin
22. Slope $= \frac{1}{3}$, passing through the origin
23. Slope $= -\frac{2}{3}$, passing through $(6, -2)$
24. Slope $= -\frac{3}{5}$, passing through $(10, -4)$
25. Passing through $(1, 2)$ and $(5, 10)$
26. Passing through $(3, 5)$ and $(8, 15)$
27. Passing through $(-3, 0)$ and $(0, 3)$
28. Passing through $(-2, 0)$ and $(0, 2)$
29. Passing through $(-3, -1)$ and $(2, 4)$
30. Passing through $(-2, -4)$ and $(1, -1)$
31. Passing through $(-3, -2)$ and $(3, 6)$
32. Passing through $(-3, 6)$ and $(3, -2)$
33. Passing through $(-3, -1)$ and $(4, -1)$
34. Passing through $(-2, -5)$ and $(6, -5)$
35. Passing through $(2, 4)$ with x-intercept $= -2$
36. Passing through $(1, -3)$ with x-intercept $= -1$
37. x-intercept $= -\frac{1}{2}$ and y-intercept $= 4$
38. x-intercept $= 4$ and y-intercept $= -2$

In Exercises 39–46, give the slope and y-intercept of each line whose equation is given. Then graph the line.

39. $y = 2x + 1$
40. $y = 3x + 2$
41. $y = -2x + 1$
42. $y = -3x + 2$
43. $y = \frac{3}{4}x - 2$
44. $y = \frac{3}{4}x - 3$
45. $y = -\frac{3}{5}x + 7$
46. $y = -\frac{2}{5}x + 6$

In Exercises 47–52, graph each equation in the rectangular coordinate system.

47. $y = -2$
48. $y = 4$
49. $x = -3$
50. $x = 5$
51. $y = 0$
52. $x = 0$

In Exercises 53–60,
  a. Rewrite the given equation in slope-intercept form.
  b. Give the slope and y-intercept.
  c. Graph the equation.

53. $3x + y - 5 = 0$
54. $4x + y - 6 = 0$
55. $2x + 3y - 18 = 0$
56. $4x + 6y + 12 = 0$
57. $8x - 4y - 12 = 0$
58. $6x - 5y - 20 = 0$
59. $3x - 9 = 0$
60. $4y + 28 = 0$

In Exercises 61–68, use the given conditions to write an equation for each line in point-slope form and slope-intercept form.

61. Passing through $(-8, -10)$ and parallel to the line whose equation is $y = -4x + 3$
62. Passing through $(-2, -7)$ and parallel to the line whose equation is $y = -5x + 4$

63. Passing through $(2, -3)$ and perpendicular to the line whose equation is $y = \frac{1}{5}x + 6$
64. Passing through $(-4, 2)$ and perpendicular to the line whose equation is $y = \frac{1}{3}x + 7$
65. Passing through $(-2, 2)$ and parallel to the line whose equation is $2x - 3y - 7 = 0$
66. Passing through $(-1, 3)$ and parallel to the line whose equation is $3x - 2y - 5 = 0$
67. Passing through $(4, -7)$ and perpendicular to the line whose equation is $x - 2y - 3 = 0$
68. Passing through $(5, -9)$ and perpendicular to the line whose equation is $x + 7y - 12 = 0$

Application Exercises

69. The scatter plot shows that from 1985 to 2001, the number of Americans participating in downhill skiing remained relatively constant. Write an equation that models the number of participants in downhill skiing, y, in millions, for this period.

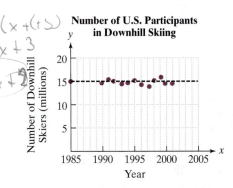

Source: National Ski Areas Association

If talk about a federal budget surplus sounded too good to be true, that's because it probably was. The Congressional Budget Office's estimates for 2010 range from a $1.2 trillion budget surplus to a $286 billion deficit. Use the information provided by the Congressional Budget Office graphs to solve Exercises 70–71.

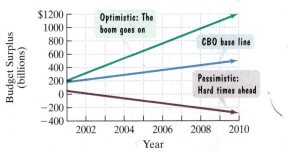

Source: Congressional Budget Office

70. Turn back a page and look at the line that indicates hard times ahead. Find the slope of this line using $(2001, 50)$ and $(2010, -286)$. Use a calculator and round to the nearest whole number. Describe what the slope represents.

71. Turn back a page and look at the line that indicates the boom goes on. Find the slope of this line using $(2001, 200)$ and $(2010, 1200)$. Use a calculator and round to the nearest whole number. Describe what the slope represents.

72. Horrified at the cost the last time you needed a prescription drug? The graph shows that the cost of the average retail prescription has been rising steadily since 1991.

Average Cost of a Retail Prescription

Years after 1991

Source: Newsweek

a. According to the graph, what is the y-intercept? Describe what this represents in this situation.

b. Use the coordinates of the two points shown to compute the slope. What does this mean about the cost of the average retail prescription?

c. Write a linear equation in slope-intercept form that models the cost of the average retail prescription, y, x years after 1991.

d. Use your model from part (c) to predict the cost of the average retail prescription in 2010.

73. For 61 years, Social Security has been a huge success. It is the primary source of income for 66% of Americans over 65 and the only thing that keeps 42% of the elderly from poverty. However, the number of workers per Social Security beneficiary has been declining steadily since 1950.

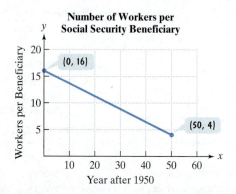

Number of Workers per Social Security Beneficiary

Year after 1950

Source: Social Security Administration

a. According to the graph, what is the y-intercept? Describe what this represents in this situation.

b. Use the coordinates of the two points shown to compute the slope. What does this mean about the number of workers per beneficiary?

c. Write a linear equation in slope-intercept form that models the number of workers per beneficiary, y, x years after 1950.

d. Use your model from part (c) to predict number of workers per beneficiary in 2010. For every 8 workers, how many beneficiaries will there be?

74. We seem to be fed up with being lectured at about our waistlines. The points in the graph show the average weight of American adults from 1990 through 2000. Also shown is a line that passes through or near the points.

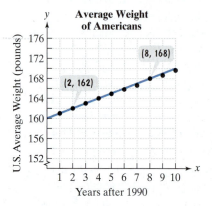

Average Weight of Americans

Years after 1990

Source: Diabetes Care

a. Use the two points whose coordinates are shown by the voice balloons to find the point-slope equation of the line that models average weight of Americans, y, in pounds, x years after 1990.

b. Write the equation in part (a) in slope-intercept form.

c. Use the slope-intercept equation to predict the average weight of Americans in 2008.

75. Films may not be getting any better, but in this era of moviegoing, the number of screens available for new films and the classics has exploded. The points in the graph show the number of screens in the United States from 1995 through 2000. Also shown is a line that passes through or near the points.

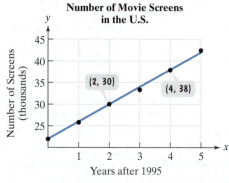

Number of Movie Screens in the U.S.

Years after 1995

Source: Motion Picture Association of America

a. Use the two points whose coordinates are shown by the voice balloons to find the point-slope equation of the line that models the number of screens, y, in thousands, x years after 1995.

b. Write the equation in part (a) in slope-intercept form.

c. Use the slope-intercept equation to predict the number of screens, in thousands, in 2008.

76. The scatter plot shows the relationship between the percentage of married women of child-bearing age using contraceptives and the births per woman in selected countries. Also shown is the regression line. Use two points on this line to write both its point-slope and slope-intercept equations. Then find the number of births per woman if 90% of married women of child-bearing age use contraceptives.

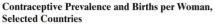
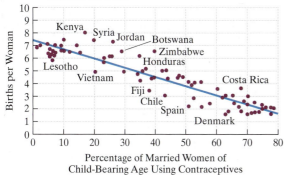

Contraceptive Prevalence and Births per Woman, Selected Countries

Source: Population Reference Bureau

77. Shown, again, is the scatter plot that indicates a relationship between the percentage of adult females in a country who are literate and the mortality of children under five. Also shown is a line that passes through or near the points. Find a linear equation that models the data by finding the slope-intercept equation of the line. Use the model to make a prediction about child mortality based on the percentage of adult females in a country who are literate.

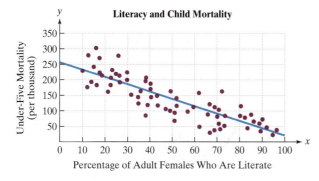

Literacy and Child Mortality

Source: United Nations

In Exercises 78–80, find a linear equation in slope-intercept form that models the given description. Describe what each variable in your model represents. Then use the model to make a prediction.

78. In 1995, the average temperature of Earth was 57.7°F and has increased at a rate of 0.01°F per year since then.

79. In 1995, 60% of U.S. adults read a newspaper and this percentage has decreased at a rate of 0.7% per year since then.

80. A computer that was purchased for $4000 is depreciating at a rate of $950 per year.

81. A business discovers a linear relationship between the number of shirts it can sell and the price per shirt. In particular, 20,000 shirts can be sold at $19 each, and 2000 of the same shirts can be sold at $55 each. Write the slope-intercept equation of the *demand line* that models the number of shirts that can be sold, y, at a price of x dollars. Then determine the number of shirts that can be sold at $50 each.

Writing in Mathematics

82. What is the slope of a line and how is it found?

83. Describe how to write the equation of a line if two points along the line are known.

84. Explain how to derive the slope-intercept form of a line's equation, $y = mx + b$, from the point-slope form

$$y - y_1 = m(x - x_1).$$

85. Explain how to graph the equation $x = 2$. Can this equation be expressed in slope-intercept form? Explain.

86. Explain how to use the general form of a line's equation to find the line's slope and y-intercept.

87. If two lines are parallel, describe the relationship between their slopes.

88. If two lines are perpendicular, describe the relationship between their slopes.

88. If you know a point on a line and you know the equation of a line perpendicular to this line, explain how to write the line's equation.

90. A formula in the form $y = mx + b$ models the cost, y, of a four-year college x years after 2003. Would you expect m to be positive, negative, or zero? Explain your answer.

91. We saw that the percentage of people satisfied with their lives remains relatively constant for all age groups. Exercise 69 showed that the number of skiers in the United States has remained relatively constant over time. Give another example of a real-world phenomenon that has remained relatively constant. Try writing an equation that models this phenomenon.

Technology Exercises

Use a graphing utility to graph each equation in Exercises 92–95. Then use the $\boxed{\text{TRACE}}$ *feature to trace along the line and find the coordinates of two points. Use these points to compute the line's slope. Check your result by using the coefficient of x in the line's equation.*

92. $y = 2x + 4$

93. $y = -3x + 6$

94. $y = -\frac{1}{2}x - 5$

95. $y = \frac{3}{4}x - 2$

96. Is there a relationship between alcohol from moderate wine consumption and heart disease death rate? The table gives data from 19 developed countries.

France

Country	A	B	C	D	E	F	G
Liters of alcohol from drinking wine, per person, per year (x)	2.5	3.9	2.9	2.4	2.9	0.8	9.1
Deaths from heart disease, per 100,000 people per year (y)	211	167	131	191	220	297	71

U.S.

Country	H	I	J	K	L	M	N	O	P	Q	R	S
(x)	0.8	0.7	7.9	1.8	1.9	0.8	6.5	1.6	5.8	1.3	1.2	2.7
(y)	211	300	107	167	266	227	86	207	115	285	199	172

Source: New York Times, December 28, 1994

a. Use the statistical menu of your graphing utility to enter the 19 ordered pairs of data items shown in the table.

b. Use the $\boxed{\text{DRAW}}$ menu and the scatter plot capability to draw a scatter plot of the data.

c. Select the linear regression option. Use your utility to obtain values for a and b for the equation of the regression line, $y = ax + b$. You may also be given a **correlation coefficient, r.** Values of r close to 1 indicate that the points can be described by a linear relationship and the regression line has a positive slope. Values of r close to -1 indicate that the points can be described by a linear relationship and the regression line has a negative slope. Values of r close to 0 indicate no linear relationship between the variables. In this case, a linear model does not accurately describe the data.

d. Use the appropriate sequence (consult your manual) to graph the regression equation on top of the points in the scatter plot.

Critical Thinking Exercises

97. Which one of the following is true?
  a. A linear equation with nonnegative slope has a graph that rises from left to right.
  b. The equations $y = 4x$ and $y = -4x$ have graphs that are perpendicular lines.
  c. The line whose equation is $5x + 6y - 30 = 0$ passes through the point $(6, 0)$ and has slope $-\frac{5}{6}$.
  d. The graph of $y = 7$ in the rectangular coordinate system is the single point $(7, 0)$.

98. Prove that the equation of a line passing through $(a, 0)$ and $(0, b)$ $(a \neq 0, b \neq 0)$ can be written in the form $\frac{x}{a} + \frac{y}{b} = 1$. Why is this called the *intercept form* of a line?

99. Use the figure shown to make the following lists.
  a. List the slopes $m_1, m_2, m_3,$ and m_4 in order of decreasing size.
  b. List the y-intercepts $b_1, b_2, b_3,$ and b_4 in order of decreasing size.

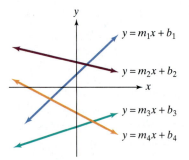

100. Excited about the success of celebrity stamps, post office officials were rumored to have put forth a plan to institute two new types of thermometers. On these new scales, $°E$ represents degrees Elvis and $°M$ represents degrees Madonna. If it is known that $40°E = 25°M$, $280°E = 125°M$, and degrees Elvis is linearly related to degrees Madonna, write an equation expressing E in terms of M.

Group Exercise

101. Group members should consult an almanac, newspaper, magazine, or the Internet to find data that lie approximately on or near a straight line. Working by hand or using a graphing utility, construct a scatter plot for the data. If working by hand, draw a line that approximately fits the data and then write its equation. If using a graphing utility, obtain the equation of the regression line. Then use the equation of the line to make a prediction about what might happen in the future. Are there circumstances that might affect the accuracy of this prediction? List some of these circumstances.

SECTION 2.2   *Distance and Midpoint Formulas; Circles*

Objectives

1. Find the distance between two points.
2. Find the midpoint of a line segment.
3. Write the standard form of a circle's equation.
4. Give the center and radius of a circle whose equation is in standard form.
5. Convert the general form of a circle's equation to standard form.

It's a good idea to know your way around a circle. Clocks, angles, maps, and compasses are based on circles. Circles occur everywhere in nature: in ripples on water, patterns on a butterfly's wings, and cross sections of trees. Some consider the circle to be the most pleasing of all shapes.

The rectangular coordinate system gives us a unique way of knowing a circle. It enables us to translate a circle's geometric definition into an algebraic equation. To do this, we must first develop a formula for the distance between any two points in rectangular coordinates.

1 Find the distance between two points.

The Distance Formula

Using the Pythagorean Theorem, we can find the distance between the two points $P_1(x_1, y_1)$ and $P_2(x_2, y_2)$ in the rectangular coordinate system. The two points are illustrated in Figure 2.14.

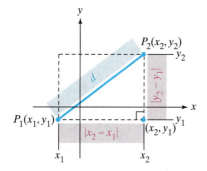

Figure 2.14

The distance that we need to find is represented by d and shown in blue. Notice that the distance between two points on the dashed horizontal line is the absolute value of the difference between the x-coordinates of the two points. This distance, $|x_2 - x_1|$, is shown in pink. Similarly, the distance between two points on the dashed vertical line is the absolute value of the difference between the y-coordinates of the two points. This distance, $|y_2 - y_1|$, is also shown in pink.

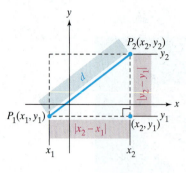

Figure 2.14, repeated

Because the dashed lines are horizontal and vertical, a right triangle is formed. Thus, we can use the Pythagorean Theorem to find distance d. By the Pythagorean Theorem,

$$d^2 = |x_2 - x_1|^2 + |y_2 - y_1|^2$$
$$d = \sqrt{|x_2 - x_1|^2 + |y_2 - y_1|^2}$$
$$d = \sqrt{(x_2 - x_1)^2 + (y_2 - y_1)^2}.$$

This result is called the **distance formula.**

The Distance Formula

The distance, d, between the points (x_1, y_1) and (x_2, y_2) in the rectangular coordinate system is

$$d = \sqrt{(x_2 - x_1)^2 + (y_2 - y_1)^2}.$$

When using the distance formula, it does not matter which point you call (x_1, y_1) and which you call (x_2, y_2).

EXAMPLE 1   Using the Distance Formula

Find the distance between $(-1, -3)$ and $(2, 3)$.

Solution   Letting $(x_1, y_1) = (-1, -3)$ and $(x_2, y_2) = (2, 3)$, we obtain

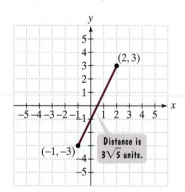

Figure 2.15 Finding the distance between two points

$$d = \sqrt{(x_2 - x_1)^2 + (y_2 - y_1)^2} \qquad \text{Use the distance formula.}$$
$$= \sqrt{[2 - (-1)]^2 + [3 - (-3)]^2} \qquad \text{Substitute the given values.}$$
$$= \sqrt{(2 + 1)^2 + (3 + 3)^2} \qquad \begin{array}{l}\text{Apply the definition of subtraction}\\ \text{within the grouping symbols.}\end{array}$$
$$= \sqrt{3^2 + 6^2} \qquad \text{Perform the resulting additions.}$$
$$= \sqrt{9 + 36} \qquad \text{Square 3 and 6.}$$
$$= \sqrt{45} \qquad \text{Add.}$$
$$= 3\sqrt{5} \approx 6.71. \qquad \sqrt{45} = \sqrt{9 \cdot 5} = \sqrt{9}\sqrt{5} = 3\sqrt{5}$$

The distance between the given points is $3\sqrt{5}$ units, or approximately 6.71 units. The situation is illustrated in Figure 2.15.

 Check Point 1   Find the distance between $(2, -2)$ and $(5, 2)$.

2 Find the midpoint of a line segment.

The Midpoint Formula

The distance formula can be used to derive a formula for finding the midpoint of a line segment between two given points. The formula is given as follows:

The Midpoint Formula

Consider a line segment whose endpoints are (x_1, y_1) and (x_2, y_2). The coordinates of the segment's midpoint are

$$\left(\frac{x_1 + x_2}{2}, \frac{y_1 + y_2}{2} \right).$$

To find the midpoint, take the average of the two x-coordinates and the average of the two y-coordinates.

EXAMPLE 2 Using the Midpoint Formula

Find the midpoint of the line segment with endpoints $(1, -6)$ and $(-8, -4)$.

Solution To find the coordinates of the midpoint, we average the coordinates of the endpoints.

$$\text{Midpoint} = \left(\frac{1 + (-8)}{2}, \frac{-6 + (-4)}{2} \right) = \left(\frac{-7}{2}, \frac{-10}{2} \right) = \left(-\frac{7}{2}, -5 \right)$$

Average the x-coordinates. Average the y-coordinates.

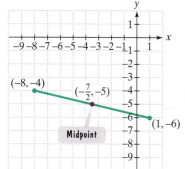

Figure 2.16 Finding a line segment's midpoint

Figure 2.16 illustrates that the point $\left(-\frac{7}{2}, -5 \right)$ is midway between the points $(1, -6)$ and $(-8, -4)$.

Check Point 2 Find the midpoint of the line segment with endpoints $(1, 2)$ and $(7, -3)$.

Circles

Our goal is to translate a circle's geometric definition into an equation. We begin with this geometric definition.

Definition of a Circle

A **circle** is the set of all points in a plane that are equidistant from a fixed point, called the **center.** The fixed distance from the circle's center to any point on the circle is called the **radius.**

Figure 2.17 is our starting point for obtaining a circle's equation. We've placed the circle into a rectangular coordinate system. The circle's center is (h, k) and its radius is r. We let (x, y) represent the coordinates of any point on the circle.

What does the geometric definition of a circle tell us about point (x, y) in Figure 2.17? The point is on the circle if and only if its distance from the center is r. We can use the distance formula to express this idea algebraically:

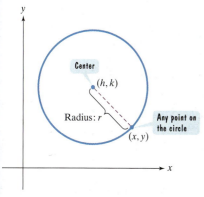

Figure 2.17 A circle centered at (h, k) with radius r

The distance between (x, y) and (h, k)

$$\sqrt{(x - h)^2 + (y - k)^2}$$

is always r.

$$= r$$

Squaring both sides of $\sqrt{(x - h)^2 + (y - k)^2} = r$ yields the *standard form of the equation of a circle*.

The Standard Form of the Equation of a Circle

The **standard form of the equation of a circle** with center (h, k) and radius r is

$$(x - h)^2 + (y - k)^2 = r^2.$$

---

3 Write the standard form of a circle's equation.

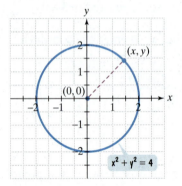

Figure 2.18 The graph of $x^2 + y^2 = 4$

EXAMPLE 3 Finding the Standard Form of a Circle's Equation

Write the standard form of the equation of the circle with center $(0, 0)$ and radius 2. Graph the circle.

Solution The center is $(0, 0)$. Because the center is represented as (h, k) in the standard form of the equation, $h = 0$ and $k = 0$. The radius is 2, so we will let $r = 2$ in the equation.

$(x - h)^2 + (y - k)^2 = r^2$ This is the standard form of a circle's equation.

$(x - 0)^2 + (y - 0)^2 = 2^2$ Substitute 0 for h, 0 for k, and 2 for r.

$x^2 + y^2 = 4$ Simplify.

The standard form of the equation of the circle is $x^2 + y^2 = 4$. Figure 2.18 shows the graph.

Check Point 3 Write the standard form of the equation of the circle with center $(0, 0)$ and radius 4.

Technology

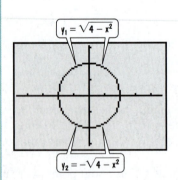

To graph a circle with a graphing utility, first solve the equation for y.

$$x^2 + y^2 = 4$$
$$y^2 = 4 - x^2$$
$$y = \pm\sqrt{4 - x^2}$$

Graph the two equations

$$y_1 = \sqrt{4 - x^2} \quad \text{and} \quad y_2 = -\sqrt{4 - x^2}$$

in the same viewing rectangle. The graph of $y_1 = \sqrt{4 - x^2}$ is the top semicircle because y is always positive. The graph of $y_2 = -\sqrt{4 - x^2}$ is the bottom semicircle because y is always negative. Use a $\boxed{\text{ZOOM SQUARE}}$ setting so that the circle looks like a circle. (Many graphing utilities have problems connecting the two semicircles because the segments directly across horizontally from the center become nearly vertical.)

Example 3 and Check Point 3 involved circles centered at the origin. The standard form of the equation of all such circles is $x^2 + y^2 = r^2$, where r is the circle's radius. Now, let's consider a circle whose center is not at the origin.

**EXAMPLE 4   Finding the Standard Form
of a Circle's Equation**

Write the standard form of the equation of the circle with center $(-2, 3)$ and radius 4.

Solution   The center is $(-2, 3)$. Because the center is represented as (h, k) in the standard form of the equation, $h = -2$ and $k = 3$. The radius is 4, so we will let $r = 4$ in the equation.

$$(x - h)^2 + (y - k)^2 = r^2 \qquad \text{\textcolor{blue}{This is the standard form of a circle's equation.}}$$
$$[x - (-2)]^2 + (y - 3)^2 = 4^2 \qquad \text{\textcolor{blue}{Substitute –2 for h, 3 for k, and 4 for r.}}$$
$$(x + 2)^2 + (y - 3)^2 = 16 \qquad \text{\textcolor{blue}{Simplify.}}$$

The standard form of the equation of the circle is $(x + 2)^2 + (y - 3)^2 = 16$.

> **Check Point 4**   Write the standard form of the equation of the circle with center $(5, -6)$ and radius 10.

④ Give the center and radius of a circle whose equation is in standard form.

**EXAMPLE 5   Using the Standard Form of a Circle's Equation
to Graph the Circle**

Find the center and radius of the circle whose equation is

$$(x - 2)^2 + (y + 4)^2 = 9$$

and graph the equation.

Solution   In order to graph the circle, we need to know its center, (h, k), and its radius, r. We can find the values for h, k, and r by comparing the given equation to the standard form of the equation of a circle.

$$(x - 2)^2 + (y + 4)^2 = 9$$
$$(x - 2)^2 + [y - (-4)]^2 = 3^2$$

This is $(x - h)^2$, with $h = 2$. This is $(y - k)^2$, with $k = -4$. This is r^2, with $r = 3$.

We see that $h = 2$, $k = -4$, and $r = 3$. Thus, the circle has center $(h, k) = (2, -4)$ and a radius of 3 units. To graph this circle, first plot the center $(2, -4)$. Because the radius is 3, you can locate at least four points on the circle by going out three units to the right, to the left, up, and down from the center.

The points three units to the right and to the left of $(2, -4)$ are $(5, -4)$ and $(-1, -4)$, respectively. The points three units up and down from $(2, -4)$ are $(2, -1)$ and $(2, -7)$, respectively.

Using these points, we obtain the graph in Figure 2.19.

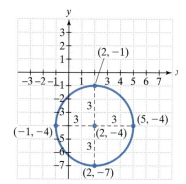

Figure 2.19 The graph of $(x - 2)^2 + (y + 4)^2 = 9$

> **Check Point 5**   Find the center and radius of the circle whose equation is
> $$(x + 3)^2 + (y - 1)^2 = 4$$
> and graph the equation.

If we square $x - 2$ and $y + 4$ in the standard form of the equation from Example 5, we obtain another form for the circle's equation.

$$(x - 2)^2 + (y + 4)^2 = 9 \qquad \text{This is the standard form of the equation from Example 5.}$$

$$x^2 - 4x + 4 + y^2 + 8y + 16 = 9 \qquad \text{Square } x - 2 \text{ and } y + 4.$$

$$x^2 + y^2 - 4x + 8y + 20 = 9 \qquad \text{Combine numerical terms and rearrange terms.}$$

$$x^2 + y^2 - 4x + 8y + 11 = 0 \qquad \text{Subtract 9 from both sides.}$$

This result suggests that an equation in the form $x^2 + y^2 + Dx + Ey + F = 0$ can represent a circle. This is called the *general form of the equation of a circle*.

> ### The General Form of the Equation of a Circle
> The **general form of the equation of a circle** is
> $$x^2 + y^2 + Dx + Ey + F = 0.$$

5 Convert the general form of a circle's equation to standard form.

We can convert the general form of the equation of a circle to the standard form $(x - h)^2 + (y - k)^2 = r^2$. We do so by completing the square on x and y. Let's see how this is done.

EXAMPLE 6 Converting the General Form of a Circle's Equation to Standard Form and Graphing the Circle

Write in standard form and graph: $x^2 + y^2 + 4x - 6y - 23 = 0$.

Solution Because we plan to complete the square on both x and y, let's rearrange terms so that x-terms are arranged in descending order, y-terms are arranged in descending order, and the constant term appears on the right.

$$x^2 + y^2 + 4x - 6y - 23 = 0 \qquad \text{This is the given equation.}$$

$$(x^2 + 4x \quad) + (y^2 - 6y \quad) = 23 \qquad \text{Rewrite in anticipation of completing the square.}$$

$$(x^2 + 4x + 4) + (y^2 - 6y + 9) = 23 + 4 + 9$$

Complete the square on x: $\frac{1}{2} \cdot 4 = 2$ and $2^2 = 4$, so add 4 to both sides. Complete the square on y: $\frac{1}{2}(-6) = -3$ and $(-3)^2 = 9$, so add 9 to both sides.

Remember that numbers added on the left side must also be added on the right side.

$$(x + 2)^2 + (y - 3)^2 = 36 \qquad \text{Factor on the left and add on the right.}$$

This last equation is in standard form. We can identify the circle's center and radius by comparing this equation to the standard form of the equation of a circle, $(x - h)^2 + (y - k)^2 = r^2$.

$$(x + 2)^2 + (y - 3)^2 = 36$$

$$[x - (-2)]^2 + (y - 3)^2 = 6^2$$

This is $(x - h)^2$, with $h = -2$. This is $(y - k)^2$, with $k = 3$. This is r^2, with $r = 6$.

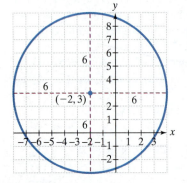

Figure 2.20 The graph of $(x + 2)^2 + (y - 3)^2 = 36$

We use the center, $(h, k) = (-2, 3)$, and the radius, $r = 6$, to graph the circle. The graph is shown in Figure 2.20.

Technology

To graph $x^2 + y^2 + 4x - 6y - 23 = 0$, rewrite the equation as a quadratic equation in y.

$$y^2 - 6y + (x^2 + 4x - 23) = 0$$

Now solve for y using the quadratic formula, with $a = 1, b = -6$, and $c = x^2 + 4x - 23$.

$$y = \frac{-b \pm \sqrt{b^2 - 4ac}}{2a} = \frac{-(-6) \pm \sqrt{(-6)^2 - 4 \cdot 1(x^2 + 4x - 23)}}{2 \cdot 1} = \frac{6 \pm \sqrt{36 - 4(x^2 + 4x - 23)}}{2}$$

Because we will enter these equations, there is no need to simplify. Enter

$$y_1 = \frac{6 + \sqrt{36 - 4(x^2 + 4x - 23)}}{2}$$

and

$$y_2 = \frac{6 - \sqrt{36 - 4(x^2 + 4x - 23)}}{2}.$$

Use a ZOOM SQUARE setting. The graph is shown on the right.

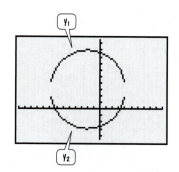

Check Point 6 Write in standard form and graph:

$$x^2 + y^2 + 4x - 4y - 1 = 0.$$

EXERCISE SET 2.2

Practice Exercises

In Exercises 1–18, find the distance between each pair of points. If necessary, round answers to two decimals places.

1. $(2, 3)$ and $(14, 8)$
2. $(5, 1)$ and $(8, 5)$
3. $(4, 1)$ and $(6, 3)$
4. $(2, 3)$ and $(3, 5)$
5. $(0, 0)$ and $(-3, 4)$
6. $(0, 0)$ and $(3, -4)$
7. $(-2, -6)$ and $(3, -4)$
8. $(-4, -1)$ and $(2, -3)$
9. $(0, -3)$ and $(4, 1)$
10. $(0, -2)$ and $(4, 3)$
11. $(3.5, 8.2)$ and $(-0.5, 6.2)$
12. $(2.6, 1.3)$ and $(1.6, -5.7)$
13. $(0, -\sqrt{3})$ and $(\sqrt{5}, 0)$
14. $(0, -\sqrt{2})$ and $(\sqrt{7}, 0)$
15. $(3\sqrt{3}, \sqrt{5})$ and $(-\sqrt{3}, 4\sqrt{5})$
16. $(2\sqrt{3}, \sqrt{6})$ and $(-\sqrt{3}, 5\sqrt{6})$
17. $\left(\frac{7}{3}, \frac{1}{5}\right)$ and $\left(\frac{1}{3}, \frac{6}{5}\right)$
18. $\left(-\frac{1}{4}, -\frac{1}{7}\right)$ and $\left(\frac{3}{4}, \frac{6}{7}\right)$

In Exercises 19–30, find the midpoint of each line segment with the given endpoints.

19. $(6, 8)$ and $(2, 4)$
20. $(10, 4)$ and $(2, 6)$
21. $(-2, -8)$ and $(-6, -2)$
22. $(-4, -7)$ and $(-1, -3)$
23. $(-3, -4)$ and $(6, -8)$
24. $(-2, -1)$ and $(-8, 6)$
25. $\left(-\frac{7}{2}, \frac{3}{2}\right)$ and $\left(-\frac{5}{2}, -\frac{11}{2}\right)$
26. $\left(-\frac{2}{5}, \frac{7}{15}\right)$ and $\left(-\frac{2}{5}, -\frac{4}{15}\right)$
27. $(8, 3\sqrt{5})$ and $(-6, 7\sqrt{5})$
28. $(7\sqrt{3}, -6)$ and $(3\sqrt{3}, -2)$
29. $(\sqrt{18}, -4)$ and $(\sqrt{2}, 4)$
30. $(\sqrt{50}, -6)$ and $(\sqrt{2}, 6)$

In Exercises 31–40, write the standard form of the equation of the circle with the given center and radius.

31. Center $(0, 0), r = 7$
32. Center $(0, 0), r = 8$
33. Center $(3, 2), r = 5$
34. Center $(2, -1), r = 4$
35. Center $(-1, 4), r = 2$
36. Center $(-3, 5), r = 3$
37. Center $(-3, -1), r = \sqrt{3}$

38. Center $(-5, -3)$, $r = \sqrt{5}$

39. Center $(-4, 0)$, $r = 10$

40. Center $(-2, 0)$, $r = 6$

In Exercises 41–48, give the center and radius of the circle described by the equation and graph each equation.

41. $x^2 + y^2 = 16$

42. $x^2 + y^2 = 49$

43. $(x - 3)^2 + (y - 1)^2 = 36$

44. $(x - 2)^2 + (y - 3)^2 = 16$

45. $(x + 3)^2 + (y - 2)^2 = 4$

46. $(x + 1)^2 + (y - 4)^2 = 25$

47. $(x + 2)^2 + (y + 2)^2 = 4$

48. $(x + 4)^2 + (y + 5)^2 = 36$

In Exercises 49–56, complete the square and write the equation in standard form. Then give the center and radius of each circle and graph the equation.

49. $x^2 + y^2 + 6x + 2y + 6 = 0$

50. $x^2 + y^2 + 8x + 4y + 16 = 0$

51. $x^2 + y^2 - 10x - 6y - 30 = 0$

52. $x^2 + y^2 - 4x - 12y - 9 = 0$

53. $x^2 + y^2 + 8x - 2y - 8 = 0$

54. $x^2 + y^2 + 12x - 6y - 4 = 0$

55. $x^2 - 2x + y^2 - 15 = 0$

56. $x^2 + y^2 - 6y - 7 = 0$

 Application Exercises

57. A rectangular coordinate system with coordinates in miles is placed on the map in the figure shown. Bangkok has coordinates $(-115, 170)$ and Phnom Penh has coordinates $(65, 70)$. How long will it take a plane averaging 400 miles per hour to fly directly from one city to the other? Round to the nearest tenth of an hour. Approximately how many minutes is the flight?

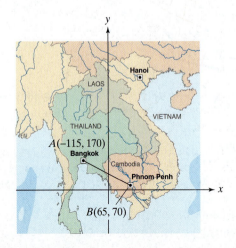

58. We refer to the driveway in the figure shown as being *circular*, meaning that it is bounded by two circles. The figure indicates that the radius of the larger circle is 52 feet and the radius of the smaller circle is 38 feet.

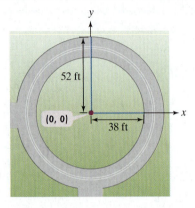

a. Use the coordinate system shown to write the equation of the smaller circle.

b. Use the coordinate system shown to write the equation of the larger circle.

59. The ferris wheel in the figure has a radius of 68 feet. The clearance between the wheel and the ground is 14 feet. The rectangular coordinate system shown has its origin on the ground directly below the center of the wheel. Use the coordinate system to write the equation of the circular wheel.

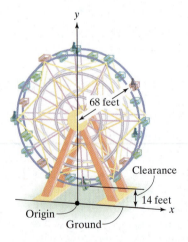

60. The circle formed by the middle lane of a circular running track can be described algebraically by $x^2 + y^2 = 4$, where all measurements are in miles. If you run around the track's middle lane twice, approximately how many miles have you covered?

 Writing in Mathematics

61. In your own words, describe how to find the distance between two points in the rectangular coordinate system.

62. In your own words, describe how to find the midpoint of a line segment if its endpoints are known.

63. What is a circle? Without using variables, describe how the definition of a circle can be used to obtain a form of its equation.

64. Give an example of a circle's equation in standard form. Describe how to find the center and radius for this circle.

65. How is the standard form of a circle's equation obtained from its general form?

66. Does $(x - 3)^2 + (y - 5)^2 = 0$ represent the equation of a circle? If not, describe the graph of this equation.

67. Does $(x - 3)^2 + (y - 5)^2 = -25$ represent the equation of a circle? What sort of set is the graph of this equation?

 Technology Exercises

In Exercises 68–70, use a graphing utility to graph each circle whose equation is given.

68. $x^2 + y^2 = 25$

69. $(y + 1)^2 = 36 - (x - 3)^2$

70. $x^2 + 10x + y^2 - 4y - 20 = 0$

 Critical Thinking Exercises

71. Which one of the following is true?

a. The equation of the circle whose center is at the origin with radius 16 is $x^2 + y^2 = 16$.

b. The graph of $(x - 3)^2 + (y + 5)^2 = 36$ is a circle with radius 6 centered at $(-3, 5)$.

c. The graph of $(x - 4) + (y + 6) = 25$ is a circle with radius 5 centered at $(4, -6)$.

d. None of the above is true.

72. Show that the points $A(1, 1 + d)$, $B(3, 3 + d)$, and $C(6, 6 + d)$ are collinear (lie along a straight line) by showing that the distance from A to B plus the distance from B to C equals the distance from A to C.

73. Prove the midpoint formula by using the following procedure.

a. Show that the distance between (x_1, y_1) and $\left(\dfrac{x_1 + x_2}{2}, \dfrac{y_1 + y_2}{2} \right)$ is equal to the distance between (x_2, y_2) and $\left(\dfrac{x_1 + x_2}{2}, \dfrac{y_1 + y_2}{2} \right)$.

b. Use the procedure from Exercise 72 and the distances from part (a) to show that the points (x_1, y_1), $\left(\dfrac{x_1 + x_2}{2}, \dfrac{y_1 + y_2}{2} \right)$, and (x_2, y_2) are collinear.

In Exercises 74–75, write the standard form and the general form of the equation of each circle.

74. Center at $(3, -5)$ and passing through the point $(-2, 1)$

75. Passing through $(-7, 2)$ and $(1, 2)$; these points are endpoints of the diameter, the line that passes through the circle's center.

76. Find the area of the donut-shaped region bounded by the graphs of $(x - 2)^2 + (y + 3)^2 = 25$ and $(x - 2)^2 + (y + 3)^2 = 36$.

77. A **tangent line** to a circle is a line that intersects the circle at exactly one point. The tangent line is perpendicular to the radius of the circle at this point of contact. Write the point-slope equation of a line tangent to the circle whose equation is $x^2 + y^2 = 25$ at the point $(3, -4)$.

SECTION 2.3 *Basics of Functions*

Objectives

1. Find the domain and range of a relation.

2. Determine whether a relation is a function.

3. Determine whether an equation represents a function.

4. Evaluate a function.

5. Find and simplify a function's difference quotient.

6. Understand and use piecewise functions.

7. Find the domain of a function.

Jerry Orbach	$34,000
Charles Shaugnessy	$31,800
Andy Richter	$29,400
Norman Schwarzkopf	$28,000
Jon Stewart	$28,000

The answer: See the above list. The question: Who are *Celebrity Jeopardy's* five all-time highest earners? The list indicates a correspondence between the five all-time highest earners and their winnings. We can write this correspondence using a set of ordered pairs:

{(Orbach, $34,000), (Shaugnessy, $31,800), (Richter, $29,400), (Schwarzkopf, $28,000), (Stewart, $28,000)}.

1 Find the domain and range of a relation.

The mathematical term for a set of ordered pairs is a *relation*.

> **Definition of a Relation**
> A **relation** is any set of ordered pairs. The set of all first components of the ordered pairs is called the **domain** of the relation, and the set of all second components is called the **range** of the relation.

EXAMPLE 1 Finding the Domain and Range of a Relation

Find the domain and range of the relation:

{(Orbach, $34,000), (Shaugnessy, $31,800), (Richter, $29,400), (Schwarzkopf, $28,000), (Stewart, $28,000)}.

Solution The domain is the set of all first components. Thus, the domain is
{Orbach, Shaugnessy, Richter, Schwarzkopf, Stewart}.

The range is the set of all second components. Thus, the range is
{$34,000, $31,800, $29,400, $28,000}.

Check Point 1 Find the domain and the range of the relation:
{(5, 12.8), (10, 16.2), (15, 18.9), (20, 20.7), (25, 21.8)}.

As you worked Check Point 1, did you wonder if there was a rule that assigned the "inputs" in the domain to the "outputs" in the range? For example, for the ordered pair (15, 18.9), how does the output 18.9 depend on the input 15? Think paid vacation days! The first number in each ordered pair is the number of years a full-time employee has been employed by a medium to large U.S. company. The second number is the average number of paid vacation days each year. Consider, for example, the ordered pair (15, 18.9).

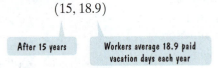

The relation in the vacation-days example can be pictured as follows:

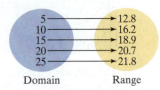

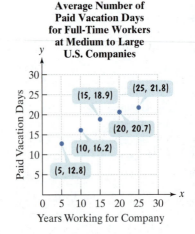

Figure 2.21 The graph of a relation showing a correspondence between years with a company and paid vacation days

Source: Bureau of Labor Statistics

A scatter plot, like the one shown in Figure 2.21, is another way to represent the relation.

2 Determine whether a relation is a function.

Jerry Orbach	$34,000
Charles Shaugnessy	$31,800
Andy Richter	$29,400
Norman Schwarzkopf	$28,000
Jon Stewart	$28,000

Functions

Shown, again, in the margin are *Celebrity Jeopardy's* five all-time highest winners and their winnings. We've used this information to define two relations. Figure 2.22(a) shows a correspondence between winners and their winnings. Figure 2.22(b) shows a correspondence between winnings and winners.

Figure 2.22(a)
Winners correspond to winnings

Figure 2.22(b)
Winnings correspond to winners

A relation in which each member of the domain corresponds to exactly one member of the range is a **function**. Can you see that the relation in Figure 2.22(a) is a function? Each winner in the domain corresponds to exactly one winning amount in the range. If we know the winner, we can be sure of the amount won. Notice that more than one element in the domain can correspond to the same element in the range. (Schwarzkopf and Stewart both won $28,000.)

Is the relation in Figure 2.22(b) a function? Does each member of the domain correspond to precisely one member of the range? This relation is not a function because there is a member of the domain that corresponds to two members of the range:

($28,000, Schwarzkopf)   ($28,000, Stewart).

The member of the domain, $28,000, corresponds to both Schwarzkopf and Stewart in the range. If we know the amount won, $28,000, we cannot be sure of the winner. Because **a function is a relation in which no two ordered pairs have the same first component and different second components,** the ordered pairs ($28,000, Schwarzkopf) and ($28,000, Stewart) are not ordered pairs of a function.

Same first component

($28,000, Schwarzkopf)   ($28,000, Stewart)

Different second components

Definition of a Function

A **function** is a correspondence from a first set, called the **domain,** to a second set, called the **range,** such that each element in the domain corresponds to *exactly one* element in the range.

Example 2 illustrates that not every correspondence between sets is a function.

EXAMPLE 2   Determining Whether a Relation is a Function

Determine whether each relation is a function:

    a. $\{(1, 6), (2, 6), (3, 8), (4, 9)\}$ **b.** $\{(6, 1), (6, 2), (8, 3), (9, 4)\}$.

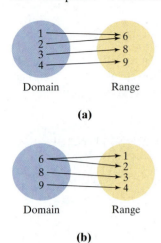

Domain   Range

(a)

Domain   Range

(b)

Figure 2.23

Study Tip

The word "range" can mean many things, from a chain of mountains to a cooking stove. For functions, it means the set of all function values. For graphing utilities, it means the setting used for the viewing rectangle. Try not to confuse these meanings.

3 Determine whether an equation represents a function.

Solution   We will make a figure for each relation that shows the domain and the range.

a. We begin with the relation $\{(1, 6), (2, 6), (3, 8), (4, 9)\}$. Figure 2.23(a) shows that every element in the domain corresponds to exactly one element in the range. The element 1 in the domain corresponds to the element 6 in the range. Furthermore, 2 corresponds to 6, 3 corresponds to 8, and 4 corresponds to 9. No two ordered pairs in the given relation have the same first component and different second components. Thus, the relation is a function.

b. We now consider the relation $\{(6, 1), (6, 2), (8, 3), (9, 4)\}$. Figure 2.23(b) shows that 6 corresponds to both 1 and 2. If any element in the domain corresponds to more than one element in the range, the relation is not a function. This relation is not a function; two ordered pairs have the same first component and different second components.

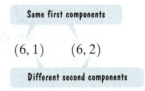

Same first components

$(6, 1) \qquad (6, 2)$

Different second components

Look at Figure 2.23 again. The fact that 1 and 2 in the domain have the same image, 6, in the range does not violate the definition of a function. **A function can have two different first components with the same second component.** By contrast, a relation is not a function when two different ordered pairs have the same first component and different second components. Thus, the relation in Example 2(b) is not a function.

Check Point 2   Determine whether each relation is a function:
a. $\{(1, 2), (3, 4), (5, 6), (5, 8)\}$
b. $\{(1, 2), (3, 4), (6, 5), (8, 5)\}$.

Functions as Equations

Functions are usually given in terms of equations rather than as sets of ordered pairs. For example, here is an equation that models paid vacation days each year as a function of years working for a company:

$$y = -0.016x^2 + 0.93x + 8.5.$$

The variable x represents years working for a company. The variable y represents the average number of vacation days each year. The variable y is a function of the variable x. For each value of x, there is one and only one value of y. The variable x is called the **independent variable** because it can be assigned any value from the domain. Thus, x can be assigned any positive integer representing the number of years working for a company. The variable y is called the **dependent variable** because its value depends on x. Paid vacation days depend on years working for a company. The value of the dependent variable, y, is calculated after selecting a value for the independent variable, x.

We have seen that not every set of ordered pairs defines a function. Similarly, not all equations with the variables x and y define a function. If an equation is solved for y and more than one value of y can be obtained for a given x, then the equation does not define y as a function of x.

EXAMPLE 3 **Determining Whether an Equation Represents a Function**

Determine whether each equation defines y as a function of x:

a. $x^2 + y = 4$ **b.** $x^2 + y^2 = 4$.

Solution   Solve each equation for y in terms of x. If two or more values of y can be obtained for a given x, the equation is not a function.

a. $x^2 + y = 4$ This is the given equation.

$x^2 + y - x^2 = 4 - x^2$ Solve for y by subtracting x^2 from both sides.

$y = 4 - x^2$ Simplify.

From this last equation we can see that for each value of x, there is one and only one value of y. For example, if $x = 1$, then $y = 4 - 1^2 = 3$. The equation defines y as a function of x.

b. $x^2 + y^2 = 4$ This given equation describes a circle.

$x^2 + y^2 - x^2 = 4 - x^2$ Isolate y^2 by subtracting x^2 from both sides.

$y^2 = 4 - x^2$ Simplify.

$y = \pm\sqrt{4 - x^2}$ Apply the square root method.

The $\pm$ in this last equation shows that for certain values of x (all values between -2 and 2), there are two values of y. For example, if $x = 1$, then $y = \pm\sqrt{4 - 1^2} = \pm\sqrt{3}$. For this reason, the equation does not define y as a function of x.

Check Point 3   Solve each equation for y and then determine whether the equation defines y as a function of x:
a. $2x + y = 6$ **b.** $x^2 + y^2 = 1$.

---

4 Evaluate a function.

Function Notation

When an equation represents a function, the function is often named by a letter such as $f, g, h, F, G,$ or H. Any letter can be used to name a function. Suppose that f names a function. Think of the domain as the set of the function's inputs and the range as the set of the function's outputs. As shown in Figure 2.24, the input is represented by x and the output by $f(x)$. The special notation **$f(x)$**, read "f of x" or "f at x," represents the **value of the function at the number x.**

Study Tip

The notation $f(x)$ does *not* mean "f times x." The notation describes the value of the function at x.

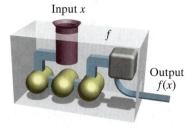

Input x

Output $f(x)$

Figure 2.24 A function as a machine with inputs and outputs

Let's make this clearer by considering a specific example. We know that the equation

$$y = -0.016x^2 + 0.93x + 8.5$$

defines y as a function of x. We'll name the function f. Now, we can apply our new function notation.

Input	Output	Equation	We read this equation as "f of x equals $-0.016x^2 + 0.93x + 8.5$."
x	$f(x)$	$f(x) = -0.016x^2 + 0.93x + 8.5$	

Suppose we are interested in finding $f(10)$, the function's output when the input is 10. To find the value of the function at 10, we substitute 10 for x. We are **evaluating the function** at 10.

$f(x) = -0.016x^2 + 0.93x + 8.5$ This is the given function.

$f(10) = -0.016(10)^2 + 0.93(10) + 8.5$ Replace each occurrence of x with 10.

$= -0.016(100) + 0.93(10) + 8.5$ Evaluate the exponential expression: $10^2 = 100$.

$= -1.6 + 9.3 + 8.5$ Perform the multiplications.

$= 16.2$ Add from left to right.

The statement $f(10) = 16.2$, read "f of 10 equals 16.2," tells us that the value of the function at 10 is 16.2. When the function's input is 10, its output is 16.2 (After 10 years, workers average 16.2 vacation days each year.) To find other function values, such as $f(15)$, $f(20)$, or $f(23)$, substitute the specified input values for x into the function's equation.

If a function is named f and x represents the independent variable, the notation $f(x)$ corresponds to the y-value for a given x. Thus,

$$f(x) = -0.016x^2 + 0.93x + 8.5 \quad \text{and} \quad y = -0.016x^2 + 0.93x + 8.5$$

define the same function. This function may be written as

$$y = f(x) = -0.016x^2 + 0.93x + 8.5.$$

Technology

Graphing utilities can be used to evaluate functions. The screens below show the evaluation of

$$f(x) = -0.016x^2 + 0.93x + 8.5$$

at 10 on a T1-83 graphing calculator. The function f is named Y_1.

```
Plot1 Plot2 Plot3
\Y1 ■ -.016X²+0.93
X+8.5■
∖Y2=
∖Y3=
∖Y4=
∖Y5=
∖Y6=
```

```
Y1(10)
            16.2
```

EXAMPLE 4   **Evaluating a Function**

If $f(x) = x^2 + 3x + 5$, evaluate:

   a. $f(2)$ **b.** $f(x + 3)$ **c.** $f(-x)$.

Solution   We substitute 2, $x + 3$, and $-x$ for x in the definition of f. When replacing x with a variable or an algebraic expression, you might find it helpful to think of the function's equation as

$$f(\boxed{x}) = \boxed{x}^2 + 3\boxed{x} + 5.$$

a. We find $f(2)$ by substituting 2 for x in the equation.

$$f(\boxed{2}) = \boxed{2}^2 + 3 \cdot \boxed{2} + 5 = 4 + 6 + 5 = 15$$

Thus, $f(2) = 15$.

b. We find $f(x + 3)$ by substituting $x + 3$ for x in the equation.

$$f(\boxed{x + 3}) = \boxed{(x + 3)}^2 + 3\boxed{(x + 3)} + 5$$

Equivalently,

$$f(x + 3) = (x + 3)^2 + 3(x + 3) + 5$$

$$= x^2 + 6x + 9 + 3x + 9 + 5$$ Square $x + 3$ using $(A + B)^2 = A^2 + 2AB + B^2$.

Distribute 3 throughout the parentheses.

$$= x^2 + 9x + 23.$$ Combine like terms.

c. We find $f(-x)$ by substituting $-x$ for x in the equation.

$$f(\;-x\;) = \;(-x)^2 + 3\;(-x)\; + 5$$

Equivalently,

$$f(-x) = (-x)^2 + 3(-x) + 5$$
$$= x^2 - 3x + 5.$$

Discovery

Using $f(x) = x^2 + 3x + 5$ and the answers in parts (b) and (c):

1. Is $f(x + 3)$ equal to $f(x) + f(3)$?
2. Is $f(-x)$ equal to $-f(x)$?

Check Point 4 If $f(x) = x^2 - 2x + 7$, evaluate:

a. $f(-5)$ **b.** $f(x + 4)$ **c.** $f(-x)$.

5 Find and simplify a function's difference quotient.

Functions and Difference Quotients

We have seen how slope can be interpreted as a rate of change. In the next section, we will be studying the average rate of change of a function. A ratio, called the *difference quotient*, plays an important role in understanding the rate at which functions change.

Definition of a Difference Quotient

The expression

$$\frac{f(x + h) - f(x)}{h}$$

for $h \neq 0$ is called the **difference quotient**.

EXAMPLE 5 Evaluating and Simplifying a Difference Quotient

If $f(x) = x^2 + 3x + 5$, find and simplify:

a. $f(x + h)$ **b.** $\dfrac{f(x + h) - f(x)}{h}, h \neq 0.$

Solution

a. We find $f(x + h)$ by replacing x with $x + h$ each time that x appears in the equation.

$$f(x) \quad = \quad x^2 \quad + \quad 3x \quad + \quad 5$$

Replace x with $x + h$.   Replace x with $x + h$.   Replace x with $x + h$.   Copy the 5. There is no x in this term.

$$f(x + h) \quad = \quad (x + h)^2 \quad + \quad 3(x + h) \quad + \quad 5$$

$$= x^2 + 2xh + h^2 + 3x + 3h \quad + \quad 5$$

b. Using our result from part (a), we obtain the following:

> This is f(x + h) from part (a).

> This is f(x) from the given equation.

$$\frac{f(x + h) - f(x)}{h} = \frac{x^2 + 2xh + h^2 + 3x + 3h + 5 - (x^2 + 3x + 5)}{h}$$

$$= \frac{x^2 + 2xh + h^2 + 3x + 3h + 5 - x^2 - 3x - 5}{h}$$

Remove parentheses and change the sign of each term in the parentheses.

$$= \frac{(x^2 - x^2) + (3x - 3x) + (5 - 5) + 2xh + h^2 + 3h}{h}$$

Group like terms.

$$= \frac{2xh + h^2 + 3h}{h}$$

Simplify.

$$= \frac{h(2x + h + 3)}{h}$$

Factor h from the numerator.

$$= 2x + h + 3, h \neq 0.$$

Cancel identical factors of h in the numerator and denominator.

Check Point 5   If $f(x) = x^2 - 7x + 3$, find and simplify:

a. $f(x + h)$ **b.** $\dfrac{f(x + h) - f(x)}{h}, h \neq 0.$

6   Understand and use piecewise functions.

Piecewise Functions

The early part of the twentieth century was the golden age of immigration in America. More than 13 million people migrated to the United States between 1900 and 1914. By 1910, foreign-born residents accounted for 15% of the total U.S. population. The graph in Figure 2.25 shows the percentage of Americans who were foreign born throughout the twentieth century.

We can model the data from 1910 through 2000 with two equations, one from 1910 through 1970, years in which the percentage was decreasing, and one from 1970 through 2000, years in which the percentage was increasing. These two trends can be approximated by the function

$$P(t) = \begin{cases} -\dfrac{11}{60}t + 15 & \text{if } 0 \leq t < 60 \\[2mm] \dfrac{1}{5}t - 8 & \text{if } 60 \leq t \leq 90 \end{cases}$$

Percentage of Americans Who Were Foreign Born in the Twentieth Century

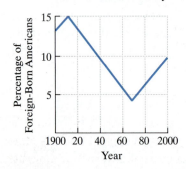

Figure 2.25 *Source*: U.S. Census Bureau

in which t represents the number of years after 1910 and $P(t)$ is the percentage of foreign-born Americans. A function that is defined by two (or more) equations over a specified domain is called a **piecewise function.**

EXAMPLE 6   Evaluating a Piecewise Function

Use the function $P(t)$, described previously, to find and interpret:

   a. $P(30)$ **b.** $P(80)$.

Solution

   a. To find $P(30)$, we let $t = 30$. Because 30 is less than 60, we use the first line of the piecewise function.

$$P(t) = -\tfrac{11}{60}t + 15 \qquad \text{This is the function's equation for } 0 \le t < 60.$$

$$P(30) = -\tfrac{11}{60} \cdot 30 + 15 \qquad \text{Replace } t \text{ with 60.}$$

$$= 9.5$$

   This means that 30 years after 1910, in 1940, 9.5% of Americans were foreign born.

   b. To find $P(80)$, we let $t = 80$. Because 80 is between 60 and 90, we use the second line of the piecewise function.

$$P(t) = \frac{1}{5}t - 8 \qquad \text{This is the function's equation for } 60 \le t \le 90.$$

$$P(80) = \tfrac{1}{5} \cdot 80 - 8 \qquad \text{Replace } t \text{ with 80.}$$

$$= 8$$

   This means that 80 years after 1910, in 1990, 8% of Americans were foreign born.

Check Point 6   If $f(x) = \begin{cases} x^2 + 3 & \text{if } x < 0 \\ 5x + 3 & \text{if } x \ge 0 \end{cases}$, find:

   a. $f(-5)$ **b.** $f(6)$.

7 Find the domain of a function.

The Domain of a Function

Let's reconsider the function that models the percentage of foreign-born Americans t years after 1910, up through and including 2000. The domain of this function is

$$\{0, \quad 1, \quad 2, \quad 3, \quad \ldots, \quad 90\}.$$

0 years after 1910 is 1910.   3 years after 1910 is 1913.   90 years after 1910 brings the domain up to the year 2000.

Functions that model data often have their domains explicitly given along with the function's equation. However, for most functions, only an equation is given, and the domain is not specified. In cases like this, the domain of f is the largest set of real numbers for which the value of $f(x)$ is a real number. For example, consider the function

$$f(x) = \frac{1}{x - 3}.$$

Because division by 0 is undefined (and not a real number), the denominator $x - 3$ cannot be 0. Thus, x cannot equal 3. The domain of the function consists of all real numbers other than 3, represented by $\{x \mid x \neq 3\}$. We say that f is not defined at 3, or $f(3)$ does not exist.

Just as the domain of a function must exclude real numbers that cause division by zero, it must also exclude real numbers that result in an even root of a negative number. For example, consider the function

$$g(x) = \sqrt{x}.$$

The equation tells us to take the square root of x. Because only nonnegative numbers have real square roots, the expression under the radical sign, x, must be greater than or equal to 0. The domain of g is $\{x \mid x \geq 0\}$, or the interval $[0, \infty)$.

Finding a Function's Domain

If a function f does not model data or verbal conditions, its domain is the largest set of real numbers for which the value of $f(x)$ is a real number. Exclude from a function's domain real numbers that cause division by zero and real numbers that result in an even root of a negative number.

EXAMPLE 7   Finding the Domain of a Function

Find the domain of each function:

a. $f(x) = x^2 - 7x$ **b.** $g(x) = \dfrac{6x}{x^2 - 9}$ **c.** $h(x) = \sqrt{3x + 12}.$

Solution

a. The function $f(x) = x^2 - 7x$ contains neither division nor an even root. The domain of f is the set of all real numbers.

b. The function $g(x) = \dfrac{6x}{x^2 - 9}$ contains division. Because division by 0 is undefined, we must exclude from the domain values of x that cause $x^2 - 9$ to be 0. Thus, x cannot equal -3 or 3. The domain of g is $\{x \mid x \neq -3, x \neq 3\}$.

c. The function $h(x) = \sqrt{3x + 12}$ contains an even root. Because only nonnegative numbers have real square roots, the quantity under the radical sign, $3x + 12$, must be greater than or equal to 0.

$$3x + 12 \geq 0$$

$$3x \geq -12$$

$$x \geq -4$$

The domain of h is $\{x \mid x \geq -4\}$, or the interval $[-4, \infty)$.

Technology

You can graph a function and often get hints about its domain. For example, $h(x) = \sqrt{3x + 12}$, or $y = \sqrt{3x + 12}$, appears only for $x \geq -4$, verifying $[-4, \infty)$ as the domain.

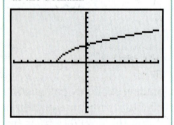

Check Point 7   Find the domain of each function:

a. $f(x) = x^2 + 3x - 17$ **b.** $g(x) = \dfrac{5x}{x^2 - 49}$

c. $h(x) = \sqrt{9x - 27}.$

EXERCISE SET 2.3

 Practice Exercises

In Exercises 1–8, determine whether each relation is a function. Give the domain and range for each relation.

1. $\{(1, 2), (3, 4), (5, 5)\}$ **2.** $\{(4, 5), (6, 7), (8, 8)\}$

3. $\{(3, 4), (3, 5), (4, 4), (4, 5)\}$

4. $\{(5, 6), (5, 7)\ (6, 6), (6, 7)\}$

5. $\{(-3, -3), (-2, -2), (-1, -1), (0, 0)\}$

6. $\{(-7, -7), (-5, -5), (-3, -3), (0, 0)\}$

7. $\{(1, 4), (1, 5), (1, 6)\}$ **8.** $\{(4, 1), (5, 1), (6, 1)\}$

In Exercises 9–20, determine whether each equation defines y as a function of x.

9. $x + y = 16$ **10.** $x + y = 25$

11. $x^2 + y = 16$ **12.** $x^2 + y = 25$

13. $x^2 + y^2 = 16$ **14.** $x^2 + y^2 = 25$

15. $x = y^2$ **16.** $4x = y^2$

17. $y = \sqrt{x + 4}$ **18.** $y = -\sqrt{x + 4}$

19. $x + y^3 = 8$ **20.** $x + y^3 = 27$

In Exercises 21–32, evaluate each function at the given values of the independent variable and simplify.

21. $f(x) = 4x + 5$
   a. $f(6)$ **b.** $f(x + 1)$ **c.** $f(-x)$

22. $f(x) = 3x + 7$
   a. $f(4)$ **b.** $f(x + 1)$ **c.** $f(-x)$

23. $g(x) = x^2 + 2x + 3$
   a. $g(-1)$ **b.** $g(x + 5)$ **c.** $g(-x)$

24. $g(x) = x^2 - 10x - 3$
   a. $g(-1)$ **b.** $g(x + 2)$ **c.** $g(-x)$

25. $h(x) = x^4 - x^2 + 1$
   a. $h(2)$ **b.** $h(-1)$
   c. $h(-x)$ **d.** $h(3a)$

26. $h(x) = x^3 - x + 1$
   a. $h(3)$ **b.** $h(-2)$
   c. $h(-x)$ **d.** $h(3a)$

27. $f(r) = \sqrt{r + 6} + 3$
   a. $f(-6)$ **b.** $f(10)$ **c.** $f(x - 6)$

28. $f(r) = \sqrt{25 - r} - 6$
   a. $f(16)$ **b.** $f(-24)$ **c.** $f(25 - 2x)$

29. $f(x) = \dfrac{4x^2 - 1}{x^2}$
   a. $f(2)$ **b.** $f(-2)$ **c.** $f(-x)$

30. $f(x) = \dfrac{4x^3 + 1}{x^3}$
   a. $f(2)$ **b.** $f(-2)$ **c.** $f(-x)$

31. $f(x) = \dfrac{x}{|x|}$
   a. $f(6)$ **b.** $f(-6)$ **c.** $f(r^2)$

32. $f(x) = \dfrac{|x + 3|}{x + 3}$
   a. $f(5)$ **b.** $f(-5)$ **c.** $f(-9 - x)$

In Exercises 33–44, find and simplify the difference quotient
$$\frac{f(x + h) - f(x)}{h}, \quad h \neq 0$$
for the given function.

33. $f(x) = 4x$ **34.** $f(x) = 7x$

35. $f(x) = 3x + 7$ **36.** $f(x) = 6x + 1$

37. $f(x) = x^2$ **38.** $f(x) = 2x^2$

39. $f(x) = x^2 - 4x + 3$ **40.** $f(x) = x^2 - 5x + 8$

41. $f(x) = 6$ **42.** $f(x) = 7$

43. $f(x) = \dfrac{1}{x}$ **44.** $f(x) = \dfrac{1}{2x}$

In Exercises 45–50, evaluate each piecewise function at the given values of the independent variable.

45. $f(x) = \begin{cases} 3x + 5 & \text{if } x < 0 \\ 4x + 7 & \text{if } x \geq 0 \end{cases}$
   a. $f(-2)$ **b.** $f(0)$ **c.** $f(3)$

46. $f(x) = \begin{cases} 6x - 1 & \text{if } x < 0 \\ 7x + 3 & \text{if } x \geq 0 \end{cases}$
   a. $f(-3)$ **b.** $f(0)$ **c.** $f(4)$

47. $g(x) = \begin{cases} x + 3 & \text{if } x \geq -3 \\ -(x + 3) & \text{if } x < -3 \end{cases}$
   a. $g(0)$ **b.** $g(-6)$ **c.** $g(-3)$

48. $g(x) = \begin{cases} x + 5 & \text{if } x \geq -5 \\ -(x + 5) & \text{if } x < -5 \end{cases}$
   a. $g(0)$ **b.** $g(-6)$ **c.** $g(-5)$

49. $h(x) = \begin{cases} \dfrac{x^2 - 9}{x - 3} & \text{if } x \neq 3 \\ 6 & \text{if } x = 3 \end{cases}$
   a. $h(5)$ **b.** $h(0)$ **c.** $h(3)$

50. $h(x) = \begin{cases} \dfrac{x^2 - 25}{x - 5} & \text{if } x \neq 5 \\ 10 & \text{if } x = 5 \end{cases}$
   a. $h(7)$ **b.** $h(0)$ **c.** $h(5)$

In Exercises 51–74, find the domain of each function.

51. $f(x) = 4x^2 - 3x + 1$ **52.** $f(x) = 8x^2 - 5x + 2$

53. $g(x) = \dfrac{3}{x - 4}$ **54.** $g(x) = \dfrac{2}{x + 5}$

55. $h(x) = \dfrac{7x}{x^2 - 16}$ **56.** $h(x) = \dfrac{12x}{x^2 - 36}$

57. $f(x) = \dfrac{2}{(x + 3)(x - 7)}$ **58.** $f(x) = \dfrac{15}{(x + 8)(x - 3)}$

59. $H(r) = \dfrac{4}{r^2 + 11r + 24}$ **60.** $H(r) = \dfrac{5}{6r^2 + r - 2}$

61. $f(t) = \dfrac{3}{t^2 + 4}$ **62.** $f(t) = \dfrac{5}{t^2 + 9}$

63. $f(x) = \sqrt{x - 3}$ **64.** $f(x) = \sqrt{x + 2}$

65. $f(x) = \dfrac{1}{\sqrt{x - 3}}$ **66.** $f(x) = \dfrac{1}{\sqrt{x + 2}}$

67. $g(x) = \sqrt{5x + 35}$ **68.** $g(x) = \sqrt{7x - 70}$

69. $f(x) = \sqrt{24 - 2x}$ **70.** $f(x) = \sqrt{84 - 6x}$

71. $f(x) = \sqrt{x^2 - 5x - 14}$ **72.** $f(x) = \sqrt{x^2 - 5x - 24}$

73. $f(x) = \dfrac{\sqrt{x - 2}}{x - 5}$ **74.** $f(x) = \dfrac{\sqrt{x - 3}}{x - 6}$

Application Exercises

75. The bar graph shows the percentage of people in the United States using the Internet by education level. Write five ordered pairs for

(education level, percentage using Internet)

as a relation. Find the domain and the range of the relation. Is this relation a function? Explain your answer.

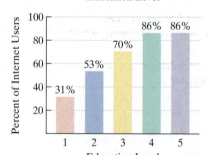

Internet Use in the U.S. by Education Level

Percent of Internet Users — Education Level
(1 = less than high school, 2 = high school graduate, 3 = some college, 4 = college graduate, 5 = advanced degree)

31%, 53%, 70%, 86%, 86%

Source: U.C.L.A. Center for Communication Policy

The table shows the ten longest-running television shows of the twentieth century. Use the information in the table to solve Exercises 76–78.

Ten Longest-Running National Network TV Series of the Twentieth Century

Program	Number of Seasons the Show Ran
"Walt Disney"	33
"60 Minutes"	33
"The Ed Sullivan Show"	24
"Gunsmoke"	20
"The Red Skelton Show"	20
"Meet the Press"	18
"What's My Line?"	18
"I've Got a Secret"	17
"Lassie"	17
"The Lawrence Welk Show"	17

Source: Nielsen Media Research

76. Consider the relation for which the domain represents the ten longest-running series and the range represents the number of seasons the series ran. Is this relation a function? Explain your answer.

77. Consider the relation for which the domain represents the number of seasons the ten longest-running series ran and the range represents the ten longest-running series. Is this relation a function? Explain your answer.

78. Use your answers from Exercises 76 and 77 to answer the following question: If the components in a function's ordered pairs are reversed, must the resulting relation also be a function?

79. The function

$$P(x) = 0.72x^2 + 9.4x + 783$$

models the gray wolf population in the United States, $P(x)$, x years after 1960. Find and interpret $P(30)$. How well does the function model the actual value shown in the bar graph?

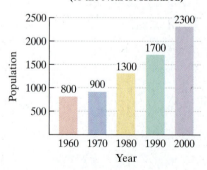

Gray Wolf Population (to the Nearest Hundred)

Population — Year

800 (1960), 900 (1970), 1300 (1980), 1700 (1990), 2300 (2000)

Source: U.S. Department of the Interior

80. As the use of the Internet increases, so has the number of computer infections from viruses. The function

$$N(x) = 0.2x^2 - 1.2x + 2$$

models the number of infections per month for every 1000 computers, $N(x)$, x years after 1990. Find and interpret $N(10)$. How well does the function model the actual value shown in the bar graph?

Computer Infection Rates

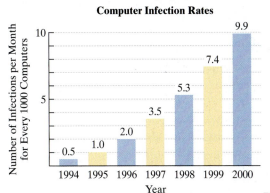

Source: Jupiter Communications

The function

$$P(t) = 6.85\sqrt{t} + 19$$

models the percentage of U.S. households online, P(t), t years after 1997. In Exercises 81–84, use this function to find and interpret the given expression. If necessary, use a calculator and round to the nearest whole percent. How well does the function model the actual data shown in the bar graph?

Percentage of U.S. Households Online

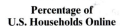

Source: Forrester Research

81. $P(0)$ **82.** $P(4)$

83. $P(3) - P(1)$ **84.** $P(2) - P(1)$

The number of lawyers in the United States can be modeled by the function

$$f(x) = \begin{cases} 6.5x + 200 & \text{if } 0 \le x < 23 \\ 26.2x - 252 & \text{if } x \ge 23 \end{cases}$$

where x represents the number of years after 1951 and f(x) represents the number of lawyers, in thousands. In Exercises 85–88, use this function to find and interpret each of the following.

85. $f(0)$ **86.** $f(10)$

87. $f(50)$ **88.** $f(60)$

During a particular year, the taxes owed, T(x), in dollars, filing separately with an adjusted gross income of x dollars is given by the piecewise function

$$T(x) = \begin{cases} 0.15x & \text{if } 0 \le x < 17{,}900 \\ 0.28(x - 17{,}900) + 2685 & \text{if } 17{,}900 \le x < 43{,}250 \\ 0.31(x - 43{,}250) + 9783 & \text{if } x \ge 43{,}250 \end{cases}$$

In Exercises 89–90, use this function to find and interpret each of the following.

89. $T(40{,}000)$ **90.** $T(70{,}000)$

In Exercises 91–94, you will be developing functions that model given conditions.

91. A company that manufactures bicycles has a fixed cost of $100,000. It costs $100 to produce each bicycle. The total cost for the company is the sum of its fixed cost and variable costs. Write the total cost, C, as a function of the number of bicycles produced. Then find and interpret $C(90)$.

92. A car was purchased for $22,500. The value of the car decreases by $3200 per year for the first six years. Write a function that describes the value of the car, V, after x years, where $0 \le x \le 7$. Then find and interpret $V(3)$.

93. You commute to work a distance of 40 miles and return on the same route at the end of the day. Your average rate on the return trip is 30 miles per hour faster than your average rate on the outgoing trip. Write the total time, T, in hours, devoted to your outgoing and return trips as a function of your rate on the outgoing trip. Then find and interpret $T(30)$. Hint:

$$\text{Time traveled} = \frac{\text{Distance traveled}}{\text{Rate of travel}}.$$

94. A chemist working on a flu vaccine needs to mix a 10% sodium-iodine solution with a 60% sodium-iodine solution to obtain a 50-milliliter mixture. Write the amount of sodium iodine in the mixture, S, in milliliters, as a function of the number of milliliters of the 10% solution used. Then find and interpret $S(30)$.

Writing in Mathematics

95. If a relation is represented by a set of ordered pairs, explain how to determine whether the relation is a function.

96. How do you determine if an equation in x and y defines y as a function of x?

97. A student in introductory algebra hears that functions are studied in subsequent algebra courses. The student asks you what a function is. Provide the student with a clear, relatively concise response.

98. Describe one advantage of using $f(x)$ rather than y in a function's equation.

99. Explain how to find the difference quotient,

$$\frac{f(x + h) - f(x)}{h}, \text{ if a function's equation is given.}$$

100. What is a piecewise function?

101. How is the domain of a function determined?

102. For people filing a single return, federal income tax is a function of adjusted gross income because for each value of adjusted gross income there is a specific tax to be paid. On the other hand, the price of a house is not a function of the lot size on which the house sits because houses on same-sized lots can sell for many different prices.

  a. Describe an everyday situation between variables that is a function.

  b. Describe an everyday situation between variables that is not a function.

Technology Exercises

Use a graphing utility to find the domain of each function in Exercises 103–105. Then verify your observation algebraically.

103. $f(x) = \sqrt{x - 1}$ **104.** $g(x) = \sqrt{2x + 6}$

105. $h(x) = \sqrt{15 - 3x}$

Critical Thinking Exercises

106. Write a function defined by an equation in x whose domain is $\{x \mid x \neq -4, x \neq 11\}$.

107. Write a function defined by an equation in x whose domain is $[-6, \infty)$.

108. Give an example of an equation that does not define y as a function of x but that does define x as a function of y.

109. If $f(x) = ax^2 + bx + c$ and $r_1 = \dfrac{-b + \sqrt{b^2 - 4ac}}{2a}$,

find $f(r_1)$ without doing any algebra and explain how you arrived at your result.

Group Exercise

110. Almanacs, newspapers, magazines, and the Internet contain bar graphs and line graphs that describe how things are changing over time. For example, the graphs in Exercises 79–82 show how various phenomena are changing over time. Find a bar or line graph showing yearly changes that you find intriguing. Describe to the group what interests you about this data. The group should select their two favorite graphs. For each graph selected:

  a. Rewrite the data so that they are presented as a relation in the form of a set of ordered pairs.

  b. Determine whether the relation in part (a) is a function. Explain why the relation is a function, or why it is not.

SECTION 2.4 *Graphs of Functions*

Objectives

1. Graph functions by plotting points.

2. Obtain information about a function from its graph.

3. Use the vertical line test to identify functions.

4. Identify intervals on which a function increases, decreases, or is constant.

5. Use graphs to locate relative maxima or minima.

6. Find a function's average rate of change.

7. Identify even or odd functions and recognize their symmetries.

8. Graph step functions.

Have you ever seen a gas-guzzling car from the 1950s, with its huge fins and overstated design? The worst year for automobile fuel efficiency was 1958, when cars averaged a dismal 12.4 miles per gallon. The function

$$f(x) = 0.0075x^2 - 0.2672x + 14.8$$

models the average number of miles per gallon for U.S. automobiles, $f(x)$, x years after 1940. If we could see the graph of the function's equation, we would get a much better idea of the relationship between time and fuel efficiency. In this section, we will learn how to use the graph of a function to obtain useful information about the function.

Graphs of Functions

A graph enables us to visualize a function's behavior. The graph shows the relationship between the function's two variables more clearly than the function's equation does. The **graph of a function** is the graph of its ordered pairs. For example, the graph of $f(x) = \sqrt{x}$ is the set of points (x, y) in the rectangular coordinate system satisfying the equation $y = \sqrt{x}$. Thus, one way to graph a function is by plotting several of its ordered pairs and drawing a line or smooth curve through them. With the function's graph, we can picture its domain on the x-axis and its range on the y-axis. Our first example illustrates how this is done.

1   Graph functions by plotting points.

EXAMPLE 1   Graphing a Function by Plotting Points

Graph $f(x) = x^2 + 1$. To do so, use integer values of x from the set $\{-3, -2, -1, 0, 1, 2, 3\}$ to obtain seven ordered pairs. Plot each ordered pair and draw a smooth curve through the points. Use the graph to specify the function's domain and range.

Solution   The graph of $f(x) = x^2 + 1$ is, by definition, the graph of $y = x^2 + 1$. We begin by setting up a partial table of coordinates.

x	$f(x) = x^2 + 1$	(x, y) or $(x, f(x))$
-3	$f(-3) = (-3)^2 + 1 = 10$	$(-3, 10)$
-2	$f(-2) = (-2)^2 + 1 = 5$	$(-2, 5)$
-1	$f(-1) = (-1)^2 + 1 = 2$	$(-1, 2)$
0	$f(0) = 0^2 + 1 = 1$	$(0, 1)$
1	$f(1) = 1^2 + 1 = 2$	$(1, 2)$
2	$f(2) = 2^2 + 1 = 5$	$(2, 5)$
3	$f(3) = 3^2 + 1 = 10$	$(3, 10)$

Now, we plot the seven points and draw a smooth curve through them, as shown in Figure 2.26. The graph of f has a cuplike shape. The points on the graph of f have x-coordinates that extend indefinitely to the left and to the right. Thus, the domain consists of all real numbers, represented by $(-\infty, \infty)$. By contrast, the points on the graph have y-coordinates that start at 1 and extend indefinitely upward. Thus, the range consists of all real numbers greater than or equal to 1, represented by $[1, \infty)$.

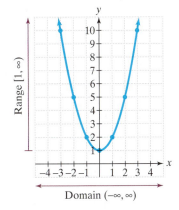

Figure 2.26 The graph of $f(x) = x^2 + 1$

Check Point 1   Graph $f(x) = x^2 - 2$, using integers from -3 to 3 for x in the partial table of coordinates. Use the graph to specify the function's domain and range.

Technology

Does your graphing utility have a $\boxed{\text{TABLE}}$ feature? If so, you can use it to create tables of coordinates for a function. You will need to enter the equation of the function and specify the starting value for x, $\boxed{\text{TblStart}}$, and the increment between successive x-values, $\boxed{\Delta\text{Tbl}}$. For the table of coordinates in Example 1, we start the table at $x = -3$ and increment by 1. Using the up- or down-arrow keys, you can scroll through the table and determine as many ordered pairs of the graph as desired.

2 Obtain information about a function from its graph.

Obtaining Information from Graphs

You can obtain information about a function from its graph. At the right or left of a graph, you will find closed dots, open dots, or arrows.

- A closed dot indicates that the graph does not extend beyond this point and the point belongs to the graph.

- An open dot indicates that the graph does not extend beyond this point and the point does not belong to the graph.

- An arrow indicates that the graph extends indefinitely in the direction in which the arrow points.

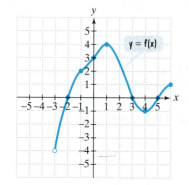

Figure 2.27

EXAMPLE 2   Obtaining Information from a Function's Graph

Use the graph of the function f, shown in Figure 2.27, to answer the following questions:

a. What are the function values $f(-1)$ and $f(1)$?

b. What is the domain of f?

c. What is the range of f?

Solution

a. Because $(-1, 2)$ is a point on the graph of f, the y-coordinate, 2, is the value of the function at the x-coordinate, -1. Thus, $f(-1) = 2$. Similarly, because $(1, 4)$ is also a point on the graph of f, this indicates that $f(1) = 4$.

b. The open dot on the left shows that $x = -3$ is not in the domain of f. By contrast, the closed dot on the right shows that $x = 6$ is in the domain of f. We determine the domain of f by noticing that the points on the graph of f have x-coordinates between -3, excluding -3, and 6, including 6. For each number x between -3 and 6, there is a point $(x, f(x))$ on the graph. Thus, the domain of f is $\{x \mid -3 < x \le 6\}$, or the interval $(-3, 6]$.

c. The points on the graph all have y-coordinates between -4, not including -4, and 4, including 4. The graph does not extend below $y = -4$ or above $y = 4$. Thus, the range of f is $\{y \mid -4 < y \le 4\}$, or the interval $(-4, 4]$.

Check Point 2 Use the graph of function f, shown below, to find $f(4)$, the domain, and the range.

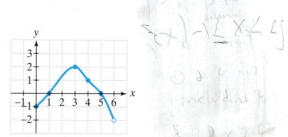

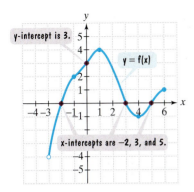

Figure 2.28 Identifying intercepts

Figure 2.28 illustrates how we can identify a graph's intercepts. To find the x-intercepts, look for the points at which the graph crosses the x-axis. There are three such points: $(-2, 0)$, $(3, 0)$, and $(5, 0)$. Thus, the x-intercepts are -2, 3, and 5. We express this in function notation by writing $f(-2) = 0$, $f(3) = 0$, and $f(5) = 0$. We say that -2, 3, and 5 are the *zeros of the function*. The **zeros of a function**, f, are the x-values for which $f(x) = 0$.

To find the y-intercept, look for the point at which the graph crosses the y-axis. This occurs at $(0, 3)$. Thus, the y-intercept is 3. We express this in function notation by writing $f(0) = 3$.

By the definition of a function, for each value of x we can have at most one value for y. What does this mean in terms of intercepts? **A function can have more than one x-intercept but at most one y-intercept.**

3 Use the vertical line test to identify functions.

The Vertical Line Test

Not every graph in the rectangular coordinate system is the graph of a function. The definition of a function specifies that no value of x can be paired with two or more different values of y. Consequently, if a graph contains two or more different points with the same first coordinate, the graph cannot represent a function. This is illustrated in Figure 2.29. Observe that points sharing a common first coordinate are vertically above or below each other.

This observation is the basis of a useful test for determining whether a graph defines y as a function of x. The test is called the **vertical line test.**

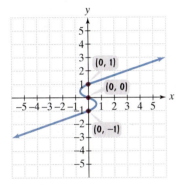

Figure 2.29 y is not a function of x because 0 is paired with three values of y, namely, $1, 0$, and -1.

The Vertical Line Test for Functions

If any vertical line intersects a graph in more than one point, the graph does not define y as a function of x.

EXAMPLE 3 **Using the Vertical Line Test**

Use the vertical line test to identify graphs in which y is a function of x.

a.
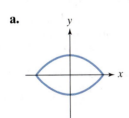

b.
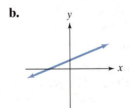

c.
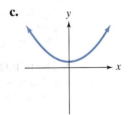

d.
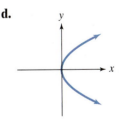

Solution y is a function of x for the graphs in **b** and **c**.

a.
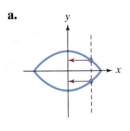

y is **not a function** of x.
Two values of y
correspond to an x-value.

b.
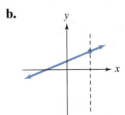

y **is a function** of x.

c.
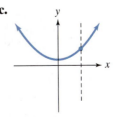

y **is a function** of x.

d.
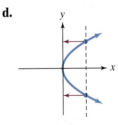

y is **not a function** of x.
Two values of y
correspond to an x-value.

Average Number of Paid Vacation Days for Full-Time Workers at Medium to Large U.S. Companies

Check Point 3 Use the vertical line test to identify graphs in which y is a function of x.

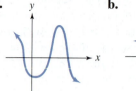

a.

b.

c.

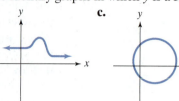

Average Number of Paid Vacation Days for Full-Time Workers at Medium to Large U.S. Companies

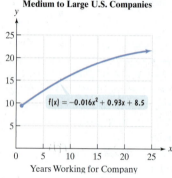

$f(x) = -0.016x^2 + 0.93x + 8.5$

Paid Vacation Days

Years Working for Company

Figure 2.30 *Source*: Bureau of Labor Statistics

EXAMPLE 4   Analyzing the Graph of a Function

The function

$$f(x) = -0.016x^2 + 0.93x + 8.5$$

models the average number of paid vacation days each year, $f(x)$, for full-time workers at medium to large U.S. companies after x years. The graph of f is shown in Figure 2.30.

a. Explain why f represents the graph of a function.
b. Use the graph to find a reasonable estimate of $f(5)$.
c. For what value of x is $f(x) = 20$?
d. Describe the general trend shown by the graph.

Solution

a. No vertical line intersects the graph of f more than once. By the vertical line test, f represents the graph of a function.

b. To find $f(5)$, or f of 5, we locate 5 on the x-axis. The figure shows the point on the graph of f for which 5 is the first coordinate. From this point, we look to the y-axis to find the corresponding y-coordinate. A reasonable estimate of the y-coordinate is 13. Thus, $f(5) \approx 13$. After 5 years, a worker can expect approximately 13 paid vacation days.

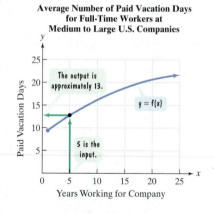

Average Number of Paid Vacation Days for Full-Time Workers at Medium to Large U.S. Companies

The output is approximately 13.

$y = f(x)$

5 is the input.

Paid Vacation Days

Years Working for Company

c. To find the value of x for which $f(x) = 20$, we locate 20 on the y-axis. The figure shows that there is one point on the graph of f for which 20 is the second coordinate. From this point, we look to the x-axis to find the corresponding x-coordinate. A reasonable estimate of the x-coordinate is 18. Thus, $f(x) = 20$ for $x \approx 18$. A worker with 20 paid vacation days has been with the company approximately 18 years.

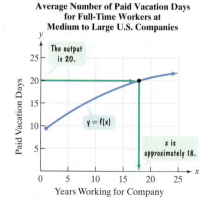

Average Number of Paid Vacation Days for Full-Time Workers at Medium to Large U.S. Companies

The output is 20.

$y = f(x)$

x is approximately 18.

Paid Vacation Days

Years Working for Company

d. The graph of f is rising from left to right. This shows that paid vacation days increase as time with the company increases. However, the rate of increase is slowing down as the graph moves to the right. This means that the increase in paid vacation days takes place more slowly the longer an employee is with the company.

Check Point 4

a. Use the graph of f in Figure 2.30 to find a reasonable estimate of $f(10)$. ⌐17

b. For what value of x is $f(x) = 15$? Round to the nearest whole number. ~8

④ Identify intervals on which a function increases, decreases, or is constant.

Increasing and Decreasing Functions

Too late for that flu shot now! It's only 8 A.M. and you're feeling lousy. Your temperature is 101°F. Fascinated by the way that algebra models the world (your author is projecting a bit here), you decide to construct graphs showing your body temperature as a function of the time of day. You decide to let x represent the number of hours after 8 A.M. and $f(x)$ your temperature at time x.

At 8 A.M. your temperature is 101°F and you are not feeling well. However, your temperature starts to decrease. It reaches normal (98.6°F) by 11 A.M. Feeling energized, you construct the graph shown on the right, indicating decreasing temperature for $\{x \mid 0 < x < 3\}$, or on the interval $(0, 3)$.

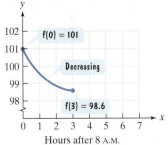

Temperature decreases on $(0, 3)$, reaching 98.6° by 11 A.M.

Did creating that first graph drain you of your energy? Your temperature starts to rise after 11 A.M. By 1 P.M., 5 hours after 8 A.M., your temperature reaches 100°F. However, you keep plotting points on your graph. At right, we can see that your temperature increases for $\{x \mid 3 < x < 5\}$, or on the interval $(3, 5)$.

The graph of f is decreasing to the left of $x = 3$ and increasing to the right of $x = 3$. Thus, your temperature 3 hours after 8 A.M. was at a relative minimum. Your relative minimum temperature was 98.6°.

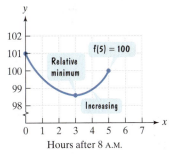

Temperature increases on $(3, 5)$.

By 3 P.M., your temperature is no worse than it was at 1 P.M.: It is still 100°F. (Of course, it's no better, either.) Your temperature remained the same, or constant, for $\{x \mid 5 < x < 7\}$, or on the interval $(5, 7)$.

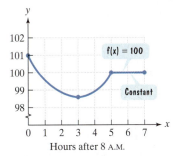

Temperature remains constant at 100° on $(5, 7)$.

The time-temperature flu scenario illustrates that a function f is increasing when its graph rises, decreasing when its graph falls, and remains constant when it neither rises nor falls. Let's now provide a more algebraic description for these intuitive concepts.

Increasing, Decreasing, and Constant Functions

1. A function is **increasing** on an open interval, I, if for any x_1 and x_2 in the interval, where $x_1 < x_2$, then $f(x_1) < f(x_2)$.
2. A function is **decreasing** on an open interval, I, if for any x_1 and x_2 in the interval, where $x_1 < x_2$, then $f(x_1) > f(x_2)$.
3. A function is **constant** on an open interval, I, if for any x_1 and x_2 in the interval, where $x_1 < x_2$, then $f(x_1) = f(x_2)$.

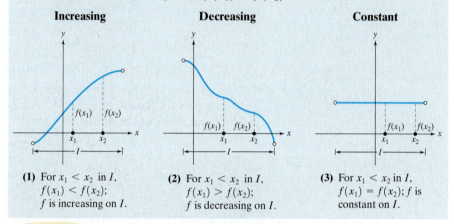

Increasing	Decreasing	Constant
(1) For $x_1 < x_2$ in I, $f(x_1) < f(x_2)$; f is increasing on I.	**(2)** For $x_1 < x_2$ in I, $f(x_1) > f(x_2)$; f is decreasing on I.	**(3)** For $x_1 < x_2$ in I, $f(x_1) = f(x_2)$; f is constant on I.

EXAMPLE 5 Intervals on Which a Function Increases, Decreases, or Is Constant

Give the intervals on which each function whose graph is shown is increasing, decreasing, or constant.

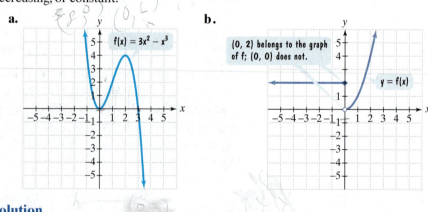

a. $f(x) = 3x^2 - x^3$

b. $(0, 2)$ belongs to the graph of f; $(0, 0)$ does not. $y = f(x)$

Solution

a. The function is decreasing on the interval $(-\infty, 0)$, increasing on the interval $(0, 2)$, and decreasing on the interval $(2, \infty)$.

b. Although the function's equations are not given, the graph indicates that the function is defined in two pieces. The part of the graph to the left of the y-axis shows that the function is constant on the interval $(-\infty, 0)$. The part to the right of the y-axis shows that the function is increasing on the interval $(0, \infty)$.

Check Point 5 Give the intervals on which the function whose graph is shown is increasing, decreasing, or constant.

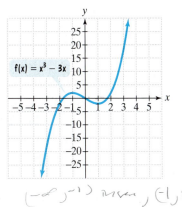

$(1, 0)$

$(-\infty, -1)$ rising, $(-1, 1)$

decreasing

5 Use graphs to locate relative maxima or minima.

Figure 2.31 Fuel efficiency of U.S. automobiles over time

Relative Maxima and Relative Minima

The points at which a function changes its increasing or decreasing behavior can be used to find the *relative maximum* or *relative minimum* values of the function. For example, consider the function with which we opened this section:

$$f(x) = 0.0075x^2 - 0.2672x + 14.8.$$

Recall that the function models the average number of miles per gallon of U.S. automobiles, $f(x)$, x years after 1940. The graph of this function is shown as a continuous curve in Figure 2.31. (It can also be shown as a series of points, each point representing a year and miles per gallon for that year.)

The graph of f is decreasing to the left of $x = 18$ and increasing to the right of $x = 18$. Thus, 18 years after 1940, in 1958, fuel efficiency was at a minimum. We say that the relative minimum fuel efficiency is $f(18)$, or approximately 12.4 miles per gallon. Mathematicians use the word "relative" to suggest that relative to an open interval about 18, the value $f(18)$ is smallest.

Definitions of Relative Maximum and Relative Minimum

1. A function value $f(a)$ is a **relative maximum** of f if there exists an open interval about a such that $f(a) > f(x)$ for all x in the open interval.

2. A function value $f(b)$ is a **relative minimum** of f if there exists an open interval about b such that $f(b) < f(x)$ for all x in the open interval.

Study Tip

The word *local* is sometimes used instead of *relative* when describing maxima or minima. If f has a relative, or local, maximum at a, $f(a)$ is greater than the values of f near a. If f has a relative, or local, minimum at b, $f(b)$ is less than the values of f near b.

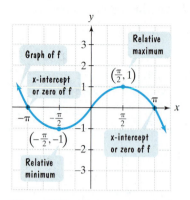

Figure 2.32 Using a graph to locate where f has a relative maximum or minimum

If the graph of a function is given, we can often visually locate the number(s) at which the function has a relative maximum or a relative minimum. For example, the graph of f in Figure 2.32 shows that:

• f has a relative maximum at $\dfrac{\pi}{2}$.

The relative maximum is $f\left(\dfrac{\pi}{2}\right) = 1$.

• f has a relative minimum at $-\dfrac{\pi}{2}$.

The relative minimum is $f\left(-\dfrac{\pi}{2}\right) = -1$.

Notice that f does not have a relative maximum or minimum at $-\pi$ and π, the x-intercepts, or zeros, of the function.

6 Find a function's average rate of change.

The Average Rate of Change of a Function

We have seen that the slope of a line can be interpreted as its rate of change. If the graph of a function is not a straight line, we speak of an **average rate of change** between any two points on its graph. To find the average rate of change, calculate the slope of the line containing the two points. This line is called a **secant line.**

The Average Rate of Change of a Function

Let $(x_1, f(x_1))$ and $(x_2, f(x_2))$ be distinct points on the graph of a function f. (See Figure 2.33.) The **average rate of change of f** from x_1 to x_2 is

$$\frac{f(x_2) - f(x_1)}{x_2 - x_1}.$$

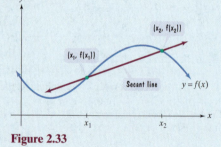

Figure 2.33

EXAMPLE 6   Finding the Average Rate of Change

Find the average rate of change of $f(x) = x^2$ from:

a. $x_1 = 0$ to $x_2 = 1$ **b.** $x_1 = 1$ to $x_2 = 2$ **c.** $x_1 = -2$ to $x_2 = 0$.

Solution

a. The average rate of change of $f(x) = x^2$ from $x_1 = 0$ to $x_2 = 1$ is

$$\frac{f(x_2) - f(x_1)}{x_2 - x_1} = \frac{f(1) - f(0)}{1 - 0} = \frac{1^2 - 0^2}{1} = 1.$$

Figure 2.34(a) shows the secant line of $f(x) = x^2$ from $x_1 = 0$ to $x_2 = 1$. The average rate of change is positive, and the function is increasing on the interval $(0, 1)$.

b. The average rate of change of $f(x) = x^2$ from $x_1 = 1$ to $x_2 = 2$ is

$$\frac{f(x_2) - f(x_1)}{x_2 - x_1} = \frac{f(2) - f(1)}{2 - 1} = \frac{2^2 - 1^2}{1} = 3.$$

Figure 2.34(b) shows the secant line of $f(x) = x^2$ from $x_1 = 1$ to $x_2 = 2$. The average rate of change is positive, and the function is increasing on the interval $(1, 2)$. Can you see that the graph rises more steeply on the interval $(1, 2)$ than on $(0, 1)$? This is because the average rate of change from $x_1 = 1$ to $x_2 = 2$ is greater than the average rate of change from $x_1 = 0$ to $x_2 = 1$.

c. The average rate of change of $f(x) = x^2$ from $x_1 = -2$ to $x_2 = 0$ is

$$\frac{f(x_2) - f(x_1)}{x_2 - x_1} = \frac{f(0) - f(-2)}{0 - (-2)} = \frac{0^2 - (-2)^2}{2} = \frac{-4}{2} = -2.$$

Figure 2.34(c) shows the secant line of $f(x) = x^2$ from $x_1 = -2$ to $x_2 = 0$. The average rate of change is negative, and the function is decreasing on the interval $(-2, 0)$.

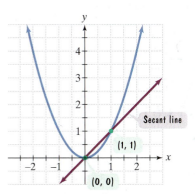

Figure 2.34(a) The secant line of $f(x) = x^2$ from $x_1 = 0$ to $x_2 = 1$

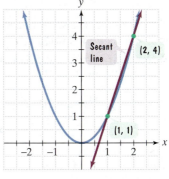

Figure 2.34(b) The secant line of $f(x) = x^2$ from $x_1 = 1$ to $x_2 = 2$

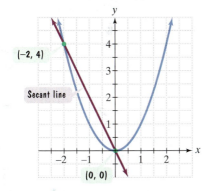

Figure 2.34(c) The secant line of $f(x) = x^2$ from $x_1 = -2$ to $x_2 = 0$

Check Point 6 Find the average rate of change of $f(x) = x^3$ from

a. $x_1 = 0$ to $x_2 = 1$

b. $x_1 = 1$ to $x_2 = 2$

c. $x_1 = -2$ to $x_2 = 0$.

Suppose we are interested in the average rate of change of f from $x_1 = x$ to $x_2 = x + h$. In this case, the average rate of change is

$$\frac{f(x_2) - f(x_1)}{x_2 - x_1} = \frac{f(x + h) - f(x)}{x + h - x} = \frac{f(x + h) - f(x)}{h}.$$

Do you recognize the last expression? It is the difference quotient that you used in the previous section to practice evaluating functions. Thus, the difference quotient gives the average rate of change of a function from x to $x + h$. In the difference quotient, h is thought of as a number very close to 0. In this way, the average rate of change can be found for a very short interval.

7 Identify even or odd functions and recognize their symmetries.

Even and Odd Functions and Symmetry

Is beauty in the eye of the beholder? Or are there certain objects (or people) that are so well balanced and proportioned that they are universally pleasing to the eye? What constitutes an attractive human face? In Figure 2.35, we've drawn lines between paired features and marked the midpoints. Notice how the features line up almost perfectly. Each half of the face is a mirror image of the other half through the white vertical line.

Did you know that graphs of some equations exhibit exactly the kind of symmetry shown by the attractive face in Figure 2.35? The word *symmetry* comes from the Greek *symmetria*, meaning "the same measure." We can identify graphs with symmetry by looking at a function's equation and determining if the function is *even* or *odd*.

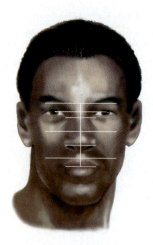

Figure 2.35 To most people, an attractive face is one in which each half is an almost perfect mirror image of the other half

---

Definition of Even and Odd Functions

The function f is an **even function** if

$$f(-x) = f(x) \quad \text{for all } x \text{ in the domain of } f.$$

The right side of the equation of an even function does not change if x is replaced with $-x$.

The function f is an **odd function** if

$$f(-x) = -f(x) \quad \text{for all } x \text{ in the domain of } f.$$

Every term in the right side of the equation of an odd function changes sign if x is replaced with $-x$.

EXAMPLE 7   Identifying Even or Odd Functions

Identify each of the following functions as even, odd, or neither:

a. $f(x) = x^3$ **b.** $g(x) = x^4 - 2x^2$ **c.** $h(x) = x^2 + 2x + 1$.

Solution   In each case, replace x with $-x$ and simplify. If the right side of the equation stays the same, the function is even. If every term on the right changes sign, the function is odd.

a. We use the given function's equation, $f(x) = x^3$, to find $f(-x)$.

Use $f(x) = x^3$.

Replace x with $-x$.   Replace x with $-x$.

$$f(-x) = (-x)^3 = (-x)(-x)(-x) = -x^3$$

There is only one term in the equation $f(x) = x^3$, and the term changed signs when we replaced x with $-x$. Because $f(-x) = -f(x)$, f is an odd function.

b. We use the given function's equation, $g(x) = x^4 - 2x^2$, to find $g(-x)$.

Use $g(x) = x^4 - 2x^2$.

> Replace x with −x.

$$g(-x) = (-x)^4 - 2(-x)^2 = (-x)(-x)(-x)(-x) - 2(-x)(-x)$$

$$= x^4 - 2x^2$$

The right side of the equation of the given function, $g(x) = x^4 - 2x^2$, did not change when we replaced x with $-x$. Because $g(-x) = g(x)$, g is an even function.

c. We use the given function's equation, $h(x) = x^2 + 2x + 1$, to find $h(-x)$.

Use $h(x) = x^2 + 2x + 1$.

> Replace x with −x.

$$h(-x) = (-x)^2 + 2(-x) + 1 = x^2 - 2x + 1$$

The right side of the equation of the given function, $h(x) = x^2 + 2x + 1$, changed when we replaced x with $-x$. Thus, $h(-x) \neq h(x)$, so h is not an even function. The sign of *each* of the three terms in the equation for $h(x)$ did not change when we replaced x with $-x$. Only the second term changed signs. Thus, $h(-x) \neq -h(x)$, so h is not an odd function. We conclude that h is neither an even nor an odd function.

Check Point 7   Determine whether each of the following functions is even, odd, or neither:

a. $f(x) = x^2 + 6$ **b.** $g(x) = 7x^3 - x$ **c.** $h(x) = x^5 + 1$.

Now, let's see what even and odd functions tell us about a function's graph. Begin with the even function $f(x) = x^2 - 4$, shown in Figure 2.36. The function is even because

$$f(-x) = (-x)^2 - 4 = x^2 - 4 = f(x).$$

Examine the pairs of points shown, such as $(3, 5)$ and $(-3, 5)$. Notice that we obtain the same y-coordinate whenever we evaluate the function at a value of x and the value of its opposite, $-x$. Like the attractive face, each half of the graph is a mirror image of the other half through the y-axis. If we were to fold the paper along the y-axis, the two halves of the graph would coincide. This causes the graph to be *symmetric with respect to the y-axis*. A graph is **symmetric with respect to the y-axis** if, for every point (x, y) on the graph, the point $(-x, y)$ is also on the graph. All even functions have graphs with this kind of symmetry.

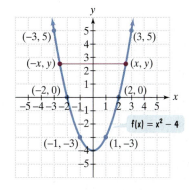

Figure 2.36 y-axis symmetry with $f(-x) = f(x)$

Even Functions and y-Axis Symmetry

The graph of an even function in which $f(-x) = f(x)$ is symmetric with respect to the y-axis.

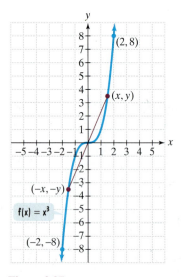

Figure 2.37 Origin symmetry with $f(-x) = -f(x)$

Now, consider the graph of the function $f(x) = x^3$. In Example 5, we saw that $f(-x) = -f(x)$, so this is an odd function. Although the graph in Figure 2.37 is not symmetric with respect to the y-axis, it is symmetric in another way. Look at the pairs of points, such as $(2, 8)$ and $(-2, -8)$. For each point (x, y) on the graph, the point $(-x, -y)$ is also on the graph. The points $(2, 8)$ and $(-2, -8)$ are reflections of one another in the origin. This means that:

- the points are the same distance from the origin, and
- the points lie on a line through the origin.

A graph is **symmetric with respect to the origin** if, for every point (x, y) on the graph, the point $(-x, -y)$ is also on the graph. Observe that the first- and third-quadrant portions of $f(x) = x^3$ are reflections of one another with respect to the origin. Notice that $f(x)$ and $f(-x)$ have opposite signs, so that $f(-x) = -f(x)$. All odd functions have graphs with origin symmetry.

Odd Functions and Origin Symmetry

The graph of an odd function in which $f(-x) = -f(x)$ is symmetric with respect to the origin.

8 Graph step functions.

Step Functions

Have you ever mailed a letter that seemed heavier than usual? Perhaps you worried that the letter would not have enough postage. Costs for mailing a letter weighing up to 5 ounces are given in Table 2.3. If your letter weighs an ounce or less, the cost is $0.37. If your letter weighs 1.05 ounces, 1.50 ounces, 1.90 ounces, or 2.00 ounces, the cost "steps" to $0.60. The cost does not take on any value between $0.37 and $0.60. If your letter weighs 2.05 ounces, 2.50 ounces, 2.90 ounces, or 3 ounces, the cost "steps" to $0.83. Cost increases are $0.23 per step.

Now, let's see what the graph of the function that models this situation looks like. Let

$$x = \text{the weight of the letter, in ounces, and}$$

$$y = f(x) = \text{the cost of mailing a letter weighing } x \text{ ounces.}$$

Table 2.3 Cost of First-Class Mail (Effective June 30, 2002)

Weight Not Over	Cost
1 ounce	$0.37
2 ounces	0.60
3 ounces	0.83
4 ounces	1.06
5 ounces	1.29

Source: U.S. Postal Service

The graph is shown in Figure 2.38. Notice how it consists of a series of steps that jump vertically 0.23 unit at each integer. The graph is constant between each pair of consecutive integers.

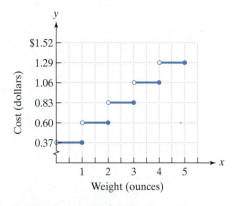

Figure 2.38

Mathematicians have defined functions that describe situations where function values graphically form discontinuous steps. One such function is called the **greatest integer function,** symbolized by int(x) or $[x]$. And what is int(x)?

int(x) = the greatest integer that is less than or equal to x.

For example,

int(1) = 1, int(1.3) = 1, int(1.5) = 1, int(1.9) = 1.

1 is the greatest integer that is less than or equal to 1, 1.3, 1.5, and 1.9.

Here are some additional examples:

int(2) = 2, int(2.3) = 2, int(2.5) = 2, int(2.9) = 2.

2 is the greatest integer that is less than or equal to 2, 2.3, 2.5, and 2.9.

Notice how we jumped from 1 to 2 in the function values for int(x). In particular,

If $1 \le x < 2$, then int(x) = 1.

If $2 \le x < 3$, then int(x) = 2.

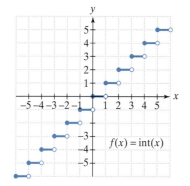

The graph of $f(x) = $ int(x) is shown in Figure 2.39. The graph of the greatest integer function jumps vertically one unit at each integer. However, the graph is constant between each pair of consecutive integers. The rightmost horizontal step shown in the graph illustrates that

If $5 \le x < 6$, then int (x) = 5.

In general,

If $n \le x < n + 1$, where n is an integer, then int(x) = n.

Figure 2.39 The graph of the greatest integer function

By contrast to the graph for the cost of first-class mail, the graph of the greatest integer function includes the point on the left of each horizontal step, but does not include the point on the right. The domain of $f(x) = $ int(x) is the set of all real numbers, $(-\infty, \infty)$. The range is the set of all integers.

Technology

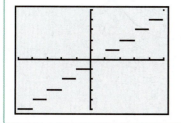

The graph of $f(x) = $ int(x), shown on the left, was obtained with a graphing utility. By graphing in "dot" mode, we can see the discontinuities at the integers. By looking at the graph, it is impossible to tell that, for each step, the point on the left is included and the point on the right is not. We must trace along the graph to obtain such information.

EXERCISE SET 2.4

Practice Exercises

Graph the function in Exercises 1–14. Use the integer values of x given to the right of the function to obtain ordered pairs. Use the graph to specify the function's domain and range.

1. $f(x) = x^2 + 2$ $x = -3, -2, -1, 0, 1, 2, 3$

2. $f(x) = x^2 - 1$ $x = -3, -2, -1, 0, 1, 2, 3$

3. $g(x) = \sqrt{x} - 1$ $x = 0, 1, 4, 9$

4. $g(x) = \sqrt{x} + 2$ $x = 0, 1, 4, 9$

5. $h(x) = \sqrt{x - 1}$ $x = 1, 2, 5, 10$

6. $h(x) = \sqrt{x + 2}$ $x = -2, -1, 2, 7$

7. $f(x) = |x| - 1$ $x = -3, -2, -1, 0, 1, 2, 3$

8. $f(x) = |x| + 1$ $x = -3, -2, -1, 0, 1, 2, 3$

9. $g(x) = |x - 1|$ $x = -3, -2, -1, 0, 1, 2, 3$

10. $g(x) = |x + 1|$ $x = -3, -2, -1, 0, 1, 2, 3$

11. $f(x) = 5$ $x = -3, -2, -1, 0, 1, 2, 3$

12. $f(x) = 3$ $x = -3, -2, -1, 0, 1, 2, 3$

13. $f(x) = x^3 - 2$ $x = -2, -1, 0, 1, 2$

14. $f(x) = x^3 + 2$ $x = -2, -1, 0, 1, 2$

*In Exercises 15–30, use the graph to determine **a.** the function's domain; **b.** the function's range; **c.** the x-intercepts, if any; **d.** the y-intercept, if any; and **e.** the function values indicated below some of the graphs.*

15.

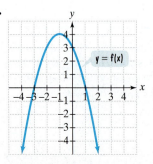

16.

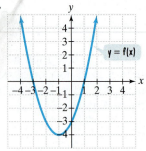

17.

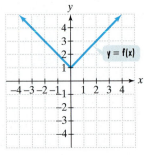

$f(-1) = ?$ $f(3) = ?$

18.

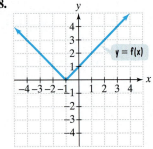

$f(-4) = ?$ $f(3) = ?$

19.

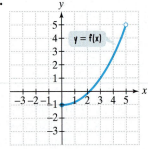

$f(3) = ?$

20.

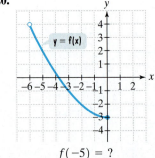

$f(-5) = ?$

21.

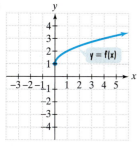

$f(4) = ?$

22.

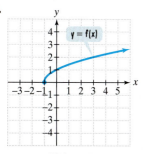

$f(3) = ?$

23.

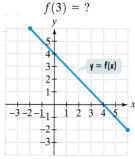

$f(-1) = ?$

24.

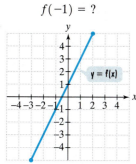

$f(-2) = ?$

25.

$f(-4) = ?$ $f(4) = ?$

26.

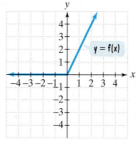

$f(-2) = ?$ $f(2) = ?$

27.

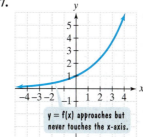

y = f(x) approaches but never touches the x-axis.

28.

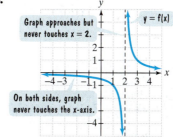

Graph approaches but never touches x = 2.

On both sides, graph never touches the x-axis.

29.

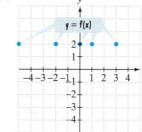

30.

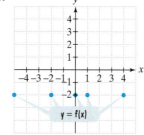

In Exercises 31–38, use the vertical line test to identify graphs in which y is a function of x.

31.

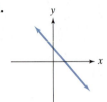

32.

33.

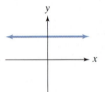

34.

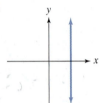

35.

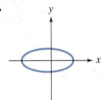

36.

37.

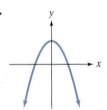

38.

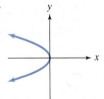

47.

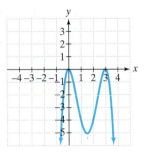

48.

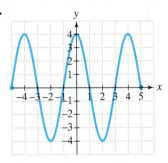

49.

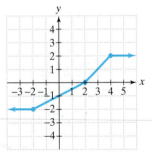

In Exercises 39–50, use the graph to determine:

   a. *intervals on which the function is increasing, if any.*

   b. *intervals on which the function is decreasing, if any.*

   c. *intervals on which the function is constant, if any.*

39. Use the graph in Exercise 15.

40. Use the graph in Exercise 16.

41. Use the graph in Exercise 21.

42. Use the graph in Exercise 22.

43. Use the graph in Exercise 23.

44. Use the graph in Exercise 24.

45. Use the graph in Exercise 25.

46. Use the graph in Exercise 26.

50.

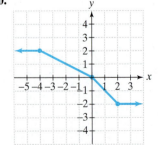

In Exercises 51–54, the graph of a function f is given. Use the graph to find:

  a. The numbers, if any, at which f has a relative maximum. What are these relative maxima?

  b. The numbers, if any, at which f has a relative minimum. What are these relative minima?

51.

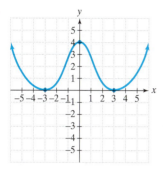

52.

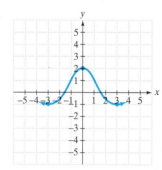

53.

$f(x) = 2x^3 + 3x^2 - 12x + 1$

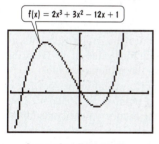

[−4, 4, 1] by [−15, 25, 5]

54.

$f(x) = 2x^3 - 15x^2 + 24x + 19$

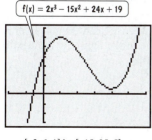

[−2, 6, 1] by [−15, 35, 5]

In Exercises 55–60, find the average rate of change of the function from x_1 to x_2.

55. $f(x) = 3x$ from $x_1 = 0$ to $x_2 = 5$

56. $f(x) = 6x$ from $x_1 = 0$ to $x_2 = 4$

57. $f(x) = x^2 + 2x$ from $x_1 = 3$ to $x_2 = 5$

58. $f(x) = x^2 - 2x$ from $x_2 = 3$ to $x_2 = 6$

59. $f(x) = \sqrt{x}$ from $x_1 = 4$ to $x_2 = 9$

60. $f(x) = \sqrt{x}$ from $x_1 = 9$ to $x_2 = 16$

In Exercises 61–72, determine whether each function is even, odd, or neither.

61. $f(x) = x^3 + x$ **62.** $f(x) = x^3 - x$

63. $g(x) = x^2 + x$,   **64.** $g(x) = x^2 - x$

65. $h(x) = x^2 - x^4$ **66.** $h(x) = 2x^2 + x^4$

67. $f(x) = x^2 - x^4 + 1$ **68.** $f(x) = 2x^2 + x^4 + 1$

69. $f(x) = \frac{1}{5}x^6 - 3x^2$ **70.** $f(x) = 2x^3 - 6x^5$

71. $f(x) = x\sqrt{1 - x^2}$ **72.** $f(x) = x^2\sqrt{1 - x^2}$

In Exercises 73–76, use possible symmetry to determine whether each graph is the graph of an even function, an odd function, or a function that is neither even nor odd.

73.

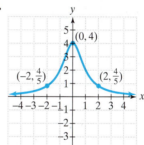

74.

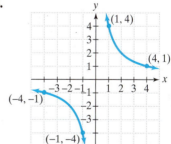

75.

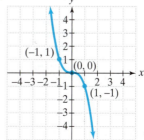

76.

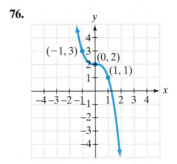

In Exercises 77–82, if $f(x) = int(x)$, find each function value.

77. $f(1.06)$

78. $f(2.99)$

79. $f(\frac{1}{3})$

80. $f(-1.5)$

81. $f(-2.3)$

82. $f(-99.001)$

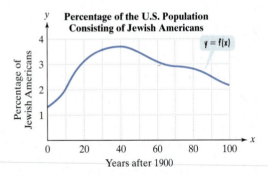

Application Exercises

The figure shows the percentage of the U.S. population made up of Jewish Americans, $f(x)$, as a function of time, x, where x is the number of years after 1900. Use the graph to solve Exercises 83–90.

Percentage of the U.S. Population Consisting of Jewish Americans

y = f(x)

Source: American Jewish Yearbook

83. Use the graph to find a reasonable estimate of $f(60)$. What does this mean in terms of the variables in this situation?

84. Use the graph to find a reasonable estimate of $f(100)$. What does this mean in terms of the variables in this situation?

85. For what value or values of x is $f(x) = 3$? Round to the nearest year. What does this mean in terms of the variables in this situation?

86. For what value or values of x is $f(x) = 2.5$? Round to the nearest year. What does this mean in terms of the variables in this situation?

87. In which year did the percentage of Jewish Americans in the U.S. population reach a maximum? What is a reasonable estimate of the percentage for that year?

88. In which year was the percentage of Jewish Americans in the U.S. population at a minimum? What is a reasonable estimate of the percentage for that year?

89. Explain why f represents the graph of a function.

90. Describe the general trend shown by the graph.

The function

$$f(x) = 0.4x^2 - 36x + 1000$$

models the number of accidents, $f(x)$, per 50 million miles driven as a function of the driver's age, x, in years, where x includes drivers from ages 16 through 74. The graph of f is shown. Use the graph of f, and possibly the equation, to solve Exercises 91–93.

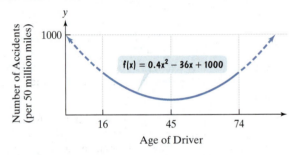

91. State the intervals on which the function is increasing and decreasing and describe what this means in terms of the variables modeled by the function.

92. For what value of x does the graph reach its lowest point? What is the minimum value of y? Describe the practical significance of this minimum value.

93. Use the graph to identify two different ages for which drivers have the same number of accidents. Use the equation for f to find the number of accidents for drivers at each of these ages.

94. Based on a study by Vance Tucker (*Scientific American*, May 1969), the power expenditure of migratory birds in flight is a function of their flying speed, x, in miles per hour, modeled by $f(x) = 0.67x^2 - 27.74x + 387$. Power expenditure, $f(x)$, is measured in calories, and migratory birds generally fly between 12 and 30 miles per hour. The graph of f is shown in the figure on the next page, with a domain of $[12, 30]$.

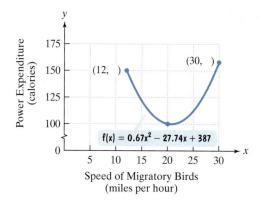

Power Expenditure (calories) vs. Speed of Migratory Birds (miles per hour)

(12,)   (30,)

$f(x) = 0.67x^2 - 27.74x + 387$

a. State the intervals on which the function is increasing and decreasing and describe what this means in terms of the variables modeled by the function.

b. For what approximate value of x does the graph reach its lowest point? What is the minimum value of y? Describe the practical significance of this minimum value.

95. The cost of a telephone call between two cities is \$0.10 for the first minute and \$0.05 for each additional minute or portion of a minute. Draw a graph of the cost, C, in dollars, of the phone call as a function of time, t, in minutes, on the interval $(0, 5]$.

96. A cargo service charges a flat fee of \$4 plus \$1 for each pound or fraction of a pound to mail a package. Let $C(x)$ represent the cost to mail a package that weighs x pounds. Graph the cost function on the interval $(0, 5]$.

97. Researchers at Yale University have suggested that levels of passion and commitment in human relations are functions of time. Based on the shapes of the graphs shown, which do you think depicts passion and which represents commitment? Explain how you arrived at your answer.

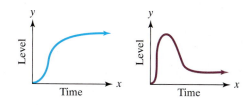

Writing in Mathematics

98. Discuss one disadvantage to using point plotting as a method for graphing functions.

99. Explain how to use a function's graph to find the function's domain and range.

100. Explain how the vertical line test is used to determine whether a graph is a function.

101. What does it mean if function f is increasing on an interval?

102. Suppose that a function f is increasing on (a, b) and decreasing on (b, c). Describe what occurs at $x = b$. What does the function value $f(b)$ represent?

103. What is a secant line?

104. What is the average rate of change of a function?

105. If you are given a function's equation, how do you determine if the function is even, odd, or neither?

106. If you are given a function's graph, how do you determine if the function is even, odd, or neither?

107. What is a step function? Give an example of an everyday situation that can be modeled using such a function. Do not use the cost-of-mail example.

108. Explain how to find int(-3.000004).

Technology Exercises

109. The function

$$f(x) = -0.00002x^3 + 0.008x^2 - 0.3x + 6.95$$

models the number of annual physician visits, $f(x)$, by a person of age x. Graph the function in a $[0, 100, 5]$ by $[0, 40, 2]$ viewing rectangle. What does the shape of the graph indicate about the relationship between one's age and the number of annual physician visits? Use the $\boxed{\text{TRACE}}$ or minimum function capability to find the coordinates of the minimum point on the graph of the function. What does this mean?

In Exercises 110–115, use a graphing utility to graph each function. Use a $[-5, 5, 1]$ by $[-5, 5, 1]$ viewing rectangle. Then find the intervals on which the function is increasing, decreasing, or constant.

110. $f(x) = x^3 - 6x^2 + 9x + 1$ **111.** $g(x) = |4 - x^2|$

112. $h(x) = |x - 2| + |x + 2|$ **113.** $f(x) = x^{1/3}(x - 4)$

114. $g(x) = x^{2/3}$ **115.** $h(x) = 2 - x^{2/5}$

116. a. Graph the functions $f(x) = x^n$ for $n = 2, 4$, and 6 in a $[-2, 2, 1]$ by $[-1, 3, 1]$ viewing rectangle.

b. Graph the functions $f(x) = x^n$ for $n = 1, 3$, and 5 in a $[-2, 2, 1]$ by $[-2, 2, 1]$ viewing rectangle.

c. If n is even, where is the graph of $f(x) = x^n$ increasing and where is it decreasing?

d. If n is odd, what can you conclude about the graph of $f(x) = x^n$ in terms of increasing or decreasing behavior?

e. Graph all six functions in a $[-1, 3, 1]$ by $[-1, 3, 1]$ viewing rectangle. What do you observe about the graphs in terms of how flat or how steep they are?

Critical Thinking Exercises

117. Which one of the following is true based on the graph of f in the figure?

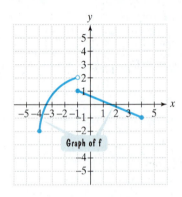

Graph of f

a. The domain of f is $[-4, 1)$ or $(1, 4]$.

b. The range of f is $[-2, 2]$.

c. $f(-1) - f(4) = 2$

d. $f(0) = 2.1$

118. Sketch the graph of f using the following properties. (More than one correct graph is possible.) f is a piecewise function that is decreasing on $(-\infty, 2)$, $f(2) = 0$, f is increasing on $(2, \infty)$, and the range of f is $[0, \infty)$.

119. Define a piecewise function on the intervals $(-\infty, 2]$, $(2, 5)$, and $[5, \infty)$ that does not "jump" at 2 or 5 such that one piece is a constant function, another piece is an increasing function, and the third piece is a decreasing function.

120. Suppose that $h(x) = \dfrac{f(x)}{g(x)}$. The function f can be even, odd, or neither. The same is true for the function g.

a. Under what conditions is h definitely an even function?

b. Under what conditions is h definitely an odd function?

121. Take another look at the cost of first-class mail and its graph (Table 2.3 and Figure 2.38 on page 226. Change the description of the heading in the left column of Table 2.3 so that the graph includes the point on the left of each horizontal step, but does not include the point on the right.

Group Exercise

122. In Exercise 97, passion and commitment are graphed over time. For this activity, you will be creating a graph of a particular experience that involved your feelings of love, anger, sadness, or any other emotion you choose. The horizontal axis should be labeled time and the vertical axis the emotion you are graphing. You will not be using your algebra skills to create your graph; however, you should try to make the graph as precise as possible. You may use negative numbers on the vertical axis, if appropriate. After each group member has created a graph, pool together all of the graphs and study them to see if there are any similarities in the graphs for a particular emotion or for all emotions.

SECTION 2.5 *Transformations of Functions*

Objectives

1. Recognize graphs of common functions.
2. Use vertical shifts to graph functions.
3. Use horizontal shifts to graph functions.
4. Use reflections to graph functions.
5. Use vertical stretching and shrinking to graph functions.
6. Graph functions involving a sequence of transformations.

Have you seen *Terminator 2*, *The Mask*, or *The Matrix*? These were among the first films to use spectacular effects in which a character or object having one shape was transformed in a fluid fashion into a quite different shape. The name for such a transformation is **morphing.** The effect allows a real actor to be seamlessly transformed into a computer-generated animation. The animation can be made to perform impossible feats before it is morphed back to the conventionally filmed image.

   Like transformed movie images, the graph of one function can be turned into the graph of a different function. To do this, we need to rely on a function's equation. Knowing that a graph is a transformation of a familiar graph makes graphing easier.

1 Recognize graphs of common functions.

Graphs of Common Functions

Table 2.4 below and on page 236 gives names to six frequently encountered functions in algebra. The table shows each function's graph and lists characteristics of the function. Study the shape of each graph and take a few minutes to verify the function's characteristics from its graph. Knowing these graphs is essential for analyzing their transformations into more complicated graphs.

Table 2.4 Algebra's Common Graphs

Constant Function	Identity Function	Standard Quadratic Function
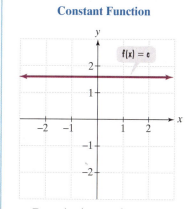	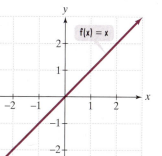	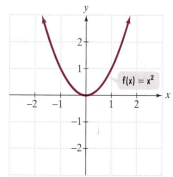
• Domain: $(-\infty, \infty)$	• Domain: $(-\infty, \infty)$	• Domain: $(-\infty, \infty)$
• Range: the single number c	• Range: $(-\infty, \infty)$	• Range: $[0, \infty)$
• Constant on $(-\infty, \infty)$	• Increasing on $(-\infty, \infty)$	• Decreasing on $(-\infty, 0)$ and increasing on $(0, \infty)$
• Even function	• Odd function	• Even function

Table 2.4   Algebra's Common Graphs (*continued*)

Standard Cubic Function	Square Root Function	Absolute Value Function
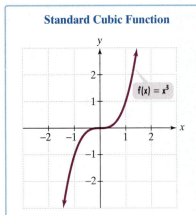	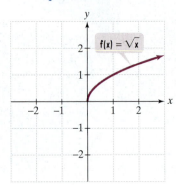	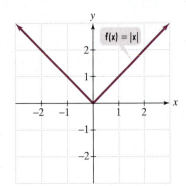
• Domain: $(-\infty, \infty)$	• Domain: $[0, \infty)$	• Domain: $(-\infty, \infty)$
• Range: $(-\infty, \infty)$	• Range: $[0, \infty)$	• Range: $[0, \infty)$
• Increasing on $(-\infty, \infty)$	• Increasing on $(0, \infty)$	• Decreasing on $(-\infty, 0)$ and increasing on $(0, \infty)$
• Odd function	• Neither even nor odd	• Even function

Discovery

The study of how changing a function's equation can affect its graph can be explored with a graphing utility. Use your graphing utility to verify the hand-drawn graphs as you read this section.

 2 Use vertical shifts to graph functions.

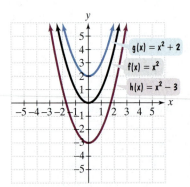

Figure 2.40 Vertical shifts

The graph of $f(x) = x^2$ can be gradually morphed into the graph of $g(x) = x^2 + 2$ by using animation to graph $f(x) = x^2 + c$ for $0 \le c \le 2$. By selecting many values for c, we can create an animated sequence in which change appears to occur continuously.

Vertical Shifts

Let's begin by looking at three graphs whose shapes are the same. Figure 2.40 shows the graphs. The black graph in the middle is the standard quadratic function, $f(x) = x^2$. Now, look at the blue graph on the top. The equation of this graph, $g(x) = x^2 + 2$, adds 2 to the right side of $f(x) = x^2$. What effect does this have on the graph of f? It shifts the graph vertically up by 2 units.

$$g(x) = x^2 + 2 = f(x) + 2$$

The graph of g shifts the graph of f up 2 units.

Finally, look at the red graph on the bottom of Figure 2.40. The equation of this graph, $h(x) = x^2 - 3$, subtracts 3 from the right side of $f(x) = x^2$. What effect does this have on the graph of f? It shifts the graph vertically down by 3 units.

$$h(x) = x^2 - 3 = f(x) - 3$$

The graph of h shifts the graph of f down 3 units.

In general, if c is positive, $y = f(x) + c$ shifts the graph of f upward c units and $y = f(x) - c$ shifts the graph of f downward c units. These are called **vertical shifts** of the graph of f.

Vertical Shifts

Let f be a function and c a positive real number.

*   The graph of $y = f(x) + c$ is the graph of $y = f(x)$ shifted c units vertically upward.
*   The graph of $y = f(x) - c$ is the graph of $y = f(x)$ shifted c units vertically downward.

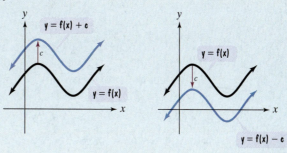

EXAMPLE 1   Vertical Shift Down

Use the graph of $f(x) = |x|$ to obtain the graph of $g(x) = |x| - 4$.

Solution   The graph of $g(x) = |x| - 4$ has the same shape as the graph of $f(x) = |x|$. However, it is shifted down vertically 4 units. We have constructed a table showing some of the coordinates for f and g. The graphs of f and g are shown in Figure 2.41.

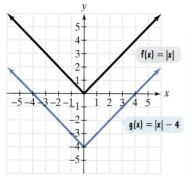

Figure 2.41

x	$y = f(x) = \|x\|$	$(x, f(x))$	$y = g(x)$ $= \|x\| - 4 = f(x) - 4$	$(x, g(x))$
-2	$\|-2\| = 2$	$(-2, 2)$	$\|-2\| - 4 = -2$	$(-2, -2)$
-1	$\|-1\| = 1$	$(-1, 1)$	$\|-1\| - 4 = -3$	$(-1, -3)$
0	$\|0\| = 0$	$(0, 0)$	$\|0\| - 4 = -4$	$(0, -4)$
1	$\|1\| = 1$	$(1, 1)$	$\|1\| - 4 = -3$	$(1, -3)$
2	$\|2\| = 2$	$(2, 2)$	$\|2\| - 4 = -2$	$(2, -2)$

Check Point 1   Use the graph of $f(x) = |x|$ to obtain the graph of $g(x) = |x| + 3$.

3 Use horizontal shifts to graph functions.

Horizontal Shifts

We return to the graph of $f(x) = x^2$, the standard quadratic function. In Figure 2.42 on the next page, the graph of function f is in the middle of the three graphs. Turn the page and verify this observation.

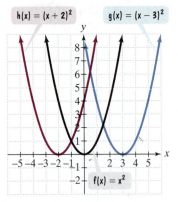

Figure 2.42 Horizontal shifts

By contrast to the vertical shift situation, this time there are graphs to the left and to the right of the graph of f. Look at the blue graph on the right. The equation of this graph, $g(x) = (x - 3)^2$, subtracts 3 from each value of x in the domain of $f(x) = x^2$. What effect does this have on the graph of f? It shifts the graph horizontally to the right by 3 units.

$$g(x) = (x - 3)^2 = f(x - 3)$$

The graph of g · · · shifts the graph of f 3 units to the right.

Now, look at the red graph on the left in Figure 2.42. The equation of this graph, $h(x) = (x + 2)^2$, adds 2 to each value of x in the domain of $f(x) = x^2$. What effect does this have on the graph of f? It shifts the graph horizontally to the left by 2 units.

$$h(x) = (x + 2)^2 = f(x + 2)$$

The graph of h · · · shifts the graph of f 2 units to the left.

In general, if c is positive, $y = f(x + c)$ shifts the graph of f to the left c units and $y = f(x - c)$ shifts the graph of f to the right c units. These are called **horizontal shifts** of the graph of f.

Study Tip

We know that positive numbers are to the right of zero on a number line and negative numbers are to the left of zero. This positive-negative orientation does not apply to horizontal shifts. A *positive* number causes a shift to the *left* and a *negative* number causes a shift to the *right*.

Horizontal Shifts

Let f be a function and c a positive real number.

- The graph of $y = f(x + c)$ is the graph of $y = f(x)$ shifted to the left c units.
- The graph of $y = f(x - c)$ is the graph of $y = f(x)$ shifted to the right c units.

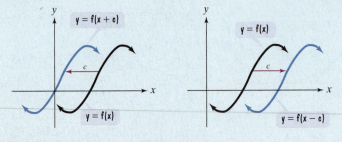

EXAMPLE 2   Horizontal Shift to the Left

Use the graph of $f(x) = \sqrt{x}$ to obtain the graph of $g(x) = \sqrt{x + 5}$.

Solution   Compare the equations for $f(x) = \sqrt{x}$ and $g(x) = \sqrt{x + 5}$. The equation for g adds 5 to each value of x in the domain of f.

$$y = g(x) = \sqrt{x + 5} = f(x + 5)$$

The graph of g · · · shifts the graph of f 5 units to the left.

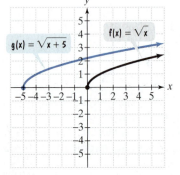

Figure 2.43 Shifting $f(x) = \sqrt{x}$ five units left

The graph of $g(x) = \sqrt{x + 5}$ has the same shape as the graph of $f(x) = \sqrt{x}$. However, it is shifted horizontally to the left 5 units. We have created tables on the next page showing some of the coordinates for f and g. As shown in Figure 2.43, every point in the graph of g is exactly 5 units to the left of a corresponding point on the graph of f.

x	$y = f(x) = \sqrt{x}$	$(x, f(x))$
0	$\sqrt{0} = 0$	$(0, 0)$
1	$\sqrt{1} = 1$	$(1, 1)$
4	$\sqrt{4} = 2$	$(4, 2)$

x	$y = g(x) = \sqrt{x + 5}$	$(x, g(x))$
-5	$\sqrt{-5 + 5} = \sqrt{0} = 0$	$(-5, 0)$
-4	$\sqrt{-4 + 5} = \sqrt{1} = 1$	$(-4, 1)$
-1	$\sqrt{-1 + 5} = \sqrt{4} = 2$	$(-1, 2)$

Check Point 2 Use the graph of $f(x) = \sqrt{x}$ to obtain the graph of $g(x) = \sqrt{x - 4}$.

Some functions can be graphed by combining horizontal and vertical shifts. These functions will be variations of a function whose equation you know how to graph, such as the standard quadratic function, the standard cubic function, the square root function, or the absolute value function.

In our next example, we will use the graph of the standard quadratic function, $f(x) = x^2$, to obtain the graph of $h(x) = (x + 1)^2 - 3$. We will graph three functions:

$$f(x) = x^2 \qquad g(x) = (x + 1)^2 \qquad h(x) = (x + 1)^2 - 3.$$

Start by graphing the standard quadratic function.

Shift the graph of f horizontally one unit to the left.

Shift the graph of g vertically down 3 units.

Discovery

Work Example 3 by first shifting the graph of $f(x) = x^2$ three units down, graphing $g(x) = x^2 - 3$. Now, shift this graph one unit left to graph $h(x) = (x + 1)^2 - 3$. Did you obtain the graph in Figure 2.44(c)? What can you conclude?

EXAMPLE 3 **Combining Horizontal and Vertical Shifts**

Use the graph of $f(x) = x^2$ to obtain the graph of $h(x) = (x + 1)^2 - 3$.

Solution

Step 1 Graph $f(x) = x^2$. The graph of the standard quadratic function is shown in Figure 2.44(a). We've identified three points on the graph.

Step 2 Graph $g(x) = (x + 1)^2$. Because we add 1 to each value of x in the domain of the standard quadratic function, $f(x) = x^2$, we shift the graph of f horizontally one unit to the left. This is shown in Figure 2.44(b). Notice that every point in the graph in Figure 2.44(b) has an x-coordinate that is one less than the x-coordinate for the corresponding point in the graph in Figure 2.44(a).

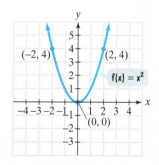

(a) The graph of $f(x) = x^2$

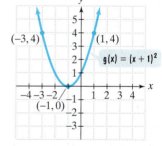

(b) The graph of $g(x) = (x + 1)^2$

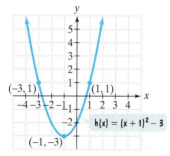

(c) The graph of $h(x) = (x + 1)^2 - 3$

Figure 2.44

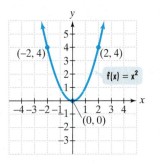

(a) The graph of $f(x) = x^2$

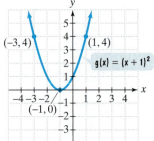

(b) The graph of $g(x) = (x + 1)^2$

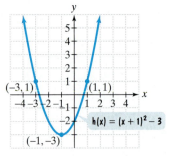

(c) The graph of $h(x) = (x + 1)^2 - 3$

Figure 2.44, repeated

Step 3   Graph $h(x) = (x + 1)^2 - 3$. Because we subtract 3, we shift the graph in Figure 2.44(b) vertically down 3 units. The graph is shown in Figure 2.44(c). Notice that every point in the graph in Figure 2.44(c) has a y-coordinate that is three less than the y-coordinate of the corresponding point in the graph in Figure 2.44(b).

Check Point 3   Use the graph of $f(x) = \sqrt{x}$ to obtain the graph of $h(x) = \sqrt{x - 1} - 2$.

④ Use reflections to graph functions.

Reflections of Graphs

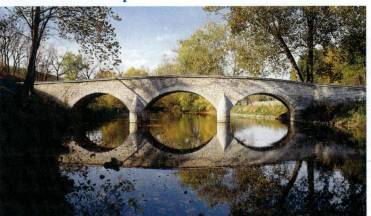

This photograph shows a reflection of an old bridge in a Maryland river. This perfect reflection occurs because the surface of the water is absolutely still. A mild breeze rippling the water's surface would distort the reflection.

Is it possible for graphs to have mirror-like qualities? Yes. Figure 2.45 shows the graphs of $f(x) = x^2$ and $g(x) = -x^2$. The graph of g is a **reflection about the x-axis** of the graph of f. In general, the graph of $y = -f(x)$ reflects the graph of f about the x-axis. Thus, the graph of g is a reflection of the graph of f about the x-axis because

$$g(x) = -x^2 = -f(x).$$

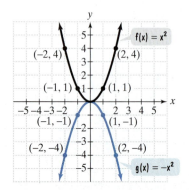

Figure 2.45 Reflections about the x-axis

Reflection about the x-Axis

The graph of $y = -f(x)$ is the graph of $y = f(x)$ reflected about the x-axis.

EXAMPLE 4   Reflection about the *x*-Axis

Use the graph of $f(x) = \sqrt{x}$ to obtain the graph of $g(x) = -\sqrt{x}$.

Solution   Compare the equations for $f(x) = \sqrt{x}$ and $g(x) = -\sqrt{x}$. The graph of g is a reflection about the *x*-axis of the graph of f because

$$g(x) = -\sqrt{x} = -f(x).$$

We have created a table showing some of the coordinates for f and g. The graphs of f and g are shown in Figure 2.46.

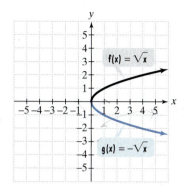

Figure 2.46 Reflecting $f(x) = \sqrt{x}$ about the *x*-axis

x	$f(x) = \sqrt{x}$	$(x, f(x))$	$g(x) = -\sqrt{x}$	$(x, g(x))$
0	$\sqrt{0} = 0$	$(0, 0)$	$-\sqrt{0} = 0$	$(0, 0)$
1	$\sqrt{1} = 1$	$(1, 1)$	$-\sqrt{1} = -1$	$(1, -1)$
4	$\sqrt{4} = 2$	$(4, 2)$	$-\sqrt{4} = -2$	$(4, -2)$

Check Point 4   Use the graph of $f(x) = |x|$ to obtain the graph of $g(x) = -|x|$.

It is also possible to reflect graphs about the *y*-axis.

Reflection about the *y*-Axis
The graph of $y = f(-x)$ is the graph of $y = f(x)$ reflected about the *y*-axis.

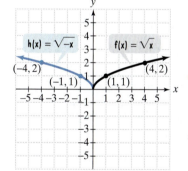

Figure 2.47 Reflecting $f(x) = \sqrt{x}$ about the *y*-axis

EXAMPLE 5   Reflection about the *y*-Axis

Use the graph of $f(x) = \sqrt{x}$ to obtain the graph of $h(x) = \sqrt{-x}$.

Solution   Compare the equations for $f(x) = \sqrt{x}$ and $h(x) = \sqrt{-x}$. The graph of h is a reflection about the *y*-axis of the graph of f because

$$h(x) = \sqrt{-x} = f(-x).$$

We have created tables showing some of the coordinates for f and h. The graphs of f and h are shown in Figure 2.47.

x	$f(x) = \sqrt{x}$	$(x, f(x))$
0	$\sqrt{0} = 0$	$(0, 0)$
1	$\sqrt{1} = 1$	$(1, 1)$
4	$\sqrt{4} = 2$	$(4, 2)$

x	$h(x) = \sqrt{-x}$	$(x, h(x))$
0	$\sqrt{-0} = \sqrt{0} = 0$	$(0, 0)$
-1	$\sqrt{-(-1)} = \sqrt{1} = 1$	$(-1, 1)$
-4	$\sqrt{-(-4)} = \sqrt{4} = 2$	$(-4, 2)$

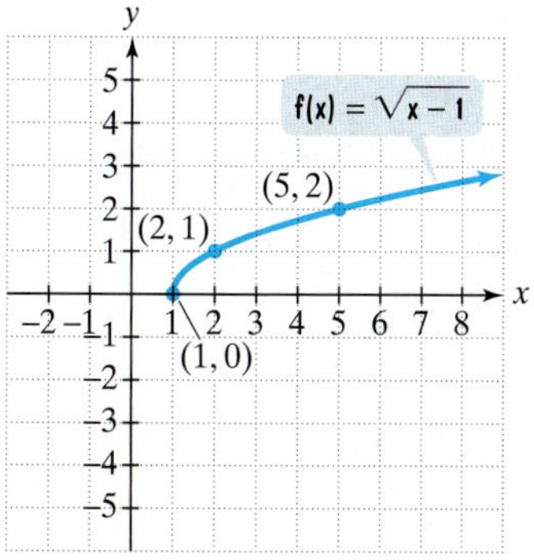

Check Point 5   Use the graph of $f(x) = \sqrt{x - 1}$ in Figure 2.48 to obtain the graph of $h(x) = \sqrt{-x - 1}$.

Figure 2.48

5 Use vertical stretching and shrinking to graph functions.

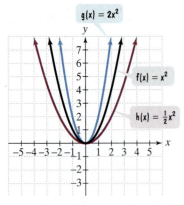

Figure 2.49 Stretching and shrinking $f(x) = x^2$

Vertical Stretching and Shrinking

Morphing does much more than move an image horizontally, vertically, or about an axis. An object having one shape is transformed into a different shape. Horizontal shifts, vertical shifts, and reflections do not change the basic shape of a graph. How can we shrink and stretch graphs, thereby altering their basic shapes?

Look at the three graphs in Figure 2.49. The black graph in the middle is the graph of the standard quadratic function, $f(x) = x^2$. Now, look at the blue graph on the top. The equation of this graph is $g(x) = 2x^2$. Thus, for each x, the y-coordinate of g is 2 times as large as the corresponding y-coordinate on the graph of f. The result is a narrower graph. We say that the graph of g is obtained by vertically *stretching* the graph of f. Now, look at the red graph on the bottom. The equation of this graph is $h(x) = \frac{1}{2}x^2$, or $h(x) = \frac{1}{2}f(x)$. Thus, for each x, the y-coordinate of h is one-half as large as the corresponding y-coordinate on the graph of f. The result is a wider graph. We say that the graph of h is obtained by vertically *shrinking* the graph of f.

These observations can be summarized as follows:

> ### Stretching and Shrinking Graphs
> Let f be a function and c a positive real number.
> - If $c > 1$, the graph of $y = cf(x)$ is the graph of $y = f(x)$ vertically stretched by multiplying each of its y-coordinates by c.
> - If $0 < c < 1$, the graph of $y = cf(x)$ is the graph of $y = f(x)$ vertically shrunk by multiplying each of its y-coordinates by c.

EXAMPLE 6   Vertically Stretching a Graph

Use the graph of $f(x) = |x|$ to obtain the graph of $g(x) = 2|x|$.

Solution   The graph of $g(x) = 2|x|$ is obtained by vertically stretching the graph of $f(x) = |x|$. We have constructed a table showing some of the coordinates for f and g. Observe that the y-coordinate on the graph of g is twice as large as the corresponding y-coordinate on the graph of f. The graphs of f and g are shown in Figure 2.50.

x	$f(x) = \lvert x\rvert$	$(x, f(x))$	$g(x) = 2\lvert x\rvert = 2f(x)$	$(x, g(x))$
-2	$\lvert -2\rvert = 2$	$(-2, 2)$	$2\lvert -2\rvert = 4$	$(-2, 4)$
-1	$\lvert -1\rvert = 1$	$(-1, 1)$	$2\lvert -1\rvert = 2$	$(-1, 2)$
0	$\lvert 0\rvert = 0$	$(0, 0)$	$2\lvert 0\rvert = 0$	$(0, 0)$
1	$\lvert 1\rvert = 1$	$(1, 1)$	$2\lvert 1\rvert = 2$	$(1, 2)$
2	$\lvert 2\rvert = 2$	$(2, 2)$	$2\lvert 2\rvert = 4$	$(2, 4)$

 Check Point 6   Use the graph of $f(x) = |x|$ to obtain the graph of $g(x) = 3|x|$.

Figure 2.50 Stretching $f(x) = |x|$

EXAMPLE 7 Vertically Shrinking a Graph

Use the graph of $f(x) = |x|$ to obtain the graph of $h(x) = \frac{1}{2}|x|$.

Solution The graph of $h(x) = \frac{1}{2}|x|$ is obtained by vertically shrinking the graph of $f(x) = |x|$. We have constructed a table showing some of the coordinates for f and h. Observe that the y-coordinate on the graph of h is one-half the corresponding y-coordinate on the graph of f. The graphs of f and h are shown in Figure 2.51.

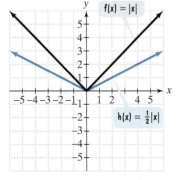

Figure 2.51 Shrinking $f(x) = |x|$

x	$f(x) = |x|$	$(x, f(x))$	$h(x) = \frac{1}{2}|x| = \frac{1}{2}f(x)$	$(x, h(x))$
-2	$|-2| = 2$	$(-2, 2)$	$\frac{1}{2}|-2| = 1$	$(-2, 1)$
-1	$|-1| = 1$	$(-1, 1)$	$\frac{1}{2}|-1| = \frac{1}{2}$	$(-1, \frac{1}{2})$
0	$|0| = 0$	$(0, 0)$	$\frac{1}{2}|0| = 0$	$(0, 0)$
1	$|1| = 1$	$(1, 1)$	$\frac{1}{2}|1| = \frac{1}{2}$	$(1, \frac{1}{2})$
2	$|2| = 2$	$(2, 2)$	$\frac{1}{2}|2| = 1$	$(2, 1)$

Check Point 7 Use the graph of $f(x) = |x|$ to obtain the graph of $h(x) = \frac{1}{4}|x|$.

6 Graph functions involving a sequence of transformations.

Sequences of Transformations

Table 2.5 summarizes the procedures for transforming the graph of $y = f(x)$.

Table 2.5 Summary of Transformations
In each case, c represents a positive real number.

To Graph:	Draw the Graph of f and:	Changes in the Equation of $y = f(x)$
Vertical shifts		
$\quad y = f(x) + c$	Raise the graph of f by c units.	c is added to $f(x)$.
$\quad y = f(x) - c$	Lower the graph of f by c units.	c is subtracted from $f(x)$.
Horizontal shifts		
$\quad y = f(x + c)$	Shift the graph of f to the left c units.	x is replaced with $x + c$.
$\quad y = f(x - c)$	Shift the graph of f to the right c units.	x is replaced with $x - c$.
Reflection about the x-axis $\quad y = -f(x)$	Reflect the graph of f about the x-axis.	$f(x)$ is multiplied by -1.
Reflection about the y-axis $\quad y = f(-x)$	Reflect the graph of f about the y-axis.	x is replaced with $-x$.
Vertical stretching or shrinking		
$\quad y = cf(x), c > 1$	Multiply each y-coordinate of $y = f(x)$ by c, vertically stretching the graph of f.	$f(x)$ is multiplied by $c, c > 1$.
$\quad y = cf(x), 0 < c < 1$	Multiply each y-coordinate of $y = f(x)$ by c, vertically shrinking the graph of f.	$f(x)$ is multiplied by $c, 0 < c < 1$.

A function involving more than one transformation can be graphed by performing transformations in the following order:

1. Horizontal shifting
2. Vertical stretching or shrinking
3. Reflecting
4. Vertical shifting

EXAMPLE 8 **Graphing Using a Sequence of Transformations**

Use the graph of $f(x) = \sqrt{x}$ to graph $g(x) = \sqrt{1 - x} + 3$.

Solution The following sequence of steps is illustrated in Figure 2.52. We begin with the graph of $f(x) = \sqrt{x}$.

Step 1   Horizontal Shifting   Graph $y = \sqrt{x + 1}$.   Because x is replaced with $x + 1$, the graph of $f(x) = \sqrt{x}$ is shifted 1 unit to the left.

Step 2   Vertical Stretching or Shrinking   Because the equation $y = \sqrt{x + 1}$ is not multiplied by a constant in $g(x) = \sqrt{1 - x} + 3$, no stretching or shrinking is involved.

Step 3   Reflecting   We are interested in graphing $y = \sqrt{1 - x} + 3$, or $y = \sqrt{-x + 1} + 3$. We have now graphed $y = \sqrt{x + 1}$. We can graph $y = \sqrt{-x + 1}$ by noting that x is replaced with $-x$. Thus, we graph $y = \sqrt{-x + 1}$ by reflecting the graph of $y = \sqrt{x + 1}$ about the y-axis.

Step 4   Vertical Shifting   We can use the graph of $y = \sqrt{1 - x}$ to get the graph of $g(x) = \sqrt{1 - x} + 3$. Because 3 is added, shift the graph of $y = \sqrt{1 - x}$ up 3 units.

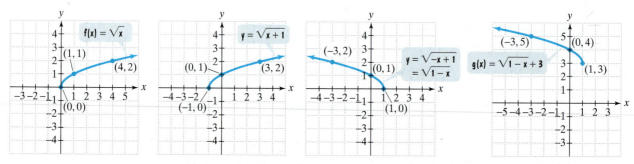

Figure 2.52 Using $f(x) = \sqrt{x}$ to graph $g(x) = \sqrt{1 - x} + 3$

Check Point 8   Use the graph of $f(x) = x^2$ to graph $g(x) = -(x - 2)^2 + 3$.

EXERCISE SET 2.5

Practice Exercises

In Exercises 1–10, begin by graphing the standard quadratic function, $f(x) = x^2$. Then use transformations of this graph to graph the given function.

1. $g(x) = x^2 - 2$

2. $g(x) = x^2 - 1$

3. $g(x) = (x - 2)^2$

4. $g(x) = (x - 1)^2$

5. $h(x) = -(x - 2)^2$

6. $h(x) = -(x - 1)^2$

7. $h(x) = (x - 2)^2 + 1$

8. $h(x) = (x - 1)^2 + 2$

9. $g(x) = 2(x - 2)^2$

10. $g(x) = \frac{1}{2}(x - 1)^2$

In Exercises 11–22, begin by graphing the square root function, $f(x) = \sqrt{x}$. Then use transformations of this graph to graph the given function.

11. $g(x) = \sqrt{x} + 2$

12. $g(x) = \sqrt{x} + 1$

13. $g(x) = \sqrt{x + 2}$

14. $g(x) = \sqrt{x + 1}$

15. $h(x) = -\sqrt{x + 2}$

16. $h(x) = -\sqrt{x + 1}$

17. $h(x) = \sqrt{-x + 2}$

18. $h(x) = \sqrt{-x + 1}$

19. $g(x) = \frac{1}{2}\sqrt{x + 2}$

20. $g(x) = 2\sqrt{x + 1}$

21. $h(x) = \sqrt{x + 2} - 2$

22. $h(x) = \sqrt{x + 1} - 1$

In Exercises 23–34, begin by graphing the absolute value function, $f(x) = |x|$. Then use transformations of this graph to graph the given function.

23. $g(x) = |x| + 4$

24. $g(x) = |x| + 3$

25. $g(x) = |x + 4|$

26. $g(x) = |x + 3|$

27. $h(x) = |x + 4| - 2$

28. $h(x) = |x + 3| - 2$

29. $h(x) = -|x + 4|$

30. $h(x) = -|x + 3|$

31. $g(x) = -|x + 4| + 1$

32. $g(x) = -|x + 4| + 2$

33. $h(x) = 2|x + 4|$

34. $h(x) = 2|x + 3|$

In Exercises 35–44, begin by graphing the standard cubic function, $f(x) = x^3$. Then use transformations of this graph to graph the given function.

35. $g(x) = x^3 - 3$

36. $g(x) = x^3 - 2$

37. $g(x) = (x - 3)^3$

38. $g(x) = (x - 2)^3$

39. $h(x) = -x^3$

40. $h(x) = -(x - 2)^3$

41. $h(x) = \frac{1}{2}x^3$

42. $h(x) = \frac{1}{4}x^3$

43. $r(x) = (x - 3)^3 + 2$

44. $r(x) = (x - 2)^3 + 1$

In Exercises 45–52, use the graph of the function f to sketch the graph of the given function g.

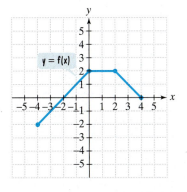

45. $g(x) = f(x) + 1$

46. $g(x) = f(x) + 2$

47. $g(x) = f(x + 1)$

48. $g(x) = f(x + 2)$

49. $g(x) = -f(x)$

50. $g(x) = \frac{1}{2}f(x)$

51. $g(x) = \frac{1}{2}f(x + 1)$

52. $g(x) = -f(x + 2)$

In Exercises 53–56, write a possible equation for the function whose graph is shown. Each graph shows a transformation of a common function.

53.

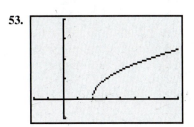

$[-2, 8, 1]$ by $[-1, 4, 1]$

54.

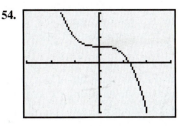

$[-3, 3, 1]$ by $[-6, 6, 1]$

55.

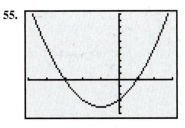

$[-5, 3, 1]$ by $[-5, 10, 1]$

56.

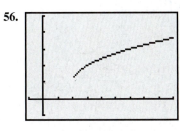

$[-1, 9, 1]$ by $[-1, 5, 1]$

Application Exercises

57. The function $f(x) = 2.9\sqrt{x} + 20.1$ models the median height, $f(x)$, in inches, of boys who are x months of age. The graph of f is shown.

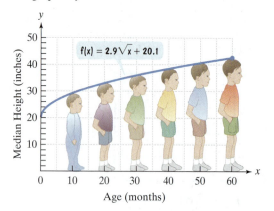

$f(x) = 2.9\sqrt{x} + 20.1$

a. Describe how the graph can be obtained using transformations of the square root function $f(x) = \sqrt{x}$.

b. According to the model, what is the median height of boys who are 48 months, or four years, old? Use a calculator and round to the nearest tenth of an inch. The actual median height for boys at 48 months is 40.8 inches. How well does the model describe the actual height?

c. Use the model to find the average rate of change, in inches per month, between birth and 10 months. Round to the nearest tenth.

d. Use the model to find the average rate of change, in inches per month, between 50 and 60 months. Round to the nearest tenth. How does this compare with your answer in part (c)? How is this difference shown by the graph?

58. The graph shows the amount of money, in billions of dollars, of new student loans from 1993 through 2000.

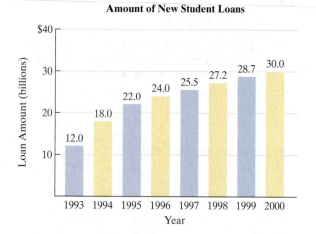

Amount of New Student Loans

Year

Source: U.S. Department of Education

The data shown can be modeled by the function $f(x) = 6.75\sqrt{x} + 12$, where $f(x)$ is the amount, in billions of dollars, of new student loans x years after 1993.

a. Describe how the graph of f can be obtained using transformations of the square root function $f(x) = \sqrt{x}$. Then sketch the graph of f over the interval $0 \le x \le 9$. If applicable, use a graphing utility to verify your hand-drawn graph.

b. According to the model, how much was loaned in 2000? Round to the nearest tenth of a billion. How well does the model describe the actual data?

c. Use the model to find the average rate of change, in billions of dollars per year, between 1993 and 1995. Round to the nearest tenth.

d. Use the model to find the average rate of change, in billions of dollars per year, between 1998 and 2000. Round to the nearest tenth. How does this compare with you answer in part (c)? How is this difference shown by your graph?

e. Rewrite the function so that it represents the amount, $f(x)$, in billions of dollars, of new student loans x years after 1995.

Writing in Mathematics

59. What must be done to a function's equation so that its graph is shifted vertically upward?

60. What must be done to a function's equation so that its graph is shifted horizontally to the right?

61. What must be done to a function's equation so that its graph is reflected about the x-axis?

62. What must be done to a function's equation so that its graph is reflected about the y-axis?

63. What must be done to a function's equation so that its graph is stretched?

Technology Exercises

64. a. Use a graphing utility to graph $f(x) = x^2 + 1$.

    b. Graph $f(x) = x^2 + 1$, $g(x) = f(2x)$, $h(x) = f(3x)$, and $k(x) = f(4x)$ in the same viewing rectangle.

    c. Describe the relationship among the graphs of f, g, h, and k, with emphasis on different values of x for points on all four graphs that give the same y-coordinate.

    d. Generalize by describing the relationship between the graph of f and the graph of g, where $g(x) = f(cx)$ for $c > 1$.

    e. Try out your generalization by sketching the graphs of $f(cx)$ for $c = 1$, $c = 2$, $c = 3$, and $c = 4$ for a function of your choice.

65. a. Use a graphing utility to graph $f(x) = x^2 + 1$.

    b. Graph $f(x) = x^2 + 1$, and $g(x) = f(\frac{1}{2}x)$, and $h(x) = f(\frac{1}{4}x)$ in the same viewing rectangle.

    c. Describe the relationship among the graphs of f, g, and h, with emphasis on different values of x for points on all three graphs that give the same y-coordinate.

    d. Generalize by describing the relationship between the graph of f and the graph of g, where $g(x) = f(cx)$ for $0 < c < 1$.

    e. Try out your generalization by sketching the graphs of $f(cx)$ for $c = 1$, and $c = \frac{1}{2}$, and $c = \frac{1}{4}$ for a function of your choice.

Critical Thinking Exercises

66. Which one of the following is true?

  a. If $f(x) = |x|$ and $g(x) = |x + 3| + 3$, then the graph of g is a translation of three units to the right and three units upward of the graph of f.

  b. If $f(x) = -\sqrt{x}$ and $g(x) = \sqrt{-x}$, then f and g have identical graphs.

  c. If $f(x) = x^2$ and $g(x) = 5(x^2 - 2)$, then the graph of g can be obtained from the graph of f by stretching f five units followed by a downward shift of two units.

  d. If $f(x) = x^3$ and $g(x) = -(x - 3)^3 - 4$, then the graph of g can be obtained from the graph of f by moving f three units to the right, reflecting in the x-axis, and then moving the resulting graph down four units.

In Exercises 67–70, functions f and g are graphed in the same rectangular coordinate system. If g is obtained from f through a sequence of transformations, find an equation for g.

67.

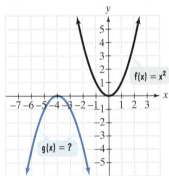

68.

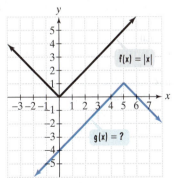

69.

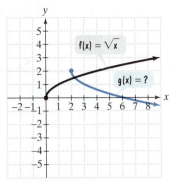

70.

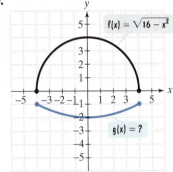

For Exercises 71–74, assume that (a, b) is a point on the graph of f. What is the corresponding point on the graph of each of the following functions?

71. $y = f(-x)$ **72.** $y = 2f(x)$

73. $y = f(x - 3)$ **74.** $y = f(x) - 3$

Group Exercise

75. This activity is a group research project on morphing and should result in a presentation made by group members to the entire class. Be sure to include morphing images that will intrigue class members. You should have no problem finding an array of fascinating images online. Also include a discussion of films using spectacular morphing effects. Rent videos of these films and show appropriate excerpts.

SECTION 2.6 Combinations of Functions; Composite Functions

Objectives

1. Combine functions arithmetically, specifying domains.

2. Form composite functions.

3. Determine domains for composite functions.

4. Write functions as compositions.

They say a fool and his money are soon parted and the rest of us just wait to be taxed. It's hard to believe that the United States was a low-tax country in the early part of the twentieth century. Figure 2.53 shows how the tax burden has grown since then. We can use the information shown to illustrate how two functions can be combined to form a new function. In this section, you will learn how to combine functions to obtain new functions.

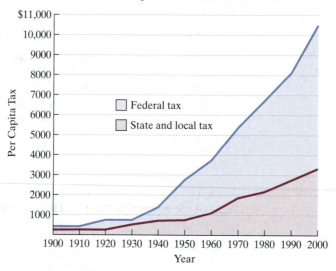

U.S. Per Capita Tax Burden in 2000 Dollars

Figure 2.53   *Source:* Tax Foundation

---

1 Combine functions arithmetically, specifying domains.

Combinations of Functions

To begin our discussion, take a look at the information shown for the year 2000. The total per capita tax burden is approximately $10,500. The per capita state and local tax is approximately $3400. The per capita federal tax is the difference between these amounts.

$$\text{Per capita federal tax} = \$10,500 - \$3400 = \$7100$$

We can think of this subtraction as the subtraction of function values. We do this by introducing the following functions:

Let $T(x)$ = total per capita tax in year x.

Let $S(x)$ = per capita state and local tax in year x.

Using Figure 2.53, we see that

$$T(2000) = \$10{,}500 \quad \text{and} \quad S(2000) = \$3400.$$

We can subtract these function values by introducing a new function, $T - S$, defined by the subtraction of $T(x)$ and $S(x)$. Thus,

$$(T - S)(x) = T(x) - S(x) = \begin{array}{l} \text{total per capita tax in year x} \\ \text{minus state and local per} \\ \text{capita tax in year x.} \end{array}$$

For example,

$$(T - S)(2000) = T(2000) - S(2000) = \$10{,}500 - \$3400 = \$7100.$$

> In 2000, the difference between total tax and state and local tax was $7100. This is the per capita federal tax.

Figure 2.53 illustrates that information involving differences of functions often appears in graphs seen in newspapers and magazines. Like numbers and algebraic expressions, two functions can be added, subtracted multiplied, or divided as long as there are numbers common to the domains of both functions. The common domain for functions T and S in Figure 2.53 is

$$\{1900, 1901, 1902, 1903, \dots, 2000\}.$$

Because functions are usually given as equations, we perform operations by carrying out these operations with the algebraic expressions that appear on the right side of the equations. For example, we can combine the following two functions using addition:

$$f(x) = 2x + 1 \quad \text{and} \quad g(x) = x^2 - 4.$$

To do so, we add the terms to the right of the equal sign for $f(x)$ to the terms to the right of the equal sign for $g(x)$. Here is how it's done:

$$
\begin{aligned}
(f + g)(x) &= f(x) + g(x) \\
&= (2x + 1) + (x^2 - 4) && \text{Add terms for f(x) and g(x).} \\
&= 2x - 3 + x^2 && \text{Combine like terms.} \\
&= x^2 + 2x - 3 && \text{Arrange terms in descending powers of x.}
\end{aligned}
$$

The name of this new function is $f + g$. Thus, the sum $f + g$ is the function defined by $(f + g)(x) = x^2 + 2x - 3$. The domain of $f + g$ consists of the numbers x that are in the domain of f and in the domain of g. Because neither f nor g contains division or even roots, the domain of each function is the set of all real numbers. Thus, the domain of $f + g$ is also the set of all real numbers.

EXAMPLE 1 Finding the Sum of Two Functions

Let $f(x) = x^2 - 3$ and $g(x) = 4x + 5$. Find:

    a. $(f + g)(x)$ **b.** $(f + g)(3)$.

Solution

a. $(f + g)(x) = f(x) + g(x) = (x^2 - 3) + (4x + 5) = x^2 + 4x + 2$. Thus, $(f + g)(x) = x^2 + 4x + 2$.

b. We find $(f + g)(3)$ by substituting 3 for x in the equation for $f + g$.

$$(f + g)(x) = x^2 + 4x + 2 \qquad \textit{This is the equation for } f + g.$$

Substitute 3 for x.

$$(f + g)(3) = 3^2 + 4 \cdot 3 + 2 = 9 + 12 + 2 = 23$$

Check Point 1 Let $f(x) = 3x^2 + 4x - 1$ and $g(x) = 2x + 7$. Find:

    a. $(f + g)(x)$ **b.** $(f + g)(4)$.

Here is a general definition for function addition:

> ### The Sum of Functions
>
> Let f and g be two functions. The **sum $f + g$** is the function defined by
> $$(f + g)(x) = f(x) + g(x).$$
> The domain of $f + g$ is the set of all real numbers that are common to the domain of f and the domain of g.

EXAMPLE 2 Adding Functions and Determining the Domain

Let $f(x) = \sqrt{x + 3}$ and $g(x) = \sqrt{x - 2}$. Find:

    a. $(f + g)(x)$ **b.** the domain of $f + g$.

Solution

a. $(f + g)(x) = f(x) + g(x) = \sqrt{x + 3} + \sqrt{x - 2}$

b. The domain of $f + g$ is the set of all real numbers that are common to the domain of f and the domain of g. Thus, we must find the domains of f and g. We will do so for f first.

    Note that $f(x) = \sqrt{x + 3}$ is a function involving the square root of $x + 3$. Because the square root of a negative quantity is not a real number, the value of $x + 3$ must be nonnegative. Thus, the domain of f is all x such that $x + 3 \geq 0$. Equivalently, the domain is $\{x | x \geq -3\}$, or $[-3, \infty)$.

    Likewise, $g(x) = \sqrt{x - 2}$ is also a square root function. Because the square root of a negative quantity is not a real number, the value of $x - 2$ must be nonnegative. Thus, the domain of g is all x such that $x - 2 \geq 0$. Equivalently, the domain is $\{x | x \geq 2\}$, or $[2, \infty)$.

    Now, we can use a number line to determine the domain of $f + g$. Figure 2.54 shows the domain of f in blue and the domain of g in red. Can you see that all real numbers greater than or equal to 2 are common to both domains? This is shown in purple on the number line. Thus, the domain of $f + g$ is $[2, \infty)$.

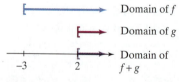

Domain of f

Domain of g

Domain of $f + g$

$-3 \qquad 2$

Figure 2.54 Finding the domain of the sum $f + g$

Technology

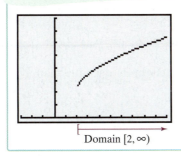

The graph on the left is the graph of

$$y = \sqrt{x + 3} + \sqrt{x - 2}$$

in a $[-3, 10, 1]$ by $[0, 8, 1]$ viewing rectangle. The graph reveals what we discovered algebraically in Example 2(b). The domain of this function is $[2, \infty)$.

Domain $[2, \infty)$

Check Point 2

Let $f(x) = \sqrt{x - 3}$ and $g(x) = \sqrt{x + 1}$. Find:

a. $(f + g)(x)$ **b.** the domain of $f + g$.

We can also combine functions using subtraction, multiplication, and division by performing operations with the algebraic expressions that appear on the right side of the equations. For example, the functions $f(x) = x + 3$ and $g(x) = x - 1$ can be combined to form the difference, product, and quotient of f and g. Here's how it's done.

Difference: f − g

$$(f - g)(x) = f(x) - g(x)$$

$$= (x + 3) - (x - 1) = x + 3 - x + 1 = 4$$

Product: fg

$$(fg)(x) = f(x) \cdot g(x)$$

$$= (x + 3)(x - 1) = x^2 + 2x - 3$$

Quotient: $\frac{f}{g}$

$$\left(\frac{f}{g}\right)(x) = \frac{f(x)}{g(x)} = \frac{x + 3}{x - 1}, \quad x \neq 1$$

Just like the domain for $f + g$, the domain for each of these functions consists of all real numbers that are common to the domains of f and g. In the case of the quotient function $\dfrac{f(x)}{g(x)}$, we must remember not to divide by 0, so we add the further restriction that $g(x) \neq 0$.

The following definitions summarize our discussion:

Definitions: Sum, Difference, Product, and Quotient of Functions

Let f and g be two functions. The **sum** $f + g$, the **difference** $f - g$, the **product** fg, and the **quotient** $\frac{f}{g}$ are functions whose domains are the set of all real numbers common to the domains of f and g, defined as follows:

1. Sum: $(f + g)(x) = f(x) + g(x)$

2. Difference: $(f - g)(x) = f(x) - g(x)$

3. Product: $(fg)(x) = f(x) \cdot g(x)$

4. Quotient: $\left(\dfrac{f}{g}\right)(x) = \dfrac{f(x)}{g(x)}$, provided $g(x) \neq 0$

EXAMPLE 3   Combining Functions

If $f(x) = 2x - 1$ and $g(x) = x^2 + x - 2$, find:

a. $(f - g)(x)$ **b.** $(fg)(x)$ **c.** $\left(\dfrac{f}{g}\right)(x)$.

Determine the domain for each function.

Solution

a. $(f - g)(x) = f(x) - g(x)$ *This is the definition of the difference $f - g$.*

$\qquad = (2x - 1) - (x^2 + x - 2)$ *Subtract $g(x)$ from $f(x)$.*

$\qquad = 2x - 1 - x^2 - x + 2$ *Perform the subtraction.*

$\qquad = -x^2 + x + 1$ *Combine like terms and arrange terms in descending powers of x.*

b. $(fg)(x) = f(x) \cdot g(x)$ *This is the definition of the product fg.*

$\qquad = (2x - 1)(x^2 + x - 2)$ *Multiply $f(x)$ and $g(x)$.*

$\qquad = 2x(x^2 + x - 2) - 1(x^2 + x - 2)$ *Multiply each term in the second factor by $2x$ and -1, respectively.*

$\qquad = 2x^3 + 2x^2 - 4x - x^2 - x + 2$ *Use the distributive property.*

$\qquad = 2x^3 + (2x^2 - x^2) + (-4x - x) + 2$ *Rearrange terms so that like terms are adjacent.*

$\qquad = 2x^3 + x^2 - 5x + 2$ *Combine like terms.*

c. $\left(\dfrac{f}{g}\right)(x) = \dfrac{f(x)}{g(x)}$ *This is the definition of the quotient $\dfrac{f}{g}$.*

$\qquad = \dfrac{2x - 1}{x^2 + x - 2}$ *Divide the algebraic expressions for $f(x)$ and $g(x)$.*

> ## Study Tip
>
> If the function $\frac{f}{g}$ can be simplified, determine the domain *before* simplifying.
>
> **EXAMPLE:**
>
> $f(x) = x^2 - 4$ and
>
> $g(x) = x - 2$
>
> $\left(\dfrac{f}{g}\right)(x) = \dfrac{x^2 - 4}{x - 2}$
>
> Domain of $\left(\dfrac{f}{g}\right)$ is $\{x | x \neq 2\}$.
>
> $= \dfrac{\overset{1}{\cancel{(x + 2)}}(x - 2)}{\cancel{(x - 2)}} = x + 2$

Because the equations for f and g do not involve division or contain even roots, the domain of both f and g is the set of all real numbers. Thus, the domain of $f - g$ and fg is the set of all real numbers. However, for $\frac{f}{g}$, the denominator cannot equal zero. We can factor the denominator as follows:

$$\left(\dfrac{f}{g}\right)(x) = \dfrac{2x - 1}{x^2 + x - 2} = \dfrac{2x - 1}{(x + 2)(x - 1)}.$$

Because $x + 2 \neq 0$, $x \neq -2$. Because $x - 1 \neq 0$, $x \neq 1$.

We see that the domain for $\frac{f}{g}$ is the set of all real numbers except -2 and 1: $\{x | x \neq -2, x \neq 1\}$.

Check Point 3   If $f(x) = x - 5$ and $g(x) = x^2 - 1$, find:

a. $(f - g)(x)$ **b.** $(fg)(x)$ **c.** $\left(\dfrac{f}{g}\right)(x)$.

Determine the domain for each function.

2   Form composite functions.

Composite Functions

There is another way of combining two functions. To help understand this new combination, suppose that your computer store is having a sale. The models that are on sale cost either $300 less than the regular price or 85% of the regular price. If x represents the computer's regular price, both discounts can be described with the following functions:

$$f(x) = x - 300 \qquad\qquad g(x) = 0.85x.$$

> The computer is on sale for $300 less than its regular price.

> The computer is on sale for 85% of its regular price.

At the store, you bargain with the salesperson. Eventually, she makes an offer you can't refuse: The sale price is 85% of the regular price followed by a $300 reduction:

$$0.85x - 300.$$

> 85% of the regular price

> followed by a $300 reduction

In terms of functions f and g, this offer can be obtained by taking the output of $g(x) = 0.85x$, namely $0.85x$, and using it as the input of f:

$$f(x) = x - 300$$

> Replace x with 0.85x, the output of g(x) = 0.85x.

$$f(0.85x) = 0.85x - 300.$$

Because $0.85x$ is $g(x)$, we can write this last equation as

$$f(g(x)) = 0.85x - 300.$$

We read this equation as "f of g of x is equal to $0.85x - 300$." We call $f(g(x))$ the *composition of the function f with g*, or a *composite function*. This composite function is written $f \circ g$. Thus,

$$(f \circ g)(x) = f(g(x)) = 0.85x - 300.$$

Like all functions, we can evaluate $f \circ g$ for a specified value of x in the function's domain. For example, here's how to find the value of this function at 1400:

$$(f \circ g)(x) = 0.85x - 300 \qquad \text{\textit{This composite function describes the offer you cannot refuse.}}$$

> Replace x with 1400.

$$(f \circ g)(1400) = 0.85(1400) - 300 = 1190 - 300 = 890.$$

This means that a computer that regularly sells for $1400 is on sale for $890 subject to both discounts.

    Before you run out to buy a new computer, let's generalize our discussion of the computer's double discount and define the composition of any two functions.

The Composition of Functions

The **composition of the function** f **with** g is denoted by $f \circ g$ and is defined by the equation

$$(f \circ g)(x) = f(g(x)).$$

The domain of the **composite function** $f \circ g$ is the set of all x such that

1. x is in the domain of g and
2. $g(x)$ is in the domain of f.

The composition of f with g, $f \circ g$, is pictured as a machine with inputs and outputs in Figure 2.55. The diagram indicates that the output of g, or $g(x)$, becomes the input for "machine" f. If $g(x)$ is not in the domain of f, it cannot be input into machine f, and so $g(x)$ must be discarded.

Inputs, x

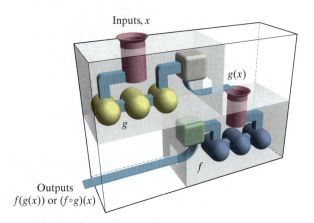

Figure 2.55 Inputting one function into a second function

Outputs
$f(g(x))$ or $(f \circ g)(x)$

$g(x)$

EXAMPLE 4 Forming Composite Functions

Given $f(x) = 3x - 4$ and $g(x) = x^2 + 6$, find:

  a. $(f \circ g)(x)$ **b.** $(g \circ f)(x)$.

Solution

a. We begin with $(f \circ g)(x)$, the composition of f with g. Because $(f \circ g)(x)$ means $f(g(x))$, we must replace each occurrence of x in the equation for f with $g(x)$.

$$f(x) = 3x - 4 \qquad \text{This is the given equation for } f.$$

Replace x with g(x).

$$(f \circ g)(x) = f(g(x)) = 3g(x) - 4$$
$$= 3(x^2 + 6) - 4 \quad \text{Because } g(x) = x^2 + 6, \text{ replace } g(x) \text{ with } x^2 + 6.$$
$$= 3x^2 + 18 - 4 \quad \text{Use the distributive property.}$$
$$= 3x^2 + 14 \qquad \text{Simplify.}$$

Thus, $(f \circ g)(x) = 3x^2 + 14$.

b. Next, we find $(g \circ f)(x)$, the composition of g with f. Because $(g \circ f)(x)$ means $g(f(x))$, we must replace each occurrence of x in the equation for g with $f(x)$.

$$g(x) = x^2 + 6 \qquad \text{\textit{This is the given equation for g.}}$$

Replace x with f(x).

$$(g \circ f)(x) = g(f(x)) = (f(x))^2 + 6$$
$$= (3x - 4)^2 + 6 \qquad \text{\textit{Because f(x) = 3x − 4, replace}}$$
$$\text{\textit{f(x) with 3x − 4.}}$$
$$= 9x^2 - 24x + 16 + 6 \qquad \text{\textit{Use (A − B)}}^2 = \text{\textit{A}}^2 − 2\text{\textit{AB}} + \text{\textit{B}}^2$$
$$\text{\textit{to square 3x − 4.}}$$
$$= 9x^2 - 24x + 22 \qquad \text{\textit{Simplify.}}$$

Thus, $(g \circ f)(x) = 9x^2 - 24x + 22$. Notice that $(f \circ g)(x)$ is not the same function as $(g \circ f)(x)$.

Check Point 4 Given $f(x) = 5x + 6$ and $g(x) = x^2 - 1$, find:

a. $(f \circ g)(x)$ **b.** $(g \circ f)(x)$.

3   Determine domains for composite functions.

We need to be careful in determining the domain for the composite function

$$(f \circ g)(x) = f(g(x)).$$

The following values must be excluded from the input x:

- If x is not in the domain of g, it must not be in the domain of $f \circ g$.
- Any x for which $g(x)$ is not in the domain of f must not be in the domain of $f \circ g$.

EXAMPLE 5   **Forming a Composite Function and Finding Its Domain**

Given $f(x) = \dfrac{2}{x - 1}$ and $g(x) = \dfrac{3}{x}$, find:

a. $(f \circ g)(x)$ **b.** the domain of $f \circ g$.

Solution

a. Because $(f \circ g)(x)$ means $f(g(x))$, we must replace x in $f(x) = \dfrac{2}{x - 1}$ with $g(x)$.

Study Tip

The procedure for simplifying complex fractions can be found in Section P.6, pages 66–67.

$$(f \circ g)(x) = f(g(x)) = \frac{2}{g(x) - 1} = \frac{2}{\frac{3}{x} - 1} = \frac{2}{\frac{3}{x} - 1} \cdot \frac{x}{x} = \frac{2x}{3 - x}$$

$$g(x) = \frac{3}{x} \qquad \text{\textit{Simplify the complex fraction by multiplying by }} \frac{x}{x}, \text{\textit{ or 1.}}$$

Thus, $(f \circ g)(x) = \dfrac{2x}{3 - x}$.

b. We determine the domain of $(f \circ g)(x)$ in two steps.

Rules for Excluding Numbers from the Domain of $(f \circ g)(x) = f(g(x))$	Applying the Rules to $f(x) = \dfrac{2}{x-1}$ and $g(x) = \dfrac{3}{x}$
If x is not in the domain of g, it must not be in the domain of $f \circ g$.	The domain of g is $\{x \mid x \neq 0\}$. Thus, 0 must be excluded from the domain of $f \circ g$.
Any x for which $g(x)$ is not in the domain of f must not be in the domain of $f \circ g$.	The domain of f is $\{x \mid x \neq 1\}$. This means we must exclude from the domain of $f \circ g$ any x for which $g(x) = 1$. $\dfrac{3}{x} = 1$ *Set g(x) equal to 1.* $3 = x$ *Multiply both sides by x.* 3 must be excluded from the domain of $f \circ g$.

The domain of $f \circ g$ is $\{x \mid x \neq 0 \text{ and } x \neq 3\}$.

Check Point 5   Given $f(x) = \dfrac{4}{x+2}$ and $g(x) = \dfrac{1}{x}$, find:

a. $(f \circ g)(x)$ **b.** the domain of $f \circ g$.

4 Write functions as compositions

Decomposing Functions

When you form a composite function, you "compose" two functions to form a new function. It is also possible to reverse this process. That is, you can "decompose" a given function and express it as a composition of two functions. Although there is more than one way to do this, there is often a "natural" selection that comes to mind first. For example, consider the function h defined by

$$h(x) = (3x^2 - 4x + 1)^5.$$

The function h takes $3x^2 - 4x + 1$ and raises it to the power 5. A natural way to write h as a composition of two functions is to raise the function $g(x) = 3x^2 - 4x + 1$ to the power 5. Thus, if we let

$$f(x) = x^5 \text{ and } g(x) = 3x^2 - 4x + 1, \text{ then}$$
$$(f \circ g)(x) = f(g(x)) = f(3x^2 - 4x + 1) = (3x^2 - 4x + 1)^5.$$

EXAMPLE 6   Writing a Function as a Composition

Express as a composition of two functions:

$$h(x) = \sqrt[3]{x^2 + 1}.$$

Solution   The function h takes $x^2 + 1$ and takes its cube root. A natural way to write h as a composition of two functions is to take the cube root of the function $g(x) = x^2 + 1$. Thus, we let

$$f(x) = \sqrt[3]{x} \text{ and } g(x) = x^2 + 1.$$

> **Study Tip**
>
> Suppose the form of function h is
>
> $h(x) = (\text{algebraic expression})^{\text{power}}$.
>
> Function h can be expressed as a composition, $f \circ g$, using
>
> $f(x) = x^{\text{power}}$
>
> $g(x) = \text{algebraic expression}$.

We can check this composition by finding $(f \circ g)(x)$. This should give the original function, namely $h(x) = \sqrt[3]{x^2 + 1}$.

$$(f \circ g)(x) = f(g(x)) = f(x^2 + 1) = \sqrt[3]{x^2 + 1} = h(x)$$

Check Point 6 Express as a composition of two functions:

$$h(x) = \sqrt{x^2 + 5}.$$

EXERCISE SET 2.6

Practice Exercises

1. If $f(x) = 2x^2 - 5$ and $g(x) = 3x + 7$, find:
   a. $(f + g)(x)$ **b.** $(f + g)(4)$.

2. If $f(x) = 3x^2 - 2x + 1$ and $g(x) = 4x - 1$, find:
   a. $(f + g)(x)$ **b.** $(f + g)(5)$.

3. If $f(x) = \sqrt{x - 6}$ and $g(x) = \sqrt{x + 2}$, find:
   a. $(f + g)(x)$ **b.** the domain of $f + g$.

4. If $f(x) = \sqrt{x - 8}$ and $g(x) = \sqrt{x + 5}$, find:
   a. $(f + g)(x)$ **b.** the domain of $f + g$.

In Exercises 5–16, find $f + g, f - g, fg,$ and $\frac{f}{g}$. Determine the domain for each function.

5. $f(x) = 2x + 3, \quad g(x) = x - 1$

6. $f(x) = 3x - 4, \quad g(x) = x + 2$

7. $f(x) = x - 5, \quad g(x) = 3x^2$

8. $f(x) = x - 6, \quad g(x) = 5x^2$

9. $f(x) = 2x^2 - x - 3, \quad g(x) = x + 1$

10. $f(x) = 6x^2 - x - 1, \quad g(x) = x - 1$

11. $f(x) = \sqrt{x}, \quad g(x) = x - 4$

12. $f(x) = \dfrac{1}{x}, \quad g(x) = x - 5$

13. $f(x) = 2 + \dfrac{1}{x}, \quad g(x) = \dfrac{1}{x}$

14. $f(x) = 6 - \dfrac{1}{x}, \quad g(x) = \dfrac{1}{x}$

15. $f(x) = \sqrt{x + 4}, \quad g(x) = \sqrt{x - 1}$

16. $f(x) = \sqrt{x + 6}, \quad g(x) = \sqrt{x - 3}$

In Exercises 17–28, find:
   a. $(f \circ g)(x)$
   b. $(g \circ f)(x)$
   c. $(f \circ g)(2)$.

17. $f(x) = 2x, \quad g(x) = x + 7$

18. $f(x) = 3x, \quad g(x) = x - 5$

19. $f(x) = x + 4, \quad g(x) = 2x + 1$

20. $f(x) = 5x + 2, \quad g(x) = 3x - 4$

21. $f(x) = 4x - 3, \quad g(x) = 5x^2 - 2$

22. $f(x) = 7x + 1, \quad g(x) = 2x^2 - 9$

23. $f(x) = x^2 + 2, \quad g(x) = x^2 - 2$

24. $f(x) = x^2 + 1, \quad g(x) = x^2 - 3$

25. $f(x) = \sqrt{x}, \quad g(x) = x - 1$

26. $f(x) = \sqrt{x}, \quad g(x) = x + 2$

27. $f(x) = 2x - 3, \quad g(x) = \dfrac{x + 3}{2}$

28. $f(x) = 6x - 3, \quad g(x) = \dfrac{x + 3}{6}$

In Exercises 29–38, find:
   a. $(f \circ g)(x)$ **b.** the domain of $f \circ g$.

29. $f(x) = \dfrac{2}{x + 3}, \quad g(x) = \dfrac{1}{x}$

30. $f(x) = \dfrac{5}{x + 4}, \quad g(x) = \dfrac{1}{x}$

31. $f(x) = \dfrac{x}{x + 1}, \quad g(x) = \dfrac{4}{x}$

32. $f(x) = \dfrac{x}{x + 5}, \quad g(x) = \dfrac{6}{x}$

33. $f(x) = \sqrt{x}, \quad g(x) = x + 3$

34. $f(x) = \sqrt{x}, \quad g(x) = x - 3$

35. $f(x) = x^2 + 4, \quad g(x) = \sqrt{1 - x}$

36. $f(x) = x^2 + 1, \quad g(x) = \sqrt{2 - x}$

37. $f(x) = 4 - x^2, \quad g(x) = \sqrt{x^2 - 4}$

38. $f(x) = 9 - x^2, \quad g(x) = \sqrt{x^2 - 9}$

In Exercises 39–46, express the given function h as a composition of two functions f and g so that $h(x) = (f \circ g)(x)$.

39. $h(x) = (3x - 1)^4$

40. $h(x) = (2x - 5)^3$

41. $h(x) = \sqrt[3]{x^2 - 9}$

42. $h(x) = \sqrt{5x^2 + 3}$

43. $h(x) = |2x - 5|$

44. $h(x) = |3x - 4|$

45. $h(x) = \dfrac{1}{2x - 3}$

46. $h(x) = \dfrac{1}{4x + 5}$

In Exercises 47–58, use the graphs of f and g to evaluate each function.

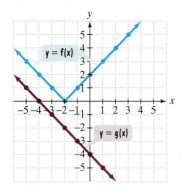

47. $(f + g)(-3)$

48. $(f + g)(-4)$

49. $(f - g)(2)$

50. $(g - f)(2)$

51. $\left(\dfrac{f}{g}\right)(-6)$

52. $\left(\dfrac{f}{g}\right)(-5)$

53. $(fg)(-4)$

54. $(fg)(-2)$

55. $(f \circ g)(2)$

56. $(f \circ g)(1)$

57. $(g \circ f)(0)$

58. $(g \circ f)(-1)$

Application Exercises

It seems that Phideau's medical bills are costing us an arm and a paw. The graph shows veterinary costs, in billions of dollars, for dogs and cats in five selected years. Let

$D(x)$ = veterinary costs, in billions of dollars, for dogs in year x

$C(x)$ = veterinary costs, in billions of dollars, for cats in year x.

Use the graph to solve Exercises 59–62.

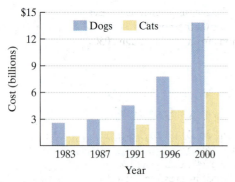

Veterinary Costs in the U.S.

Source: American Veterinary Medical Association

59. Find an estimate of $(D + C)(2000)$. What does this mean in terms of the variables in this situation?

60. Find an estimate of $(D - C)(2000)$. What does this mean in terms of the variables in this situation?

61. Using the information shown in the graph, what is the domain of $D + C$?

62. Using the information shown in the graph, what is the domain of $D - C$?

Consider the following functions:

$f(x)$ = *population of the world's more developed regions in year x*

$g(x)$ = *population of the world's less developed regions in year x*

$h(x)$ = *total world population in year x.*

Use these functions and the graph shown to answer Exercises 63–66.

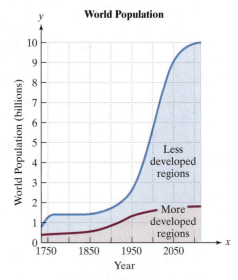

World Population

Source: Population Reference Bureau

63. What does the function $f + g$ represent?

64. What does the function $h - g$ represent?

65. Use the graph to estimate $(f + g)(2000)$.

66. Use the graph to estimate $(h - g)(2000)$.

67. A company that sells radios has a yearly fixed cost of $600,000. It costs the company $45 to produce each radio. Each radio will sell for $65. The company's costs and revenue are modeled by the following functions:

$C(x) = 600,000 + 45x$ This function models the company's costs.

$R(x) = 65x.$ This function models the company's revenue.

Find and interpret $(R - C)(20,000)$, $(R - C)(30,000)$ and $(R - C)(40,000)$.

68. A department store has two locations in a city. From 1998 through 2002, the profits for each of the store's two branches are modeled by the functions $f(x) = -0.44x + 13.62$ and $g(x) = 0.51x + 11.14$. In each model, x represents the number of years after 1998 and f and g represent the profit, in millions of dollars.
   a. What is the slope of f? Describe what this means.
   b. What is the slope of g? Describe what this means.
   c. Find $f + g$. What is the slope of this function? What does this mean?

69. The regular price of a computer is x dollars. Let $f(x) = x - 400$ and $g(x) = 0.75x$.
   a. Describe what the functions f and g model in terms of the price of the computer.
   b. Find $(f \circ g)(x)$ and describe what this models in terms of the price of the computer.
   c. Repeat part (b) for $(g \circ f)(x)$.
   d. Which composite function models the greater discount on the computer, $f \circ g$ or $g \circ f$? Explain.

70. The regular price of a pair of jeans is x dollars. Let $f(x) = x - 5$ and $g(x) = 0.6x$.
   a. Describe what functions f and g model in terms of the price of the jeans.
   b. Find $(f \circ g)(x)$ and describe what this models in terms of the price of the jeans.
   c. Repeat part (b) for $(g \circ f)(x)$.
   d. Which composite function models the greater discount on the jeans, $f \circ g$ or $g \circ f$? Explain.

Writing in Mathematics

71. If equations for functions f and g are given, explain how to find $f + g$.

72. If the equations of two functions are given, explain how to obtain the quotient function and its domain.

73. If equations for functions f and g are given, describe two ways to find $(f - g)(3)$.

74. Explain how to use the graphs in Figure 2.53 on page 248 to estimate the per capita federal tax for any one of the years shown on the horizontal axis.

75. Describe a procedure for finding $(f \circ g)(x)$. What is the name of this function?

76. Describe the values of x that must be excluded from the domain of $(f \circ g)(x)$.

Technology Exercises

77. The function $f(t) = -0.14t^2 + 0.51t + 31.6$ models the U.S. population ages 65 and older, $f(t)$, in millions, t years after 1990. The function $g(t) = 0.54t^2 + 12.64t + 107.1$ models the total yearly cost of Medicare, $g(t)$, in billions of dollars, t years after 1990. Graph the function $\frac{g}{f}$ in a $[0, 15, 1]$ by $[0, 60, 1]$ viewing rectangle. What does the shape of the graph indicate about the per capita costs of Medicare for the U.S. population ages 65 and over with increasing time?

78. Graph $y_1 = x^2 - 2x, y_2 = x$, and $y_3 = y_1 \div y_2$ in the same $[-10, 10, 1]$ by $[-10, 10, 1]$ viewing rectangle. Then use the $\boxed{\text{TRACE}}$ feature to trace along y_3. What happens at $x = 0$? Explain why this occurs.

79. Graph $y_1 = x^2 - 4$, $y_2 = \sqrt{4 - x^2}$, and $y_3 = y_2^2 - 4$ in the same $[-5, 5, 1]$ by $[-5, 5, 1]$ viewing rectangle. If y_1 represents f and y_2 represents g, use the graph of y_3 to find the domain of $f \circ g$. Then verify your observation algebraically.

Critical Thinking Exercises

80. Which one of the following is true?
   a. If $f(x) = x^2 - 4$ and $g(x) = \sqrt{x^2 - 4}$, then $(f \circ g)(x) = -x^2$ and $(f \circ g)(5) = -25$.
   b. There can never be two functions f and g, where $f \neq g$, for which $(f \circ g)(x) = (g \circ f)(x)$.
   c. If $f(7) = 5$ and $g(4) = 7$ then $(f \circ g)(4) = 35$.
   d. If $f(x) = \sqrt{x}$ and $g(x) = 2x - 1$, then $(f \circ g)(5) = g(2)$.

81. Prove that if f and g are even functions, then fg is also an even function.

82. Define two functions f and g so that $f \circ g = g \circ f$.

83. Use the graphs given in Exercises 63–66 to create a graph that shows the population, in billions, of less developed regions from 1950 through 2050.

Group Exercise

84. Consult an almanac, newspaper, magazine, or the Internet to find data displayed in a graph in the style of Figure 2.53 on page 248. Using the two graphs that group members find most interesting, introduce two functions that are related to the graphs. Then write and solve a problem involving function subtraction for each selected graph. If you are not sure where to begin, reread page 248–249 or look at Exercises 63–66 in this exercise set.

SECTION 2.7 *Inverse Functions*

Objectives

1. Verify inverse functions.
2. Find the inverse of a function.
3. Use the horizontal line test to determine if a function has an inverse function.
4. Use the graph of a one-to-one function to graph its inverse function.

In most societies, women say they prefer to marry men who are older than themselves, whereas men say they prefer women who are younger. Evolutionary psychologists attribute these preferences to female concern with a partner's material resources and male concern with a partner's fertility (*Source:* David M. Buss, *Psychological Inquiry*, 6, 1–30). When the man is considerably older than the woman, people rarely comment. However, when the woman is older, as in the relationship between actors Susan Sarandon and Tim Robbins, people take notice.

Figure 2.56 shows the preferred age in a mate in five selected countries. We can focus on the data for the women and define a function.

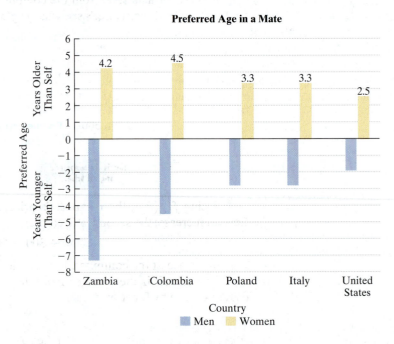

Figure 2.56

Source: Carole Wade and Carol Tavris, P*sychology Sixth Edition*, Prentice Hall, 2000

Let the domain of the function be the set of the five countries shown in the graph. Let the range be the set of the average number of years women in each of the respective countries prefer men who are older than themselves. The function can be written as follows:

$$f:\{(\text{Zambia}, 4.2), (\text{Colombia}, 4.5), (\text{Poland}, 3.3), (\text{Italy}, 3.3), (\text{U.S.}, 2.5)\}.$$

Now let's "undo" f by interchanging the first and second components in each of its ordered pairs. Switching the inputs and outputs of f, we obtain the following relation:

> **Same first component**

Undoing f:$\{(4.2, \text{Zambia}), (4.5, \text{Colombia}), (3.3, \text{Poland}), (3.3, \text{Italy}), (2.5, \text{U.S.})\}$.

> **Different second components**

Can you see that this relation is not a function? Two of its ordered pairs have the same first component and different second components. This violates the definition of a function.

If a function f is a set of ordered pairs, (x, y), then the changes produced by f can be "undone" by reversing the components of all the ordered pairs. The resulting relation, (y, x), may or may not be a function. In this section, we will develop these ideas by studying functions whose compositions have a special "undoing" relationship.

Inverse Functions

Here are two functions that describe situations related to the price of a computer, x:
$$f(x) = x - 300 \qquad g(x) = x + 300.$$

Function f subtracts \$300 from the computer's price and function g adds \$300 to the computer's price. Let's see what $f(g(x))$ does. Put $g(x)$ into f:

$$f(x) = x - 300 \qquad \text{This is the given equation for } f.$$

> **Replace x with g(x).**

$$f(g(x)) = g(x) - 300$$
$$= x + 300 - 300 \qquad \text{Because } g(x) = x + 300, \text{ replace } g(x) \text{ with } x + 300.$$
$$= x. \qquad \text{This is the computer's original price.}$$

Using $f(g(x))$, the computer's price, x, went through two changes: the first, an increase; the second, a decrease:

$$x + 300 - 300.$$

The final price of the computer, x, is identical to its starting price, x.

In general, if the changes made to x by function g are undone by the changes made by function f, then
$$f(g(x)) = x.$$

Assume, also, that this "undoing" takes place in the other direction:
$$g(f(x)) = x.$$

Under these conditions, we say that each function is the *inverse function* of the other. The fact that g is the inverse of f is expressed by renaming g as f^{-1}, read "f-inverse." For example, the inverse functions
$$f(x) = x - 300 \qquad g(x) = x + 300$$

are usually named as follows:
$$f(x) = x - 300 \qquad f^{-1}(x) = x + 300.$$

With these ideas in mind, we present the formal definition of the inverse of a function:

Definition of the Inverse of a Function

Let f and g be two functions such that

$$f(g(x)) = x \qquad \text{for every } x \text{ in the domain of } g$$

and

$$g(f(x)) = x \qquad \text{for every } x \text{ in the domain of } f.$$

The function g is the **inverse of the function** f, and is denoted by f^{-1} (read "f-inverse"). Thus, $f(f^{-1}(x)) = x$ and $f^{-1}(f(x)) = x$. The domain of f is equal to the range of f^{-1}, and vice versa.

1 Verify inverse functions.

EXAMPLE 1 Verifying Inverse Functions

Show that each function is an inverse of the other:

$$f(x) = 5x \qquad \text{and} \qquad g(x) = \frac{x}{5}.$$

Solution To show that f and g are inverses of each other, we must show that $f(g(x)) = x$ and $g(f(x)) = x$. We begin with $f(g(x))$.

$$f(x) = 5x \qquad \text{\textcolor{blue}{This is the given equation for f.}}$$

Replace x with g(x).

$$f(g(x)) = 5g(x) = 5\left(\frac{x}{5}\right) = x$$

Next, we find $g(f(x))$.

$$g(x) = \frac{x}{5} \qquad \text{\textcolor{blue}{This is the given equation for g.}}$$

Replace x with f(x).

$$g(f(x)) = \frac{f(x)}{5} = \frac{5x}{5} = x$$

Because g is the inverse of f (and vice versa), we can use inverse notation and write

$$f(x) = 5x \qquad \text{and} \qquad f^{-1}(x) = \frac{x}{5}.$$

Notice how f^{-1} undoes the change produced by f: f changes x by multiplying by 5 and f^{-1} undoes this by dividing by 5.

Check Point 1 Show that each function is an inverse of the other:

$$f(x) = 7x \qquad \text{and} \qquad g(x) = \frac{x}{7}.$$

EXAMPLE 2   Verifying Inverse Functions

Show that each function is an inverse of the other:

$$f(x) = 3x + 2 \quad \text{and} \quad g(x) = \frac{x - 2}{3}.$$

Solution   To show that f and g are inverses of each other, we must show that $f(g(x)) = x$ and $g(f(x)) = x$. We begin with $f(g(x))$.

$$f(x) = 3x + 2 \quad \text{\textit{This is the equation for f.}}$$

Replace x with $g(x)$.

$$f(g(x)) = 3g(x) + 2 = 3\left(\frac{x - 2}{3}\right) + 2 = x - 2 + 2 = x$$

Next, we find $g(f(x))$.

$$g(x) = \frac{x - 2}{3} \quad \text{\textit{This is the equation for g.}}$$

Replace x with $f(x)$.

$$g(f(x)) = \frac{f(x) - 2}{3} = \frac{(3x + 2) - 2}{3} = \frac{3x}{3} = x$$

Because g is the inverse of f (and vice versa), we can use inverse notation and write

$$f(x) = 3x + 2 \quad \text{and} \quad f^{-1}(x) = \frac{x - 2}{3}.$$

Notice how f^{-1} undoes the changes produced by f: f changes x by *multiplying* by 3 and *adding* 2, and f^{-1} undoes this by *subtracting* 2 and *dividing* by 3. This "undoing" process is illustrated in Figure 2.57.

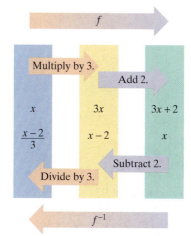

Figure 2.57 f^{-1} undoes the changes produced by f.

Check Point 2   Show that each function is an inverse of the other:

$$f(x) = 4x - 7 \quad \text{and} \quad g(x) = \frac{x + 7}{4}.$$

2 Find the inverse of a function.

Finding the Inverse of a Function

The definition of the inverse of a function tells us that the domain of f is equal to the range of f^{-1}, and vice versa. This means that if the function f is the set of ordered pairs (x, y), then the inverse of f is the set of ordered pairs (y, x). If a function is defined by an equation, we can obtain the equation for f^{-1}, the inverse of f, by interchanging the role of x and y in the equation for the function f.

Finding the Inverse of a Function

The equation for the inverse of a function f can be found as follows:

1. Replace $f(x)$ with y in the equation for $f(x)$.
2. Interchange x and y.
3. Solve for y. If this equation does not define y as a function of x, the function f does not have an inverse function and this procedure ends. If this equation does define y as a function of x, the function f has an inverse function.
4. If f has an inverse function, replace y in step 3 by $f^{-1}(x)$. We can verify our result by showing that $f(f^{-1}(x)) = x$ and $f^{-1}(f(x)) = x$.

Study Tip

The procedure for finding a function's inverse uses a *switch-and-solve* strategy. Switch x and y, then solve for y.

EXAMPLE 3   Finding the Inverse of a Function

Find the inverse of $f(x) = 7x - 5$.

Solution

Step 1   **Replace $f(x)$ with y:**

$$y = 7x - 5$$

Step 2   **Interchange x and y:**

$$x = 7y - 5 \quad \text{This is the inverse function.}$$

Step 3   **Solve for y:**

$$x + 5 = 7y \qquad \text{Add 5 to both sides.}$$

$$\frac{x + 5}{7} = y \qquad \text{Divide both sides by 7.}$$

Discovery

In Example 3, we found that if $f(x) = 7x - 5$, then

$$f^{-1}(x) = \frac{x + 5}{7}.$$

Verify this result by showing that

$$f(f^{-1}(x)) = x$$

and

$$f^{-1}(f(x)) = x.$$

Step 4   **Replace y with $f^{-1}(x)$:**

$$f^{-1}(x) = \frac{x + 5}{7} \qquad \text{The equation is written with } f^{-1} \text{ on the left.}$$

Thus, the inverse of $f(x) = 7x - 5$ is $f^{-1}(x) = \dfrac{x + 5}{7}$.

The inverse function, f^{-1}, undoes the changes produced by f. f changes x by multiplying by 7 and subtracting 5. f^{-1} undoes this by adding 5 and dividing by 7.

Check Point 3   Find the inverse of $f(x) = 2x + 7$.

EXAMPLE 4   Finding the Equation of the Inverse

Find the inverse of $f(x) = x^3 + 1$.

Solution

Step 1   **Replace $f(x)$ with y:** $y = x^3 + 1$

Step 2   **Interchange x and y:** $x = y^3 + 1$

Step 3   **Solve for y:** $x - 1 = y^3$

$$\sqrt[3]{x - 1} = \sqrt[3]{y^3}$$

$$\sqrt[3]{x - 1} = y$$

Step 4   **Replace y with $f^{-1}(x)$:** $f^{-1}(x) = \sqrt[3]{x - 1}$.

Thus, the inverse of $f(x) = x^3 + 1$ is $f^{-1}(x) = \sqrt[3]{x - 1}$.

Check Point 4   Find the inverse of $f(x) = 4x^3 - 1$.

3 Use the horizontal line test to determine if a function has an inverse function.

The Horizontal Line Test and One-to-One Functions

Let's see what happens if we try to find the inverse of the standard quadratic function, $f(x) = x^2$.

Step 1 **Replace $f(x)$ with y:** $y = x^2$.

Step 2 **Interchange x and y:** $x = y^2$.

Step 3 **Solve for y:** We apply the square root method to solve $y^2 = x$ for y.

We obtain

$$y = \pm\sqrt{x}.$$

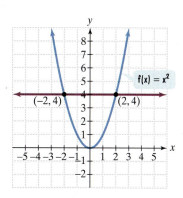

Figure 2.58 The horizontal line intersects the graph twice.

The $\pm$ in this last equation shows that for certain values of x (all positive real numbers), there are two values of y. Because this equation does not represent y as a function of x, the standard quadratic function does not have an inverse function.

Can we look at the graph of a function and tell if it represents a function with an inverse? Yes. The graph of the standard quadratic function is shown in Figure 2.58. Four units above the x-axis, a horizontal line is drawn. This line intersects the graph at two of its points, $(-2, 4)$ and $(2, 4)$. Because inverse functions have ordered pairs with the coordinates reversed, let's see what happens if we reverse these coordinates. We obtain $(4, -2)$ and $(4, 2)$. A function provides exactly one output for each input. However, the input 4 is associated with two outputs, -2 and 2. The points $(4, -2)$ and $(4, 2)$ do not define a function.

If any horizontal line, such as the one in Figure 2.58, intersects a graph at two or more points, these points will not define a function when their coordinates are reversed. This suggests the **horizontal line test** for inverse functions:

Discovery

How might you restrict the domain of $f(x) = x^2$, graphed in Figure 2.58, so that the remaining portion of the graph passes the horizontal line test?

The Horizontal Line Test For Inverse Functions

A function f has an inverse that is a function, f^{-1}, if there is no horizontal line that intersects the graph of the function f at more than one point.

EXAMPLE 5   Applying the Horizontal Line Test

Which of the following graphs represent functions that have inverse functions?

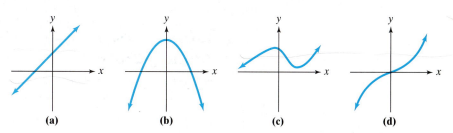

Solution   Can you see that horizontal lines can be drawn in parts (b) and (c) that intersect the graphs more than once? This is illustrated in the figure at the top of the next page. These graphs do not pass the horizontal line test. The graphs in parts (b) and (c) are not the graphs of functions with inverse functions. By contrast, no horizontal line can be drawn in parts (a) and (d) that intersect the graphs more than once. These graphs pass the horizontal line test. Thus, the graphs parts (a) and (d) represent functions that have inverse functions.

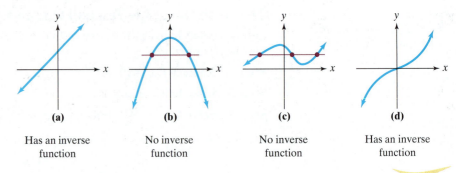

(a)	(b)	(c)	(d)
Has an inverse function	No inverse function	No inverse function	Has an inverse function

Check Point 5 Which of the following graphs represent functions that have inverse functions?

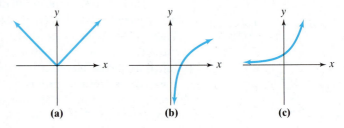

(a)   (b)   (c)

A function passes the horizontal line test when no two different ordered pairs have the same second component. This means that if $x_1 \neq x_2$, then $f(x_1) \neq f(x_2)$. Such a function is called a **one-to-one function.** Thus, a one-to-one function is a function in which no two different ordered pairs have the same second component. Only one-to-one functions have inverse functions. Any function that passes the horizontal line test is a one-to-one function. Any one-to-one function has a graph that passes the horizontal line test.

4 Use the graph of a one-to-one function to graph its inverse function.

Graphs of f and f^{-1}

There is a relationship between the graph of a one-to-one function, f, and its inverse, f^{-1}. Because inverse functions have ordered pairs with the coordinates reversed, if the point (a, b) is on the graph of f, then the point (b, a) is on the graph of f^{-1}. The points (a, b) and (b, a) are symmetric with respect to the line $y = x$. Thus, **the graph of f^{-1} is a reflection of the graph of f about the line $y = x$.** This is illustrated in Figure 2.59.

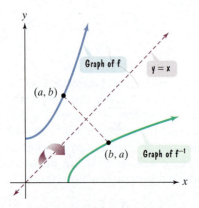

Figure 2.59 The graph of f^{-1} is a reflection of the graph of f about $y = x$.

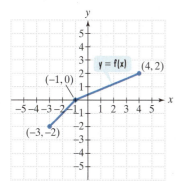

Figure 2.60

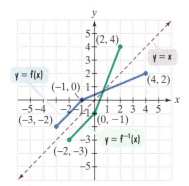

Figure 2.61 The graphs of f and f^{-1}

EXAMPLE 6 Graphing the Inverse Function

Use the graph of f in Figure 2.60 to draw the graph of its inverse function.

Solution We begin by noting that no horizontal line intersects the graph of f at more than one point, so f does have an inverse function. Because the points $(-3, -2)$, $(-1, 0)$, and $(4, 2)$ are on the graph of f, the graph of the inverse function, f^{-1}, has points with these ordered pairs reversed. Thus, $(-2, -3)$, $(0, -1)$, and $(2, 4)$ are on the graph of f^{-1}. We can use these points to graph f^{-1}. The graph of f^{-1} is shown in Figure 2.61. Note that the graph of f^{-1} is the reflection of the graph of f about the line $y = x$.

Check Point 6 Use the graph of f in the figure below to draw the graph of its inverse function.

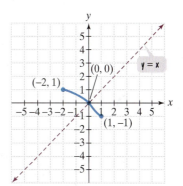

EXERCISE SET 2.7

Practice Exercises

In Exercises 1–10, find $f(g(x))$ and $g(f(x))$ and determine whether each pair of functions f and g are inverses of each other.

1. $f(x) = 4x$ and $g(x) = \dfrac{x}{4}$

2. $f(x) = 6x$ and $g(x) = \dfrac{x}{6}$

3. $f(x) = 3x + 8$ and $g(x) = \dfrac{x - 8}{3}$

4. $f(x) = 4x + 9$ and $g(x) = \dfrac{x - 9}{4}$

5. $f(x) = 5x - 9$ and $g(x) = \dfrac{x + 5}{9}$

6. $f(x) = 3x - 7$ and $g(x) = \dfrac{x + 3}{7}$

7. $f(x) = \dfrac{3}{x - 4}$ and $g(x) = \dfrac{3}{x} + 4$

8. $f(x) = \dfrac{2}{x - 5}$ and $g(x) = \dfrac{2}{x} + 5$

9. $f(x) = -x$ and $g(x) = -x$

10. $f(x) = \sqrt[3]{x - 4}$ and $g(x) = x^3 + 4$

The functions in Exercises 11–30 are all one-to-one. For each function:

a. *Find an equation for $f^{-1}(x)$, the inverse function.*

b. *Verify that your equation is correct by showing that $f(f^{-1}(x)) = x$ and $f^{-1}(f(x)) = x$.*

11. $f(x) = x + 3$

12. $f(x) = x + 5$

13. $f(x) = 2x$

14. $f(x) = 4x$

15. $f(x) = 2x + 3$

16. $f(x) = 3x - 1$

17. $f(x) = x^3 + 2$

18. $f(x) = x^3 - 1$

19. $f(x) = (x + 2)^3$

20. $f(x) = (x - 1)^3$

21. $f(x) = \dfrac{1}{x}$

22. $f(x) = \dfrac{2}{x}$

23. $f(x) = \sqrt{x}$

24. $f(x) = \sqrt[3]{x}$

25. $f(x) = x^2 + 1$, for $x \geq 0$

26. $f(x) = x^2 - 1$, for $x \geq 0$

27. $f(x) = \dfrac{2x + 1}{x - 3}$

28. $f(x) = \dfrac{2x - 3}{x + 1}$

29. $f(x) = \sqrt[3]{x - 4} + 3$

30. $f(x) = x^{3/5}$

Which graphs in Exercises 31–36 represent functions that have inverse functions?

31.

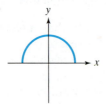

32.

33.

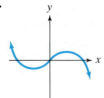

34.

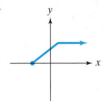

35.

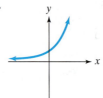

36.

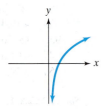

In Exercises 37–40, use the graph of f to draw the graph of its inverse function.

37.

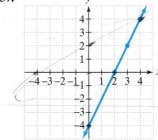

38.

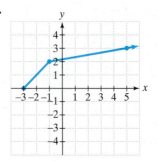

39.

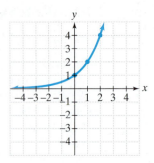

40.

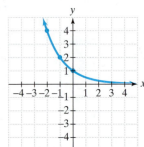

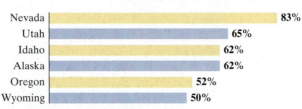

Application Exercises

41. Refer of Figure 2.56 on page 260. Recall that the bar graphs in the figure show the preferred age in a mate in five selected countries.

    a. Consider a function, f, whose domain is the set of the five countries shown in the graph. Let the range be the set of the average number of years men in each of the respective countries prefer women who are younger than themselves. (You will need to use the graph to estimate these values. Assume that the bars for Poland and Italy have the same length. Round to the nearest tenth of a year.) Write function f as a set of ordered pairs.

    b. Write the relation that is the inverse of f as a set of ordered pairs. Is this relation a function? Explain your answer.

42. The bar graph shows the percentage of land owned by the federal government in western states in which the government owns at least half of the land.

Percentage of Land Owned by the Federal Government

Nevada	83%
Utah	65%
Idaho	62%
Alaska	62%
Oregon	52%
Wyoming	50%

Source: Bureau of Land Management

a. Consider a function, f, whose domain is the set of six states shown. Let the range be the percentage of land owned by the federal government in each of the respective states. Write the function f as a set of ordered pairs.

b. Write the relation that is the inverse of f as a set of ordered pairs. Is this relation a function? Explain your answer.

43. The graph represents the probability of two people in the same room sharing a birthday as a function of the number of people in the room. Call the function f.

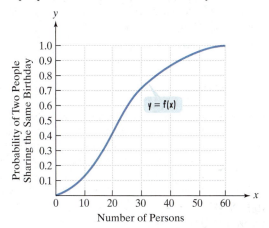

a. Explain why f has an inverse that is a function.

b. Describe in practical terms the meaning of $f^{-1}(0.25)$, $f^{-1}(0.5)$, and $f^{-1}(0.7)$.

44. The graph shows the average age at which women in the United States marry for the first time over a 110-year period.

Source: U.S. Census Bureau

a. Does this graph have an inverse that is a function? What does this mean about the average age at which U.S. women marry during the period shown?

b. Identify two or more years in which U.S. women married for the first time at the same average age. What is a reasonable estimate of this average age?

45. The formula

$$y = f(x) = \frac{9}{5}x + 32$$

is used to convert from x degrees Celsius to y degrees Fahrenheit. The formula

$$y = g(x) = \frac{5}{9}(x - 32)$$

is used to convert from x degrees Fahrenheit to y degrees Celsius. Show that f and g are inverse functions.

46. One yardstick for measuring how steadily—if slowly—athletic performance improved is the mile run. In 1923, the record for the mile was a comparatively sleepy 4 minutes, 10.4 seconds. In 1954, Roger Bannister of Britain cracked the 4-minute mark, coming in at 3 minutes, 59.4 seconds. In the half-century since, about 0.3 second per year has been shaved off Bannister's record.

Mile Records			
1886	4:12.3	1958	3:54.5
1923	4:10.4	1966	3:51.3
1933	4:07.6	1979	3:48.9
1945	4:01.3	1985	3:46.3
1954	3:59.4	1999	3:43.1

Source: U.S.A. Track and Field

a. Consider the following information:

- In 1954, the record was 3 minutes, 59.4 seconds, or 239.4 seconds.
- The record has decreased by 0.3 second per year since then.

Use this information to write a function, f, that models the mile record, $f(x)$, in seconds, x years after 1954.

b. Find the inverse of the mile-record function. Describe what each variable in the inverse function represents.

c. According to the inverse model, how many years after 1954 will someone run a 3-minute, or 180-second, mile? In which year will this occur?

Writing in Mathematics

47. Explain how to determine if two functions are inverses of each other.

48. Describe how to find the inverse of a one-to-one function.

49. What is the horizontal line test and what does it indicate?

50. Describe how to use the graph of a one-to-one function to draw the graph of its inverse function.

51. How can a graphing utility be used to visually determine if two functions are inverses of each other?

Technology Exercises

In Exercises 52–60, use a graphing utility to graph the function. Use the graph to determine whether the function has an inverse that is a function (that is, whether the function is one-to-one).

52. $f(x) = x^2 - 1$

53. $f(x) = \sqrt[3]{2 - x}$

54. $f(x) = \dfrac{x^3}{2}$

55. $f(x) = \dfrac{x^4}{4}$

56. $f(x) = \text{int}(x - 2)$

57. $f(x) = |x - 2|$

58. $f(x) = (x - 1)^3$

59. $f(x) = -\sqrt{16 - x^2}$

60. $f(x) = x^3 + x + 1$

In Exercises 61–63, use a graphing utility to graph f and g in the same viewing rectangle. In addition, graph the line $y = x$ and visually determine if f and g are inverses.

61. $f(x) = 4x + 4, \ g(x) = 0.25x - 1$

62. $f(x) = \dfrac{1}{x} + 2, \ g(x) = \dfrac{1}{x - 2}$

63. $f(x) = \sqrt[3]{x} - 2, \ g(x) = (x + 2)^3$

Critical Thinking Exercises

64. Which one of the following is true?

    a. The inverse of $\{(1, 4), (2, 7)\}$ is $\{(2, 7), (1, 4)\}$.

    b. The function $f(x) = 5$ is one-to-one.

    c. If $f(x) = 3x$, then $f^{-1}(x) = \dfrac{1}{3x}$.

    d. The domain of f is the same as the range of f^{-1}.

65. If $f(x) = 3x$ and $g(x) = x + 5$, find $(f \circ g)^{-1}(x)$ and $(g^{-1} \circ f^{-1})(x)$.

66. Show that

$$f(x) = \frac{3x - 2}{5x - 3}$$

is its own inverse.

67. *Freedom 7* was the spacecraft that carried the first American into space in 1961. Total flight time was 15 minutes, and the spacecraft reached a maximum height of 116 miles. Consider a function, s, that expresses *Freedom 7*'s height, $s(t)$, in miles, after t minutes. Is s a one-to-one function? Explain your answer.

68. If $f(2) = 6$, find x satisfying $8 + f^{-1}(x - 1) = 10$.

Group Exercise

69. In Tom Stoppard's play *Arcadia*, the characters dream and talk about mathematics, including ideas involving graphing, composite functions, symmetry, and lack of symmetry in things that are tangled, mysterious, and unpredictable. Group members should read the play. Present a report on the ideas discussed by the characters that are related to concepts that we studied in this chapter. Bring in a copy of the play and read appropriate excerpts.

CHAPTER SUMMARY, REVIEW, AND TEST

Summary

DEFINITIONS AND CONCEPTS	EXAMPLES
2.1 Lines and Slope	
a. The slope, m, of the line through (x_1, y_1) and (x_2, y_2) is $m = \dfrac{y_2 - y_1}{x_2 - x_1}$.	Ex. 1, p.177
b. Equations of lines include point-slope form, $y - y_1 = m(x - x_1)$, slope-intercept form, $y = mx + b$, and general form, $Ax + By + C = 0$. The equation of a horizontal line is $y = b$; a vertical line is $x = a$.	Ex. 2 & 3, p.179; Ex. 5 & 6, p.182
c. Parallel lines have equal slopes. Perpendicular lines have slopes that are negative reciprocals.	Ex. 8 & 9, p.184–185
2.2 Distance and Midpoint Formulas; Circles	
a. The distance, d, between the points (x_1, y_1) and (x_2, y_2) is given by $d = \sqrt{(x_2 - x_1)^2 + (y_2 - y_1)^2}$.	Ex. 1, p.194
b. The midpoint of the line segment whose endpoints are (x_1, y_1) and (x_2, y_2) is the point with coordinates $\left(\dfrac{x_1 + x_2}{2}, \dfrac{y_1 + y_2}{2} \right)$.	Ex. 2, p.195

c. The standard form of the equation of a circle with center (h, k) and radius r is $(x - h)^2 + (y - k)^2 = r^2$.

d. The general form of the equation of a circle is $x^2 + y^2 + Dx + Ey + F = 0$.

e. To convert from the general form to the standard form of a circle's equation, complete the square on x and y.

2.3 Basics of Functions

a. A relation is any set of ordered pairs. The set of first components is the domain and the set of second components is the range.

b. A function is a correspondence from a first set, called the domain, to a second set, called the range, such that each element in the domain corresponds to exactly one element in the range. If any element in a relation's domain corresponds to more than one element in the range, the relation is not a function.

c. Functions are usually given in terms of equations involving x and y, in which x is the independent variable and y is the dependent variable. If an equation is solved for y and more than one value of y can be obtained for a given x, then the equation does not define y as a function of x. If an equation defines a function, the value of the function at x, $f(x)$, often replaces y.

d. The difference quotient is

$$\frac{f(x + h) - f(x)}{h}, h \neq 0.$$

e. If a function f does not model data or verbal conditions, its domain is the largest set of real numbers for which the value of $f(x)$ is a real number. Exclude from the function's domain real numbers that cause division by zero and real numbers that result in an even root of a negative number.

2.4 Graphs of Functions

a. The graph of a function is the graph of its ordered pairs.

b. The vertical line test for functions: If any vertical line intersects a graph in more than one point, the graph does not define y as a function of x.

c. A function is increasing on intervals where its graph rises, decreasing on intervals where it falls, and constant on intervals where it neither rises nor falls. Precise definitions are given in the box on page 220.

d. If the graph of a function is given, we can often visually locate the number(s) at which the function has a relative maximum or relative minimum. Precise definitions are given in the box on page 221.

e. The average rate of change of f from x_1 to x_2 is

$$\frac{f(x_2) - f(x_1)}{x_2 - x_1}.$$

f. The graph of an even function in which $f(-x) = f(x)$ is symmetric with respect to the y-axis. The graph of an odd function in which $f(-x) = -f(x)$ is symmetric with respect to the origin.

g. The graph of $f(x) = \text{int}(x)$, where $\text{int}(x)$ is the greatest integer that is less than or equal to x, has function values that form discontinuous steps, shown in Figure 2.39 on page 227. If $n \leq x < n + 1$, where n is an integer, then $\text{int}(x) = n$.

2.5 Transformations of Functions

a. Table 2.4 on pages 235–236 shows the graphs of the constant function, $f(x) = c$, the identity function, $f(x) = x$, the standard quadratic function, $f(x) = x^2$, the standard cubic function, $f(x) = x^3$, the square root function, $f(x) = \sqrt{x}$, and the absolute value function, $f(x) = |x|$. The table also lists characteristics of each function.

b. Table 2.5 on page 243 summarizes how to graph a function using vertical shifts, $y = f(x) \pm c$, horizontal shifts, $y = f(x \pm c)$, reflections about the x-axis, $y = -f(x)$, reflections about the y-axis, $y = f(-x)$, vertical stretching, $y = cf(x)$, $c > 1$, and vertical shrinking, $y = cf(x)$, $0 < c < 1$.

c. A function involving more than one transformation can be graphed in the following order: (1) horizontal shifting; (2) vertical stretching or shrinking; (3) reflecting; (4) vertical shifting.

2.6 Combinations of Functions; Composite and Inverse Functions

a. When functions are given as equations, they can be added, subtracted, multiplied, or divided by performing operations with the algebraic expressions that appear on the right side of the equations. Definitions for the sum $f + g$, the difference $f - g$, the product fg, and the quotient $\dfrac{f}{g}$ functions are given in the box on page 251.

Ex. 1, p. 250;
Ex. 2, p. 250;
Ex. 3, p. 252

b. The composition of functions f and g, $f \circ g$, is defined by $(f \circ g)(x) = f(g(x))$. The domain of the composite function $f \circ g$ is given in the box on page 256. This composite function is obtained by replacing each occurrence of x in the equation for f with $g(x)$.

Ex. 4, p. 254;
Ex. 5, p. 255

2.7 Inverse Functions

a. If $f(g(x)) = x$ and $g(f(x)) = x$, function g is the inverse of function f, denoted f^{-1} and read "f inverse." Thus, to show that f and g are inverses of each other, one must show $f(g(x)) = x$ and $g(f(x)) = x$.

Ex. 1, p. 262;
Ex. 2, p. 263

b. The procedure for finding a function's inverse uses a switch-and-solve strategy. Switch x and y, then solve for y. The procedure is given in the box on page 263.

Ex. 3 & 4, p. 264

c. The horizontal line test for inverse functions: A function f has an inverse that is a function, f^{-1}, if there is no horizontal line that intersects the graph of the function f at more than one point.

Ex. 5, p. 265

d. A one-to-one function is one in which no two different ordered pairs have the same second component. Only one-to-one functions have inverse functions.

e. If the point (a, b) is on the graph of f, then the point (b, a) is on the graph of f^{-1}. The graph of f^{-1} is a reflection of the graph of f about the line $y = x$.

Ex. 6, p. 267

Review Exercises

2.1

In Exercises 1–4, find the slope of the line passing through each pair of points or state that the slope is undefined. Then indicate whether the line through the points rises, falls, is horizontal, or is vertical.

1. $(3, 2)$ and $(5, 1)$
2. $(-1, -2)$ and $(-3, -4)$
3. $\left(-3, \frac{1}{4}\right)$ and $\left(6, \frac{1}{4}\right)$
4. $(-2, 5)$ and $(-2, 10)$

In Exercises 5–6, use the given conditions to write an equation for each line in point-slope form and slope-intercept form.

5. Passing through $(-3, 2)$ with slope -6
6. Passing through $(1, 6)$ and $(-1, 2)$

In Exercises 7–10, give the slope and y-intercept of each line whose equation is given. Then graph the line.

7. $y = \frac{2}{5}x - 1$
8. $y = -4x + 5$
9. $2x + 3y + 6 = 0$
10. $2y - 8 = 0$

11. Corporations in the United States are doing quite well, thank you. The scatter plot in the next column shows corporate profits, in billions of dollars, from 1990 through 2000. Also shown is a line that passes through or near the points.

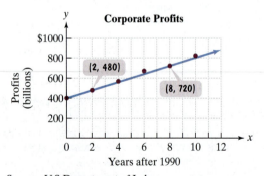

Corporate Profits

Source: U.S Department of Labor

a. Use the two points whose coordinates are shown by the voice balloons to find the point-slope equation of the line that models corporate profits, y, in billions of dollars, x years after 1990.

b. Write the equation in part (a) in slope-intercept form.

c. Use the linear model to predict corporate profits in 2010.

12. The scatter plot on the next page shows the number of minutes each that 16 people exercise per week and the number of headaches per month each person experiences.

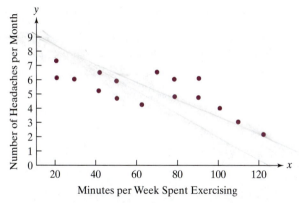

Number of Headaches per Month

Minutes per Week Spent Exercising

a. Draw a line that fits the data so that the spread of the data points around the line is as small as possible.

b. Use the coordinates of two points along your line to write its point-slope and slope-intercept equations.

c. Use the equation in part (b) to predict the number of headaches per month for a person exercising 130 minutes per week.

In Exercises 13–14, use the given conditions to write an equation for each line in point-slope form and slope-intercept form.

13. Passing through $(4, -7)$ and parallel to the line whose equation is $3x + y - 9 = 0$

14. Passing through $(-3, 6)$ and perpendicular to the line whose equation is $y = \frac{1}{3}x + 4$

2.2

In Exercises 15–16, find the distance between each pair of points. If necessary, round answers to two decimal places.

15. $(-2, -3)$ and $(3, 9)$ **16.** $(-4, 3)$ and $(-2, 5)$

In Exercises 17–18, find the midpoint of each line segment with the given endpoints.

17. $(2, 6)$ and $(-12, 4)$ **18.** $(4, -6)$ and $(-15, 2)$

In Exercises 19–20, write the standard form of the equation of the circle with the given center and radius.

19. Center $(0, 0)$, $r = 3$ **20.** Center $(-2, 4)$, $r = 6$

In Exercises 21–23, give the center and radius of each circle and graph its equation.

21. $x^2 + y^2 = 1$ **22.** $(x + 2)^2 + (y - 3)^2 = 9$

23. $x^2 + y^2 - 4x + 2y - 4 = 0$

2.3

In Exercises 24–26, determine whether each relation is a function. Give the domain and range for each relation.

24. $\{(2, 7), (3, 7), (5, 7)\}$ **25.** $\{(1, 10)\ (2, 500), (13, \pi)\}$

26. $\{(12, 13), (14, 15), (12, 19)\}$

In Exercises 27–29, determine whether each equation defines y as a function of x.

27. $2x + y = 8$ **28.** $3x^2 + y = 14$

29. $2x + y^2 = 6$

In Exercises 30–33, evaluate each function at the given values of the independent variable and simplify.

30. $f(x) = 5 - 7x$
    a. $f(4)$ **b.** $f(x + 3)$ **c.** $f(-x)$

31. $g(x) = 3x^2 - 5x + 2$
    a. $g(0)$ **b.** $g(-2)$
    c. $g(x - 1)$ **d.** $g(-x)$

32. $g(x) = \begin{cases} \sqrt{x - 4} & \text{if } x \geq 4 \\ 4 - x & \text{if } x < 4 \end{cases}$
    a. $g(13)$ **b.** $g(0)$ **c.** $g(-3)$

33. $f(x) = \begin{cases} \dfrac{x^2 - 1}{x - 1} & \text{if } x \neq 1 \\ 12 & \text{if } x = 1 \end{cases}$
    a. $f(-2)$ **b.** $f(1)$ **c.** $f(2)$

In Exercises 34–35, find and simplify the difference quotient

$$\frac{f(x + h) - f(x)}{h}, \quad h \neq 0$$

for the given function.

34. $f(x) = 8x - 11$ **35.** $f(x) = x^2 - 13x + 5$

In Exercises 36–40, find the domain of each function.

36. $f(x) = x^2 + 6x - 3$ **37.** $g(x) = \dfrac{4}{x - 7}$

38. $h(x) = \sqrt{8 - 2x}$ **39.** $f(x) = \dfrac{x}{x^2 - 1}$

40. $g(x) = \dfrac{\sqrt{x - 2}}{x - 5}$

2.4

Graph the functions in Exercises 41–42. Use the integer values of x given to the right of the function to obtain the ordered pairs. Use the graph to specify the function's domain and range.

41. $f(x) = x^2 - 4x + 4$ $x = -1, 0, 1, 2, 3, 4$

42. $f(x) = |2 - x|$ $x = -1, 0, 1, 2, 3, 4$

In Exercises 43–45, use the graph to determine **a.** *the function's domain;* **b.** *the function's range;* **c.** *the x-intercepts, if any;* **d.** *the y-intercept, if any;* **e.** *intervals on which the function is increasing, decreasing, or constant; and* **f.** *the function values indicated below the graphs.*

43.

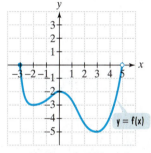

$y = f(x)$

$f(-2) = ?\ \ f(3) = ?$

44.

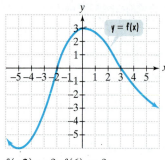

$$f(-2) = ? \quad f(6) = ?$$

45.

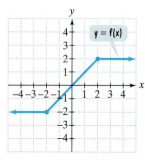

$$f(-9) = ? \quad f(14) = ?$$

In Exercises 46–47, find:

  a. *The numbers, if any, at which f has a relative maximum. What are these relative maxima?*
  b. *The numbers, if any, at which f has a relative minimum. What are these relative minima?*

46. Use the graph in Exercise 43.

47. Use the graph in Exercise 44.

In Exercises 48–51, use the vertical line test to identify graphs in which y is a function of x.

48.

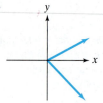

49.

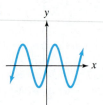

50.

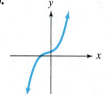

51.

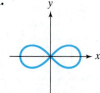

52. Find the average rate of change of $f(x) = x^2 - 4x$ from $x_1 = 5$ to $x_2 = 9$.

53. The graph shows annual spending per uniformed member of the U.S. military in inflation-adjusted dollars. Find the average rate of change of spending per year from 1955 through 2000. Round to the nearest dollar per year.

Source: Center for Strategic and Budgetary Assessments

In Exercises 54–56, determine whether each function is even, odd, or neither. State each function's symmetry. If you are using a graphing utility, graph the function and verify its possible symmetry.

54. $f(x) = x^3 - 5x$

55. $f(x) = x^4 - 2x^2 + 1$

56. $f(x) = 2x\sqrt{1 - x^2}$

57. The graph shows the height, in meters, of a vulture in terms of its time, in seconds, in flight.

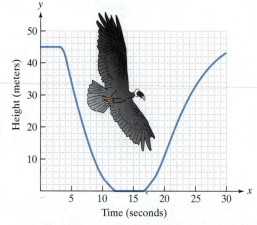

  a. Is the vulture's height a function of time? Use the graph to explain why or why not.
  b. On which interval is the function decreasing? Describe what this means in practical terms.
  c. On which intervals is the function constant? What does this mean for each of these intervals?
  d. On which interval is the function increasing? What does this mean?

58. A cargo service charges a flat fee of $5 plus $1.50 for each pound or fraction of a pound. Graph shipping cost, $C(x)$, in dollars, as a function of weight, x, in pounds, for $0 < x \leq 5$.

2.5

In Exercises 59–61, begin by graphing the standard quadratic function, $f(x) = x^2$. Then use transformations of this graph to graph the given function.

59. $g(x) = x^2 + 2$ **60.** $h(x) = (x + 2)^2$

61. $r(x) = -(x + 1)^2$

In Exercises 62–64, begin by graphing the square root function, $f(x) = \sqrt{x}$. Then use transformations of this graph to graph the given function.

62. $g(x) = \sqrt{x + 3}$ **63.** $h(x) = \sqrt{3 - x}$

64. $r(x) = 2\sqrt{x + 2}$

In Exercises 65–67, begin by graphing the absolute value function, $f(x) = |x|$. Then use transformations of this graph to graph the given function.

65. $g(x) = |x + 2| - 3$ **66.** $h(x) = -|x - 1| + 1$

67. $r(x) = \frac{1}{2}|x + 2|$

In Exercises 68–70, begin by graphing the standard cubic function, $f(x) = x^3$. Then use transformations of this graph to graph the given function.

68. $g(x) = \frac{1}{2}(x - 1)^3$ **69.** $h(x) = -(x + 1)^3$

70. $r(x) = \frac{1}{4}x^3 - 1$

In Exercises 71–73, use the graph of the function f to sketch the graph of the given function g.

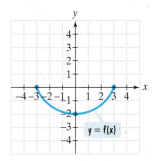

71. $g(x) = f(x + 2) + 3$ **72.** $g(x) = \frac{1}{2}f(x - 1)$

73. $g(x) = -2 + 2f(x + 2)$

2.6

In Exercises 74–76, find $f + g$, $f - g$, fg, and $\frac{f}{g}$. Determine the domain for each function.

74. $f(x) = 3x - 1$, $g(x) = x - 5$

75. $f(x) = x^2 + x + 1$, $g(x) = x^2 - 1$

76. $f(x) = \sqrt{x + 7}$, $g(x) = \sqrt{x - 2}$

In Exercises 77–78, find **a.** $(f \circ g)(x)$; **b.** $(g \circ f)(x)$; **c.** $(f \circ g)(3)$.

77. $f(x) = x^2 + 3$, $g(x) = 4x - 1$

78. $f(x) = \sqrt{x}$, $g(x) = x + 1$

In Exercises 79–80, find **a.** $(f \circ g)(x)$; **b.** the domain of $(f \circ g)$.

79. $f(x) = \dfrac{x + 1}{x - 2}$, $g(x) = \dfrac{1}{x}$

80. $f(x) = \sqrt{x - 1}$, $g(x) = x + 3$

In Exercises 81–82, express the given function h as a composition of two functions f and g so that $h(x) = (f \circ g)(x)$.

81. $h(x) = (x^2 + 2x - 1)^4$ **82.** $h(x) = \sqrt[3]{7x + 4}$

2.7

In Exercises 83–84, find $f\left(g(x)\right)$ and $g\left(f(x)\right)$ and determine whether each pair of functions f and g are inverses of each other.

83. $f(x) = \dfrac{3}{5}x + \dfrac{1}{2}$ and $g(x) = \dfrac{5}{3}x - 2$

84. $f(x) = 2 - 5x$ and $g(x) = \dfrac{2 - x}{5}$

The functions in Exercises 85–87 are all one-to-one. For each function:
   a. Find an equation for $f^{-1}(x)$, the inverse function.
   b. Verify that your equation is correct by showing that $f(f^{-1}(x)) = x$ and $f^{-1}(f(x)) = x$.

85. $f(x) = 4x - 3$ **86.** $f(x) = \sqrt{x + 2}$

87. $f(x) = 8x^3 + 1$

Which graphs in Exercises 88–91 represent functions that have inverse functions?

88.

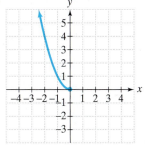

89.

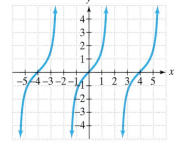

90.

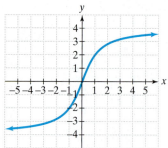

91.

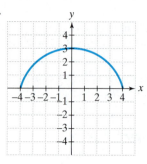

92. Use the graph of f in the figure shown to draw the graph of its inverse function.

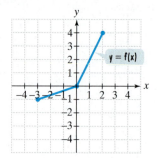

Chapter 2 Test

In Exercises 1–2, use the given conditions to write an equation for each line in point-slope form and slope-intercept form.

1. Passing through $(2, 1)$ and $(-1, -8)$

2. Passing through $(-4, 6)$ and perpendicular to the line whose equation is $y = -\frac{1}{4}x + 5$

3. Strong demand plus higher fuel and labor costs are driving up the price of flying. The graph shows the national averages for one-way fares. Also shown is a line that models that data.

National Averages for One-Way Airline Fares: Business Travel

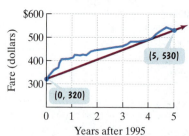

Source: American Express

a. Use the two points whose coordinates are shown by the voice balloons to write the slope-intercept equation of the line that models the average one-way fare, y, in dollars, x years after 1995.

b. According to the model, what will the national average for one-way fares be in 2008?

4. Give the center and radius of the circle whose equation is $x^2 + y^2 + 4x - 6y - 3 = 0$ and graph the equation.

5. List by letter all relations that are not functions.

a. $\{(7, 5), (8, 5), (9, 5)\}$

b. $\{(5, 7), (5, 8), (5, 9)\}$

c.

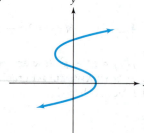

d. $x^2 + y^2 = 100$

e.

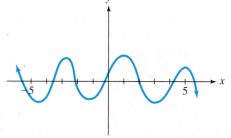

6. If $f(x) = x^2 - 2x + 5$, find $f(x - 1)$ and simplify.

7. If $g(x) = \begin{cases} \sqrt{x - 3} & \text{if } x \geq 3 \\ 3 - x & \text{if } x < 3 \end{cases}$, find $g(-1)$ and $g(7)$.

8. If $f(x) = \sqrt{12 - 3x}$, find the domain of f.

9. If $f(x) = x^2 + 11x - 7$, find and simplify the difference quotient $\dfrac{f(x + h) - f(x)}{h}$.

10. Use the graph of function f to answer the following questions.

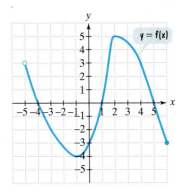

a. What is $f(4) - f(-3)$?
b. What is the domain of f?
c. What is the range of f?
d. On which interval or intervals is f increasing?
e. On which interval or intervals is f decreasing?
f. For what number does f have a relative maximum? What is the relative maximum?
g. For what number does f have a relative minimum? What is the relative minimum?
h. What are the x-intercepts?
i. What is the y-intercept?

11. Find the average rate of change of $f(x) = 3x^2 - 5$ from $x_1 = 6$ to $x_2 = 10$.

12. Determine whether $f(x) = x^4 - x^2$ is even, odd, or neither. Use your answer to explain why the graph in the figure shown cannot be the graph of f.

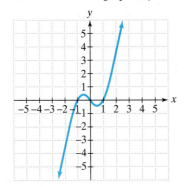

13. The figure at the top of the next column shows how the graph of $h(x) = -2(x - 3)^2$ is obtained from the graph of $f(x) = x^2$. Describe this process, using the graph of g in your description.

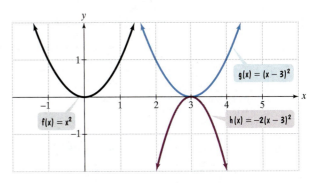

14. Begin by graphing the absolute value function, $f(x) = |x|$. Then use transformations of this graph to graph $g(x) = \frac{1}{2}|x + 1| + 3$.

If $f(x) = x^2 + 3x - 4$ and $g(x) = 5x - 2$, find each function or function value in Exercises 15–19.

15. $(f - g)(x)$

16. $\left(\dfrac{f}{g}\right)(x)$ and its domain

17. $(f \circ g)(x)$

18. $(g \circ f)(x)$

19. $f(g(2))$

20. If $f(x) = \dfrac{7}{x - 4}$ and $g(x) = \dfrac{2}{x}$, find $(f \circ g)(x)$ and the domain of $f \circ g$.

21. Express $h(x) = (2x + 13)^7$ as a composition of two functions f and g so that $h(x) = (f \circ g)(x)$.

22. If $f(x) = \sqrt{x - 2}$, find the equation for $f^{-1}(x)$. Then verify that your equation is correct by showing that $f(f^{-1}(x)) = x$ and $f^{-1}(f(x)) = x$.

23. A function f models the amount given to charity as a function of income. The graph of f is shown in the figure.

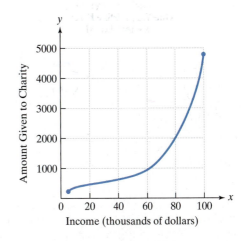

a. Explain why f has an inverse that is a function.
b. Find $f(80)$.
c. Describe in practical terms the meaning of $f^{-1}(2000)$.

24. Use a graphing utility to graph $f(x) = \dfrac{x^3}{3} + x^2 - 15x + 3$

in a $[-10, 10, 1]$ by $[-30, 70, 10]$ viewing rectangle. Use the graph to answer the following questions.

a. Is f one-to-one? Explain.

b. Is f even, odd, or neither? Explain.

c. What is the range of f?

d. On which interval or intervals is f increasing?

e. On which interval or intervals is f decreasing?

f. For what number does f have a relative maximum? What is the relative maximum?

g. For what number does f have a relative minimum? What is the relative minimum?

Cumulative Review Exercises (Chapters P–2)

Simplify each expression in Exercises 1 and 2.

1. $\dfrac{4x^2y}{2x^5y^{-3}}$

2. $\dfrac{5}{4\sqrt{2}}$

3. Factor: $x^3 - 4x^2 + 2x - 8$.

In Exercises 4 and 5, perform the operations and simplify.

4. $\dfrac{x-3}{x+4} + \dfrac{x}{x-2}$

5. $\dfrac{4 + \dfrac{2}{x}}{4 - \dfrac{2}{x}}$

Solve each equation in Exercises 6–9.

6. $(x + 3)(x - 4) = 8$

7. $3(4x - 1) = 4 - 6(x - 3)$

8. $\sqrt{x + 2} = x$

9. $x^{2/3} - x^{1/3} - 6 = 0$

Solve each inequality in Exercises 10 and 11. Express the answer in interval notation.

10. $\dfrac{x}{2} - 3 \le \dfrac{x}{4} + 2$

11. $\dfrac{x + 3}{x - 2} \le 2$

12. Write the point-slope form and the slope-intercept form of the line passing through $(-2, 5)$ and perpendicular to the line whose equation is $y = -\frac{1}{4}x + \frac{1}{3}$.

13. Graph $f(x) = \sqrt{x}$ and then use transformations of this graph to graph $g(x) = \sqrt{x - 3} + 4$ in the same rectangular coordinate system.

14. If $f(x) = 2 + \sqrt{x - 3}$, find the equation for $f^{-1}(x)$.

15. If $f(x) = 3 - x^2$, find $\dfrac{f(x + h) - f(x)}{h}$ and simplify.

16. Solve for c: $A = \dfrac{cd}{c + d}$.

17. You invested $6000 in two accounts paying 7% and 9% annual interest, respectively. At the end of the year, the total interest from these investments was $510. How much was invested at each rate?

18. For a summer sales job, you are choosing between two pay arrangements: a weekly salary of $200 plus 5% commission on sales, or a straight 15% commission. For how many dollars of sales will the earnings be the same regardless of the pay arrangement?

19. The length of a rectangular garden is 2 feet more than twice its width. If 22 feet of fencing is needed to enclose the garden, what are its dimensions?

20. On the first five tests you have scores of $61, 95, 71, 83,$ and 80. The last test, a final exam, counts as two grades. What score do you need on the final in order to have an average score of 80?

Polynomial and Rational Functions

Chapter 3

There is a function that models the age in human years, $H(x)$, of a dog that is x years old:

$$H(x) = -0.001618x^4 + 0.077326x^3$$
$$-1.2367x^2 + 11.460x + 2.914.$$

The function contains variables to powers that are whole numbers and is an example of a **polynomial function.** In this chapter, we study polynomial functions and functions that consist of quotients of polynomials, called **rational functions.**

One of the joys of your life is your dog, your very special buddy. Lately, however, you've noticed that your companion is slowing down a bit. He's now 8 years old and you wonder how this translates into human years. You remember something about every year of a dog's life being equal to seven years for a human. Is there a more accurate description?

SECTION 3.1 Quadratic Functions

Objectives

1. Recognize characteristics of parabolas.
2. Graph parabolas.
3. Solve problems involving minimizing or maximizing quadratic functions.

The Food Stamp Program is the first line of defense against hunger for millions of American families. The program provides benefits for eligible participants to purchase approved food items at approved food stores. Over half of all participants are children; one out of six is a low-income older adult. The function

$$f(x) = -0.5x^2 + 4x + 19$$

models the number of people, $f(x)$, in millions, receiving food stamps x years after 1990. For example, to find the number of food stamp recipients in 2000, substitute 10 for x because 2000 is 10 years after 1990:

$$f(10) = -0.5(10)^2 + 4(10) + 19 = 9.$$

Thus, in 2000, there were 9 million food stamp recipients.

The function $f(x) = -0.5x^2 + 4x + 19$ is an example of a *quadratic function*. A **quadratic function** is any function of the form

$$f(x) = ax^2 + bx + c$$

where a, b, and c are real numbers and $a \neq 0$. A quadratic function is a polynomial function whose highest power is 2. In this section, we will study quadratic functions and their graphs.

1 Recognize characteristics of parabolas.

Graphs of Quadratic Functions

The graph of any quadratic function is called a **parabola.** Parabolas are shaped like cups, as shown in Figure 3.1. If the coefficient of x^2 (the value of a in $ax^2 + bx + c$) is positive, the parabola opens upward. If the coefficient of x^2 is negative, the graph opens downward. The **vertex** (or turning point) of the parabola is the minimum point on the graph when it opens upward, and the maximum point on the graph when it opens downward.

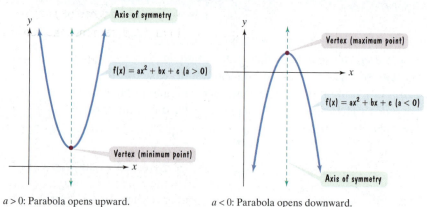

Figure 3.1 Characteristics of parabolas

$a > 0$: Parabola opens upward. $a < 0$: Parabola opens downward.

Look at the unusual image of the word "mirror" shown below. The artist, Scott Kim, has created the image so that the two halves of the whole are mirror images of each other. A parabola shares this kind of symmetry, in which a line through the vertex divides the figure in half. Parabolas are symmetric with respect to this line, called the **axis of symmetry.** The movements of gymnasts, divers, and swimmers can approximate this symmetry. If a parabola is folded along its axis of symmetry, the two halves match exactly.

2 Graph parabolas.

Graphing Quadratic Functions in Standard Form

In Section 2.5, we applied a series of transformations to the graph of $f(x) = x^2$. The graph of this function is a parabola. The vertex for this parabola is $(0, 0)$. In Figure 3.2(a), the graph of $f(x) = ax^2$ for $a > 0$ is shown in black; it opens *upward*. In Figure 3.2(b), the graph of $f(x) = ax^2$ for $a < 0$ is shown in black; it opens *downward*.

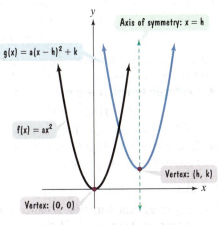

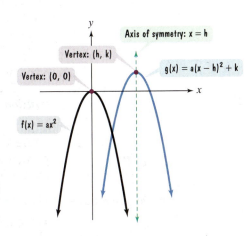

Figure 3.2 Transformations of $f(x) = ax^2$

(a) $a > 0$: Parabola opens upward.

(b) $a < 0$: Parabola opens downward.

Figure 3.2 also shows the graphs of $g(x) = a(x - h)^2 + k$ in blue. Compare these graphs to those of $f(x) = ax^2$. Observe that h determines the horizontal shift and k determines the vertical shift of the graph of $f(x) = ax^2$:

$$g(x) = a(x - h)^2 + k.$$

If $h > 0$, it shifts the graph of $f(x) = ax^2$ h units to the right.

If $k > 0$, it shifts the graph of $y = a(x-h)^2$ k units up.

Consequently, the vertex $(0, 0)$ on the black graph of $f(x) = ax^2$ moves to the point (h, k) on the blue graph of $g(x) = a(x - h)^2 + k$. The axis of symmetry is the vertical line whose equation is $x = h$.

The form of the expression for g is convenient because it immediately identifies the vertex of the parabola as (h, k). This is the **standard form** of a quadratic function.

> **The Standard Form of a Quadratic Function**
>
> The quadratic function
>
> $$f(x) = a(x - h)^2 + k, \qquad a \neq 0$$
>
> is in **standard form**. The graph of f is a parabola whose vertex is the point (h, k). The parabola is symmetric with respect to the line $x = h$. If $a > 0$, the parabola opens upward; if $a < 0$, the parabola opens downward.

The sign of a in $f(x) = a(x - h)^2 + k$ determines whether the parabola opens upward or downward. Furthermore, if $|a|$ is small, the parabola opens more widely than if $|a|$ is large. Here is a general procedure for graphing parabolas whose equations are in standard form:

> **Graphing Quadratic Functions with Equations in Standard Form**
>
> To graph $f(x) = a(x - h)^2 + k$,
>
> 1. Determine whether the parabola opens upward or downward. If $a > 0$, it opens upward. If $a < 0$, it opens downward.
> 2. Determine the vertex of the parabola. The vertex is (h, k).
> 3. Find any x-intercepts by replacing $f(x)$ with 0. Solve the resulting quadratic equation for x.
> 4. Find the y-intercept by replacing x with 0.
> 5. Plot the intercepts and vertex. Connect these points with a smooth curve that is shaped like a cup. Draw a dashed vertical line for the axis of symmetry.

EXAMPLE 1   **Graphing a Quadratic Function in Standard Form**

Graph the quadratic function $f(x) = -2(x - 3)^2 + 8$.

Solution   We can graph this function by following the steps in the preceding box. We begin by identifying values for a, h, and k.

Standard form $f(x) = a(x - h)^2 + k$

$a = -2$ $h = 3$ $k = 8$

Given equation $f(x) = -2(x - 3)^2 + 8$

Step 1   **Determine how the parabola opens.**   Note that a, the coefficient of x^2, is -2. Thus, $a < 0$; this negative value tells us that the parabola opens downward.

Step 2   **Find the vertex.**   The vertex of the parabola is (h, k). Because $h = 3$ and $k = 8$, the parabola's vertex is $(3, 8)$.

Step 3 Find the x-intercepts. Replace $f(x)$ with 0 in $f(x) = -2(x - 3)^2 + 8$.

$$0 = -2(x - 3)^2 + 8 \qquad \text{Find x-intercepts, setting f(x) equal to 0.}$$

$$2(x - 3)^2 = 8 \qquad \text{Solve for x. Add } 2(x - 3)^2 \text{ to both sides of the equation.}$$

$$(x - 3)^2 = 4 \qquad \text{Divide both sides by 2.}$$

$$(x - 3) = \pm\sqrt{4} \qquad \text{Apply the square root method. If } (x - c)^2 = d, \text{ then } x - c = \pm\sqrt{d}.$$

$$x - 3 = -2 \quad \text{or} \quad x - 3 = 2 \qquad \text{Express as two separate equations.}$$

$$x = 1 \quad \text{or} \quad x = 5 \qquad \text{Add 3 to both sides in each equation.}$$

The x-intercepts are 1 and 5. The parabola passes through $(1, 0)$ and $(5, 0)$.

Step 4 Find the y-intercept. Replace x with 0 in $f(x) = -2(x - 3)^2 + 8$.

$$f(0) = -2(0 - 3)^2 + 8 = -2(-3)^2 + 8 = -2(9) + 8 = -10$$

The y-intercept is -10. The parabola passes through $(0, -10)$.

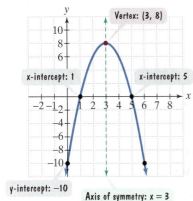

Vertex: (3, 8)

x-intercept: 1   x-intercept: 5

y-intercept: −10

Axis of symmetry: x = 3

Figure 3.3 The graph of $f(x) = -2(x - 3)^2 + 8$

Step 5 Graph the parabola. With a vertex at $(3, 8)$, x-intercepts at 1 and 5, and a y-intercept at -10, the graph of f is shown in Figure 3.3. The axis of symmetry is the vertical line whose equation is $x = 3$.

Check Point 1 Graph the quadratic function $f(x) = -(x - 1)^2 + 4$.

EXAMPLE 2 Graphing a Quadratic Function in Standard Form

Graph the quadratic function $f(x) = (x + 3)^2 + 1$.

Solution We begin by finding values for $a, h,$ and k.

Standard form $\qquad f(x) = a(x - h)^2 + k$

Given equation $\qquad f(x) = (x + 3)^2 + 1$

$$\text{or} \quad f(x) = 1(x - (-3))^2 + 1$$

$a = 1 \qquad h = -3 \qquad k = 1$

Step 1 Determine how the parabola opens. Note that a, the coefficient of x^2, is 1. Thus, $a > 0$; this positive value tells us that the parabola opens upward.

Step 2 Find the vertex. The vertex of the parabola is (h, k). Because $h = -3$ and $k = 1$, the parabola's vertex is $(-3, 1)$.

Step 3 Find the x-intercepts. Replace $f(x)$ with 0 in $f(x) = (x + 3)^2 + 1$. Because the vertex is $(-3, 1)$, which lies above the x-axis, and the parabola opens upward, it appears that this parabola has no x-intercepts. We can verify this observation algebraically.

$$0 = (x + 3)^2 + 1 \quad \text{Find possible x-intercepts, setting f(x) equal to 0.}$$

$$-1 = (x + 3)^2 \quad \text{Solve for x. Subtract 1 from both sides.}$$

$$x + 3 = \pm\sqrt{-1} \quad \text{Apply the square root method.}$$

$$x + 3 = \pm i \quad \text{Recall that } \sqrt{-1} = i, \text{ an imaginary number.}$$

$$x = -3 \pm i \quad \text{Subtract 3 from both sides.}$$

Because this equation has no real solutions, the parabola has no x-intercepts.

Step 4 Find the y-intercept. Replace x with 0 in $f(x) = (x + 3)^2 + 1$.

$$f(0) = (0 + 3)^2 + 1 = 3^2 + 1 = 9 + 1 = 10$$

The y-intercept is 10. The parabola passes through $(0, 10)$.

Step 5 Graph the parabola. With a vertex at $(-3, 1)$, no x-intercepts, and a y-intercept at 10, the graph of f is shown in Figure 3.4. The axis of symmetry is the vertical line whose equation is $x = -3$.

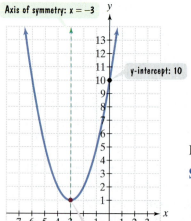

Figure 3.4 The graph of $g(x) = (x + 3)^2 + 1$

Check Point 2 Graph the quadratic function $f(x) = (x - 2)^2 + 1$.

Graphing Quadratic Functions in the Form $f(x) = ax^2 + bx + c$

Quadratic functions are frequently expressed in the form $f(x) = ax^2 + bx + c$. How can we identify the vertex of a parabola whose equation is in this form? By completing the square, we can find a way to describe the vertex in terms of a and b.

$$f(x) = ax^2 + bx + c$$

$$= a\left(x^2 + \frac{b}{a}x\right) + c \quad \text{Factor out a from } ax^2 + bx.$$

$$= a\left(x^2 + \frac{b}{a}x + \frac{b^2}{4a^2}\right) + c - a\left(\frac{b^2}{4a^2}\right)$$

Complete the square by adding the square of half the coefficient of x.

By completing the square, we added $a \cdot \frac{b^2}{4a^2}$. To avoid changing the function's equation, we must subtract this term.

$$= a\left(x + \frac{b}{2a}\right)^2 + c - \frac{b^2}{4a} \quad \text{Write the trinomial as the square of a binomial and simplify the constant term.}$$

Compare this form of the equation with a quadratic function's standard form.

Standard form $\qquad f(x) = a(x - h)^2 + k$

$$h = -\frac{b}{2a} \qquad k = c - \frac{b^2}{4a}$$

Equation under discussion $\quad f(x) = a\left(x - \left(-\frac{b}{2a}\right)\right)^2 + c - \frac{b^2}{4a}$

The important part of this observation is that h, the x-coordinate of the vertex, is $-\dfrac{b}{2a}$. The y-coordinate can be found by evaluating the function at $-\dfrac{b}{2a}$.

> ### The Vertex of a Parabola Whose Equation Is $f(x) = ax^2 + bx + c$
> Consider the parabola defined by the quadratic function
> $f(x) = ax^2 + bx + c$. The parabola's vertex is $\left(-\dfrac{b}{2a}, f\left(-\dfrac{b}{2a}\right)\right)$.

We can apply our five-step procedure and graph parabolas in the form $f(x) = ax^2 + bx + c$. The only step that is different is how we determine the vertex.

EXAMPLE 3 Graphing a Quadratic Function in the Form $f(x) = ax^2 + bx + c$

Graph the quadratic function $\quad f(x) = -x^2 + 4x - 1$.
Solution

Step 1 Determine how the parabola opens. Note that a, the coefficient of x^2, is -1. Thus, $a < 0$; this negative value tells us that the parabola opens downward.

Step 2 Find the vertex. We know that the x-coordinate of the vertex is $x = -\dfrac{b}{2a}$. We identify a, b, and c in $f(x) = ax^2 + bx + c$.

$$f(x) = -x^2 + 4x - 1$$

$$a = -1 \qquad b = 4 \qquad c = -1$$

Substitute the values of a and b into the equation for the x-coordinate:

$$x = -\frac{b}{2a} = -\frac{4}{2(-1)} = \frac{-4}{-2} = 2.$$

The x-coordinate of the vertex is 2. We substitute 2 for x in $f(x) = -x^2 + 4x - 1$, the equation of the function, to find the y-coordinate:

$$f(2) = -2^2 + 4 \cdot 2 - 1 = -4 + 8 - 1 = 3.$$

The vertex is $(2, 3)$.

Step 3   Find the x-intercepts.   Replace $f(x)$ with 0 in $f(x) = -x^2 + 4x - 1$. We obtain $0 = -x^2 + 4x - 1$ or $-x^2 + 4x - 1 = 0$. This equation cannot be solved by factoring. We will use the quadratic formula to solve it.

$$a = -1, \qquad b = 4, \qquad c = -1$$

$$x = \frac{-b \pm \sqrt{b^2 - 4ac}}{2a} = \frac{-4 \pm \sqrt{4^2 - 4(-1)(-1)}}{2(-1)} = \frac{-4 \pm \sqrt{16 - 4}}{-2}$$

$$x = \frac{-4 - \sqrt{12}}{-2} \approx 3.7 \quad \text{or} \quad x = \frac{-4 + \sqrt{12}}{-2} \approx 0.3$$

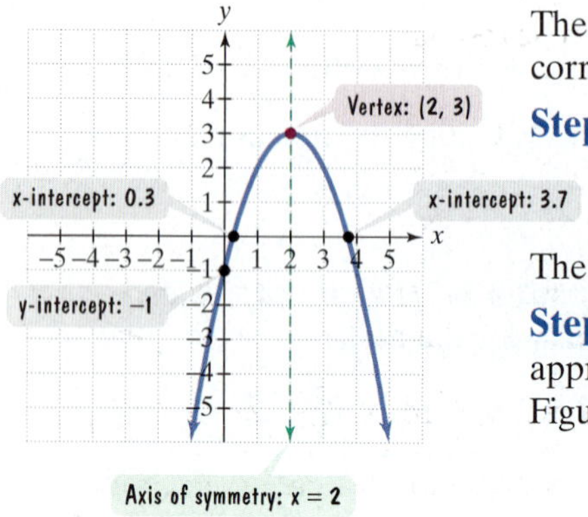

The x-intercepts are approximately 0.3 and 3.7. The parabola passes through the corresponding points, which we approximate as $(0.3, 0)$ and $(3.7, 0)$.

Step 4   Find the y-intercept.   Replace x with 0 in $f(x) = -x^2 + 4x - 1$.

$$f(0) = -0^2 + 4 \cdot 0 - 1 = -1$$

The y-intercept is -1. The parabola passes through $(0, -1)$.

Step 5   Graph the parabola.   With a vertex at $(2, 3)$, x-intercepts at approximately 0.3 and 3.7, and a y-intercept at -1, the graph of f is shown in Figure 3.5. The axis of symmetry is the vertical line whose equation is $x = 2$.

Figure 3.5 The graph of $f(x) = -x^2 + 4x - 1$

Check Point 3   Graph the quadratic function $f(x) = x^2 - 2x - 3$.

3   Solve problems involving minimizing or maximizing quadratic functions.

Applications of Quadratic Functions

When did the maximum number of Americans participate in the food stamp program? What is the age of a driver having the least number of car accidents? How do people launching fireworks know when they should explode to be viewed at the greatest possible height? The answers to these questions involve finding the maximum or minimum value of quadratic functions.

Consider the quadratic function $f(x) = ax^2 + bx + c$. If $a > 0$, the parabola opens upward and the vertex is its lowest point. If $a < 0$, the parabola opens downward and the vertex is its highest point. The x-coordinate of the vertex is $-\dfrac{b}{2a}$. Thus, we can find the minimum or maximum value of f by evaluating the quadratic function at $x = -\dfrac{b}{2a}$.

> **Minimum and Maximum: Quadratic Functions**
> Consider $f(x) = ax^2 + bx + c$.
>
> **1.** If $a > 0$, then f has a minimum that occurs at $x = -\dfrac{b}{2a}$.
> This minimum value is $f\left(-\dfrac{b}{2a}\right)$.
>
> **2.** If $a < 0$, then f has a maximum that occurs at $x = -\dfrac{b}{2a}$.
> This maximum value is $f\left(-\dfrac{b}{2a}\right)$.

EXAMPLE 4 An Application: The Food Stamp Program

The function

$$f(x) = -0.5x^2 + 4x + 19$$

models the number of people, $f(x)$, in millions, receiving food stamps x years after 1990. (*Source: New York Times*) In which year was this number at a maximum? How many food stamp recipients were there for that year?

Solution The quadratic function is in the form $f(x) = ax^2 + bx + c$ with $a = -0.5$ and $b = 4$. Because $a < 0$, the function has a maximum value that occurs at $x = -\dfrac{b}{2a}$.

$$x = -\frac{b}{2a} = -\frac{4}{2(-0.5)} = \frac{-4}{-1} = 4$$

This means that the number of people receiving food stamps was at a maximum 4 years after 1990, in 1994. The number of recipients, in millions, for that year was

$$f(4) = -0.5(4)^2 + 4(4) + 19 = -8 + 16 + 19 = 27.$$

In 1994, the number of people receiving food stamps reached a maximum of 27 million.

Technology

The graph of the function modeling the millions of food stamp recipients

$$f(x) = -0.5x^2 + 4x + 19$$

is shown in a $[0, 10, 1]$ by $[0, 35, 5]$ viewing rectangle. The maximum function feature verifies that 4 years after 1990, the number of recipients reached a maximum of 27 million. Notice that x gives the location of the maximum and y gives the maximum value.

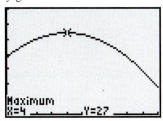

Check Point 4 The function $f(x) = 0.4x^2 - 36x + 1000$ models the number of accidents, $f(x)$, per 50 million miles driven, in terms of a driver's age, x, in years, where $16 \le x \le 74$. What is the age of a driver having the least number of car accidents? What is the minimum number of car accidents per 50 million miles driven?

Visualizing Irritability by Age

The quadratic function

$$P(x) = -0.05x^2 + 4.2x - 26$$

models the percentage of coffee drinkers, $P(x)$, who are x years old who become irritable if they do not have coffee at their regular time. Figure 3.6 shows the graph of the function. The vertex reveals that 62.2% of 42-year-old coffee drinkers become irritable. This is the maximum percentage for any age, x, in the function's domain.

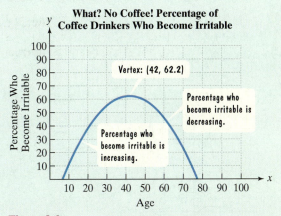

Figure 3.6

Source: LMK Associates

Technology

We've come a long way from the small nation of "embattled farmers" who launched the American Revolution. In the early days of our Republic, 95% of the population was involved in farming. The graph in Figure 3.7 shows the number of farms in the United States from 1850 through 2010 (projected). Because the graph is shaped like a cup, with an increasing number of farms from 1850 to 1910 and a decreasing number of farms from 1910 to 2010, a quadratic function is an appropriate model for the data. You can use the statistical menu of a graphing utility to enter the data in Figure 3.7. We entered the data using (number of decades after 1850, millions of U.S. farms). The data are shown to the right of Figure 3.7.

Number of U. S. Farms, 1850–2010

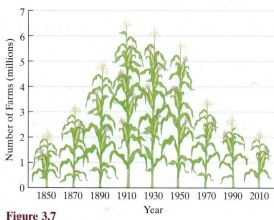

Data:
(0, 2.3), (2, 3.3), (4, 5.1),
(6, 6.7), (8, 6.4), (10, 5.8),
(12, 3.6), (14, 2.9), (16, 2.3)

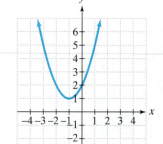

Figure 3.7

Source: U.S. Bureau of the Census

Upon entering the QUADratic REGression program, we obtain the results shown in the screen. Thus, the quadratic function of best fit is

$$f(x) = -0.064x^2 + 0.99x + 2.2$$

where x represents the number of decades after 1850 and $f(x)$ represents the number of U.S. farms, in millions.

EXERCISE SET 3.1

Practice Exercises

In Exercises 1–4, the graph of a quadratic function is given. Write the function's equation, selecting from the following options.

$$f(x) = (x + 1)^2 - 1 \qquad g(x) = (x + 1)^2 + 1$$
$$h(x) = (x - 1)^2 + 1 \qquad j(x) = (x - 1)^2 - 1$$

1.

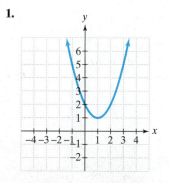

2.

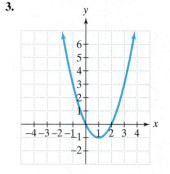

3.

4.

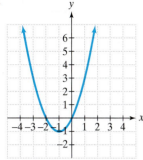

8.

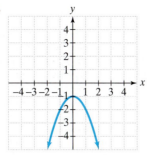

In Exercises 5–8, the graph of a quadratic function is given. Write the function's equation, selecting from the following options.

$$f(x) = x^2 + 2x + 1 \qquad g(x) = x^2 - 2x + 1$$
$$h(x) = x^2 - 1 \qquad j(x) = -x^2 - 1$$

In Exercises 9–16, find the coordinates of the vertex for the parabola defined by the given quadratic function.

9. $f(x) = 2(x - 3)^2 + 1$ **10.** $f(x) = -3(x - 2)^2 + 12$

11. $f(x) = -2(x + 1)^2 + 5$ **12.** $f(x) = -2(x + 4)^2 - 8$

13. $f(x) = 2x^2 - 8x + 3$ **14.** $f(x) = 3x^2 - 12x + 1$

15. $f(x) = -x^2 - 2x + 8$ **16.** $f(x) = -2x^2 + 8x - 1$

5.

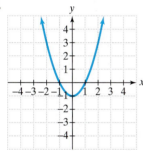

In Exercises 17–34, use the vertex and intercepts to sketch the graph of each quadratic function. Give the equation of the parabola's axis of symmetry. Use the graph to determine the function's domain and range.

17. $f(x) = (x - 4)^2 - 1$ **18.** $f(x) = (x - 1)^2 - 2$

19. $f(x) = (x - 1)^2 + 2$ **20.** $f(x) = (x - 3)^2 + 2$

21. $y - 1 = (x - 3)^2$ **22.** $y - 3 = (x - 1)^2$

23. $f(x) = 2(x + 2)^2 - 1$ **24.** $f(x) = \frac{5}{4} - \left(x - \frac{1}{2}\right)^2$

25. $f(x) = 4 - (x - 1)^2$ **26.** $f(x) = 1 - (x - 3)^2$

6.

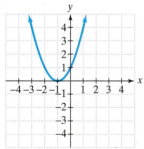

27. $f(x) = x^2 - 2x - 3$ **28.** $f(x) = x^2 - 2x - 15$

29. $f(x) = x^2 + 3x - 10$ **30.** $f(x) = 2x^2 - 7x - 4$

31. $f(x) = 2x - x^2 + 3$ **32.** $f(x) = 5 - 4x - x^2$

33. $f(x) = 2x - x^2 - 2$ **34.** $f(x) = 6 - 4x + x^2$

7.

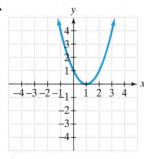

In Exercises 35–40, determine, without graphing, whether the given quadratic function has a minimum value or a maximum value. Then find the coordinates of the minimum or the maximum point.

35. $f(x) = 3x^2 - 12x - 1$ **36.** $f(x) = 2x^2 - 8x - 3$

37. $f(x) = -4x^2 + 8x - 3$ **38.** $f(x) = -2x^2 - 12x + 3$

39. $f(x) = 5x^2 - 5x$ **40.** $f(x) = 6x^2 - 6x$

Application Exercises

Cigarette Consumption per U.S. Adult

Source: U.S. Department of Health and Human Services

41. Please see the graph above. The function

$$f(x) = -3.1x^2 + 51.4x + 4024.5$$

models the average annual per capita consumption of cigarettes, $f(x)$, by Americans 18 and older x years after 1960. According to this model, in which year did cigarette consumption per capita reach a maximum? What was the consumption for that year? Does this accurately model what actually occurred as shown by the graph above?

42. The function

$$f(x) = 104.5x^2 - 1501.5x + 6016$$

models the death rate per year per 100,000 males, $f(x)$, for U.S. men who average x hours of sleep each night. How many hours of sleep, to the nearest tenth of an hour, corresponds to the minimum death rate? What is this minimum death rate, to the nearest whole number?

43. Fireworks are launched into the air. The quadratic function

$$s(t) = -16t^2 + 200t + 4$$

models the fireworks' height, $s(t)$, in feet, t seconds after they are launched. When should the fireworks explode so that they go off at the greatest height? What is that height?

44. A football is thrown by a quarterback to a receiver 40 yards away. The quadratic function

$$s(t) = -0.025t^2 + t + 5$$

models the football's height above the ground, $s(t)$, in feet, when it is t yards from the quarterback. How many yards from the quarterback does the football reach its greatest height? What is that height?

In the United States, HCV, or hepatitis C virus, is four times as widespread as HIV. Few of the nation's three to four million carriers have any idea they are infected. The graph shows the projected mortality, in thousands, from the virus. Use the information in the graph to solve Exercises 45–46.

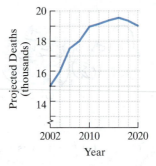

Projected Hepatitis C Deaths in the United States

Source: *American Journal of Public Health*

45. Why is a quadratic function an appropriate model for the data shown in the graph?

46. Suppose that a quadratic function is used to model the data shown with ordered pairs representing (number of years after 2002, thousands of hepatitis C deaths). Determine, without obtaining an actual quadratic function that models the data, the approximate coordinates of the vertex for the function's graph. Describe what this means in practical terms. Use the word "maximum" in your description.

47. You have 120 feet of fencing to enclose a rectangular plot that borders on a river. If you do not fence the side along the river, find the length and width of the plot that will maximize the area. What is the largest area that can be enclosed?

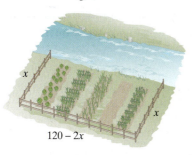

$120 - 2x$

48. The figure shown indicates that you have 100 yards of fencing to enclose a rectangular area. Find the dimensions of the rectangle that maximize the enclosed area. What is the maximum area?

$50 - x$ x

49. A rain gutter is made from sheets of aluminum that are 20 inches wide. As shown in the figure, the edges are turned up to form right angles. Determine the depth of the gutter that will maximize its cross-sectional area and allow the greatest amount of water to flow.

Flat sheet
20 inches
wide

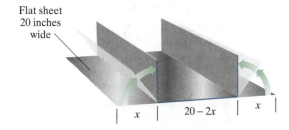

x $20 - 2x$ x

50. Hunky Beef, a local sandwich store, has a fixed weekly cost of $525.00, and variable costs for making a roast beef sandwich are $0.55.

  a. Let x represent the number of roast beef sandwiches made and sold each week. Write the weekly cost function, C, for Hunky Beef. (*Hint:* The cost function is the sum of fixed and variable costs.)

  b. The function $R(x) = -0.001x^2 + 3x$ describes the money, in dollars, that Hunky Beef takes in each week from the sale of x roast beef sandwiches. Use this revenue function and the cost function from part (a) to write the stores weekly profit function, P. (*Hint:* The profit function is the difference between revenue and cost functions.)

  c. Use the store's profit function to determine the number of roast beef sandwiches it should make and sell each week to maximize profit. What is the maximum weekly profit?

Writing in Mathematics

51. What is a quadratic function?

52. What is a parabola? Describe its shape.

53. Explain how to decide whether a parabola opens upward or downward.

54. Describe how to find a parabola's vertex if its equation is expressed in standard form. Give an example.

55. Describe how to find a parabola's vertex if its equation is in the form $f(x) = ax^2 + bx + c$. Use $f(x) = x^2 - 6x + 8$ as an example.

56. A parabola that opens upward has its vertex at $(1, 2)$. Describe as much as you can about the parabola based on this information. Include in your discussion the number of x-intercepts (if any) for the parabola.

57. The quadratic function

$$f(x) = -0.018x^2 + 1.93x - 25.34$$

describes the miles per gallon, $f(x)$, of a Ford Taurus driven at x miles per hour. Suppose that you own a Ford Taurus. Describe how you can use this function to save money.

Technology Exercises

58. Use a graphing utility to verify any five of your hand-drawn graphs in Exercises 17–34.

59. **a.** Use a graphing utility to graph $y = 2x^2 - 82x + 720$ in a standard viewing rectangle. What do you observe?

  b. Find the coordinates of the vertex for the given quadratic function.

  c. The answer to part (b) is $(20.5, -120.5)$. Because the leading coefficient of the given function (2) is positive, the vertex is a minimum point on the graph. Use this fact to help find a viewing rectangle that will give a relatively complete picture of the parabola. With an axis of symmetry at $x = 20.5$, the setting for x should extend past this, so try Xmin = 0 and Xmax = 30. The setting for y should include (and probably go below) the y-coordinate of the graph's minimum point, so try Ymin = -130. Experiment with Ymax until your utility shows the parabola's major features.

  d. In general, explain how knowing the coordinates of a parabola's vertex can help determine a reasonable viewing rectangle on a graphing utility for obtaining a complete picture of the parabola.

In Exercises 60–63, find the vertex for each parabola. Then determine a reasonable viewing rectangle on your graphing utility and use it to graph the quadratic function.

60. $y = -0.25x^2 + 40x$

61. $y = -4x^2 + 20x + 160$

62. $y = 5x^2 + 40x + 600$

63. $y = 0.01x^2 + 0.6x + 100$

64. The function $y = 0.011x^2 - 0.097x + 4.1$ models the number of people in the United States, y, in millions, holding more than one job x years after 1970. Use a graphing utility to graph the function in a $[0, 20, 1]$ by $[3, 6, 1]$ viewing rectangle. $\boxed{\text{TRACE}}$ along the curve or use your utility's minimum value feature to approximate the coordinates of the parabola's vertex. Describe what this represents in practical terms.

65. The following data show fuel efficiency, in miles per gallon, for all U.S. automobiles in the indicated year.

x (Years after 1940)	y (Average Number of Miles per Gallon for U.S. Automobiles)
1940: 0	14.8
1950: 10	13.9
1960: 20	13.4
1970: 30	13.5
1980: 40	15.9
1990: 50	20.2
1998: 58	21.8

Source: U.S. Department of Transportation.

a. Use a graphing utility to draw a scatter plot of the data. Explain why a quadratic function is appropriate for modeling these data.

b. Use the quadratic regression feature to find the quadratic function that best fits the data.

c. Use the model in part (b) to determine the worst year for automobile fuel efficiency. What was the average number of miles per gallon for that year?

d. Use a graphing utility to draw a scatter plot of the data and graph the quadratic function of best fit on the scatter plot.

Critical Thinking Exercises

66. Which one of the following is true?

a. No quadratic functions have a range of $(-\infty, \infty)$.

b. The vertex of the parabola described by $f(x) = 2(x - 5)^2 - 1$ is $(5, 1)$.

c. The graph of $f(x) = -2(x + 4)^2 - 8$ has one y-intercept and two x-intercepts.

d. The maximum value of y for the quadratic function $f(x) = -x^2 + x + 1$ is 1.

67. What explanations can you offer for your answer to Exercise 41? Use a graphing utility to graph f. Do you agree with the long-term predictions made by the graph? Explain.

In Exercises 68–69, find the axis of symmetry for each parabola whose equation is given. Use the axis of symmetry to find a second point on the parabola whose y-coordinate is the same as the given point.

68. $f(x) = 3(x + 2)^2 - 5;\ (-1, -2)$

69. $f(x) = (x - 3)^2 + 2;\ (6, 11)$

70. A rancher has 1000 feet of fencing to construct six corrals, as shown in the figure. Find the dimensions that maximize the enclosed area. What is the maximum area?

Group Exercise

71. Each group member should consult an almanac, newspaper, magazine, or the Internet to find data that can be modeled by a quadratic function. Group members should select the two sets of data that are most interesting and relevant. For each data set selected:

a. Use the quadratic regression feature of a graphing utility to find the quadratic function that best fits the data.

b. Use the equation of the quadratic function to make a prediction from the data. What circumstances might affect the accuracy of your prediction?

c. Use the equation of the quadratic function to write and solve a problem involving maximizing or minimizing the function.

SECTION 3.2 *Polynomial Functions and Their Graphs*

Objectives

1. Recognize characteristics of graphs of polynomial functions.

2. Determine end behavior.

3. Use factoring to find zeros of polynomial functions.

4. Identify the multiplicity of a zero.

5. Understand the relationship between degree and turning points.

6. Graph polynomial functions.

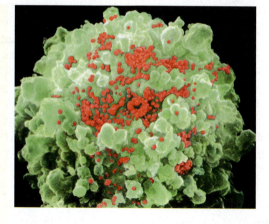

Magnified 6000 times, this color-scanned image shows a T-lymphocyte blood cell (green) infected with the HIV virus (red). Depletion of the number of T-cells causes destruction of the immune system.

In 1980, U.S. doctors diagnosed 41 cases of a rare form of cancer, Kaposi's sarcoma, that involved skin lesions, pneumonia, and severe immunological deficiencies. All cases involved gay men ranging in age from 26 to 51. By the end of 2000, approximately 775,000 Americans, straight and gay, male and female, old and young, were infected with the HIV virus.

Modeling AIDS-related data and making predictions about the epidemic's havoc is serious business. Changing circumstances and unforeseen events have resulted in models that are not particularly useful over long periods of time. For example, the function

$$f(x) = -143x^3 + 1810x^2 - 187x + 2331$$

models the number of AIDS cases diagnosed in the United States x years after 1983. The model was obtained using cases diagnosed from 1983 through 1991. Figure 3.8 shows the graph of f from 1983 through 1991 in a $[0, 8, 1]$ by $[0, 50{,}000, 5000]$ viewing rectangle. The function used to describe the number of new AIDS cases in the United States over a limited period of time is an example of a *polynomial function*.

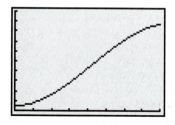

Figure 3.8 The graph of a function modeling the number of new AIDS cases in the U.S. from 1983 through 1991

Definition of a Polynomial Function

Let n be a nonnegative integer and let $a_n, a_{n-1}, \ldots, a_2, a_1, a_0$, be real numbers, with $a_n \neq 0$. The function defined by

$$f(x) = a_n x^n + a_{n-1} x^{n-1} + \cdots + a_2 x^2 + a_1 x + a_0$$

is called a **polynomial function of x of degree n.** The number a_n, the coefficient of the variable to the highest power, is called the **leading coefficient.**

A constant function $f(x) = c$, where $c \neq 0$, is a polynomial function of degree 0. A linear function $f(x) = mx + b$, where $m \neq 0$, is a polynomial function of degree 1. A quadratic function $f(x) = ax^2 + bx + c$, where $a \neq 0$, is a polynomial function of degree 2. In this section, we focus on polynomial functions of degree 3 or higher.

1 Recognize characteristics of graphs of polynomial functions.

Smooth, Continuous Graphs

Polynomial functions of degree 2 or less have graphs that are either parabolas or lines. We can graph such functions by plotting points. We can also graph polynomial functions of degree 3 or higher by plotting points. However, the process is rather tedious: Many points must be plotted. It may be easier to use a graphing utility for such functions. Regardless of the graphing method you use, you will find an ability to recognize the basic features of polynomial functions helpful. For example, they may help you choose an appropriate viewing rectangle for a graphing utility.

Two important features of the graphs of polynomial functions are that they are *smooth* and *continuous*. By **smooth,** we mean that the graph contains only rounded curves with no sharp corners. By **continuous,** we mean that the graph has no breaks and can be drawn without lifting your pencil from the rectangular coordinate system. These ideas are illustrated in Figure 3.9.

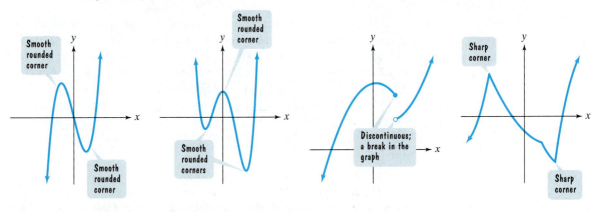

Figure 3.9 Recognizing graphs of polynomial functions

2 Determine end behavior.

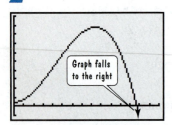

Figure 3.10 By extending the viewing rectangle, y is eventually negative and the function no longer models the number of AIDS cases.

End Behavior of Polynomial Functions

Figure 3.10 shows the graph of the function

$$f(x) = -143x^3 + 1810x^2 - 187x + 2331,$$

which models U.S. AIDS cases from 1983 through 1991. Look what happens to the graph when we extend the year up through 1998 with a $[0, 15, 1]$ by $[-5000, 50,000, 5000]$ viewing rectangle. By year 13 (1996), the values of y are negative and the function no longer models AIDS cases. We've added an arrow to the graph at the far right to emphasize that it continues to decrease without bound. It is this far-right *end behavior* of the graph that makes it inappropriate for modeling AIDS cases into the future.

The behavior of a graph of a function to the far left or the far right is called its **end behavior.** Although the graph of a polynomial function may have intervals where it increases or decreases, the graph will eventually rise or fall without bound as it moves far to the left or far to the right.

How can you determine whether the graph of a polynomial function goes up or down at each end? The end behavior of a polynomial function

$$f(x) = a_n x^n + a_{n-1} x^{n-1} + \cdots + a_1 x + a_0$$

depends upon the leading term $a_n x^n$. In particular, the sign of the leading coefficient, a_n, and the degree, n, of the polynomial function reveal its end behavior. In terms of end behavior, only the term of highest degree counts, summarized by the **Leading Coefficient Test.**

The Leading Coefficient Test

As x increases or decreases without bound, the graph of the polynomial function

$$f(x) = a_n x^n + a_{n-1} x^{n-1} + a_{n-2} x^{n-2} + \cdots + a_1 x + a_0 \quad (a_n \neq 0)$$

eventually rises or falls. In particular,

1. For n odd:

2. For n even:

If the leading coefficient is positive, the graph falls to the left and rises to the right.	If the leading coefficient is negative, the graph rises to the left and falls to the right.	If the leading coefficient is positive, the graph rises to the left and to the right.	If the leading coefficient is negative, the graph falls to the left and to the right.
$a_n > 0$	$a_n < 0$	$a_n > 0$	$a_n < 0$

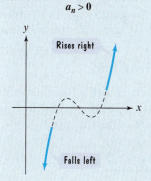
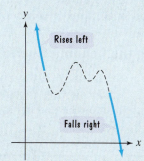
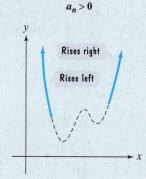
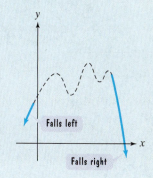

Study Tip

Odd-degree polynomial functions have graphs with opposite behavior at each end. Even-degree polynomial functions have graphs with the same behavior at each end.

EXAMPLE 1   Using the Leading Coefficient Test

Use the Leading Coefficient Test to determine the end behavior of the graph of

$$f(x) = x^3 + 3x^2 - x - 3.$$

Solution   Because the degree is odd ($n = 3$) and the leading coefficient, 1, is positive, the graph falls to the left and rises to the right, as shown in Figure 3.11.

Check Point 1   Use the Leading Coefficient Test to determine the end behavior of the graph of $f(x) = x^4 - 4x^2$.

Figure 3.11 The graph of $f(x) = x^3 + 3x^2 - x - 3$

EXAMPLE 2   Using the Leading Coefficient Test

Use end behavior to explain why

$$f(x) = -143x^3 + 1810x^2 - 187x + 2331$$

is only an appropriate model for AIDS cases for a limited time period.

Solution   Because the degree is odd ($n = 3$) and the leading coefficient, -143, is negative, the graph rises to the left and falls to the right. The fact that it falls to the right indicates at some point the number of AIDS cases will be negative, an impossibility. If a function has a graph that decreases without bound over time, it will not be capable of modeling nonnegative phenomena over long time periods.

Check Point 2 The polynomial function

$$f(x) = -0.27x^3 + 9.2x^2 - 102.9x + 400$$

models the ratio of students to computers, $f(x)$, in U.S. public schools x years after 1980. Use end behavior to determine whether this function could be an appropriate model for computers in the classroom well into the twenty-first century. Explain your answer.

If you use a graphing utility to graph a polynomial function, it is important to select a viewing rectangle that accurately reveals the graph's end behavior. If the viewing rectangle is too small, it may not accurately show the end behavior.

EXAMPLE 3 Using the Leading Coefficient Test

The graph of $f(x) = -x^4 + 8x^3 + 4x^2 + 2$ was obtained with a graphing utility using a $[-8, 8, 1]$ by $[-10, 10, 1]$ viewing rectangle. The graph is shown in Figure 3.12(a). Does the graph show the end behavior of the function?

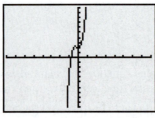
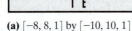
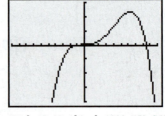

Figure 3.12   **(a)** $[-8, 8, 1]$ by $[-10, 10, 1]$ **(b)** $[-10, 10, 1]$ by $[-1000, 750, 250]$

Solution Note that the degree is even $(n = 4)$ and the leading coefficient, -1, is negative. Even-degree polynomial functions have graphs with the same behavior at each end. The Leading Coefficient Test indicates that the graph should fall to the left and fall to the right. The graph in Figure 3.12(a) is falling to the left, but it is not falling to the right. Therefore, the graph is not complete enough to show end behavior. A more complete graph of the function is shown in a larger viewing rectangle in Figure 3.12(b).

Check Point 3 The graph of $f(x) = x^3 + 13x^2 + 10x - 4$ is shown in a standard viewing rectangle in Figure 3.13. Use the Leading Coefficient Test to determine whether the graph shows the end behavior of the function. Explain your answer.

Figure 3.13

3 Use factoring to find zeros of polynomial functions.

Zeros of Polynomial Functions

If f is a polynomial function, then the values of x for which $f(x)$ is equal to 0 are called the **zeros** of f. These values of x are the **roots**, or **solutions**, of the polynomial equation $f(x) = 0$. Each real root of the polynomial equation appears as an x-intercept of the graph of the polynomial function.

EXAMPLE 4 **Finding Zeros of a Polynomial Function**

Find all zeros of $f(x) = x^3 + 3x^2 - x - 3$.

Solution By definition, the zeros are the values of x for which $f(x)$ is equal to 0. Thus, we set $f(x)$ equal to 0:

$$f(x) = x^3 + 3x^2 - x - 3 = 0.$$

We solve the polynomial equation $x^3 + 3x^2 - x - 3 = 0$ for x as follows:

$$x^3 + 3x^2 - x - 3 = 0 \qquad \text{This is the equation needed to find the}$$
$$\text{function's zeros.}$$

$$x^2(x + 3) - 1(x + 3) = 0 \qquad \text{Factor } x^2 \text{ from the first two terms and } -1 \text{ from}$$
$$\text{the last two terms.}$$

$$(x + 3)(x^2 - 1) = 0 \qquad \text{A common factor of } x + 3 \text{ is factored from the}$$
$$\text{expression.}$$

$$x + 3 = 0 \qquad \text{or} \quad x^2 - 1 = 0 \qquad \text{Set each factor equal to 0.}$$
$$x = -3 \qquad\qquad x^2 = 1 \qquad \text{Solve for } x.$$
$$x = \pm 1 \qquad \text{Remember that if } x^2 = d, \text{ then } x = \pm\sqrt{d}.$$

The zeros of f are $-3, -1$, and 1. The graph of f in Figure 3.14 shows that each zero is an x-intercept.

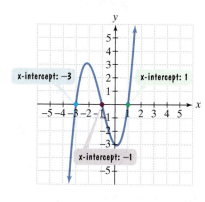

x-intercept: −3

x-intercept: 1

x-intercept: −1

Figure 3.14 The real zeros of $f(x) = x^3 + 3x^2 - x - 3$, namely $-3, -1$, and 1, are the x-intercepts for the graph of f.

Check Point 4 Find all zeros of $f(x) = x^3 + 2x^2 - 4x - 8$.

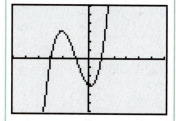
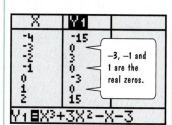
EXAMPLE 5 **Finding Zeros of a Polynomial Function**

Find all zeros of $f(x) = -x^4 + 4x^3 - 4x^2$.

Solution We find the zeros of f by setting $f(x)$ equal to 0 and solving the resulting equation.

$$-x^4 + 4x^3 - 4x^2 = 0 \qquad \text{We now have a polynomial equation.}$$
$$x^4 - 4x^3 + 4x^2 = 0 \qquad \text{Multiply both sides by } -1. \text{ This step is optional.}$$
$$x^2(x^2 - 4x + 4) = 0 \qquad \text{Factor out } x^2.$$
$$x^2(x - 2)^2 = 0 \qquad \text{Factor completely.}$$
$$x^2 = 0 \quad \text{or} \quad (x - 2)^2 = 0 \qquad \text{Set each factor equal to 0.}$$
$$x = 0 \qquad\qquad x = 2 \qquad \text{Solve for } x.$$

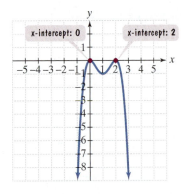

Figure 3.15 The zeros of $f(x) = -x^4 + 4x^3 - 4x^2$, namely 0 and 2, are the x-intercepts for the graph of f.

The zeros of $f(x) = -x^4 + 4x^3 - 4x^2$ are 0 and 2. The graph of f, shown in Figure 3.15, has x-intercepts at 0 and 2.

Check Point 5 Find all zeros of $f(x) = x^4 - 4x^2$.

We can use the results of factoring to express a polynomial as a product of factors. For instance, in Example 5, we can use our factoring to express the function's equation as follows:

$$f(x) = -x^4 + 4x^3 - 4x^2 = -(x^4 - 4x^3 + 4x^2) = -x^2(x - 2)^2.$$

The factor x occurs twice: $x^2 = x \cdot x$.

The factor $(x - 2)$ occurs twice: $(x - 2)^2 = (x - 2)(x - 2)$.

4 Identify the multiplicity of a zero.

Notice that each factor occurs twice. In factoring the equation for the polynomial function f, if the same factor $x - r$ occurs k times, but not $k + 1$ times, we call r a **repeated zero with multiplicity k.** For the polynomial function

$$f(x) = -x^2(x - 2)^2,$$

0 and 2 are both repeated zeros with multiplicity 2. For the polynomial function

$$f(x) = 4(x - 5)(x + 2)^3\left(x - \tfrac{1}{4}\right)^4,$$

5 is a zero with multiplicity 1, -2 is a repeated zero with multiplicity 3, and $\tfrac{1}{4}$ is a repeated zero with multiplicity 4.

The multiplicity of a zero tells us if the graph of a polynomial function touches the x-axis at the zero and turns around, or crosses the x-axis at the zero. For example, look again at the graph of $f(x) = -x^4 + 4x^3 - 4x^2$ in Figure 3.15. Each zero, 0 and 2, is a repeated zero with multiplicity 2. The graph of f touches, but does not cross, the x-axis at each of these zeros of even multiplicity. By contrast, a graph crosses the x-axis at zeros of odd multiplicity.

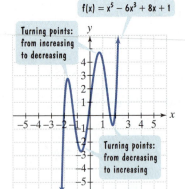

Figure 3.16 Graph with four turning points

5 Understand the relationship between degree and turning points.

Multiplicity and x-Intercepts

If r is a zero of even multiplicity, then the graph **touches** the x-axis and turns around at r. If r is a zero of odd multiplicity, then the graph **crosses** the x-axis at r. Regardless of whether a zero is even or odd, graphs tend to flatten out at zeros with multiplicity greater than one.

Turning Points of Polynomial Functions

The graph of $f(x) = x^5 - 6x^3 + 8x + 1$ is shown in Figure 3.16. The graph has four smooth **turning points.** At each turning point, the graph changes direction from increasing to decreasing or vice versa. The function value at a turning point is either a relative maximum of f or a relative minimum of f. The given equation has 5 as its greatest exponent and is therefore a polynomial function of degree 5. Notice that the graph has four turning points. In general, **if f is a polynomial of degree n, then the graph of f has at most $n - 1$ turning points.**

6 Graph polynomial functions.

A Strategy for Graphing Polynomial Functions

Here's a general strategy for graphing a polynomial function. A graphing utility is a valuable complement to this strategy. Some of the steps listed in the following box will help you to select a viewing rectangle that shows the important parts of the graph.

Graphing a Polynomial Function

$$f(x) = a_n x^n + a_{n-1} x^{n-1} + a_{n-2} x^{n-2} + \cdots + a_1 x + a_0, \quad a_n \neq 0$$

1. Use the Leading Coefficient Test to determine the graph's end behavior.
2. Find x-intercepts by setting $f(x) = 0$ and solving the resulting polynomial equation. If there is an x-intercept at r as a result of $(x - r)^k$ in the complete factorization of $f(x)$, then
   a. If k is even, the graph touches the x-axis at r and turns around.
   b. If k is odd, the graph crosses the x-axis at r.
   c. If $k > 1$, the graph flattens out at $(r, 0)$.
3. Find the y-intercept by computing $f(0)$.
4. Use symmetry, if applicable, to help draw the graph:
   a. y-axis symmetry: $f(-x) = f(x)$
   b. Origin symmetry: $f(-x) = -f(x)$
5. Use the fact that the maximum number of turning points of the graph is $n - 1$ to check whether it is drawn correctly.

EXAMPLE 6 Graphing a Polynomial Function

Graph: $f(x) = x^4 - 2x^2 + 1$.

Solution

Step 1 Determine end behavior. Because the degree is even ($n = 4$) and the leading coefficient, 1, is positive, the graph rises to the left and rises to the right.

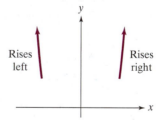

Rises left   Rises right

Step 2 Find x-intercepts (zeros of the function) by setting $f(x) = 0$.

$$x^4 - 2x^2 + 1 = 0$$
$$(x^2 - 1)(x^2 - 1) = 0 \quad \text{Factor.}$$
$$(x + 1)(x - 1)(x + 1)(x - 1) = 0 \quad \text{Factor completely.}$$
$$(x + 1)^2(x - 1)^2 = 0 \quad \text{Express the factoring in a more compact notation.}$$
$$(x + 1)^2 = 0 \quad \text{or} \quad (x - 1)^2 = 0 \quad \text{Set each factor equal to 0.}$$
$$x = -1 \qquad\qquad x = 1 \quad \text{Solve for x.}$$

We see that -1 and 1 are both repeated zeros with multiplicity 2. Because of the even multiplicity, the graph touches the x-axis at -1 and 1 and turns around. Furthermore, the graph tends to flatten out at these zeros with multiplicity greater than one.

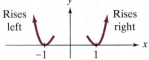

Step 3 Find the y-intercept by computing $f(0)$. We use $f(x) = x^4 - 2x^2 + 1$ and compute $f(0)$.

$$f(0) = 0^4 - 2 \cdot 0^2 + 1 = 1$$

There is a y-intercept at 1, so the graph passes through $(0, 1)$:

Step 4 Use possible symmetry to help draw the graph. Our partial graph suggests y-axis symmetry. Let's verify this by finding $f(-x)$.

$$f(x) \quad = x^4 \quad - \quad 2x^2 \quad + \quad 1$$

Replace x with $-x$.

$$f(-x) = (-x)^4 - 2(-x)^2 + 1 = x^4 - 2x^2 + 1$$

Because $f(-x) = f(x)$, the graph of f is symmetric with respect to the y-axis. Figure 3.17 shows the graph of $f(x) = x^4 - 2x^2 + 1$.

Step 5 Use the fact that the maximum number of turning points of the graph is $n - 1$ to check whether it is drawn correctly. Because $n = 4$, the maximum number of turning points is $4 - 1$, or 3. Because the graph in Figure 3.17 has three turning points, we have not violated the maximum number possible.

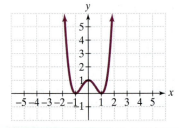

Figure 3.17 The graph of $f(x) = x^4 - 2x^2 + 1$

Check Point 6 Use the five-step strategy to graph $f(x) = x^3 - 3x^2$.

EXERCISE SET 3.2

 Practice Exercises

In Exercises 1–10, determine which functions are polynomial functions. For those that are, identify the degree.

1. $f(x) = 5x^2 + 6x^3$

2. $f(x) = 7x^2 + 9x^4$

3. $g(x) = 7x^5 - \pi x^3 + \frac{1}{5}x$

4. $g(x) = 6x^7 + \pi x^5 + \frac{2}{3}x$

5. $h(x) = 7x^3 + 2x^2 + \frac{1}{x}$

6. $h(x) = 8x^3 - x^2 + \frac{2}{x}$

7. $f(x) = x^{1/2} - 3x^2 + 5$

8. $f(x) = x^{1/3} - 4x^2 + 7$

9. $f(x) = \frac{x^2 + 7}{x^3}$

10. $f(x) = \frac{x^2 + 7}{3}$

In Exercises 11–14, identify which graphs are not those of polynomial functions.

11.

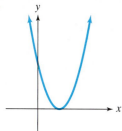

12.

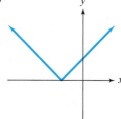

13.

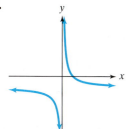

14.

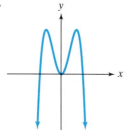

In Exercises 15–20, use the Leading Coefficient Test to determine the end behavior of the graph of the given polynomial function. Then use this end behavior to match the polynomial function with its graph. [The graphs are labeled (a) through (f).]

15. $f(x) = -x^4 + x^2$ **16.** $f(x) = x^3 - 4x^2$

17. $f(x) = (x - 3)^2$

18. $f(x) = -x^3 - x^2 + 5x - 3$

19. $f(x) = x - 3$ **20.** $f(x) = (x + 1)^2(x - 1)^2$

(a)

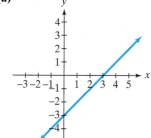

(b)

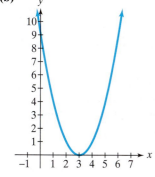

(c)

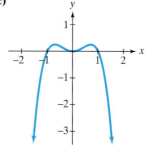

(d)

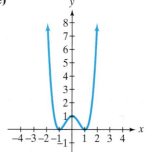

(e)

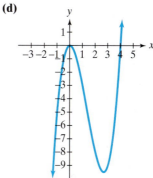

(f)

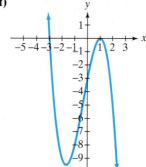

In Exercises 21–26, use the Leading Coefficient Test to determine the end behavior of the graph of the polynomial function.

21. $f(x) = 5x^3 + 7x^2 - x + 9$

22. $f(x) = 11x^3 - 6x^2 + x + 3$

23. $f(x) = 5x^4 + 7x^2 - x + 9$

24. $f(x) = 11x^4 - 6x^2 + x + 3$

25. $f(x) = -5x^4 + 7x^2 - x + 9$

26. $f(x) = -11x^4 - 6x^2 + x + 3$

In Exercises 27–34, find the zeros for each polynomial function and give the multiplicity for each zero. State whether the graph crosses the x-axis, or touches the x-axis and turns around, at each zero.

27. $f(x) = 2(x - 5)(x + 4)^2$

28. $f(x) = 3(x + 5)(x + 2)^2$

29. $f(x) = 4(x - 3)(x + 6)^3$

30. $f(x) = -3(x + \frac{1}{2})(x - 4)^3$

31. $f(x) = x^3 - 2x^2 + x$

32. $f(x) = x^3 + 4x^2 + 4x$

33. $f(x) = x^3 + 7x^2 - 4x - 28$

34. $f(x) = x^3 + 5x^2 - 9x - 45$

In Exercises 35–50,

   a. *Use the Leading Coefficient Test to determine the graph's end behavior.*
   b. *Find x-intercepts by setting $f(x) = 0$ and solving the resulting polynomial equation. State whether the graph crosses the x-axis, or touches the x-axis and turns around, at each intercept.*
   c. *Find the y-intercept by setting x equal to 0 and computing $f(0)$.*
   d. *Determine whether the graph has y-axis symmetry, origin symmetry, or neither.*
   e. *If necessary, find a few additional points and graph the function. Use the fact that the maximum number of turning points of the graph is $n - 1$ to check whether it is drawn correctly.*

35. $f(x) = x^3 + 2x^2 - x - 2$

36. $f(x) = x^3 + x^2 - 4x - 4$

37. $f(x) = x^4 - 9x^2$ **38.** $f(x) = x^4 - x^2$

39. $f(x) = -x^4 + 16x^2$ **40.** $f(x) = -x^4 + 4x^2$

41. $f(x) = x^4 - 2x^3 + x^2$ **42.** $f(x) = x^4 - 6x^3 + 9x^2$

43. $f(x) = -2x^4 + 4x^3$ **44.** $f(x) = -2x^4 + 2x^3$

45. $f(x) = 6x^3 - 9x - x^5$ **46.** $f(x) = 6x - x^3 - x^5$

47. $f(x) = 3x^2 - x^3$ **48.** $f(x) = \frac{1}{2} - \frac{1}{2}x^4$

49. $f(x) = -3(x - 1)^2(x^2 - 4)$

50. $f(x) = -2(x - 4)^2(x^2 - 25)$

 Application Exercises

51. A herd of 100 elk is introduced to a small island. The number of elk, $N(t)$, after t years is described by the polynomial function $N(t) = -t^4 + 21t^2 + 100$.

   a. Use the Leading Coefficient Test to determine the graph's end behavior to the right. What does this mean about what will eventually happen to the elk population?
   b. Graph the function.
   c. Graph only the portion of the function that serves as a realistic model for the elk population over time. When does the population become extinct?

52. The common cold is caused by a rhinovirus. After x days of invasion by the viral particles, the number of particles in our bodies, $f(x)$, in billions, can be modeled by the polynomial function

$$f(x) = -0.75x^4 + 3x^3 + 5.$$

Use the Leading Coefficient Test to determine the graph's end behavior to the right. What does this mean about the number of viral particles in our bodies over time?

53. The polynomial function

$$f(x) = -0.87x^3 + 0.35x^2 + 81.62x + 7684.94$$

models the number of thefts, $f(x)$, in thousands, in the United States x years after 1987. Will this function be useful in modeling the number of thefts over an extended period of time? Explain your answer.

54. The graphs show the percentage of husbands and wives with one or more children who said their marriage was going well "all the time" at various stages in their relationships.

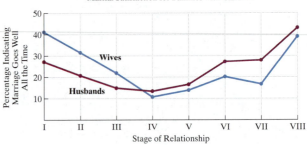

Marital Satisfaction for Families with Children

Source: Rollins, B., & Feldman, H. (1970), Marital satisfaction over the family life cycle. *Journal of Marriage and the Family,* 32, 20–28.

Stage   I: Beginning families
Stage   II: Child-bearing families
Stage   III: Families with preschool children
Stage   IV: Families with school-age children

Stage   V: Families with teenagers

Stage   VI: Families with adult children leaving home

Stage VII: Families in the middle years

Stage VIII: Aging families

a. Between which stages was marital satisfaction for wives decreasing?

b. Between which stages was marital satisfaction for wives increasing?

c. How many turning points (from decreasing to increasing or from increasing to decreasing) are shown in the graph for wives?

d. Suppose that a polynomial function is used to model the data shown in the graph for wives using

(stage in the relationship, percentage indicating that the marriage was going well all the time).

Use the number of turning points to determine the degree of the polynomial function of best fit.

e. For the model in part (d), should the leading coefficient of the polynomial function be positive or negative? Explain your answer.

55. The bar graph below shows women's earnings as a percentage of men's from 1970 through 2000. Suppose that a polynomial function is used to model the data shown using

(number of years after 1970, women's earnings as a percentage of men's).

Determine the degree of the polynomial function of best fit. Should the leading coefficient be positive or negative? Explain your answer.

The Wage Gap in the United States

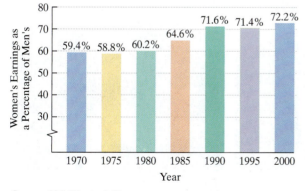

Source: U.S. Women's Bureau

Writing in Mathematics

56. What is a polynomial function?

57. What do we mean when we describe the graph of a polynomial function as smooth and continuous?

58. What is meant by the end behavior of a polynomial function?

59. Explain how to use the Leading Coefficient Test to determine the end behavior of a polynomial function.

60. Why is a third-degree polynomial function with a negative leading coefficient not appropriate for modeling non-negative real-world phenomena over a long period of time?

61. What are the zeros of a polynomial function and how are they found?

62. Explain the relationship between the multiplicity of a zero and whether or not the graph crosses or touches the *x*-axis at that zero.

63. Explain the relationship between the degree of a polynomial function and the number of turning points on its graph.

64. Can the graph of a polynomial function have no *x*-intercepts? Explain.

65. Can the graph of a polynomial function have no *y*-intercept? Explain.

66. Describe a strategy for graphing a polynomial function. In your description, mention intercepts, the polynomial's degree, and turning points.

67. In a favorable habitat and without natural predators, a population of reindeer is introduced to an island preserve. The reindeer population, $f(t)$, t years after their introduction is modeled by the polynomial function $f(t) = -0.125t^5 + 3.125t^4 + 4000$. Discuss the growth and decline of the reindeer population. Describe the factors that might contribute to this population model.

68. The graphs shown in Exercise 54 indicate that marital satisfaction tends to be greatest at the beginning and at the end of the stages in the relationship, with a decline occurring in the middle. What explanations can you offer for this trend?

Technology Exercises

69. Use a graphing utility to verify any five of the graphs that you drew by hand in Exercises 35–50.

Write a polynomial function that imitates the end behavior of each graph in Exercises 70–73. The dashed portions of the graphs indicate that you should focus only on imitating the left and right behavior of the graph and can be flexible about what occurs between the left and right ends. Then use your graphing utility to graph the polynomial function and verify that you imitated the end behavior shown in the given graph.

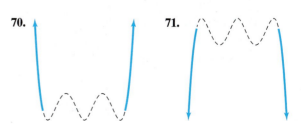

72.

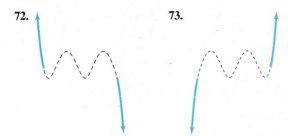

73.

In Exercises 74–77, use a graphing utility with a viewing rectangle large enough to show end behavior to graph each polynomial function.

74. $f(x) = x^3 + 13x^2 + 10x - 4$
75. $f(x) = -2x^3 + 6x^2 + 3x - 1$
76. $f(x) = -x^4 + 8x^3 + 4x^2 + 2$
77. $f(x) = -x^5 + 5x^4 - 6x^3 + 2x + 20$

In Exercises 78–79, use a graphing utility to graph f and g in the same viewing rectangle. Then use the ⟨ZOOM OUT⟩ *feature to show that f and g have identical end behavior.*

78. $f(x) = x^3 - 6x + 1, \quad g(x) = x^3$
79. $f(x) = -x^4 + 2x^3 - 6x, \quad g(x) = -x^4$

Critical Thinking Exercises

80. Which one of the following is true?
  a. If $f(x) = -x^3 + 4x$, then the graph of f falls to the left and to the right.
  b. A mathematical model that is a polynomial of degree n whose leading term is $a_n x^n$, n odd and $a_n < 0$, is ideally

suited to describe nonnegative phenomena over unlimited periods of time.
  c. There is more than one third-degree polynomial function with the same three x-intercepts.
  d. The graph of a function with origin symmetry can rise to the left and to the right.

Use the descriptions in Exercises 81–82 to write an equation of a polynomial function with the given characteristics. Use a graphing utility to graph your function to see if you are correct. If not, modify the function's equation and repeat this process.

81. Crosses the x-axis at $-4, 0$, and 3; lies above the x-axis between -4 and 0; lies below the x-axis between 0 and 3
82. Touches the x-axis at 0 and crosses the x-axis at 2; lies below the x-axis between 0 and 2

Group Exercise

83. This exercise is based on the group's work in Exercise 71 of Exercise Set 3.1. For the two data sets that the group selected:
  a. Use the polynomial regression feature of a graphing utility to find the third-degree polynomial function that best fits the data.
  b. Use this function to repeat the predictions that you made with the quadratic function. How do these predictions compare with those that you obtained previously?
  c. For each data set, describe whether the quadratic function or the third-degree function is a better fit. Use a graphing utility, a scatter plot of the data, and the function of best fit drawn on the scatter plot to help determine which function is the better fit.

SECTION 3.3 *Dividing Polynomials; Remainder and Factor Theorems*

Objectives

1. Use long division to divide polynomials.
2. Use synthetic division to divide polynomials.
3. Evaluate a polynomial using the Remainder Theorem.
4. Use the Factor Theorem to solve a polynomial equation.

For those of you who are dog lovers, you might still be thinking of the polynomial function that models the age in human years, $H(x)$, of a dog that is x years old, namely

$$H(x) = -0.001618x^4 + 0.077326x^3 - 1.2367x^2 + 11.460x + 2.914.$$

Suppose that you are in your twenties, say 25. What is Fido's equivalent age? To answer this question, we must substitute 25 for $H(x)$ and solve the resulting polynomial equation for x:

$$25 = -0.001618x^4 + 0.077326x^3 - 1.2367x^2 + 11.460x + 2.914.$$

How can we solve such an equation? You might begin by subtracting 25 from both sides to obtain zero on one side. But then what? The factoring that we used in the previous section will not work in this situation.

In Sections 3.4 and 3.5, we will present techniques for solving certain kinds of polynomial equations. These techniques will further enhance your ability to manipulate algebraically the polynomial functions that model your world. Because these techniques are based on understanding polynomial division, in this section we look at two methods for dividing polynomials.

① Use long division to divide polynomials.

Long Division of Polynomials and the Division Algorithm

We begin by looking at division by a polynomial containing more than one term, such as

$$x + 3 \overline{)x^2 + 10x + 21.}$$

Divisor has two terms.   Dividend has three terms.

When a divisor has more than one term, the four steps used to divide whole numbers—**divide**, **multiply**, **subtract**, **bring down the next term**—form the repetitive procedure for polynomial long division.

EXAMPLE 1   Long Division of Polynomials

Divide $x^2 + 10x + 21$ by $x + 3$.

Solution   The following steps illustrate how polynomial division is very similar to numerical division.

$$x + 3 \overline{)x^2 + 10x + 21}$$

Arrange the terms of the dividend ($x^2 + 10x + 21$) and the divisor ($x + 3$) in descending powers of x.

$$x + 3 \overline{\smash{)}\,x^2 + 10x + 21} \quad\overset{\textstyle x}{}$$

Divide x^2 (the first term in the dividend) by x (the first term in the divisor): $\dfrac{x^2}{x} = x$. Align like terms.

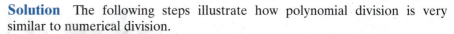

$$\begin{array}{r} x \\ x + 3 \overline{)x^2 + 10x + 21} \\ \underline{x^2 + 3x} \end{array}$$

times   equals

Multiply each term in the divisor x + 3 by x, aligning terms of the product under like terms in the dividend.

$$\begin{array}{r} x \\ x + 3 \overline{)x^2 + 10x + 21} \\ \underline{\ominus x^2 + \ominus 3x} \\ 7x \end{array}$$

Change signs of the polynomial being subtracted.

Subtract $x^2 + 3x$ from $x^2 + 10x$ by changing the sign of each term in the lower expression and adding.

$$\begin{array}{r} x \\ x + 3 \overline{)x^2 + 10x + 21} \\ \underline{x^2 + 3x} \downarrow \\ 7x + 21 \end{array}$$

Bring down 21 from the original dividend and add algebraically to form a new dividend.

$$x + 3 \overline{)x^2 + 10x + 21}$$ with quotient $x + 7$
$$\underline{x^2 + 3x} \downarrow$$
$$7x + 21$$

Find the second term of the quotient. **Divide** the first term of $7x + 21$ by x, the first term of the divisor: $\dfrac{7x}{x} = 7.$

times

$$x + 3 \overline{)x^2 + 10x + 21}$$ with quotient $x + 7$
$$\underline{x^2 + 3x}$$
$$7x + 21$$
$$\underline{7x + 21}$$
$$0 \quad \text{Remainder}$$

equals

Multiply the divisor x + 3 by 7, aligning under like terms in the new dividend. Then **subtract** to obtain the remainder of 0.

The quotient is $x + 7$. Because the remainder is 0, we can conclude that $x + 3$ is a factor of $x^2 + 10x + 21$ and

$$\frac{x^2 + 10x + 21}{x + 3} = x + 7.$$

Check Point 1 Divide $x^2 + 14x + 45$ by $x + 9$.

Before considering additional examples, let's summarize the general procedure for dividing one polynomial by another.

Long Division of Polynomials

1. **Arrange the terms** of both the dividend and the divisor in descending powers of the variable.

2. **Divide** the first term in the dividend by the first term in the divisor. The result is the first term of the quotient.

3. **Multiply** every term in the divisor by the first term in the quotient. Write the resulting product beneath the dividend with like terms lined up.

4. **Subtract** the product from the dividend.

5. **Bring down** the next term in the original dividend and write it next to the remainder to form a new dividend.

6. Use this new expression as the dividend and repeat this process until the remainder can no longer be divided. This will occur when the degree of the remainder (the highest exponent on a variable in the remainder) is less than the degree of the divisor.

In our next long division, we will obtain a nonzero remainder.

EXAMPLE 2 Long Division of Polynomials

Divide $4 - 5x - x^2 + 6x^3$ by $3x - 2$.

Solution We begin by writing the divisor and dividend in descending powers of x.

Is It Hot in Here, or Is It Just Me?

In the 1980s, a rising trend in global surface temperature was observed, and the term "global warming" was coined. Scientists are more convinced than ever that burning coal, oil, and gas results in a buildup of gases and particles that trap heat and raise the planet's temperature. The average increase in global surface temperature, $T(x)$, in degrees Centigrade, x years after 1980 can be modeled by the polynomial function

$$T(x) = \frac{21}{5,000,000} x^3$$
$$- \frac{127}{1,000,000} x^2 + \frac{1293}{50,000} x.$$

Use your graphing utility to graph the function in a $[0, 60, 3]$ by $[0, 2, 0.1]$ viewing rectangle. (Place parentheses around each fractional coefficient when you enter the equation.) What do you observe about global warming through the year 2040?

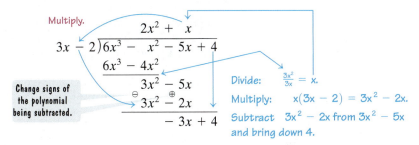

Now we divide $3x^2$ by $3x$ to obtain x, multiply x and the divisor, and subtract.

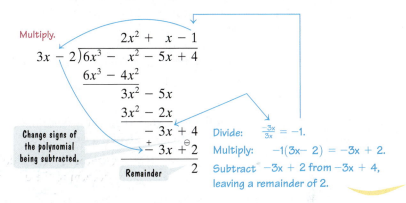

Now we divide $-3x$ by $3x$ to obtain -1, multiply -1 and the divisor, and subtract.

In Example 2, the quotient is $2x^2 + x - 1$ and the remainder is 2. When there is a nonzero remainder, as in this example, list the quotient, plus the remainder above the divisor. Thus,

$$\underbrace{\frac{6x^3 - x^2 - 5x + 4}{3x - 2}}_{\substack{\text{Dividend} \\ \text{Divisor}}} = 2x^2 + x - 1 + \underbrace{\frac{2}{3x - 2}}_{\substack{\text{Remainder} \\ \text{Divisor}}}.$$

Multiplying both sides of this equation by $3x - 2$ results in the following equation:

$$6x^3 - x^2 - 5x + 4 = (3x - 2)(2x^2 + x - 1) + 2.$$

Dividend   Divisor   Quotient   Remainder

Polynomial long division is checked by multiplying the divisor with the quotient and then adding the remainder. This should give the dividend. The process illustrates the **Division Algorithm.**

> **The Division Algorithm**
>
> If $f(x)$ and $d(x)$ are polynomials, with $d(x) \neq 0$, and the degree of $d(x)$ is less than or equal to the degree of $f(x)$, then there exist unique polynomials $q(x)$ and $r(x)$ such that
>
> $$f(x) \quad = \quad d(x) \quad \cdot \quad q(x) \quad + \quad r(x).$$
>
> Dividend   Divisor   Quotient   Remainder
>
> The remainder, $r(x)$, equals 0 or it is of degree less than the degree of $d(x)$. If $r(x) = 0$, we say that $d(x)$ **divides evenly** into $f(x)$ and that $d(x)$ and $q(x)$ are **factors** of $f(x)$.

Check Point 2   Divide $7 - 11x - 3x^2 + 2x^3$ by $x - 3$. Use the remainder to express your result in fractional form.

If a power of x is missing in either a dividend or a divisor, add that power of x with a coefficient of 0 and then divide. In this way, like terms will be aligned as you carry out the long division.

EXAMPLE 3   Long Division of Polynomials

Divide $6x^4 + 5x^3 + 3x - 5$ by $3x^2 - 2x$.

Solution   We write the dividend, $6x^4 + 5x^3 + 3x - 5$, as $6x^4 + 5x^3 + 0x^2 + 3x - 5$ to keep all like terms aligned.

$$
\begin{array}{r}
2x^2 + 3x + 2 \\
3x^2 - 2x \overline{)6x^4 + 5x^3 + 0x^2 + 3x - 5} \\
\ominus\; 6x^4 \oplus 4x^3 \\
\hline
9x^3 + 0x^2 \\
\ominus\; 9x^3 \oplus 6x^2 \\
\hline
6x^2 + 3x \\
\ominus\; 6x^2 \oplus 4x \\
\hline
7x - 5
\end{array}
$$

Multiply.

$2x^2(3x^2 - 2x)$

$3x(3x^2 - 2x)$

$2(3x^2 - 2x)$

Remainder

The division process is finished because the degree of $7x - 5$, which is 1, is less than the degree of the divisor $3x^2 - 2x$, which is 2. The answer is

$$\frac{6x^4 + 5x^3 + 3x - 5}{3x^2 - 2x} = 2x^2 + 3x + 2 + \frac{7x - 5}{3x^2 - 2x}.$$

Check Point 3   Divide $2x^4 + 3x^3 - 7x - 10$ by $x^2 - 2x$.

2 Use synthetic division to divide polynomials.

Dividing Polynomials Using Synthetic Division

We can use **synthetic division** to divide polynomials if the divisor is of the form $x - c$. This method provides a quotient more quickly than long division. Let's compare the two methods showing $x^3 + 4x^2 - 5x + 5$ divided by $x - 3$.

Long Division

$$
\begin{array}{r}
x^2 + 7x + 16 \\
x - 3 \overline{)x^3 + 4x^2 - 5x + 5} \\
\ominus x^3 \overset{\oplus}{-} 3x^2 \\
\hline
7x^2 - 5x \\
\ominus 7x^2 \overset{\oplus}{-} 21x \\
\hline
16x + 5 \\
\ominus 16x \overset{\oplus}{-} 48 \\
\hline
53
\end{array}
$$

Quotient

Dividend

Divisor
$x - c$;
$c = 3$

Remainder

Synthetic Division

$$
\begin{array}{r|rrrr}
3 & 1 & 4 & -5 & 5 \\
& & 3 & 21 & 48 \\
\hline
& 1 & 7 & 16 & 53
\end{array}
$$

Notice the relationship between the polynomials in the long division process and the numbers that appear in synthetic division.

These are the coefficients of the dividend $x^3 + 4x^2 - 5x + 5$.

The divisor is $x - 3$. This is 3, or c, in $x - c$.

$$
\begin{array}{r|rrrr}
3 & 1 & 4 & -5 & 5 \\
& & 3 & 21 & 48 \\
\hline
& 1 & 7 & 16 & 53
\end{array}
$$

These are the coefficients of the quotient $x^2 + 7x + 16$.

This is the remainder.

Now let's look at the steps involved in synthetic division.

Synthetic Division

To divide a polynomial by $x - c$,

Example

1. Arrange polynomials in descending powers, with a 0 coefficient for any missing term.

$$x - 3 \overline{)x^3 + 4x^2 - 5x + 5}$$

2. Write c for the divisor, $x - c$. To the right, write the coefficients of the dividend.

$$
\begin{array}{r|rrrr}
3 & 1 & 4 & -5 & 5
\end{array}
$$

3. Write the leading coefficient of the dividend on the bottom row.

$$
\begin{array}{r|rrrr}
3 & 1 & 4 & -5 & 5 \\
& \downarrow & & & \\
\hline
& 1
\end{array}
$$

Bring down 1.

4. Multiply c (in this case, 3) times the value just written on the bottom row. Write the product in the next column in the second row.

$$
\begin{array}{r|rrrr}
3 & 1 & 4 & -5 & 5 \\
& & 3 & & \\
\hline
& 1
\end{array}
$$

Multiply by 3: $3 \cdot 1 = 3$.

5. Add the values in this new column, writing the sum in the bottom row.

$$
\begin{array}{r|rrrr}
3 & 1 & 4 & -5 & 5 \\
& & 3 & & \\
\hline
& 1 & 7
\end{array}
$$

Add.

6. Repeat this series of multiplications and additions until all columns are filled in.

$$\begin{array}{r|rrrr} 3 & 1 & 4 & -5 & 5 \\ & & 3 & 21 & \\ \hline & 1 & 7 & 16 & \end{array}$$ Add.

Multiply by 3: $3 \cdot 7 = 21$.

$$\begin{array}{r|rrrr} 3 & 1 & 4 & -5 & 5 \\ & & 3 & 21 & 48 \\ \hline & 1 & 7 & 16 & 53 \end{array}$$ Add.

Multiply by 3: $3 \cdot 16 = 48$.

7. Use the numbers in the last row to write the quotient, plus the remainder above the divisor. **The degree of the first term of the quotient is one less than the degree of the first term of the dividend.** The final value in this row is the remainder.

Written from the last row of the synthetic division

$$x - 3 \overline{\smash{\big)}\, x^3 + 4x^2 - 5x + 5} \quad 1x^2 + 7x + 16 + \dfrac{53}{x-3}$$

EXAMPLE 4   Using Synthetic Division

Use synthetic division to divide $5x^3 + 6x + 8$ by $x + 2$.

Solution The divisor must be in the form $x - c$. Thus, we write $x + 2$ as $x - (-2)$. This means that $c = -2$. Writing a 0 coefficient for the missing x^2-term in the dividend, we can express the division as follows:

$$x - (-2) \overline{\smash{\big)}\, 5x^3 + 0x^2 + 6x + 8}.$$

Now we are ready to set up the problem so that we can use synthetic division.

Use the coefficients of the dividend $5x^3 + 0x^2 + 6x + 8$ in descending powers of x.

This is c in $x-(-2)$.

$$\begin{array}{r|rrrr} -2 & 5 & 0 & 6 & 8 \end{array}$$

We begin the synthetic division process by bringing down 5. This is followed by a series of multiplications and additions.

1. Bring down 5.

$$\begin{array}{r|rrrr} -2 & 5 & 0 & 6 & 8 \\ & & & & \\ \hline & 5 & & & \end{array}$$

2. Multiply: $-2(5) = -10$.

$$\begin{array}{r|rrrr} -2 & 5 & 0 & 6 & 8 \\ & & -10 & & \\ \hline & 5 & & & \end{array}$$

Multiply by -2.

3. Add: $0 + (-10) = -10$.

$$\begin{array}{r|rrrr} -2 & 5 & 0 & 6 & 8 \\ & & -10 & & \\ \hline & 5 & -10 & & \end{array}$$ Add.

4. Multiply: $-2(-10) = 20$.

$$\begin{array}{r|rrrr} -2 & 5 & 0 & 6 & 8 \\ & & -10 & 20 & \\ \hline & 5 & -10 & & \end{array}$$

Multiply by -2.

5. Add: $6 + 20 = 26$.

$$\begin{array}{r|rrrr} -2 & 5 & 0 & 6 & 8 \\ & & -10 & 20 & \\ \hline & 5 & -10 & 26 & \end{array}$$ Add.

6. Multiply: $-2(26) = -52$.

$$\begin{array}{r|rrrr} -2 & 5 & 0 & 6 & 8 \\ & & -10 & 20 & -52 \\ \hline & 5 & -10 & 26 & \end{array}$$

Multiply by -2.

7. Add: $8 + (-52) = -44$.

$$\begin{array}{r|rrrr} -2 & 5 & 0 & 6 & 8 \\ & & -10 & 20 & -52 \\ \hline & 5 & -10 & 26 & -44 \end{array}$$ Add.

The numbers in the last row represent the coefficients of the quotient and the remainder. The degree of the first term of the quotient is one less than that of the dividend. Because the degree of the dividend, $5x^3 + 6x + 8$, is 3, the degree of the quotient is 2. This means that the 5 in the last row represents $5x^2$.

$$
\begin{array}{r|rrrr}
-2 & 5 & 0 & 6 & 8 \\
   &   & -10 & 20 & -52 \\
\hline
   & 5 & -10 & 26 & -44
\end{array}
$$

The quotient is $5x^2 - 10x + 26$.   The remainder is -44.

Thus,

$$
x + 2 \overline{)5x^3 + 6x + 8} \quad = \quad 5x^2 - 10x + 26 - \frac{44}{x + 2}
$$

Check Point 4   Use synthetic division to divide

$$x^3 - 7x - 6 \text{ by } x + 2.$$

3   Evaluate a polynomial using the Remainder Theorem.

The Remainder Theorem

Let's consider the Division Algorithm when the dividend, $f(x)$, is divided by $x - c$. In this case, the remainder must be a constant because its degree is less than one, the degree of $x - c$.

$$f(x) = (x - c)q(x) + r$$

Dividend   Divisor   Quotient   The remainder, r, is a constant when dividing by x - c.

Now let's evaluate f at c.

$$f(c) = (c - c)q(c) + r \qquad \text{Find } f(c) \text{ by letting } x = c \text{ in } f(x) = (x - c)q(x) + r.$$
This will give an expression for r.

$$f(c) = 0 \cdot q(c) + r \qquad c - c = 0$$

$$f(c) = r \qquad 0 \cdot q(c) = 0 \text{ and } 0 + r = r.$$

What does this last equation mean? If a polynomial is divided by $x - c$, the remainder is the value of the polynomial at c. This result is called the **Remainder Theorem.**

The Remainder Theorem

If the polynomial $f(x)$ is divided by $x - c$, then the remainder is $f(c)$.

Example 5 shows how we can use the Remainder Theorem to evaluate a polynomial function at 2. Rather than substituting 2 for x, we divide the function by $x - 2$. The remainder is $f(2)$.

EXAMPLE 5 **Using the Remainder Theorem to Evaluate a Polynomial Function**

Given $f(x) = x^3 - 4x^2 + 5x + 3$, use the Remainder Theorem to find $f(2)$.

Solution By the Remainder Theorem, if $f(x)$ is divided by $x - 2$, then the remainder is $f(2)$. We'll use synthetic division to divide.

$$\begin{array}{r|rrrr} 2 & 1 & -4 & 5 & 3 \\ & & 2 & -4 & 2 \\ \hline & 1 & -2 & 1 & 5 \end{array} \quad \text{Remainder}$$

The remainder, 5, is the value of $f(2)$. Thus, $f(2) = 5$. We can verify that this is correct by evaluating $f(2)$ directly. Using $f(x) = x^3 - 4x^2 + 5x + 3$, we obtain

$$f(2) = 2^3 - 4 \cdot 2^2 + 5 \cdot 2 + 3 = 8 - 16 + 10 + 3 = 5.$$

Check Point 5 Given $f(x) = 3x^3 + 4x^2 - 5x + 3$, use the Remainder Theorem to find $f(-4)$.

4 Use the Factor Theorem to solve a polynomial equation.

The Factor Theorem

Let's look again at the Division Algorithm when the divisor is of the form $x - c$.

$$f(x) = (x - c)q(x) + r$$

Dividend   Divisor   Quotient   Constant remainder

By the Remainder Theorem, the remainder r is $f(c)$, so we can substitute $f(c)$ for r:

$$f(x) = (x - c)q(x) + f(c).$$

Notice that if $f(c) = 0$, then

$$f(x) = (x - c)q(x)$$

so that $x - c$ is a factor of $f(x)$. This means that for the polynomial function $f(x)$, if $f(c) = 0$, then $x - c$ is a factor of $f(x)$.

Let's reverse directions and see what happens if $x - c$ is a factor of $f(x)$. This means that

$$f(x) = (x - c)q(x).$$

If we replace x with c, we obtain

$$f(c) = (c - c)q(c) = 0.$$

Thus, if $x - c$ is a factor of $f(x)$, then $f(c) = 0$.

We have proved a result known as the **Factor Theorem.**

The Factor Theorem

Let $f(x)$ be a polynomial.

a. If $f(c) = 0$, then $x - c$ is a factor of $f(x)$.

b. If $x - c$ is a factor of $f(x)$, then $f(c) = 0$.

The example that follows shows how the Factor Theorem can be used to solve a polynomial equation.

EXAMPLE 6   Using the Factor Theorem

Solve the equation $2x^3 - 3x^2 - 11x + 6 = 0$ given that 3 is a zero of $f(x) = 2x^3 - 3x^2 - 11x + 6$.

Solution   We are given that $f(3) = 0$. The Factor Theorem tells us that $x - 3$ is a factor of $f(x)$. We'll use synthetic division to divide $f(x)$ by $x - 3$.

$$\underline{3}\begin{array}{rrrr} 2 & -3 & -11 & 6 \\ & 6 & 9 & -6 \\ \hline 2 & 3 & -2 & 0 \end{array}$$

$$\begin{array}{r} 2x^2 + 3x - 2 \\ x - 3\overline{)2x^3 - 3x^2 - 11x + 6} \end{array}$$

> The remainder, 0, verifies that x − 3 is a factor of $2x^3 - 3x^2 - 11x + 6$.

Equivalently,

$$2x^3 - 3x^2 - 11x + 6 = (x - 3)(2x^2 + 3x - 2).$$

Now we can solve the polynomial equation.

$$2x^3 - 3x^2 - 11x + 6 = 0 \qquad \text{This is the given equation.}$$
$$(x - 3)(2x^2 + 3x - 2) = 0 \qquad \text{Factor using the result from the synthetic division.}$$
$$(x - 3)(2x - 1)(x + 2) = 0 \qquad \text{Factor the trinomial.}$$
$$x - 3 = 0 \quad \text{or} \quad 2x - 1 = 0 \quad \text{or} \quad x + 2 = 0 \qquad \text{Set each factor equal to 0.}$$
$$x = 3 \qquad\qquad x = \tfrac{1}{2} \qquad\qquad x = -2 \qquad \text{Solve for x.}$$

The solution set is $\left\{-2, \tfrac{1}{2}, 3\right\}$.

Technology

Because the solution set of

$$2x^3 - 3x^2 - 11x + 6 = 0$$

is $\left\{-2, \tfrac{1}{2}, 3\right\}$, this implies that the polynomial function

$$f(x) = 2x^3 - 3x^2 - 11x + 6$$

has x-intercepts (or zeros) at $-2, \tfrac{1}{2}$, and 3. This is verified by the graph of f.

x-intercept: $\tfrac{1}{2}$

x-intercept: -2 x-intercept: 3

$[-10, 10, 1]$ by $[-15, 15, 1]$

Based on the Factor Theorem, the following statements are useful in solving polynomial equations:

1. If $f(x)$ is divided by $x - c$ and the remainder is zero, then c is a zero of f and c is a root of the polynomial equation $f(x) = 0$.
2. If $f(x)$ is divided by $x - c$ and the remainder is zero, then $x - c$ is a factor of $f(x)$.

Check Point 6   Solve the equation $15x^3 + 14x^2 - 3x - 2 = 0$ given that -1 is a zero of $f(x) = 15x^3 + 14x^2 - 3x - 2$.

EXERCISE SET 3.3

Practice Exercises

In Exercises 1–16, divide using long division. State the quotient, $q(x)$, and the remainder, $r(x)$.

1. $(x^2 + 8x + 15) \div (x + 5)$
2. $(x^2 + 3x - 10) \div (x - 2)$
3. $(x^3 + 5x^2 + 7x + 2) \div (x + 2)$
4. $(x^3 - 2x^2 - 5x + 6) \div (x - 3)$

5. $(6x^3 + 7x^2 + 12x - 5) \div (3x - 1)$
6. $(6x^3 + 17x^2 + 27x + 20) \div (3x + 4)$
7. $(12x^2 + x - 4) \div (3x - 2)$
8. $(4x^2 - 8x + 6) \div (2x - 1)$
9. $\dfrac{2x^3 + 7x^2 + 9x - 20}{x + 3}$
10. $\dfrac{3x^2 - 2x + 5}{x - 3}$
11. $\dfrac{4x^4 - 4x^2 + 6x}{x - 4}$
12. $\dfrac{x^4 - 81}{x - 3}$

13. $\dfrac{6x^3 + 13x^2 - 11x - 15}{3x^2 - x - 3}$

14. $\dfrac{x^4 + 2x^3 - 4x^2 - 5x - 6}{x^2 + x - 2}$

15. $\dfrac{18x^4 + 9x^3 + 3x^2}{3x^2 + 1}$ **16.** $\dfrac{2x^5 - 8x^4 + 2x^3 + x^2}{2x^3 + 1}$

In Exercises 17–32, divide using synthetic division.

17. $(2x^2 + x - 10) \div (x - 2)$

18. $(x^2 + x - 2) \div (x - 1)$

19. $(3x^2 + 7x - 20) \div (x + 5)$

20. $(5x^2 - 12x - 8) \div (x + 3)$

21. $(4x^3 - 3x^2 + 3x - 1) \div (x - 1)$

22. $(5x^3 - 6x^2 + 3x + 11) \div (x - 2)$

23. $(6x^5 - 2x^3 + 4x^2 - 3x + 1) \div (x - 2)$

24. $(x^5 + 4x^4 - 3x^2 + 2x + 3) \div (x - 3)$

25. $(x^2 - 5x - 5x^3 + x^4) \div (5 + x)$

26. $(x^2 - 6x - 6x^3 + x^4) \div (6 + x)$

27. $\dfrac{x^5 + x^3 - 2}{x - 1}$ **28.** $\dfrac{x^7 + x^5 - 10x^3 + 12}{x + 2}$

29. $\dfrac{x^4 - 256}{x - 4}$ **30.** $\dfrac{x^7 - 128}{x - 2}$

31. $\dfrac{2x^5 - 3x^4 + x^3 - x^2 + 2x - 1}{x + 2}$

32. $\dfrac{x^5 - 2x^4 - x^3 + 3x^2 - x + 1}{x - 2}$

33. Given $f(x) = 2x^3 - 11x^2 + 7x - 5$, use the Remainder Theorem to find $f(4)$.

34. Given $f(x) = x^3 - 7x^2 + 5x - 6$, use the Remainder Theorem to find $f(3)$.

35. Given $f(x) = 7x^4 - 3x^3 + 6x + 9$, use the Remainder Theorem to find $f(-5)$.

36. Given $f(x) = 3x^4 + 6x^3 - 2x + 4$, use the Remainder Theorem to find $f(-4)$.

37. Use synthetic division to divide $f(x) = x^3 - 4x^2 + x + 6$ by $x + 1$. Use the result to find all zeros of f.

38. Use synthetic division to divide $f(x) = x^3 - 2x^2 - x + 2$ by $x + 1$. Use the result to find all zeros of f.

39. Solve the equation $2x^3 - 5x^2 + x + 2 = 0$ given that 2 is a zero of $f(x) = 2x^3 - 5x^2 + x + 2$.

40. Solve the equation $2x^3 - 3x^2 - 11x + 6 = 0$ given that -2 is a zero of $f(x) = 2x^3 - 3x^2 - 11x + 6$.

41. Solve the equation $12x^3 + 16x^2 - 5x - 3 = 0$ given that $-\frac{3}{2}$ is a root.

42. Solve the equation $3x^3 + 7x^2 - 22x - 8 = 0$ given that $-\frac{1}{3}$ is a root.

Application Exercises

43. A rectangle with length $2x + 5$ inches has an area of $2x^4 + 15x^3 + 7x^2 - 135x - 225$ square inches. Write a polynomial that represents its width.

44. If you travel a distance of $x^3 + 3x^2 + 5x + 3$ miles at a rate of $x + 1$ miles per hour, write a polynomial that represents the number of hours you traveled.

During the 1980s, the controversial economist Arthur Laffer promoted the idea that tax increases lead to a reduction in government revenue. Called supply-side economics, the theory uses function such as

$$f(x) = \frac{80x - 8000}{x - 110}, \quad 30 \le x \le 100.$$

This function models the government tax revenue, $f(x)$, in tens of billions of dollars, in terms of the tax rate, x. The graph of the function is shown. It illustrates tax revenue decreasing quite dramatically as the tax rate increases. At a tax rate of (gasp) 100%, the government takes all our money and no one has an incentive to work. With no income earned, zero dollars in tax revenue is generated.

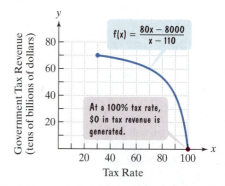

Use function f and its graph to solve Exercises 45–46.

45. a. Find and interpret $f(30)$. Identify the solution as a point on the graph of the function.

   b. Rewrite the function by using long division to perform

   $$(80x - 8000) \div (x - 110).$$

   Then use this new form of the function to find $f(30)$. Do you obtain the same answer as you did in part (a)?

   c. Is f a polynomial function? Explain your answer.

46. a. Find and interpret $f(40)$. Identify the solution as a point on the graph of the function.

   b. Rewrite the function by using long division to perform

   $$(80x - 8000) \div (x - 110).$$

   Then use this new form of the function to find $f(40)$. Do you obtain the same answer as you did in part (a)?

   c. Is f a polynomial function? Explain your answer.

Writing in Mathematics

47. Explain how to perform long division of polynomials. Use $2x^3 - 3x^2 - 11x + 7$ divided by $x - 3$ in your explanation.

48. In your own words, state the Division Algorithm.

49. How can the Division Algorithm be used to check the quotient and remainder in a long division problem?

50. Explain how to perform synthetic division. Use the division problem in Exercise 47 to support your explanation.

51. State the Remainder Theorem.

52. Explain how the Remainder Theorem can be used to find $f(-6)$ if $f(x) = x^4 + 7x^3 + 8x^2 + 11x + 5$. What advantage is there to using the Remainder Theorem in this situation rather than evaluating $f(-6)$ directly?

53. How can the Factor Theorem be used to determine if $x - 1$ is a factor of $x^3 - 2x^2 - 11x + 12$?

54. If you know that -2 is a zero of

$$f(x) = x^3 + 7x^2 + 4x - 12,$$

explain how to solve the equation

$$x^3 + 7x^2 + 4x - 12 = 0.$$

55. The idea of supply-side economics (see Exercises 45–46) is that an increase in the tax rate may actually reduce government revenue. What explanation can you offer for this theory?

 **Technology Exercises**

In Exercises 56–59, use a graphing utility to determine if the division has been performed correctly Graph the function on each side of the equation in the same viewing rectangle. If the graphs do not coincide, correct the expression on the right side by using polynomial long division. Then verify your correction using the graphing utility.

56. $(6x^2 + 16x + 8) \div (3x + 2) = 2x + 4, x \neq -\frac{2}{3}$

57. $\dfrac{x^4 + 6x^3 + 6x^2 - 10x - 3}{x^2 + 2x - 3} = x^2 + 4x + 1,$
$x \neq -3, x \neq 1$

58. $\dfrac{2x^3 - 3x^2 - 3x + 4}{x - 1} = 2x^2 - x + 4, x \neq 1$

59. $\dfrac{3x^4 + 4x^3 - 32x^2 - 5x - 20}{x + 4} = 3x^3 + 8x^2 - 5,$
$x \neq -4$

 Critical Thinking Exercises

60. Which one of the following is true?
   a. If a trinomial in x of degree 6 is divided by a trinomial in x of degree 3, the degree of the quotient is 2.
   b. Synthetic division could not be used to find the quotient of $10x^3 - 6x^2 + 4x - 1$ and $x - \frac{1}{2}$.
   c. Any problem that can be done by synthetic division can also be done by the method for long division of polynomials.
   d. If a polynomial long-division problem results in a remainder that is a whole number, then the divisor is a factor of the dividend.

61. Find k so that $4x + 3$ is a factor of
$$20x^3 + 23x^2 - 10x + k.$$

62. When $2x^2 - 7x + 9$ is divided by a polynomial, the quotient is $2x - 3$ and the remainder is 3. Find the polynomial.

63. Find the quotient of $x^{3n} + 1$ and $x^n + 1$.

64. Synthetic division is a process for dividing a polynomial by $x - c$. The coefficient of x is 1. How might synthetic division be used if you are dividing by $2x - 4$?

SECTION 3.4 *Zeros of Polynomial Functions*

Objectives

1. Use the Rational Zero Theorem to find possible rational zeros.

2. Find zeros of a polynomial function.

3. Solve polynomial equations.

4. Use Descartes's Rule of Signs.

A moth has moved into your closet. She appeared in your bedroom at night, but somehow her relatively stout body escaped your clutches. Within a few weeks, swarms of moths in your tattered wardrobe suggest that Mama Moth was in the family way. There must be at least 200 critters nesting in every crevice of your clothing.

Two hundred plus moth-tykes from one female moth; is this possible? Indeed it is. The number of eggs, N, in a female moth is a function of her abdominal width, W, in millimeters, modeled by

$$N(W) = 14W^3 - 17W^2 - 16W + 34$$

for $1.5 \leq W \leq 3.5$. Because there are 200 moths feasting on your favorite sweaters, Mama's abdominal width can be estimated by finding the roots of the polynomial equation

$$14W^3 - 17W^2 - 16W + 34 = 200.$$

With mathematics present even in your quickly disappearing attire, we move from rags to polynomial equations. The process of solving such equations begins with listing possibilities for Mama Moth's abdominal width. To do this, we turn to a theorem that plays an important role in finding zeros of polynomial functions.

1 Use the Rational Zero Theorem to find possible rational zeros.

The Rational Zero Theorem

The Rational Zero Theorem gives a list of all possible rational zeros of a polynomial function. Equivalently, the theorem gives all possible rational roots of a polynomial equation. Not every number in the list will be a zero of the function, but every rational zero of the polynomial function will appear somewhere in the list.

> **The Rational Zero Theorem**
>
> If $f(x) = a_n x^n + a_{n-1} x^{n-1} + \cdots + a_1 x + a_0$ has *integer* coefficients and $\dfrac{p}{q}$
>
> (where $\dfrac{p}{q}$ is reduced) is a rational zero, then p is a factor of the constant term, a_0, and q is a factor of the leading coefficient, a_n.

You can explore the "why" behind the Rational Zero Theorem in Exercise 66 of Exercise Set 3.4. For now, let's see if we can figure out what the theorem tells us about possible rational zeros. In order to use the theorem, list all the integers that are factors of the constant term, a_0. Then list all the integers that are factors of the leading coefficient, a_n. Finally list all possible rational zeros:

$$\text{Possible rational zeros} = \frac{\text{Factors of the constant term}}{\text{Factors of the leading coefficient}}.$$

EXAMPLE 1 Using the Rational Zero Theorem

List all possible rational zeros of $f(x) = -x^4 + 3x^2 + 4$.

Solution The constant term is 4. We list all of its factors: $\pm 1, \pm 2, \pm 4$. The leading coefficient is -1. Its factors are ± 1.

| Factors of the constant term: | $\pm 1, \quad \pm 2, \quad \pm 4$ |
| Factors of the leading coefficient: | ± 1 |

Because

$$\text{Possible rational zeros} = \frac{\text{Factors of the constant term}}{\text{Factors of the leading coefficient}},$$

we must take each number in the first row, $\pm 1, \pm 2, \pm 4$, and divide by each number in the second row, ± 1.

Study Tip

Always keep in mind the relationship among zeros, roots, and x-intercepts. The zeros of function f are the roots, or solutions, of the equation $f(x) = 0$. Furthermore, the real zeros, or real roots, are the x-intercepts of the graph of f.

$$\text{Possible rational zeros} = \frac{\text{Factors of } 4}{\text{Factors of } -1} = \frac{\pm 1, \pm 2, \pm 4}{\pm 1} = \pm 1, \quad \pm 2, \quad \pm 4$$

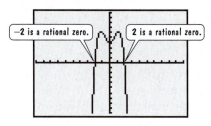

There are six possible rational zeros. The graph of $f(x) = -x^4 + 3x^2 + 4$ is shown in Figure 3.18. The x-intercepts are -2 and 2. Thus, -2 and 2 are the actual rational zeros.

| -2 is a rational zero. | | 2 is a rational zero. |

Figure 3.18 The graph of $f(x) = -x^4 + 3x^2 + 4$ shows that -2 and 2 are rational zeros.

Check Point 1   List all possible rational zeros of

$$f(x) = x^3 + 2x^2 - 5x - 6.$$

EXAMPLE 2   Using the Rational Zero Theorem

List all possible rational zeros of $f(x) = 15x^3 + 14x^2 - 3x - 2$.

Solution   The constant term is -2 and the leading coefficient is 15.

$$\text{Possible rational zeros} = \frac{\text{Factors of the constant term, } -2}{\text{Factors of the leading coefficient, } 15}$$
$$= \frac{\pm 1, \pm 2}{\pm 1, \pm 3, \pm 5, \pm 15}$$
$$= \pm 1, \quad \pm 2, \quad \pm \tfrac{1}{3}, \quad \pm \tfrac{2}{3}, \quad \pm \tfrac{1}{5}, \quad \pm \tfrac{2}{5}, \quad \pm \tfrac{1}{15}, \quad \pm \tfrac{2}{15}$$

| Divide ± 1 and ± 2 by ± 1. | Divide ± 1 and ± 2 by ± 3. | Divide ± 1 and ± 2 by ± 5. | Divide ± 1 and ± 2 by ± 15. |

There are 16 possible rational zeros. The actual solution set of

$$15x^3 + 14x^2 - 3x - 2 = 0$$

is $\left\{-1, -\tfrac{1}{3}, \tfrac{2}{5}\right\}$, which contains 3 of the 16 possible zeros.

Check Point 2   List all possible rational zeros of

$$f(x) = 4x^5 + 12x^4 - x - 3.$$

2   Find zeros of a polynomial function.

How do we determine which (if any) of the possible rational zeros are rational zeros of the polynomial function? To find the first rational zero, we can use a trial-and-error process involving synthetic division: If $f(x)$ is divided by $x - c$ and the remainder is zero, then c is a zero of f. After we identify the first rational zero, we use the result of the synthetic division to factor the original polynomial. Then we set each factor equal to zero to identify any additional rational zeros.

EXAMPLE 3 Finding Zeros of a Polynomial Function

Find all rational zeros of $f(x) = x^3 + 2x^2 - 5x - 6$.

Solution We begin by listing all possible rational zeros.

Possible rational zeros

$$= \frac{\text{Factors of the constant term, } -6}{\text{Factors of the leading coefficient, } 1} = \frac{\pm 1, \pm 2, \pm 3, \pm 6}{\pm 1} = \pm 1, \pm 2, \pm 3, \pm 6$$

> Divide the eight numbers in the numerator by ±1.

Now we will use synthetic division to see if we can find a rational zero among the possible rational zeros $\pm 1, \pm 2, \pm 3, \pm 6$. Keep in mind that if $f(x)$ is divided by $x - c$ and the remainder is zero, then c is a zero of f. Let's start by testing 1. If 1 is not a rational zero, then we will test other possible rational zeros.

Test 1

Possible rational zero

Coefficients of $f(x) = x^3 + 2x^2 - 5x - 6$

$$
\begin{array}{r|rrrr}
1 & 1 & 2 & -5 & -6 \\
  &   & 1 & 3 & -2 \\
\hline
  & 1 & 3 & -2 & -8
\end{array}
$$

> The nonzero remainder shows that 1 is not a zero.

Test 2

Possible rational zero

Coefficients of $f(x) = x^3 + 2x^2 - 5x - 6$

$$
\begin{array}{r|rrrr}
2 & 1 & 2 & -5 & -6 \\
  &   & 2 & 8 & 6 \\
\hline
  & 1 & 4 & 3 & 0
\end{array}
$$

> The zero remainder shows that 2 is a zero.

The zero remainder tells us that 2 is a zero of the polynomial function $f(x) = x^3 + 2x^2 - 5x - 6$. Equivalently, 2 is a solution, or root, of the polynomial equation $x^3 + 2x^2 - 5x - 6 = 0$. Thus, $x - 2$ is a factor of the polynomial.

$$x^3 + 2x^2 - 5x - 6 = 0 \qquad \text{Finding the zeros of } f(x) = x^3 + 2x^2 - 5x - 6 \text{ is}$$
the same as finding the roots of this equation.

$$(x - 2)(x^2 + 4x + 3) = 0 \qquad \text{Factor using the result from the synthetic division.}$$

$$(x - 2)(x + 3)(x + 1) = 0 \qquad \text{Factor completely.}$$

$$x - 2 = 0 \quad \text{or} \quad x + 3 = 0 \quad \text{or} \quad x + 1 = 0 \qquad \text{Set each factor equal to zero.}$$

$$x = 2 \qquad\qquad x = -3 \qquad\qquad x = -1 \qquad \text{Solve for x.}$$

The solution set is $\{-3, -1, 2\}$. The rational zeros of f are $-3, -1$, and 2.

Check Point 3 Find all rational zeros of

$$f(x) = x^3 + 8x^2 + 11x - 20.$$

Our work in Example 3 involved solving a third-degree equation. We found one factor by synthetic division and then factored the remaining quadratic factor. If the degree of a polynomial function or equation is 4 or higher, it is often necessary to find more than one linear factor by synthetic division.

One way to speed up the process of finding the first zero is to graph the function. Any x-intercept is a zero.

3 Solve polynomial equations.

EXAMPLE 4 Solving a Polynomial Equation

Solve: $x^4 - 6x^2 - 8x + 24 = 0$.

Solution Recall that we refer to the *zeros* of a polynomial function and the *roots* of a polynomial equation. Because we are given an equation, we will use the word "roots," rather than "zeros," in the solution process. We begin by listing all possible rational roots.

Possible rational roots

$$= \frac{\text{Factors of the constant term, 24}}{\text{Factors of the leading coefficient, 1}}$$

$$= \frac{\pm1, \pm2, \pm3, \pm4, \pm6, \pm8, \pm12, \pm24}{\pm1} = \pm1, \pm2, \pm3, \pm4, \pm6, \pm8, \pm12, \pm24$$

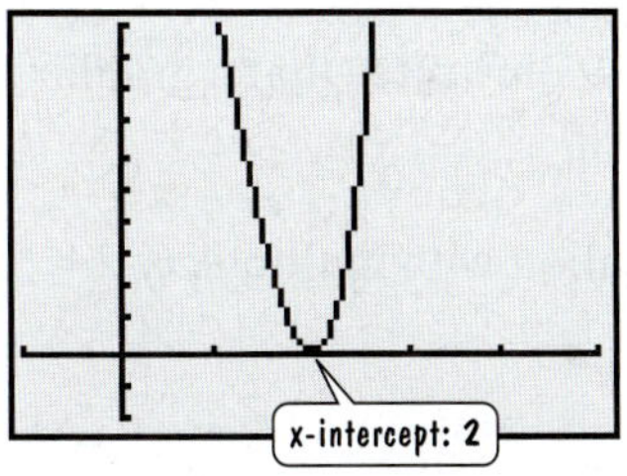

The graph of $f(x) = x^4 - 6x^2 - 8x + 24$ is shown in Figure 3.19. Because the x-intercept is 2, we will test 2 by synthetic division and show that it is a root of the given equation.

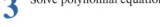
x-intercept: 2

$$\begin{array}{r|rrrrr} 2 & 1 & 0 & -6 & -8 & 24 \\ & & 2 & 4 & -4 & -24 \\ \hline & 1 & 2 & -2 & -12 & 0 \end{array}$$

Careful!
$x^4 - 6x^2 - 8x + 24 = x^4 + 0x^3 - 6x^2 - 8x + 24$

The zero remainder indicates that 2 is a root of $x^4 - 6x^2 - 8x + 24 = 0$.

Figure 3.19 The graph of $f(x) = x^4 - 6x^2 - 8x + 24$ in a $[-1, 5, 1]$ by $[-2, 10, 1]$ viewing rectangle

Now we can rewrite the given equation in factored form.

$$x^4 - 6x^2 - 8x + 24 = 0 \quad \text{This is the given equation.}$$

$$(x - 2)(x^3 + 2x^2 - 2x - 12) = 0 \quad \text{This is the result obtained from the synthetic division.}$$

$$x - 2 = 0 \quad \text{or} \quad x^3 + 2x^2 - 2x - 12 = 0 \quad \text{Set each factor equal to 0.}$$

We can use the same approach to look for rational roots of the polynomial equation $x^3 + 2x^2 - 2x - 12 = 0$, listing all possible rational roots. However, take a second look at the graph in Figure 3.19. Because the graph turns around at 2, this means that 2 is a root of even multiplicity. Thus, 2 must also be a root of $x^3 + 2x^2 - 2x - 12 = 0$, confirmed by the following synthetic division.

$$\begin{array}{r|rrrr} 2 & 1 & 2 & -2 & -12 \\ & & 2 & 8 & 12 \\ \hline & 1 & 4 & 6 & 0 \end{array}$$

These are the coefficients of $x^3 + 2x^2 - 2x - 12 = 0$.

The zero remainder indicates that 2 is a root of $x^3 + 2x^2 - 2x - 12 = 0$.

Now we can solve the original equation as follows:

$$x^4 - 6x^2 - 8x + 24 = 0 \quad \text{This is the given equation.}$$

$$(x - 2)(x^3 + 2x^2 - 2x - 12) = 0 \quad \text{This factorization is obtained from the first synthetic division.}$$

$$(x - 2)(x - 2)(x^2 + 4x + 6) = 0 \quad \text{This factorization is obtained from the second synthetic division.}$$

$$x - 2 = 0 \quad \text{or} \quad x - 2 = 0 \quad \text{or} \quad x^2 + 4x + 6 = 0 \quad \text{Set each factor equal to 0.}$$

$$x = 2 \qquad\qquad x = 2 \qquad\qquad x^2 + 4x + 6 = 0 \quad \text{Solve.}$$

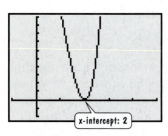

Figure 3.19, repeated

We can use the quadratic formula to solve $x^2 + 4x + 6 = 0$.

$$x = \frac{-b \pm \sqrt{b^2 - 4ac}}{2a}$$

We use the quadratic formula because $x^2 + 4x + 6$ cannot be factored.

$$= \frac{-4 \pm \sqrt{4^2 - 4(1)(6)}}{2(1)}$$

Let $a = 1$, $b = 4$, and $c = 6$.

$$= \frac{-4 \pm \sqrt{-8}}{2}$$

Multiply and subtract under the radical.

$$= \frac{-4 \pm 2i\sqrt{2}}{2}$$

$\sqrt{-8} = \sqrt{4(2)(-1)} = 2i\sqrt{2}$

$$= -2 \pm i\sqrt{2}$$

Simplify.

The solution set of the original equation, $x^4 - 6x^2 - 8x + 24 = 0$, is $\{2, -2 - i\sqrt{2}, -2 + i\sqrt{2}\}$. The graph in Figure 3.19 illustrates that a graphing utility does not reveal the two imaginary roots.

In Example 4, 2 is a repeated root of the equation with multiplicity 2. The example illustrates two general properties:

> **Properties of Polynomial Equations**
>
> **1.** If a polynomial equation is of degree n, then counting multiple roots separately, the equation has n roots.
>
> **2.** If $a + bi$ is a root of a polynomial equation ($b \neq 0$), then the complex imaginary number $a - bi$ is also a root. Complex imaginary roots, if they exist, occur in conjugate pairs.

These ideas will be developed in more detail in the next section.

Check Point 4 Solve: $x^4 - 6x^3 + 22x^2 - 30x + 13 = 0$.

4 Use Descartes's Rule of Signs.

An equation can have as many true [positive] roots as it contains changes of sign, from plus to minus or from minus to plus. ... René Descartes (1596–1650) in *La Géométrie* (1637)

Descartes's Rule of Signs

Because an nth-degree polynomial equation might have roots that are imaginary numbers, we should note that such an equation can have *at most n* real roots. **Descartes's Rule of Signs** provides even more specific information about the number of real zeros that a polynomial can have. The rule is based on considering *variations in sign* between consecutive coefficients. For example, the function

$$f(x) = 3x^7 - 2x^5 - x^4 + 7x^2 + x - 3$$

has three sign changes.

> **Descartes's Rule of Signs**
>
> Let $f(x) = a_n x^n + a_{n-1} x^{n-1} + \cdots + a_2 x^2 + a_1 x + a_0$ be a polynomial with real coefficients.
>
> **1.** The number of *positive real zeros* of f is either equal to the number of sign changes of $f(x)$ or is less than that number by an even integer. If

there is only one variation in sign, there is exactly one positive real zero.

2. The number of *negative real zeros* of f is either equal to the number of sign changes of $f(-x)$ or is less than that number by an even integer. If $f(-x)$ has only one variation in sign, then f has exactly one negative real zero.

EXAMPLE 5   Using Descartes's Rule of Signs

Determine the possible number of positive and negative real zeros of $f(x) = x^3 + 2x^2 + 5x + 4$.

Solution

1. To find possibilities for positive real zeros, count the number of sign changes in the equation for $f(x)$. Because all the terms are positive, there are no variations in sign. Thus, there are no positive real zeros.

2. To find possibilities for negative real zeros, count the number of sign changes in the equation for $f(-x)$. We obtain this equation by replacing x with $-x$ in the given function.

$$f(x) = x^3 + 2x^2 + 5x + 4 \quad \text{This is the given polynomial function.}$$

Replace x with −x.

$$f(-x) = (-x)^3 + 2(-x)^2 + 5(-x) + 4$$
$$= -x^3 + 2x^2 - 5x + 4$$

Now count the sign changes.

$$f(-x) = -x^3 + 2x^2 - 5x + 4$$
$$\quad\quad\quad 1 \quad\quad 2 \quad\quad 3$$

There are three variations in sign. The number of negative real zeros of f is either equal to the number of sign changes, 3, or is less than this number by an even integer. This means that there are either 3 negative real zeros or $3 - 2 = 1$ negative real zero.

What do the results of Example 5 mean in terms of solving
$$x^3 + 2x^2 + 5x + 4 = 0?$$

Without using Descartes's Rule of Signs, we list possible rational roots as follows:

Possible rational roots

$$= \frac{\text{Factors of the constant term, 4}}{\text{Factors of the leading coefficient, 1}} = \frac{\pm 1, \pm 2, \pm 4}{\pm 1} = \pm 1, \ \pm 2, \ \pm 4.$$

However, Descartes's Rule of Signs informed us that $f(x) = x^3 + 2x^2 + 5x + 4$ has no positive real zeros. Thus, the polynomial equation $x^3 + 2x^2 + 5x + 4 = 0$ has no positive real roots. This means that we can eliminate the positive numbers from our list of possible rational roots. Possible rational roots include only $-1, -2,$ and -4. We can use synthetic division to test the three possible rational roots. Our test on two of the three rational roots is shown on the next page.

$$\begin{array}{r|rrrr} -1 & 1 & 2 & 5 & 4 \\ & & -1 & -1 & -4 \\ \hline & 1 & 1 & 4 & 0 \end{array}$$

The zero remainder shows that −1 is a root.

$$\begin{array}{r|rrrr} -2 & 1 & 2 & 5 & 4 \\ & & -2 & 0 & -10 \\ \hline & 1 & 0 & 5 & -6 \end{array}$$

The nonzero remainder shows that −2 is not a root.

By solving the equation $x^3 + 2x^2 + 5x + 4 = 0$, you will find that this equation of degree 3 has three roots. One root is -1, and the other two roots are imaginary numbers in a conjugate pair. Verify this by completing the solution process.

 Check Point 5 Determine the possible number of positive and negative real zeros of $f(x) = x^4 - 14x^3 + 71x^2 - 154x + 120$.

EXERCISE SET 3.4

 ## Practice Exercises

In Exercises 1–8, use the Rational Zero Theorem to list all possible rational zeros for each given function.

1. $f(x) = x^3 + x^2 - 4x - 4$

2. $f(x) = x^3 + 3x^2 - 6x - 8$

3. $f(x) = 3x^4 - 11x^3 - x^2 + 19x + 6$

4. $f(x) = 2x^4 + 3x^3 - 11x^2 - 9x + 15$

5. $f(x) = 4x^4 - x^3 + 5x^2 - 2x - 6$

6. $f(x) = 3x^4 - 11x^3 - 3x^2 - 6x + 8$

7. $f(x) = x^5 - x^4 - 7x^3 + 7x^2 - 12x - 12$

8. $f(x) = 4x^5 - 8x^4 - x + 2$

In Exercises 9–14,
   a. *List all possible rational zeros.*
   b. *Use synthetic division to test the possible rational zeros and find an actual zero.*
   c. *Use the zero from part (b) to find all the zeros of the polynomial function.*

9. $f(x) = x^3 + x^2 - 4x - 4$

10. $f(x) = x^3 - 2x^2 - 11x + 12$

11. $f(x) = 2x^3 - 3x^2 - 11x + 6$

12. $f(x) = 2x^3 - 5x^2 + x + 2$

13. $f(x) = 3x^3 + 7x^2 - 22x - 8$

14. $f(x) = 3x^3 + 8x^2 - 15x + 4$

In Exercises 15–22,
   a. *List all possible rational roots.*
   b. *Use synthetic division to test the possible rational roots and find an actual root.*
   c. *Use the root from part (b) and solve the equation.*

15. $x^3 - 2x^2 - 11x + 12 = 0$

16. $x^3 - 2x^2 - 7x - 4 = 0$

17. $x^3 - 10x - 12 = 0$

18. $x^3 - 5x^2 + 17x - 13 = 0$

19. $6x^3 + 25x^2 - 24x + 5 = 0$

20. $2x^3 - 5x^2 - 6x + 4 = 0$

21. $x^4 - 2x^3 - 5x^2 + 8x + 4 = 0$

22. $x^4 - 2x^2 - 16x - 15 = 0$

In Exercises 23–28, use Descartes's Rule of Signs to determine the possible number of positive and negative real zeros for each given function.

23. $f(x) = x^3 + 2x^2 + 5x + 4$

24. $f(x) = x^3 + 7x^2 + x + 7$

25. $f(x) = 5x^3 - 3x^2 + 3x - 1$

26. $f(x) = -2x^3 + x^2 - x + 7$

27. $f(x) = 2x^4 - 5x^3 - x^2 - 6x + 4$

28. $f(x) = 4x^4 - x^3 + 5x^2 - 2x - 6$

In Exercises 29–42, find all zeros of the polynomial function or solve the given polynomial equation. Use the Rational Zero Theorem, Descartes's Rule of Signs, and possibly the graph of the polynomial function shown by a graphing utility as an aid in obtaining the first zero or the first root.

29. $f(x) = x^3 - 4x^2 - 7x + 10$

30. $f(x) = x^3 + 12x^2 + 21x + 10$

31. $2x^3 - x^2 - 9x - 4 = 0$

32. $3x^3 - 8x^2 - 8x + 8 = 0$

33. $f(x) = x^4 - 2x^3 + x^2 + 12x + 8$

34. $f(x) = x^4 - 4x^3 - x^2 + 14x + 10$

35. $x^4 - 3x^3 - 20x^2 - 24x - 8 = 0$

36. $x^4 - x^3 + 2x^2 - 4x - 8 = 0$

37. $f(x) = 3x^4 - 11x^3 - x^2 + 19x + 6$

38. $f(x) = 2x^4 + 3x^3 - 11x^2 - 9x + 15$

39. $4x^4 - x^3 + 5x^2 - 2x - 6 = 0$

40. $3x^4 - 11x^3 - 3x^2 - 6x + 8 = 0$

41. $2x^5 + 7x^4 - 18x^2 - 8x + 8 = 0$

42. $4x^5 + 12x^4 - 41x^3 - 99x^2 + 10x + 24 = 0$

Application Exercises

The graphs are based on a study of the percentage of professional works completed in each age decade of life by 738 people who lived to be at least 79. Use the graphs to solve Exercises 43–44.

Age Trends in Professional Productivity

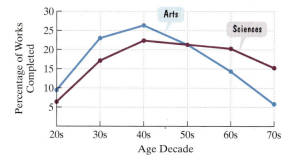

Source: Dennis, W. (1966), Creative productivity between the ages of 20 and 80 years. *Journal of Gerontology*, 21, 1–8.

43. Suppose that a polynomial function f is used to model the data shown in the graph for the arts using

(age decade, percentage of works completed).

a. Use the graph to solve the polynomial equation $f(x) = 27$. Describe what this means in terms of an age decade and productivity.

b. Describe the degree and the leading coefficient of a function f that can be used to model the data in the graph.

44. Suppose that a polynomial function g is used to model the data shown in the graph for the sciences using

(age decade, percentage of works completed).

a. Use the graph to solve the polynomial equation $g(x) = 20$. Find only the meaningful value of x and then describe what this means in terms of an age decade and productivity.

b. Describe the degree and the leading coefficient of a function g that can be used to model the data in the graph.

45. The number of eggs, N, in a female moth is a function of her abdominal width, W, in millimeters, modeled by $N = 14W^3 - 17W^2 - 16W + 34$, for $1.5 \leq W \leq 3.5$. What is the abdominal width when there are 211 eggs?

46. The concentration of a drug, $f(x)$, in parts per million, in a patient's blood x hours after the drug is administered is given by the function

$$f(x) = -x^4 + 12x^3 - 58x^2 + 132x.$$

How many hours after the drug is administered will it be eliminated from the bloodstream?

47. The width of a rectangular box is twice the height and the length is 7 inches more than the height. If the volume is 72 cubic inches, find the dimensions of the box.

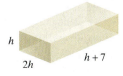

48. A box with an open top is formed by cutting squares out of the corners of a rectangular piece of cardboard 10 inches by 8 inches and then folding up the sides. If x represents the length of the side of the square cut from each corner of the rectangle, what size square must be cut if the volume of the box is to be 48 cubic inches?

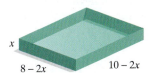

Writing in Mathematics

49. Describe how to find the possible rational zeros of a polynomial function.

50. Describe how to use Descartes's Rule of Signs to determine the possible number of positive real zeros of a polynomial function.

51. Describe how to use Descartes's Rule of Signs to determine the possible number of negative roots of a polynomial equation.

52. Why must every polynomial equation of degree 3 have at least one real root?

53. Explain why the equation $x^4 + 6x^2 + 2 = 0$ has no rational roots.

54. Suppose $\frac{3}{4}$ is a root of a polynomial equation. What does this tell us about the leading coefficient and the constant term in the equation?

55. Use the graphs for Exercises 43–44 to describe one similarity and one difference between age trends in professional productivity in the arts and the sciences.

Technology Exercises

The equations in Exercises 56–59 have real roots that are rational. Use the Rational Zero Theorem to list all possible rational roots. Then graph the polynomial function in the given viewing rectangle to determine which possible rational roots are actual roots of the equation.

56. $2x^3 - 15x^2 + 22x + 15 = 0$; $[-1, 6, 1]$ by $[-50, 50, 1]$

57. $6x^3 - 19x^2 + 16x - 4 = 0$; $[0, 2, 1]$ by $[-3, 2, 1]$

58. $2x^4 + 7x^3 - 4x^2 - 27x - 18 = 0$; $[-4, 3, 1]$ by $[-45, 45, 1]$

59. $4x^4 + 4x^3 + 7x^2 - x - 2 = 0$; $[-2, 2, 1]$ by $[-5, 5, 1]$

60. Use Descartes's Rule of Signs to determine the possible number of positive and negative real zeros of $f(x) = 3x^4 + 5x^2 + 2$. What does this mean in terms of the graph of f? Verify your result by using a graphing utility to graph f.

61. Use Descartes's Rule of Signs to determine the possible number of positive and negative real zeros of $f(x) = x^5 - x^4 + x^3 - x^2 + x - 8$. Verify your result by using a graphing utility to graph f.

62. Determine a number of polynomial functions of odd degree and graph each function. Is it possible for the graph to have no real zeros? Explain. Try doing the same thing for polynomial functions of even degree. Now is it possible to have no real zeros?

Critical Thinking Exercises

63. Which one of the following is true?
  a. The equation $x^3 + 5x^2 + 6x + 1 = 0$ has one positive real root.
  b. Descartes's Rule of Signs gives the exact number of

positive and negative real roots for a polynomial equation.
  c. Every polynomial equation of degree 3 has at least one rational root.
  d. None of the above is true.

64. Give an example of a polynomial equation that has no real roots. Describe how you obtained the equation.

65. If the volume of the solid shown in the figure is 208 cubic inches, find the value of x.

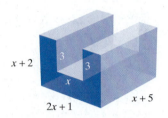

66. In this exercise, we lead you through the steps involved in the proof of the Rational Zero Theorem. Consider the polynomial equation

$$a_n x^n + a_{n-1} x^{n-1} + a_{n-2} x^{n-2} + \cdots + a_1 x + a_0 = 0$$

where $\frac{p}{q}$ is a rational root reduced to lowest terms.

  a. Substitute $\frac{p}{q}$ for x in the equation and show that the equation can be written as

$$a_n p^n + a_{n-1} p^{n-1} q$$
$$+ a_{n-2} p^{n-2} q^2 + \cdots + a_1 p q^{n-1} = -a_0 q^n.$$

  b. Why is p a factor of the left side of the equation?

  c. Because p divides the left side, it must also divide the right side. However, because $\frac{p}{q}$ is reduced to lowest terms, p cannot divide q. Thus, p and q have no common factors other than -1 and 1. Because p does divide the right side and it is not a factor of q^n, what can you conclude?

  d. Rewrite the equation from part (a) with all terms containing q on the left and the term that does not have a factor of q on the right. Use an argument that parallels parts (b) and (c) to conclude that q is a factor of a_n.

SECTION 3.5 *More On Zeros of Polynomial Functions*

Objectives

1. Find bounds for the roots of a polynomial equation.
2. Approximate real zeros.
3. Use conjugate roots to solve a polynomial equation.
4. Use the Linear Factorization Theorem to factor a polynomial.
5. Find polynomials with given zeros.

You stole my formula!

Tartaglia's Secret Formula for One Solution of $x^3 + mx = n$

$$x = \sqrt[3]{\sqrt{\left(\frac{n}{2}\right)^2 + \left(\frac{m}{3}\right)^3} + \frac{n}{2}}$$
$$-\sqrt[3]{\sqrt{\left(\frac{n}{2}\right)^2 + \left(\frac{m}{3}\right)^3} - \frac{n}{2}}$$

Popularizers of mathematics are sharing bizarre stories that are giving math a secure place in popular culture. One episode, able to compete with the wildest fare served up by television talk shows and the tabloids, involves three Italian mathematicians and, of all things, zeros of polynomial functions.

Tartaglia (1499–1557), poor and starving, has found a formula that gives a root for a third-degree polynomial equation. Cardano (1501–1576) begs Tartaglia to reveal the secret formula, wheedling it from him with the promise he will find the impoverished Tartaglia a patron. Then Cardano publishes his famous work *Ars Magna*, in which he presents Tartaglia's formula as his own. Cardano uses his most talented student, Ferrari (1522–1565), who derived a formula for a root of a fourth-degree polynomial equation, to falsely accuse Tartaglia of plagiarism. The dispute becomes violent and Tartaglia is fortunate to escape alive.

The noise from this "You Stole My Formula" episode is quieted by the work of French mathematician Evariste Galois (1811–1832). Galois proved that there is no general formula for finding roots of polynomial equations of degree 5 or higher. There are, of course, methods for finding roots. In this section, we continue our study of methods for finding zeros of polynomial functions.

1 Find bounds for the roots of a polynomial equation.

Upper and Lower Bounds for Roots

The **Upper and Lower Bound Theorem** helps us rule out many of a polynomial equations possible rational roots. Figure 3.20 illustrates that a is a **lower bound** and b is an **upper bound** for the roots of $f(x) = 0$ because every real root c of the equation satisfies $a \leq c \leq b$.

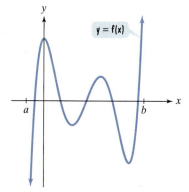

Figure 3.20 b is an upper bound and a is a lower bound for the real roots of $f(x) = 0$.

The Upper and Lower Bound Theorem

Let $f(x)$ be a polynomial with real coefficients and a positive leading coefficient, and let a and b be nonzero real numbers.

1. Divide $f(x)$ by $x - b$ (where $b > 0$) using synthetic division. If the last row containing the quotient and remainder has no negative numbers, then b is an **upper bound** for the real roots of $f(x) = 0$.
2. Divide $f(x)$ by $x - a$ (where $a < 0$) using synthetic division. If the last row containing the quotient and remainder has numbers that alternate in sign (zero entries count as positive or negative), then a is a **lower bound** for the real roots of $f(x) = 0$.

EXAMPLE 1   Finding Bounds for the Roots

Show that all the real roots of the equation $8x^3 + 10x^2 - 39x + 9 = 0$ lie between -3 and 2.

Solution   We begin by showing that 2 is an upper bound for the real roots. Divide the polynomial by $x - 2$. If all the numbers in the bottom row of the synthetic division are nonnegative, then 2 is an upper bound.

$$
\begin{array}{r|rrrr}
2 & 8 & 10 & -39 & 9 \\
  &   & 16 & 52 & 26 \\
\hline
  & 8 & 26 & 13 & 35
\end{array}
$$

> All numbers in this row are nonnegative.

The nonnegative entries in the last row verify that 2 is an upper bound. Next, we show that -3 is a lower bound for the real roots. Divide the polynomial by $x - (-3)$, or $x + 3$. If the numbers in the bottom row of the synthetic division alternate in sign, then -3 is a lower bound. Remember that the number zero can be considered positive or negative.

$$
\begin{array}{r|rrrr}
-3 & 8 & 10 & -39 & 9 \\
   &   & -24 & 42 & -9 \\
\hline
   & 8 & -14 & 3   & 0
\end{array}
$$

> Counting 0 as negative, the signs alternate:
> $+, -, +, -.$

By the Upper and Lower Bound Theorem, the alternating signs in the last row indicate that -3 is a lower bound for the roots. (The zero remainder indicates that -3 is also a root.)

Check Point 1   Show that all the real roots of the equation $2x^3 + 11x^2 - 7x - 6 = 0$ lie between -7 and 2.

How might the Upper and Lower Bound Theorem be helpful in solving a polynomial equation? Consider the equation

$$x^4 + 3x^3 - 27x^2 + 3x - 28 = 0.$$

With a leading coefficient of 1 and a constant term of -28, the possible rational roots are

$$\pm 1, \quad \pm 2, \quad \pm 4, \quad \pm 7, \quad \pm 14, \quad \pm 28.$$

We begin testing for an actual root using synthetic division. The following divisions indicate that 1 and 2 are not roots because of the nonzero remainders. However, something interesting happens when testing 4.

$$
\begin{array}{r|rrrrr}
1 & 1 & 3 & -27 & 3   & -28 \\
  &   & 1 & 4   & -23 & -20 \\
\hline
  & 1 & 4 & -23 & -20 & -48
\end{array}
\qquad
\begin{array}{r|rrrrr}
2 & 1 & 3 & -27 & 3   & -28 \\
  &   & 2 & 10 & -34 & -62 \\
\hline
  & 1 & 5 & -17 & -31 & -90
\end{array}
$$

$$
\begin{array}{r|rrrrr}
4 & 1 & 3 & -27 & 3 & -28 \\
  &   & 4 & 28 & 4 & 28 \\
\hline
  & 1 & 7 & 1   & 7 & 0
\end{array}
$$

> 4 is a root of the equation because the remainder is 0.

Nonnegative numbers

> 4 is an upper bound for the roots of the equation.

Notice that 4 is both a root and an upper bound for the roots. Should you take the time to use synthetic division and test 7, 14, and 28? There is no need to do this because all three numbers exceed 4, the upper bound for the roots. Thus, 7, 14, and 28 cannot be roots of the equation.

Technology

The Upper and Lower Bound Theorem and your knowledge of polynomial functions can help you to find a reasonable range setting when using your graphing utility. Consider

$$f(x) = x^4 + 3x^3 - 27x^2 + 3x - 28.$$

Based on our discussion, 4 is a zero and an upper bound for the zeros. We can also use synthetic division to show that -7 is a zero and a lower bound for the zeros. We can use these lower and upper bounds to determine Xmin and Xmax. We'll go one unit to the left and to the right of these bounds and use $[-8, 5, 1]$. Now, how do we determine Ymin and Ymax? Let's see what kinds of values of y we obtain when we evaluate the function between -8 and 5. Using synthetic division, direct substitution, or the table feature of some graphing utilities, we have $f(-6) = -370$, $f(-5) = -468$, $f(0) = -28$, and $f(3) = -100$. These evaluations suggest that we can use -500 for Ymin and 100 for Ymax. The graph of $f(x) = x^4 + 3x^3 - 27x^2 + 3x - 28$ is shown in a $[-8, 5, 1]$ by $[-500, 100, 20]$ viewing rectangle in Figure 3.21. Because the degree is even ($n = 4$) and the leading coefficient, 1, is positive, the graph should rise to the left and to the right. This is precisely what occurs in Figure 3.21. Our work in obtaining this complete graph is an excellent illustration of the fact that technology complements human knowledge and is not intended to replace it.

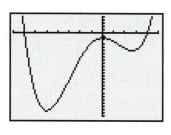

Figure 3.21

2 Approximate real zeros.

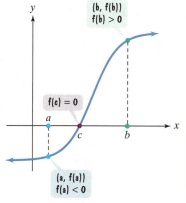

Figure 3.22 The graph must cross the x-axis at some value between a and b.

The Intermediate Value Theorem

We can find decimal approximations for real zeros of polynomial functions using a graphing utility. The **Intermediate Value Theorem** tells us of the existence of real zeros and how to approximate them. The idea behind the theorem is illustrated in Figure 3.22. The figure shows that if $(a, f(a))$ lies below the x-axis and $(b, f(b))$ lies above the x-axis, the smooth, continuous graph of a polynomial function f must cross the x-axis at some value c between a and b. This value is a real zero for the function.

These observations are summarized in the **Intermediate Value Theorem.**

The Intermediate Value Theorem for Polynomials

Let f be a polynomial function with real coefficients. If $f(a)$ and $f(b)$ have opposite signs, then there is at least one value of c between a and b for which $f(c) = 0$. Equivalently, the equation $f(x) = 0$ has at least one real root between a and b.

EXAMPLE 2 Approximating a Real Zero

a. Show that the polynomial function $f(x) = x^3 - 2x - 5$ has a real zero between 2 and 3.

b. Use the Intermediate Value Theorem to find an approximation for this real zero to the nearest tenth.

Solution

a. Let us evaluate $f(x)$ at 2 and 3. If $f(2)$ and $f(3)$ have opposite signs, then there is a real zero between 2 and 3. Using $f(x) = x^3 - 2x - 5$, we obtain

$$f(2) = 2^3 - 2 \cdot 2 - 5 = 8 - 4 - 5 = -1$$

> f(2) is negative.

and

$$f(3) = 3^3 - 2 \cdot 3 - 5 = 27 - 6 - 5 = 16.$$

> f(3) is positive.

This sign change shows that the polynomial function has a real zero between 2 and 3.

b. A numerical approach is to evaluate f at successive tenths between 2 and 3, looking for a sign change. This sign change will place the real zero between a pair of successive tenths.

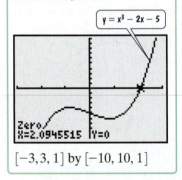
x	$f(x) = x^3 - 2x - 5$	
2	$f(2) = 2^3 - 2(2) - 5$	$= -1$
2.1	$f(2.1) = (2.1)^3 - 2(2.1) - 5 = 0.061$	

> Sign change

The sign change indicates that f has a real zero between 2 and 2.1. We now follow a similar procedure to locate the real zero between successive hundredths. We divide the interval $[2, 2.1]$ into ten equal subintervals. Then we evaluate f at each endpoint and look for a sign change.

x	$f(x) = x^3 - 2x - 5$	x	$f(x) = x^3 - 2x - 5$
2.00	$f(2.00) = -1$	2.06	$f(2.06) = -0.378184$
2.01	$f(2.01) = -0.899399$	2.07	$f(2.07) = -0.270257$
2.02	$f(2.02) = -0.797592$	2.08	$f(2.08) = -0.161088$
2.03	$f(2.03) = -0.694573$	2.09	$f(2.09) = -0.050671$
2.04	$f(2.04) = -0.590336$	2.1	$f(2.1) = 0.061$
2.05	$f(2.05) = -0.484875$		

> Sign change

The sign change indicates that f has a real zero between 2.09 and 2.1. Correct to the nearest tenth, the zero is 2.1.

> **Check Point 2** Show that the polynomial function $f(x) = 3x^3 - 10x + 9$ has a real zero between -3 and -2.

3 Use conjugate roots to solve a polynomial equation.

The Fundamental Theorem of Algebra

We have seen that if a polynomial equation is of degree n, then counting multiple roots separately, the equation has n roots. Some of these roots may be imaginary numbers—that is, nonreal complex numbers—that occur in conjugate pairs, such as $2 + i$ and $2 - i$.

EXAMPLE 3   **Using Conjugate Roots to Solve a Polynomial Equation**

Solve $x^4 - 4x^3 + 3x^2 + 8x - 10 = 0$ given that $2 + i$ is a root.

Solution   The degree of the given equation is 4. This means that there are four roots. One of the roots is $2 + i$. Because imaginary roots come in conjugate pairs, we know that $2 - i$ is a second root. By the Factor Theorem, both

$$[x - (2 + i)] \text{ and } [x - (2 - i)]$$

are factors of the given polynomial. We multiply these known factors as follows:

$$[x - (2 + i)][x - (2 - i)]$$

F   O   I   L

$$= x^2 - x(2 - i) - x(2 + i) + (2 + i)(2 - i)$$ Multiply using the FOIL method.

$$= x^2 - 2x + ix - 2x - ix + (4 - i^2)$$ Continue multiplying.

$$= x^2 - 2x + ix - 2x - ix + [4 - (-1)]$$ Simplify using $i^2 - 1$.

$$= x^2 - 4x + 5$$ Combine like terms.

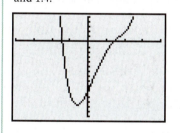

Technology

The graph of

$f(x) =$

$x^4 - 4x^3 + 3x^2 + 8x - 10$

is shown in a $[-4, 4, 1]$ by $[-15, 5, 1]$ viewing rectangle. The real roots of $f(x) = 0$, the equation in Example 3, are $-\sqrt{2}$ and $\sqrt{2}$. These appear as x-intercepts at approximately -1.4 and 1.4.

At this point we have only two of the four possible roots, $2 + i$ and $2 - i$. We can find the other two roots by factoring the given equation. The FOIL multiplication shows that $x^2 - 4x + 5$ is one of the factors. We can find the other factor(s) by dividing $x^2 - 4x + 5$ into the polynomial on the left side of the given equation.

$$
\begin{array}{r}
x^2 \qquad\quad - 2 \\
x^2 - 4x + 5 \overline{\smash{)}\, x^4 - 4x^3 + 3x^2 + 8x - 10} \\
\ominus x^4 - \oplus 4x^3 + \ominus 5x^2 \\
\hline
-2x^2 + 8x - 10 \\
\oplus \qquad \ominus \qquad \oplus \\
-2x^2 + 8x - 10 \\
\hline
0
\end{array}
$$

The zero remainder confirms that $x^2 - 4x + 5$ is a factor.

We can now solve the given equation.

$$x^4 - 4x^3 + 3x^2 + 8x - 10 = 0$$ This is the original equation.

$$(x^2 - 4x + 5)(x^2 - 2) = 0$$ Factor using the result of the polynomial long division.

$$x^2 - 4x + 5 = 0 \text{ or } x^2 - 2 = 0$$ Set each factor equal to 0.

$$x = 2 \pm i \qquad x = \pm\sqrt{2}$$ Solve for x. We know the roots of the first equation, $x^2 - 4x + 5 = 0$, by our previous analysis.

The solution set is $\{2 \pm i, \pm\sqrt{2}\}$.

Check Point 3   Solve $x^4 - 8x^3 + 64x - 105 = 0$ given that $2 - i$ is a root.

The fact that a polynomial equation of degree n has n roots is a consequence of a theorem proved in 1799 by a 22-year-old student named Carl Friedrich Gauss in his doctoral dissertation. His result is called the **Fundamental Theorem of Algebra.**

The Fundamental Theorem of Algebra

If $f(x)$ is a polynomial of degree n, where $n \geq 1$, then the equation $f(x) = 0$ has at least one complex root.

Suppose, for example, that $f(x) = 0$ represents a polynomial equation of degree n. By the Fundamental Theorem of Algebra, we know that this equation has at least one complex root; we'll call it c_1. By the Factor Theorem, we know that $x - c_1$ is a factor of $f(x)$. Therefore, we obtain

$$(x - c_1)q_1(x) = 0 \qquad \text{\textit{The degree of the polynomial } } q_1(x) \text{ \textit{is} } n - 1.$$
$$x - c_1 = 0 \quad \text{or} \quad q_1(x) = 0. \qquad \text{\textit{Set each factor equal to 0.}}$$

If the degree of $q_1(x)$ is at least 1, by the Fundamental Theorem of Algebra the equation $q_1(x) = 0$ has at least one complex root. We'll call it c_2. The Factor Theorem gives us

$$q_1(x) = 0 \qquad \text{\textit{The degree of } } q_1(x) \text{ \textit{is} } n - 1.$$
$$(x - c_2)q_2(x) = 0 \qquad \text{\textit{The degree of } } q_2(x) \text{ \textit{is} } n - 2.$$
$$x - c_2 = 0 \quad \text{or} \quad q_2(x) = 0. \qquad \text{\textit{Set each factor equal to 0.}}$$

Let's see what we have up to this point, and then continue the process.

$$f(x) = 0 \qquad \text{\textit{This is the original polynomial equation of degree } n.}$$
$$(x - c_1)q_1(x) = 0 \qquad \text{\textit{This is the result from our first application of the Fundamental Theorem.}}$$
$$(x - c_1)(x - c_2)q_2(x) = 0 \qquad \text{\textit{This is the result from our second application of the Fundamental Theorem.}}$$

By continuing this process, we will obtain the product of n linear factors. Setting each of these linear factors equal to zero results in n complex roots. Thus, if $f(x)$ is a polynomial of degree n, where $n \geq 1$, then $f(x) = 0$ has exactly n roots, where roots are counted according to their multiplicity.

4 Use the Linear Factorization Theorem to factor a polynomial.

The Linear Factorization Theorem

In Example 3, we found that $x^4 - 4x^3 + 3x^2 + 8x - 10 = 0$ has $\{2 \pm i, \pm \sqrt{2}\}$ as a solution set. The polynomial can be factored over the complex nonreal numbers as follows:

$$f(x) = x^4 - 4x^3 + 3x^2 + 8x - 10 \qquad \text{\textit{These are the four zeros.}}$$

$$= [x - (2 + i)][x - (2 - i)](x + \sqrt{2})(x - \sqrt{2}).$$

These are four linear factors.

This fourth-degree polynomial has four linear factors. Just as an nth-degree polynomial equation has n roots, an nth-degree polynomial has n linear factors. This is formally stated as the **Linear Factorization Theorem.**

The Linear Factorization Theorem

If $f(x) = a_n x^n + a_{n-1} x^{n-1} + \cdots + a_1 x + a_0$, where $n \geq 1$ and $a_n \neq 0$, then

$$f(x) = a_n(x - c_1)(x - c_2) \cdots (x - c_n),$$

where $c_1, c_2, \ldots, c_n$ are complex numbers (possibly real and not necessarily distinct). In words: An nth-degree polynomial can be expressed as the product of a nonzero constant and n linear factors.

The Linear Factorization Theorem involves factors somewhat different than those you are used to seeing. For example, the polynomial $x^2 - 3$ is irreducible over the rational numbers. However, it can be factored over the real numbers as follows:

$$x^2 - 3 = (x + \sqrt{3})(x - \sqrt{3}).$$ Use $a^2 - b^2 = (a + b)(a - b)$ with $a = x$ and $b = \sqrt{3}$.

The polynomial $x^2 + 1$ is irreducible over the real numbers, but reducible over the complex imaginary numbers:

$$x^2 + 1 = (x + i)(x - i).$$

Study Tip

The sum of squares, irreducible over the real numbers, can be factored over the imaginary numbers as

$a^2 + b^2 = (a + bi)(a - bi).$

EXAMPLE 4   Factoring a Polynomial

Factor $x^4 - 3x^2 - 28$:

a. As the product of factors that are irreducible over the rational numbers.
b. As the product of factors that are irreducible over the real numbers.
c. In completely factored form involving complex imaginary numbers.

Solution

a. $x^4 - 3x^2 - 28 = (x^2 - 7)(x^2 + 4)$ Both quadratic factors are irreducible over the rational numbers.

b. $= (x + \sqrt{7})(x - \sqrt{7})(x^2 + 4)$ The third factor is still irreducible over the real numbers.

c. $= (x + \sqrt{7})(x - \sqrt{7})(x + 2i)(x - 2i)$ This is the completely factored form using complex imaginary numbers.

 Check Point 4   Factor $x^4 - 4x^2 - 5$ as the product of factors that are irreducible over **a.** the rational numbers; **b.** the real numbers; **c.** the complex imaginary numbers.

5 Find polynomials with given zeros.

Reversing Things: Finding Polynomials when the Zeros Are Given

Many of our problems involving polynomial functions and polynomial equations dealt with the process of finding zeros and roots. The Linear Factorization Theorem enables us to reverse this process, finding a polynomial function when the zeros are given.

EXAMPLE 5   Finding a Polynomial Function with Given Zeros

Find a fourth-degree polynomial function $f(x)$ with real coefficients that has $-2, 2,$ and i as zeros and such that $f(3) = -150$.

Solution   Because i is a zero and the polynomial has real coefficients, the conjugate, $-i$, must also be a zero. We can now use the Linear Factorization Theorem.

Technology

The graph of $f(x) = -3x^4 + 9x^2 + 12$, shown in a $[-3, 3, 1]$ by $[-200, 20, 20]$ viewing rectangle, verifies that -2 and 2 are real zeros. By tracing along the curve, we can check that $f(3) = -150$.

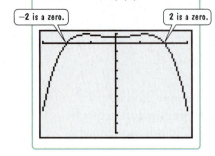

-2 is a zero.   2 is a zero.

$$f(x) = a_n (x - c_1)(x - c_2)(x - c_3)(x - c_4)$$ This is the linear factorization for a fourth-degree polynomial.

$$= a_n(x + 2)(x - 2)(x - i)(x + i)$$ Use the given zeros: $c_1 = -2, c_2 = 2, c_3 = i$, and, from above, $c_4 = -i$.

$$= a_n (x^2 - 4)(x^2 + 1)$$ Multiply.

$$f(x) = a_n (x^4 - 3x^2 - 4)$$ Complete the multiplication.

$$f(3) = a_n (3^4 - 3 \cdot 3^2 - 4) = -150$$ To find a_n, use the fact that $f(3) = -150$.

$$a_n(81 - 27 - 4) = -150$$ Solve for a_n.

$$50a_n = -150$$ Simplify: $81 - 27 - 4 = 50$

$$a_n = -3$$ Divide both sides by 50.

Substituting -3 for a_n in the formula for $f(x)$, we obtain

$$f(x) = -3 (x^4 - 3x^2 - 4).$$

Equivalently,

$$f(x) = -3x^4 + 9x^2 + 12.$$

Check Point 5 Find a third-degree polynomial function $f(x)$ with real coefficients that has -3 and i as zeros and such that $f(1) = 8$.

EXERCISE SET 3.5

Practice Exercises

Use the Upper and Lower Bound Theorem to solve Exercises 1–4.

1. Show that all the real roots of the equation $x^4 - 5x^3 + 11x^2 + 33x - 18 = 0$ lie between -4 and 7.

2. Show that all the real roots of the equation $x^4 + 11x^3 - 12x^2 + 6 = 0$ lie between -13 and 1.

3. Show that all the real roots of the equation $2x^3 + 5x^2 - 8x - 7 = 0$ lie between -4 and 2.

4. Show that all the real roots of the equation $2x^5 - 13x^3 + 2x - 5 = 0$ lie between -3 and 3.

5. Consider the equation $x^4 + 3x^3 + 2x^2 - 5x + 12 = 0$.
   a. List all possible rational roots.
   b. Determine whether 1 is a root using synthetic division. What two conclusions can you draw?
   c. Based on part (b), what possible rational roots can you eliminate?
   d. Determine whether -3 is a root using synthetic division. What two conclusions can you draw?
   e. Based on part (d), what possible rational roots can you eliminate?

6. Consider the equation
   $2x^5 + 5x^4 - 8x^3 - 14x^2 + 6x + 9 = 0$.
   a. List all possible rational roots.
   b. Determine whether $\frac{3}{2}$ is a root using synthetic division. What two conclusions can you draw?
   c. Based on part (b), what possible rational roots can you eliminate?

d. Determine whether -3 is a root using synthetic division. What two conclusions can you draw?
e. Based on part (d), what possible rational roots can you eliminate?

In Exercises 7–14, show that each polynomial has a real zero between the given integers. Then use the Intermediate Value Theorem to find an approximation for this zero to the nearest tenth. If applicable, use a graphing utility's zero feature to verify your answer.

7. $f(x) = x^3 - x - 1$; between 1 and 2

8. $f(x) = x^3 - 4x^2 + 2$; between 0 and 1

9. $f(x) = 2x^4 - 4x^2 + 1$; between -1 and 0

10. $f(x) = x^4 + 6x^3 - 18x^2$; between 2 and 3

11. $f(x) = x^3 + x^2 - 2x + 1$; between -3 and -2

12. $f(x) = x^5 - x^3 - 1$; between 1 and 2

13. $f(x) = 3x^3 - 10x + 9$; between -3 and -2

14. $f(x) = 3x^3 - 8x^2 + x + 2$; between 2 and 3

In Exercises 15–22, use the given root to find the solution set of the polynomial equation.

15. $x^3 - 2x^2 + 4x - 8 = 0$; $-2i$

16. $x^4 + 13x^2 + 36 = 0$; $3i$

17. $3x^3 - 7x^2 + 8x - 2 = 0$; $1 + i$

18. $x^3 - 7x^2 + 16x - 10 = 0$; $3 + i$

19. $x^4 - 6x^2 + 25 = 0$; $2 - i$

20. $x^4 - x^3 - 9x^2 + 29x - 60 = 0$; $1 + 2i$

21. $x^4 - 8x^3 + 64x - 105 = 0$; $2 - i$

22. $4x^4 - 28x^3 + 129x^2 - 130x + 125 = 0$; $3 - 4i$

In Exercises 23–28, factor each polynomial:

    a. *as the product of factors that are irreducible over the rational numbers.*

    b. *as the product of factors that are irreducible over the real numbers.*

    c. *in completely factored form involving complex nonreal, or imaginary, numbers.*

23. $x^4 - x^2 - 20$ **24.** $x^4 + 6x^2 - 27$

25. $x^4 + x^2 - 6$ **26.** $x^4 - 9x^2 - 22$

27. $x^4 - 2x^3 + x^2 - 8x - 12$
    (*Hint*: One factor is $x^2 + 4$.)

28. $x^4 - 4x^3 + 14x^2 - 36x + 45$
    (*Hint*: One factor is $x^2 + 9$.)

In Exercises 29–36, find an nth-degree polynomial function with real coefficients satisfying the given conditions. If you are using a graphing utility, use it to graph the function and verify the real zeros and the given function value.

29. $n = 3$; 1 and $5i$ are zeros; $f(-1) = -104$

30. $n = 3$; 4 and $2i$ are zeros; $f(-1) = -50$

31. $n = 3$; -5 and $4 + 3i$ are zeros; $f(2) = 91$

32. $n = 3$; 6 and $-5 + 2i$ are zeros; $f(2) = -636$

33. $n = 4$; i and $3i$ are zeros; $f(-1) = 20$

34. $n = 4$; $-2, -\frac{1}{2}$, and i are zeros; $f(1) = 18$

35. $n = 4$; $-2, 5$, and $3 + 2i$ are zeros; $f(1) = -96$

36. $n = 4$; $-4, \frac{1}{3}$, and $2 + 3i$ are zeros; $f(1) = 100$

In Exercises 37–44, find all the zeros of the function and write the polynomial as a product of linear factors.

37. $f(x) = x^3 - x^2 + 25x - 25$

38. $f(x) = x^3 - 10x^2 + 33x - 34$

39. $f(x) = x^3 - 8x^2 + 25x - 26$

40. $f(x) = x^3 - 8x^2 + 17x - 4$

41. $f(x) = x^4 + 37x^2 + 36$

42. $f(x) = x^4 + 8x^3 + 9x^2 - 10x + 100$

43. $f(x) = 16x^4 + 36x^3 + 16x^2 + x - 30$

44. $f(x) = 2x^4 - x^3 + 7x^2 - 4x - 4$

Application Exercises

We have seen that the polynomial function
    $H(x) = -0.001618x^4 + 0.077326x^3 - 1.2367x^2 + 11.460x + 2.914$ *models the age in human years, $H(x)$, of a dog that is x years old, where $x \geq 1$. Although the coefficients make it difficult to solve equations algebraically using this function, a graph of the function makes approximate solutions possible.*

Use the graph shown to solve Exercises 45–46. Round all answers to the nearest year.

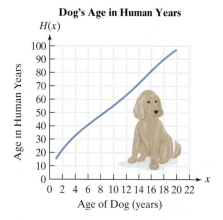

Dog's Age in Human Years

Source: U.C. Davis

45. If you are 25, what is the equivalent age for dogs?

46. If you are 90, what is the equivalent age for dogs?

47. Set up an equation to answer the question in either Exercise 45 or 46. Bring all terms to one side and obtain zero on the other side. What are some of the difficulties involved in solving this equation? Explain how the Intermediate Value Theorem can be used to verify the approximate solution that you obtained from the graph.

The United States has more people in prison, as well as more people in prison per capita, than any other western industrialized nation. The bar graph shows the number of inmates in U.S. state and federal prisons in seven selected years from 1985 through 2000.

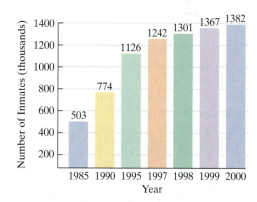

Source: U.S. Justice Department

The data in the graph can be modeled by

    a linear function, $f(x) = 61.3x + 495$

    a quadratic function,
         $g(x) = -0.131x^2 + 63.27x + 491.6$

    a third-degree polynomial function,
         $h(x) = -0.219x^3 + 4.885x^2 + 35.14x + 503.14.$

For each of these functions, x represents the number of years after 1985 and the function value represents the number of inmates, in thousands. Use this information to solve Exercises 48–49.

48. The graph indicates that in 2000, there were 1382 thousand inmates. Substitute 1382 for $f(x)$ and $g(x)$ in the linear and quadratic models. Then solve each resulting equation to find how many years after 1985, to the nearest tenth of a year, inmate population was 1382 thousand. How well do the linear and quadratic functions serve as a model for 2000?

49. The graph indicates that in 2000, there were 1382 thousand inmates. Substitute 1382 for $h(x)$ in the third-degree model. Set the resulting equation equal to 0 and show that it has a real root between 14 and 15. Then use the Intermediate Value Theorem or a graphing utility's zero feature to find an approximation, to the nearest tenth, for this root. How well does the third-degree polynomial function serve as a model for 2000?

Writing in Mathematics

50. When testing a number using synthetic division, how do you know if it is an upper bound for the real roots?

51. When testing a number using synthetic division, how do you know if it is a lower bound for the real roots?

52. How do you show that a polynomial function has a real zero between two given numbers?

53. How does the linear factorization of $f(x)$, that is,
$$f(x) = a_n(x - c_1)(x - c_2)\cdots(x - c_n),$$
show that a polynomial equation of degree n has n roots?

Technology Exercises

54. Show that -1 is a lower bound of $f(x) = x^3 - 53x^2 + 103x - 51$. Show that 60 is an upper bound. Use this information and a graphing utility to draw a relatively complete graph of f.

In Exercises 55–56, use a graphing utility to determine upper and lower bounds for the zeros of f. Does synthetic division verify your observations?

55. $f(x) = 2x^3 + x^2 - 14x - 7$

56. $f(x) = 2x^4 - 7x^3 - 5x^2 + 28x - 12$

57. The function $f(x) = -0.00002x^3 + 0.008x^2 - 0.3x + 6.95$ models the number of annual physician visits, $f(x)$, by a person of age x.

a. Graph the function for meaningful values of x and discuss what the graph reveals in terms of the variables described by the model.

b. Use the zero or root feature of your graphing utility to find the age, to the nearest year, for the group that averages 13.43 annual physician visits.

c. Verify part (b) using the graph of f.

Use a graphing utility to obtain a complete graph for each polynomial function in Exercises 58–61. Then determine the number of real zeros and the number of nonreal complex zeros for each function.

58. $f(x) = x^3 - 6x - 9$

59. $f(x) = 3x^5 - 2x^4 + 6x^3 - 4x^2 - 24x + 16$

60. $f(x) = 3x^4 + 4x^3 - 7x^2 - 2x - 3$

61. $f(x) = x^6 - 64$

Critical Thinking Exercises

In Exercises 62–65, what is the smallest degree that each polynomial could have?

62.

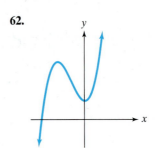

63.

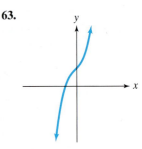

64.

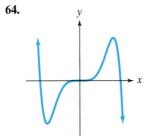

65.

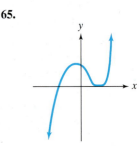

66. Explain why nonreal complex zeros are gained or lost in pairs in terms of graphs of polynomial functions.

67. Explain why a polynomial function of degree 20 cannot cross the x-axis exactly once.

68. Give an example of a function that is not subject to the Intermediate Value Theorem.

Group Exercise

69. The graph at the top of the next page shows costs for private and public four-year colleges projected through the year 2017. According to these projections, your daughter's college education at a private four-year school could cost about $250,000. This activity involves forming and using models from these data. Group members should begin by deciding whether to work with data for private or public colleges.

Cost of a Four-Year College

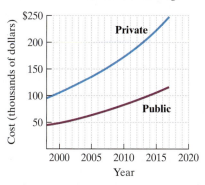

Source: U.S. Department of Education

a. Let $x = 0$ correspond to 1998, $x = 1$ to 1999, $x = 2$ to 2000, and so on up through $x = 19$ for 2017. Group members should use the chosen line graph to obtain a reasonable estimate for the cost of an education, y, in thousands of dollars, for each x.

b. Use the regression feature of a graphing utility to model the data for the cost of a four-year college x years after 1998 using a linear function, a quadratic function, and a third-degree polynomial function.

c. Use these functions to write and solve problems similar to Exercises 48 and 49.

d. Use these functions to make predictions well into the future. Which function, if any, seems to be most reasonable in its predicted cost? Of course, these are only predictions, subject to unforeseeable events. What events might render each of these models relatively useless over long periods of time?

SECTION 3.6 *Rational Functions and Their Graphs*

Objectives

1. Find the domain of rational functions.

2. Use arrow notation.

3. Identify vertical asymptotes.

4. Identify horizontal asymptotes.

5. Graph rational functions.

6. Identify slant asymptotes.

7. Solve applied problems involving rational functions.

Technology is now promising to bring light, fast, and beautiful wheelchairs to millions of disabled people. The cost of manufacturing these radically different wheelchairs can be modeled by rational functions. In this section, we will see how graphs of these functions illustrate that low prices are possible with high production levels, urgently needed in this situation. There are more than half a billion people with disabilities in developing countries; an estimated 20 million need wheelchairs right now.

1 Find the domain of rational functions.

Rational Functions

Rational functions are quotients of polynomial functions. This means that rational functions can be expressed as

$$f(x) = \frac{p(x)}{q(x)} \ .$$

where p and q are polynomial functions and $q(x) \neq 0$. The **domain** of a rational function is the set of all real numbers except the x-values that make the denominator zero. For example, the domain of the rational function

$$f(x) = \frac{x^2 + 7x + 9}{x(x - 2)(x + 5)}$$

This is $p(x)$.

This is $q(x)$.

is the set of all real numbers except 0, 2, and −5.

EXAMPLE 1   Finding the Domain of a Rational Function

Find the domain of each rational function:

a. $f(x) = \dfrac{x^2 - 9}{x - 3}$ **b.** $g(x) = \dfrac{x}{x^2 - 9}$ **c.** $h(x) = \dfrac{x + 3}{x^2 + 9}$.

Solution   Rational functions contain division. Because division by 0 is undefined, we must exclude from the domain of each function values of x that cause the polynomial function in the denominator to be 0.

a. The denominator of $f(x) = \dfrac{x^2 - 9}{x - 3}$ is 0 if $x = 3$. Thus, x cannot equal 3.

The domain of f consists of all real numbers except 3. We can express the domain in set-builder or interval notation:

$$\text{Domain of } f = \{x | x \neq 3\}$$
$$\text{Domain of } f = (-\infty, 3) \text{ or } (3, \infty).$$

b. The denominator of $g(x) = \dfrac{x}{x^2 - 9}$ is 0 if $x = -3$ or $x = 3$. Thus, the domain of g consists of all real numbers except −3 and 3. We can express the domain in set-builder or interval notation:

$$\text{Domain of } g = \{x | x \neq -3, x \neq 3\}$$
$$\text{Domain of } g = (-\infty, -3) \text{ or } (-3, 3) \text{ or } (3, \infty).$$

c. No real numbers cause the denominator of $h(x) = \dfrac{x + 3}{x^2 + 9}$ to equal 0. The domain of h consists of all real numbers.

$$\text{Domain of } h = (-\infty, \infty)$$

> **Study Tip**
>
> Because the domain of a rational function is the set of all real numbers except those for which the denominator is 0, you can identify such numbers by setting the denominator equal to 0 and solving for x. Exclude the resulting real values of x from the domain.

Check Point 1   Find the domain of each rational function:

a. $f(x) = \dfrac{x^2 - 25}{x - 5}$ **b.** $g(x) = \dfrac{x}{x^2 - 25}$ **c.** $h(x) = \dfrac{x + 5}{x^2 + 25}$.

(Ask your professor if a particular notation is preferred.)

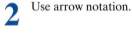 **Use arrow notation.**

The most basic rational function is the **reciprocal function,** defined by $f(x) = \dfrac{1}{x}$. The denominator of the reciprocal function is zero when $x = 0$, so the domain of f is the set of all real numbers except 0.

Let's look at the behavior of f near the excluded value 0. We start by evaluating $f(x)$ to the left of 0.

x approaches 0 from the left.

x	−1	−0.5	−0.1	−0.01	−0.001
$f(x) = \dfrac{1}{x}$	−1	−2	−10	−100	−1000

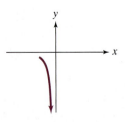

Mathematically, we say that "x approaches 0 from the left." From the table and the accompanying graph, on the bottom of the previous page, it appears that as x approaches 0 from the left, the function values, $f(x)$, decrease without bound. We say that "$f(x)$ approaches negative infinity." We use a special arrow notation to describe this situation symbolically:

$$\text{As } x \to 0^-, \ f(x) \to -\infty.$$

As x approaches 0 from the left, f(x) approaches negative infinity (that is, the graph falls).

Observe that the minus $(-)$ superscript on the 0 $(x \to 0^-)$ is read "from the left."

Next, we evaluate $f(x)$ to the right of 0.

x approaches 0 from the right.

x	0.001	0.01	0.1	0.5	1
$f(x) = \dfrac{1}{x}$	1000	100	10	2	1

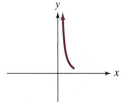

Mathematically, we say that "x approaches 0 from the right." From the table and the accompanying graph, it appears that as x approaches 0 from the right, the function values, $f(x)$, increase without bound. We say that "$f(x)$ approaches infinity." We again use a special arrow notation to describe this situation symbolically:

$$\text{As } x \to 0^+, \ f(x) \to \infty.$$

As x approaches 0 from the right, f(x) approaches infinity (that is, the graph rises).

Observe that the plus $(+)$ superscript on the 0 $(x \to 0^+)$ is read "from the right."

Now let's see what happens to the function values, $f(x)$, as x gets farther away from the origin. The following tables suggest what happens to $f(x)$ as x increases or decreases without bound.

x increases without bound:

x	1	10	100	1000
$f(x) = \dfrac{1}{x}$	1	0.1	0.01	0.001

x decreases without bound:

x	-1	-10	-100	-1000
$f(x) = \dfrac{1}{x}$	-1	-0.1	-0.01	-0.001

Figure 3.23 illustrates the end behavior of $f(x) = \dfrac{1}{x}$ as x increases or decreases without bound. The function values, $f(x)$, are getting progressively closer to 0. This means that as x increases or decreases without bound, the graph of f is approaching the horizontal line $y = 0$ (that is, the x-axis). We use the arrow notation to describe this situation:

$$\text{As } x \to \infty, f(x) \to 0 \quad \text{and} \quad \text{as } x \to -\infty, f(x) \to 0.$$

As x approaches infinity (that is, increases without bound), f(x) approaches 0.

As x approaches negative infinity (that is, decreases without bound), f(x) approaches 0.

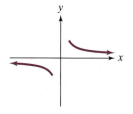

Figure 3.23 $f(x)$ approaches 0 as x increases or decreases without bound

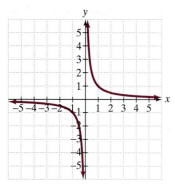

Figure 3.24 The graph of the reciprocal function $f(x) = \dfrac{1}{x}$

Thus, as x approaches infinity $(x \to \infty)$ or as x approaches negative infinity $(x \to -\infty)$, the function values are approaching zero: $f(x) \to 0$.

The graph of the reciprocal function $f(x) = \dfrac{1}{x}$ is shown in Figure 3.24. Unlike the graph of a polynomial function, the graph of the reciprocal function has a break in it and is composed of two distinct branches.

The arrow notation used throughout our discussion of the reciprocal function is summarized in the following box:

Arrow Notation

Symbol	Meaning
$x \to a^+$	x approaches a from the right.
$x \to a^-$	x approaches a from the left.
$x \to \infty$	x approaches infinity; that is, x increases without bound.
$x \to -\infty$	x approaches negative infinity; that is, x decreases without bound.

3 Identify vertical asymptotes.

Vertical Asymptotes of Rational Functions

Look again at the graph of $f(x) = \dfrac{1}{x}$. The curve approaches, but does not touch, the y-axis. The y-axis, or $x = 0$, is said to be a *vertical asymptote* of the graph. A rational function may have no vertical asymptotes, one vertical asymptote, or several vertical asymptotes. The graph of a rational function never intersects a vertical asymptote. We will use dashed lines to show asymptotes.

Definition of a Vertical Asymptote

The line $x = a$ is a **vertical asymptote** of the graph of a function f if $f(x)$ increases or decreases without bound as x approaches a.

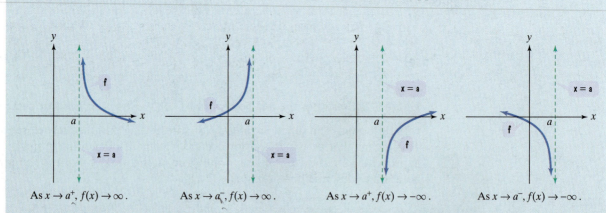

As $x \to a^+$, $f(x) \to \infty$.   As $x \to a^-$, $f(x) \to \infty$.   As $x \to a^+$, $f(x) \to -\infty$.   As $x \to a^-$, $f(x) \to -\infty$.

Thus, as x approaches a from either the left or the right, $f(x) \to \infty$ or $f(x) \to -\infty$.

If the graph of a rational function has vertical asymptotes, they can be located using the following theorem:

> **Locating Vertical Asymptotes**
>
> If $f(x) = \dfrac{p(x)}{q(x)}$ is a rational function in which $p(x)$ and $q(x)$ have no common factors and a is a zero of $q(x)$, the denominator, then $x = a$ is a vertical asymptote of the graph of f.

EXAMPLE 2 **Finding the Vertical Asymptotes of a Rational Function**

Find the vertical asymptotes, if any, of the graph of each rational function:

a. $f(x) = \dfrac{x}{x^2 - 9}$ **b.** $g(x) = \dfrac{x + 3}{x^2 - 9}$ **c.** $h(x) = \dfrac{x + 3}{x^2 + 9}$.

Solution Factoring is usually helpful in identifying zeros of denominators.

a.
$$f(x) = \frac{x}{x^2 - 9} = \frac{x}{(x + 3)(x - 3)}$$

> This factor is 0 if $x = -3$.   This factor is 0 if $x = 3$.

There are no common factors in the numerator and the denominator. The zeros of the denominator are -3 and 3. Thus, the lines $x = -3$ and $x = 3$ are the vertical asymptotes for the graph of f.

b. We will use factoring to see if there are common factors.

$$g(x) = \frac{x + 3}{x^2 - 9} = \frac{(x + 3)}{(x + 3)(x - 3)} = \frac{1}{x - 3}$$

> There is a common factor, $x + 3$, so simplify.   This denominator is 0 if $x = 3$.

The only zero of the denominator of $g(x)$ in simplified form is 3. Thus, the line $x = 3$ is the only vertical asymptote of the graph of g.

c. We cannot factor the denominator of $h(x)$ over the real numbers.

$$h(x) = \frac{x + 3}{x^2 + 9}$$

> No real numbers make this denominator 0.

The denominator has no real zeros. Thus, the graph of h has no vertical asymptotes.

Check Point 2 Find the vertical asymptotes, if any, of the graph of each rational function:

a. $f(x) = \dfrac{x}{x^2 - 1}$ **b.** $g(x) = \dfrac{x - 1}{x^2 - 1}$ **c.** $h(x) = \dfrac{x - 1}{x^2 + 1}$.

A value where the denominator of a function is zero does not necessarily result in a vertical asymptote. There is a hole corresponding to $x = a$, and not a vertical asymptote, in the graph of a function under the following conditions: The value a causes the denominator to be zero, but there is a reduced form of the functions equation in which a does not cause the denominator to be zero.

Consider, for example, the function

$$f(x) = \frac{x^2 - 9}{x - 3}.$$

Because the denominator is zero when $x = 3$, the functions domain is all real numbers except 3. However, there is a reduced form of the equation in which 3 does not cause the denominator to be zero:

$$f(x) = \frac{x^2 - 9}{x - 3} = \frac{(x + 3)(x - 3)}{x - 3} = x + 3, \quad x \neq 3$$

Denominator is zero at $x = 3$.

In this reduced form, 3 does not result in a zero denominator.

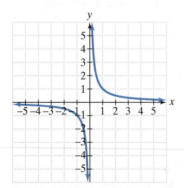

$f(x) = \frac{x^2 - 9}{x - 3}$

Hole corresponding to $x = 3$

Figure 3.25 A graph with a hole corresponding to the denominator's zero

Figure 3.25 shows that the graph has a hole corresponding to $x = 3$. Graphing utilities do not show this feature of the graph.

Horizontal Asymptotes of Rational Functions

4 Identify horizontal asymptotes.

Figure 3.24 shows the graph of the reciprocal function $f(x) = \frac{1}{x}$. As $x \to \infty$ and as $x \to -\infty$, the function values are approaching 0: $f(x) \to 0$. The line $y = 0$ (that is, the x-axis) is a *horizontal asymptote* of the graph. Many, but not all, rational functions have horizontal asymptotes.

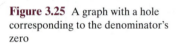

Figure 3.24 The graph of the reciprocal function $f(x) = \frac{1}{x}$, repeated

Definition of a Horizontal Asymptote

The line $y = b$ is a **horizontal asymptote** of the graph of a function f if $f(x)$ approaches b as x increases or decreases without bound.

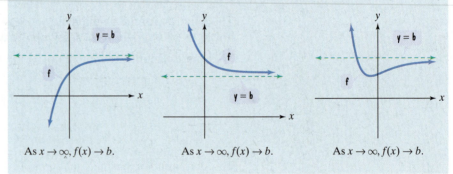

As $x \to \infty, f(x) \to b.$ As $x \to \infty, f(x) \to b.$ As $x \to \infty, f(x) \to b.$

Recall that a rational function may have several vertical asymptotes. By contrast, it can have at most one horizontal asymptote. Although a graph can never intersect a vertical asymptote, it may cross its horizontal asymptote.

If the graph of a rational function has a horizontal asymptote, it can be located using the following theorem:

Locating Horizontal Asymptotes

Let f be the rational function given by

$$f(x) = \frac{a_n x^n + a_{n-1} x^{n-1} + \cdots + a_1 x + a_0}{b_m x^m + b_{m-1} x^{m-1} + \cdots + b_1 x + b_0}, \quad a_n \neq 0, b_m \neq 0.$$

The degree of the numerator is n. The degree of the denominator is m.

1. If $n < m$, the x-axis, or $y = 0$, is the horizontal asymptote of the graph of f.

2. If $n = m$, the line $y = \dfrac{a_n}{b_m}$ is the horizontal asymptote of the graph of f.

3. If $n > m$, the graph of f has no horizontal asymptote.

EXAMPLE 3   **Finding the Horizontal Asymptote of a Rational Function**

Find the horizontal asymptote, if any, of the graph of each rational function:

a. $f(x) = \dfrac{4x}{2x^2 + 1}$ **b.** $g(x) = \dfrac{4x^2}{2x^2 + 1}$ **c.** $h(x) = \dfrac{4x^3}{2x^2 + 1}$.

Solution

a. $f(x) = \dfrac{4x}{2x^2 + 1}$

The degree of the numerator, 1, is less than the degree of the denominator, 2. Thus, the graph of f has the x-axis as a horizontal asymptote [see Figure 3.26(a)]. The equation of the horizontal asymptote is $y = 0$.

b. $g(x) = \dfrac{4x^2}{2x^2 + 1}$

The degree of the numerator, 2, is equal to the degree of the denominator, 2. The leading coefficients of the numerator and denominator, 4 and 2, are used to obtain the equation of the horizontal asymptote. The equation of the horizontal asymptote is $y = \frac{4}{2}$ or $y = 2$ [see Figure 3.26(b)].

c. $h(x) = \dfrac{4x^3}{2x^2 + 1}$

The degree of the numerator, 3, is greater than the degree of the denominator, 2. Thus, the graph of h has no horizontal asymptote [see Figure 3.26(c)].

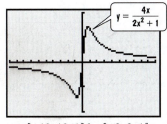

$[-10, 10, 1]$ by $[-2, 2, 1]$

(a) The horizontal asymptote of the graph is $y = 0$.

Figure 3.26

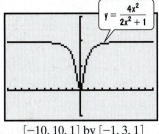

$[-10, 10, 1]$ by $[-1, 3, 1]$

(b) The horizontal asymptote of the graph is $y = 2$.

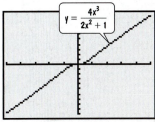

$[-5, 5, 1]$ by $[-10, 10, 1]$

(c) The graph has no horizontal asymptote.

Check Point 3 Find the horizontal asymptote, if any, of the graph of each rational function:

a. $f(x) = \dfrac{9x^2}{3x^2 + 1}$ **b.** $g(x) = \dfrac{9x}{3x^2 + 1}$ **c.** $h(x) = \dfrac{9x^3}{3x^2 + 1}.$

5 Graph rational functions.

Graphing Rational Functions

Here are some suggestions for graphing rational functions:

> **Strategy for Graphing a Rational Function**
>
> Suppose that
> $$f(x) = \frac{p(x)}{q(x)},$$
> where p and q are polynomial functions with no common factors.
>
> **1.** Determine whether the graph of f has symmetry.
> $$f(-x) = f(x): y\text{-axis symmetry}$$
> $$f(-x) = -f(x): \text{origin symmetry}$$
> **2.** Find the y-intercept (if there is one) by evaluating $f(0)$.
> **3.** Find the x-intercepts (if there are any) by solving the equation $p(x) = 0$.
> **4.** Find any vertical asymptote(s) by solving the equation $q(x) = 0$.
> **5.** Find the horizontal asymptote (if there is one) using the rule for determining the horizontal asymptote of a rational function.
> **6.** Plot at least one point between and beyond each x-intercept and vertical asymptote.
> **7.** Use the information obtained previously to graph the function between and beyond the vertical asymptotes.

EXAMPLE 4 Graphing a Rational Function

Graph: $f(x) = \dfrac{2x}{x - 1}.$

Solution

Step 1 Determine symmetry.

$$f(-x) = \frac{2(-x)}{-x - 1} = \frac{-2x}{-x - 1} = \frac{2x}{x + 1}$$

Because $f(-x)$ does not equal $f(x)$ or $-f(x)$, the graph has neither y-axis nor origin symmetry.

Step 2 Find the y-intercept. Evaluate $f(0)$.

$$f(0) = \frac{2 \cdot 0}{0 - 1} = \frac{0}{-1} = 0$$

The y-intercept is 0, and so the graph passes through the origin.

The function to be graphed,

$$f(x) = \frac{2x}{x-1},\ \text{repeated}$$

Step 3 **Find x-intercept(s).** This is done by solving $p(x) = 0$.

$$2x = 0 \quad \textcolor{blue}{\text{Set the numerator equal to 0.}}$$
$$x = 0$$

There is only one x-intercept. This verifies that the graph passes through the origin.

Step 4 **Find the vertical asymptote(s).** Solve $q(x) = 0$, thereby finding zeros of the denominator.

$$x - 1 = 0 \quad \textcolor{blue}{\text{Set the denominator equal to 0.}}$$
$$x = 1$$

The equation of the vertical asymptote is $x = 1$.

Step 5 **Find the horizontal asymptote.** Because the numerator and denominator have the same degree, the leading coefficients of the numerator and denominator, 2 and 1, are used to obtain the equation of the horizontal asymptote.

$$y = \frac{2}{1} = 2.$$

The equation of the horizontal asymptote is $y = 2$.

Step 6 **Plot points between and beyond each x-intercept and vertical asymptote.** With an x-intercept at 0 and a vertical asymptote at $x = 1$, we evaluate the function at $-2, -1, \frac{1}{2}, 2,$ and 4.

x	-2	-1	$\frac{1}{2}$	2	4
$f(x) = \dfrac{2x}{x-1}$	$\dfrac{4}{3}$	1	-2	4	$\dfrac{8}{3}$

Figure 3.27 shows these points, the y-intercept, the x-intercept, and the asymptotes.

Step 7 **Graph the function.** The graph of $f(x) = \dfrac{2x}{x-1}$ is shown in Figure 3.28.

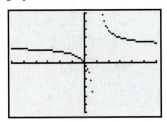
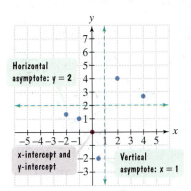

Figure 3.27 Preparing to graph the rational function $f(x) = \dfrac{2x}{x-1}$

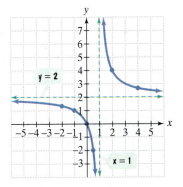

Figure 3.28 The graph of $f(x) = \dfrac{2x}{x-1}$

Check Point 4 Graph: $f(x) = \dfrac{3x}{x-2}.$

EXAMPLE 5   Graphing a Rational Function

Graph: $f(x) = \dfrac{3x^2}{x^2 - 4}$.

Solution

Step 1   Determine symmetry: $f(-x) = \dfrac{3(-x)^2}{(-x)^2 - 4} = \dfrac{3x^2}{x^2 - 4} = f(x)$: The graph of f is symmetric with respect to the y-axis.

Step 2   Find the y-intercept: $f(0) = \dfrac{3 \cdot 0^2}{0^2 - 4} = \dfrac{0}{-4} = 0$: The y-intercept is 0.

Step 3   Find the x-intercept: $3x^2 = 0$, so $x = 0$: The x-intercept is 0.

Step 4   Find the vertical asymptotes: Set $q(x) = 0$.

$$x^2 - 4 = 0 \qquad \text{Set the denominator equal to 0.}$$
$$x^2 = 4$$
$$x = \pm 2$$

The vertical asymptotes are $x = -2$ and $x = 2$.

Step 5   Find the horizontal asymptote: The horizontal asymptote is $y = \frac{3}{1} = 3$.

Step 6   Plot points between and beyond the x-intercept and the vertical asymptotes.   With an x-intercept at 0 and vertical asymptotes at $x = -2$ and $x = 2$, we evaluate the function at $-3, -1, 1, 3,$ and 4.

x	−3	−1	1	3	4
$f(x) = \dfrac{3x^2}{x^2 - 4}$	$\dfrac{27}{5}$	−1	−1	$\dfrac{27}{5}$	4

Figure 3.29 shows these points, the y-intercept, the x-intercept, and the asymptotes.

Step 7   Graph the function.   The graph of $f(x) = \dfrac{3x^2}{x^2 - 4}$ is shown in Figure 3.30. The y-axis symmetry is now obvious.

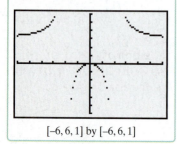
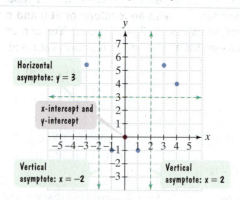

Figure 3.29 Preparing to graph $f(x) = \dfrac{3x^2}{x^2 - 4}$

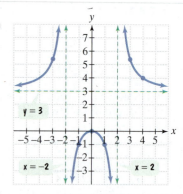

Figure 3.30 The graph of $f(x) = \dfrac{3x^2}{x^2 - 4}$

Check Point 5 Graph: $f(x) = \dfrac{2x^2}{x^2 - 9}$.

Example 6 illustrates that not every rational function has vertical and horizontal asymptotes.

EXAMPLE 6 Graphing a Rational Function

Graph: $f(x) = \dfrac{x^4}{x^2 + 1}$.

Solution

Step 1 Determine symmetry: $f(-x) = \dfrac{(-x)^4}{(-x)^2 + 1} = \dfrac{x^4}{x^2 + 1} = f(x)$:
The graph of f is symmetric with respect to the y-axis.

Step 2 Find the y-intercept: $f(0) = \dfrac{0^4}{0^2 + 1} = \dfrac{0}{1} = 0$: The y-intercept is 0.

Step 3 Find the x-intercept: $x^4 = 0$, so $x = 0$: The x-intercept is 0.

Step 4 Find the vertical asymptote: Set $q(x) = 0$.

$$x^2 + 1 = 0 \qquad \textit{Set the denominator equal to 0.}$$

$$x^2 = -1$$

Although this equation has imaginary roots ($x = \pm i$), there are no real roots. Thus, there is no vertical asymptote.

Step 5 Find the horizontal asymptote: Because the degree of the numerator, 4, is greater than the degree of the denominator, 2, there is no horizontal asymptote.

Step 6 Plot points between and beyond the x-intercept and the vertical asymptotes. With an x-intercept at 0 and no vertical asymptotes, let's look at function values at $-2, -1, 1,$ and 2. You can evaluate the function at 1 and 2. Use y-axis symmetry to obtain function values at -1 and -2:

$$f(-1) = f(1) \text{ and } f(-2) = f(2).$$

x	-2	-1	1	2
$f(x) = \dfrac{x^4}{x^2 + 1}$	$\dfrac{16}{5}$	$\dfrac{1}{2}$	$\dfrac{1}{2}$	$\dfrac{16}{5}$

Step 7 Graph the function. Figure 3.31 shows the graph of f using the points obtained from the table and y-axis symmetry. Notice that as x approaches infinity or negative infinity ($x \to \infty$ or $x \to -\infty$), the function values, $f(x)$, are getting larger without bound $[f(x) \to \infty]$.

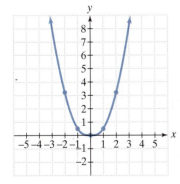

Figure 3.31 The graph of $f(x) = \dfrac{x^4}{x^2 + 1}$

Check Point 6 Graph: $f(x) = \dfrac{x^4}{x^2 + 2}$.

6 Identify slant asymptotes.

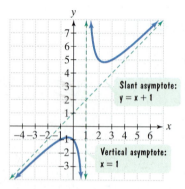

Figure 3.32 The graph of $f(x) = \dfrac{x^2 + 1}{x - 1}$ with a slant asymptote

Slant Asymptotes

Examine the graph of

$$f(x) = \frac{x^2 + 1}{x - 1}$$

shown in Figure 3.32. Note that the degree of the numerator, 2, is greater than the degree of the denominator, 1. Thus, the graph of this function has no horizontal asymptote. However, the graph has a **slant asymptote,** $y = x + 1$.

  The graph of a rational function has a slant asymptote if the degree of the **numerator is one more than the degree of the denominator.** The equation of the slant asymptote can be found by division. For example, to find the slant asymptote for the graph of $f(x) = \dfrac{x^2 + 1}{x - 1}$, divide $x - 1$ into $x^2 + 1$:

$$\underline{1}\ \begin{array}{rrr} 1 & 0 & 1 \\ & 1 & 1 \\ \hline 1 & 1 & 2 \end{array} \qquad\qquad \begin{array}{r} 1x + 1 + \dfrac{2}{x-1} \\ x-1\overline{)x^2 + 0x + 1} \end{array}$$

Remainder

Observe that

$$f(x) = \frac{x^2 + 1}{x - 1} = \underbrace{x + 1}_{\substack{\text{Slant asymptote:} \\ y = x + 1}} + \frac{2}{x - 1}$$

If $|x| \to \infty$, the value of $\dfrac{2}{x - 1}$ is approximately 0. Thus, when $|x|$ is large, the function is very close to $y = x + 1 + 0$. This means that as $x \to \infty$ or as $x \to -\infty$, the graph of f gets closer and closer to the line whose equation is $y = x + 1$. The line $y = x + 1$ is a slant asymptote of the graph.

  In general, if $f(x) = \dfrac{p(x)}{q(x)}, p$ and q have no common factors, and the degree of p is one greater than the degree of q, find the slant asymptote by dividing $q(x)$ into $p(x)$. The division will take the form

$$\frac{p(x)}{q(x)} = \underbrace{mx + b}_{\substack{\text{Slant asymptote:} \\ y = mx + b}} + \frac{\text{remainder}}{q(x)}.$$

The equation of the slant asymptote is obtained by dropping the term with the remainder. Thus, the equation of the slant asymptote is $y = mx + b$.

EXAMPLE 7 **Finding the Slant Asymptote of a Rational Function**

Find the slant asymptote of $f(x) = \dfrac{x^2 - 4x - 5}{x - 3}$.

Solution Because the degree of the numerator, 2, is exactly one more than the degree of the denominator, 1, and $x - 3$ is not a factor of $x^2 - 4x - 5$, the graph of f has a slant asymptote. To find the equation of the slant asymptote, divide $x - 3$ into $x^2 - 4x - 5$:

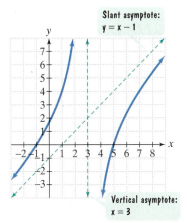

Figure 3.33 The graph of $f(x) = \dfrac{x^2 - 4x - 5}{x - 3}$

$$\underline{3|} \quad 1 \quad -4 \quad -5$$
$$ 3 \quad -3$$
$$\overline{\quad 1 \quad -1 \quad -8}$$

$1x - 1 - \dfrac{8}{x - 3}$

$x - 3 \overline{)x^2 - 4x - 5}$

Drop the remainder term and you'll have the equation of the slant asymptote.

Remainder

The equation of the slant asymptote is $y = x - 1$. Using our strategy for graphing rational functions, the graph of $f(x) = \dfrac{x^2 - 4x - 5}{x - 3}$ is shown in Figure 3.33.

Check Point 7 Find the slant asymptote of $f(x) = \dfrac{2x^2 - 5x + 7}{x - 2}$.

7 Solve applied problems involving rational functions.

Applications

There are numerous examples of asymptotic behavior in functions that describe real-world phenomena. Let's consider an example from the business world. The **cost function,** C, for a business is the sum of its fixed and variable costs:

$$C(x) = (\text{fixed cost}) + cx.$$
Cost per unit times the number of units produced, x

The **average cost** per unit for a company to produce x units is the sum of its fixed and variable costs divided by the number of units produced. The **average cost function** is a rational function that is denoted by $\bar{C}$. Thus,

$$\bar{C}(x) = \dfrac{(\text{fixed cost}) + cx}{x}.$$
Cost of producing x units: fixed plus variable costs
Number of units produced

EXAMPLE 8 **Average Cost of Producing a Wheelchair**

A company is planning to manufacture wheelchairs that are light, fast, and beautiful. Fixed monthly cost will be \$500,000, and it will cost \$400 to produce each radically innovative chair.

a. Write the cost function, C, of producing x wheelchairs.

b. Write the average cost function, $\bar{C}$, of producing x wheelchairs.

c. Find and interpret $\bar{C}(1000)$, $\bar{C}(10,000)$, and $\bar{C}(100,000)$.

d. What is the horizontal asymptote for the average cost function, $\bar{C}$? Describe what this represents for the company.

Solution

a. The cost function of producing x wheelchairs, C, is the sum of the fixed cost and the variable cost.

> Fixed cost is $500,000.

> Variable cost: $400 for each wheelchair produced

$$C(x) = 500{,}000 + 400x$$

b. The average cost function of producing x wheelchairs, $\bar{C}$, is the sum of fixed and variable costs divided by the number of wheelchairs produced.

$$\bar{C}(x) = \frac{500{,}000 + 400x}{x} \quad \text{or} \quad \bar{C}(x) = \frac{400x + 500{,}000}{x}$$

c. We evaluate $\bar{C}$ at 1000, 10,000, and 100,000, interpreting the results.

$$\bar{C}(1000) = \frac{400(1000) + 500{,}000}{1000} = 900$$

The average cost per wheelchair of producing 1000 wheelchairs per month is $900.

$$\bar{C}(10{,}000) = \frac{400(10{,}000) + 500{,}000}{10{,}000} = 450$$

The average cost per wheelchair of producing 10,000 wheelchairs per month is $450.

$$\bar{C}(100{,}000) = \frac{400(100{,}000) + 500{,}000}{100{,}000} = 405$$

The average cost per wheelchair of producing 100,000 wheelchairs per month is $405. Notice that with higher production levels, the cost of producing each wheelchair decreases.

d. We developed the average cost function

$$\bar{C}(x) = \frac{400x + 500{,}000}{x}$$

in which the degree of the numerator, 1, is equal to the degree of the denominator, 1. The leading coefficients of the numerator and denominator, 400 and 1, are used to obtain the equation of the horizontal asymptote. The equation of the horizontal asymptote is

$$y = \frac{400}{1} \quad \text{or} \quad y = 400.$$

The horizontal asymptote is shown in Figure 3.34. This means that the more wheelchairs produced per month, the closer the average cost per wheelchair for the company comes to $400. The least possible cost per wheelchair is approaching $400. Competitively low prices take place with high production levels, posing a major problem for small businesses.

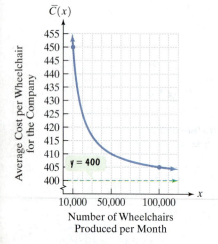

Figure 3.34 As production level increases, the average cost per wheelchair approaches $400.

> **Check Point 8** The time: the not-too-distant future. A new company is hoping to replace traditional computers and two-dimensional monitors with its virtual reality system. The fixed monthly cost will be $600,000, and it will cost $500 to produce each system.
>
> **a.** Write the cost function, C, of producing x virtual reality systems.
> **b.** Write the average cost function, $\bar{C}$, of producing x virtual reality systems.
> **c.** Find and interpret $\bar{C}(1000)$, $\bar{C}(10{,}000)$, and $\bar{C}(100{,}000)$.
> **d.** What is the horizontal asymptote for the average cost function, $\bar{C}$? Describe what this represents for the company.

EXERCISE SET 3.6

Practice Exercises

In Exercises 1–8, find the domain of each rational function.

1. $f(x) = \dfrac{5x}{x-4}$ **2.** $f(x) = \dfrac{7x}{x-8}$

3. $g(x) = \dfrac{3x^2}{(x-5)(x+4)}$ **4.** $g(x) = \dfrac{2x^2}{(x-2)(x+6)}$

5. $h(x) = \dfrac{x+7}{x^2-49}$ **6.** $h(x) = \dfrac{x+8}{x^2-64}$

7. $f(x) = \dfrac{x+7}{x^2+49}$ **8.** $f(x) = \dfrac{x+8}{x^2+64}$

Use the graph of the rational function in the figure shown to complete each statement in Exercises 9–14.

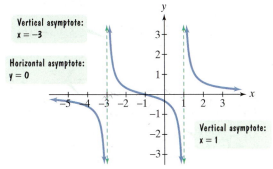

9. As $x \to -3^-$, $f(x) \to$ _____.
10. As $x \to -3^+$, $f(x) \to$ _____.
11. As $x \to 1^-$, $f(x) \to$ _____.
12. As $x \to 1^+$, $f(x) \to$ _____.
13. As $x \to -\infty$, $f(x) \to$ _____.
14. As $x \to \infty$, $f(x) \to$ _____.

Use the graph of the rational function in the figure shown to complete each statement in Exercises 15–20.

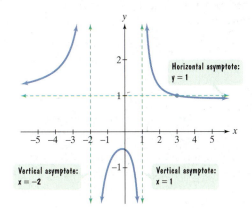

15. As $x \to 1^+$, $f(x) \to$ _____.

16. As $x \to 1^-$, $f(x) \to$ _____.
17. As $x \to -2^+$, $f(x) \to$ _____.
18. As $x \to -2^-$, $f(x) \to$ _____.
19. As $x \to \infty$, $f(x) \to$ _____.
20. As $x \to -\infty$, $f(x) \to$ _____.

In Exercises 21–28, find the vertical asymptotes, if any, of the graph of each rational function.

21. $f(x) = \dfrac{x}{x+4}$ **22.** $f(x) = \dfrac{x}{x-3}$

23. $g(x) = \dfrac{x+3}{x(x+4)}$ **24.** $g(x) = \dfrac{x+3}{x(x-3)}$

25. $h(x) = \dfrac{x}{x(x+4)}$ **26.** $h(x) = \dfrac{x}{x(x-3)}$

27. $r(x) = \dfrac{x}{x^2+4}$ **28.** $r(x) = \dfrac{x}{x^2+3}$

In Exercises 29–36, find the horizontal asymptote, if any, of the graph of each rational function.

29. $f(x) = \dfrac{12x}{3x^2+1}$ **30.** $f(x) = \dfrac{15x}{3x^2+1}$

31. $g(x) = \dfrac{12x^2}{3x^2+1}$ **32.** $g(x) = \dfrac{15x^2}{3x^2+1}$

33. $h(x) = \dfrac{12x^3}{3x^2+1}$ **34.** $h(x) = \dfrac{15x^3}{3x^2+1}$

35. $f(x) = \dfrac{-2x+1}{3x+5}$ **36.** $f(x) = \dfrac{-3x+7}{5x-2}$

In Exercises 37–58, follow the seven steps on page 342 to graph each rational function.

37. $f(x) = \dfrac{4x}{x-2}$ **38.** $f(x) = \dfrac{3x}{x-1}$

39. $f(x) = \dfrac{2x}{x^2-4}$ **40.** $f(x) = \dfrac{4x}{x^2-1}$

41. $f(x) = \dfrac{2x^2}{x^2-1}$ **42.** $f(x) = \dfrac{4x^2}{x^2-9}$

43. $f(x) = \dfrac{-x}{x+1}$ **44.** $f(x) = \dfrac{-3x}{x+2}$

45. $f(x) = -\dfrac{1}{x^2-4}$ **46.** $f(x) = -\dfrac{2}{x^2-1}$

47. $f(x) = \dfrac{2}{x^2+x-2}$ **48.** $f(x) = \dfrac{-2}{x^2-x-2}$

49. $f(x) = \dfrac{2x^2}{x^2+4}$ **50.** $f(x) = \dfrac{4x^2}{x^2+1}$

51. $f(x) = \dfrac{x+2}{x^2+x-6}$ **52.** $f(x) = \dfrac{x-4}{x^2-x-6}$

53. $f(x) = \dfrac{x^4}{x^2+2}$ **54.** $f(x) = \dfrac{2x^4}{x^2+1}$

55. $f(x) = \dfrac{x^2 + x - 12}{x^2 - 4}$

56. $f(x) = \dfrac{x^2}{x^2 + x - 6}$

57. $f(x) = \dfrac{3x^2 + x - 4}{2x^2 - 5x}$

58. $f(x) = \dfrac{x^2 - 4x + 3}{(x + 1)^2}$

In Exercises 59–66, **a.** *Find the slant asymptote of the graph of each rational function and* **b.** *Follow the seven-step strategy and use the slant asymptote to graph each rational function.*

59. $f(x) = \dfrac{x^2 - 1}{x}$

60. $f(x) = \dfrac{x^2 - 4}{x}$

61. $f(x) = \dfrac{x^2 + 1}{x}$

62. $f(x) = \dfrac{x^2 + 4}{x}$

63. $f(x) = \dfrac{x^2 + x - 6}{x - 3}$

64. $f(x) = \dfrac{x^2 - x + 1}{x - 1}$

65. $f(x) = \dfrac{x^3 + 1}{x^2 + 2x}$

66. $f(x) = \dfrac{x^3 - 1}{x^2 - 9}$

Application Exercises

67. A company is planning to manufacture mountain bikes. Fixed monthly cost will be $100,000 and it will cost $100 to produce each bicycle.

   a. Write the cost function, C, of producing x mountain bikes.

   b. Write the average cost function, $\bar{C}$, of producing x mountain bikes.

   c. Find and interpret $\bar{C}(500)$, $\bar{C}(1000)$, $\bar{C}(2000)$, and $\bar{C}(4000)$.

   d. What is the horizontal asymptote for the function, $\bar{C}$? Describe what this means in practical terms.

68. A company that manufactures running shoes has a fixed monthly cost of $300,000. It costs $30 to produce each pair of shoes.

   a. Write the cost function, C, of producing x pairs of shoes.

   b. Write the average cost function, $\bar{C}$, of producing x pairs of shoes.

   c. Find and interpret $\bar{C}(1000)$, $\bar{C}(10,000)$, and $\bar{C}(100,000)$.

   d. What is the horizontal asymptote for the average cost function, $\bar{C}$? Describe what this represents for the company.

69. Textbook sales at college stores have increased during the past two decades. The function

$$B(x) = 190.9x + 2413.99$$

models textbook sales, $B(x)$, in millions of dollars, x years after 1985. College enrollment has also increased. The function

$$E(x) = 0.234x + 12.54$$

models total college enrollment, $E(x)$, in millions, x years after 1985.

   a. Write a rational function that models the average amount of money spent on textbooks per college student, $M(x)$, in dollars per student, x years after 1985. The graph of M is shown in the figure.

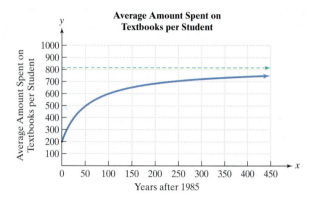

Average Amount Spent on Textbooks per Student

   b. Predict the average amount of money that will be spent on textbooks per college student in 2004. How is this shown on the graph of M?

   c. What is the horizontal asymptote for the function that models the average amount of money spent on textbooks per college student? Describe what this represents in practical terms.

70. The rational function

$$C(x) = \frac{130x}{100 - x}, \quad 0 \le x < 100,$$

describes the cost, $C(x)$, in millions of dollars, to inoculate $x\%$ of the population against a particular strain of flu.

   a. Find and interpret $C(20)$, $C(40)$, $C(60)$, $C(80)$, and $C(90)$.

   b. What is the equation of the vertical asymptote? What does this mean in terms of the variables in the function?

   c. Graph the function.

Among all deaths from a particular disease, the percentage that are smoking related (21–39 cigarettes per day) is a function of the disease's **incidence ratio.** *The incidence ratio describes the number of times more likely smokers are than nonsmokers to die from the disease. The following table shows the incidence ratios for heart disease and lung cancer for two age groups.*

Incidence Ratios

	Heart Disease	**Lung Cancer**
Ages 55–64	1.9	10
Ages 65–74	1.7	9

Source: Alexander M. Walker, *Observations and Inference*, page 20.

For example, the incidence ratio of 9 in the table means that smokers between the ages of 65 and 74 are 9 times more likely than nonsmokers in the same group to die from lung cancer. The rational function

$$P(x) = \frac{100(x - 1)}{x}$$

models the percentage of smoking-related deaths among all deaths from a disease, P(x), in terms of the disease's incidence ratio, x. The graph of the rational function is shown. Use this function to solve Exercises 71–74.

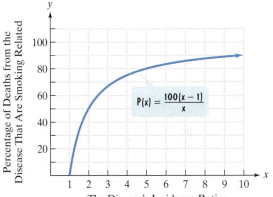

The Disease's Incidence Ratio:
The number of times more likely smokers are than nonsmokers to die from the disease

71. Find P(10). Describe what this means in terms of the incidence ratio, 10, given in the table. Identify your solution as a point on the graph.

72. Find P(9). Round to the nearest percent. Describe what this means in terms of the incidence ratio, 9, given in the table. Identify your solution as a point on the graph.

73. What is the horizontal asymptote of the graph? Describe what this means about the percentage of deaths caused by smoking with increasing incidence ratios.

74. According to the model and its graph, is there a disease for which all deaths are caused by smoking? Explain your answer.

75. Rational functions are often used to model how much we remember over time. In an experiment on memory, students in a language class are asked to memorize 40 vocabulary words in Latin, a language with which the students are not familiar. After studying the words for one day, the class is tested each day thereafter to see how many words they remember. The class average is taken and the results are graphed below.

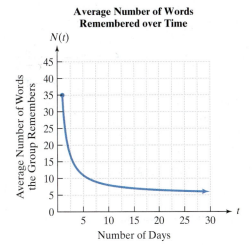

Average Number of Words Remembered over Time

Number of Days

a. Use the graph to find a reasonable estimate of the number of Latin words remembered after 1 day, 5 days, and 15 days.

b. The function that models the number of Latin words remembered by the students after t days is given by

$$N(t) = \frac{5t + 30}{t}, \quad \text{where } t \geq 1.$$

Find N(1), N(5), and N(15), comparing these values with your estimates from part (a).

c. What does the graph indicate about the number of Latin words remembered by the group over time?

d. Use the function in part (b) to find the horizontal asymptote for the graph. Describe what this horizontal asymptote means in terms of the variables modeled in this situation.

Writing in Mathematics

76. What is a rational function?

77. Use everyday language to describe the graph of a rational function f such that as $x \to -\infty, f(x) \to 3$.

78. Use everyday language to describe the behavior of a graph near its vertical asymptote if $f(x) \to \infty$ as $x \to -2^-$ and $f(x) \to -\infty$ as $x \to -2^+$.

79. If you are given the equation of a rational function, explain how to find the vertical asymptotes, if any, of the function's graph.

80. If you are given the equation of a rational function, explain how to find the horizontal asymptote, if any, of the function's graph.

81. Describe how to graph a rational function.

82. If you are given the equation of a rational function, how can you tell if the graph has a slant asymptote? If it does, how do you find its equation?

83. Is every rational function a polynomial function? Why or why not? Does a true statement result if the two adjectives *rational* and *polynomial* are reversed? Explain.

84. Although your friend has a family history of heart disease, he smokes, on average, 25 cigarettes per day. He sees the table showing incidence ratios for heart disease (see Exercises 71–74) and feels comfortable that they are less than 2, compared to 9 and 10 for lung cancer. He claims that all family deaths have been from heart disease, and decides not to give up smoking. Use the given function and its graph to describe some additional information not given in the table that might influence his decision.

Technology Exercises

85. Use a graphing utility to verify any five of your hand-drawn graphs in Exercises 37–66.

86. Use a graphing utility to verify your hand-drawn graph in Exercise 70.

87. Use a graphing utility to graph $y = \dfrac{1}{x}$, $y = \dfrac{1}{x^3}$, and $\dfrac{1}{x^5}$ in the same viewing rectangle. For odd values of n, how does changing n affect the graph of $y = \dfrac{1}{x^n}$?

88. Use a graphing utility to graph $y = \dfrac{1}{x^2}$, $y = \dfrac{1}{x^4}$, and $y = \dfrac{1}{x^6}$ in the same viewing rectangle. For even values of n, how does changing n affect the graph of $y = \dfrac{1}{x^n}$?

89. Use a graphing utility to graph
$$f(x) = \frac{x^2 - 4x + 3}{x - 2} \quad \text{and} \quad g(x) = \frac{x^2 - 5x + 6}{x - 2}.$$
What differences do you observe between the graph of f and g? How do you account for these differences?

90. The rational function
$$f(x) = \frac{27,725(x - 14)}{x^2 + 9} - 5x$$

models the number of arrests, $f(x)$, per 100,000 drivers, for driving under the influence of alcohol, as a function of a driver's age, x.

a. Graph the function in a $[0, 70, 5]$ by $[0, 400, 20]$ viewing rectangle.

b. Describe the trend shown by the graph.

c. Use the ZOOM and TRACE features or the maximum function feature of your graphing utility to find the age that corresponds to the greatest number of arrests. How many arrests, per 100,000 drivers, are there for this age group?

Critical Thinking Exercises

91. Which one of the following is true?

a. The graph of a rational function cannot have both a vertical and a horizontal asymptote.

b. It is not possible to have a rational function whose graph has no y-intercept.

c. The graph of a rational function can have three horizontal asymptotes.

d. The graph of a rational function can never cross a vertical asymptote.

92. Which one of the following is true?

a. The function $f(x) = \dfrac{1}{\sqrt{x - 3}}$ is a rational function.

b. The x-axis is a horizontal asymptote for the graph of $f(x) = \dfrac{4x - 1}{x + 3}$.

c. The number of televisions that a company can produce per week after t weeks of production is given by
$$N(t) = \frac{3000t^2 + 30,000t}{t^2 + 10t + 25}.$$
Using this model, the company will eventually be able to produce 30,000 televisions in a single week.

d. None of the given statements is true.

In Exercises 93–96, write the equation of a rational function $f(x) = \dfrac{p(x)}{q(x)}$ having the indicated properties, in which the degrees of p and q are as small as possible. More than one correct function may be possible. Graph your function using a graphing utility to verify that it has the required properties.

93. f has a vertical asymptote given by $x = 3$, a horizontal asymptote $y = 0$, y-intercept at –1, and no x-intercept.

94. f has vertical asymptotes given by $x = -2$ and $x = 2$, a horizontal asymptote $y = 2$, y-intercept at $\frac{9}{2}$, x-intercepts at -3 and 3, and y-axis symmetry.

95. f has a vertical asymptote given by $x = 1$, a slant asymptote whose equation is $y = x$, y-intercept at 2, and x-intercepts at -1 and 2.

96. f has no vertical, horizontal, or slant asymptotes, and no x-intercepts.

Group Exercise

97. Group members form the sales team for a company that makes computer video games. It has been determined that the rational function

$$f(x) = \frac{200x}{x^2 + 100}$$

models the monthly sales, $f(x)$, in thousands of games, of a new video game as a function of the number of months, x, after the game is introduced. The figure shows the graph of the function. What are the team's recommendations to the company in terms of how long the video game should be on the market before another new video game is introduced? What other factors might members want to take into account in terms of the recommendations? What

will eventually happen to sales, and how is this indicated by the graph? What does this have to do with a horizontal asymptote? What could the company do to change the behavior of this function and continue generating sales? Would this be cost effective?

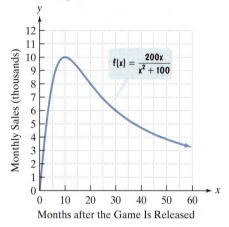

Monthly Sales of a New Video Game

$f(x) = \dfrac{200x}{x^2 + 100}$

Months after the Game Is Released

SECTION 3.7 *Modeling Using Variation*

Objectives

1. Solve direct variation problems.

2. Solve inverse variation problems.

3. Solve combined variation problems.

4. Solve problems involving joint variation.

Have you ever wondered how telecommunication companies estimate the number of phone calls expected per day between two cities? The formula

$$N = \frac{400P_1P_2}{d^2}$$

shows that the daily number of phone calls, N, increases as the populations of the cities, P_1 and P_2, in thousands, increase and decreases as the distance, d, between the cities increases.

   Certain formulas occur so frequently in applied situations that they are given special names. Variation formulas show how one quantity changes in relation to other quantities. Quantities can vary *directly*, *inversely*, or *jointly*. In this section, we look at situations that can be modeled by each of these kinds of variation. And think of this: The next time you get one of those "all-circuits-are-busy" messages, you will be able to use a variation formula to estimate how many other callers you're competing with for those precious 8-cent minutes.

1 Solve direct variation problems.

Direct Variation

Because light travels faster than sound, during a thunderstorm we see lightning before we hear thunder. The formula

$$d = 1080t$$

describes the distance, in feet, of the storm's center if it takes t seconds to hear thunder after seeing lightning. Thus,

If $t = 1$, $d = 1080 \cdot 1 = 1080$. If it takes 1 second to hear thunder, the storm's center is 1080 feet away.

If $t = 2$, $d = 1080 \cdot 2 = 2160$. If it takes 2 seconds to hear thunder, the storm's center is 2160 feet away.

If $t = 3$, $d = 1080 \cdot 3 = 3240$. If it takes 3 seconds to hear thunder, the storm's center is 3240 feet away.

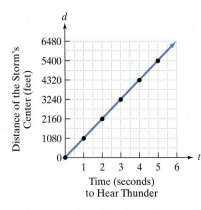

The graph of $d = 1080t$. Distance to a storm's center varies directly as the time it takes to hear thunder.

As the formula $d = 1080t$ illustrates, the distance to the storm's center is a constant multiple of how long it takes to hear the thunder. When the time is doubled, the storm's distance is doubled; when the time is tripled, the storm's distance is tripled; and so on. Because of this, the distance is said to *vary directly* as the time. The *equation of variation* is

$$d = 1080t.$$

Generalizing, we obtain the following statement:

Direct Variation

If a situation is described by an equation in the form

$$y = kx$$

where k is a nonzero constant, we say that **y varies directly as x** or **y is directly proportional to x.** The number k is called the **constant of variation** or the **constant of proportionality.**

EXAMPLE 1 Writing a Direct Variation Equation

A person's salary, S, varies directly as the number of hours worked, h.

a. Write an equation that expresses this relationship.

b. Margarita earns $18 per hour. Substitute 18 for k, the constant of variation, in the equation in part (a) and write the equation for Margarita's salary.

Solution

a. We know that y varies directly as x is expressed as

$$y = kx.$$

By changing letters, we can write an equation that describes the following English statement: Salary, S, varies directly as the number of hours worked, h.

$$S = kh$$

We can also read this as "salary is directly proportional to hours worked."

b. Substituting 18 for k in $S = kh$, the direct variation equation, gives

$$S = 18h.$$

This equation describes Margarita's salary in terms of the number of hours she works. For example, if she works 10 hours, we can substitute 10 for h and determine her salary:

$$S = 18(10) = 180.$$

Her salary for working 10 hours is $180. Notice that, as the number of hours worked increases, the salary increases.

 Check Point 1 A person's hair length, L, in inches, varies directly as the number of years it has been growing, N.

a. Write an equation that expresses this relationship.

b. The longest moustache on record was grown by Kalyan Sain of India. His moustache grew 4 inches each year. Substitute 4 for k, the constant of variation, in the equation in part (a) and write the equation for the length of Sain's moustache.

c. Sain grew his moustache for 17 years. Substitute 17 for N in the equation from part (b) and find its length.

In Example 1 and CheckPoint 1, the constants of variation, or proportionality, were given. If the constant of variation is not given, we can find it by substituting given values in the variation formula and solving for k. Example 2 shows how this is done.

EXAMPLE 2   Finding k, the Constant of Variation

Height, H, varies directly as foot length, F.

a. Write an equation that expresses this relationship.

b. Photographs of large footprints were published in 1951. Some speculated that these footprints were made by the Abominable Snowman. Each footprint was 23 inches long. The Abominable Snowman's height was determined to be 154.1 inches. (This is 12 feet, 10.1 inches, so it might not be a pleasant experience to run into this critter on a mellow hike through the woods!) Use $H = 154.1$ and $F = 23$ to find the constant of variation.

Solution

a. We know that y varies directly as x is expressed as

$$y = kx.$$

By changing letters, we can write an equation that describes the following English statement: Height, H, varies directly as foot length, F.

$$H = kF$$

> Equivalently, height is directly proportional to foot length.

b. The Abominable Snowman's height is 154.1 inches, and foot length is 23 inches. Substitute 154.1 for H and 23 for F in the direct variation equation.

$$H = kF$$

$$154.1 = k \cdot 23$$

Solve for k, the constant of variation, by dividing both sides of the equation by 23:

$$\frac{154.1}{23} = \frac{k \cdot 23}{23}$$

$$6.7 = k.$$

> Remember that 6.7 is also called the **constant of proportionality.**

Thus, the constant of variation is 6.7.

In Example 2, now that we know the constant of variation ($k = 6.7$), we can rewrite $H = kF$ using this constant. The equation of variation is

$$H = 6.7F.$$

We can use this equation to find other values. For example, if your foot length is 10 inches, your height is

$$H = 6.7(10) = 67,$$

or approximately 67 inches.

 The weight, W, of an aluminum canoe varies directly as its length, L.
   a. Write an equation that expresses this relationship.
   b. A 6-foot canoe weighs 75 pounds. Substitute 75 for W and 6 for L in the equation from part (a) and find k, the constant of variation.
   c. Substitute the value of k into your equation in part (a) and write the equation that describes the weight of this type of canoe in terms of its length.
   d. Use the equation from part (c) to find the weight of a 16-foot canoe of this type.

Our work up to this point provides a step-by-step procedure for solving variation problems. This procedure applies to direct variation problems as well as to the other kinds of variation problems that we will discuss.

Solving Variation Problems

1. Write an equation that describes the given English statement.

2. Substitute the given pair of values into the equation in step 1 and find the value of k.

3. Substitute the value of k into the equation in step 1.

4. Use the equation from step 3 to answer the problem's question.

EXAMPLE 3   Solving a Direct Variation Problem

The amount of garbage, G, varies directly as the population, P. Allegheny County, Pennsylvania, has a population of 1.3 million and creates 26 million pounds of garbage each week. Find the weekly garbage produced by New York City with a population of 7.3 million.

Solution

Step 1 Write an equation. We know that y varies directly as x is expressed as

$$y = kx.$$

By changing letters, we can write an equation that describes the following English statement: Garbage production, G, varies directly as the population, P.

$$G = kP$$

> Equivalently, garbage production is directly proportional to the population.

Step 2 Use the given values to find k. Allegheny County has a population of 1.3 million and creates 26 million pounds of garbage weekly. Substitute 26 for G and 1.3 for P in the direct variation equation. Then solve for k.

$$G = kP \qquad \text{This is the direct variation equation.}$$

$$26 = k \cdot 1.3 \qquad G = 26 \text{ and } P = 1.3.$$

$$\frac{26}{1.3} = \frac{k \cdot 1.3}{1.3} \qquad \text{Divide both sides by 1.3.}$$

$$20 = k \qquad \text{Simplify.}$$

Step 3 Substitute the value of k into the equation.

$$G = kP \qquad \text{Use the direct variation equation from step 1.}$$

$$G = 20P \qquad \text{Replace } k, \text{ the constant of variation, with 20.}$$

Step 4 Answer the problems question. New York City has a population of 7.3 million. To find its weekly garbage production, substitute 7.3 for P in $G = 20P$ and solve for G.

$$G = 20P \qquad \text{Use the equation from step 3.}$$

$$G = 20(7.3) \qquad \text{Substitute 7.3 for } P.$$

$$G = 146$$

The weekly garbage produced by New York City weighs approximately 146 million pounds.

Check Point 3 The pressure, P, of water on an object below the surface varies directly as its distance, D, below the surface. If a submarine experiences a pressure of 25 pounds per square inch 60 feet below the surface, how much pressure will it experience 330 feet below the surface?

The direct variation equation $y = kx$, or $f(x) = kx$, is a linear function. If $k > 0$, then the slope of the line is positive. Consequently, as x increases, y also increases.

A direct variation situation can involve variables to higher powers. For example, y can vary directly as x^2 $(y = kx^2)$ or as x^3 $(y = kx^3)$.

> ### Direct Variation with Powers
>
> **y varies directly as the nth power of x** if there exists some nonzero constant k such that
>
> $$y = kx^n.$$
>
> We also say that **y is directly proportional to the nth power of x.**

Direct variation with powers is modeled by polynomial functions. In our next example, the graph of the variation equation is the familiar parabola.

EXAMPLE 4 Solving a Direct Variation Problem

The distance, s, that a body falls from rest varies directly as the square of the time, t, of the fall. If skydivers fall 64 feet in 2 seconds, how far will they fall in 4.5 seconds?

Solution

Step 1 Write an equation. We know that y varies directly as the square of x is expressed as

$$y = kx^2.$$

By changing letters, we can write an equation that describes the following English statement: Distance, s, varies directly as the square of time, t, of the fall.

$$s = kt^2$$

> Equivalently, distance is directly proportional to the square of time.

Step 2 Use the given values to find k. Skydivers fall 64 feet in 2 seconds. Substitute 64 for s and 2 for t in the direct variation equation. Then solve for k.

$$s = kt^2 \quad \text{This is the direct variation equation.}$$
$$64 = k \cdot 2^2 \quad s = 64 \text{ and } t = 2.$$
$$64 = 4k \quad \text{Simplify.}$$
$$\frac{64}{4} = \frac{4k}{4} \quad \text{Divide both sides by 4.}$$
$$16 = k \quad \text{Simplify.}$$

Step 3 Substitute the value of k into the equation.

$$s = kt^2 \quad \text{Use the direct variation equation from step 1.}$$
$$s = 16t^2 \quad \text{Replace } k, \text{ the constant of variation, with 16.}$$

Step 4 Answer the problem's question. How far will the skydivers fall in 4.5 seconds? Substitute 4.5 for t in $s = 16t^2$ and solve for s.

$$s = 16(4.5)^2 = 16(20.25) = 324$$

Thus, in 4.5 seconds, skydivers will fall 324 feet.

We can express the variation equation from Example 4 in function notation, writing

$$s(t) = 16t^2.$$

The distance that a body falls from rest is a function of the time, t, of the fall. The parabola that is the graph of this quadratic function is shown in Figure 3.35. The graph increases rapidly from left to right, showing the effects of the acceleration of gravity.

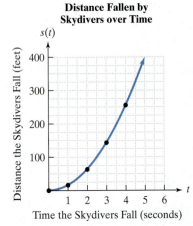

Distance Fallen by Skydivers over Time

Distance the Skydivers Fall (feet) / Time the Skydivers Fall (seconds)

Figure 3.35 The graph of $s(t) = 16t^2$

The distance required to stop a car varies directly as the square of its speed. If 200 feet are required to stop a car traveling 60 miles per hour, how many feet are required to stop a car traveling 100 miles per hour?

2 Solve inverse variation problems.

Inverse Variation

The distance from Atlanta, Georgia, to Orlando, Florida, is 450 miles. The time that it takes to drive from Atlanta to Orlando depends on the rate at which one drives and is given by

$$\text{Time} = \frac{450}{\text{Rate}}.$$

For example, if you average 45 miles per hour, the time for the drive is

$$\text{Time} = \frac{450}{45} = 10,$$

or 10 hours. If you ignore speed limits and average 75 miles per hour, the time for the drive is

$$\text{Time} = \frac{450}{75} = 6,$$

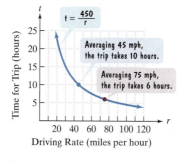

Figure 3.36

or 6 hours. As your rate (or speed) increases, the time for the trip decreases and vice versa. This is illustrated in Figure 3.36.

We can express the time for the Atlanta–Orlando trip using t for time and r for rate:

$$t = \frac{450}{r}.$$

This equation is an example of an *inverse variation* equation. Time, t, *varies inversely* as rate, r. When two quantities vary inversely, and the constant of variation is positive, such as 450, one quantity increases as the other decreases, and vice versa.

Generalizing, we obtain the following statement:

> **Inverse Variation**
>
> If a situation is described by an equation in the form
>
> $$y = \frac{k}{x}$$
>
> where k is a nonzero constant, we say that **y varies inversely as x** or **y is inversely proportional to x**. The number k is called the **constant of variation**.

Notice that the inverse variation equation

$$y = \frac{k}{x}, \quad \text{or} \quad f(x) = \frac{k}{x},$$

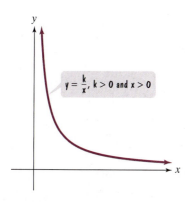

Figure 3.37 The graph of the inverse variation equation

is a rational function. For $k > 0$ and $x > 0$, the graph of the function takes on the shape shown in Figure 3.37.

We use the same procedure to solve inverse variation problems as we did to solve direct variation problems. Example 5 illustrates this procedure.

EXAMPLE 5   Solving an Inverse Variation Problem

When you use a spray can and press the valve at the top, you decrease the pressure of the gas in the can. This decrease of pressure causes the volume of the gas in the can to increase. Because the gas needs more room than is provided in the can, it expands in spray form through the small hole near the valve. In general, if the temperature is constant, the pressure, P, of a gas in a container varies inversely as the volume, V, of the container. The pressure of a gas sample in a container whose volume is 8 cubic inches is 12 pounds per square inch. If the sample expands to a volume of 22 cubic inches, what is the new pressure of the gas?

Doubling the pressure halves the volume.

Solution

Step 1   Write an equation. We know that y varies inversely as x is expressed as

$$y = \frac{k}{x}.$$

By changing letters, we can write an equation that describes the following English statement: The pressure, P, of a gas in a container varies inversely as the volume, V.

$$P = \frac{k}{V}$$

Equivalently, pressure is inversely proportional to volume.

Step 2   Use the given values to find k. The pressure of a gas sample in a container whose volume is 8 cubic inches is 12 pounds per square inch. Substitute 12 for P and 8 for V in the inverse variation equation. Then solve for k.

$$P = \frac{k}{V} \qquad \text{This is the inverse variation equation.}$$

$$12 = \frac{k}{8} \qquad P = 12 \text{ and } V = 8.$$

$$12 \cdot 8 = \frac{k}{8} \cdot 8 \qquad \text{Multiply both sides by 8.}$$

$$96 = k \qquad \text{Simplify.}$$

Step 3   Substitute the value of k into the equation.

$$P = \frac{k}{V} \qquad \text{Use the inverse variation equation from step 1.}$$

$$P = \frac{96}{V} \qquad \text{Replace } k, \text{ the constant of variation, with 96.}$$

Step 4   Answer the problem's question. We need to find the pressure when the volume expands to 22 cubic inches. Substitute 22 for V and solve for P.

$$P = \frac{96}{V} = \frac{96}{22} = 4\frac{4}{11}$$

Crude-Oil Price per Barrel

Source: U.S. Department of Commerce

3 Solve combined variation problems

When the volume is 22 cubic inches, the pressure of the gas is $4\frac{4}{11}$ pounds per square inch.

 Check Point 5 The price, P, of oil varies inversely as the supply, S. An OPEC nation sells oil for $19.50 per barrel when its daily production level is 4 million barrels. At what price will it sell oil if the daily production level is decreased to 3 million barrels?

Combined Variation

In a **combined variation** situation, direct and inverse variation occur at the same time. For example, as the advertising budget, A, of a company increases, its monthly sales, S, also increase. Monthly sales vary directly as the advertising budget:

$$S = kA.$$

By contrast, as the price of the company's product, P, increases, its monthly sales, S, decrease. Monthly sales vary inversely as the price of the product:

$$S = \frac{k}{P}.$$

We can combine these two variation equations into one combined equation:

$$S = \frac{kA}{P}.$$

The following example illustrates the application of combined variation.

EXAMPLE 6 Solving a Combined Variation Problem

The owners of Rollerblades Now determine that the monthly sales, S, of its skates vary directly as its advertising budget, A, and inversely as the price of the skates, P. When $60,000 is spent on advertising and the price of the skates is $40, the monthly sales are 12,000 pairs of rollerblades.

a. Write an equation of variation that describes this situation.

b. Determine monthly sales if the amount of the advertising budget is increased to $70,000.

Solution

a. Write an equation.

$$S = \frac{kA}{P}$$ Translate "sales vary directly as the advertising budget and inversely as the skates' price."

Use the given values to find k.

$$12{,}000 = \frac{k(60{,}000)}{40}$$ When $60,000 is spent on advertising ($A = 60,000$) and the price is $40 ($P = 40$), monthly sales are 12,000 units ($S = 12,000$).

$$12{,}000 = k \cdot 1500$$ Divide 60,000 by 40.

$$\frac{12{,}000}{1500} = \frac{k \cdot 1500}{1500}$$ Divide both sides of the equation by 1500.

$$8 = k$$ Simplify.

Using $k = 8$ and $S = \dfrac{kA}{P}$, the equation of variation that describes monthly sales is

$$S = \frac{8A}{P}.$$

b. The advertising budget is increased to $70,000, so $A = 70,000$. The skates' price is still $40, so $P = 40$.

$$S = \frac{8A}{P} \qquad \text{\color{blue}This is the combined variation equation from part (a).}$$

$$S = \frac{8(70,000)}{40} \qquad \text{\color{blue}Substitute 70,000 for A and 40 for P.}$$

$$S = 14,000 \qquad \text{\color{blue}Simplify.}$$

With a $70,000 advertising budget and $40 price, the company can expect to sell 14,000 pairs of rollerblades in a month (up from 12,000).

Check Point 6 The number of minutes needed to solve an exercise set of variation problems varies directly as the number of problems and inversely as the number of people working to solve the problems. It takes 4 people 32 minutes to solve 16 problems. How many minutes will it take 8 people to solve 24 problems?

4 Solve problems involving joint variation.

Joint Variation

Joint variation is a variation in which a variable varies directly as the product of two or more other variables. Thus, the equation $y = kxz$ is read "y varies jointly as x and z."

Joint variation plays a critical role in Isaac Newton's formula for gravitation:

$$F = G\frac{m_1 m_2}{d^2}.$$

The formula states that the force of gravitation, F, between two bodies varies jointly as the product of their masses, m_1 and m_2, and inversely as the square of the distance between them, d. (G is the gravitational constant.) The formula indicates that gravitational force exists between any two objects in the universe, increasing as the distance between the bodies decreases. One practical result is that the pull of the moon on the oceans is greater on the side of the Earth closer to the moon. This gravitational imbalance is what produces tides.

EXAMPLE 7 Modeling Centrifugal Force

The centrifugal force, C, of a body moving in a circle varies jointly with the radius of the circular path, r, and the body's mass, m, and inversely with the square of the time, t, it takes to move about one full circle. A 6-gram body moving in a circle with radius 100 centimeters at a rate of 1 revolution in 2 seconds has a centrifugal force of 6000 dynes. Find the centrifugal force of an 18-gram body moving in a circle with radius 100 centimeters at a rate of 1 revolution in 3 seconds.

Solution

$$C = \frac{krm}{t^2} \qquad \text{\color{blue}Translate "Centrifugal force, C, varies jointly with radius, r, and mass, m, and inversely with the square of time, t."}$$

$$6000 = \frac{k(100)(6)}{2^2}$$ If $r = 100$, $m = 6$, and $t = 2$, then $C = 6000$.

$$40 = k$$ Solve for k.

$$C = \frac{40rm}{t^2}$$ Substitute 40 for k in the model for centrifugal force.

$$= \frac{40(100)(18)}{3^2}$$ Find C when $r = 100$, $m = 18$, and $t = 3$.

$$= 8000$$

The centrifugal force is 8000 dynes.

Check Point 7 The volume of a cone, V, varies jointly as its height, h, and the square of its radius, r. A cone with a radius measuring 6 feet and a height measuring 10 feet has a volume of 120π cubic feet. Find the volume of a cone having a radius of 12 feet and a height of 2 feet.

EXERCISE SET 3.7

Practice Exercises

In Exercises 1–12, write an equation that expresses each relationship. Use k as the constant of variation.

1. g varies directly as h.

2. v varies directly as r.

3. a is directly proportional to the square of b.

4. s is directly proportional to the cube of v.

5. r varies inversely as t.

6. w varies inversely as l.

7. a is inversely proportional to the cube of b.

8. y is inversely proportional to the square root of x.

9. r varies directly as s and inversely as v.

10. a varies directly as d and inversely as g.

11. s varies jointly as g and the square of t.

12. V varies jointly as h and the square of r.

In Exercises 13–22, determine the constant of variation for each stated condition.

13. y varies directly as x, and $y = 75$ when $x = 3$.

14. y varies directly as x, and $y = 55$ when $x = 11$.

15. y varies directly as x^2, and $y = 45$ when $x = 3$.

16. y varies directly as x^2, and $y = 72$ when $x = 6$.

17. W varies inversely as r, and $W = 500$ when $r = 10$.

18. T varies inversely as n, and $T = 7$ when $n = 12$.

19. A varies directly as B and inversely as C, and $A = 9$ when $B = 12$ and $C = 4$.

20. D varies directly as E and inversely as F, and $D = 6$ when $E = 12$ and $F = 10$.

21. a varies jointly as b and c, and $a = 72$ when $b = 18$ and $c = 2$.

22. z varies jointly as w and y, and $z = 38$ when $w = 38$ and $y = 2$.

Use the four-step procedure for solving variation problems given on page 356 to solve Exercises 23–30.

23. y varies directly as x. $y = 35$ when $x = 5$. Find y when $x = 12$.

24. y varies directly as x. $y = 55$ when $x = 5$. Find y when $x = 13$.

25. y varies inversely as x. $y = 10$ when $x = 5$. Find y when $x = 2$.

26. y varies inversely as x. $y = 5$ when $x = 3$. Find y when $x = 9$.

27. y is directly proportional to x and inversely proportional to the square of z. $y = 20$ when $x = 50$ and $z = 5$. Find y when $x = 3$ and $z = 6$.

28. a is directly proportional to b and inversely proportional to the square of c. $a = 7$ when $b = 9$ and $c = 6$. Find a when $b = 4$ and $c = 8$.

29. y varies jointly as x and z. $y = 25$ when $x = 2$ and $z = 5$. Find y when $x = 8$ and $z = 12$.

30. C varies jointly as A and T. $C = 175$ when $A = 2100$ and $T = 4$. Find C when $A = 2400$ and $T = 6$.

Application Exercises

31. A person's fingernail growth, G, in inches, varies directly as the number of weeks it has been growing, W.

a. Write an equation that expresses this relationship.

b. Fingernails grow at a rate of about 0.02 inch per week. Substitute 0.02 for k, the constant of variation, in the equation in part (a) and write the equation for fingernail growth.

c. Substitute 52 for W to determine your fingernail length at the end of one year if for some bizarre reason you decided not to cut them and they did not break.

32. A person's wages, W, vary directly as the number of hours worked, h.

a. Write an equation that expresses this relationship.

b. For a 40-hour work week, Gloria earned $1400. Substitute 1400 for W and 40 for h in the equation from part (a) and find k, the constant of variation.

c. Substitute the value of k into your equation in part (a) and write the equation that describes Gloria's wages in terms of the number of hours she works.

d. Use the equation from part (c) to find Gloria's wages for 25 hours of work.

Use the four-step procedure for solving variation problems given on page 356 to solve Exercises 33–49.

33. The cost, C, of an airplane ticket varies directly as the number of miles, M, in the trip. A 3000-mile trip costs $400. What is the cost of a 450-mile trip?

34. An object's weight on the moon, M, varies directly as its weight on Earth, E. A person who weighs 55 kilograms on Earth weights 8.8 kilograms on the moon. What is the moon weight of a person who weighs 90 kilograms on Earth?

35. The Mach number is a measurement of speed named after the man who suggested it, Ernst Mach (1838–1916). The speed of an aircraft is directly proportional to its Mach number. Shown here are two aircraft. Use the figures for the Concorde to determine the Blackbird's speed.

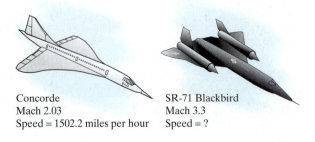

Concorde
Mach 2.03
Speed = 1502.2 miles per hour

SR-71 Blackbird
Mach 3.3
Speed = ?

36. Do you still own records, or are you strictly a CD person? Record owners claim that the quality of sound on good vinyl surpasses that of a CD, although this is up for debate. This, however, is not debatable: The number of revolutions a record makes as it is being played is directly proportional to the time that it is on the turntable. A record that lasted 3 minutes made 135 revolutions. If a record takes 2.4 minutes to play, how many revolutions does it make?

37. If all men had identical body types, their weight would vary directly as the cube of their height. Shown is Robert Wadlow, who reached a record height of 8 feet 11 inches (107 inches) before his death at age 22. If a man who is 5 feet 10 inches tall (70 inches) with the same body type as Mr. Wadlow weighs 170 pounds, what was Robert Wadlow's weight shortly before his death?

38. The distance that an object falls varies directly as the square of the time it has been falling. An object falls 144 feet in 3 seconds. Find how far it will fall in 7 seconds.

39. The time that it takes you to get to campus varies inversely as your driving rate. Averaging 20 miles per hour in terrible traffic, it takes you 1.5 hours to get to campus. How long would the trip take averaging 60 miles per hour?

40. The weight that can be supported by a 2-inch by 4-inch piece of pine (called a 2-by-4) varies inversely as its length. A 10-foot 2-by-4 can support 500 pounds. What weight can be supported by a 5-foot 2-by-4?

41. The volume of a gas in a container at a constant temperature is inversely proportional to the pressure. If the volume is 32 cubic centimeters at a pressure of 8 pounds, find the pressure when the volume is 40 cubic centimeters.

42. The current in a circuit is inversely proportional to the resistance. The current is 20 amperes when the resistance is 5 ohms. Find the current for a resistance of 16 ohms.

43. A person's body-mass index is used to assess levels of fatness, with an index from 20 to 26 considered in the desirable range. The index varies directly as one's weight, in pounds, and inversely as one's height, in inches. A person who weighs 150 pounds and is 70 inches tall has an index of 21. What is the body-mass index of a person who weighs 240 pounds and is 74 inches tall? Because the index is rounded to the nearest whole number, do so and then determine if this person's level of fatness is in the desirable range.

44. The volume of a gas varies directly as its temperature and inversely as its pressure. At a temperature of 100 Kelvin and a pressure of 15 kilograms per square meter, the gas occupies a volume of 20 cubic meters. Find the volume at a temperature of 150 Kelvin and a pressure of 30 kilograms per square meter.

45. The intensity of illumination on a surface varies inversely as the square of the distance of the light source from the surface. The illumination from a source is 25 foot-candles at a distance of 4 feet. What is the illumination when the distance is 6 feet?

46. The gravitational force with which Earth attracts an object varies inversely with the square of the distance from the center of Earth. A gravitational force of 0.4 pound acts on an object 8000 miles from Earth's center. Find the force of attraction on an object 6000 miles from the center of Earth.

47. Kinetic energy varies jointly as the mass and the square of the velocity. A mass of 8 grams and velocity of 3 centimeters per second has a kinetic energy of 36 ergs. Find the kinetic energy for a mass of 4 grams and velocity of 6 centimeters per second.

48. The electrical resistance of a wire varies directly as its length and inversely as the square of its diameter. A wire of 720 feet with $\frac{1}{4}$-inch diameter has a resistance of $1\frac{1}{2}$ ohms. Find the resistance for 960 feet of the same kind of wire if its diameter is doubled.

49. The average number of phone calls between two cities in a day varies jointly as the product of their populations and inversely as the square of the distance between them. The population of Minneapolis is 2538 thousand and the population of Cincinnati is 1818 thousand. Separated by 108 miles, the average number of telephone calls per day between the two cities is 158,233. Find the average number of telephone calls per day between Orlando, Florida (population 1225 thousand) and Seattle, Washington (population 2970 thousand), two cities that are 3403 miles apart.

Writing in Mathematics

50. What does it mean if two quantities vary directly?

51. In your own words, explain how to solve a variation problem.

52. What does it mean if two quantities vary inversely?

53. Explain what is meant by combined variation. Give an example with your explanation.

54. Explain what is meant by joint variation. Give an example with your explanation.

In Exercises 55–56, describe in words the variation shown by the given equation.

55. $z = \dfrac{k\sqrt{x}}{y^2}$ **56.** $z = kx^2\sqrt{y}$

57. We have seen that the daily number of phone calls between two cities varies jointly as their populations and inversely as the square of the distance between them. This model, used by telecommunication companies to estimate the line capacities needed between various cities, is called the *gravity model*. Compare the model to Newton's formula for gravitation on page 362 and describe why the name *gravity model* is appropriate.

Technology Exercise

58. Use a graphing utility to graph any three of the variation equations in Exercises 33–42. Then TRACE along each curve and identify the point that corresponds to the problem's solution.

Critical Thinking Exercises

59. In a hurricane, the wind pressure varies directly as the square of the wind velocity. If wind pressure is a measure of a hurricane's destructive capacity, what happens to this destructive power when the wind speed doubles?

60. The illumination from a light source varies inversely as the square of the distance from the light source. If you raise a lamp from 15 inches to 30 inches over your desk, what happens to the illumination?

61. The heat generated by a stove element varies directly as the square of the voltage and inversely as the resistance. If the voltage remains constant, what needs to be done to triple the amount of heat generated?

62. Galileo's telescope brought about revolutionary changes in astronomy. A comparable leap in our ability to observe the universe took place as a result of the Hubble Space Telescope. The space telescope can see stars and galaxies whose brightness is $\frac{1}{50}$ of the faintest objects now observable using ground-based telescopes. Use the fact that the brightness of a point source, such as a star, varies inversely as the square of its distance from an observer to show that the space telescope can see about seven times farther than a ground-based telescope.

Group Exercise

63. Begin by deciding on a product that interests the group because you are now in charge of advertising this product. Members were told that the demand for the product varies directly as the amount spent on advertising and inversely as the price of the product. However, as more money is spent on advertising, the price of your product rises. Under what conditions would members recommend an increased expense in advertising? Once you've determined what your product is, write formulas for the given conditions and experiment with hypothetical numbers. What other factors might you take into consideration in terms of your recommendation? How do these factor affect the demand for your product?

CHAPTER SUMMARY, REVIEW, AND TEST

Summary

DEFINITIONS AND CONCEPTS	EXAMPLES

3.1 Quadratic Functions

a. A quadratic function is of the form $f(x) = ax^2 + bx + c, a \neq 0$.

b. The standard form of a quadratic function is $f(x) = a(x - h)^2 + k, a \neq 0$.

c. The graph of a quadratic function is a parabola. The vertex is (h, k) or $\left(-\dfrac{b}{2a}, f\left(-\dfrac{b}{2a}\right)\right)$.

    A procedure for graphing a quadratic function is given in the box on page 282.

d. See the box on page 286 for minimum or maximum values of quadratic functions.

    Ex. 1, p. 282;
Ex. 2, p. 283;
Ex. 3, p. 285
Ex. 4, p. 287

3.2 Polynomial Functions and Their Graphs

a. Polynomial Function of x of Degree n: $f(x) = a_n x^n + a_{n-1} x^{n-1} + \cdots + a_2 x^2 + a_1 x + a_0$, $a_n \neq 0$

b. The graphs of polynomial functions are smooth and continuous.

c. The end behavior of the graph of a polynomial function depends on the leading term, given by the Leading Coefficient Test in the box on page 295.

d. The values of x for which $f(x)$ is equal to 0 are the zeros of the polynomial function f. These values are the roots, or solutions, of the polynomial equation $f(x) = 0$.

e. If $x - r$ occurs k times in a polynomial function's factorization, r is a repeated zero with multiplicity k. If k is even, the graph touches the x-axis at r; if odd, it crosses the x-axis at r.

f. If f is a polynomial of degree n, the graph of f has at most $n - 1$ turning points.

g. A strategy for graphing a polynomial function is given in the box on page 299.

    Ex. 1&2, p. 295;
Ex. 3, p. 296
Ex. 4&5, p. 297

    Ex. 6, p. 299

3.3 Dividing Polynomials; Remainder and Factor Theorems

a. Long division of polynomials is performed by dividing, multiplying, subtracting, bringing down the next term, and repeating this process until the degree of the remainder is less than the degree of the divisor. The details are given in the box on page 306.

b. The Division Algorithm: $f(x) = d(x)q(x) + r(x)$. The dividend is the product of the divisor and the quotient plus the remainder.

c. Synthetic division is used to divide a polynomial by $x - c$. The details are given in the box on pages 309–310.

d. The Remainder Theorem: If polynomial $f(x)$ is divided by $x - c$, then the remainder is $f(c)$.

e. The Factor Theorem: If $x - c$ is a factor of a polynomial function $f(x)$, c is a zero of f and a root of $f(x) = 0$. If c is a zero of f or a root of $f(x) = 0$, then $x - c$ is a factor of $f(x)$.

    Ex. 1, p. 305;
Ex. 2, p. 306;
Ex. 3, p. 308

    Ex. 4, p. 310

    Ex. 5, p. 312

    Ex. 6, p. 313

DEFINITIONS AND CONCEPTS	EXAMPLES

3.4 Zeros of Polynomial Functions

a. The Rational Zero Theorem states that possible rational zeros of a polynomial function =

$$\frac{\text{Factors of the constant term}}{\text{Factors of the leading coefficient}}.$$ The theorem is stated in the box on page 316.

Ex. 1, p. 316;
Ex. 2, p. 317;
Ex. 3, p. 318;
Ex. 4, p. 319

b. Descartes's Rule of Signs: The number of positive real zeros of f equals the number of sign changes of $f(x)$ or is less than that number by an even integer. The number of negative real zeros of f applies a similar statement to $f(-x)$.

Ex. 5, p. 321

3.5 More on Zeros of Polynomial Functions

a. The Upper and Lower Bound Theorem: The number $b > 0$ is an upper bound for the real roots of $f(x) = 0$ if synthetic division of $f(x)$ by $x - b$ results in no negative numbers. The $a < 0$ is a lower bound if synthetic division by $x - a$ results in numbers that alternate in sign, number counting zero entries as positive or negative.

Ex. 1, p. 326

b. The Intermediate Value Theorem: If $f(a)$ and $f(b)$ have opposite signs, there is at least one value of c between a and b for which $f(c) = 0$.

Ex. 2, p. 327

c. Number of roots: If $f(x)$ is a polynomial of degree $n \geq 1$, then, counting multiple roots separately, the equation $f(x) = 0$ has n roots.

d. If $a + bi$ is a root of $f(x) = 0$, then $a - bi$ is also a root.

Ex. 3, p. 329

e. The Linear Factorization Theorem: An nth-degree polynomial can be expressed as the product of n linear factors. Thus, $f(x) = a_n (x - c_1)(x - c_2)\cdots(x - c_n)$.

Ex. 4, p. 331;
Ex. 5, p. 331

3.6 Rational Functions and Their Graphs

a. Rational function: $f(x) = \dfrac{p(x)}{q(x)}$; p and q are polynomial functions and $q(x) \neq 0$. The domain of f is the set of all real numbers excluding values of x that make $q(x)$ zero.

Ex. 1, p. 336

b. Arrow notation is summarized in the box on page 338.

c. The line $x = a$ is a vertical asymptote of the graph of f if $f(x)$ increases or decreases without bound as x approaches a. Vertical asymptotes are identified using the location theorem in the box on page 339.

Ex. 2, p. 339

d. The line $y = b$ is a horizontal asymptote of the graph of f if $f(x)$ approaches b as x increases or decreases without bound. Horizontal asymptotes are identified using the location theorem in the box on page 341.

Ex. 3, p. 341

e. A strategy for graphing rational functions is given in the box on page 342.

Ex. 4, p. 342;
Ex. 5, p. 344;
Ex. 6, p. 345

f. The graph of a rational function has a slant asymptote when the degree of the numerator is one more than the degree of the denominator. The equation of the slant asymptote is found using division and dropping the remainder term.

Ex. 7, p. 347

3.7 Modeling Using Variation

a.

English Statement	Equation	
y varies directly as x.	$y = kx$	Ex. 3, p. 356
y is directly proportional to x.		
y varies directly as x^n.	$y = kx^n$	Ex. 4, p. 358
y is directly proportional to x^n.		

DEFINITIONS AND CONCEPTS		EXAMPLES
y varies inversely as x. y is inversely proportional to x.	$y = \dfrac{k}{x}$	Ex. 5, p. 360; Ex. 6, p. 361
y varies inversely as x^n. y is inversely proportional to x^n.	$y = \dfrac{k}{x^n}$	
y varies jointly as x and z.	$y = kxz$	Ex. 7, p. 362

b. A procedure for solving variation problems is given in the box on page 356.

Review Exercises

3.1

In Exercises 1–4, use the vertex and intercepts to sketch the graph of each quadratic function. Give the equation for the parabola's axis of symmetry.

1. $f(x) = -2(x - 1)^2 + 3$ **2.** $f(x) = (x + 4)^2 - 2$

3. $f(x) = -x^2 + 2x + 3$ **4.** $f(x) = 2x^2 - 4x - 6$

5. A person standing close to the edge on the top of an 80-foot building throws a ball vertically upward with an initial velocity of 64 feet per second. The function $s(t) = -16t^2 + 64t + 80$ describes the ball's height above the ground, $s(t)$, in feet, t seconds after it is thrown. After how many seconds does the ball reach its maximum height? What is the maximum height?

6. Suppose that a quadratic function is used to model the data shown in the graph using

(number of years after 1960, divorce rate per 1000 population).

U.S. Divorce Rate

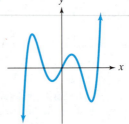

Source: U.S. Department of Health and Human Services

Determine, without obtaining an actual quadratic function that models the data, the approximate coordinates of the vertex for the function's graph. Describe what this means in practical terms. Use the word "maximum" in your description.

7. A field bordering a straight stream is to be enclosed. The side bordering the stream is not to be fenced. If 1000 yards of fencing material is to be used, what are the dimensions of the largest rectangular field that can be fenced? What is the maximum area?

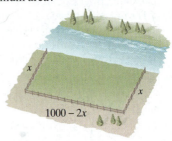

3.2

In Exercises 8–11, use the Leading Coefficient Test to determine the end behavior of the graph of the given polynomial function. Then use this end behavior to match the polynomial function with its graph. [The graphs are labeled (a) through (d).]

8. $f(x) = -x^3 + 12x^2 - x$

9. $f(x) = x^6 - 6x^4 + 9x^2$

10. $f(x) = x^5 - 5x^3 + 4x$

11. $f(x) = -x^4 + 1$

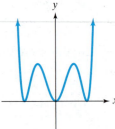

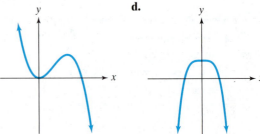

12. The function $f(x) = -0.0013x^3 + 0.78x^2 - 1.43x + 18.1$ models the percentage of U.S. families below the poverty level, $f(x)$, x years after 1960. Use end behavior to explain why the model is valid only for a limited period of time.

13. Despite a combination of drugs used to inhibit the growth of the HIV virus, a patient dies as a result of the virus overwhelming his body. Could the function $N(t) = -\frac{3}{4}t^4 + 3t^3 + 5$ model the number of viral particles, $N(t)$, in billions, in this patient's body over time? Use the graph's end behavior to the right to answer the question. Explain your answer.

In Exercises 14–15, find the zeros for each polynomial function and give the multiplicity of each zero. State whether the graph crosses or touches the x-axis at each zero.

14. $f(x) = -2(x - 1)(x + 2)^2(x + 5)^3$

15. $f(x) = x^3 - 5x^2 - 25x + 125$

In Exercises 16–21,
   a. *Use the Leading Coefficient Test to determine the graph's end behavior.*
   b. *Determine whether the graph has y-axis symmetry, origin symmetry, or neither.*
   c. *Graph the function.*

16. $f(x) = x^3 - x^2 - 9x + 9$ **17.** $f(x) = 4x - x^3$
18. $f(x) = 2x^3 + 3x^2 - 8x - 12$ **19.** $f(x) = -x^4 + 25x^2$
20. $f(x) = -x^4 + 6x^3 - 9x^2$ **21.** $f(x) = 3x^4 - 15x^3$

3.3

In Exercises 22–24, divide using long division.

22. $\left(4x^3 - 3x^2 - 2x + 1\right) \div (x + 1)$
23. $\left(10x^3 - 26x^2 + 17x - 13\right) \div (5x - 3)$
24. $\left(4x^4 + 6x^3 + 3x - 1\right) \div \left(2x^2 + 1\right)$

In Exercises 25–26, divide using synthetic division.

25. $\left(3x^4 + 11x^3 - 20x^2 + 7x + 35\right) \div (x + 5)$
26. $\left(3x^4 - 2x^2 - 10x\right) \div (x - 2)$
27. Given $f(x) = 2x^3 - 7x^2 + 9x - 3$, use the Remainder Theorem to find $f(-13)$.
28. Use synthetic division to divide $f(x) = 2x^3 + x^2 - 13x + 6$ by $x - 2$. Use the result to find all zeros of f.
29. Solve the equation $x^3 - 17x + 4 = 0$ given that 4 is a root.

3.4

In Exercises 30–31, use the Rational Zero Theorem to list all possible rational zeros for each given function.

30. $f(x) = x^4 - 6x^3 + 14x^2 - 14x + 5$
31. $f(x) = 3x^5 - 2x^4 - 15x^3 + 10x^2 + 12x - 8$

In Exercises 32–33, use Descartes's Rule of Signs to determine the possible number of positive and negative real zeros for each given function.

32. $f(x) = 3x^4 - 2x^3 - 8x + 5$

33. $f(x) = 2x^5 - 3x^3 - 5x^2 + 3x - 1$

34. Use Descartes's Rule of Signs to explain why $2x^4 + 6x^2 + 8 = 0$ has no real roots.

For Exercises 35–40,
   a. *List all possible rational roots or rational zeros.*
   b. *Use Descartes's Rule of Signs to determine the possible number of positive and negative real roots or real zeros.*
   c. *Use synthetic division to test the possible rational roots or zeros and find an actual root or zero.*
   d. *Use the root or zero from part (c) to find all the zeros or roots.*

35. $f(x) = x^3 + 3x^2 - 4$
36. $f(x) = 6x^3 + x^2 - 4x + 1$
37. $8x^3 - 36x^2 + 46x - 15 = 0$
38. $x^4 - x^3 - 7x^2 + x + 6 = 0$
39. $4x^4 + 7x^2 - 2 = 0$
40. $f(x) = 2x^4 + x^3 - 9x^2 - 4x + 4$

3.5

41. Show that all real roots of the equation
$$2x^4 - 7x^3 - 5x^2 + 28x - 12 = 0$$
lie between -2 and 6. Use this result to list all possible rational roots.

42. Consider the equation $2x^4 - x^3 - 5x^2 + 10x + 12 = 0$.
   a. List all possible rational roots.
   b. Determine whether 2 is a root using synthetic division. In terms of bounds, what can you conclude?
   c. Determine whether -2 is a root using synthetic division. In terms of bounds, what can you conclude?
   d. Use the results of parts (b) and (c) to discard some of the possible rational roots from part (a). Now what are the possible rational roots?

In Exercises 43–44, show that the polynomial has a zero between the given integers. Then use the Intermediate Value Theorem to find an approximation for this zero to the nearest tenth. If applicable, use a graphing utility's zero feature to verify your answer.

43. $f(x) = x^3 - 2x - 1$; between 1 and 2
44. $f(x) = 3x^3 + 2x^2 - 8x + 7$; between -3 and -2

In Exercises 45–47, use the given root to find the solution set of the polynomial equation.

45. $4x^3 - 47x^2 + 232x + 61 = 0$; $6 + 5i$
46. $x^4 - 4x^3 + 16x^2 - 24x + 20 = 0$; $1 - 3i$
47. $2x^4 - 17x^3 + 137x^2 - 57x - 65 = 0$; $4 + 7i$

In Exercises 48–50, find an nth-degree polynomial function with real coefficients satisfying the given conditions. If you are using a graphing utility, graph the function and verify the real zeros and the given function value.

48. $n = 3$; 2 and $2 - 3i$ are zeros; $f(1) = -10$

49. $n = 4$; i is a zero; -3 is a zero of multiplicity 2; $f(-1) = 16$

50. $n = 4$; $-2, 3$, and $1 + 3i$ are zeros; $f(2) = -40$

In Exercises 51–52, find all the zeros of each polynomial function and write the polynomial as a product of linear factors.

51. $f(x) = 2x^4 + 3x^3 + 3x - 2$

52. $g(x) = x^4 - 6x^3 + x^2 + 24x + 16$

In Exercises 53–56, graphs of fifth-degree polynomial functions are shown. In each case, specify the number of real zeros and the number of imaginary zeros. Indicate whether there are any real zeros with multiplicity other than 1.

53.
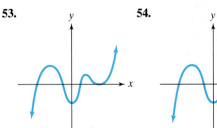

54.
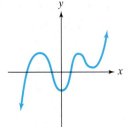

55.   **56.**
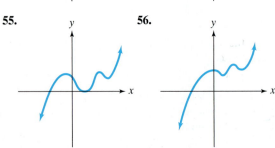

3.6

In Exercises 57–64, find the vertical asymptotes, if any, the horizontal asymptote, if there is one, and the slant asymptote, if there is one, of the graph of each rational function. Then graph the rational function.

57. $f(x) = \dfrac{2x}{x^2 - 9}$

58. $g(x) = \dfrac{2x - 4}{x + 3}$

59. $h(x) = \dfrac{x^2 - 3x - 4}{x^2 - x - 6}$

60. $r(x) = \dfrac{x^2 + 4x + 3}{(x + 2)^2}$

61. $y = \dfrac{x^2}{x + 1}$

62. $y = \dfrac{x^2 + 2x - 3}{x - 3}$

63. $f(x) = \dfrac{-2x^3}{x^2 + 1}$

64. $g(x) = \dfrac{4x^2 - 16x + 16}{2x - 3}$

65. A company is planning to manufacture affordable graphing calculators. Fixed monthly cost will be $50,000, and it will cost $25 to produce each calculator.

    a. Write the cost function, C, of producing x graphing calculators.

    b. Write the average cost function, $\bar{C}$, of producing x graphing calculators.

    c. Find and interpret $\bar{C}(50)$, $\bar{C}(100)$, $\bar{C}(1000)$, and $\bar{C}(100,000)$.

d. What is the horizontal asymptote for this function, and what does it represent?

66. In Palo Alto, California, a government agency ordered computer-related companies to contribute to a monetary pool to clean up underground water supplies. (The companies had stored toxic chemicals in leaking underground containers.) The rational function

$$C(x) = \frac{200x}{100 - x}$$

models the cost, $C(x)$, in tens of thousands of dollars, for removing x percent of the contaminant's.

    a. Find and interpret $C(90) - C(50)$.

    b. What is the equation for the vertical asymptote?

    What does this mean in terms of the variables given by the function?

Exercises 67–68 involve rational functions that model the given situations. In each case, find the horizontal asymptote as $x \to \infty$ and then describe what this means in practical terms.

67. $f(x) = \dfrac{150x + 120}{0.05x + 1}$; the number of bass, $f(x)$, after x months in a lake that was stocked with 120 bass

68. $P(x) = \dfrac{72,900}{100x^2 + 729}$; the percentage, $P(x)$, of people in the United States with x years of education who are unemployed

69. The function $p(x) = 1.96x + 3.14$ models the number of nonviolent prisoners, $p(x)$, in thousands, in New York State prisons x years after 1980. The function $q(x) = 3.04x + 21.79$ models the total number of prisoners, $q(x)$, in thousands, in New York State prisons x years after 1980.

    a. Write a function that models the fraction of nonviolent prisoners in New York prisons x years after 1980.

    b. What is the equation of the horizontal asymptote associated with the function in part (a)? Describe what this means about the percentage, to the nearest tenth of a percent, of nonviolent prisoners in New York prisons over time.

    c. Use your equation in part (b) to explain why, in 1998, New York implemented a strategy where more nonviolent offenders are granted parole and more violent offenders are denied parole.

3.7

Solve the variation problems in Exercises 70–75.

70. An electric bill varies directly as the amount of electricity used. The bill for 1400 kilowatts of electricity is $98. What is the bill for 2200 kilowatts of electricity?

71. The distance that a body falls from rest is directly proportional to the square of the time of the fall. If skydivers fall 144 feet in 3 seconds, how far will they fall in 10 seconds?

72. The time it takes to drive a certain distance is inversely proportional to the rate of travel. If it takes 4 hours at 50 miles per hour to drive the distance, how long will it take at 40 miles per hour?

73. The loudness of a stereo speaker, measured in decibels, varies inversely as the square of your distance from the speaker. When you are 8 feet from the speaker, the loudness is 28 decibels. What is the loudness when you are 4 feet from the speaker?

74. The time required to assemble computers varies directly as the number of computers assembled and inversely as the number of workers. If 30 computers can be assembled by 6 workers in 10 hours, how long would it take 5 workers to assemble 40 computers?

75. The volume of a pyramid varies jointly as its height and the area of its base. A pyramid with a height of 15 feet and a base with an area of 35 square feet has a volume of 175 cubic feet. Find the volume of a pyramid with a height of 20 feet and a base with an area of 120 square feet.

Chapter 3 Test

In Exercises 1–2, use the vertex and intercepts to sketch the graph of each quadratic function. Give the equation for the parabola's axis of symmetry.

1. $f(x) = (x + 1)^2 + 4$ **2.** $f(x) = x^2 - 2x - 3$

3. Determine, without graphing, whether the quadratic function $f(x) = -2x^2 + 12x - 16$ has a minimum value or a maximum value. Then find the coordinates of the minimum or the maximum point.

4. The function $f(x) = -x^2 + 46x - 360$ models the daily profit, $f(x)$, in hundreds of dollars, for a company that manufactures x computers daily. How many computers should be manufactured each day to maximize profit? What is the maximum daily profit?

5. Consider the function $f(x) = x^3 - 5x^2 - 4x + 20$.

   a. Use factoring to find all zeros of f.

   b. Use the Leading Coefficient Test and the zeros of f to graph the function.

6. Use end behavior to explain why the graph cannot be the graph of $f(x) = x^5 - x$. Then use intercepts to explain why the graph cannot represent $f(x) = x^5 - x$.

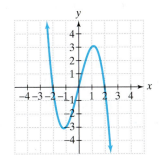

7. The graph of $f(x) = 6x^3 - 19x^2 + 16x - 4$ is shown in the figure at the top of the next column.

   a. Based on the graph of f, find the root of the equation $6x^3 - 19x^2 + 16x - 4 = 0$ that is an integer.

   b. Use synthetic division to find the other two roots of $6x^3 - 19x^2 + 16x - 4 = 0$.

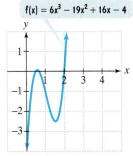

$f(x) = 6x^3 - 19x^2 + 16x - 4$

8. Use the Rational Zero Theorem to list all possible rational zeros of $f(x) = 2x^3 + 11x^2 - 7x - 6$.

9. Use Descartes's Rule of Signs to determine the possible number of positive and negative real zeros of $f(x) = 3x^5 - 2x^4 - 2x^2 + x - 1$.

10. Solve: $x^3 + 6x^2 - x - 30 = 0$.

11. Consider the function whose equation is given by $f(x) = 2x^4 - x^3 - 13x^2 + 5x + 15$.

   a. List all possible rational zeros.

   b. Use the graph of f in the figure shown and synthetic division to find all zeros of the function.

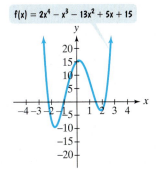

$f(x) = 2x^4 - x^3 - 13x^2 + 5x + 15$

12. Use the graph of $f(x) = 3x^4 + 4x^3 - 7x^2 - 2x - 3$ in the figure shown on the next page to find the smallest positive integer that is an upper bound and the largest negative integer that is a lower bound for the real roots of

$$3x^4 + 4x^3 - 7x^2 - 2x - 3 = 0.$$

Then use synthetic division to show that all the real roots of the equation lie between these integers.

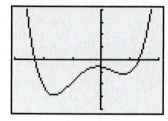

$[-3, 2, 1]$ by $[-4, 4, 1]$

13. Solve $x^4 - 7x^3 + 18x^2 - 22x + 12 = 0$ given that $1 - i$ is a root.

14. Use the graph of $f(x) = x^3 + 3x^2 - 4$ in the figure shown to factor $x^3 + 3x^2 - 4$.

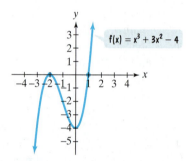

$f(x) = x^3 + 3x^2 - 4$

In Exercises 15–18, find the domain of each rational function and graph the function.

15. $f(x) = \dfrac{x}{x^2 - 16}$

16. $f(x) = \dfrac{x^2 - 9}{x - 2}$

17. $f(x) = \dfrac{x + 1}{x^2 + 2x - 3}$

18. $f(x) = \dfrac{4x^2}{x^2 + 3}$

19. Rational functions can be used to model learning. Many of these functions model the proportion of correct responses as a function of the number of trials of a particular task. One such model, called a learning curve, is

$$f(x) = \frac{0.9x - 0.4}{0.9x + 0.1},$$

where $f(x)$ is the proportion of correct responses after x trials. If $f(x) = 0$, there are no correct responses. If $f(x) = 1$, all responses are correct. The graph of the rational function is shown.

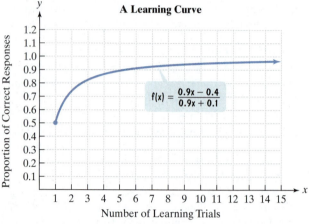

A Learning Curve

$f(x) = \dfrac{0.9x - 0.4}{0.9x + 0.1}$

Proportion of Correct Responses

Number of Learning Trials

a. According to the graph, what proportion of responses are correct after 5 learning trials?

b. According to the graph, how many learning trials are necessary for 0.95 of the responses to be correct?

c. Use the function's equation to write the equation of the horizontal asymptote. What does this mean in terms of the variables modeled by the learning curve?

20. The intensity of light received at a source varies inversely as the square of the distance from the source. A particular light has an intensity of 20 foot-candles at 15 feet. What is the light's intensity at 10 feet?

Cumulative Review Exercises (Chapters P–3)

Simplify each expression in Exercises 1–3.

1. $\dfrac{1}{2 - \sqrt{3}}$

2. $3(x^2 - 3x + 1) - 2(3x^2 + x - 4)$

3. $3\sqrt{8} + 5\sqrt{50} - 4\sqrt{32}$

4. Factor completely: $x^7 - x^5$.

Solve each equation in Exercises 5–8.

5. $|2x - 1| = 3$

6. $3x^2 - 5x + 1 = 0$

7. $9 + \dfrac{3}{x} = \dfrac{2}{x^2}$

8. $x^3 + 2x^2 - 5x - 6 = 0$

Solve each inequality in Exercises 9–10. Express the answer in interval notation.

9. $|2x - 5| > 3$

10. $3x^2 > 2x + 5$

11. Give the center and radius. Then graph the equation:
$x^2 + y^2 - 2x + 4y - 4 = 0$.

12. Solve for t: $V = C(1 - t)$.

13. If $f(x) = \sqrt{45 - 9x}$, find the domain of f.

If $f(x) = x^2 + 2x - 5$ and $g(x) = 4x - 1$, find each function or function value in Exercises 14–16.

14. $(f - g)(x)$

15. $(f \circ g)(x)$

16. $g(f(-3))$

17. Consider the function $f(x) = x^3 - 4x^2 - x + 4$.
  a. Use factoring to find all zeros of f.
  b. Use the Leading Coefficient Test and the zeros of f to graph the function.

Graph each function in Exercises 18–20.

18. $f(x) = x^2 + 2x - 8$

19. $f(x) = x^2(x - 3)$

20. $f(x) = \dfrac{x - 1}{x - 2}$

Exponential and Logarithmic Functions

Chapter 4

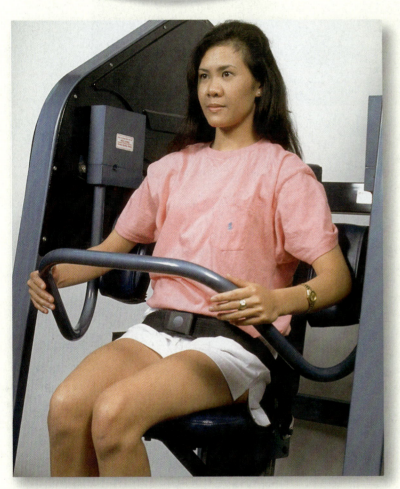

What went wrong on the space shuttle *Challenger*? Will population growth lead to a future without comfort or individual choice? Can I put aside a small amount of money and have millions for early retirement? Why did I feel I was walking too slowly on my visit to New York City? Why are people in California at far greater risk from drunk drivers than from earthquakes? What is the difference between earthquakes measuring 6 and 7 on the Richter scale? And what can I hope to accomplish in weightlifting?

The functions that you will be learning about in this chapter will provide you with the mathematics for answering these questions. You will see how these remarkable functions enable us to predict the future and rediscover the past.

You've recently taken up weightlifting, recording the maximum number of pounds you can lift at the end of each week. At first your weight limit increases rapidly, but now you notice that this growth is beginning to level off. You wonder about a function that would serve as a mathematical model to predict the number of pounds you can lift as you continue the sport.

SECTION 4.1   Exponential Functions

Objectives

1. Evaluate exponential functions.
2. Graph exponential functions.
3. Evaluate functions with base e.
4. Use compound interest formulas.

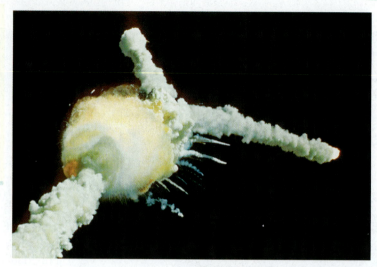

The space shuttle *Challenger* exploded approximately 73 seconds into flight on January 28, 1986. The tragedy involved damage to O-rings, which were used to seal the connections between different sections of the shuttle engines. The number of O-rings damaged increases dramatically as temperature falls.

The function

$$f(x) = 13.49(0.967)^x - 1$$

models the number of O-rings expected to fail when the temperature is $x°$F. Can you see how this function is different from polynomial functions? The variable x is in the exponent. Functions whose equations contain a variable in the exponent are called **exponential functions.** Many real-life situations, including population growth, growth of epidemics, radioactive decay, and other changes that involve rapid increase or decrease, can be described using exponential functions.

> **Definition of the Exponential Function**
>
> The **exponential function f with base b** is defined by
> $$f(x) = b^x \quad \text{or} \quad y = b^x$$
> where b is a positive constant other than 1 ($b > 0$ and $b \neq 1$) and x is any real number.

Here are some examples of exponential functions:

$$f(x) = 2^x \qquad g(x) = 10^x \qquad h(x) = 3^{x+1}.$$

Base is 2.   Base is 10.   Base is 3.

Each of these functions has a constant base and a variable exponent. By contrast, the following functions are not exponential:

$$F(x) = x^2 \qquad G(x) = 1^x \qquad H(x) = x^x.$$

Variable is the base and not the exponent.   The base of an exponential function must be a positive constant other than 1.   Variable is both the base and the exponent.

Why is $G(x) = 1^x$ not classified as an exponential function? The number 1 raised to any power is 1. Thus, the function G can be written as $G(x) = 1$, which is a constant function.

1 Evaluate exponential functions.

You will need a calculator to evaluate exponential expressions. Most scientific calculators have a $\boxed{y^x}$ key. Graphing calculators have a $\boxed{\wedge}$ key. To evaluate expressions of the form b^x, enter the base b, press $\boxed{y^x}$ or $\boxed{\wedge}$, enter the exponent x, and finally press $\boxed{=}$ or $\boxed{\text{ENTER}}$.

EXAMPLE 1 Evaluating an Exponential Function

The exponential function $f(x) = 13.49(0.967)^x - 1$ describes the number of O-rings expected to fail, $f(x)$, when the temperature is $x°$F. On the morning the *Challenger* was launched, the temperature was 31°F, colder than any previous experience. Find the number of O-rings expected to fail at this temperature.

Solution Because the temperature was 31°F, substitute 31 for x and evaluate the function.

$$f(x) = 13.49(0.967)^x - 1 \quad \text{This is the given function.}$$
$$f(31) = 13.49(0.967)^{31} - 1 \quad \text{Substitute 31 for x.}$$

Use a scientific or graphing calculator to evaluate $f(31)$. Press the following keys on your calculator to do this:

Scientific calculator: 13.49 $\boxed{\times}$ .967 $\boxed{y^x}$ 31 $\boxed{-}$ 1 $\boxed{=}$

Graphing calculator: 13.49 $\boxed{\times}$ .967 $\boxed{\wedge}$ 31 $\boxed{-}$ 1 $\boxed{\text{ENTER}}$.

The display should be approximately 3.7668627.

$$f(31) = 13.49(0.967)^{31} - 1 \approx 3.8 \approx 4$$

Thus, four O-rings are expected to fail at a temperature of 31°F.

Check Point 1 Use the function in Example 1 to find the number of O-rings expected to fail at a temperature of 60°F. Round to the nearest whole number.

2 Graph exponential functions.

Graphing Exponential Functions

We are familiar with expressions involving b^x, where x is a rational number. For example,

$$b^{1.7} = b^{17/10} = \sqrt[10]{b^{17}} \quad \text{and} \quad b^{1.73} = b^{173/100} = \sqrt[100]{b^{173}}.$$

However, note that the definition of $f(x) = b^x$ includes all real numbers for the domain x. You may wonder what b^x means when x is an irrational number, such as $b^{\sqrt{3}}$ or b^π. Using closer and closer approximations for $\sqrt{3}$ ($\sqrt{3} \approx 1.73205$), we can think of $b^{\sqrt{3}}$ as the value that has the successively closer approximations

$$b^{1.7}, b^{1.73}, b^{1.732}, b^{1.73205}, \ldots.$$

In this way, we can graph the exponential function with no holes, or points of discontinuity, at the irrational domain values.

EXAMPLE 2   Graphing an Exponential Function

Graph: $f(x) = 2^x$.

Solution   We begin by setting up a table of coordinates.

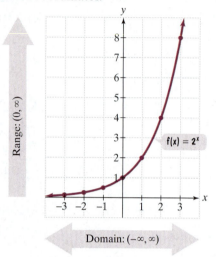

x	$f(x) = 2^x$
-3	$f(-3) = 2^{-3} = \frac{1}{8}$
-2	$f(-2) = 2^{-2} = \frac{1}{4}$
-1	$f(-1) = 2^{-1} = \frac{1}{2}$
0	$f(0) = 2^0 = 1$
1	$f(1) = 2^1 = 2$
2	$f(2) = 2^2 = 4$
3	$f(3) = 2^3 = 8$

Range: $(0, \infty)$

Domain: $(-\infty, \infty)$

Figure 4.1 The graph of $f(x) = 2^x$

We plot these points, connecting them with a continuous curve. Figure 4.1 shows the graph of $f(x) = 2^x$. Observe that the graph approaches, but never touches, the negative portion of the x-axis. Thus, the x-axis is a horizontal asymptote. The range is the set of all positive real numbers. Although we used integers for x in our table of coordinates, you can use a calculator to find additional points. For example, $f(0.3) = 2^{0.3} \approx 1.231$, $f(0.95) = 2^{0.95} \approx 1.932$. The points $(0.3, 1.231)$ and $(0.95, 1.932)$ approximately fit the graph.

Check Point 2   Graph: $f(x) = 3^x$.

Four exponential functions have been graphed in Figure 4.2. Compare the black and green graphs, where $b > 1$, to those in blue and red, where $b < 1$. When $b > 1$, the value of y increases as the value of x increases. When $b < 1$, the value of y decreases as the value of x increases. Notice that all four graphs pass through $(0, 1)$.

Study Tip

The graph of $y = \left(\frac{1}{2}\right)^x$, meaning $y = 2^{-x}$, is the graph of $y = 2^x$ reflected about the y-axis.

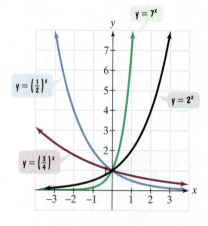

Figure 4.2 Graphs of four exponential functions

The graphs on the previous page illustrate the following general characteristics of exponential functions:

Characteristics of Exponential Functions of the Form $f(x) = b^x$

1. The domain of $f(x) = b^x$ consists of all real numbers. The range of $f(x) = b^x$ consists of all positive real numbers.
2. The graphs of all exponential functions of the form $f(x) = b^x$ pass through the point $(0, 1)$ because $f(0) = b^0 = 1 \, (b \neq 0)$. The y-intercept is 1.
3. If $b > 1$, $f(x) = b^x$ has a graph that goes up to the right and is an increasing function. The greater the value of b, the steeper the increase.
4. If $0 < b < 1$, $f(x) = b^x$ has a graph that goes down to the right and is a decreasing function. The smaller the value of b, the steeper the decrease.
5. $f(x) = b^x$ is one-to-one and has an inverse that is a function.
6. The graph of $f(x) = b^x$ approaches, but does not cross, the x-axis. The x-axis is a horizontal asymptote.

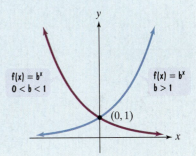

Transformations of Exponential Functions The graphs of exponential functions can be translated vertically or horizontally, reflected, stretched, or shrunk. We use the ideas of Section 2.5 to do so, as summarized in Table 4.1.

Table 4.1 Transformations Involving Exponential Functions
In each case, c represents a positive real number.

Transformation	Equation	Description
Vertical translation	$g(x) = b^x + c$	• Shifts the graph of $f(x) = b^x$ upward c units.
	$g(x) = b^x - c$	• Shifts the graph of $f(x) = b^x$ downward c units.
Horizontal translation	$g(x) = b^{x+c}$	• Shifts the graph of $f(x) = b^x$ to the left c units.
	$g(x) = b^{x-c}$	• Shifts the graph of $f(x) = b^x$ to the right c units.
Reflecting	$g(x) = -b^x$	• Reflects the graph of $f(x) = b^x$ about the x-axis.
	$g(x) = b^{-x}$	• Reflects the graph of $f(x) = b^x$ about the y-axis.
Vertical stretching or shrinking	$g(x) = cb^x$	• Stretches the graph of $f(x) = b^x$ if $c > 1$. • Shrinks the graph of $f(x) = b^x$ if $0 < c < 1$.

Using the information in Table 4.1 and a table of coordinates, you will obtain relatively accurate graphs that can be verified using a graphing utility.

EXAMPLE 3   Transformations Involving Exponential Functions

Use the graph of $f(x) = 3^x$ to obtain the graph of $g(x) = 3^{x+1}$.

Solution   Examine Table 4.1. Note that the function $g(x) = 3^{x+1}$ has the general form $g(x) = b^{x+c}$, where $c = 1$. Thus, we graph $g(x) = 3^{x+1}$ by shifting the graph of $f(x) = 3^x$ *one* unit to the *left*. We construct a table showing some of the coordinates for f and g, selecting integers from -2 to 2 for x. The graphs of f and g are shown in Figure 4.3.

x	$f(x) = 3^x$	$g(x) = 3^{x+1}$
-2	$f(-2) = 3^{-2} = \frac{1}{9}$	$g(-2) = 3^{-2+1} = 3^{-1} = \frac{1}{3}$
-1	$f(-1) = 3^{-1} = \frac{1}{3}$	$g(-1) = 3^{-1+1} = 3^0 = 1$
0	$f(0) = 3^0 = 1$	$g(0) = 3^{0+1} = 3^1 = 3$
1	$f(1) = 3^1 = 3$	$g(1) = 3^{1+1} = 3^2 = 9$
2	$f(2) = 3^2 = 9$	$g(2) = 3^{2+1} = 3^3 = 27$

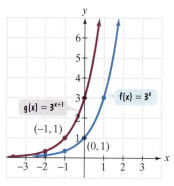

Figure 4.3 The graph of $g(x) = 3^{x+1}$ is the graph of $f(x) = 3^x$ shifted one unit to the left.

Check Point 3   Use the graph of $f(x) = 3^x$ to obtain the graph of $g(x) = 3^{x-1}$.

If an exponential function is translated upward or downward, the horizontal asymptote is shifted by the amount of the vertical shift.

EXAMPLE 4   Transformations Involving Exponential Functions

Use the graph of $f(x) = 2^x$ to obtain the graph of $g(x) = 2^x - 3$.

Solution   The function $g(x) = 2^x - 3$ has the general form $g(x) = b^x - c$, where $c = 3$. Thus, we graph $g(x) = 2^x - 3$ by shifting the graph of $f(x) = 2^x$ *down three* units. We construct a table showing some of the coordinates for f and g, selecting integers from -2 to 2 for x.

The graphs of f and g are shown in Figure 4.4. Notice that the horizontal asymptote for f, the x-axis, is shifted down three units for the horizontal asymptote for g. As a result, $y = -3$ is the horizontal asymptote for g.

x	$f(x) = 2^x$	$g(x) = 2^x - 3$
-2	$f(-2) = 2^{-2} = \frac{1}{4}$	$g(-2) = 2^{-2} - 3 = \frac{1}{4} - 3 = -2\frac{3}{4}$
-1	$f(-1) = 2^{-1} = \frac{1}{2}$	$g(-1) = 2^{-1} - 3 = \frac{1}{2} - 3 = -2\frac{1}{2}$
0	$f(0) = 2^0 = 1$	$g(0) = 2^0 - 3 = 1 - 3 = -2$
1	$f(1) = 2^1 = 2$	$g(1) = 2^1 - 3 = 2 - 3 = -1$
2	$f(2) = 2^2 = 4$	$g(2) = 2^2 - 3 = 4 - 3 = 1$

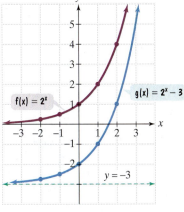

Figure 4.4 The graph of $g(x) = 2^x - 3$ is the graph of $f(x) = 2^x$ shifted down three units.

Check Point 4   Use the graph of $f(x) = 2^x$ to obtain the graph of $g(x) = 2^x + 1$.

3   Evaluate functions with base e.

The Natural Base e

An irrational number, symbolized by the letter e, appears as the base in many applied exponential functions. This irrational number is approximately equal to 2.72. More accurately,

$$e \approx 2.71828 \ldots .$$

The number e is called the **natural base.** The function $f(x) = e^x$ is called the **natural exponential function.**

Use a scientific or graphing calculator with an $\boxed{e^x}$ key to evaluate e to various powers. For example, to find e^2, press the following keys on most calculators:

Scientific calculator: 2 $\boxed{e^x}$

Graphing calculator: $\boxed{e^x}$ 2 $\boxed{\text{ENTER}}$.

The display should be approximately 7.389.

$$e^2 \approx 7.389$$

The number e lies between 2 and 3. Because $2^2 = 4$ and $3^2 = 9$, it makes sense that e^2, approximately 7.389, lies between 4 and 9.

Because $2 < e < 3$, the graph of $y = e^x$ lies between the graphs of $y = 2^x$ and $y = 3^x$, shown in Figure 4.5.

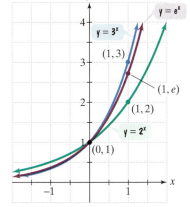

Figure 4.5 Graphs of three exponential functions

EXAMPLE 5   World Population

In a report entitled *Resources and Man*, the U.S. National Academy of Sciences concluded that a world population of 10 billion "is close to (if not above) the maximum that an intensely managed world might hope to support with some degree of comfort and individual choice." At the time the report was issued in 1969, world population was approximately 3.6 billion, with a growth rate of 2% per year. The function

$$f(x) = 3.6e^{0.02x}$$

describes world population, $f(x)$, in billions, x years after 1969. Use the function to find world population in the year 2020. Is there cause for alarm?

World Population, in Billions

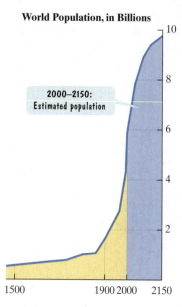

2000–2150:
Estimated population

Source: U.N. Population Division

Solution Because 2020 is 51 years after 1969, we substitute 51 for x in $f(x) = 3.6e^{0.02x}$:

$$f(51) = 3.6e^{0.02(51)}.$$

Perform this computation on your calculator.

Scientific calculator: 3.6 $\boxed{\times}$ $\boxed{(}$ .02 $\boxed{\times}$ 51 $\boxed{)}$ $\boxed{e^x}$ $\boxed{=}$

Graphing calculator: 3.6 $\boxed{\times}$ $\boxed{e^x}$ $\boxed{(}$ .02 $\boxed{\times}$ 51 $\boxed{)}$ $\boxed{\text{ENTER}}$

The display should be approximately 9.9835012. Thus,

$$f(51) = 3.6e^{0.02(51)} \approx 9.98.$$

This indicates that world population in the year 2020 will be approximately 9.98 billion. Because this number is quite close to 10 billion, the given function suggests that there may be cause for alarm.

World population in 2000 was approximately 6 billion, but the growth rate was no longer 2%. It had slowed down to 1.3%. Using this current growth rate, exponential functions now predict a world population of 7.8 billion in the year 2020. Experts think the population may stabilize at 10 billion after 2200 if the deceleration in growth rate continues.

Check Point 5 The function $f(x) = 6e^{0.013x}$ describes world population, $f(x)$, in billions, x years after 2000 subject to a growth rate of 1.3% annually. Use the function to find world population in 2050.

4 Use compound interest formulas.

Compound Interest

We all want a wonderful life with fulfilling work, good health, and loving relationships. And let's be honest: Financial security wouldn't hurt! Achieving this goal depends on understanding how money in savings accounts grows in remarkable ways as a result of *compound interest*. **Compound interest** is interest computed on your original investment as well as on any accumulated interest.

Suppose a sum of money, called the **principal**, P, is invested at an annual percentage rate r, in decimal form, compounded once per year. Because the interest is added to the principal at year's end, the accumulated value, A, is

$$A = P + Pr = P(1 + r).$$

The accumulated amount of money follows this pattern of multiplying the previous principal by $(1 + r)$ for each successive year, as indicated in Table 4.2.

Table 4.2

Time in Years	Accumulated Value after Each Compounding
0	$A = P$
1	$A = P(1 + r)$
2	$A = P(1 + r)(1 + r) = P(1 + r)^2$
3	$A = P(1 + r)^2(1 + r) = P(1 + r)^3$
4	$A = P(1 + r)^3(1 + r) = P(1 + r)^4$
⋮	⋮
t	$A = P(1 + r)^t$

This formula gives the balance, A, that a principal, P, is worth after t years at interest rate r, compounded once a year.

n	$\left(1 + \dfrac{1}{n}\right)^n$
1	2
2	2.25
5	2.48832
10	2.59374246
100	2.704813829
1000	2.716923932
10,000	2.718145927
100,000	2.718268237
1,000,000	2.718280469
1,000,000,000	2.718281827

As n takes on increasingly large values, the expression $\left(1 + \dfrac{1}{n}\right)^n$ approaches e.

Most savings institutions have plans in which interest is paid more than once a year. If compound interest is paid twice a year, the compounding period is six months. We say that the interest is **compounded semiannually.** When compound interest is paid four times a year, the compounding period is three months and the interest is said to be **compounded quarterly.** Some plans allow for monthly compounding or daily compounding.

In general, when compound interest is paid n times a year, we say that there are n **compounding periods per year.** The formula $A = P(1 + r)^t$ can be adjusted to take into account the number of compounding periods in a year. If there are n compounding periods per year, the formula becomes

$$A = P\left(1 + \frac{r}{n}\right)^{nt}.$$

Some banks use **continuous compounding,** where the number of compounding periods increases infinitely (compounding interest every trillionth of a second, every quadrillionth of a second, etc.). As n, the number of compounding periods in a year, increases without bound, the expression $\left(1 + \dfrac{1}{n}\right)^n$ approaches e. As a result, the formula for continuous compounding is $A = Pe^{rt}$. Although continuous compounding sounds terrific, it yields only a fraction of a percent more interest over a year than daily compounding.

Formulas for Compound Interest

After t years, the balance, A, in an account with principal P and annual interest rate r (in decimal form) is given by the following formulas:

1. For n compoundings per year: $A = P\left(1 + \dfrac{r}{n}\right)^{nt}$

2. For continuous compounding: $A = Pe^{rt}$.

EXAMPLE 6   Choosing between Investments

You want to invest \$8000 for 6 years, and you have a choice between two accounts. The first pays 7% per year, compounded monthly. The second pays 6.85% per year, compounded continuously. Which is the better investment?

Solution   The better investment is the one with the greater balance in the account after 6 years. Let's begin with the account with monthly compounding. We use the compound interest model with $P = 8000$, $r = 7\% = 0.07$, $n = 12$ (monthly compounding means 12 compoundings per year), and $t = 6$.

$$A = P\left(1 + \frac{r}{n}\right)^{nt} = 8000\left(1 + \frac{0.07}{12}\right)^{12 \cdot 6} \approx 12{,}160.84$$

The balance in this account after 6 years is \$12,160.84. For the second investment option, we use the model for continuous compounding with $P = 8000$, $r = 6.85\% = 0.0685$, and $t = 6$.

$$A = Pe^{rt} = 8000e^{0.0685(6)} \approx 12{,}066.60$$

Technology

The graphs illustrate that as x increases, $\left(1 + \dfrac{1}{x}\right)^x$ approaches e.

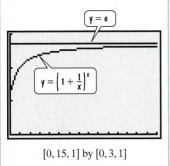

[0, 15, 1] by [0, 3, 1]

The balance in this account after 6 years is $12,066.60, slightly less than the previous amount. Thus, the better investment is the 7% monthly compounding option.

Check Point 6

A sum of $10,000 is invested at an annual rate of 8%. Find the balance in the account after 5 years subject to **a.** quarterly compounding and **b.** continuous compounding.

EXERCISE SET 4.1

Practice Exercises

In Exercises 1–10, approximate each number using a calculator. Round your answer to three decimal places.

1. $2^{3.4}$ **2.** $3^{2.4}$ **3.** $3^{\sqrt{5}}$ **4.** $5^{\sqrt{3}}$ **5.** $4^{-1.5}$

6. $6^{-1.2}$ **7.** $e^{2.3}$ **8.** $e^{3.4}$ **9.** $e^{-0.95}$ **10.** $e^{-0.75}$

In Exercises 11–18, graph each function by making a table of coordinates. If applicable, use a graphing utility to confirm your hand-drawn graph.

11. $f(x) = 4^x$ **12.** $f(x) = 5^x$

13. $g(x) = \left(\frac{3}{2}\right)^x$ **14.** $g(x) = \left(\frac{4}{3}\right)^x$

15. $h(x) = \left(\frac{1}{2}\right)^x$ **16.** $h(x) = \left(\frac{1}{3}\right)^x$

17. $f(x) = (0.6)^x$ **18.** $f(x) = (0.8)^x$

In Exercises 19–24, the graph of an exponential function is given. Select the function for each graph from the following options:

$$f(x) = 3^x, g(x) = 3^{x-1}, h(x) = 3^x - 1,$$
$$F(x) = -3^x, G(x) = 3^{-x}, H(x) = -3^{-x}.$$

19.

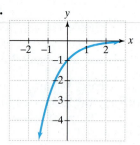

20.

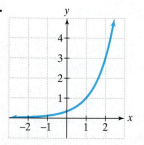

21.

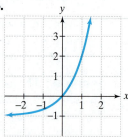

22.

23.

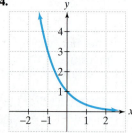

24.

In Exercises 25–34, begin by graphing $f(x) = 2^x$. Then use transformations of this graph and a table of coordinates to graph the given function. If applicable, use a graphing utility to confirm your hand-drawn graphs.

25. $g(x) = 2^{x+1}$

26. $g(x) = 2^{x+2}$

27. $g(x) = 2^x - 1$

28. $g(x) = 2^x + 2$

29. $h(x) = 2^{x+1} - 1$

30. $h(x) = 2^{x+2} - 1$

31. $g(x) = -2^x$

32. $g(x) = 2^{-x}$

33. $g(x) = 2 \cdot 2^x$

34. $g(x) = \frac{1}{2} \cdot 2^x$

In Exercises 35–40, graph functions f and g in the same rectangular coordinate system. If applicable, use a graphing utility to confirm your hand-drawn graphs.

35. $f(x) = 3^x$ and $g(x) = 3^{-x}$

36. $f(x) = 3^x$ and $g(x) = -3^x$

37. $f(x) = 3^x$ and $g(x) = \frac{1}{3} \cdot 3^x$

38. $f(x) = 3^x$ and $g(x) = 3 \cdot 3^x$

39. $f(x) = (\frac{1}{2})^x$ and $g(x) = (\frac{1}{2})^{x-1} + 1$

40. $f(x) = (\frac{1}{2})^x$ and $g(x) = (\frac{1}{2})^{x-1} + 2$

Use the compound interest formulas $A = P\left(1 + \dfrac{r}{n}\right)^{nt}$ and $A = Pe^{rt}$ to solve Exercises 41–44. Round answers to the nearest cent.

41. Find the accumulated value of an investment of $10,000 for 5 years at an interest rate of 5.5% if the money is **a.** compounded semiannually; **b.** compounded quarterly; **c.** compounded monthly; **d.** compounded continuously.

42. Find the accumulated value of an investment of $5000 for 10 years at an interest rate of 6.5% if the money is **a.** compounded semiannually; **b.** compounded quarterly; **c.** compounded monthly; **d.** compounded continuously.

43. Suppose that you have $12,000 to invest. Which investment yields the greatest return over 3 years: 7% compounded monthly or 6.85% compounded continuously?

44. Suppose that you have $6000 to invest. Which investment yields the greatest return over 4 years: 8.25% compounded quarterly or 8.3% compounded semiannually?

Application Exercises

Use a calculator with a $\boxed{y^x}$ key or a $\boxed{\wedge}$ key to solve Exercises 45–52.

45. The exponential function $f(x) = 67.38(1.026)^x$ describes the population of Mexico, $f(x)$, in millions, x years after 1980.

   a. Substitute 0 for x and, without using a calculator, find Mexico's population in 1980.

   b. Substitute 27 for x and use your calculator to find Mexico's population in the year 2007 as predicted by this function.

   c. Find Mexico's population in the year 2034 as predicted by this function.

   d. Find Mexico's population in the year 2061 as predicted by this function.

   e. What appears to be happening to Mexico's population every 27 years?

46. The 1986 explosion at the Chernobyl nuclear power plant in the former Soviet Union sent about 1000 kilograms of radioactive cesium-137 into the atmosphere. The function $f(x) = 1000(0.5)^{x/30}$ describes the amount, $f(x)$, in kilograms, of cesium-137 remaining in Chernobyl x years after 1986. If even 100 kilograms of cesium-137 remain in Chernobyl's atmosphere, the area is considered unsafe for human habitation. Find $f(80)$ and determine if Chernobyl will be safe for human habitation by 2066.

It is 8:00 P.M. and West Side Story is scheduled to begin. When the curtain does not go up, a rumor begins to spread through the 400-member audience: The lead roles of Tony and Maria might be understudied by Anthony Hopkins and Jodie Foster. The function

$$f(x) = \frac{400}{1 + 399(0.67)^x}$$

models the number of people in the audience, f (x), who have heard the rumor x minutes after 8:00. Use this function to solve Exercises 47–48.

47. Evaluate $f(10)$ and describe what this means in practical terms.

48. Evaluate $f(20)$ and describe what this means in practical terms.

The formula $S = C(1 + r)^t$ models inflation, where C = the value today, r = the annual inflation rate, and S = the inflated value t years from now. Use this formula to solve Exercises 49–50.

49. If the inflation rate is 6%, how much will a house now worth $65,000 be worth in 10 years?

50. If the inflation rate is 3%, how much will a house now worth $110,000 be worth in 5 years?

51. A decimal approximation for $\sqrt{3}$ is 1.7320508. Use a calculator to find $2^{1.7}, 2^{1.73}, 2^{1.732}, 2^{1.73205}$, and $2^{1.7320508}$. Now find $2^{\sqrt{3}}$. What do you observe?

52. A decimal approximation for π is 3.141593. Use a calculator to find $2^3, 2^{3.1}, 2^{3.14}, 2^{3.141}, 2^{3.1415}, 2^{3.14159}$, and $2^{3.141593}$. Now find 2^π. What do you observe?

The graph on the next page shows the number of Americans enrolled in HMOs, in millions, from 1992 through 2000. The data can be modeled by the exponential function

$$f(x) = 36.1e^{0.113x},$$

which describes enrollment in HMOs, f (x), in millions, x years after 1992. Use this function to solve Exercises 53–54.

Enrollment in HMOs

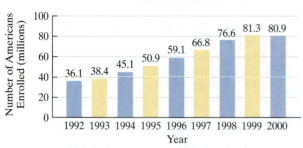

Source: Department of Health and Human Services

53. According to the model, how many Americans will be enrolled in HMOs in the year 2006? Round to the nearest tenth of a million.

54. According to the model, how many Americans will be enrolled in HMOs in the year 2008? Round to the nearest tenth of a million.

55. In college, we study large volumes of information—information that, unfortunately, we do not often retain for very long. The function

$$f(x) = 80e^{-0.5x} + 20$$

describes the percentage of information, $f(x)$, that a particular person remembers x weeks after learning the information.
   a. Substitute 0 for x and, without using a calculator, find the percentage of information remembered at the moment it is first learned.
   b. Substitute 1 for x and find the percentage of information that is remembered after 1 week.
   c. Find the percentage of information that is remembered after 4 weeks.
   d. Find the percentage of information that is remembered after one year (52 weeks).

56. In 1626, Peter Minuit convinced the Wappinger Indians to sell him Manhattan Island for $24. If the Native Americans had put the $24 into a bank account paying 5% interest, how much would the investment be worth in the year 2000 if interest were compounded
   a. monthly?   **b.** continuously?

57. The function

$$N(t) = \frac{30{,}000}{1 + 20e^{-1.5t}}$$

describes the number of people, $N(t)$, who become ill with influenza t weeks after its initial outbreak in a town with 30,000 inhabitants. The horizontal asymptote in the graph at the top of the next column indicates that there is a limit to the epidemic's growth.
   a. How many people became ill with the flu when the epidemic began? (When the epidemic began, $t = 0$.)
   b. How many people were ill by the end of the third week?
   c. Why can't the spread of an epidemic simply grow indefinitely? What does the horizontal asymptote shown in the graph indicate about the limiting size of the population that becomes ill?

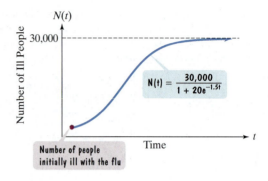

$$N(t) = \frac{30{,}000}{1 + 20e^{-1.5t}}$$

Number of people initially ill with the flu

Writing in Mathematics

58. What is an exponential function?

59. What is the natural exponential function?

60. Use a calculator to evaluate $\left(1 + \dfrac{1}{x}\right)^x$ for $x = 10, 100, 1000, 10{,}000, 100{,}000,$ and $1{,}000{,}000$. Describe what happens to the expression as x increases.

61. Write an example similar to Example 6 on page 358 in which continuous compounding at a slightly lower yearly interest rate is a better investment than compounding n times per year.

62. Describe how you could use the graph of $f(x) = 2^x$ to obtain a decimal approximation for $\sqrt{2}$.

63. The exponential function $y = 2^x$ is one-to-one and has an inverse function. Try finding the inverse function by exchanging x and y and solving for y. Describe the difficulty that you encounter in this process. What is needed to overcome this problem?

64. In 2000, world population was approximately 6 billion with an annual growth rate of 1.3%. Discuss two factors that would cause this growth rate to slow down over the next ten years.

Technology Exercises

65. Graph $y = 13.49(0.967)^x - 1$, the function for the number of O-rings expected to fail at $x°F$, in a $[0, 90, 10]$ by $[0, 20, 5]$ viewing rectangle. If NASA engineers had used this function and its graph, is it likely they would have allowed the *Challenger* to be launched when the temperature was 31°F? Explain.

66. You have $10,000 to invest. One bank pays 5% interest compounded quarterly and the other pays 4.5% interest compounded monthly.
   a. Use the formula for compound interest to write a function for the balance in each account at any time t.
   b. Use a graphing utility to graph both functions in an appropriate viewing rectangle. Based on the graphs, which bank offers the better return on your money?

67. a. Graph $y = e^x$ and $y = 1 + x + \dfrac{x^2}{2}$ in the same viewing rectangle.

b. Graph $y = e^x$ and $y = 1 + x + \dfrac{x^2}{2} + \dfrac{x^3}{6}$ in the same viewing rectangle.

c. Graph $y = e^x$ and $y = 1 + x + \dfrac{x^2}{2} + \dfrac{x^3}{6} + \dfrac{x^4}{24}$ in the same viewing rectangle.

d. Describe what you observe in parts (a)–(c). Try generalizing this observation.

Critical Thinking Exercises

68. Which one of the following is true?

a. As the number of compounding periods increases on a fixed investment, the amount of money in the account over a fixed interval of time will increase without bound.

b. The functions $f(x) = 3^{-x}$ and $g(x) = -3^x$ have the same graph.

c. $e = 2.718$

d. The functions $f(x) = \left(\frac{1}{3}\right)^x$ and $g(x) = 3^{-x}$ have the same graph.

69. The graphs labeled (a)–(d) in the figure represent $y = 3^x$, $y = 5^x$, $y = \left(\frac{1}{3}\right)^x$, and $y = \left(\frac{1}{5}\right)^x$, but not necessarily in that order. Which is which? Describe the process that enables you to make this decision.

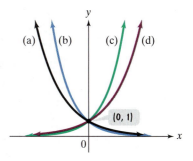

70. Graph $f(x) = 2^x$ and its inverse function in the same rectangular coordinate system.

71. The *hyperbolic cosine* and *hyperbolic sine* functions are defined by

$$\cosh x = \frac{e^x + e^{-x}}{2} \quad \text{and} \quad \sinh x = \frac{e^x - e^{-x}}{2}.$$

Prove that $(\cosh x)^2 - (\sinh x)^2 = 1$.

SECTION 4.2 Logarithmic Functions

Objectives

1. Change from logarithmic to exponential form.

2. Change from exponential to logarithmic form.

3. Evaluate logarithms.

4. Use basic logarithmic properties.

5. Graph logarithmic functions.

6. Find the domain of a logarithmic function.

7. Use common logarithms.

8. Use natural logarithms.

The earthquake that ripped through northern California on October 17, 1989, measured 7.1 on the Richter scale, killed more than 60 people, and injured more than 2400. Shown here is San Francisco's Marina district, where shock waves tossed houses off their foundations and into the street.

A higher measure on the Richter scale is more devastating than it seems because for each increase in one unit on the scale, there is a tenfold increase in the intensity of an earthquake. In this section, our focus is on the inverse of the exponential function, called the logarithmic function. The logarithmic function will help you to understand diverse phenomena, including earthquake intensity, human memory, and the pace of life in large cities.

Study Tip

In case you need to review inverse functions, they are discussed in Section 2.7 on pages 260–267. The horizontal line test appears on page 265.

The Definition of Logarithmic Functions

No horizontal line can be drawn that intersects the graph of an exponential function at more than one point. This means that the exponential function is one-to-one and has an inverse. The inverse function of the exponential function with base b is called the *logarithmic function with base b*.

> ### Definition of the Logarithmic Function
>
> For $x > 0$ and $b > 0$, $b \neq 1$,
>
> $$y = \log_b x \text{ is equivalent to } b^y = x.$$
>
> The function $f(x) = \log_b x$ is the **logarithmic function with base b**.

Study Tip

The inverse function of $y = b^x$ is $x = b^y$. Logarithms give us a way to express this inverse function for y in terms of x.

The equations

$$y = \log_b x \quad \text{and} \quad b^y = x$$

are different ways of expressing the same thing. The first equation is in **logarithmic form** and the second equivalent equation is in **exponential form.** Notice that a **logarithm, y, is an exponent.** You should learn the location of the base and exponent in each form.

> ### Location of Base and Exponent in Exponential and Logarithmic Forms
>
> Exponent
>
> Logarithmic Form: $y = \log_b x$ Exponential Form: $b^y = x$
>
> Base   Base

1 Change from logarithmic to exponential form.

EXAMPLE 1 Changing from Logarithmic to Exponential Form

Write each equation in its equivalent exponential form:

a. $2 = \log_5 x$ **b.** $3 = \log_b 64$ **c.** $\log_3 7 = y$.

Solution   We use the fact that $y = \log_b x$ means $b^y = x$.

a. $2 = \log_5 x$ means $5^2 = x$.   **b.** $3 = \log_b 64$ means $b^3 = 64$.

Logarithms are exponents.   Logarithms are exponents.

c. $\log_3 7 = y$ or $y = \log_3 7$ means $3^y = 7$.

Check Point 1   Write each equation in its equivalent exponential form:

a. $3 = \log_7 x$ **b.** $2 = \log_b 25$ **c.** $\log_4 26 = y$.

2 Change from exponential to logarithmic form.

EXAMPLE 2 Changing from Exponential to Logarithmic Form

Write each equation in its equivalent logarithmic form:

a. $12^2 = x$ **b.** $b^3 = 8$ **c.** $e^y = 9$.

Solution We use the fact that $b^y = x$ means $y = \log_b x$.

a. $12^2 = x$ means $2 = \log_{12} x$.   **b.** $b^3 = 8$ means $3 = \log_b 8$.

> Exponents are logarithms.   Exponents are logarithms.

c. $e^y = 9$ means $y = \log_e 9$.

Check Point 2   Write each equation in its equivalent logarithmic form:

a. $2^5 = x$ **b.** $b^3 = 27$ **c.** $e^y = 33$.

3   Evaluate logarithms.

Remembering that logarithms are exponents makes it possible to evaluate some logarithms by inspection. The logarithm of x with base b, $\log_b x$, is the exponent to which b must be raised to get x. For example, suppose we want to evaluate $\log_2 32$. We ask, 2 to what power gives 32? Because $2^5 = 32$, $\log_2 32 = 5$.

EXAMPLE 3   Evaluating Logarithms

Evaluate:

a. $\log_2 16$ **b.** $\log_3 9$ **c.** $\log_{25} 5$.

Solution

Logarithmic Expression	Question Needed for Evaluation	Logarithmic Expression Evaluated
a. $\log_2 16$	2 to what power gives 16?	$\log_2 16 = 4$ because $2^4 = 16$.
b. $\log_3 9$	3 to what power gives 9?	$\log_3 9 = 2$ because $3^2 = 9$.
c. $\log_{25} 5$	25 to what power gives 5?	$\log_{25} 5 = \frac{1}{2}$ because $25^{1/2} = \sqrt{25} = 5$.

Check Point 3   Evaluate:

a. $\log_{10} 100$ **b.** $\log_3 3$ **c.** $\log_{36} 6$.

4   Use basic logarithmic properties.

Basic Logarithmic Properties

Because logarithms are exponents, they have properties that can be verified using properties of exponents.

Basic Logarithmic Properties Involving One

1. $\log_b b = 1$ because 1 is the exponent to which b must be raised to obtain b. $\left(b^1 = b\right)$

2. $\log_b 1 = 0$ because 0 is the exponent to which b must be raised to obtain 1. $\left(b^0 = 1\right)$

EXAMPLE 4   Using Properties of Logarithms

Evaluate:

a. $\log_7 7$ **b.** $\log_5 1$.

Solution

a. Because $\log_b b = 1$, we conclude $\log_7 7 = 1$.
b. Because $\log_b 1 = 0$, we conclude $\log_5 1 = 0$.

Check Point 4 Evaluate:

a. $\log_9 9$ **b.** $\log_8 1$.

The inverse of the exponential function is the logarithmic function. Thus, if $f(x) = b^x$, then $f^{-1}(x) = \log_b x$. In Chapter 2, we saw how inverse functions "undo" one another. In particular,

$$f(f^{-1}(x)) = x \text{ and } f^{-1}(f(x)) = x.$$

Applying these relationships to exponential and logarithmic functions, we obtain the following **inverse properties of logarithms:**

Inverse Properties of Logarithms

For $b > 0$ and $b \neq 1$,

$$\log_b b^x = x$$ The logarithm with base b of b raised to a power equals that power.

$$b^{\log_b x} = x$$ b raised to the logarithm with base b of a number equals that number.

EXAMPLE 5   Using Inverse Properties of Logarithms

Evaluate:

a. $\log_4 4^5$ **b.** $6^{\log_6 9}$.

Solution

a. Because $\log_b b^x = x$, we conclude $\log_4 4^5 = 5$.
b. Because $b^{\log_b x} = x$, we conclude $6^{\log_6 9} = 9$.

Check Point 5 Evaluate:

a. $\log_7 7^8$ **b.** $3^{\log_3 17}$.

5 Graph logarithmic functions.

Graphs of Logarithmic Functions

How do we graph logarithmic functions? We use the fact that the logarithmic function is the inverse of the exponential function. This means that the logarithmic function reverses the coordinates of the exponential function. It also means that the graph of the logarithmic function is a reflection of the graph of the exponential function about the line $y = x$.

EXAMPLE 6 **Graphs of Exponential and Logarithmic Functions**

Graph $f(x) = 2^x$ and $g(x) = \log_2 x$ in the same rectangular coordinate system.

Solution We first set up a table of coordinates for $f(x) = 2^x$. Reversing, these coordinates gives the coordinates for the inverse function $g(x) = \log_2 x$.

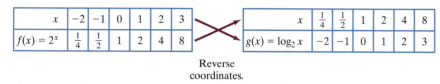

x	-2	-1	0	1	2	3
$f(x) = 2^x$	$\frac{1}{4}$	$\frac{1}{2}$	1	2	4	8

x	$\frac{1}{4}$	$\frac{1}{2}$	1	2	4	8
$g(x) = \log_2 x$	-2	-1	0	1	2	3

Reverse coordinates.

We now plot the ordered pairs in both tables, connecting them with smooth curves. Figure 4.6 shows the graphs of $f(x) = 2^x$ and its inverse function $g(x) = \log_2 x$. The graph of the inverse can also be drawn by reflecting the graph of $f(x) = 2^x$ about the line $y = x$.

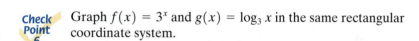

Figure 4.6 The graphs of $f(x) = 2^x$ and its inverse function

Check Point 6 Graph $f(x) = 3^x$ and $g(x) = \log_3 x$ in the same rectangular coordinate system.

Figure 4.7 illustrates the relationship between the graph of the exponential function, shown in blue, and its inverse, the logarithmic function, shown in red, for bases greater than 1 and for bases between 0 and 1.

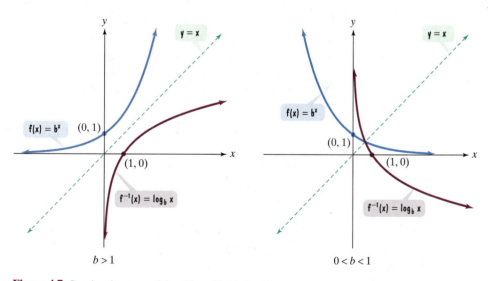

Figure 4.7 Graphs of exponential and logarithmic functions

Discovery

Verify each of the four characteristics in the box for the red graphs in Figure 4.7.

Characteristics of the Graphs of Logarithmic Functions of the Form $f(x) = \log_b x$

- The x-intercept is 1. There is no y-intercept.
- The y-axis is a vertical asymptote.
- If $b > 1$, the function is increasing. If $0 < b < 1$, the function is decreasing.
- The graph is smooth and continuous. It has no sharp corners or gaps.

The graphs of logarithmic functions can be translated vertically or horizontally, reflected, stretched, or shrunk. We use the ideas of Section 2.5 to do so, as summarized in Table 4.3.

Table 4.3 Transformations Involving Logarithmic Functions
In each case, c represents a positive real number.

Transformation	Equation	Description
Vertical translation	$g(x) = \log_b x + c$	• Shifts the graph of $f(x) = \log_b x$ upward c units.
	$g(x) = \log_b x - c$	• Shifts the graph of $f(x) = \log_b x$ downward c units.
Horizontal translation	$g(x) = \log_b (x + c)$	• Shifts the graph of $f(x) = \log_b x$ to the left c units. Vertical asymptote: $x = -c$.
	$g(x) = \log_b (x - c)$	• Shifts the graph of $f(x) = \log_b x$ to the right c units. Vertical asymptote: $x = c$.
Reflecting	$g(x) = -\log_b x$	• Reflects the graph of $f(x) = \log_b x$ about the x-axis.
	$g(x) = \log_b (-x)$	• Reflects the graph of $f(x) = \log_b x$ about the y-axis.
Vertical stretching or shrinking	$g(x) = c \log_b x$	• Stretches the graph of $f(x) = \log_b x$ if $c > 1$. • Shrinks the graph of $f(x) = \log_b x$ if $0 < c < 1$.

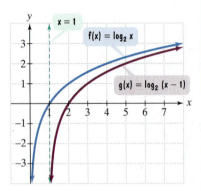

Figure 4.8 Shifting $f(x) = \log_2 x$ one unit to the right

For example, Figure 4.8 illustrates that the graph of $g(x) = \log_2 (x - 1)$ is the graph of $f(x) = \log_2 x$ shifted one unit to the right. If a logarithmic function is translated to the left or to the right, both the x-intercept and the vertical asymptote are shifted by the amount of the horizontal shift. In Figure 4.8, the x-intercept of f is 1. Because g is shifted one unit to the right, its x-intercept is 2. Also observe that the vertical asymptote for f, the y-axis, is shifted one unit to the right for the vertical asymptote for g. Thus, $x = 1$ is the vertical asymptote for g.

Here are some other examples of transformations of graphs of logarithmic functions:

- The graph of $g(x) = 3 + \log_4 x$ is the graph of $f(x) = \log_4 x$ shifted up three units, shown in Figure 4.9.
- The graph of $h(x) = -\log_2 x$ is the graph of $f(x) = \log_2 x$ reflected about the x-axis, shown in Figure 4.10.
- The graph of $r(x) = \log_2 (-x)$ is the graph of $f(x) = \log_2 x$ reflected about the y-axis, shown in Figure 4.11.

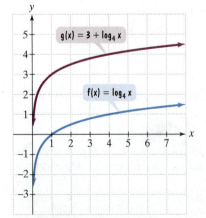

Figure 4.9 Shifting vertically up three units

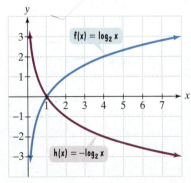

Figure 4.10 Reflection about the x-axis

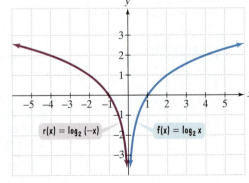

Figure 4.11 Reflection about the y-axis

6 Find the domain of a logarithmic function.

The Domain of a Logarithmic Function

In Section 4.1, we learned that the domain of an exponential function of the form $f(x) = b^x$ includes all real numbers and its range is the set of positive real numbers. Because the logarithmic function reverses the domain and the range of the exponential function, the **domain of a logarithmic function of the form $f(x) = \log_b x$ is the set of all positive real numbers.** Thus, $\log_2 8$ is defined because the value of x in the logarithmic expression, 8, is greater than zero and therefore is included in the domain of the logarithmic function $f(x) = \log_2 x$. However, $\log_2 0$ and $\log_2 (-8)$ are not defined because 0 and -8 are not positive real numbers and therefore are excluded from the domain of the logarithmic function $f(x) = \log_2 x$. In general, the domain of $f(x) = \log_b (x + c)$ consists of all x for which $x + c > 0$.

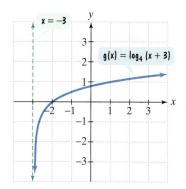

Figure 4.12 The domain of $g(x) = \log_4 (x + 3)$ is $(-3, \infty)$.

EXAMPLE 7   Finding the Domain of a Logarithmic Function

Find the domain of $g(x) = \log_4 (x + 3)$.

Solution   The domain of g consists of all x for which $x + 3 > 0$. Solving this inequality for x, we obtain $x > -3$. Thus, the domain of g is $(-3, \infty)$. This is illustrated in Figure 4.12. The vertical asymptote is $x = -3$, and all points on the graph of g have x-coordinates that are greater than -3.

Check Point 7   Find the domain of $h(x) = \log_4 (x - 5)$.

7 Use common logarithms.

Common Logarithms

The logarithmic function with base 10 is called the **common logarithmic function.** The function $f(x) = \log_{10} x$ is usually expressed as $f(x) = \log x$. A calculator with a LOG key can be used to evaluate common logarithms. Here are some examples:

Logarithm	Most Scientific Calculator Keystrokes	Most Graphing Calculator Keystrokes	Display (or Approximate Display)
$\log 1000$	1000 LOG	LOG 1000 ENTER	3
$\log \dfrac{5}{2}$	(5 ÷ 2) LOG	LOG (5 ÷ 2) ENTER	0.39794
$\dfrac{\log 5}{\log 2}$	5 LOG ÷ 2 LOG =	LOG 5 ÷ LOG 2 ENTER	2.32193
$\log (-3)$	3 +/- LOG	LOG (−) 3 ENTER	ERROR

The error message given by many calculators for log (-3) is a reminder that the domain of every logarithmic function, including the common logarithmic function, is the set of positive real numbers.

Many real-life phenomena start with rapid growth, and then the growth begins to level off. This type of behavior can be modeled by logarithmic functions.

EXAMPLE 8 Modeling Height of Children

The percentage of adult height attained by a boy who is x years old can be modeled by

$$f(x) = 29 + 48.8 \log (x + 1)$$

where x represents the boy's age and $f(x)$ represents the percentage of his adult height. Approximately what percent of his adult height is a boy at age eight?

Solution We substitute the boy's age, 8, for x and evaluate the function.

$f(x) = 29 + 48.8 \log (x + 1)$ This is the given function.

$f(8) = 29 + 48.8 \log (8 + 1)$ Substitute 8 for x.

$= 29 + 48.8 \log 9$ Graphing calculator keystrokes:

≈ 76 29 $+$ 48.8 $\boxed{\text{LOG}}$ 9 $\boxed{\text{ENTER}}$

Thus, an 8-year-old boy is approximately 76% of his adult height.

Check Point 8 Use the function in Example 8 to answer this question: Approximately what percent of his adult height is a boy at age 10?

The basic properties of logarithms that were listed earlier in this section can be applied to common logarithms.

Properties of Common Logarithms

General Properties

1. $\log_b 1 = 0$
2. $\log_b b = 1$
3. $\log_b b^x = x$
4. $b^{\log_b x} = x$

Common Logarithm Properties

1. $\log 1 = 0$
2. $\log 10 = 1$
3. $\log 10^x = x$
4. $10^{\log x} = x$

The property $\log 10^x = x$ can be used to evaluate common logarithms involving powers of 10. For example,

$$\log 100 = \log 10^2 = 2, \quad \log 1000 = \log 10^3 = 3, \quad \text{and} \quad \log 10^{7.1} = 7.1.$$

EXAMPLE 9 Earthquake Intensity

The magnitude, R, on the Richter scale of an earthquake of intensity I is given by

$$R = \log \frac{I}{I_0}$$

where I_0 is the intensity of a barely felt zero-level earthquake. The earthquake that destroyed San Francisco in 1906 was $10^{8.3}$ times as intense as a zero-level earthquake. What was its magnitude on the Richter scale?

Solution Because the earthquake was $10^{8.3}$ times as intense as a zero-level earthquake, the intensity, I, is $10^{8.3}I_0$.

$$R = \log \frac{I}{I_0}$$ This is the formula for magnitude on the Richter scale.

$$R = \log \frac{10^{8.3}I_0}{I_0}$$ Substitute $10^{8.3}I_0$ for I.

$$= \log 10^{8.3}$$ Simplify.

$$= 8.3$$ Use the property $\log 10^x = x$.

San Francisco's 1906 earthquake registered 8.3 on the Richter scale.

Check Point 9 Use the formula in Example 9 to solve this problem. If an earthquake is 10,000 times as intense as a zero-level quake ($I = 10{,}000I_0$), what is its magnitude on the Richter scale?

8 Use natural logarithms.

Natural Logarithms

The logarithmic function with base e is called the **natural logarithmic function.** The function $f(x) = \log_e x$ is usually expressed as $f(x) = \ln x$, read "el en of x." A calculator with an $\boxed{\text{LN}}$ key can be used to evaluate natural logarithms.

Like the domain of all logarithmic functions, the domain of the natural logarithmic function is the set of all positive real numbers. Thus, the domain of $f(x) = \ln(x + c)$ consists of all x for which $x + c > 0$.

EXAMPLE 10 Finding Domains of Natural Logarithmic Functions

Find the domain of each function:
    a. $f(x) = \ln(3 - x)$ **b.** $g(x) = \ln(x - 3)^2$.

Solution

    a. The domain of f consists of all x for which $3 - x > 0$. Solving this inequality for x, we obtain $x < 3$. Thus, the domain of f is $\{x \mid x < 3\}$, or $(-\infty, 3)$. This is verified by the graph in Figure 4.13.

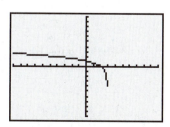

Figure 4.13 The domain of $f(x) = \ln(3 - x)$ is $(-\infty, 3)$.

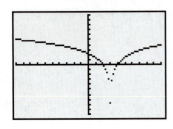

Figure 4.14 3 is excluded from the domain of $g(x) = \ln(x-3)^2$.

b. The domain of g consists of all x for which $(x-3)^2 > 0$. It follows that the domain of g is the set of all real numbers except 3. Thus, the domain of g is $\{x|x \neq 3\}$, or, in interval notation, $(-\infty, 3)$ or $(3, \infty)$. This is shown by the graph in Figure 4.14. To make it more obvious that 3 is excluded from the domain, we changed the $\boxed{\text{MODE}}$ to Dot.

Check Point 10 Find the domain of each function:

a. $f(x) = \ln(4-x)$ **b.** $g(x) = \ln x^2$.

The basic properties of logarithms that were listed earlier in this section can be applied to natural logarithms.

Properties of Natural Logarithms

General Properties	Natural Logarithm Properties
1. $\log_b 1 = 0$	**1.** $\ln 1 = 0$
2. $\log_b b = 1$	**2.** $\ln e = 1$
3. $\log_b b^x = x$	**3.** $\ln e^x = x$
4. $b^{\log_b x} = x$	**4.** $e^{\ln x} = x$

The property $\ln e^x = x$ can be used to evaluate natural logarithms involving powers of e. For example,

$$\ln e^2 = 2, \quad \ln e^3 = 3, \quad \ln e^{7.1} = 7.1, \quad \text{and} \quad \ln \frac{1}{e} = \ln e^{-1} = -1.$$

EXAMPLE 11 **Using Inverse Properties**

Use inverse properties to simplify:

a. $\ln e^{7x}$ **b.** $e^{\ln 4x^2}$.

Solution

a. Because $\ln e^x = x$, we conclude that $\ln e^{7x} = 7x$.

b. Because $e^{\ln x} = x$, we conclude $e^{\ln 4x^2} = 4x^2$.

Check Point 11 Use inverse properties to simplify:

a. $\ln e^{25x}$ **b.** $e^{\ln \sqrt{x}}$.

EXAMPLE 12 **Walking Speed and City Population**

As the population of a city increases, the pace of life also increases. The formula

$$W = 0.35 \ln P + 2.74$$

models average walking speed, W, in feet per second, for a resident of a city whose population is P thousand. Find the average walking speed for people living in New York City with a population of 7323 thousand.

Solution   We use the formula and substitute 7323 for P, the population in thousands.

$$W = 0.35 \ln P + 2.74 \qquad \text{This is the given formula.}$$

$$W = 0.35 \ln 7323 + 2.74 \qquad \text{Substitute 7323 for } P.$$

$$\approx 5.9 \qquad \text{Graphing calculator keystrokes:}$$
$$\text{0.35 } \boxed{\text{LN}} \text{ 7323 } \boxed{+} \text{ 2.74 } \boxed{\text{ENTER}}.$$

The average walking speed in New York City is approximately 5.9 feet per second.

Check Point 12   Use the formula $W = 0.35 \ln P + 2.74$ to find the average walking speed in Jackson, Mississippi, with a population of 197 thousand.

EXERCISE SET 4.2

Practice Exercises

In Exercises 1–8, write each equation in its equivalent exponential form.

1. $4 = \log_2 16$

2. $6 = \log_2 64$

3. $2 = \log_3 x$

4. $2 = \log_9 x$

5. $5 = \log_b 32$

6. $3 = \log_b 27$

7. $\log_6 216 = y$

8. $\log_5 125 = y$

In Exercises 9–20, write each equation in its equivalent logarithmic form.

9. $2^3 = 8$

10. $5^4 = 625$

11. $2^{-4} = \frac{1}{16}$

12. $5^{-3} = \frac{1}{125}$

13. $\sqrt[3]{8} = 2$

14. $\sqrt[3]{64} = 4$

15. $13^2 = x$

16. $15^2 = x$

17. $b^3 = 1000$

18. $b^3 = 343$

19. $7^y = 200$

20. $8^y = 300$

In Exercises 21–38, evaluate each expression without using a calculator.

21. $\log_4 16$

22. $\log_7 49$

23. $\log_2 64$

24. $\log_3 27$

25. $\log_7 \sqrt{7}$

26. $\log_6 \sqrt{6}$

27. $\log_2 \frac{1}{8}$

28. $\log_3 \frac{1}{9}$

29. $\log_{64} 8$

30. $\log_{81} 9$

31. $\log_5 5$

32. $\log_{11} 11$

33. $\log_4 1$

34. $\log_6 1$

35. $\log_5 5^7$

36. $\log_4 4^6$

37. $8^{\log_8 19}$

38. $7^{\log_7 23}$

39. Graph $f(x) = 4^x$ and $g(x) = \log_4 x$ in the same rectangular coordinate system.

40. Graph $f(x) = 5^x$ and $g(x) = \log_5 x$ in the same rectangular coordinate system.

41. Graph $f(x) = \left(\frac{1}{2}\right)^x$ and $g(x) = \log_{1/2} x$ in the same rectangular coordinate system.

42. Graph $f(x) = \left(\frac{1}{4}\right)^x$ and $g(x) = \log_{1/4} x$ in the same rectangular coordinate system.

In Exercises 43–48, the graph of a logarithmic function is given. Select the function for each graph from the following options:

$$f(x) = \log_3 x, \, g(x) = \log_3 (x - 1), \, h(x) = \log_3 x - 1,$$

$$F(x) = -\log_3 x, \, G(x) = \log_3 (-x), \, H(x) = 1 - \log_3 x.$$

43.

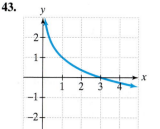

44.

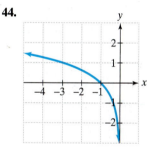

45.

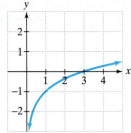

46.

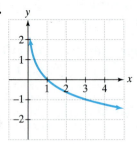

47.

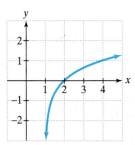

48.

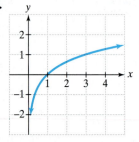

In Exercises 49–54, begin by graphing $f(x) = \log_2 x$. Then use transformations of this graph to graph the given function. What is the graph's x-intercept? What is the vertical asymptote?

49. $g(x) = \log_2 (x + 1)$ **50.** $g(x) = \log_2 (x + 2)$

51. $h(x) = 1 + \log_2 x$ **52.** $h(x) = 2 + \log_2 x$

53. $g(x) = \frac{1}{2} \log_2 x$ **54.** $g(x) = -2 \log_2 x$

In Exercises 55–60, find the domain of each logarithmic function.

55. $f(x) = \log_5 (x + 4)$ **56.** $f(x) = \log_5 (x + 6)$

57. $f(x) = \log (2 - x)$ **58.** $f(x) = \log (7 - x)$

59. $f(x) = \ln (x - 2)^2$ **60.** $f(x) = \ln (x - 7)^2$

In Exercises 61–74, evaluate each expression without using a calculator.

61. $\log 100$ **62.** $\log 1000$ **63.** $\log 10^7$

64. $\log 10^8$ **65.** $10^{\log 33}$ **66.** $10^{\log 53}$

67. $\ln 1$ **68.** $\ln e$ **69.** $\ln e^6$

70. $\ln e^7$ **71.** $\ln \dfrac{1}{e^6}$ **72.** $\ln \dfrac{1}{e^7}$

73. $e^{\ln 125}$ **74.** $e^{\ln 300}$

In Exercises 75–80, use inverse properties of logarithms to simplify each expression.

75. $\ln e^{9x}$ **76.** $\ln e^{13x}$ **77.** $e^{\ln 5x^2}$

78. $e^{\ln 7x^2}$ **79.** $10^{\log \sqrt{x}}$ **80.** $10^{\log \sqrt[3]{x}}$

Application Exercises

The percentage of adult height attained by a girl who is x years old can be modeled by

$$f(x) = 62 + 35 \log (x - 4)$$

where x represents the girl's age (from 5 to 15) and $f(x)$ represents the percentage of her adult height. Use the function to solve Exercises 81–82.

81. Approximately what percent of her adult height is a girl at age 13?

82. Approximately what percent of her adult height is a girl at age ten?

83. The annual amount that we spend to attend sporting events can be modeled by

$$f(x) = 2.05 + 1.3 \ln x$$

where x represents the number of years after 1984 and $f(x)$ represents the total annual expenditures for admission to spectator sports, in billions of dollars. In 2000, approximately how much was spent on admission to spectator sports?

84. The percentage of U.S. households with cable television can be modeled by

$$f(x) = 18.32 + 15.94 \ln x$$

where x represents the number of years after 1979 and $f(x)$ represents the percentage of U.S. households with cable television. What percentage of U.S. households had cable television in 1990?

The loudness level of a sound, D, in decibels, is given by the formula

$$D = 10 \log (10^{12} I)$$

where I is the intensity of the sound, in watts per meter^2. Decibel levels range from 0, a barely audible sound, to 160, a sound resulting in a ruptured eardrum. Use the formula to solve Exercises 85–86.

85. The sound of a blue whale can be heard 500 miles away, reaching an intensity of 6.3×10^6 watts per meter^2. Determine the decibel level of this sound. At close range, can the sound of a blue whale rupture the human eardrum?

86. What is the decibel level of a normal conversation, 3.2×10^{-6} watt per meter^2?

87. Students in a psychology class took a final examination. As part of an experiment to see how much of the course content they remembered over time, they took equivalent forms of the exam in monthly intervals thereafter. The average score for the group, $f(t)$, after t months was modeled by the function

$$f(t) = 88 - 15 \ln (t + 1), \quad 0 \le t \le 12.$$

a. What was the average score on the original exam?

b. What was the average score after 2 months? 4 months? 6 months? 8 months? 10 months? one year?

c. Sketch the graph of f (either by hand or with a graphing utility). Describe what the graph indicates in terms of the material retained by the students.

Writing in Mathematics

88. Describe the relationship between an equation in logarithmic form and an equivalent equation in exponential form.

89. What question can be asked to help evaluate $\log_3 81$?

90. Explain why the logarithm of 1 with base b is 0.

91. Describe the following property using words: $\log_b b^x = x$.

92. Explain how to use the graph of $f(x) = 2^x$ to obtain the graph of $g(x) = \log_2 x$.

93. Explain how to find the domain of a logarithmic function.

94. New York City is one of the world's great walking cities. Use the formula in Example 12 on page 394 to describe what frequently happens to tourists exploring the city by foot.

95. Logarithmic models are well suited to phenomena in which growth is initially rapid but then begins to level off. Describe something that is changing over time that can be modeled using a logarithmic function.

96. Suppose that a girl is 4′ 6″ at age 10. Explain how to use the function in Exercises 81–82 to determine how tall she can expect to be as an adult.

Technology Exercises

In Exercises 97–100, graph f and g in the same viewing rectangle. Then describe the relationship of the graph of g to the graph of f.

97. $f(x) = \ln x, g(x) = \ln (x + 3)$

98. $f(x) = \ln x, g(x) = \ln x + 3$

99. $f(x) = \log x, g(x) = -\log x$

100. $f(x) = \log x, g(x) = \log (x - 2) + 1$

101. Students in a mathematics class took a final examination. They took equivalent forms of the exam in monthly intervals thereafter. The average score, $f(t)$, for the group after t months was modeled by the human memory function $f(t) = 75 - 10 \log (t + 1)$, where $0 \le t \le 12$.

Use a graphing utility to graph the function. Then determine how many months will elapse before the average score falls below 65.

102. Graph f and g in the same viewing rectangle.

a. $f(x) = \ln (3x), g(x) = \ln 3 + \ln x$

b. $f(x) = \log (5x^2), g(x) = \log 5 + \log x^2$

c. $f(x) = \ln (2x^3), g(x) = \ln 2 + \ln x^3$

d. Describe what you observe in parts (a)–(c). Generalize this observation by writing an equivalent expression for $\log_b (MN)$, where $M > 0$ and $N > 0$.

e. Complete this statement: The logarithm of a product is equal to _____.

103. Graph each of the following functions in the same viewing rectangle and then place the functions in order from the one that increases most slowly to the one that increases most rapidly.

$$y = x, y = \sqrt{x}, y = e^x, y = \ln x, y = x^x, y = x^2$$

Critical Thinking Exercises

104. Which one of the following is true?

a. $\dfrac{\log_2 8}{\log_2 4} = \dfrac{8}{4}$

b. $\log (-100) = -2$

c. The domain of $f(x) = \log_2 x$ is $(-\infty, \infty)$.

d. $\log_b x$ is the exponent to which b must be raised to obtain x.

105. Without using a calculator, find the exact value of

$$\frac{\log_3 81 - \log_\pi 1}{\log_{2\sqrt{2}} 8 - \log 0.001}.$$

106. Solve for x: $\log_4[\log_3(\log_2 x)] = 0$.

107. Without using a calculator, determine which is the greater number: $\log_4 60$ or $\log_3 40$.

Group Exercise

108. This group exercise involves exploring the way we grow. Group members should create a graph for the function that models the percentage of adult height attained by a boy who is x years old, $f(x) = 29 + 48.8 \log (x + 1)$. Let $x = 1, 2, 3, \dots, 12$, find function values, and connect the resulting points with a smooth curve. Then create a function that models the percentage of adult height attained by a girl who is x years old, $g(x) = 62 + 35 \log (x - 4)$. Let $x = 5, 6, 7, \dots, 15$, find function values, and connect the resulting points with a smooth curve. Group members should then discuss similarities and differences in the growth patterns for boys and girls based on the graphs.

SECTION 4.3 Properties of Logarithms

Objectives

1. Use the product rule.
2. Use the quotient rule.
3. Use the power rule.
4. Expand logarithmic expressions.
5. Condense logarithmic expressions.
6. Use the change-of-base property.

We all learn new things in different ways. In this section, we consider important properties of logarithms. What would be the most effective way for you to learn about these properties? Would it be helpful to use your graphing utility and discover one of these properties for yourself? To do so, work Exercise 102 in Exercise Set 4.2 before continuing. Would the properties become more meaningful if you could see exactly where they come from? If so, you will find details of the proofs of many of these properties in the appendix. The remainder of our work in this chapter will be based on the properties of logarithms that you learn in this section.

1 Use the product rule.

The Product Rule

Properties of exponents correspond to properties of logarithms. For example, when we multiply with the same base, we add exponents:

$$b^m \cdot b^n = b^{m+n}.$$

This property of exponents, coupled with an awareness that a logarithm is an exponent, suggests the following property, called the **product rule:**

Discovery

We know that log 100,000 = 5. Show that you get the same result by writing 100,000 as 1000 · 100 and then using the product rule. Then verify the product rule by using other numbers whose logarithms are easy to find.

The Product Rule

Let b, M, and N be positive real numbers with $b \neq 1$.

$$\log_b (MN) = \log_b M + \log_b N$$

The logarithm of a product is the sum of the logarithms.

When we use the product rule to write a single logarithm as the sum of two logarithms, we say that we are **expanding a logarithmic expression.** For example, we can use the product rule to expand $\ln(4x)$:

$$\ln (7x) = \ln 7 + \ln x.$$

The logarithm of a product   is   the sum of the logarithms.

EXAMPLE 1   Using the Product Rule

Use the product rule to expand each logarithmic expression:

    a. $\log_4 (7 \cdot 5)$ **b.** $\log (10x)$.

Solution

    a. $\log_4 (7 \cdot 5) = \log_4 7 + \log_4 5$ The logarithm of a product is the sum of the logarithms.

    b. $\log (10x) = \log 10 + \log x$ The logarithm of a product is the sum of the logarithms. These are common logarithms with base 10 understood.

                $= 1 + \log x$ Because $\log_b b = 1$, then $\log 10 = 1$.

> **Check Point 1**   Use the product rule to expand each logarithmic expression:
>
>   **a.** $\log_6 (7 \cdot 11)$ **b.** $\log (100x)$.

2   Use the quotient rule.

The Quotient Rule

When we divide with the same base, we subtract exponents:

$$\frac{b^m}{b^n} = b^{m-n}.$$

This property suggests the following property of logarithms, called the **quotient rule:**

Discovery

We know that $\log_2 16 = 4$. Show that you get the same result by writing 16 as $\dfrac{32}{2}$ and then using the quotient rule. Then verify the quotient rule using other numbers whose logarithms are easy to find.

> **The Quotient Rule**
>
> Let b, M, and N be positive real numbers with $b \neq 1$.
>
> $$\log_b \left(\frac{M}{N} \right) = \log_b M - \log_b N$$
>
> The logarithm of a quotient is the difference of the logarithms.

When we use the quotient rule to write a single logarithm as the difference of two logarithms, we say that we are **expanding a logarithmic expression.** For example, we can use the quotient rule to expand $\log \dfrac{x}{2}$:

$$\log \frac{x}{2} = \log x - \log 2.$$

The logarithm of a quotient   is   the difference of the logarithms.

EXAMPLE 2   Using the Quotient Rule

Use the quotient rule to expand each logarithmic expression:

    a. $\log_7 \left(\dfrac{19}{x} \right)$ **b.** $\ln \left(\dfrac{e^3}{7} \right)$.

Solution

a. $\log_7\left(\dfrac{19}{x}\right) = \log_7 19 - \log_7 x$ The logarithm of a quotient is the difference of the logarithms.

b. $\ln\left(\dfrac{e^3}{7}\right) = \ln e^3 - \ln 7$ The logarithm of a quotient is the difference of the logarithms. These are natural logarithms with base e understood.

$\qquad\qquad = 3 - \ln 7$ Because $\ln e^x = x$, then $\ln e^3 = 3$.

Check Point 2   Use the quotient rule to expand each logarithmic expression:

a. $\log_8\left(\dfrac{23}{x}\right)$ **b.** $\ln\left(\dfrac{e^5}{11}\right)$.

3   Use the power rule.

The Power Rule

When an exponential expression is raised to a power, we multiply exponents:

$$(b^m)^n = b^{mn}.$$

This property suggests the following property of logarithms, called the **power rule:**

> **The Power Rule**
>
> Let b and M be positive real numbers with $b \neq 1$, and let p be any real number.
>
> $$\log_b M^p = p \log_b M$$
>
> The logarithm of a number with an exponent is the product of the exponent and the logarithm of that number.

When we use the power rule to "pull the exponent to the front," we say that we are **expanding a logarithmic expression.** For example, we can use the power rule to expand $\ln x^2$:

$$\ln x^2 = 2 \ln x.$$

The logarithm of a number with an exponent   is   the product of the exponent and the logarithm of that number.

Figure 4.15 shows the graphs of $y = \ln x^2$ and $y = 2 \ln x$. Are $\ln x^2$ and $2 \ln x$ the same? The graphs illustrate that $y = \ln x^2$ and $y = 2 \ln x$ have different domains. The graphs are only the same if $x > 0$. Thus, we should write

$$\ln x^2 = 2 \ln x \text{ for } x > 0.$$

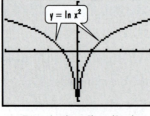

Domain: $(-\infty, 0)$ or $(0, \infty)$

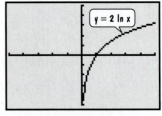

Domain: $(0, \infty)$

Figure 4.15 $\ln x^2$ and $2 \ln x$ have different domains.

Study Tip

The graphs show

$$y_1 = \ln(x + 3)$$

and

$$y_2 = \ln x + \ln 3.$$

The graphs are not the same. The graph of y_1 is the graph of the natural logarithmic function shifted 3 units to the left. By contrast, the graph of y_2 is the graph of the natural logarithmic function shifted upward by $\ln 3$, or about 1.1 units. Thus we see that

$$\ln(x + 3) \neq \ln x + \ln 3.$$

In general,

$$\log_b(M + N) \neq \log_b M + \log_b N.$$

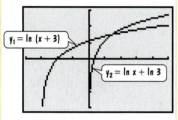

Try to avoid the following errors:

INCORRECT

$$\log_b(M + N) = \log_b M + \log_b N$$

$$\log_b(M - N) = \log_b M - \log_b N$$

$$\log_b(M \cdot N) = \log_b M \cdot \log_b N$$

$$\log_b\left(\frac{M}{N}\right) = \frac{\log_b M}{\log_b N}$$

$$\frac{\log_b M}{\log_b N} = \log_b M - \log_b N$$

4 Expand logarithmic expressions.

When expanding a logarithmic expression, you might want to determine whether the rewriting has changed the domain of the expression. For the rest of this section, assume that all variable and variable expressions represent positive numbers.

EXAMPLE 3 Using the Power Rule

Use the power rule to expand each logarithmic expression:

a. $\log_5 7^4$ **b.** $\ln \sqrt{x}$.

Solution

a. $\log_5 7^4 = 4 \log_5 7$ The logarithm of a number with an exponent is the exponent times the logarithm of the number.

b. $\ln \sqrt{x} = \ln x^{1/2}$ Rewrite the radical using a rational exponent.

$\quad\quad = \frac{1}{2} \ln x$ Use the power rule to bring the exponent to the front.

Check Point 3 Use the power rule to expand each logarithmic expression:

a. $\log_6 3^9$ **b.** $\ln \sqrt[3]{x}$.

Expanding Logarithmic Expressions

It is sometimes necessary to use more than one property of logarithms when you expand a logarithmic expression. Properties for expanding logarithmic expressions are as follows:

Properties for Expanding Logarithmic Expressions

For $M > 0$ and $N > 0$:

1. $\log_b(MN) = \log_b M + \log_b N$ Product rule

2. $\log_b\left(\frac{M}{N}\right) = \log_b M - \log_b N$ Quotient rule

3. $\log_b M^p = p \log_b M$ Power rule

EXAMPLE 4 Expanding Logarithmic Expressions

Use logarithmic properties to expand each expression as much as possible:

a. $\log_b(x^2 \sqrt{y})$ **b.** $\log_6\left(\dfrac{\sqrt[3]{x}}{36y^4}\right)$.

Solution We will have to use two or more of the properties for expanding logarithms in each part of this example.

a. $\log_b(x^2 \sqrt{y}) = \log_b(x^2 y^{1/2})$ Use exponential notation.

$\quad\quad = \log_b x^2 + \log_b y^{1/2}$ Use the product rule.

$\quad\quad = 2 \log_b x + \frac{1}{2} \log_b y$ Use the power rule.

b. $\log_6\left(\dfrac{\sqrt[3]{x}}{36y^4}\right) = \log_6\left(\dfrac{x^{1/3}}{36y^4}\right)$ *Use exponential notation.*

$$= \log_6 x^{1/3} - \log_6(36y^4)$$ *Use the quotient rule.*

$$= \log_6 x^{1/3} - (\log_6 36 + \log_6 y^4)$$ *Use the product rule on $\log_6(36y^4)$.*

$$= \frac{1}{3}\log_6 x - (\log_6 36 + 4\log_6 y)$$ *Use the power rule.*

$$= \frac{1}{3}\log_6 x - \log_6 36 - 4\log_6 y$$ *Apply the distributive property.*

$$= \frac{1}{3}\log_6 x - 2 - 4\log_6 y$$ *$\log_6 36 = 2$ because 2 is the power to which we must raise 6 to get 36. $(6^2 = 36)$*

Check Point 4   Use logarithmic properties to expand each expression as much as possible:

a. $\log_b(x^4\sqrt[3]{y})$ **b.** $\log_5 \dfrac{\sqrt{x}}{25y^3}$.

5 Condense logarithmic expressions.

Condensing Logarithmic Expressions

To **condense a logarithmic expression,** we write the sum or difference of two or more logarithmic expressions as a single logarithmic expression. We use the properties of logarithms to do so.

Study Tip

These properties are the same as those in the box on page 401. The only difference is that we've reversed the sides in each property from the previous box.

Properties for Condensing Logarithmic Expressions

For $M > 0$ and $N > 0$:

1. $\log_b M + \log_b N = \log_b(MN)$ *Product rule*

2. $\log_b M - \log_b N = \log_b\left(\dfrac{M}{N}\right)$ *Quotient rule*

3. $p\log_b M = \log_b M^p$ *Power rule*

EXAMPLE 5   Condensing Logarithmic Expressions

Write as a single logarithm:

a. $\log_4 2 + \log_4 32$ **b.** $\log(4x - 3) - \log x$.

Solution

a. $\log_4 2 + \log_4 32 = \log_4(2 \cdot 32)$ *Use the product rule.*

$$= \log_4 64$$ *We now have a single logarithm. However, we can simplify.*

$$= 3$$ *$\log_4 64 = 3$ because $4^3 = 64$.*

b. $\log(4x - 3) - \log x = \log\left(\dfrac{4x - 3}{x}\right)$ *Use the quotient rule.*

Check Point 5

Write as a single logarithm:

a. $\log 25 + \log 4$ **b.** $\log (7x + 6) - \log x$.

Coefficients of logarithms must be 1 before you can condense them using the product and quotient rules. For example, to condense

$$2 \ln x + \ln (x + 1),$$

the coefficient of the first term must be 1. We use the power rule to rewrite the coefficient as an exponent:

1. Use the power rule to make the number in front an exponent.

$$2 \ln x + \ln (x + 1) = \ln x^2 + \ln (x + 1) = \ln [x^2(x + 1)].$$

2. Use the product rule. The sum of logarithms with coefficients 1 is the logarithm of the product.

EXAMPLE 6 Condensing Logarithmic Expressions

Write as a single logarithm:
a. $\frac{1}{2} \log x + 4 \log (x - 1)$ **b.** $3 \ln (x + 7) - \ln x$
c. $4 \log_b x - 2 \log_b 6 + \frac{1}{2} \log_b y$.

Solution

a. $\frac{1}{2} \log x + 4 \log (x - 1)$

$= \log x^{1/2} + \log (x - 1)^4$ Use the power rule so that all coefficients are 1.

$= \log [x^{1/2} (x - 1)^4]$ Use the product rule. The condensation can be expressed as $\log [\sqrt{x} (x - 1)^4]$.

b. $3 \ln (x + 7) - \ln x$

$= \ln (x + 7)^3 - \ln x$ Use the power rule so that all coefficients are 1.

$= \ln \left[\dfrac{(x + 7)^3}{x} \right]$ Use the quotient rule.

c. $4 \log_b x - 2 \log_b 6 + \frac{1}{2} \log_b y$

$= \log_b x^4 - \log_b 6^2 + \log_b y^{1/2}$ Use the power rule so that all coefficients are 1.

$= (\log_b x^4 - \log_b 36) + \log_b y^{1/2}$ This optional step emphasizes the order of operations.

$= \log_b \left(\dfrac{x^4}{36} \right) + \log_b y^{1/2}$ Use the quotient rule.

$= \log_b \left(\dfrac{x^4}{36} \cdot y^{1/2} \right)$ or $\log \left(\dfrac{x^4 \sqrt{y}}{36} \right)$ Use the product rule.

Check Point 6

Write as a single logarithm:

a. $2 \ln x + \frac{1}{3} \ln (x + 5)$ **b.** $2 \log (x - 3) - \log x$
c. $\frac{1}{4} \log_b x - 2 \log_b 5 + 10 \log_b y$.

6 Use the change-of-base property.

The Change-of-Base Property

We have seen that calculators give the values of both common logarithms (base 10) and natural logarithms (base e). To find a logarithm with any other base, we can use the following change-of-base property:

The Change-of-Base Property

For any logarithmic bases a and b, and any positive number M,

$$\log_b M = \frac{\log_a M}{\log_a b}.$$

The logarithm of M with base b is equal to the logarithm of M with any new base divided by the logarithm of b with that new base.

In the change-of-base property, base b is the base of the original logarithm. Base a is a new base that we introduce. Thus, the change-of-base property allows us to change from base b to *any* new base a, as long as the newly introduced base is a positive number not equal to 1.

The change-of-base property is used to write a logarithm in terms of quantities that can be evaluated with a calculator. Because calculators contain keys for common (base 10) and natural (base e) logarithms, we will frequently introduce base 10 or base e.

Change-of-Base Property

$$\log_b M = \frac{\log_a M}{\log_a b}$$

a is the new introduced base.

Introducing Common Logarithms

$$\log_b M = \frac{\log_{10} M}{\log_{10} b}$$

10 is the new introduced base.

Introducing Natural Logarithms

$$\log_b M = \frac{\log_e M}{\log_e b}$$

e is the new introduced base.

Using the notations for common logarithms and natural logarithms, we have the following results:

The Change-of-Base Property: Introducing Common and Natural Logarithms

Introducing Common Logarithms

$$\log_b M = \frac{\log M}{\log b}$$

Introducing Natural Logarithms

$$\log_b M = \frac{\ln M}{\ln b}$$

EXAMPLE 7 **Changing Base to Common Logarithms**

Use common logarithms to evaluate $\log_5 140$.

Solution Because $\log_b M = \dfrac{\log M}{\log b}$,

$$\log_5 140 = \frac{\log 140}{\log 5}$$

$$\approx 3.07.$$

Use a calculator: 140 [LOG] [÷] 5 [LOG] [=] or [LOG] 140 [÷] [LOG] 5 [ENTER].

This means that $\log_5 140 \approx 3.07$.

Discovery

Find a reasonable estimate of $\log_5 140$ to the nearest whole number. 5 to what power is 140? Compare your estimate to the value obtained in Example 7.

Check Point 7 Use common logarithms to evaluate $\log_7 2506$.

EXAMPLE 8 Changing Base to Natural Logarithms

Use natural logarithms to evaluate $\log_5 140$.

Solution Because $\log_b M = \dfrac{\ln M}{\ln b}$,

$$\log_5 140 = \frac{\ln 140}{\ln 5}$$
$$\approx 3.07.$$

Use a calculator: 140 [LN] [÷] 5 [LN] [=] or [LN] 140 [÷] [LN] 5 [ENTER].

We have again shown that $\log_5 140 \approx 3.07$.

 Check Point 8 Use natural logarithms to evaluate $\log_7 2506$.

We can use the change-of-base property to graph logarithmic functions with bases other than 10 or e on a graphing utility. For example, Figure 4.16 shows the graphs of

$$y = \log_2 x \quad \text{and} \quad y = \log_{20} x$$

in a [0, 10, 1] by [−3, 3, 1] viewing rectangle. Because $\log_2 x = \dfrac{\ln x}{\ln 2}$ and $\log_{20} x = \dfrac{\ln x}{\ln 20}$, the functions can be entered as

$$y_1 = [LN] \ x \ [÷] [LN] \ 2$$
$$\text{and} \ y_2 = [LN] \ x \ [÷] [LN] \ 20.$$

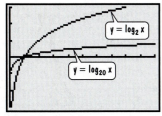

Figure 4.16 Using the change-of-base property to graph logarithmic functions

EXERCISE SET 4.3

Practice Exercises

In Exercises 1–40, use properties of logarithms to expand each logarithmic expression as much as possible. Where possible, evaluate logarithmic expressions without using a calculator.

1. $\log_5 (7 \cdot 3)$

2. $\log_8 (13 \cdot 7)$

3. $\log_7 (7x)$

4. $\log_9 (9x)$

5. $\log (1000x)$

6. $\log (10{,}000x)$

7. $\log_7 \left(\dfrac{7}{x}\right)$

8. $\log_9 \left(\dfrac{9}{x}\right)$

9. $\log \left(\dfrac{x}{100}\right)$

10. $\log \left(\dfrac{x}{1000}\right)$

11. $\log_4 \left(\dfrac{64}{y}\right)$

12. $\log_5 \left(\dfrac{125}{y}\right)$

13. $\ln \left(\dfrac{e^2}{5}\right)$

14. $\ln \left(\dfrac{e^4}{8}\right)$

15. $\log_b x^3$

16. $\log_b x^7$

17. $\log N^{-6}$

18. $\log M^{-8}$

19. $\ln \sqrt[5]{x}$

20. $\ln \sqrt[7]{x}$

21. $\log_b (x^2 y)$

22. $\log_b (xy^3)$

23. $\log_4 \left(\dfrac{\sqrt{x}}{64}\right)$

24. $\log_5 \left(\dfrac{\sqrt{x}}{25}\right)$

25. $\log_6 \left(\dfrac{36}{\sqrt{x+1}}\right)$

26. $\log_8 \left(\dfrac{64}{\sqrt{x+1}}\right)$

27. $\log_b \left(\dfrac{x^2 y}{z^2}\right)$

28. $\log_b \left(\dfrac{x^3 y}{z^2}\right)$

29. $\log \sqrt{100x}$

30. $\ln \sqrt{ex}$

31. $\log \sqrt[3]{\dfrac{x}{y}}$

32. $\log \sqrt[5]{\dfrac{x}{y}}$

33. $\log_b \left(\dfrac{\sqrt{x}\,y^3}{z^3} \right)$

34. $\log_b \left(\dfrac{\sqrt[3]{x}\,y^4}{z^5} \right)$

35. $\log_5 \sqrt[3]{\dfrac{x^2 y}{25}}$

36. $\log_2 \sqrt[5]{\dfrac{x y^4}{16}}$

37. $\ln \left[\dfrac{x^3 \sqrt{x^2 + 1}}{(x + 1)^4} \right]$

38. $\ln \left[\dfrac{x^4 \sqrt{x^2 + 3}}{(x + 3)^5} \right]$

39. $\log \left[\dfrac{10 x^2 \sqrt[3]{1 - x}}{7(x + 1)^2} \right]$

40. $\log \left[\dfrac{100 x^3 \sqrt[3]{5 - x}}{3(x + 7)^2} \right]$

In Exercises 41–70, use properties of logarithms to condense each logarithmic expression. Write the expression as a single logarithm whose coefficient is 1. Where possible, evaluate logarithmic expressions.

41. $\log 5 + \log 2$

42. $\log 250 + \log 4$

43. $\ln x + \ln 7$

44. $\ln x + \ln 3$

45. $\log_2 96 - \log_2 3$

46. $\log_3 405 - \log_3 5$

47. $\log (2x + 5) - \log x$

48. $\log (3x + 7) - \log x$

49. $\log x + 3 \log y$

50. $\log x + 7 \log y$

51. $\frac{1}{2} \ln x + \ln y$

52. $\frac{1}{3} \ln x + \ln y$

53. $2 \log_b x + 3 \log_b y$

54. $5 \log_b x + 6 \log_b y$

55. $5 \ln x - 2 \ln y$

56. $7 \ln x - 3 \ln y$

57. $3 \ln x - \frac{1}{3} \ln y$

58. $2 \ln x - \frac{1}{2} \ln y$

59. $4 \ln (x + 6) - 3 \ln x$

60. $8 \ln (x + 9) - 4 \ln x$

61. $3 \ln x + 5 \ln y - 6 \ln z$

62. $4 \ln x + 7 \ln y - 3 \ln z$

63. $\frac{1}{2} (\log x + \log y)$

64. $\frac{1}{3} (\log_4 x - \log_4 y)$

65. $\frac{1}{2} (\log_5 x + \log_5 y) - 2 \log_5 (x + 1)$

66. $\frac{1}{3} (\log_4 x - \log_4 y) + 2 \log_4 (x + 1)$

67. $\frac{1}{3} [2 \ln (x + 5) - \ln x - \ln (x^2 - 4)]$

68. $\frac{1}{3} [5 \ln (x + 6) - \ln x - \ln (x^2 - 25)]$

69. $\log x + \log 7 + \log (x^2 - 1) - \log (x + 1)$

70. $\log x + \log 15 + \log (x^2 - 4) - \log (x + 2)$

In Exercises 71–78, use common logarithms or natural logarithms and a calculator to evaluate to four decimal places.

71. $\log_5 13$

72. $\log_6 17$

73. $\log_{14} 87.5$

74. $\log_{16} 57.2$

75. $\log_{0.1} 17$

76. $\log_{0.3} 19$

77. $\log_\pi 63$

78. $\log_\pi 400$

In Exercises 79–82, use a graphing utility and the change-of-base property to graph each function.

79. $y = \log_3 x$

80. $y = \log_{15} x$

81. $y = \log_2 (x + 2)$

82. $y = \log_3 (x - 2)$

Application Exercises

83. The loudness level of a sound can be expressed by comparing the sound's intensity to the intensity of a sound barely audible to the human ear. The formula

$$D = 10(\log I - \log I_0)$$

describes the loudness level of a sound, D, in decibels, where I is the intensity of the sound, in watts per meter^2, and I_0 is the intensity of a sound barely audible to the human ear.

a. Express the formula so that the expression in parentheses is written as a single logarithm.

b. Use the form of the formula from part (a) to answer this question: If a sound has an intensity 100 times the intensity of a softer sound, how much larger on the decibel scale is the loudness level of the more intense sound?

84. The formula

$$t = \frac{1}{c} \left[\ln A - \ln (A - N) \right]$$

describes the time, t, in weeks, that it takes to achieve mastery of a portion of a task, where A is the maximum learning possible, N is the portion of the learning that is to be achieved, and c is a constant used to measure an individual's learning style.

a. Express the formula so that the expression in brackets is written as a single logarithm.

b. The formula is also used to determine how long it will take chimpanzees and apes to master a task. For example, a typical chimpanzee learning sign language can master a maximum of 65 signs. Use the form of the formula from part (a) to answer this question: How many weeks will it take a chimpanzee to master 30 signs if c for that chimp is 0.03?

Writing in Mathematics

85. Describe the product rule for logarithms and give an example.

86. Describe the quotient rule for logarithms and give an example.

87. Describe the power rule for logarithms and give an example.

88. Without showing the details, explain how to condense $\ln x - 2 \ln (x + 1)$.

89. Describe the change-of-base property and give an example.

90. Explain how to use your calculator to find $\log_{14} 283$.

91. You overhear a student talking about a property of logarithms in which division becomes subtraction. Explain what the student means by this.

92. Find $\ln 2$ using a calculator. Then calculate each of the following: $1 - \frac{1}{2}$; $1 - \frac{1}{2} + \frac{1}{3}$; $1 - \frac{1}{2} + \frac{1}{3} - \frac{1}{4}$; $1 - \frac{1}{2} + \frac{1}{3} - \frac{1}{4} + \frac{1}{5}$; .... Describe what you observe.

Technology Exercises

93. a. Use a graphing utility (and the change-of-base property) to graph $y = \log_3 x$.

   b. Graph $y = 2 + \log_3 x$, $y = \log_3 (x + 2)$, and $y = -\log_3 x$ in the same viewing rectangle as $y = \log_3 x$. Then describe the change or changes that need to be made to the graph of $y = \log_3 x$ to obtain each of these three graphs.

94. Graph $y = \log x$, $y = \log (10x)$, and $y = \log (0.1x)$ in the same viewing rectangle. Describe the relationship among the three graphs. What logarithmic property accounts for this relationship?

95. Use a graphing utility and the change-of-base property to graph $y = \log_3 x$, $y = \log_{25} x$, and $y = \log_{100} x$ in the same viewing rectangle.

   a. Which graph is on the top in the interval $(0, 1)$? Which is on the bottom?

   b. Which graph is on the top in the interval $(1, \infty)$? Which is on the bottom?

   c. Generalize by writing a statement about which graph is on top, which is on the bottom, and in which intervals, using $y = \log_b x$ where $b > 1$.

Disprove each statement in Exercises 96–100 by
   a. *letting y equal a positive constant of your choice.*
   b. *using a graphing utility to graph the function on each side of the equal sign. The two functions should have different graphs, showing that the equation is not true in general.*

96. $\log(x + y) = \log x + \log y$ **97.** $\log \dfrac{x}{y} = \dfrac{\log x}{\log y}$

98. $\ln(x - y) = \ln x - \ln y$ **99.** $\ln(xy) = (\ln x)(\ln y)$

100. $\dfrac{\ln x}{\ln y} = \ln x - \ln y$

Critical Thinking Exercises

101. Which one of the following is true?

   a. $\dfrac{\log_7 49}{\log_7 7} = \log_7 49 - \log_7 7$

   b. $\log_b(x^3 + y^3) = 3 \log_b x + 3 \log_b y$

   c. $\log_b(xy)^5 = (\log_b x + \log_b y)^5$

   d. $\ln \sqrt{2} = \dfrac{\ln 2}{2}$

102. Use the change-of-base property to prove that

$$\log e = \frac{1}{\ln 10}.$$

103. If $\log 3 = A$ and $\log 7 = B$, find $\log_7 9$ in terms of A and B.

104. Write as a single term that does not contain a logarithm:

$$e^{\ln 8x^5 - \ln 2x^2}.$$

105. If $f(x) = \log_b x$, show that

$$\frac{f(x + h) - f(x)}{h} = \log_b \left(1 + \frac{h}{x}\right)^{1/h}, h \neq 0.$$

SECTION 4.4   *Exponential and Logarithmic Equations*

Objectives

1. Solve exponential equations.

2. Solve logarithmic equations.

3. Solve applied problems involving exponential and logarithmic equations.

Is an early retirement awaiting you?

You inherited $30,000. You'd like to put aside $25,000 and eventually have over half a million dollars for early retirement. Is this possible? In this section, you will see how techniques for solving equations with variable exponents provide an answer to the question.

1 Solve exponential equations.

Exponential Equations

An **exponential equation** is an equation containing a variable in an exponent. Examples of exponential equations include

$$3^x = 81, \quad 4^x = 15, \quad \text{and} \quad 40e^{0.6x} = 240.$$

Each side of the first equation can be expressed with the same base. Can you see that we can rewrite

$$3^x = 81 \quad \text{as} \quad 3^x = 3^4?$$

All exponential functions are one-to-one—that is, if b is a positive number other than 1 and $b^M = b^N$, then $M = N$. Because we have expressed $3^x = 81$ as $3^x = 3^4$, we conclude that $x = 4$. The equation's solution set is {4}.

Most exponential equations cannot be rewritten so that each side has the same base. Logarithms are extremely useful in solving such equations. The solution begins with isolating the exponential expression and taking the natural logarithm on both sides. Why can we do this? All logarithmic relations are functions. Thus, if M and N are positive real numbers and $M = N$, then $\log_b M = \log_b N$.

Using Natural Logarithms to Solve Exponential Equations

1. Isolate the exponential expression.
2. Take the natural logarithm on both sides of the equation.
3. Simplify using one of the following properties:

$$\ln b^x = x \ln b \quad \text{or} \quad \ln e^x = x.$$

4. Solve for the variable.

EXAMPLE 1   Solving an Exponential Equation

Solve: $4^x = 15$.

Solution   Because the exponential expression, 4^x, is already isolated on the left, we begin by taking the natural logarithm on both sides of the equation.

Discovery

The base that is used when taking the logarithm on both sides of an equation can be any base at all. Solve $4^x = 15$ by taking the common logarithm on both sides. Solve again, this time taking the logarithm with base 4 on both sides. Use the change-of-base property to show that the solutions are the same as the one obtained in Example 1.

$4^x = 15$	This is the given equation.
$\ln 4^x = \ln 15$	Take the natural logarithm on both sides.
$x \ln 4 = \ln 15$	Use the power rule and bring the variable exponent to the front: $\ln b^x = x \ln b$.
$x = \dfrac{\ln 15}{\ln 4}$	Solve for x by dividing both sides by $\ln 4$.

We now have an exact value for x. We use the exact value for x in the equation's solution set. Thus, the equation's solution is $\dfrac{\ln 15}{\ln 4}$ and the solution set is $\left\{ \dfrac{\ln 15}{\ln 4} \right\}$. We can obtain a decimal approximation by using a calculator: $x \approx 1.95$. Because $4^2 = 16$, it seems reasonable that the solution to $4^x = 15$ is approximately 1.95

Check Point 1 Solve: $5^x = 134$. Find the solution set and then use a calculator to obtain a decimal approximation to two decimal places for the solution.

EXAMPLE 2   Solving an Exponential Equation

Solve: $40e^{0.6x} = 240$.

Solution   We begin by dividing both sides by 40 to isolate the exponential expression, $e^{0.6x}$. Then we take the natural logarithm on both sides of the equation.

$40e^{0.6x} = 240$	This is the given equation.
$e^{0.6x} = 6$	Isolate the exponential factor by dividing both sides by 40.
$\ln e^{0.6x} = \ln 6$	Take the natural logarithm on both sides.
$0.6x = \ln 6$	Use the inverse property $\ln e^x = x$ on the left.
$x = \dfrac{\ln 6}{0.6} \approx 2.99$	Divide both sides by 0.6.

Thus, the solution of the equation is $\dfrac{\ln 6}{0.6} \approx 2.99$. Try checking this approximate solution in the original equation to verify that $\left\{\dfrac{\ln 6}{0.6}\right\}$ is the solution set.

Check Point 2 Solve: $7e^{2x} = 63$. Find the solution set and then use a calculator to obtain a decimal approximation to two decimal places for the solution.

EXAMPLE 3   Solving an Exponential Equation

Solve: $5^{4x-7} - 3 = 10$

Solution   We begin by adding 3 to both sides to isolate the exponential expression, 5^{4x-7}. Then we take the natural logarithm on both sides of the equation.

$5^{4x-7} - 3 = 10$	This is the given equation.
$5^{4x-7} = 13$	Add 3 to both sides.
$\ln 5^{4x-7} = \ln 13$	Take the natural logarithm on both sides.
$(4x - 7)\ln 5 = \ln 13$	Use the power rule to bring the exponent to the front: $\ln M^p = p\ln M$.
$4x\ln 5 - 7\ln 5 = \ln 13$	Use the distributive property and distribute $\ln 5$ to both terms in parentheses.
$4x\ln 5 = \ln 13 + 7\ln 5$	Isolate the variable term by adding $7\ln 5$ to both sides.
$x = \dfrac{\ln 13 + 7\ln 5}{4\ln 5}$	Isolate x by dividing both sides by $4\ln 5$.

The solution set is $\left\{\dfrac{\ln 13 + 7\ln 5}{4\ln 5}\right\}$. The solution is approximately 2.15.

Check Point 3 Solve: $6^{3x-4} - 7 = 2081$. Find the solution set and then use a calculator to obtain a decimal approximation to two decimal places for the solution.

EXAMPLE 4 Solving an Exponential Equation

Solve: $e^{2x} - 4e^x + 3 = 0$.

Solution   The given equation is quadratic in form. If $t = e^x$, the equation can be expressed as $t^2 - 4t + 3 = 0$. Because this equation can be solved by factoring, we factor to isolate the exponential term.

$$e^{2x} - 4e^x + 3 = 0 \qquad \text{This is the given equation.}$$

$$(e^x - 3)(e^x - 1) = 0 \qquad \text{Factor on the left. Notice that if } t = e^x,$$
$$t^2 - 4t + 3 = (t - 3)(t - 1).$$

$$e^x - 3 = 0 \quad \text{or} \quad e^x - 1 = 0 \qquad \text{Set each factor equal to 0.}$$

$$e^x = 3 \qquad\qquad e^x = 1 \qquad \text{Solve for } e^x.$$

$$\ln e^x = \ln 3 \qquad\qquad x = 0 \qquad \text{Take the natural logarithm on both sides of the first equation. The equation on the right can be solved by inspection.}$$

$$x = \ln 3 \qquad\qquad\qquad\qquad\quad \ln e^x = x$$

The solution set is $\{0, \ln 3\}$. The solutions are 0 and approximately 1.10.

Technology

Shown below is the graph of $y = e^{2x} - 4e^x + 3$. There are two x-intercepts, one at 0 and one at approximately 1.10. These intercepts verify our algebraic solution.

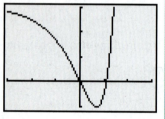

Check Point 4 Solve: $e^{2x} - 8e^x + 7 = 0$. Find the solution set and then use a calculator to obtain a decimal approximation to two decimal places, if necessary, for the solutions.

2 Solve logarithmic equations.

Logarithmic Equations

A **logarithmic equation** is an equation containing a variable in a logarithmic expression. Examples of logarithmic equations include

$$\log_4 (x + 3) = 2 \quad \text{and} \quad \ln (2x) = 3.$$

If a logarithmic equation is in the form $\log_b x = c$, we can solve the equation by rewriting it in its equivalent exponential form $b^c = x$. Example 5 illustrates how this is done.

EXAMPLE 5 Solving a Logarithmic Equation

Solve: $\log_4 (x + 3) = 2$.

Solution   We first rewrite the equation as an equivalent equation in exponential form using the fact that $\log_b x = c$ means $b^c = x$.

$$\log_4(x + 3) = 2 \qquad \text{means} \qquad 4^2 = x + 3$$

Logarithms are exponents.

Technology

The graphs of
$y_1 = \log_4(x + 3)$ and $y_2 = 2$
have an intersection point whose x-coordinate is 13. This verifies that {13} is the solution set for $\log_4(x + 3) = 2$.

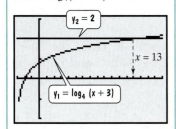

$[-3, 17, 1]$ by $[-2, 3, 1]$

Note:
Because

$$\log_b x = \frac{\ln x}{\ln b}$$

(change-of-base property), we entered y_1 using

$$y_1 = \frac{\ln(x + 3)}{\ln 4}.$$

Now we solve the equivalent equation for x.

$4^2 = x + 3$	This is the equation equivalent to $\log_4(x + 3) = 2$.
$16 = x + 3$	Square 4.
$13 = x$	Subtract 3 from both sides.

Check 13:

$$\log_4(x + 3) = 2 \qquad \text{This is the given logarithmic equation.}$$
$$\log_4(13 + 3) \overset{?}{=} 2 \qquad \text{Substitute 13 for x.}$$
$$\log_4 16 \overset{?}{=} 2$$
$$2 = 2 \checkmark \qquad \log_4 16 = 2 \text{ because } 4^2 = 16.$$

This true statement indicates that the solution set is {13}.

Check Point 5 Solve: $\log_2(x - 4) = 3$.

Logarithmic expressions are defined only for logarithms of positive real numbers. Always check proposed solutions of a logarithmic equation in the original equation. Exclude from the solution set any proposed solution that produces the logarithm of a negative number or the logarithm of 0.

To rewrite the logarithmic equation $\log_b x = c$ in the equivalent exponential form $b^c = x$, we need a single logarithm whose coefficient is one. It is sometimes necessary to use properties of logarithms to condense logarithms into a single logarithm. In the next example, we use the product rule for logarithms to obtain a single logarithmic expression on the left side.

EXAMPLE 6 Using the Product Rule to Solve a Logarithmic Equation

Solve: $\log_2 x + \log_2(x - 7) = 3$.

Solution

$\log_2 x + \log_2(x - 7) = 3$	This is the given equation.
$\log_2[x(x - 7)] = 3$	Use the product rule to obtain a single logarithm: $\log_b M + \log_b N = \log_b(MN)$.
$2^3 = x(x - 7)$	$\log_b x = c$ means $b^c = x$.
$8 = x^2 - 7x$	Apply the distributive property on the right and evaluate 2^3 on the left.
$0 = x^2 - 7x - 8$	Set the equation equal to 0.
$0 = (x - 8)(x + 1)$	Factor.
$x - 8 = 0$ or $x + 1 = 0$	Set each factor equal to 0.
$x = 8 \qquad\qquad x = -1$	Solve for x.

Check 8:

$$\log_2 x + \log_2(x - 7) = 3$$
$$\log_2 8 + \log_2(8 - 7) \overset{?}{=} 3$$
$$\log_2 8 + \log_2 1 \overset{?}{=} 3$$
$$3 + 0 \overset{?}{=} 3$$
$$3 = 3 \checkmark$$

Check −1:

$$\log_2 x + \log_2(x - 7) = 3$$
$$\log_2(-1) + \log_2(-1 - 7) \overset{?}{=} 3$$

The number −1 does not check.
Negative numbers do not have logarithms.

The solution set is {8}.

Check Point 6 Solve: $\log x + \log(x - 3) = 1$.

Equations involving natural logarithms can be solved using the inverse property $e^{\ln x} = x$. For example, to solve

$$\ln x = 5$$

we write both sides of the equation as exponents on base e:

$$e^{\ln x} = e^5.$$

This is called **exponentiating both sides** of the equation. Using the inverse property $e^{\ln x} = x$, we simplify the left side of the equation and obtain the solution:

$$x = e^5.$$

EXAMPLE 7 Solving an Equation with a Natural Logarithm

Solve: $3 \ln(2x) = 12$.

Solution

$3 \ln(2x) = 12$	This is the given equation.
$\ln(2x) = 4$	Divide both sides by 3.
$e^{\ln(2x)} = e^4$	Exponentiate both sides.
$2x = e^4$	Use the inverse property to simplify the left side: $e^{\ln x} = x$.
$x = \dfrac{e^4}{2} \approx 27.30$	Divide both sides by 2.

Check $\dfrac{e^4}{2}$:

$3 \ln(2x) = 12$	This is the given logarithmic equation.
$3 \ln\left[2\left(\dfrac{e^4}{2}\right)\right] \stackrel{?}{=} 12$	Substitute $\dfrac{e^4}{2}$ for x.
$3 \ln e^4 \stackrel{?}{=} 12$	Simplify: $\dfrac{\cancel{2}}{1} \cdot \dfrac{e^4}{\cancel{2}} = e^4$.
$3 \cdot 4 \stackrel{?}{=} 12$	Because $\ln e^x = x$, we conclude $\ln e^4 = 4$.
$12 = 12$ ✓	

This true statement indicates that the solution set is $\left\{\dfrac{e^4}{2}\right\}$.

Check Point 7 Solve: $4 \ln 3x = 8$.

3 Solve applied problems involving exponential and logarithmic equations.

Applications

Our first applied example provides a mathematical perspective on the old slogan "Alcohol and driving don't mix." In California, where 38% of fatal traffic crashes involve drinking drivers, it is illegal to drive with a blood alcohol concentration of 0.08 or higher. At these levels, drivers may be arrested and charged with driving under the influence.

EXAMPLE 8   Alcohol and Risk of a Car Accident

Medical research indicates that the risk of having a car accident increases exponentially as the concentration of alcohol in the blood increases. The risk is modeled by

$$R = 6e^{12.77x}$$

where x is the blood alcohol concentration and R, given as a percent, is the risk of having a car accident. What blood alcohol concentration corresponds to a 17% risk of a car accident?

Solution   For a risk of 17%, we let $R = 17$ in the equation and solve for x, the blood alcohol concentration.

$R = 6e^{12.77x}$	This is the given equation.
$6e^{12.77x} = 17$	Substitute 17 for R and (optional) reverse the two sides of the equation.
$e^{12.77x} = \dfrac{17}{6}$	Isolate the exponential factor by dividing both sides by 6.
$\ln e^{12.77x} = \ln\left(\dfrac{17}{6}\right)$	Take the natural logarithm on both sides.
$12.77x = \ln\left(\dfrac{17}{6}\right)$	Use the inverse property $\ln e^x = x$ on the left.
$x = \dfrac{\ln\left(\dfrac{17}{6}\right)}{12.77} \approx 0.08$	Divide both sides by 12.77.

For a blood alcohol concentration of 0.08, the risk of a car accident is 17%. In many states, it is illegal to drive at this blood alcohol concentration.

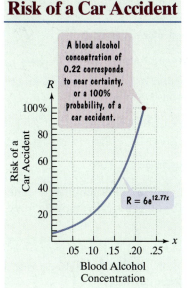

Visualizing the Relationship between Blood Alcohol Concentration and the Risk of a Car Accident

A blood alcohol concentration of 0.22 corresponds to near certainty, or a 100% probability, of a car accident.

$R = 6e^{12.77x}$

Blood Alcohol Concentration

Check Point 8   Use the formula in Example 8 to answer this question: What blood alcohol concentration corresponds to a 7% risk of a car accident? (In many states, drivers under the age of 21 can lose their license for driving at this level.)

Suppose that you inherit $30,000. Is it possible to invest $25,000 and have over half a million dollars for early retirement? Our next example illustrates the power of compound interest.

EXAMPLE 9   Revisiting the Formula for Compound Interest

The formula

$$A = P\left(1 + \frac{r}{n}\right)^{nt}$$

describes the accumulated value, A, of a sum of money, P, the principal, after t years at annual percentage rate r (in decimal form) compounded n times a year. How long will it take $25,000 to grow to $500,000 at 9% annual interest compounded monthly?

Solution

$$A = P\left(1 + \frac{r}{n}\right)^{nt}$$　This is the given formula.

$$500{,}000 = 25{,}000\left(1 + \frac{0.09}{12}\right)^{12t}$$　A (the desired accumulated value) = $500,000,
P (the principal) = $25,000,
r (the interest rate) = 9% = 0.09, and n = 12
(monthly compounding).

Our goal is to solve the equation for t. Let's reverse the two sides of the equation and then simplify within parentheses.

$$25{,}000\left(1 + \frac{0.09}{12}\right)^{12t} = 500{,}000$$　Reverse the two sides of the previous equation.

$$25{,}000(1 + 0.0075)^{12t} = 500{,}000$$　Divide within parentheses: $\dfrac{0.09}{12} = 0.0075$.

$$25{,}000(1.0075)^{12t} = 500{,}000$$　Add within parentheses.

$$(1.0075)^{12t} = 20$$　Divide both sides by 25,000.

$$\ln(1.0075)^{12t} = \ln 20$$　Take the natural logarithm on both sides.

$$12t\ln(1.0075) = \ln 20$$　Use the power rule to bring the exponent to the front: $\ln M^p = p\ln M$.

$$t = \frac{\ln 20}{12\ln 1.0075}$$　Solve for t, dividing both sides by 12 ln 1.0075.

$$\approx 33.4$$　Use a calculator.

After approximately 33.4 years, the $25,000 will grow to an accumulated value of $500,000. If you set aside the money at age 20, you can begin enjoying a life of leisure at about age 53.

Check Point 9　How long, to the nearest tenth of a year, will it take $1000 to grow to $3600 at 8% annual interest compounded quarterly?

> ## Playing Doubles: Interest Rates and Doubling Time
>
> One way to calculate what your savings will be worth at some point in the future is to consider doubling time. Shown below is how long it takes for your money to double at different annual interest rates subject to continuous compounding.
>
> | Annual Interest Rate | Years to Double |
> |---|---|
> | 5% | 13.9 years |
> | 7% | 9.9 years |
> | 9% | 7.7 years |
> | 11% | 6.3 years |
>
> Of course, the first problem is collecting some money to invest. The second problem is finding a reasonably safe investment with a return of 9% or more.

Yogi Berra, catcher and renowned hitter for the New York Yankees (1946–1963), said it best: "Prediction is very hard, especially when it's about the future." At the start of the twenty-first century, we are plagued by questions about the environment. Will we run out of gas? How hot will it get? Will there be neighborhoods where the air is pristine? Can we make garbage disappear? Will there be any wilderness left? Which wild animals will become extinct? These concerns have led to the growth of the environmental industry in the United States.

EXAMPLE 10　The Growth of the Environmental Industry

The formula

$$N = 461.87 + 299.4\ln x$$

models the thousands of workers, N, in the environmental industry in the United States x years after 1979. By which year will there be 1,500,000, or 1500 thousand, U.S. workers in the environmental industry?

Solution   We substitute 1500 for N and solve for x, the number of years after 1979.

$$N = 461.87 + 299.4 \ln x \quad \text{This is the given formula.}$$

$$461.87 + 299.4 \ln x = 1500 \qquad \text{Substitute 1500 for N and reverse the two sides of the equation.}$$

Our goal is to isolate $\ln x$. We can then find x by exponentiating both sides of the equation, using the inverse property $e^{\ln x} = x$.

$$299.4 \ln x = 1038.13 \qquad \text{Subtract 461.87 from both sides.}$$

$$\ln x = \frac{1038.13}{299.4} \qquad \text{Divide both sides by 299.4.}$$

$$e^{\ln x} = e^{1038.13/299.4} \qquad \text{Exponentiate both sides.}$$

$$x = e^{1038.13/299.4} \qquad e^{\ln x} = x$$

$$\approx 32 \qquad \text{Use a calculator.}$$

Approximately 32 years after 1979, in the year 2011, there will be 1.5 million U.S. workers in the environmental industry.

Check Point 10   Use the formula in Example 10 to find by which year there will be two million, or 2000 thousand, U.S. workers in the environmental industry.

EXERCISE SET 4.4

Practice Exercises

Solve each exponential equation in Exercises 1–26. Express the solution set in terms of natural logarithms. Then use a calculator to obtain a decimal approximation, correct to two decimal places, for the solution.

1. $10^x = 3.91$

2. $10^x = 8.07$

3. $e^x = 5.7$

4. $e^x = 0.83$

5. $5^x = 17$

6. $19^x = 143$

7. $5e^x = 23$

8. $9e^x = 107$

9. $3e^{5x} = 1977$

10. $4e^{7x} = 10{,}273$

11. $e^{1-5x} = 793$

12. $e^{1-8x} = 7957$

13. $e^{5x-3} - 2 = 10{,}476$

14. $e^{4x-5} - 7 = 11{,}243$

15. $7^{x+2} = 410$

16. $5^{x-3} = 137$

17. $7^{0.3x} = 813$

18. $3^{x/7} = 0.2$

19. $5^{2x+3} = 3^{x-1}$

20. $7^{2x+1} = 3^{x+2}$

21. $e^{2x} - 3e^x + 2 = 0$

22. $e^{2x} - 2e^x - 3 = 0$

23. $e^{4x} + 5e^{2x} - 24 = 0$

24. $e^{4x} - 3e^{2x} - 18 = 0$

25. $3^{2x} + 3^x - 2 = 0$

26. $2^{2x} + 2^x - 12 = 0$

Solve each logarithmic equation in Exercises 27–44. Be sure to reject any value of x that produces the logarithm of a negative number or the logarithm of 0.

27. $\log_3 x = 4$

28. $\log_5 x = 3$

29. $\log_4(x + 5) = 3$

30. $\log_5(x - 7) = 2$

31. $\log_3(x - 4) = -3$

32. $\log_7(x + 2) = -2$

33. $\log_4(3x + 2) = 3$

34. $\log_2(4x + 1) = 5$

35. $\log_5 x + \log_5(4x - 1) = 1$

36. $\log_6(x + 5) + \log_6 x = 2$

37. $\log_3(x - 5) + \log_3(x + 3) = 2$

38. $\log_2(x - 1) + \log_2(x + 1) = 3$

39. $\log_2(x + 2) - \log_2(x - 5) = 3$

40. $\log_4(x + 2) - \log_4(x - 1) = 1$

41. $2 \log_3(x + 4) = \log_3 9 + 2$

42. $3 \log_2(x - 1) = 5 - \log_2 4$

43. $\log_2(x - 6) + \log_2(x - 4) - \log_2 x = 2$

44. $\log_2(x - 3) + \log_2 x - \log_2(x + 2) = 2$

Exercises 45–52 involve equations with natural logarithms. Solve each equation by isolating the natural logarithm and exponentiating both sides. Express the answer in terms of e. Then use a calculator to obtain a decimal approximation, correct to two decimal places, for the solution.

45. $\ln x = 2$

46. $\ln x = 3$

47. $5 \ln(2x) = 20$

48. $6 \ln(2x) = 30$

49. $6 + 2 \ln x = 5$

50. $7 + 3 \ln x = 6$

51. $\ln \sqrt{x + 3} = 1$

52. $\ln \sqrt{x + 4} = 1$

Application Exercises

Use the formula $R = 6e^{12.77x}$, where x is the blood alcohol concentration and R, given as a percent, is the risk of having a car accident, to solve Exercises 53–54.

53. What blood alcohol concentration corresponds to a 25% risk of a car accident?

54. What blood alcohol concentration corresponds to a 50% risk of a car accident?

55. The formula $A = 18.9e^{0.0055t}$ models the population of New York State, A, in millions, t years after 2000.

    a. What was the population of New York in 2000?

    b. When will the population of New York reach 19.6 million?

56. The formula $A = 15.9e^{0.0235t}$ models the population of Florida, A, in millions, t years after 2000.

    a. What was the population of Florida in 2000?

    b. When will the population of Florida reach 17.5 million?

In Exercises 57–60, complete the table for a savings account subjected to n compoundings yearly $\left[A = P\left(1 + \dfrac{r}{n}\right)^{nt} \right]$. Round answers to one decimal place.

	Amount Invested	Number of Compounding Periods	Annual Interest Rate	Accumulated Amount	Time t in Years
57.	$12,500	4	5.75%	$20,000	
58.	$7250	12	6.5%	$15,000	
59.	$1000	360		$1400	2
60.	$5000	360		$9000	4

In Exercises 61–64, complete the table for a savings account subjected to continuous compounding $(A = Pe^{rt})$. Round answers to one decimal place.

	Amount Invested	Annual Interest Rate	Accumulated Amount	Time t in Years
61.	$8000	8%	Double the amount invested	
62.	$8000		$12,000	2
63.	$2350		Triple the amount invested	7
64.	$17,425	4.25%	$25,000	

65. The function $f(x) = 15{,}557 + 5259 \ln x$ models the average cost of a new car, $f(x)$, in dollars, x years after 1989. When was the average cost of a new car $25,000?

66. The function $f(x) = 68.41 + 1.75 \ln x$ models the life expectancy, $f(x)$, in years, for African-American females born x years after 1969. In which birth year was life expectancy 73.7 years? Round to the nearest year.

The function $P(x) = 95 - 30\log_2 x$ models the percentage, $P(x)$, of students who could recall the important features of a classroom lecture as a function of time, where x represents the number of days that have elapsed since the lecture was given. The figure shows the graph of the function. Use this information to solve Exercises 67–68. Round answers to one decimal place.

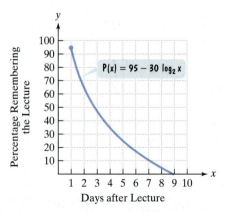

67. After how many days do only half the students recall the important features of the classroom lecture? (Let $P(x) = 50$ and solve for x.) Locate the point on the graph that conveys this information.

68. After how many days have all students forgotten the important features of the classroom lecture? (Let $P(x) = 0$ and solve for x.) Locate the point on the graph on the previous page that conveys this information.

The pH of a solution ranges from 0 to 14. An acid solution has a pH less than 7. Pure water is neutral and has a pH of 7. Normal, unpolluted rain has a pH of about 5.6. The pH of a solution is given by

$$pH = -\log x$$

where x represents the concentration of the hydrogen ions in the solution, in moles per liter. Use the formula to solve Exercises 69–70.

69. An environmental concern involves the destructive effects of acid rain. The most acidic rainfall ever had a pH of 2.4. What was the hydrogen ion concentration? Express the answer as a power of 10, and then round to the nearest thousandth.

70. The figure shows very acidic rain in the northeast United States. What is the hydrogen ion concentration of rainfall with a pH of 4.2? Express the answer as a power of 10, and then round to the nearest hundred-thousandth.

Acid Rain over Canada and the United States

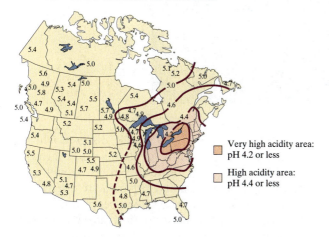

Very high acidity area: pH 4.2 or less

High acidity area: pH 4.4 or less

Source: National Atmospheric Program

Writing in Mathematics

71. Explain how to solve an exponential equation. Use $3^x = 140$ in your explanation.

72. Explain how to solve a logarithmic equation. Use $\log_3(x - 1) = 4$ in your explanation.

73. In many states, a 17% risk of a car accident with a blood alcohol concentration of 0.08 is the lowest level for charging a motorist with driving under the influence. Do you agree with the 17% risk as a cutoff percentage, or do you feel that the percentage should be lower or higher?

Explain your answer. What blood alcohol concentration corresponds to what you believe is an appropriate percentage?

74. Have you purchased a new or used car recently? If so, describe if the function in Exercise 65 accurately models what you paid for your car. If there is a big difference between the figure given by the formula and the amount that you paid, how can you explain this difference?

Technology Exercises

In Exercises 75–82, use your graphing utility to graph each side of the equation in the same viewing rectangle. Then use the x-coordinate of the intersection point to find the equation's solution set. Verify this value by direct substitution into the equation.

75. $2^{x+1} = 8$ **76.** $3^{x+1} = 9$

77. $\log_3(4x - 7) = 2$ **78.** $\log_3(3x - 2) = 2$

79. $\log(x + 3) + \log x = 1$ **80.** $\log(x - 15) + \log x = 2$

81. $3^x = 2x + 3$ **82.** $5^x = 3x + 4$

Hurricanes are one of nature's most destructive forces. These low-pressure areas often have diameters of over 500 miles. The function $f(x) = 0.48 \ln(x + 1) + 27$ models the barometric air pressure, $f(x)$, in inches of mercury, at a distance of x miles from the eye of a hurricane. Use this function to solve Exercises 83–84.

83. Graph the function in a $[0, 500, 50]$ by $[27, 30, 1]$ viewing rectangle. What does the shape of the graph indicate about barometric air pressure as the distance from the eye increases?

84. Use an equation to answer this question: How far from the eye of a hurricane is the barometric air pressure 29 inches of mercury? Use the TRACE and ZOOM features or the intersect command of your graphing utility to verify your answer.

85. The function $P(t) = 145e^{-0.092t}$ models a runner's pulse, $P(t)$, in beats per minute, t minutes after a race, where $0 \le t \le 15$. Graph the function using a graphing utility. TRACE along the graph and determine after how many minutes the runner's pulse will be 70 beats per minute. Round to the nearest tenth of a minute. Verify your observation algebraically.

86. The function $W(t) = 2600(1 - 0.51e^{-0.075t})^3$ models the weight, $W(t)$, in kilograms, of a female African elephant at age t years. (1 kilogram ≈ 2.2 pounds) Use a graphing utility to graph the function. Then TRACE along the curve to estimate the age of an adult female elephant weighing 1800 kilograms.

Critical Thinking Exercises

87. Which one of the following is true?

a. If $\log(x + 3) = 2$, then $e^2 = x + 3$.

b. If $\log(7x + 3) - \log(2x + 5) = 4$, then in exponential form $10^4 = (7x + 3) - (2x + 5)$.

c. If $x = \dfrac{1}{k} \ln y$, then $y = e^{kx}$.

d. Examples of exponential equations include $10^x = 5.71$, $e^x = 0.72$, and $x^{10} = 5.71$.

88. If $4000 is deposited into an account paying 3% interest compounded annually and at the same time $2000 is deposited into an account paying 5% interest compounded annually, after how long will the two accounts have the same balance?

Solve each equation in Exercises 89–91. Check each proposed solution by direct substitution or with a graphing utility.

89. $(\ln x)^2 = \ln x^2$

90. $(\log x)(2 \log x + 1) = 6$

91. $\ln(\ln x) = 0$

Group Exercise

92. Research applications of logarithmic functions as mathematical models and plan a seminar based on your group's research. Each group member should research one of the following areas or any other area of interest: pH (acidity of solutions), intensity of sound (decibels), brightness of stars, consumption of natural resources, human memory, progress over time in a sport, profit over time. For the area that you select, explain how logarithmic functions are used and provide examples.

SECTION 4.5 *Modeling with Exponential and Logarithmic Functions*

Objectives

1. Model exponential growth and decay.

2. Use logistic growth models.

3. Model data with exponential and logarithmic functions.

4. Express an exponential model in base e.

The most casual cruise on the Internet shows how people disagree when it comes to making predictions about the effects of the world's growing population. Some argue that there is a recent slowdown in the growth rate, economies remain robust, and famines in Biafra and Ethiopia are aberrations rather than signs of the future. Others say that the 6 billion people on Earth is twice as many as can be supported in middle-class comfort, and the world is running out of arable land and fresh water. Debates about entities that are growing exponentially can be approached mathematically: We can create functions that model data and use these functions to make predictions. In this section we will show you how this is done.

1 Model exponential growth and decay.

Exponential Growth and Decay

One of algebra's many applications is to predict the behavior of variables. This can be done with *exponential growth* and *decay models*. With exponential growth or decay, quantities grow or decay at a rate directly proportional to their size. Populations that are growing exponentially grow extremely rapidly as they get larger because there are more adults to have offspring. For example, the **growth rate** for world population is 1.3%, or 0.013. This means that each year world population is 1.3% more than what it was in the previous year. In 2001, world population was approximately 6.2 billion. Thus, we compute the world population in 2002 as follows:

6.2 billion + 1.3% of 6.2 billion = 6.2 + (0.013)(6.2) = 6.2806.

This computation suggests that 6.2806 billion people will populate the world in 2002. The 0.0806 billion represents an increase of 80.6 million people from 2001 to 2002, the equivalent of the population of Germany. Using 1.3% as the annual growth rate, world population for 2003 is found in a similar manner:

6.2806 + 1.3% of 6.2806 = 6.2806 + (0.013)(6.2806) ≈ 6.3622.

This computation suggests that approximately 6.3622 billion people will populate the world in 2003.

The explosive growth of world population may remind you of the growth of money in an account subject to compound interest. Just as the growth rate for world population is multiplied by the population plus any increase in the population, a compound interest rate is multiplied by your original investment plus any accumulated interest. The balance in an account subject to continuous compounding and world population are special cases of an *exponential growth model*.

Study Tip

You have seen the formula for exponential growth before, but with different letters. It is the formula for compound interest with continous compounding.

$$A = Pe^{rt}$$

| Amount at time t | Principal is the original amount. | Interest rate is the growth rate. |

$$A = A_o e^{kt}$$

Exponential Growth and Decay Models

The mathematical model for **exponential growth** or **decay** is given by

$$f(t) = A_0 e^{kt} \quad \text{or} \quad A = A_0 e^{kt}.$$

- **If $k > 0$, the function models the amount, or size, of a *growing* entity.** A_0 is the original amount, or size, of the growing entity at time $t = 0$, A is the amount at time t, and k is a constant representing the growth rate.
- **If $k < 0$, the function models the amount, or size, of a *decaying* entity.** A_0 is the original amount, or size, of the decaying entity at time $t = 0$, A is the amount at time t, and k is a constant representing the decay rate.

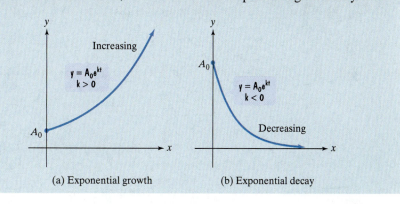

(a) Exponential growth   (b) Exponential decay

Sometimes we need to use given data to determine k, the rate of growth or decay. After we compute the value of k, we can use the formula $A = A_0 e^{kt}$ to make predictions. This idea is illustrated in our first two examples.

EXAMPLE 1   Modeling the Growth of the Minimum Wage

The graph in Figure 4.17 shows the growth of the minimum wage from 1970 through 2000. In 1970, the minimum wage was $1.60 per hour. By 2000, it had grown to $5.15 per hour.

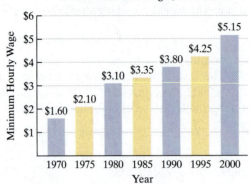

Federal Minimum Wages, 1970–2000

Figure 4.17 *Source:* U.S. Employment Standards Administration

a. Find the exponential growth function that models the data for 1970 through 2000.

b. By which year will the minimum wage reach $7.50 per hour?

Solution

a. We use the exponential growth model

$$A = A_0 e^{kt}$$

in which t is the number of years after 1970. This means that 1970 corresponds to $t = 0$. At that time the minimum wage was $1.60, so we substitute 1.6 for A_0 in the growth model:

$$A = 1.6 e^{kt}.$$

We are given that $5.15 is the minimum wage in 2000. Because 2000 is 30 years after 1970, when $t = 30$ the value of A is 5.15. Substituting these numbers into the growth model will enable us to find k, the growth rate. We know that $k > 0$ because the problem involves growth.

$A = 1.6 e^{kt}$	Use the growth model, $A = A_0 e^{kt}$, with $A_0 = 1.6$.
$5.15 = 1.6 e^{k \cdot 30}$	When $t = 30$, $A = 5.15$. Substitute these numbers into the model.
$e^{30k} = \dfrac{5.15}{1.6}$	Isolate the exponential factor by dividing both sides by 1.6. We also reversed the sides.
$\ln e^{30k} = \ln \dfrac{5.15}{1.6}$	Take the natural logarithm on both sides.
$30k = \ln \dfrac{5.15}{1.6}$	Simplify the left side using $\ln e^x = x$.
$k = \dfrac{\ln \dfrac{5.15}{1.6}}{30} \approx 0.039$	Divide both sides by 30 and solve for k.

We substitute 0.039 for k in the growth model to obtain the exponential growth function for the minimum wage. It is

$$A = 1.6 e^{0.039t}$$

where t is measured in years after 1970.

b. To find the year in which the minimum wage will reach $7.50 per hour, we substitute 7.5 for A in the model from part (a) and solve for t.

$A = 1.6e^{0.039t}$ *This is the model from part (a).*

$7.5 = 1.6e^{0.039t}$ *Substitute 7.5 for A.*

$e^{0.039t} = \dfrac{7.5}{1.6}$ *Divide both sides by 1.6. We also reversed the sides.*

$\ln e^{0.039t} = \ln \dfrac{7.5}{1.6}$ *Take the natural logarithm on both sides.*

$0.039t = \ln \dfrac{7.5}{1.6}$ *Simplify on the left using $\ln e^x = x$.*

$t = \dfrac{\ln \dfrac{7.5}{1.6}}{0.039} \approx 40$ *Solve for t by dividing both sides by 0.039.*

Because 40 is the number of years after 1970, the model indicates that the minimum wage will reach $7.50 by 1970 + 40, or in the year 2010.

Check Point 1 In 1990, the population of Africa was 643 million and by 2000 it had grown to 813 million.

a. Use the exponential growth model $A = A_0e^{kt}$, in which t is the number of years after 1990, to find the exponential growth function that models the data.

b. By which year will Africa's population reach 2000 million, or two billion?

Lying with Statistics

Benjamin Disraeli, Queen Victoria's prime minister, stated that there are "lies, damned lies, and statistics." The problem is not that data lie, but rather that liars use data. For example, the data in Example 1 create the impression that wages are on the rise and workers are better off each year. The graph in Figure 4.18 is more effective in creating an accurate picture. Why? It is adjusted for inflation and measured in constant 1996 dollars. Something else to think about: In predicting a minimum wage of $7.50 by 2010, are we using the best possible model for the data? We return to this issue in the exercise set.

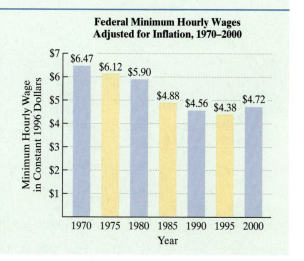

Federal Minimum Hourly Wages
Adjusted for Inflation, 1970–2000

Figure 4.18
Source: U.S. Employment Standards Administration

Carbon Dating and Artistic Development

The artistic community was electrified by the discovery in 1995 of spectacular cave paintings in a limestone cavern in France. Carbon dating of the charcoal from the site showed that the images, created by artists of remarkable talent, were 30,000 years old, making them the oldest cave paintings ever found. The artists seemed to have used the cavern's natural contours to heighten a sense of perspective. The quality of the painting suggests that the art of early humans did not mature steadily from primitive to sophisticated in any simple linear fashion.

Our next example involves exponential decay and its use in determining the age of fossils and artifacts. The method is based on considering the percentage of carbon-14 remaining in the fossil or artifact. Carbon-14 decays exponentially with a *half-life* of approximately 5715 years. The **half-life** of a substance is the time required for half of a given sample to disintegrate. Thus, after 5715 years a given amount of carbon-14 will have decayed to half the original amount. Carbon dating is useful for artifacts or fossils up to 80,000 years old. Older objects do not have enough carbon-14 left to date age accurately.

EXAMPLE 2 Carbon-14 Dating: The Dead Sea Scrolls

a. Use the fact that after 5715 years a given amount of carbon-14 will have decayed to half the original amount to find the exponential decay model for carbon-14.

b. In 1947, earthenware jars containing what are known as the Dead Sea Scrolls were found by an Arab Bedouin herdsman. Analysis indicated that the scroll wrappings contained 76% of their original carbon-14. Estimate the age of the Dead Sea Scrolls.

Solution   We begin with the exponential decay model $A = A_0 e^{kt}$. We know that $k < 0$ because the problem involves the decay of carbon-14. After 5715 years ($t = 5715$), the amount of carbon-14 present, A, is half the original amount A_0. Thus, we can substitute $\dfrac{A_0}{2}$ for A in the exponential decay model. This will enable us to find k, the decay rate.

a.

$$A = A_0 e^{kt}$$
Begin with the exponential decay model .

$$\frac{A_0}{2} = A_0 e^{k \cdot 5715}$$
After 5715 years ($t = 5715$), $A = \dfrac{A_0}{2}$ (because the amount present, A, is half the original amount, A_0).

$$\frac{1}{2} = e^{5715k}$$
Divide both sides of the equation by A_0.

$$\ln \frac{1}{2} = \ln e^{5715k}$$
Take the natural logarithm on both sides.

$$\ln \frac{1}{2} = 5715k$$
Simplify the right side using $\ln e^x = x$.

$$k = \frac{\ln \dfrac{1}{2}}{5715} \approx -0.000121$$
Divide both sides by 5715 and solve for k.

Substituting for k in the decay model, $A = A_0 e^{kt}$, the model for carbon-14 is

$$A = A_0 e^{-0.000121t}.$$

b.

$$A = A_0 e^{-0.000121t}$$
This is the decay model for carbon-14.

$$0.76 A_0 = A_0 e^{-0.000121t}$$
A, the amount present, is 76% of the original amount, so $A = 0.76 A_0$.

$$0.76 = e^{-0.000121t}$$
Divide both sides of the equation by A_0.

$$\ln 0.76 = \ln e^{-0.000121t}$$
Take the natural logarithm on both sides.

$$\ln 0.76 = \ln e^{-0.000121t}$$ We've repeated this equation from the bottom of the previous page.

$$\ln 0.76 = -0.000121t$$ Simplify the right side using $\ln e^x = x$.

$$t = \frac{\ln 0.76}{-0.000121} \approx 2268$$ Divide both sides by -0.000121 and solve for t.

The Dead Sea Scrolls are approximately 2268 years old plus the number of years between 1947 and the current year.

Check Point 2   Strontium-90 is a waste product from nuclear reactors. As a consequence of fallout from atmospheric nuclear tests, we all have a measurable amount of strontium-90 in our bones.
   a. Use the fact that after 28 years a given amount of strontium-90 will have decayed to half the original amount to find the exponential decay model for strontium-90.
   b. Suppose that a nuclear accident occurs and releases 60 grams of strontium-90 into the atmosphere. How long will it take for strontium-90 to decay to a level of 10 grams?

2 Use logistic growth models.

Logistic Growth Models

From population growth to the spread of an epidemic, nothing on Earth can grow exponentially indefinitely. Growth is always limited. This is shown in Figure 4.19 by the horizontal asymptote. The **logistic growth model** is an exponential function used to model situations in which growth is limited.

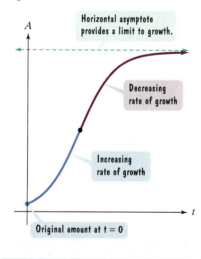

A

Horizontal asymptote provides a limit to growth.

Decreasing rate of growth

Increasing rate of growth

Original amount at t = 0

t

Figure 4.19 The logistic growth curve has a horizontal asymptote that limits the growth of A over time.

Logistic Growth Model

The mathematical model for limited logistic growth is given by

$$f(t) = \frac{c}{1 + ae^{-bt}} \quad \text{or} \quad A = \frac{c}{1 + ae^{-bt}}$$

where a, b, and c are constants, with $c > 0$ and $b > 0$.

As time increases $(t \to \infty)$, the expression ae^{-bt} in the model approaches 0, and A gets closer and closer to c. This means that $y = c$ is a horizontal asymptote for the graph of the function. Thus, the value of A can never exceed c and c represents the limiting size that A can attain.

EXAMPLE 3 Modeling the Spread of the Flu

The function

$$f(t) = \frac{30{,}000}{1 + 20e^{-1.5t}}$$

describes the number of people, $f(t)$, who have become ill with influenza t weeks after its initial outbreak in a town with 30,000 inhabitants.

a. How many people became ill with the flu when the epidemic began?
b. How many people were ill by the end of the fourth week?
c. What is the limiting size of $f(t)$, the population that becomes ill?

Solution

a. The time at the beginning of the flu epidemic is $t = 0$. Thus, we can find the number of people who were ill at the beginning of the epidemic by substituting 0 for t.

$$f(t) = \frac{30{,}000}{1 + 20e^{-1.5t}} \qquad \text{This is the given logistic growth function.}$$

$$f(0) = \frac{30{,}000}{1 + 20e^{-1.5(0)}} \qquad \text{When the epidemic began, } t = 0.$$

$$= \frac{30{,}000}{1 + 20} \qquad e^{-1.5(0)} = e^{0} = 1$$

$$\approx 1429$$

Approximately 1429 people were ill when the epidemic began.

b. We find the number of people who were ill at the end of the fourth week by substituting 4 for t in the logistic growth function.

$$f(t) = \frac{30{,}000}{1 + 20e^{-1.5t}} \qquad \text{Use the given logistic growth function.}$$

$$f(4) = \frac{30{,}000}{1 + 20e^{-1.5(4)}} \qquad \text{To find the number of people ill by the end of week four, let } t = 4.$$

$$= 28{,}583 \qquad \text{Use a calculator.}$$

Approximately 28,583 people were ill by the end of the fourth week. Compared with the number of people who were ill initially, 1429, this illustrates the virulence of the epidemic.

c. Recall that in the logistic growth model, $f(t) = \dfrac{c}{1 + ae^{-bt}}$, the constant c represents the limiting size that $f(t)$ can attain. Thus, the number in the numerator, 30,000, is the limiting size of the population that becomes ill.

Technology

The graph of the logistic growth function for the flu epidemic

$$y = \frac{30{,}000}{1 + 20e^{-1.5x}}$$

can be obtained using a graphing utility. We started x at 0 and ended at 10. This takes us to week 10. (In Example 3, we found that by week 4 approximately 28,583 people were ill.) We also know that 30,000 is the limiting size, so we took values of y up to 30,000. Using a $[0, 10, 1]$ by $[0, 30{,}000, 3000]$ viewing rectangle, the graph of the logistic growth function is shown below.

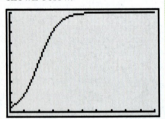

Check Point 3

In a learning theory project, psychologists discovered that

$$f(t) = \frac{0.8}{1 + e^{-0.2t}}$$

is a model for describing the proportion of correct responses, $f(t)$, after t learning trials.

Check Point 3 continued

a. Find the proportion of correct responses prior to learning trials taking place.

b. Find the proportion of correct responses after 10 learning trials.

c. What is the limiting size of $f(t)$, the proportion of correct responses, as continued learning trials take place?

3 Model data with exponential and logarithmic functions.

The Art of Modeling

Throughout this chapter, we have been working with models that were given. However, we can create functions that model data by observing patterns in scatter plots. Figure 4.20 shows scatter plots for data that are exponential or logarithmic.

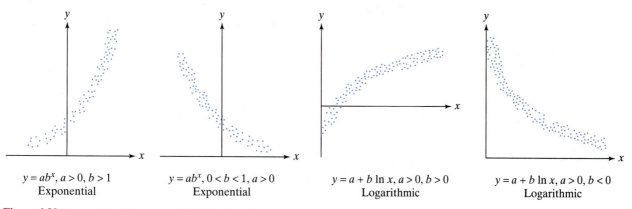

$y = ab^x, a > 0, b > 1$
Exponential

$y = ab^x, 0 < b < 1, a > 0$
Exponential

$y = a + b \ln x, a > 0, b > 0$
Logarithmic

$y = a + b \ln x, a > 0, b < 0$
Logarithmic

Figure 4.20 Scatter plots for exponential or logarithmic models

Graphing utilities can be used to find the equation of a function that is derived from data. For example, earlier in the chapter we encountered a function that modeled the size of a city and the average walking speed, in feet per second, of pedestrians. The function was derived from the data in Table 4.4. The scatter plot is shown in Figure 4.21.

Table 4.4

x, Population (thousands)	y, Walking Speed (feet per second)
5.5	3.3
14	3.7
71	4.3
138	4.4
342	4.8

Source: Mark and Helen Bornstein, "The Pace of Life"

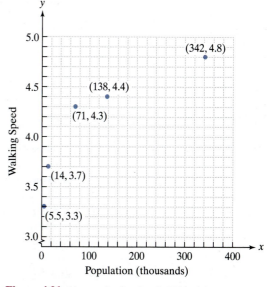

Figure 4.21 Scatter plot for data in Table 4.4

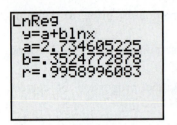

Figure 4.22 A logarithmic model for the data in Table 4.4

Because the data in this scatter plot increase rapidly at first and then begin to level off a bit, the shape suggests that a logarithmic model might be a good choice. A graphing utility fits the data in Table 4.4 to a logarithmic model of the form $y = a + b \ln x$ by using the Natural Logarithmic REGression (LnReg) option (see Figure 4.22). From the figure, we see that the logarithmic model of the data, with numbers rounded to three decimal places, is

$$y = 2.735 + 0.352 \ln x.$$

The number r that appears in Figure 4.22 is called the **correlation coefficient** and is a measure of how well the model fits the data. The value of r is such that $-1 \le r \le 1$. A positive r means that as the x-values increase, so do the y-values. A negative r means that as the x-values increase, the y-values decrease. **The closer that r is to -1 or 1, the better the model fits the data.** Because r is approximately 0.996, the model

$$y = 2.735 + 0.352 \ln x$$

fits the data very well.

Now let's look at data whose scatter plot suggests an exponential model. The data in Table 4.5 indicate world population for six years. The scatter plot is shown in Figure 4.23.

Table 4.5

x, Year	y, World Population (billions)
1950	2.6
1960	3.1
1970	3.7
1980	4.5
1989	5.3
2001	6.2

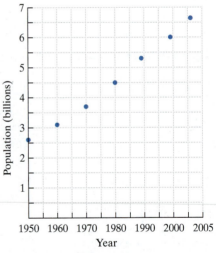

Figure 4.23 A scatter plot for data in Table 4.5

Because the data in this scatter plot have a rapidly increasing pattern, the shape suggests that an exponential model might be a good choice. (You might also want to try a linear model.) If you select the exponential option, you will use a graphing utility's Exponential REGression option. With this feature, a graphing utility fits the data to an exponential model of the form $y = ab^x$.

When computing an exponential model of the form $y = ab^x$, many graphing utilities rewrite the equation using logarithms. Because the domain of the logarithmic function is the set of positive numbers, **zero must not be a value for x** when using such utilities. What does this mean in terms of our data for world population that starts in the year 1950? We must start values of x after 0. Thus, we'll assign x to represent the number of years after 1949.

This gives us the data shown in Table 4.6. Using the Exponential REGression option, we obtain the equation in Figure 4.24.

Table 4.6

x, Numbers of Years after 1949		y, World Population (billions)
1	(1950)	2.6
11	(1960)	3.1
21	(1970)	3.7
31	(1980)	4.5
40	(1989)	5.3
52	(2001)	6.2

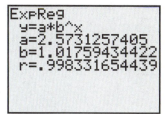

Figure 4.24 An exponential model for the data in Table 4.6

From Figure 4.24, we see that the exponential model of the data for world population, y, in billions, x years after 1949, with numbers rounded to three decimal places, is

$$y = 2.573(1.018)^x.$$

The correlation coefficient, r, is close to 1, indicating that the model fits the data very well.

Because $b = e^{\ln b}$, we can rewrite any model in the form $y = ab^x$ in terms of base e.

4 Express an exponential model in base e.

> **Expressing an Exponential Model in Base e**
>
> $y = ab^x$ is equivalent to $y = ae^{(\ln b)\cdot x}$.

EXAMPLE 4   Rewriting an Exponential Model in Base e

Rewrite $y = 2.573(1.018)^x$ in terms of base e.

Solution

$y = ab^x$ is equivalent to $y = ae^{(\ln b)\cdot x}$.

$y = 2.573(1.018)^x$ is equivalent to $y = 2.573e^{(\ln 1.018)\cdot x}$.

Using $\ln 1.018 \approx 0.018$, the exponential growth model for world population, y, in billions, x years after 1949 is

$$y = 2.573e^{0.018x}.$$

In Example 4, we can replace y with A and x with t so that the model has the same letters as those in the exponential growth model $A = A_0e^{kt}$.

$A = A_0 e^{kt}$ This is the exponential growth model.

$A = 2.573e^{0.018t}$ This is the model for world population.

The value of k, 0.018, indicates a growth rate of 1.8%. Although this is an excellent model for the data, we must be careful about making projections about world population using this growth function. Why? World population growth rate is now 1.3%, not 1.8%, so our model will overestimate future populations.

Check Point 4 Rewrite $y = 4(7.8)^x$ in terms of base e. Express the answer in terms of a natural logarithm, and then round to three decimal places.

When using a graphing utility to model data, begin with a scatter plot, drawn either by hand or with the graphing utility, to obtain a general picture for the shape of the data. It might be difficult to determine which model best fits the data—linear, logarithmic, exponential, quadratic, or something else. If necessary, use your graphing utility to fit several models to the data. The best model is the one that yields the value r, the correlation coefficient, closest to 1 or -1. Finding a proper fit for data can be almost as much art as it is mathematics. In this era of technology, the process of creating models that best fit data is one that involves more decision making than computation.

EXERCISE SET 4.5

Practice and Application Exercises

The exponential growth model $A = 203e^{0.011t}$ describes the population of the United States, A, in millions, t years after 1970. Use this model to solve Exercises 1–4.

1. What was the population of the United States in 1970?
2. By what percentage is the population of the United States increasing each year?
3. When will the U.S. population be 300 million?
4. When will the U.S. population be 350 million?

India is currently one of the world's fastest-growing countries. By 2040, the population of India will be larger than the population of China; by 2050, nearly one-third of the world's population will live in these two countries alone. The exponential growth model $A = 574e^{0.026t}$ describes the population of India, A, in millions, t years after 1974. Use this model to solve Exercises 5–8.

5. By what percentage is the population of India increasing each year?
6. What was the population of India in 1974?
7. When will India's population be 1624 million?
8. When will India's population be 2732 million?
9. Low interest rates, easy credit, and strong demand from new immigrants have driven up the average sales price of new one-family houses in the United States. In 1995, the average sales price was $158,700 and by 2000 it had increased to $207,200.
   a. Use the exponential growth model $A = A_0e^{kt}$, in which t is the number of years after 1995, to find the exponential growth function that models the data.
   b. According to your model, by which year will the average sales price of a new one-family house reach $300,000?

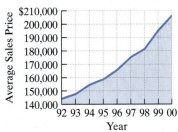

Average Sales Prices of New One-Family Houses

Source: U.S. Census Bureau

About the size of New Jersey, Israel has seen its population soar to more than 6 million since it was established. With the help of U.S. aid, the country now has a diversified economy rivaling those of other developed Western nations. By contrast, the Palestinians, living under Israeli occupation and a corrupt regime, endure bleak conditions. The graphs show that by 2050, Palestinians in the West Bank, Gaza Strip, and East Jerusalem will outnumber Israelis. Exercises 10–12 involve the projected growth of these two populations.

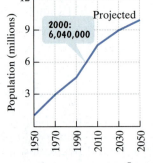

Population of Israel

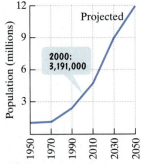

Palestinian Population in West Bank, Gaza, and East Jerusalem

Source: Newsweek

10. In 2000, the population of the Palestinians in the West Bank, Gaza Strip, and East Jerusalem was approximately 3.2 million and by 2050 it is projected to grow to 12 million. Use the exponential growth model $A = A_0e^{kt}$, in which t is the number of years after 2000, to find the exponential growth function that models the data.

11. In 2000, the population of Israel was approximately 6.04 million and by 2050 it is projected to grow to 10 million. Use the exponential growth model $A = A_0e^{kt}$, in which t is the number of years after 2000, to find an exponential growth function that models the data.

12. Use the growth models in Exercises 10 and 11 to determine the year in which the two populations will be the same.

An artifact originally had 16 grams of carbon-14 present. The decay model $A = 16e^{-0.000121t}$ describes the amount of carbon-14 present, A, in grams, after t years. Use this model to solve Exercises 13–14.

13. How many grams of carbon-14 will be present after 5715 years?

14. How many grams of carbon-14 will be present after 11,430 years?

15. The half-life of the radioactive element krypton-91 is 10 seconds. If 16 grams of krypton-91 are initially present, how many grams are present after 10 seconds? 20 seconds? 30 seconds? 40 seconds? 50 seconds?

16. The half-life of the radioactive element plutonium-239 is 25,000 years. If 16 grams of plutonium-239 are initially present how many grams are present after 25,000 years? 50,000 years? 75,000 years? 100,000 years? 125,000 years?

Use the exponential decay model for carbon-14, $A = A_0e^{-0.000121t}$, to solve Exercises 17–18.

17. Prehistoric cave paintings were discovered in a cave in France. The paint contained 15% of the original carbon-14. Estimate the age of the paintings.

18. Skeletons were found at a construction site in San Francisco in 1989. The skeletons contained 88% of the expected amount of carbon-14 found in a living person. In 1989, how old were the skeletons?

19. The August 1978 issue of *National Geographic* described the 1964 find of dinosaur bones of a newly discovered dinosaur weighing 170 pounds, measuring 9 feet, with a 6-inch claw on one toe of each hind foot. The age of the dinosaur was estimated using potassium-40 dating of rocks surrounding the bones.

  a. Potassium-40 decays exponentially with a half-life of approximately 1.31 billion years. Use the fact that after 1.31 billion years a given amount of potassium-40 will have decayed to half the original amount to show that the decay model for potassium-40 is given by $A = A_0e^{-0.52912t}$, where t is in billions of years.

  b. Analysis of the rocks surrounding the dinosaur bones indicated that 94.5% of the original amount of potassium-40 was still present. Let $A = 0.945A_0$ in the model in part (a) and estimate the age of the bones of the dinosaur.

20. A bird species in danger of extinction has a population that is decreasing exponentially $(A = A_0e^{kt})$. Five years ago the population was at 1400 and today only 1000 of the birds are alive. Once the population drops below 100, the situation will be irreversible. When will this happen?

21. Use the exponential growth model, $A = A_0e^{kt}$, to show that the time it takes a population to double (to grow from A_0 to $2A_0$) is given by $t = \dfrac{\ln 2}{k}$.

22. Use the exponential growth model, $A = A_0e^{kt}$, to show that the time it takes a population to triple (to grow from A_0 to $3A_0$) is given by $t = \dfrac{\ln 3}{k}$.

Use the formula $t = \dfrac{\ln 2}{k}$ that gives the time for a population with a growth rate k to double to solve Exercises 23–24. Express each answer to the nearest whole year.

23. China is growing at a rate of 1.1% per year. How long will it take China to double its population?

24. Japan is growing at a rate of 0.3% per year. How long will it take Japan to double its population?

25. The logistic growth function

$$f(t) = \frac{100,000}{1 + 5000e^{-t}}$$

describes the number of people, $f(t)$, who have become ill with influenza t weeks after its initial outbreak in a particular community.

  a. How many people became ill with the flu when the epidemic began?

  b. How many people were ill by the end of the fourth week?

  c. What is the limiting size of the population that becomes ill?

26. The logistic growth function

$$f(t) = \frac{500}{1 + 83.3e^{-0.162t}}$$

describes the population, $f(t)$, of an endangered species of birds t years after they are introduced to a nonthreatening habitat.

  a. How many birds were initially introduced to the habitat?

  b. How many birds are expected in the habitat after 10 years?

  c. What is the limiting size of the bird population that the habitat will sustain?

The logistic growth function

$$P(x) = \frac{90}{1 + 271e^{-0.122x}}$$

models the percentage, $P(x)$, of Americans who are x years old with some coronary heart disease. Use the function to solve Exercises 27–30.

27. What percentage of 20-year-olds have some coronary heart disease?

28. What percentage of 80-year-olds have some coronary heart disease?

29. At what age is the percentage of some coronary heart disease 50%?

30. At what age is the percentage of some coronary heart disease 70%?

In Exercises 31–34, rewrite the equation in terms of base e. Express the answer in terms of a natural logarithm, and then round to three decimal places.

31. $y = 100(4.6)^x$ 32. $y = 1000(7.3)^x$

33. $y = 2.5(0.7)^x$ 34. $y = 4.5(0.6)^x$

Writing in Mathematics

35. Nigeria has a growth rate of 0.031 or 3.1%. Describe what this means.

36 How can you tell if an exponential model describes exponential growth or exponential decay?

37. Suppose that a population that is growing exponentially increases from 800,000 people in 2003 to 1,000,000 people in 2006. Without showing the details, describe how to obtain the exponential growth function that models the data.

38. What is the half-life of a substance?

39. Describe a difference between exponential growth and logistic growth.

40. Describe the shape of a scatter plot that suggests modeling the data with an exponential function.

41. Based on the graphs in Exercises 10–12, for which population is an exponential growth function a better model? Explain your answer.

42. You take up weightlifting and record the maximum number of pounds you can lift at the end of each week. You start off with rapid growth in terms of the weight you can lift from week to week, but then the growth begins to level off. Describe how to obtain a function that models the number of pounds you can lift at the end of each week. How can you use this function to predict what might happen if you continue the sport?

43. Would you prefer that your salary be modeled exponentially or logarithmically? Explain your answer.

44. One problem with all exponential growth models is that nothing can grow exponentially forever. Describe factors that might limit the size of a population.

Technology Exercises

In Example 1 on page 420–421, we used two data points and an exponential function to model federal minimum wages that were not adjusted for inflation from 1970 through 2000. The data are shown again in the table. Use all seven data points to solve Exercises 45–49.

x, Number of Years after 1969	y, Federal Minimum Wage
1	1.60
6	2.10
11	3.10
16	3.35
21	3.80
26	4.25
31	5.15

45. Use your graphing utility's Exponential REGression option to obtain a model of the form $y = ab^x$ that fits the data. How well does the correlation coefficient, r, indicate that the model fits the data?

46. Use your graphing utility's Logarithmic REGression option to obtain a model of the form $y = a + b \ln x$ that fits the data. How well does the correlation coefficient, r, indicate that the model fits the data?

47. Use your graphing utility's Linear REGression option to obtain a model of the form $y = ax + b$ that fits the data. How well does the correlation coefficient, r, indicate that the model fits the data?

48. Use your graphing utility's Power REGression option to obtain a model of the form $y = ax^b$ that fits the data. How well does the correlation coefficient, r, indicate that the model fits the data?

49. Use the value of r in Exercises 45–48 to select the model of best fit. Use this model to predict by which year the minimum wage will reach $7.50. How does this answer compare to the year we found in Example 1, namely 2010? If you obtained a different year, how do you account for this difference?

50. In Exercises 27–30, you worked with the logistic growth function

$$P(x) = \frac{90}{1 + 271e^{-0.122x}}$$

which models the percentage, $P(x)$, of Americans who are x years old with some coronary heart disease. Use your graphing utility to graph the function in a $[0, 100, 10]$ by $[0, 100, 10]$ viewing rectangle. Describe as specifically as possible what the logistic curve indicates about aging and the percentage of Americans with coronary heart disease.

In Exercises 51–52, use a graphing utility to find the model that best fits the given data. Then use the model to make a reasonable prediction for a value that exceeds those shown on the graph's horizontal axis.

51.

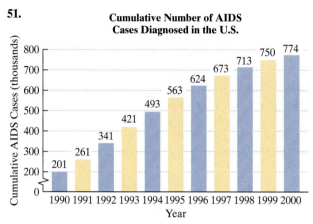

Cumulative Number of AIDS Cases Diagnosed in the U.S.

Source: Centers for Disease Control

52.

Percentage of Miscarriages, by Age

Source: *Time*

Critical Thinking Exercises

53. The World Health Organization makes predictions about the number of AIDS cases based on a compromise between a linear model and an exponential growth model. Explain why the World Health Organization does this.

54. Over a period of time, a hot object cools to the temperature of the surrounding air. This is described mathematically by

$$T = C + (T_0 - C)e^{-kt},$$

where t is the time it takes for an object to cool from temperature T_0 to temperature T, C is the surrounding air temperature, and k is a positive constant that is associated with the cooling object. A cake removed from the oven has a temperature of 210°F and is left to cool in a room that has a temperature of 70°F. After 30 minutes, the temperature of the cake is 140°F. What is the temperature of the cake after 40 minutes?

Group Exercise

55. This activity is intended for three or four people who would like to take up weightlifting. Each person in the group should record the maximum number of pounds that he or she can lift at the end of each week for the first 10 consecutive weeks. Use the Logarithmic REGression option of a graphing utility to obtain a model showing the amount of weight that group members can lift from week 1 through week 10. Graph each of the models in the same viewing rectangle to observe similarities and differences among weight-growth patterns of each member. Use the functions to predict the amount of weight that group members will be able to lift in the future. If the group continues to work out together, check the accuracy of these predictions.

56. Each group member should consult an almanac, newspaper, magazine, or the Internet to find data that can be modeled by exponential or logarithmic functions. Group members should select the two sets of data that are most interesting and relevant. For each data set selected, find a model that best fits the data. Each group member should make one prediction based on the model and then discuss a consequence of this prediction. What factors might change the accuracy of each prediction?

CHAPTER SUMMARY, REVIEW, AND TEST

Summary

DEFINITIONS AND CONCEPTS	EXAMPLES

4.1 Exponential Functions

a. The exponential function with base b is defined by $f(x) = b^x$, where $b > 0$ and $b \neq 1$. — Ex. 1, p. 375

b. Characteristics of exponential functions and graphs for $0 < b < 1$ and $b > 1$ are shown in the box on page 377. — Ex. 2, p. 376

c. Transformations involving exponential functions are summarized in Table 4.1 on page 377. — Exs. 3 & 4, p. 378

d. The natural exponential function $f(x) = e^x$. The irrational number e is called the natural base, where $e \approx 2.7183$. — Ex. 5, p. 379

e. Formulas for compound interest: After t years, the balance, A, in an account with principal P and annual interest rate r (in decimal form) is given by one of the following formulas: — Ex. 6, p. 381

   1. For n compoundings per year: $A = P\left(1 + \dfrac{r}{n}\right)^{nt}$

   2. For continuous compounding: $A = Pe^{rt}$

4.2 Logarithmic Functions

a. Definition of the logarithmic function: For $x > 0$ and $b > 0$, $b \neq 1$, $y = \log_b x$ is equivalent to $b^y = x$. The function $f(x) = \log_b x$ is the logarithmic function with base b. This function is the inverse function of the exponential function with base b. — Ex. 1, p. 386; Ex. 2, p. 386; Ex. 3, p. 387

b. Graphs of logarithmic functions for $b > 1$ and $0 < b < 1$ are shown in Figure 4.7 on page 389. Characteristics of the graphs are summarized in the box that follows the figure. — Ex. 6, p. 389

c. Transformations involving logarithmic functions are summarized in Table 4.3 on page 390.

d. The domain of a logarithmic function of the form $f(x) = \log_b x$ is the set of all positive real numbers. The domain of $f(x) = \log_b(x + c)$ consists of all x for which $x + c > 0$. — Ex. 7, p. 391; Ex. 10, p. 391

e. Common and natural logarithms: $f(x) = \log x$ means $f(x) = \log_{10} x$ and is the common logarithmic function. $f(x) = \ln x$ means $f(x) = \log_e x$ and is the natural logarithmic function. — Ex. 8, p. 392; Ex. 9, p. 392

f. Basic Logarithmic Properties

Base b $(b > 0, b \neq 1)$	Base 10 (Common Logarithms)	Base e (Natural Logarithms)	
$\log_b 1 = 0$	$\log 1 = 0$	$\ln 1 = 0$	Ex. 4, p. 388
$\log_b b = 1$	$\log 10 = 1$	$\ln e = 1$	Ex. 5, p. 388
$\log_b b^x = x$	$\log 10^x = 1$	$\ln e^x = x$	Ex. 11, p. 394
$b^{\log_b x} = x$	$10^{\log x} = x$	$e^{\ln x} = x$	Ex. 12, p. 394

4.3 Properties of Logarithms

a. *The Product Rule:* $\log_b(MN) = \log_b M + \log_b N$ — Ex. 1, p. 399

b. *The Quotient Rule:* $\log_b\left(\dfrac{M}{N}\right) = \log_b M - \log_b N$ — Ex. 2, p. 399

c. *The Power Rule:* $\log_b M^p = p \log_b M$ — Ex. 3, p. 401

d. *The Change-of-Base Property:*

The General Property	Introducing Common Logarithms	Introducing Natural Logarithms	
$\log_b M = \dfrac{\log_a M}{\log_a b}$	$\log_b M = \dfrac{\log M}{\log b}$	$\log_b M = \dfrac{\ln M}{\ln b}$	Ex. 7, p. 404 Ex. 8, p. 405

DEFINITIONS AND CONCEPTS

EXAMPLES

4.4 Exponential and Logarithmic Equations

a. An exponential equation is an equation containing a variable in an exponent. The solution procedure involves isolating the exponential expression and taking the natural logarithm on both sides. The box on page 408 provides the details.

Ex.1, p. 408;
Ex.2, p. 409;
Ex.3, p. 409;
Ex.4, p. 410

b. A logarithmic equation is an equation containing a variable in a logarithmic expression. Logarithmic equations in the form $\log_b x = c$ can be solved by rewriting as $b^c = x$.

Ex.5, p. 410

c. When checking logarithmic equations, reject proposed solutions that produce the logarithm of a negative number or the logarithm of 0 in the original equation.

Ex.6, p. 411

d. Equations involving natural logarithms are solved by isolating the natural logarithm with coefficient 1 on one side and exponentiating both sides. Simplify using $e^{\ln x} = x$.

Ex.7, p. 412

4.5 Modeling with Exponential and Logarithmic Functions

a. Exponential growth and decay models are given by $A = A_0 e^{kt}$ in which t represents time, A_0 is the amount present at $t = 0$, and A is the amount present at time t. If $k > 0$, the model describes growth and k is the growth rate. If $k < 0$, the model describes decay and k is the decay rate.

Ex.1, p. 420;
Ex.2, p. 422

b. The logistic growth model, given by $A = \dfrac{c}{1 + ae^{-bt}}$, describes situations in which growth is limited. $y = c$ is a horizontal asymptote for the graph, and growth, A, can never exceed c.

Ex.3, p. 424

c. Scatter plots for exponential and logarithmic models are shown in Figure 4.20 on page 425. When using a graphing utility to model data, the closer that the correlation coefficient, r, is to -1 or 1, the better the model fits the data.

d. Expressing an Exponential Model in Base e: $y = ab^x$ is equivalent to $y = ae^{(\ln b) \cdot x}$.

Ex.4, p. 427

Review Exercises

4.1

In Exercises 1–4, the graph of an exponential function is given. Select the function for each graph from the following options:

$$f(x) = 4^x, g(x) = 4^{-x},$$

$$h(x) = -4^{-x}, r(x) = -4^{-x} + 3.$$

1.

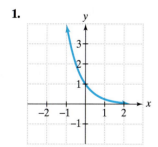

2.

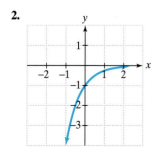

3.

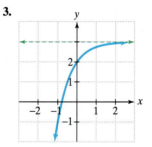

4.

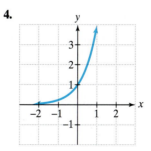

In Exercises 5–8, sketch by hand the graphs of the two functions in the same rectangular coordinate system. Use a table of coordinates to sketch the first function and transformations of this function with a table of coordinates to graph the second function.

5. $f(x) = 2^x$ and $g(x) = 2^{x-1}$

6. $f(x) = 3^x$ and $g(x) = 3^x - 1$

7. $f(x) = 3^x$ and $g(x) = -3^x$

8. $f(x) = \left(\frac{1}{2}\right)^x$ and $g(x) = \left(\frac{1}{2}\right)^{-x}$

Use the compound interest formulas to solve Exercises 9–10.

9. Suppose that you have $5000 to invest. Which investment yields the greater return over 5 years: 5.5% compounded semiannually or 5.25% compounded monthly?

10. Suppose that you have $14,000 to invest. Which investment yields the greater return over 10 years: 7% compounded monthly or 6.85% compounded continuously?

11. A cup of coffee is taken out of a microwave oven and placed in a room. The temperature, T, in degrees Fahrenheit, of the coffee after t minutes is modeled by the function $T = 70 + 130e^{-0.04855t}$. The graph of the function is shown in the figure.

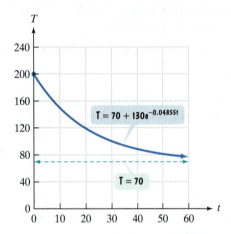

Use the graph to answer each of the following questions.

a. What was the temperature of the coffee when it was first taken out of the microwave?

b. What is a reasonable estimate of the temperature of the coffee after 20 minutes? Use your calculator to verify this estimate.

c. What is the limit of the temperature to which the coffee will cool? What does this tell you about the temperature of the room?

4.2

In Exercises 12–14, write each equation in its equivalent exponential form.

12. $\frac{1}{2} = \log_{49} 7$ **13.** $3 = \log_4 x$ **14.** $\log_3 81 = y$

In Exercises 15–17, write each equation in its equivalent logarithmic form.

15. $6^3 = 216$ **16.** $b^4 = 625$ **17.** $13^y = 874$

In Exercises 18–25, evaluate each expression without using a calculator. If evaluation is not possible, state the reason.

18. $\log_4 64$ **19.** $\log_5 \frac{1}{25}$ **20.** $\log_3 (-9)$

21. $\log_{16} 4$ **22.** $\log_{17} 17$ **23.** $\log_3 3^8$

24. $\ln e^5$ **25.** $\log_3 (\log_8 8)$

26. Graph $f(x) = 2^x$ and $g(x) = \log_2 x$ in the same rectangular coordinate system.

27. Graph $f(x) = \left(\frac{1}{3}\right)^x$ and $g(x) = \log_{1/3} x$ in the same rectangular coordinate system.

In Exercises 28–31, the graph of a logarithmic function is given. Select the function for each graph from the following options:

$$f(x) = \log x, \, g(x) = \log(-x),$$
$$h(x) = \log(2 - x), \, r(x) = 1 + \log(2 - x).$$

28.

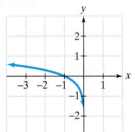

29.

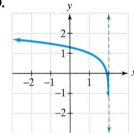

30.

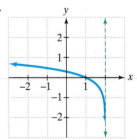

31.

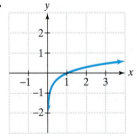

In Exercises 32–34, begin by graphing $f(x) = \log_2 x$. Then use transformations of this graph to graph the given function. What is the graph's x-intercept? What is the vertical asymptote?

32. $g(x) = \log_2(x - 2)$ **33.** $h(x) = -1 + \log_2 x$

34. $r(x) = \log_2 (-x)$

In Exercises 35–37, find the domain of each logarithmic function.

35. $f(x) = \log_8 (x + 5)$ **36.** $f(x) = \log (3 - x)$

37. $f(x) = \ln (x - 1)^2$

In Exercises 38–40, use inverse properties of logarithms to simplify each expression.

38. $\ln e^{6x}$ **39.** $e^{\ln \sqrt{x}}$ **40.** $10^{\log 4x^2}$

41. On the Richter scale, the magnitude, R, of an earthquake of intensity I is given by $R = \log \dfrac{I}{I_0}$, where I_0 is the intensity of a barely felt zero-level earthquake. If the intensity of an earthquake is $1000 I_0$, what is its magnitude on the Richter scale?

42. Students in a psychology class took a final examination. As part of an experiment to see how much of the course content they remembered over time, they took equivalent forms of the exam in monthly intervals thereafter. The average score, $f(t)$, for the group after t months is modeled by the function $f(t) = 76 - 18 \log (t + 1)$, where $0 \le t \le 12$.

   a. What was the average score when the exam was first given?

   b. What was the average score after 2 months? 4 months? 6 months? 8 months? one year?

   c. Use the results from parts (a) and (b) to graph f. Describe what the shape of the graph indicates in terms of the material retained by the students.

43. The formula

$$t = \frac{1}{c} \ln \left(\frac{A}{A - N} \right)$$

describes the time, t, in weeks, that it takes to achieve mastery of a portion of a task. In the formula, A represents maximum learning possible, N is the portion of the learning that is to be achieved, and c is a constant used to measure an individual's learning style. A 50-year-old man decides to start running as a way to maintain good health. He feels that the maximum rate he could ever hope to achieve is 12 miles per hour. How many weeks will it take before the man can run 5 miles per hour if $c = 0.06$ for this person?

4.3

In Exercises 44–47, use properties of logarithms to expand each logarithmic expression as much as possible. Where possible, evaluate logarithmic expressions without using a calculator.

44. $\log_6 (36x^3)$ **45.** $\log_4 \left(\dfrac{\sqrt{x}}{64} \right)$

46. $\log_2 \left(\dfrac{xy^2}{64} \right)$ **47.** $\ln \sqrt[3]{\dfrac{x}{e}}$

In Exercises 48–51, use properties of logarithms to condense each logarithmic expression. Write the expression as a single logarithm whose coefficient is 1.

48. $\log_b 7 + \log_b 3$ **49.** $\log 3 - 3 \log x$

50. $3 \ln x + 4 \ln y$ **51.** $\frac{1}{2} \ln x - \ln y$

In Exercises 52–53, use common logarithms or natural logarithms and a calculator to evaluate to four decimal places.

52. $\log_6 72{,}348$ **53.** $\log_4 0.863$

4.4

Solve each exponential equation in Exercises 54–58. Express the answer in terms of natural logarithms. Then use a calculator to obtain a decimal approximation, correct to two decimal places, for the solution.

54. $8^x = 12{,}143$ **55.** $9e^{5x} = 1269$

56. $e^{12-5x} - 7 = 123$ **57.** $5^{4x+2} = 37{,}500$

58. $e^{2x} - e^x - 6 = 0$

Solve each logarithmic equation in Exercises 59–63.

59. $\log_4 (3x - 5) = 3$

60. $\log_2 (x + 3) + \log_2 (x - 3) = 4$

61. $\log_3 (x - 1) - \log_3 (x + 2) = 2$

62. $\ln x = -1$ **63.** $3 + 4 \ln (2x) = 15$

64. The formula $A = 10.1 e^{0.005t}$ models the population of Los Angeles, California, A, in millions, t years after 1992. If the growth rate continues into the future, when will the population reach 13 million?

65. The amount of carbon dioxide in the atmosphere, measured in parts per million, has been increasing as a result of the burning of oil and coal. The buildup of gases and particles traps heat and raises the planet's temperature, a phenomenon called the *greenhouse effect*. Carbon dioxide accounts for about half of the warming. The function $f(t) = 364(1.005)^t$ projects carbon dioxide concentration, $f(t)$, in parts per million, t years after 2000. Using the projections given by the function, when will the carbon dioxide concentration be double the preindustrial level of 280 parts per million?

66. The formula $\overline{C}(x) = 15{,}557 + 5259 \ln x$ models the average cost of a new car, $\overline{C}(x)$, x years after 1989. When will the average cost of a new car reach \$30,000?

67. Use the formula for compound interest with n compoundings each year to solve this problem. How long, to the nearest tenth of a year, will it take \$12,500 to grow to \$20,000 at 6.5% annual interest compounded quarterly?

Use the formula for continuous compounding to solve Exercises 68–69.

68. How long, to the nearest tenth of a year, will it take \$50,000 to triple in value at 7.5% annual interest compounded continuously?

69. What interest rate is required for an investment subject to continuous compounding to triple in 5 years?

4.5

70. According to the U.S. Bureau of the Census, in 1990 there were 22.4 million residents of Hispanic origin living in the United States. By 2000, the number had increased to 35.3 million. The exponential growth function $A = 22.4e^{kt}$ describes the U.S. Hispanic population, A, in millions, t years after 1990.

a. Find k, correct to three decimal places.

b. Use the resulting model to project the Hispanic resident population in 2010.

c. In which year will the Hispanic resident population reach 60 million?

71. Use the exponential decay model for carbon-14, $A = A_0e^{-0.000121t}$, to solve this exercise. Prehistoric cave paintings were discovered in the Lascaux cave in France. The paint contained 15% of the original carbon-14. Estimate the age of the paintings at the time of the discovery.

72. The function

$$f(t) = \frac{500,000}{1 + 2499e^{-0.92t}}$$

models the number of people, $f(t)$, in a city who have become ill with influenza t weeks after its initial outbreak.

a. How many people became ill with the flu when the epidemic began?

b. How many people were ill by the end of the sixth week?

c. What is the limiting size of $f(t)$, the population that becomes ill?

In Exercises 73–74, rewrite the equation in terms of base e. Express the answer in terms of a natural logarithm, and then round to three decimal places.

73. $y = 73(2.6)^x$

74. $y = 6.5(0.43)^x$

75. The figure shows world population projections through the year 2150. The data are from the United Nations Family Planning Program and are based on optimistic or pessimistic expectations for successful control of human population growth. Suppose that you are interested in modeling these data using exponential, logarithmic, linear, and quadratic functions. Which function would you use to model each of the projections? Explain your choices. For the choice corresponding to a quadratic model, would your formula involve one with a positive or negative leading coefficient? Explain.

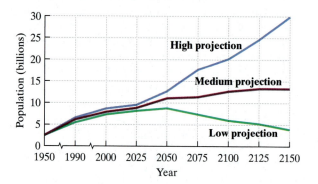

76. The figure shows the number of people in the United States age 65 and over, with projected figures for the year 2010 and beyond.

U. S. Population Age 65 and Over

Source: U.S. Bureau of the Census

Let x represent the number of years after 1899 and let y represent the U.S. population, in millions, age 65 and over. Use your graphing utility to find the model that best fits the data in the bar graph. Then use the model to find the projected U.S. population age 65 and over in 2050.

Chapter 4 Test

1. Graph $f(x) = 2^x$ and $g(x) = 2^{x+1}$ in the same rectangular coordinate system.
2. Graph $f(x) = \log_2 x$ and $g(x) = \log_2(x - 1)$ in the same rectangular coordinate system.
3. Write in exponential form: $\log_5 125 = 3$.
4. Write in logarithmic form: $\sqrt{36} = 6$.
5. Find the domain of $f(x) = \ln(3 - x)$.

In Exercises 6–7, use properties of logarithms to expand each logarithmic expression as much as possible. Where possible, evaluate logarithmic expressions without using a calculator.

6. $\log_4(64x^5)$
7. $\log_3\left(\dfrac{\sqrt[3]{x}}{81}\right)$

In Exercises 8–9, write each expression as a single logarithm.

8. $6 \log x + 2 \log y$
9. $\ln 7 - 3 \ln x$

10. Use a calculator to evaluate $\log_{15} 71$ to four decimal places.

In Exercises 11–16, solve each equation.

11. $5^x = 1.4$
12. $400e^{0.005x} = 1600$
13. $e^{2x} - 6e^x + 5 = 0$
14. $\log_6(4x - 1) = 3$
15. $\log x + \log(x + 15) = 2$
16. $2\ln(3x) = 8$

17. Suppose you have \$3000 to invest. Which investment yields the greater return over 10 years: 6.5% compounded semiannually or 6% compounded continuously? How much more (to the nearest dollar) is yielded by the better investment?

18. On the decibel scale, the loudness of a sound, D, in decibels, is given by $D = 10 \log \dfrac{I}{I_0}$, where I is the intensity of the sound, in watts per meter^2, and I_0 is the intensity of a sound barely audible to the human ear. If the intensity of a sound is $10^{12}I_0$, what is its loudness in decibels? (Such a sound is potentially damaging to the ear.)

19. The function
$$P(t) = 89.18e^{-0.004t}$$
models the percentage, $P(t)$, of married men in the United States who were employed t years after 1959.
   a. What percentage of married men were employed in 1959?
   b. Is the percentage of married men who are employed increasing or decreasing? Explain.
   c. In what year were 77% of U.S. married men employed?

20. The 1990 population of Europe was 509 million; in 2000, it was 729 million. Write the exponential growth function that describes the population of Europe, in millions, t years after 1990.

21. Use the exponential decay model for carbon-14, $A = A_0 e^{-0.000121t}$, to solve this exercise. Bones of a prehistoric man were discovered and contained 5% of the original amount of carbon-14. How long ago did the man die?

22. The logistic growth function
$$f(t) = \frac{140}{1 + 9e^{-0.165t}}$$
describes the population, $f(t)$, of an endangered species of elk t years after they were introduced to a nonthreatening habitat.
   a. How many elk were initially introduced to the habitat?
   b. How many elk are expected in the habitat after 10 years?
   c. What is the limiting size of the elk population that the habitat will sustain?

Cumulative Review Exercises (Chapters 1–4)

Solve each equation in Exercises 1–5.

1. $|3x - 4| = 2$
2. $\sqrt{2x - 5} - \sqrt{x - 3} = 1$
3. $x^4 + x^3 - 3x^2 - x + 2 = 0$
4. $e^{5x} - 32 = 96$
5. $\log_2(x + 5) + \log_2(x - 1) = 4$

Solve each inequality in Exercises 6–7. Express the answer in interval notation.

6. $14 - 5x \geq -6$
7. $|2x - 4| \leq 2$

8. Write the point-slope form and the slope-intercept form of the line passing through $(1, 3)$ and $(3, -3)$.
9. If $f(x) = x^2$ and $g(x) = x + 2$, find $(f \circ g)(x)$ and $(g \circ f)(x)$.
10. If $f(x) = 2x - 7$, find $f^{-1}(x)$.
11. Divide $x^3 + 5x^2 + 3x - 10$ by $x + 2$.
12. Use the Rational Zero Theorem to list all possible rational zeros for $f(x) = 4x^3 - 7x - 3$.

13. The value of y varies directly as the square of x. If $x = 3$ when $y = 12$, find y when $x = 15$.
14. Solve $x^3 - 4x^2 + 6x - 4 = 0$ given that $1 + i$ is a root.

In Exercises 15–18, graph each equation.

15. $(x - 3)^2 + (y + 2)^2 = 4$
16. $f(x) = (x - 2)^2 - 1$
17. $f(x) = \dfrac{x^2 - 1}{x^2 - 4}$
18. $f(x) = (x - 2)^2(x + 1)$

19. You are paid time-and-a-half for each hour worked over 40 hours a week. Last week you worked 50 hours and earned \$660. What is your normal hourly salary?

20. The function $F(t) = 1 - k \ln(t + 1)$ models the fraction of people, $F(t)$, who remember all the words in a list of nonsense words t hours after memorizing the list. After 3 hours, only half the people could remember all the words. Determine the value of k and then predict the fraction of people in the group who will remember all the words after 6 hours. Round to three decimal places and then express the fraction with a denominator of 1000.

Systems of Equations and Inequalities

Chapter 5

Most things in life depend on many variables. Temperature and precipitation are two variables that have a critical effect on whether regions are forests, grasslands, or deserts. Airlines deal with numerous variables during weather disruptions at large connecting airports. They must solve the problem of putting their operation back together again to minimize the cost of the disruption and passenger inconvenience. In this chapter, forests, grasslands, and airline service are viewed in the same way—situations with several variables. You will learn methods for modeling and solving problems in these situations.

A major weather disruption delayed your flight for hours, but you finally made it. You are in Yosemite National Park in California, surrounded by evergreen forests, alpine meadows, and sheer walls of granite. Soaring cliffs, plunging waterfalls, gigantic trees, rugged canyons, mountains and valleys stand in stark contrast to the angry chaos at the airport. This is so different from where you live and attend college, a region in which grasslands predominate.

SECTION 5.1   *Systems of Linear Equations in Two Variables*

Objectives

1. Decide whether an ordered pair is a solution of a linear system.

2. Solve linear systems by substitution.

3. Solve linear systems by addition.

4. Identify systems that do not have exactly one ordered-pair solution.

5. Solve problems using systems of linear equations.

Key West residents Brian Goss (left), George Wallace, and Michael Mooney (right) hold on to each other as they battle 90 mph winds along Houseboat Row in Key West, Fla., on Friday, Sept. 25, 1998. The three had sought shelter behind a Key West hotel as Hurricane Georges descended on the Florida Keys, but were forced to seek other shelter when the storm conditions became too rough. Hundreds of people were killed by the storm when it swept through the Caribbean.

Problems ranging from scheduling airline flights to controlling traffic flow to routing phone calls over the nation's communication network often require solutions in a matter of moments. The solution to these real-world problems can involve solving thousands of equations having thousands of variables. AT&T's domestic long-distance network involves 800,000 variables! Meteorologists describing atmospheric conditions surrounding a hurricane must solve problems involving thousands of equations rapidly and efficiently. The difference between a two-hour warning and a two-day warning is a life-and-death issue for thousands of people in the path of one of nature's most destructive forces.

Although we will not be solving 800,000 equations with 800,000 variables, we will turn our attention to two equations with two variables, such as

$$2x - 3y = -4$$
$$2x + y = 4.$$

The methods that we consider for solving such problems provide the foundation for solving far more complex systems with many variables.

1 Decide whether an ordered pair is a solution of a linear system.

Systems of Linear Equations and Their Solutions

We have seen that all equations in the form $Ax + By = C$ are straight lines when graphed. Two such equations, such as those listed above, are called a **system of linear equations** or a **linear system. A solution to a system of linear equations** is an ordered pair that satisfies all equations in the system. For example, (3, 4) satisfies the system

$$x + y = 7 \qquad (3 + 4 \text{ is, indeed, } 7.)$$
$$x - y = -1. \qquad (3 - 4 \text{ is, indeed, } -1.)$$

Thus, $(3, 4)$ satisfies both equations and is a solution of the system. The solution can be described by saying that $x = 3$ and $y = 4$. The solution can also be described using set notation. The solution set to the system is $\{(3, 4)\}$—that is, the set consisting of the ordered pair $(3, 4)$.

A system of linear equations can have exactly one solution, no solution, or infinitely many solutions. We begin with systems that have exactly one solution.

EXAMPLE 1 Determining Whether an Ordered Pair Is a Solution of a System

Determine whether $(4, -1)$ is a solution of the system

$$x + 2y = 2$$
$$x - 2y = 6.$$

Solution Because 4 is the x-coordinate and -1 is the y-coordinate of $(4, -1)$, we replace x with 4 and y with -1.

$$
\begin{array}{ll}
x + 2y = 2 & \qquad x - 2y = 6 \\
4 + 2(-1) \stackrel{?}{=} 2 & \qquad 4 - 2(-1) \stackrel{?}{=} 6 \\
4 + (-2) \stackrel{?}{=} 2 & \qquad 4 - (-2) \stackrel{?}{=} 6 \\
2 = 2, \text{ true} & \qquad 4 + 2 \stackrel{?}{=} 6 \\
& \qquad 6 = 6, \text{ true}
\end{array}
$$

The pair $(4, -1)$ satisfies both equations: It makes each equation true. Thus, the pair is a solution of the system. The solution set to the system is $\{(4, -1)\}$.

Study Tip

When solving linear systems by graphing, neatly drawn graphs are essential for determining points of intersection.

- Use rectangular coordinate graph paper.
- Use a ruler or straightedge.
- Use a pencil with a sharp point.

The solution of a system of linear equations can be found by graphing both of the equations in the same rectangular coordinate system. For a system with one solution, the **coordinates of the point of intersection give the system's solution.** For example, the system in Example 1,

$$x + 2y = 2$$
$$x - 2y = 6$$

is graphed in Figure 5.1. The solution of the system, $(4, -1)$, corresponds to the point of intersection of the lines.

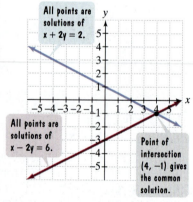

Figure 5.1 Visualizing a system's solution

 Check Point 1 Determine whether $(1, 2)$ is a solution of the system

$$2x - 3y = -4$$
$$2x + y = 4.$$

2 Solve linear systems by substitution.

Eliminating a Variable Using the Substitution Method

Finding the solution to a linear system by graphing equations may not be easy to do. For example, a solution of $\left(-\frac{2}{3}, \frac{157}{29}\right)$ would be difficult to "see" as an intersection point on a graph.

Let's consider a method that does not depend on finding a system's solution visually: the substitution method. This method involves converting the system to one equation in one variable by an appropriate substitution.

EXAMPLE 2 Solving a System by Substitution

Solve by the substitution method:

$$y = -x - 1$$
$$4x - 3y = 24.$$

Solution

Step 1 Solve either of the equations for one variable in terms of the other. This step has already been done for us. The first equation, $y = -x - 1$, has y solved in terms of x.

Step 2 Substitute the expression from step 1 into the other equation. We substitute the expression $-x - 1$ for y in the other equation:

$$y = \boxed{-x - 1} \qquad 4x - 3\boxed{y} = 24 \qquad \text{Substitute } -x - 1 \text{ for } y.$$

This gives us an equation in one variable, namely

$$4x - 3(-x - 1) = 24.$$

The variable y has been eliminated.

Step 3 Solve the resulting equation containing one variable.

$$
\begin{aligned}
4x - 3(-x - 1) &= 24 && \text{This is the equation containing one variable.} \\
4x + 3x + 3 &= 24 && \text{Apply the distributive property.} \\
7x + 3 &= 24 && \text{Combine like terms.} \\
7x &= 21 && \text{Subtract 3 from both sides.} \\
x &= 3 && \text{Divide both sides by 7.}
\end{aligned}
$$

Step 4 Back-substitute the obtained value into one of the original equations. We now know that the x-coordinate of the solution is 3. To find the y-coordinate, we back-substitute the x-value into either original equation. We will use

$$y = -x - 1.$$

Substitute 3 for x.

$$y = -3 - 1 = -4$$

With $x = 3$ and $y = -4$, the proposed solution is $(3, -4)$.

Step 5 Check the proposed solution in both of the system's given equations. Replace x with 3 and y with -4.

$$
\begin{array}{ll}
y = -x - 1 & 4x - 3y = 24 \\
-4 \stackrel{?}{=} -3 - 1 & 4(3) - 3(-4) \stackrel{?}{=} 24 \\
-4 = -4, \text{ true} & 12 + 12 \stackrel{?}{=} 24 \\
& 24 = 24, \text{ true}
\end{array}
$$

The pair $(3, -4)$ satisfies both equations. The system's solution set is $\{(3, -4)\}$.

Technology

A graphing utility can be used to solve the system in Example 2. Graph each equation and use the intersection feature. The utility displays the solution $(3, -4)$ as $x = 3, y = -4$.

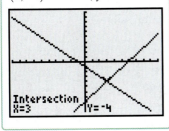

Intersection
X=3   Y=-4

Check Point 2 Solve by the substitution method:

$$y = 5x - 13$$
$$2x + 3y = 12.$$

Before considering additional examples, let's summarize the steps used in the substitution method.

Study Tip

In step 1, you can choose which variable to isolate in which equation. If possible, solve for a variable whose coefficient is 1 or −1 to avoid working with fractions.

Solving Linear Systems by Substitution

1. Solve either of the equations for one variable in terms of the other. (If one of the equations is already in this form, you can skip this step.)
2. Substitute the expression found in step 1 into the other equation. This will result in an equation in one variable.
3. Solve the equation containing one variable.
4. Back-substitute the value found in step 3 into one of the original equations. Simplify and find the value of the remaining variable.
5. Check the proposed solution in both of the system's given equations.

EXAMPLE 3 Solving a System by Substitution

Solve by the substitution method:

$$5x - 4y = 9$$
$$x - 2y = -3.$$

Solution

Step 1 Solve either of the equations for one variable in terms of the other. We begin by isolating one of the variables in either of the equations. By solving for x in the second equation, which has a coefficient of 1, we can avoid fractions.

$$x - 2y = -3 \qquad \text{This is the second equation in the given system.}$$
$$x = 2y - 3 \qquad \text{Solve for x by adding 2y to both sides.}$$

Step 2 Substitute the expression from step 1 into the other equation. We substitute $2y - 3$ for x in the first equation.

$$x = \boxed{2y - 3} \qquad 5\boxed{x} - 4y = 9$$

This gives us an equation in one variable, namely

$$5(2y - 3) - 4y = 9.$$

The variable x has been eliminated.

Step 3 Solve the resulting equation containing one variable.

$$5(2y - 3) - 4y = 9 \qquad \text{This is the equation containing one variable.}$$
$$10y - 15 - 4y = 9 \qquad \text{Apply the distributive property.}$$
$$6y - 15 = 9 \qquad \text{Combine like terms.}$$
$$6y = 24 \qquad \text{Add 15 to both sides.}$$
$$y = 4 \qquad \text{Divide both sides by 6.}$$

Step 4   Back-substitute the obtained value into one of the original equations. We back-substitute 4 for y into one of the original equations to find x. Let's use both equations to show that we obtain the same value for x in either case.

Using the first equation:	Using the second equation:
$5x - 4y = 9$	$x - 2y = -3$
$5x - 4(4) = 9$	$x - 2(4) = -3$
$5x - 16 = 9$	$x - 8 = -3$
$5x = 25$	$x = 5$
$x = 5$	

With $x = 5$ and $y = 4$, the proposed solution is $(5, 4)$.

Step 5   Check.   Take a moment to show that $(5, 4)$ satisfies both given equations. The solution set is $\{(5, 4)\}$.

Check Point 3   Solve by the substitution method:

$$3x + 2y = -1$$
$$x - y = 3.$$

3 Solve linear systems by addition.

Eliminating a Variable Using the Addition Method

The substitution method is most useful if one of the given equations has an isolated variable. A second, and frequently the easiest, method for solving a linear system is the addition method. Like the substitution method, the addition method involves eliminating a variable and ultimately solving an equation containing only one variable. However, this time we eliminate a variable by adding the equations.

For example, consider the following system of linear equations:

$$3x - 4y = 11$$
$$-3x + 2y = -7.$$

When we add these two equations, the x-terms are eliminated. This occurs because the coefficients of the x-terms, 3 and -3, are opposites (additive inverses) of each other:

$$3x - 4y = 11$$
$$-3x + 2y = -7$$

Add: $-2y = 4$ The sum is an equation in one variable.

$y = -2$ Solve for y, dividing both sides by -2.

Now we can back-substitute -2 for y into one of the original equations to find x. It does not matter which equation you use; you will obtain the same value for x in either case. If we use either equation, we can show that $x = 1$ and the solution $(1, -2)$ satisfies both equations in the system.

When we use the addition method, we want to obtain two equations whose sum is an equation containing only one variable. The key step is to obtain, for one of the variables, coefficients that differ only in sign. To do this, we may need to multiply one or both equations by some nonzero number so that the coefficients of one of the variables, x or y, become opposites. Then when the two equations are added, this variable is eliminated.

EXAMPLE 4   Solving a System by the Addition Method

Solve by the addition method:

$$3x + 2y = 48$$
$$9x - 8y = -24.$$

Solution   We must rewrite one or both equations in equivalent forms so that the coefficients of the same variable (either x or y) are opposites of each other. Consider the terms in x in each equation, that is, $3x$ and $9x$. To eliminate x, we can multiply each term of the first equation by -3 and then add the equations.

$$3x + 2y = 48 \quad \underrightarrow{\text{Multiply by } -3.} \quad -9x - 6y = -144$$

$$9x - 8y = -24 \quad \underrightarrow{\text{No change}} \quad 9x - 8y = -24$$

$$\text{Add:} \qquad -14y = -168$$

$$y = 12 \qquad \text{Solve for } y, \text{ dividing both sides by } -14.$$

Thus, $y = 12$. We back-substitute this value into either one of the given equations. We'll use the first one.

$$3x + 2y = 48 \qquad \text{This the first equation in the given system.}$$
$$3x + 2(12) = 48 \qquad \text{Substitute 12 for } y.$$
$$3x + 24 = 48 \qquad \text{Multiply.}$$
$$3x = 24 \qquad \text{Subtract 24 from both sides.}$$
$$x = 8 \qquad \text{Divide both sides by 3.}$$

The solution $(8, 12)$ can be shown to satisfy both equations in the system. Consequently, the solution set is $\{(8, 12)\}$.

Solving Linear Systems by Addition

1. If necessary, rewrite both equations in the form $Ax + By = C$.
2. If necessary, multiply either equation or both equations by appropriate nonzero numbers so that the sum of the x-coefficients or the sum of the y-coefficients is 0.
3. Add the equations in step 2. The sum is an equation in one variable.
4. Solve the equation in one variable.
5. Back-substitute the value obtained in step 4 into either of the given equations and solve for the other variable.
6. Check the solution in both of the original equations.

Check Point 4   Solve by the addition method:

$$4x + 5y = 3$$
$$2x - 3y = 7.$$

Some linear systems have solutions that are not integers. If the value of one variable turns out to be a "messy" fraction, back-substitution might lead to cumbersome arithmetic. If this happens, you can return to the original system and use addition to find the value of the other variable.

> **EXAMPLE 5**   **Solving a System by the Addition Method**

Solve by the addition method:

$$2x = 7y - 17$$
$$5y = 17 - 3x.$$

Solution

Step 1   Rewrite both equations in the form $Ax + By = C$.   We first arrange the system so that variable terms appear on the left and constants appear on the right. We obtain

$$2x - 7y = -17 \quad \text{Subtract 7y from both sides of the first equation.}$$
$$3x + 5y = 17. \quad \text{Add 3x to both sides of the second equation.}$$

Step 2   If necessary, multiply either equation or both equations by appropriate numbers so that the sum of the x-coefficients or the sum of the y-coefficients is 0.   We can eliminate x or y. Let's eliminate x by multiplying the first equation by 3 and the second equation by -2.

$$2x - 7y = -17 \quad \underrightarrow{\text{Multiply by 3.}} \quad 3 \cdot 2x - 3 \cdot 7y = 3(-17) \longrightarrow 6x - 21y = -51$$
$$3x + 5y = 17 \quad \underrightarrow{\text{Multiply by } -2.} \quad -2 \cdot 3x + (-2) \cdot 5y = -2(17) \longrightarrow -6x - 10y = -34$$

Steps 3 and 4   Add the equations and solve the equation in one variable.

$$\begin{array}{r} 6x - 21y = -51 \\ -6x - 10y = -34 \\ \hline \end{array}$$
$$\text{Add:} \quad -31y = -85$$
$$\frac{-31y}{-31} = \frac{-85}{-31} \quad \text{Divide both sides by } -31.$$
$$y = \frac{85}{31} \quad \text{Simplify.}$$

Step 5   Back-substitute and find the value for the other variable.   Back-substitution of $\frac{85}{31}$ for y into either of the given equations results in cumbersome arithmetic. Instead, let's use the addition method on the given system in the form $Ax + By = C$ to find the value for x. Thus, we eliminate y by multiplying the first equation by 5 and the second equation by 7.

$$2x - 7y = -17 \quad \underrightarrow{\text{Multiply by 5.}} \quad 10x - 35y = -85$$
$$3x + 5y = 17 \quad \underrightarrow{\text{Multiply by 7.}} \quad 21x + 35y = 119$$
$$\text{Add:} \quad 31x = 34$$
$$x = \tfrac{34}{31} \quad \text{Divide both sides by 31.}$$

Step 6   Check.   For this system, a calculator is helpful in showing the solution $\left(\frac{34}{31}, \frac{85}{31}\right)$ satisfies both equations. Consequently, the solution set is $\left\{\left(\frac{34}{31}, \frac{85}{31}\right)\right\}$.

> **Check Point 5**   Solve by the addition method:
>
> $$4x = 5 + 2y$$
> $$3y = 4 - 2x.$$

4 Identify systems that do not have exactly one ordered-pair solution.

Linear Systems Having No Solution or Infinitely Many Solutions

We have seen that a system of linear equations in two variables represents a pair of lines. The lines either intersect, are parallel, or are identical. Thus, there are three possibilities for the number of solutions to a system of two linear equations.

The Number of Solutions to a System of Two Linear Equations

The number of solutions to a system of two linear equations in two variables is given by one of the following. (See Figure 5.2.)

Number of Solutions	What This Means Graphically
Exactly one ordered-pair solution	The two lines intersect at one point.
No solution	The two lines are parallel.
Infinitely many solutions	The two lines are identical.

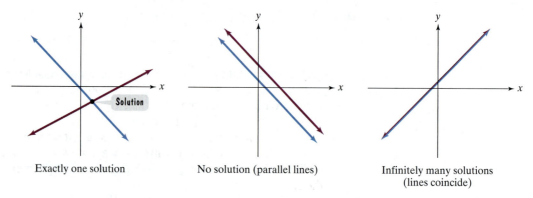

Exactly one solution   No solution (parallel lines)   Infinitely many solutions (lines coincide)

Figure 5.2 Possible graphs for a system of two linear equations in two variables

A linear system with no solution is called an **inconsistent system.** If you attempt to solve such a system by substitution or addition, you will eliminate both variables. A false statement such as $0 = 17$ will be the result.

EXAMPLE 6   A System with No Solution

Solve the system:

$$4x + 6y = 12$$
$$6x + 9y = 12.$$

Solution   Because no variable is isolated, we will use the addition method. To obtain coefficients of x that differ only in sign, we multiply the first equation by 3 and multiply the second equation by -2.

$$4x + 6y = 12 \xrightarrow{\text{Multiply by 3.}} 12x + 18y = 36$$
$$6x + 9y = 12 \xrightarrow{\text{Multiply by } -2.} \underline{-12x - 18y = -24}$$
$$\text{Add:} \qquad 0 = 12$$

There are no values of x and y for which $0 = 12$.

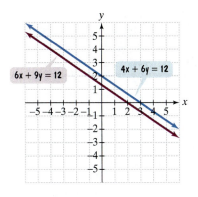

Figure 5.3 The graph of an inconsistent system

The false statement $0 = 12$ indicates that the system is inconsistent and has no solution. The solution set is the empty set, $\varnothing$.

The lines corresponding to the two equations in Example 6 are shown in Figure 5.3. The lines are parallel and have no point of intersection.

Discovery

Show that the graphs of $4x + 6y = 12$ and $6x + 9y = 12$ must be parallel lines by solving each equation for y. What is the slope and y-intercept for each line? What does this mean? If a linear system is inconsistent, what must be true about the slopes and y-intercepts for the system's graphs?

Check Point 6   Solve the system:

$$x + 2y = 4$$
$$3x + 6y = 13.$$

A linear system that has at least one solution is called a **consistent system.** Lines that intersect and lines that coincide both represent consistent systems. If the lines coincide, then the consistent system has infinitely many solutions, represented by every point on the line.

The equations in a linear system with infinitely many solutions are called **dependent.** If you attempt to solve such a system by substitution or addition, you will eliminate both variables. However, a true statement such as $0 = 0$ will be the result.

EXAMPLE 7   A System with Infinitely Many Solutions

Solve the system:

$$y = 3 - 2x$$
$$4x + 2y = 6.$$

Solution   Because the variable y is isolated in the first equation, we can use the substitution method. We substitute the expression for y in the other equation.

$$y = \boxed{3 - 2x} \quad 4x + 2\boxed{y} = 6 \qquad \text{Substitute } 3 - 2x \text{ for } y.$$

$$4x + 2y \quad = 6 \qquad \text{This is the second equation in the given system.}$$

$$4x + 2(3 - 2x) = 6 \qquad \text{Substitute } 3 - 2x \text{ for } y.$$

$$4x + 6 - 4x = 6 \qquad \text{Apply the distributive property.}$$

$$6 = 6 \qquad \text{Simplify. This statement is true for all values of } x \text{ and } y.$$

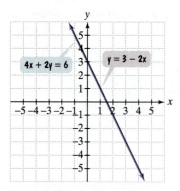

Figure 5.4 The graph of a system with infinitely many solutions

In our final step, both variables have been eliminated, and the resulting statement $6 = 6$ is true. This true statement indicates that the system has infinitely many solutions. The solution set consists of all points (x, y) lying on either of the coinciding lines, $y = 3 - 2x$ or $4x + 2y = 6$, as shown in Figure 5.4.

We express the solution set for the system in one of two equivalent ways:

$$\{(x, y) \mid y = 3 - 2x\} \quad \text{The set of all ordered pairs } (x, y) \text{ such that } y = 3 - 2x$$

$$\text{or} \quad \{(x, y) \mid 4x + 2y = 6\} \quad \text{The set of all ordered pairs } (x, y) \text{ such that } 4x + 2y = 6$$

Check Point 7   Solve the system:

$$y = 4x - 4$$
$$8x - 2y = 8.$$

5 Solve problems using systems of linear equations.

Applications

As a young entrepreneur, did you ever try selling lemonade in your front yard? Suppose that you charged 55¢ for each cup and you sold 45 cups. Your **revenue** is your income from selling these 45 units, or $\$0.55(45) = \$24.75$. Your *revenue function* from selling x cups is

$$R(x) = 0.55x.$$

This is the unit price: 55¢ for each cup.   This is the number of units sold.

For any business, the **revenue function,** R, is the money generated by selling x units of the product:

$$R(x) = px.$$

Price per unit x units sold

Back to selling lemonade and energizing the neighborhood with white sugar: Is your revenue for the afternoon also your profit? No. We need to consider the cost of the business. You estimate that the lemons, white sugar, and bottled water cost 5¢ per cup. Furthermore, mommy dearest is charging you a $\$10$ rental fee for use of your (her?) front yard. In Chapter 3, we saw that the **cost function,** C, for any business is the sum of its fixed and variable costs. Thus, your cost function for selling x cups of lemonade is

$$C(x) = 10 + 0.05x.$$

This is your $\$10$ fixed cost.   This is your variable cost: 5¢ for each cup produced.

Profit

What does every entrepreneur, from a kid selling lemonade to Bill Gates, want to do? Generate profit, of course. The profit function, P, generated after producing and selling x units of a product is the difference between the revenue function, R, and the cost function, C. The profit function for the lemonade business in Figure 5.5 is

$$P(x) = R(x) - C(x)$$
$$= 0.55x - (10 + 0.05x)$$
$$= 0.50x - 10.$$

The graph of this profit function is shown below. The red portion lies below the x-axis and shows a loss when fewer than 20 units are sold. The lemonade business is "in the red." The black portion lies above the x-axis and shows a gain when more than 20 units are sold. The lemonade business is "in the black."

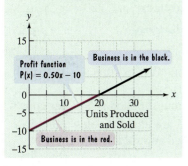

Figure 5.5 shows the graphs of the revenue and cost functions for the lemonade business. Similar graphs and models apply no matter how small or large a business venture may be.

$$R(x) = 0.55x \qquad C(x) = 10 + 0.05x$$

Revenue is 55¢ times the number of cups sold.

Cost is $10 plus 5¢ times the number of cups produced.

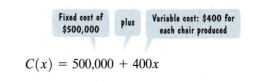

Figure 5.5

The lines intersect at the point (20, 11). This means that when 20 cups are produced and sold, both cost and revenue are $11. In business, this point of intersection is called the **break-even point.** At the break-even point, the money coming in is equal to the money going out. Can you see what happens for x-values less than 20? The red cost graph is above the blue revenue graph. The cost is greater than the revenue and the business is losing money. Thus, if you sell fewer than 20 cups of lemonade, the result is a *loss*. By contrast, look at what happens for x-values greater than 20. The blue revenue graph is above the red cost graph. The revenue is greater than the cost and the business is making money. Thus, if you sell more than 20 cups of lemonade, the result is a *gain*.

EXAMPLE 8 Finding a Break-Even Point

Technology is now promising to bring light, fast, and beautiful wheelchairs to millions of disabled people. A company is planning to manufacture these radically different wheelchairs. Fixed cost will be $500,000 and it will cost $400 to produce each wheelchair. Each wheelchair will be sold for $600.

a. Write the cost function, C, of producing x wheelchairs.

b. Write the revenue function, R, from the sale of x wheelchairs.

c. Determine the break-even point. Describe what this means.

Solution

a. The cost function is the sum of the fixed cost and variable cost.

Fixed cost of $500,000 — plus — Variable cost: $400 for each chair produced

$$C(x) = 500,000 + 400x$$

Technology

The graphs of the wheelchair company's cost and revenue functions

$$C(x) = 500,000 + 400x$$

and $R(x) = 600x$ are shown in a

$$[0, 5000, 1000]$$ by
$$[0, 3,000,000, 1,000,000]$$

viewing rectangle. The intersection point confirms that the company breaks even by producing and selling 2500 wheelchairs.

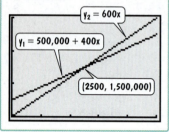

b. The revenue function is the money generated from the sale of x wheelchairs.

Revenue per chair, $600, times   the number of chairs sold

$$R(x) = 600x$$

c. The break-even point occurs where the graphs of C and R intersect. Thus, we find this point by solving the system

$$C(x) = 500,000 + 400x \qquad y = 500,000 + 400x$$
$$R(x) = 600x \qquad \text{or} \qquad y = 600x.$$

Using substitution, we can substitute $600x$ for y in the first equation.

$$600x = 500,000 + 400x \qquad \text{Substitute } 600x \text{ for } y.$$
$$200x = 500,000 \qquad \text{Subtract } 400x \text{ from both sides.}$$
$$x = 2500 \qquad \text{Divide both sides by } 200.$$

Back-substituting 2500 for x in either of the system's equations (or functions), we obtain

$$R(2500) = 600(2500) = 1,500,000.$$

We used $R(x) = 600x$.

The break-even point is (2500, 1,500,000). This means that the company will break even if it produces and sells 2500 wheelchairs. At this level, the money coming in is equal to the money going out: $1,500,000.

Check Point 8 A company that manufactures running shoes has a fixed cost of $300,000. Additionally, it costs $30 to produce each pair of shoes. They are sold at $80 per pair.

a. Write the cost function, C, of producing x pairs of running shoes.

b. Write the revenue function, R, from the sale of x pairs of running shoes.

c. Determine the break-even point. Describe what this means.

An important application of systems of equations arises in connection with supply and demand. As the price of a product increases, the demand for that product decreases. However, at higher prices suppliers are willing to produce greater quantities of the product.

EXAMPLE 9   Supply and Demand Models

A chain of video stores specializes in cult films. The weekly demand and supply models for *The Rocky Horror Picture Show* are given by

$$N = -13p + 760 \qquad \text{Demand model}$$
$$N = 2p + 430 \qquad \text{Supply model}$$

in which p is the price of the video and N is the number of copies of the video sold or supplied each week to the chain of stores.

a. How many copies of the video can be sold and supplied at $18 per copy?

b. Find the price at which supply and demand are equal. At this price, how many copies of *Rocky Horror* can be supplied and sold each week?

Solution

a. To find how many copies of the video can be sold and supplied at $18 per copy, we substitute 18 for p in the demand and supply models.

Demand Model	**Supply Model**
$N = -13p + 760$	$N = 2p + 430$

Substitute 18 for p.   Substitute 18 for p.

$$N = -13 \cdot 18 + 760 = 526 \qquad N = 2 \cdot 18 + 430 = 466$$

At $18 per video, the chain can sell 526 copies of *Rocky Horror* in a week. The manufacturer is willing to supply 466 copies per week. This will result in a shortage of copies of the video. Under these conditions, the retail chain is likely to raise the price of the video.

b. We can find the price at which supply and demand are equal by solving the demand-supply linear system. We will use substitution, substituting $-13p + 760$ for N in the second equation.

$$N = \boxed{-13p + 760} \qquad \boxed{N} = 2p + 430 \qquad \text{Substitute } -13p + 760 \text{ for N.}$$

$$-13p + 760 = 2p + 430 \qquad \text{The resulting equation contains only one variable.}$$

$$-15p + 760 = 430 \qquad \text{Subtract 2p from both sides.}$$

$$-15p = -330 \qquad \text{Subtract 760 from both sides.}$$

$$p = 22 \qquad \text{Divide both sides by } -15.$$

The price at which supply and demand are equal is $22 per video. To find the value of N, the number of videos supplied and sold weekly at this price, we back-substitute 22 for p into either the demand or the supply model. We'll use both models to make sure we get the same number in each case.

Figure 5.6 Priced at $22 per video, 474 copies of the video can be supplied and sold weekly.

Demand Model	**Supply Model**
$N = -13p + 760$	$N = 2p + 430$

Substitute 22 for p.   Substitute 22 for p.

$$N = -13 \cdot 22 + 760 = 474 \qquad N = 2 \cdot 22 + 430 = 474$$

At a price of $22 per video, 474 units of the video can be supplied and sold weekly. The intersection point, $(22, 474)$, is shown in Figure 5.6.

Check Point 9 The demand for a product is modeled by $N = -20p + 1000$ and the supply for the product by $N = 5p + 250$. In these models, p is the price of the product and N is the number supplied or sold weekly. At what price will supply equal demand? At that price, how many units of the product will be supplied and sold each week?

EXERCISE SET 5.1

Practice Exercises

In Exercises 1–4, determine whether the given ordered pair is a solution of the system.

1. $(2, 3)$
$x + 3y = 11$
$x - 5y = -13$

2. $(-3, 5)$
$9x + 7y = 8$
$8x - 9y = -69$

3. $(2, 5)$
$2x + 3y = 17$
$x + 4y = 16$

4. $(8, 5)$
$5x - 4y = 20$
$3y = 2x + 1$

In Exercises 5–18, solve each system by the substitution method.

5. $x + y = 4$
$y = 3x$

6. $x + y = 6$
$y = 2x$

7. $x + 3y = 8$
$y = 2x - 9$

8. $2x - 3y = -13$
$y = 2x + 7$

9. $x = 4y - 2$
$x = 6y + 8$

10. $x = 3y + 7$
$x = 2y - 1$

11. $5x + 2y = 0$
$x - 3y = 0$

12. $4x + 3y = 0$
$2x - y = 0$

13. $2x + 5y = -4$
$3x - y = 11$

14. $2x + 5y = 1$
$-x + 6y = 8$

15. $2x - 3y = 8 - 2x$
$3x + 4y = x + 3y + 14$

16. $3x - 4y = x - y + 4$
$2x + 6y = 5y - 4$

17. $y = \frac{1}{3}x + \frac{2}{3}$
$y = \frac{5}{7}x - 2$

18. $y = -\frac{1}{2}x + 2$
$y = \frac{3}{4}x + 7$

In Exercises 19–30, solve each system by the addition method.

19. $x + y = 1$
$x - y = 3$

20. $x + y = 6$
$x - y = -2$

21. $2x + 3y = 6$
$2x - 3y = 6$

22. $3x + 2y = 14$
$3x - 2y = 10$

23. $x + 2y = 2$
$-4x + 3y = 25$

24. $2x - 7y = 2$
$3x + y = -20$

25. $4x + 3y = 15$
$2x - 5y = 1$

26. $3x - 7y = 13$
$6x + 5y = 7$

27. $3x - 4y = 11$
$2x + 3y = -4$

28. $2x + 3y = -16$
$5x - 10y = 30$

29. $3x = 4y + 1$
$3y = 1 - 4x$

30. $5x = 6y + 40$
$2y = 8 - 3x$

In Exercises 31–42, solve by the method of your choice. Identify systems with no solution and systems with infinitely many solutions, using set notation to express their solution sets.

31. $x = 9 - 2y$
$x + 2y = 13$

32. $6x + 2y = 7$
$y = 2 - 3x$

33. $y = 3x - 5$
$21x - 35 = 7y$

34. $9x - 3y = 12$
$y = 3x - 4$

35. $3x - 2y = -5$
$4x + y = 8$

36. $2x + 5y = -4$
$3x - y = 11$

37. $x + 3y = 2$
$3x + 9y = 6$

38. $4x - 2y = 2$
$2x - y = 1$

39. $\frac{x}{4} - \frac{y}{4} = -1$
$x + 4y = -9$

40. $\frac{x}{6} - \frac{y}{2} = \frac{1}{3}$
$x + 2y = -3$

41. $2x = 3y + 4$
$4x = 3 - 5y$

42. $4x = 3y + 8$
$2x = -14 + 5y$

In Exercises 43–46, let x represent one number and let y represent the other number. Use the given conditions to write a system of equations. Solve the system and find the numbers.

43. The sum of two numbers is 7. If one number is subtracted from the other, their difference is −1. Find the numbers.

44. The sum of two numbers is 2. If one number is subtracted from the other, their difference is 8. Find the numbers.

45. Three times a first number decreased by a second number is 1. The first number increased by twice the second number is 12. Find the numbers.

46. The sum of three times a first number and twice a second number is 8. If the second number is subtracted from twice the first number, the result is 3. Find the numbers.

Application Exercises

Exercises 47–50 describe a number of business ventures. For each exercise,

a. *Write the cost function, C.*
b. *Write the revenue function, R.*
c. *Determine the break-even point. Describe what this means.*

47. A company that manufactures small canoes has a fixed cost of $18,000. It costs $20 to produce each canoe. The selling price is $80 per canoe. (In solving this exercise, let x represent the number of canoes produced and sold.)

48. A company that manufactures bicycles has a fixed cost of $100,000. It costs $100 to produce each bicycle. The selling price is $300 per bike. (In solving this exercise, let x represent the number of bicycles produced and sold.)

49. You invest in a new play. The cost includes an overhead of $30,000, plus production costs of $2500 per performance. A sold-out performance brings in $3125. (In solving this exercise, let x represent the number of sold-out performances.)

50. You invested $30,000 and started a business writing greeting cards. Supplies cost 2¢ per card and you are selling each card for 50¢. (In solving this exercise, let x represent the number of cards produced and sold.)

51. At a price of p dollars per ticket, the number of tickets to a rock concert that can be sold is given by the demand model $N = -25p + 7500$. At a price of p dollars per ticket, the number of tickets that the concert's promoters are willing to make available is given by the supply model $N = 5p + 6000$.

  a. How many tickets can be sold and supplied for $40 per ticket?

  b. Find the ticket price at which supply and demand are equal. At this price, how many tickets will be supplied and sold?

52. The weekly demand and supply models for a particular brand of scientific calculator for a chain of stores are given by the demand model $N = -53p + 1600$, and the supply model $N = 75p + 320$. In these models, p is the price of the calculator and N is the number of calculators sold or supplied each week to the stores.

  a. How many calculators can be sold and supplied at $12 per calculator?

  b. Find the price at which supply and demand are equal. At this price, how many calculators of this type can be supplied and sold each week?

53. In the United States, deaths from car accidents, per 100,000 persons, are decreasing at a faster rate than deaths from gunfire, shown by the blue and red lines that model the data points in the figure.

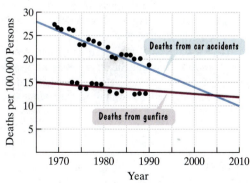

Annual Deaths in the U.S. from Car Accidents and Gunfire

Source: Journal of the American Medical Association

The function $y = -0.4x + 28$ models deaths from car accidents, y, per 100,000 persons, x years after 1965. The function $0.07x + y = 15$ models deaths from gunfire, y, per 100,000 persons, x years after 1965. Use these models to project when the number of deaths from gunfire will equal the number of deaths from car accidents. Round to the nearest year. How many annual deaths, per 100,000 persons, will there be from gunfire and from car accidents at that time? Describe how this is illustrated by the lines in the figure shown.

54. The June 7, 1999 issue of *Newsweek* presented statistics showing progress African Americans have made in education, health, and finance. Infant mortality for African Americans is decreasing at a faster rate than it is for whites, shown by the graphs below. Infant mortality for African Americans can be modeled by $M = -0.41x + 22$ and for whites by $M = -0.18x + 10$. In both models, x is the number of years after 1980 and M is infant mortality, measured in deaths per 1000 live births. Use these models to project when infant mortality for African Americans and whites will be the same. What is infant mortality rate for both groups at that time?

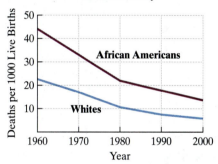

Infant Mortality

Source: National Center for Health Statistics

The graphs show average weekly earnings of full-time wage and salary workers 25 and older, by educational attainment. Exercises 55–56 involve the information in these graphs.

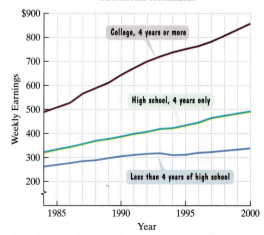

Average Weekly Earnings by Educational Attainment

Source: U.S. Bureau of Labor Statistics

55. In 1985, college graduates averaged $508 in weekly earnings. This amount has increased by approximately $25 in weekly earnings per year. By contrast, in 1985, high school graduates averaged $345 in weekly earnings. This amount has only increased by approximately $9 in weekly earnings per year.

a. Write a function that models weekly earnings, E, for college graduates x years after 1985.

b. Write a function that models weekly earnings, E, for high school graduates x years after 1985.

c. How many years after 1985 will college graduates be earning twice as much per week as high school graduates? In which year will this occur? What will be the weekly earnings for each group at that time?

56. In 1985, college graduates averaged $508 in weekly earnings. This amount has increased by approximately $25 in weekly earnings per year. By contrast, in 1985, people with less than four years of high school averaged $270 in weekly earnings. This amount has only increased by approximately $4 in weekly earnings per year.

a. Write a function that models weekly earnings, E, for college graduates x years after 1985.

b. Write a function that models weekly earnings, E, for people with less than four years of high school x years after 1985.

c. How many years after 1985 will college graduates be earning three times as much per week as people with less than four years of high school? (Round to the nearest whole number.) In which year will this occur? What will be the weekly earnings for each group at that time?

Use a system of linear equations to solve Exercises 57–67.

The graph shows the calories in some favorite fast foods. Use the information in Exercises 57-58 to find the exact caloric content of the specified foods.

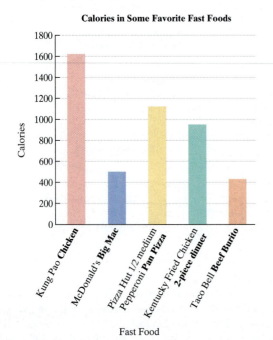

Calories in Some Favorite Fast Foods

Fast Food

Source: Center for Science in the Public Interest

57. One pan pizza and two beef burritos provide 1980 calories. Two pan pizzas and one beef burrito provide 2670 calories. Find the caloric content of each item.

58. One Kung Pao chicken and two Big Macs provide 2620 calories. Two Kung Pao chickens and one Big Mac provide 3740 calories. Find the caloric content of each item.

59. Cholesterol intake should be limited to 300 mg or less each day. One serving of scrambled eggs from McDonalds and one Double Beef Whopper from Burger King exceed this intake by 241 mg. Two servings of scrambled eggs and three Double Beef Whoppers provide 1257 mg of cholesterol. Determine the cholesterol content in each item.

60. Two medium eggs and three cups of ice cream contain 701 milligrams of cholesterol. One medium egg and one cup of ice cream exceed the suggested daily cholesterol intake of 300 milligrams by 25 milligrams. Determine the cholesterol content in each item.

61. A hotel has 200 rooms. Those with kitchen facilities rent for $100 per night and those without kitchen facilities rent for $80 per night. On a night when the hotel was completely occupied, revenues were $17,000. How many of each type of room does the hotel have?

62. In a new development, 50 one- and two-bedroom condominiums were sold. Each one-bedroom condominium sold for $120 thousand and each two bedroom condominium sold for $150 thousand. If sales totaled $7050 thousand, how many of each type of unit was sold?

63. A rectangular lot whose perimeter is 360 feet is fenced along three sides. An expensive fencing along the lot's length costs $20 per foot, and an inexpensive fencing along the two side widths costs only $8 per foot. The total cost of the fencing along the three sides comes to $3280. What are the lot's dimensions?

64. A rectangular lot whose perimeter is 320 feet is fenced along three sides. An expensive fencing along the lot's length costs $16 per foot, and an inexpensive fencing along the two side widths costs only $5 per foot. The total cost of the fencing along the three sides comes to $2140. What are the lot's dimensions?

65. When a crew rows with the current, it travels 16 miles in 2 hours. Against the current, the crew rows 8 miles in 2 hours. Let $x =$ the crew's rowing rate in still water and let $y =$ the rate of the current. The following chart summarizes this information:

	Rate	×	Time	=	Distance
Rowing with current	$x + y$		2		16
Rowing against current	$x - y$		2		8

Find the rate of rowing in still water and the rate of the current.

66. When an airplane flies with the wind, it travels 800 miles in 4 hours. Against the wind, it takes 5 hours to cover the same distance. Find the plane's rate in still air and the rate of the wind.

67. Find the measures of the angles marked $x°$ and $y°$ in the figure.

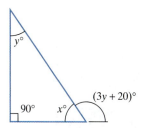

Writing in Mathematics

68. What is a system of linear equations? Provide an example with your description.

69. What is the solution to a system of linear equations?

70. Explain how to solve a system of equations using the substitution method. Use $y = 3 - 3x$ and $3x + 4y = 6$ to illustrate your explanation.

71. Explain how to solve a system of equations using the addition method. Use $3x + 5y = -2$ and $2x + 3y = 0$ to illustrate your explanation.

72. When is it easier to use the addition method rather than the substitution method to solve a system of equations?

73. When using the addition or substitution method, how can you tell if a system of linear equations has infinitely many solutions? What is the relationship between the graphs of the two equations?

74. When using the addition or substitution method, how can you tell if a system of linear equations has no solution? What is the relationship between the graphs of the two equations?

75. Describe the break-even point for a business.

76. The law of supply and demand states that, in a free market economy, a commodity tends to be sold at its equilibrium price. At this price, the amount that the seller will supply is the same amount that the consumer will buy. Explain how systems of equations can be used to determine the equilibrium price.

77. The function $y = 0.94x + 5.64$ models annual U.S. consumption of chicken, y, in pounds per person, x years after 1950. The function $0.74x + y = 146.76$ models annual U.S. consumption of red meat, y, in pounds per person, x years after 1950. What is the most efficient method for solving this system? What does the solution mean in terms of the variables in the functions? (It is not necessary to solve the system.)

78. In Exercise 77, find the slope of each model. Describe what this means in terms of the rate of change of chicken

consumption and the rate of change of red meat consumption. Why must the graphs have an intersection point? What happens to the right of the intersection point?

Technology Exercises

79. Verify your solutions to any five exercises in Exercises 5–42 by using a graphing utility to graph the two equations in the system in the same viewing rectangle. Then use the intersection feature to display the solution.

80. Some graphing utilities can give the solution to a linear system of equations. (Consult your manual for details.) This capability is usually accessed with the ⌈SIMULT⌉ (simultaneous equations) feature. First, you will enter 2, for two equations in two variables. With each equation in $Ax + By = C$ form, you will then enter the coefficients for x and y and the constant term, one equation at a time. After entering all six numbers, press ⌈SOLVE⌉. The solution will be displayed on the screen. (The x-value may be displayed as $x_1 =$ and the y-value as $x_2 =$.) Use this capability to verify the solution to any five of the exercises you solved in the practice exercises of this exercise set. Describe what happens when you use your graphing utility on a system with no solution or infinitely many solutions.

Critical Thinking Exercises

81. Write a system of equations having $\{(-2, 7)\}$ as a solution set. (More than one system is possible.)

82. Solve the system for x and y in terms of $a_1, b_1, c_1, a_2, b_2,$ and c_2:

$$a_1x + b_1y = c_1$$
$$a_2x + b_2y = c_2.$$

83. Two identical twins can only be recognized by the characteristic that one always tells the truth and the other always lies. One twin tells you of a lucky number pair: "When I multiply my first lucky number by 3 and my second lucky number by 6, the addition of the resulting numbers produces a sum of 12. When I add my first lucky number and twice my second lucky number, the sum is 5." Which twin is talking?

84. A marching band has 52 members, and there are 24 in the pom-pom squad. They wish to form several hexagons and squares like those diagrammed below. Can it be done with no people left over?

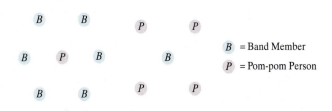

Group Exercise

85. The group should write four different word problems that can be solved using a system of linear equations in two variables. All of the problems should be on different topics. The group should turn in the four problems and their algebraic solutions.

SECTION 5.2   *Systems of Linear Equations in Three Variables*

Objectives

1. Verify the solution of a system of linear equations in three variables.

2. Solve systems of linear equations in three variables.

3. Solve problems using systems in three variables.

All animals sleep, but the length of time they sleep varies widely: Cattle sleep for only a few minutes at a time. We humans seem to need more sleep than other animals. Without enough sleep, we have difficulty concentrating, make mistakes in routine tasks, lose energy, and feel bad-tempered. There is a relationship between hours of sleep and death rate per year per 100,000 people. How many hours of sleep will put you in the group with the minimum death rate? In this section, we will answer this question by solving a system of linear equations with more than two variables.

1 Verify the solution of a system of linear equations in three variables.

Systems of Linear Equations in Three Variables and Their Solutions

An equation such as $x + 2y - 3z = 9$ is called a **linear equation in three variables.** In general, any equation of the form

$$Ax + By + Cz = D$$

where $A, B, C,$ and D are real numbers such that $A, B,$ and C are not all 0, is a linear equation in the variables $x, y,$ and $z.$ The graph of this linear equation in three variables is a plane in three-dimensional space.

The process of solving a system of three linear equations in three variables is geometrically equivalent to finding the point of intersection (assuming that there is one) of three planes in space (see Figure 5.7). A **solution** to a system of linear equations in three variables is an ordered triple of real numbers that satisfies all equations of the system. The **solution set** of the system is the set of all its solutions.

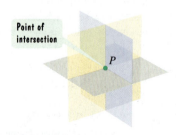

Point of intersection

P

Figure 5.7

EXAMPLE 1   **Determining Whether an Ordered Triple Satisfies a System**

Show that the ordered triple $(-1, 2, -2)$ is a solution of the system:

$$x + 2y - 3z = 9$$
$$2x - y + 2z = -8$$
$$-x + 3y - 4z = 15.$$

Solution   Because -1 is the x-coordinate, 2 is the y-coordinate, and -2 is the z-coordinate of $(-1, 2, -2)$, we replace x with -1, y with 2, and z with -2 in each of the three equations.

$$x + 2y - 3z = 9$$
$$-1 + 2(2) - 3(-2) \stackrel{?}{=} 9$$
$$-1 + 4 + 6 \stackrel{?}{=} 9$$
$$9 = 9, \text{ true}$$

$$2x - y + 2z = -8$$
$$2(-1) - 2 + 2(-2) \stackrel{?}{=} -8$$
$$-2 - 2 - 4 \stackrel{?}{=} -8$$
$$-8 = -8, \text{ true}$$

$$-x + 3y - 4z = 15$$
$$-(-1) + 3(2) - 4(-2) \stackrel{?}{=} 15$$
$$1 + 6 + 8 \stackrel{?}{=} 15$$
$$15 = 15, \text{ true}$$

The ordered triple $(-1, 2, -2)$ satisfies the three equations: It makes each equation true. Thus, the ordered triple is a solution of the system.

Check Point 1   Show that the ordered triple $(-1, -4, 5)$ is a solution of the system:

$$x - 2y + 3z = 22$$
$$2x - 3y - z = 5$$
$$3x + y - 5z = -32.$$

2 Solve systems of linear equations in three variables.

Solving Systems of Linear Equations in Three Variables by Eliminating Variables

The method for solving a system of linear equations in three variables is similar to that used on systems of linear equations in two variables. We use addition to eliminate any variable, reducing the system to two equations in two variables. Once we obtain a system of two equations in two variables, we use addition or substitution to eliminate a variable. The result is a single equation in one variable. We solve this equation to get the value of the remaining variable. Other variable values are found by back-substitution.

Solving Linear Systems in Three Variables by Eliminating Variables

1. **Reduce the system to two equations in two variables.** This is usually accomplished by taking two different pairs of equations and using the addition method to eliminate the same variable from each pair.

2. **Solve the resulting system of two equations in two variables using addition or substitution.** The result is an equation in one variable that gives the value of that variable.

3. **Back-substitute the value of the variable found in step 2 into either of the equations in two variables to find the value of the second variable.**

4. **Use the values of the two variables from steps 2 and 3 to find the value of the third variable by back-substituting into one of the original equations.**

5. **Check the proposed solution in each of the original equations.**

EXAMPLE 2 Solving a System in Three Variables

Solve the system:

$$5x - 2y - 4z = 3 \qquad \text{Equation 1}$$
$$3x + 3y + 2z = -3 \qquad \text{Equation 2}$$
$$-2x + 5y + 3z = 3. \qquad \text{Equation 3}$$

Solution   There are many ways to proceed. Because our initial goal is to reduce the system to two equations in two variables, **the central idea is to take two different pairs of equations and eliminate the same variable from each pair.**

Step 1   **Reduce the system to two equations in two variables.**   We choose any two equations and use the addition method to eliminate a variable. Let's eliminate z using Equations 1 and 2. We do so by multiplying Equation 2 by 2. Then we add equations.

$$
\begin{array}{lll}
(\text{Equation 1}) \;\; 5x - 2y - 4z = 3 & \xrightarrow{\text{No change}} & 5x - 2y - 4z = 3 \\
(\text{Equation 2}) \;\; 3x + 3y + 2z = -3 & \xrightarrow{\text{Multiply by 2.}} & 6x + 6y + 4z = -6 \\
& \text{Add:} & 11x + 4y \quad\;\; = -3 \;\; \text{Equation 4}
\end{array}
$$

Now we must eliminate the *same* variable using another pair of equations. We can eliminate z from Equations 2 and 3. First, we multiply Equation 2 by -3. Next, we multiply Equation 3 by 2. Finally, we add equations.

$$
\begin{array}{lll}
(\text{Equation 2}) \quad 3x + 3y + 2z = -3 & \xrightarrow{\text{Multiply by } -3.} & -9x - 9y - 6z = 9 \\
(\text{Equation 3}) \;\; -2x + 5y + 3z = 3 & \xrightarrow{\text{Multiply by 2.}} & -4x + 10y + 6z = 6 \\
& \text{Add:} & -13x + \;\; y \qquad\;\; = 15 \;\; \text{Equation 5}
\end{array}
$$

Equations 4 and 5 give us a system of two equations in two variables.

Step 2   **Solve the resulting system of two equations in two variables.**   We will use the addition method to solve Equations 4 and 5 for x and y. To do so, we multiply Equation 5 on both sides by -4 and add this to Equation 4.

$$
\begin{array}{lll}
(\text{Equation 4}) \;\; 11x + 4y = -3 & \xrightarrow{\text{No change}} & 11x + 4y = -3 \\
(\text{Equation 5}) \;\; -13x + y = 15 & \xrightarrow{\text{Multiply by } -4.} & 52x - 4y = -60 \\
& \text{Add:} & 63x \qquad\;\; = -63 \\
& & \quad x \qquad\;\; = -1 \quad \text{Divide both sides by 63.}
\end{array}
$$

Step 3   **Use back-substitution in one of the equations in two variables to find the value of the second variable.**   We back-substitute -1 for x in either Equation 4 or 5 to find the value of y.

$$-13x + y = 15 \qquad \text{Equation 5}$$
$$-13(-1) + y = 15 \qquad \text{Substitute } -1 \text{ for } x.$$
$$13 + y = 15 \qquad \text{Multiply.}$$
$$y = 2 \qquad \text{Subtract 13 from both sides.}$$

Step 4   Back-substitute the values found for two variables into one of the original equations to find the value of the third variable.   We can now use any one of the original equations and back-substitute the values of x and y to find the value for z. We will use Equation 2.

$$3x + 3y + 2z = -3 \quad \text{Equation 2}$$

$$3(-1) + 3(2) + 2z = -3 \quad \text{Substitute } -1 \text{ for } x \text{ and } 2 \text{ for } y.$$

$$3 + 2z = -3 \quad \begin{array}{l}\text{Multiply and then add:} \\ 3(-1) + 3(2) = -3 + 6 = 3.\end{array}$$

$$2z = -6 \quad \text{Subtract 3 from both sides.}$$

$$z = -3 \quad \text{Divide both sides by 2.}$$

With $x = -1$, $y = 2$, and $z = -3$, the proposed solution is the ordered triple $(-1, 2, -3)$.

Step 5   Check.   Check the proposed solution, $(-1, 2, -3)$, by substituting the values for x, y, and z into each of the three original equations. These substitutions yield three true statements. Thus, the solution set is $\{(-1, 2, -3)\}$.

Check Point 2   Solve the system:

$$\begin{aligned} x + 4y - z &= 20 \\ 3x + 2y + z &= 8 \\ 2x - 3y + 2z &= -16. \end{aligned}$$

In some examples, one of the variables is already eliminated from a given equation. In this case, the same variable should be eliminated from the other two equations, thereby making it possible to omit one of the elimination steps. We illustrate this idea in Example 3.

EXAMPLE 3   Solving a System of Equations with a Missing Term

Solve the system:

$$\begin{aligned} x + z &= 8 \quad \text{Equation 1} \\ x + y + 2z &= 17 \quad \text{Equation 2} \\ x + 2y + z &= 16. \quad \text{Equation 3} \end{aligned}$$

Solution

Step 1   Reduce the system to two equations in two variables.   Because Equation 1 contains only x and z, we could omit one of the elimination steps by eliminating y using Equations 2 and 3. This will give us two equations in x and z. To eliminate y using Equations 2 and 3, we multiply Equation 2 by -2 and add Equation 3.

(Equation 2) $x + y + 2z = 17$ $\xrightarrow{\text{Multiply by } -2.}$ $-2x - 2y - 4z = -34$

(Equation 3) $x + 2y + z = 16$ $\xrightarrow{\text{No change}}$ $\underline{x + 2y + z = 16}$

Add: $-x - 3z = -18$ Equation 4

Equation 4 and the given Equation 1 provide us with a system of two equations in two variables.

Step 2 Solve the resulting system of two equations in two variables. We will solve Equations 1 and 4 for x and z.

$x + \qquad z = 8$ Equation 1

$x + y + 2z = 17$ Equation 2

$x + 2y + z = 16$ Equation 3

The system we are solving, repeated

$$
\begin{array}{rl}
x + z = 8 & \text{Equation 1} \\
-x - 3z = -18 & \text{Equation 4} \\
\hline
\text{Add:} \quad -2z = -10 & \\
z = 5 & \text{Divide both sides by } -2.
\end{array}
$$

Step 3 Use back-substitution in one of the equations in two variables to find the value of the second variable. To find x, we back-substitute 5 for z in either Equation 1 or 4. We will use Equation 1.

$$
\begin{array}{rl}
x + z = 8 & \text{Equation 1} \\
x + 5 = 8 & \text{Substitute 5 for z.} \\
x = 3 & \text{Subtract 5 from both sides.}
\end{array}
$$

Step 4 Back-substitute the values found for two variables into one of the original equations to find the value of the third variable. To find y, we back-substitute 3 for x and 5 for z into Equation 2 or 3. We can't use Equation 1 because y is missing in this equation. We will use Equation 2.

$$
\begin{array}{rl}
x + y + 2z = 17 & \text{Equation 2} \\
3 + y + 2(5) = 17 & \text{Substitute 3 for x and 5 for z.} \\
y + 13 = 17 & \text{Multiply and add.} \\
y = 4 & \text{Subtract 13 from both sides.}
\end{array}
$$

We found that $z = 5$, $x = 3$, and $y = 4$. Thus, the proposed solution is the ordered triple $(3, 4, 5)$.

Step 5 Check. Substituting 3 for x, 4 for y, and 5 for z into each of the three original equations yields three true statements. Consequently, the solution set is $\{(3, 4, 5)\}$.

Check Point 3 Solve the system:

$$
\begin{array}{rl}
2y - z = 7 \\
x + 2y + z = 17 \\
2x - 3y + 2z = -1.
\end{array}
$$

A system of linear equations in three variables represents three planes. The three planes need not intersect at one point. The planes may have no common point of intersection and represent an inconsistent system with no solution. By contrast, the planes may coincide or intersect along a line. In these cases, the planes have infinitely many points in common and represent systems with infinitely many solutions. Systems of linear equations in three variables that are inconsistent or that contain dependent equations will be discussed in Chapter 9.

3 Solve problems using systems in three variables.

Applications

Systems of equations may allow us to find models for data without using a graphing utility. Three data points that do not lie on or near a line determine the graph of a quadratic function of the form $y = ax^2 + bx + c$, $a \neq 0$. Quadratic functions often model situations in which values of y are decreasing and then increasing, suggesting the cuplike shape of a parabola.

EXAMPLE 4 Modeling Data Relating Sleep and Death Rate

In a study relating sleep and death rate, the following data were obtained. Use the function $y = ax^2 + bx + c$ to model the data.

x (Average Number of Hours of Sleep)	y (Death Rate per Year per 100,000 Males)
4	1682
7	626
9	967

Solution We need to find values for a, b, and c in $y = ax^2 + bx + c$. We can do so by solving a system of three linear equations in a, b, and c. We obtain the three equations by using the values of x and y from the data as follows:

$$y = ax^2 + bx + c \qquad \text{Use the quadratic function to model the data.}$$

When x = 4, y = 1682: $1682 = a \cdot 4^2 + b \cdot 4 + c$ or $16a + 4b + c = 1682$
When x = 7, y = 626: $626 = a \cdot 7^2 + b \cdot 7 + c$ or $49a + 7b + c = 626$
When x = 9, y = 967: $967 = a \cdot 9^2 + b \cdot 9 + c$ or $81a + 9b + c = 967.$

The easiest way to solve this system is to eliminate c from two pairs of equations, obtaining two equations in a and b. Solving this system gives $a = 104.5$, $b = -1501.5$, and $c = 6016$. We now substitute the values for a, b, and c into $y = ax^2 + bx + c$. The function that models the given data is

$$y = 104.5x^2 - 1501.5x + 6016.$$

We can use the model that we obtained in Example 4 to find the death rate of males who average, say, 6 hours of sleep. First, write the model in function notation:

$$f(x) = 104.5x^2 - 1501.5x + 6016.$$

Substitute 6 for x:

$$f(6) = 104.5(6)^2 - 1501.5(6) + 6016 = 769.$$

According to the model, the death rate for males who average 6 hours of sleep is 769 deaths per 100,000 males.

Technology

The graph of

$$y = 104.5x^2 - 1501.5x + 6016$$

is displayed in a [3, 12, 1] by [500, 2000, 100] viewing rectangle. The minimum function feature shows that the lowest point on the graph, the vertex, is approximately (7.2, 622.5). Men who average 7.2 hours of sleep are in the group with the lowest death rate, approximately 622.5 per 100,000.

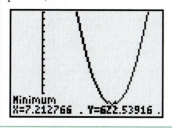

Check Point 4 Find the quadratic function $y = ax^2 + bx + c$ whose graph passes through the points $(1, 4), (2, 1),$ and $(3, 4)$.

EXERCISE SET 5.2

Practice Exercises

In Exercises 1–4, determine if the given ordered triple is a solution of the system.

1. $\begin{aligned} x + y + z &= 4 \\ x - 2y - z &= 1 \\ 2x - y - 2z &= -1 \end{aligned}$
$(2, -1, 3)$

2. $\begin{aligned} x + y + z &= 0 \\ x + 2y - 3z &= 5 \\ 3x + 4y + 2z &= -1 \end{aligned}$
$(5, -3, -2)$

3. $\begin{aligned} x - 2y &= 2 \\ 2x + 3y &= 11 \\ y - 4z &= -7 \end{aligned}$
$(4, 1, 2)$

4. $\begin{aligned} x - 2z &= -5 \\ y - 3z &= -3 \\ 2x - z &= -4 \end{aligned}$
$(-1, 3, 2)$

Solve each system in Exercises 5–18.

5. $\begin{aligned} x + y + 2z &= 11 \\ x + y + 3z &= 14 \\ x + 2y - z &= 5 \end{aligned}$

6. $\begin{aligned} 2x + y - 2z &= -1 \\ 3x - 3y - z &= 5 \\ x - 2y + 3z &= 6 \end{aligned}$

7. $\begin{aligned} 4x - y + 2z &= 11 \\ x + 2y - z &= -1 \\ 2x + 2y - 3z &= -1 \end{aligned}$

8. $\begin{aligned} x - y + 3z &= 8 \\ 3x + y - 2z &= -2 \\ 2x + 4y + z &= 0 \end{aligned}$

9. $\begin{aligned} 3x + 5y + 2z &= 0 \\ 12x - 15y + 4z &= 12 \\ 6x - 25y - 8z &= 8 \end{aligned}$

10. $\begin{aligned} 2x + 3y + 7z &= 13 \\ 3x + 2y - 5z &= -22 \\ 5x + 7y - 3z &= -28 \end{aligned}$

11. $\begin{aligned} 2x - 4y + 3z &= 17 \\ x + 2y - z &= 0 \\ 4x - y - z &= 6 \end{aligned}$

12. $\begin{aligned} x + z &= 3 \\ x + 2y - z &= 1 \\ 2x - y + z &= 3 \end{aligned}$

13. $\begin{aligned} 2x + y &= 2 \\ x + y - z &= 4 \\ 3x + 2y + z &= 0 \end{aligned}$

14. $\begin{aligned} x + 3y + 5z &= 20 \\ y - 4z &= -16 \\ 3x - 2y + 9z &= 36 \end{aligned}$

15. $\begin{aligned} x + y &= -4 \\ y - z &= 1 \\ 2x + y + 3z &= -21 \end{aligned}$

16. $\begin{aligned} x + y &= 4 \\ x + z &= 4 \\ y + z &= 4 \end{aligned}$

17. $\begin{aligned} 3(2x + y) + 5z &= -1 \\ 2(x - 3y + 4z) &= -9 \\ 4(1 + x) &= -3(z - 3y) \end{aligned}$

18. $\begin{aligned} 7z - 3 &= 2(x - 3y) \\ 5y + 3z - 7 &= 4x \\ 4 + 5z &= 3(2x - y) \end{aligned}$

In Exercises 19–20, let x represent the first number, y the second number, and z the third number. Use the given conditions to write a system of equations. Solve the system and find the numbers.

19. The sum of three numbers is 16. The sum of twice the first number, 3 times the second number, and 4 times the third number is 46. The difference between 5 times the first number and the second number is 31. Find the three numbers.

20. The following is known about three numbers: Three times the first number plus the second number plus twice the third number is 5. If 3 times the second number is subtracted from the sum of the first number and 3 times the third number, the result is 2. If the third number is subtracted from 2 times the first number and 3 times the second number, the result is 1. Find the numbers.

In Exercises 21–24, find the quadratic function $y = ax^2 + bx + c$ whose graph passes through the given points.

21. $(-1, 6), (1, 4), (2, 9)$

22. $(-2, 7), (1, -2), (2, 3)$

23. $(-1, -4), (1, -2), (2, 5)$

24. $(1, 3), (3, -1), (4, 0)$

Application Exercises

25. The bar graph shows that the number of gays discharged from the military decreased from 1998 to 1999 and increased from 1999 to 2000.

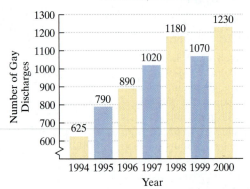

Number of Gay Discharges from the Military Under "Don't Ask, Don't Tell"

Source: New York Times

a. Write the data for 1998, 1999, and 2000 as ordered pairs (x, y), where x is the number of years after 1998 and y is the number of gay discharges from the military.

b. The three data points in part (a) can be modeled by the quadratic function $y = ax^2 + bx + c$. Substitute each ordered pair into this function, one ordered pair at a time, and write a system of three linear equations in three variables that can be used to find values for a, b, and c.

c. Solve the system in part (b). Then write the quadratic function that models the data for 1998 through 2000.

26. The bar graph shows that the percentage of the U.S. population that was foreign-born decreased between 1940 and 1970 and then increased between 1970 and 2000.

Percentage of U.S. Population That Was Foreign-Born, 1900-2000

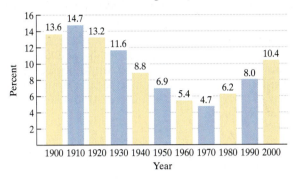

Source: U.S. Census Bureau

a. Write the data for 1940, 1970, and 2000 as ordered pairs (x, y), where x is the number of years after 1940 and y is the percentage of the U.S. population that was foreign-born in that year.

b. The three data points in part (a) can be modeled by the quadratic function $y = ax^2 + bx + c$. Substitute each ordered pair into this function, one ordered pair at a time, and write a system of linear equations in three variables that can be used to find values for a, b, and c.

c. Solve the system in part (b). Then write the quadratic function that models the data for 1940 through 2000.

27. You throw a ball straight up from a rooftop. The ball misses the rooftop on its way down and eventually strikes the ground. A mathematical model can be used to describe the relationship for the ball's height above the ground, y, after x seconds. Consider the following data:

x, seconds after the ball is thrown	y, ball's height, in feet, above the ground
1	224
3	176
4	104

a. Find the quadratic function $y = ax^2 + bx + c$ whose graph passes through the given points.

b. Use the function in part (a) to find the value for y when $x = 5$. Describe what this means.

28. A mathematical model can be used to describe the relationship between the number of feet a car travels once the brakes are applied, y, and the number of seconds the car is in motion after the brakes are applied, x. A research firm collects the following data:

x, seconds in motion after brakes are applied	y, feet car travels once the brakes are applied
1	46
2	84
3	114

a. Find the quadratic function $y = ax^2 + bx + c$ whose graph passes through the given points.

b. Use the function in part (a) to find the value for y when $x = 6$. Describe what this means.

Use a system of linear equations in three variables to solve Exercises 29–35.

29. In current U.S. dollars, John D. Rockefeller's 1913 fortune of $900 million would be worth about $189 billion. The bar graph shows that Rockefeller is the wealthiest among the world's five richest people of all time. The combined estimated wealth, in current billions of U.S. dollars, of Andrew Carnegie, Cornelius Vanderbilt, and Bill Gates is $256 billion. The difference between Carnegie's estimated wealth and Vanderbilt's is $4 billion. The difference between Vanderbilt's estimated wealth and Gate's is $36 billion. Find the estimated wealth, in current billions of U.S. dollars, of Carnegie, Vanderbilt, and Gates.

The Richest People of All Time Estimated Wealth, in Current Billions of U.S. Dollars

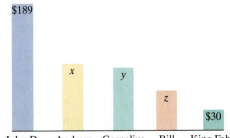

Source: Scholastic Book of World Records

30. The circle graph indicates computers in use for the United States and the rest of the world. The percentage of the world's computers in Europe and Japan combined is 13% less than the percentage of the world's computers in the United States. If the percentage of the world's computers in Europe is doubled, it is only 3% more than the percentage of the world's computers in the United States. Find the percentage of the world's computers in the United States, Europe, and Japan.

Percentage of the World's Computers: U.S. and the World

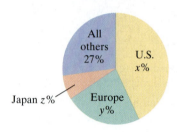

Source: Jupiter Communications

31. At a college production of *Evita*, 400 tickets were sold. The ticket prices were $8, $10, and $12, and the total income from ticket sales was $3700. How many tickets of each type were sold if the combined number of $8 and $10 tickets sold was 7 times the number of $12 tickets sold?

32. A certain brand of razor blades comes in packages of 6, 12, and 24 blades, costing $2, $3, and $4 per package, respectively. A store sold 12 packages containing a total of 162 razor blades and took in $35. How many packages of each type were sold?

33. A person invested $6700 for one year, part at 8%, part at 10%, and the remainder at 12%. The total annual income from these investments was $716. The amount of money invested at 12% was $300 more than the amount invested at 8% and 10% combined. Find the amount invested at each rate.

34. A person invested $17,000 for one year, part at 10%, part at 12%, and the remainder at 15%. The total annual income from these investments was $2110. The amount of money invested at 12% was $1000 less than the amount invested at 10% and 15% combined. Find the amount invested at each rate.

35. Find the measures of the angles marked $x°$, $y°$, and $z°$ in the following triangle.

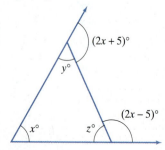

Writing in Mathematics

36. What is a system of linear equations in three variables?

37. How do you determine whether a given ordered triple is a solution of a system in three variables?

38. Describe in general terms how to solve a system in three variables.

39. AIDS is taking a deadly toll on southern Africa. Describe how to use the techniques that you learned in this section to obtain a model for African life span using projections with AIDS. Let x represent the number of years after 1985 and let y represent African life span in that year.

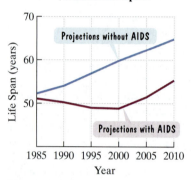

Source: United Nations

Technology Exercises

40. Does your graphing utility have a feature that allows you to solve linear systems by entering coefficients and constant terms? If so, use this feature to verify the solutions to any five exercises that you worked by hand from Exercises 5–16.

41. Verify your results in Exercises 21–24 by using a graphing utility to graph the resulting parabola. Trace along the curve and convince yourself that the three points given in the exercise lie on the parabola.

42. Some graphing utilities will do three-dimensional graphing. For example, on the TI-92, press MODE, go to GRAPH, press the arrow to the right, select 3D, then ENTER. When you display the Y = screen, you will see the equations are functions of x and y. Thus, you must solve each of a linear system's equations for z before entering the equation. For example,

$$x + y + z = 19$$

is solved for z, giving

$$z = 19 - x - y.$$

(Consult your manual.) If your utility does three-dimensional graphing, graph five of the systems in Exercises 5–16 and trace along the planes to find their common point of intersection.

Critical Thinking Exercises

43. Describe how the system

$$x + y - z - 2w = -8$$
$$x - 2y + 3z + w = 18$$
$$2x + 2y + 2z - 2w = 10$$
$$2x + y - z + w = 3$$

could be solved. Is it likely that in the near future a graphing utility will be available to provide a geometric solution (using intersecting graphs) to this system? Explain.

44. A modernistic painting consists of triangles, rectangles, and pentagons, all drawn so as to not overlap or share sides. Within each rectangle are drawn 2 red roses, and each pentagon contains 5 carnations. How many triangles, rectangles, and pentagons appear in the painting if the painting contains a total of 40 geometric figures, 153 sides of geometric figures, and 72 flowers?

Group Exercise

45. Group members should develop appropriate functions that model each of the projections shown in Exercise 39.

SECTION 5.3 *Partial Fractions*

Objective

1. Find the partial fraction decomposition of a rational expression.

The rising and setting of the sun suggest the obvious: Things change over time. Calculus is the study of rates of change, allowing the motion of the rising sun to be measured by "freezing the frame" at one instant in time. If you are given a function, calculus reveals its rate of change at any "frozen" instant. In this section, you will learn an algebraic technique used in calculus to find a function if its rate of change is known.

The Idea behind Partial Fraction Decomposition

Systems of linear equations can be used to reverse the process of adding and subtracting rational expressions—for example,

$$\frac{3}{x-4} - \frac{2}{x+2} = \frac{3(x+2) - 2(x-4)}{(x-4)(x+2)}$$

$$= \frac{3x + 6 - 2x + 8}{(x-4)(x+2)} = \frac{x + 14}{(x-4)(x+2)}.$$

In order to reverse this process, we must show that

$$\frac{x + 14}{(x - 4)(x + 2)} = \frac{3}{x - 4} - \frac{2}{x + 2} \quad \text{or} \quad \frac{3}{x - 4} + \frac{-2}{x + 2}.$$

Each of the two fractions on the right is called a **partial fraction.** The sum of these fractions is called the **partial fraction decomposition** of the rational expression on the left-hand side.

Partial fraction decompositions can be written for rational expressions of the form $\frac{P(x)}{Q(x)}$, where P and Q have no common factors and the highest power in the numerator is less than the highest power in the denominator. In this section, we will show you how to write the partial fraction decompositions for each of the following rational expressions:

$$\frac{9x^2 - 9x + 6}{(2x - 1)(x + 2)(x - 2)}$$

$P(x) = 9x^2 - 9x + 6$; highest power = 2

$Q(x) = (2x - 1)(x + 2)(x - 2)$; multiplying factors, highest power = 3

$$\frac{5x^3 - 3x^2 + 7x - 3}{\left(x^2 + 1\right)^2}.$$

$P(x) = 5x^3 - 3x^2 + 7x - 3$; highest power = 3

$Q(x) = (x^2 + 1)^2$; squaring this expression, highest power = 4

1 Find the partial fraction decomposition of a rational expression.

The Steps in Partial Fraction Decomposition

The partial fraction decomposition of a rational expression depends on the factors of the denominator. We consider four cases involving different kinds of factors in the denominator.

Case 1: The Partial Fraction Decomposition of a Rational Expression with Distinct Linear Factors in the Denominator   If the denominator has a linear factor of the form $ax + b$, then the partial fraction decomposition will contain a term of the form

$$\frac{A}{ax + b}.$$

Constant

Linear factor

Each distinct linear factor in the denominator produces a partial fraction of the form *constant over linear factor*. For example,

$$\frac{9x^2 - 9x + 6}{(2x - 1)(x + 2)(x - 2)} = \frac{A}{2x - 1} + \frac{B}{x + 2} + \frac{C}{x - 2}.$$

We write a constant over each linear factor in the denominator.

The form of the partial fraction decomposition for a rational expression with distinct linear factors in the denominator is

$$\frac{P(x)}{(a_1 x + b_1)(a_2 x + b_2)(a_3 x + b_3) \cdots (a_n x + b_n)}$$

$$= \frac{A_1}{a_1 x + b_1} + \frac{A_2}{a_2 x + b_2} + \frac{A_3}{a_3 x + b_3} + \cdots + \frac{A_n}{a_n x + b_n}.$$

EXAMPLE 1 Partial Fraction Decomposition with Distinct Linear Factors

Find the partial fraction decomposition of

$$\frac{x + 14}{(x - 4)(x + 2)}.$$

Solution We begin by setting up the partial fraction decomposition with the unknown constants. Write a constant over each of the two distinct linear factors in the denominator.

$$\frac{x + 14}{(x - 4)(x + 2)} = \frac{A}{x - 4} + \frac{B}{x + 2}$$

Our goal is to find A and B. We do this by multiplying both sides of the equation by the least common denominator, $(x - 4)(x + 2)$.

$$(x - 4)(x + 2)\frac{x + 14}{(x - 4)(x + 2)} = (x - 4)(x + 2)\left(\frac{A}{x - 4} + \frac{B}{x + 2}\right)$$

We use the distributive property on the right side.

$$(x - 4)(x + 2)\frac{x + 14}{(x - 4)(x + 2)}$$
$$= (x - 4)(x + 2)\frac{A}{(x - 4)} + (x - 4)(x + 2)\frac{B}{(x + 2)}$$

Dividing out common factors in numerators and denominators, we obtain

$$x + 14 = A(x + 2) + B(x - 4).$$

To find values for A and B that make both sides equal, we'll express the sides in exactly the same form by writing the variable x-terms and then writing the constant terms. Apply the distributive property on the right side.

$$x + 14 = Ax + 2A + Bx - 4B$$
$$x + 14 = Ax + Bx + 2A - 4B$$
$$1x + 14 = (A + B)x + (2A - 4B)$$

As shown by the arrows, if two polynomials are equal, coefficients of like powers of x must be equal $(A + B = 1)$ and their constant terms must be equal $(2A - 4B = 14)$. Consequently, A and B satisfy the following two equations:

$$A + B = 1$$
$$2A - 4B = 14.$$

We can use the addition method to solve this linear system in two variables. By multiplying the first equation by -2 and adding equations, we obtain $A = 3$ and $B = -2$. Thus,

$$\frac{x + 14}{(x - 4)(x + 2)} = \frac{A}{x - 4} + \frac{B}{x + 2} = \frac{3}{x - 4} + \frac{-2}{x + 2}\left(\text{or } \frac{3}{x - 4} - \frac{2}{x + 2}\right).$$

Steps in Partial Fraction Decomposition

1. Set up the partial fraction decomposition with the unknown constants A, B, C, etc., in the numerator of the decomposition.

2. Multiply both sides of the resulting equation by the least common denominator.

3. Simplify the right-hand side of the equation.

4. Write both sides in descending powers, equate coefficients of like powers of x, and equate constant terms.

5. Solve the resulting linear system for A, B, C, etc.

6. Substitute the values for A, B, C, etc., into the equation in step 1 and write the partial fraction decomposition.

 Check Point 1   Find the partial fraction decomposition of $\dfrac{5x - 1}{(x - 3)(x + 4)}$.

Case 2:   **The Partial Fraction Decomposition of a Rational Expression with Linear Factors in the Denominator, Some of Which Are Repeated**
Suppose that $(ax + b)^n$ is a factor of the denominator. This means that the linear factor $ax + b$ is repeated n times. When this occurs, the partial fraction decomposition will contain the following sum of n fractions:

$$\frac{P(x)}{(ax + b)^n} = \frac{A_1}{ax + b} + \frac{A_2}{(ax + b)^2} + \frac{A_3}{(ax + b)^3} + \cdots + \frac{A_n}{(ax + b)^n}.$$

Include one fraction with a constant numerator for each power of $ax + b$.

EXAMPLE 2   **Partial Fraction Decomposition with Repeated Linear Factors**

Find the partial fraction decomposition of $\dfrac{x - 18}{x(x - 3)^2}$.

Solution

Step 1   **Set up the partial fraction decomposition with the unknown constants.** Because the linear factor $x - 3$ is repeated twice, we must include one fraction with a constant numerator for each power of $x - 3$.

$$\frac{x - 18}{x(x - 3)^2} = \frac{A}{x} + \frac{B}{x - 3} + \frac{C}{(x - 3)^2}$$

Step 2   **Multiply both sides of the resulting equation by the least common denominator.** We clear fractions, multiplying both sides by $x(x - 3)^2$, the least common denominator.

$$x(x - 3)^2 \left[\frac{x - 18}{x(x - 3)^2} \right] = x(x - 3)^2 \left[\frac{A}{x} + \frac{B}{x - 3} + \frac{C}{(x - 3)^2} \right]$$

We use the distributive property on the right side.

$$x(x-3)^2 \cdot \frac{x-18}{x(x-3)^2}$$

$$= x(x-3)^2 \cdot \frac{A}{x} + x(x-3)^2 \cdot \frac{B}{(x-3)} + x(x-3)^2 \cdot \frac{C}{(x-3)^2}$$

Dividing out common factors in numerators and denominators, we obtain

$$x - 18 = A(x-3)^2 + Bx(x-3) + Cx.$$

Step 3 Simplify the right side of the equation. Square $x - 3$. Then apply the distributive property.

$$x - 18 = A(x^2 - 6x + 9) + Bx(x-3) + Cx \qquad \text{Square } x - 3 \text{ using}$$
$$(A - B)^2 = A^2 - 2AB + B^2.$$

$$x - 18 = Ax^2 - 6Ax + 9A + Bx^2 - 3Bx + Cx \qquad \text{Apply the distributive property.}$$

Step 4 Write both sides in descending powers, equate coefficients of like powers of x, and equate constant terms. The left side, $x - 18$, is in descending powers of x: $x - 18x^0$. We will write the right side in descending powers of x.

$$x - 18 = Ax^2 + Bx^2 - 6Ax - 3Bx + Cx + 9A$$

Express both sides in the same form.

$$0x^2 + 1x - 18 = (A + B)x^2 + (-6A - 3B + C)x + 9A$$

Equating coefficients of like powers of x and constant terms results in the following system of linear equations:

$$A + B = 0$$
$$-6A - 3B + C = 1$$
$$9A = -18.$$

Step 5 Solve the resulting system for $A, B,$ and C. Dividing both sides of the last equation by 9, we obtain $A = -2$. Substituting -2 for A in the first equation, $A + B = 0$, gives $-2 + B = 0$ or $B = 2$. We find C by substituting -2 for A and 2 for B in the middle equation, $-6A - 3B + C = 1$. We obtain $C = -5$.

Step 6 Substitute the values of A, B, and C and write the partial fraction decomposition. With $A = -2$, $B = 2$, and $C = -5$, the required partial fraction decomposition is

$$\frac{x-18}{x(x-3)^2} = \frac{A}{x} + \frac{B}{x-3} + \frac{C}{(x-3)^2} = -\frac{2}{x} + \frac{2}{x-3} - \frac{5}{(x-3)^2}.$$

Check Point 2 Find the partial fraction decomposition of $\dfrac{x+2}{x(x-1)^2}$.

Case 3: The Partial Fraction Decomposition of a Rational Expression with Prime, Nonrepeated Quadratic Factors in the Denominator Suppose that $ax^2 + bx + c$ is a factor of the denominator and that this quadratic factor cannot be factored into linear factors with real coefficients. Under these conditions, the partial fraction decomposition will contain a term of the form

$$\frac{Ax + B}{ax^2 + bx + c}. \qquad \begin{array}{l}\text{Linear numerator}\\[4pt]\text{Quadratic factor}\end{array}$$

Each distinct prime quadratic factor in the denominator produces a partial fraction of the form *linear numerator over quadratic factor*. For example,

$$\frac{3x^2 + 17x + 14}{(x - 2)(x^2 + 2x + 4)} = \frac{A}{x - 2} + \frac{Bx + C}{x^2 + 2x + 4}.$$

> We write a constant over the linear factor in the denominator.

> We write a linear numerator over the prime quadratic factor in the denominator.

Our next example illustrates how a linear system in three variables is used to determine values for A, B, and C.

EXAMPLE 3 Partial Fraction Decomposition

Find the partial fraction decomposition of

$$\frac{3x^2 + 17x + 14}{(x - 2)(x^2 + 2x + 4)}.$$

Solution

Step 1 Set up the partial fraction decomposition with the unknown constants. We put a constant (A) over the linear factor and a linear expression ($Bx + C$) over the prime quadratic factor.

$$\frac{3x^2 + 17x + 14}{(x - 2)(x^2 + 2x + 4)} = \frac{A}{x - 2} + \frac{Bx + C}{x^2 + 2x + 4}$$

Step 2 Multiply both sides of the resulting equation by the least common denominator. We clear fractions, multiplying both sides by $(x - 2)$ $(x^2 + 2x + 4)$, the least common denominator.

$$(x - 2)(x^2 + 2x + 4)\left[\frac{3x^2 + 17x + 14}{(x - 2)(x^2 + 2x + 4)}\right] = (x - 2)(x^2 + 2x + 4)\left[\frac{A}{x - 2} + \frac{Bx + C}{x^2 + 2x + 4}\right]$$

We use the distributive property on the right side.

$$(x - 2)(x^2 + 2x + 4) \cdot \frac{3x^2 + 17x + 14}{(x - 2)(x^2 + 2x + 4)}$$

$$= (x - 2)(x^2 + 2x + 4) \cdot \frac{A}{x - 2} + (x - 2)(x^2 + 2x + 4) \cdot \frac{Bx + C}{x^2 + 2x + 4}$$

Dividing out common factors in numerators and denominators, we obtain

$$3x^2 + 17x + 14 = A(x^2 + 2x + 4) + (Bx + C)(x - 2).$$

Step 3 Simplify the right side of the equation. We multiply on the right side by distributing A over each term in parentheses and multiplying $(Bx + C)(x - 2)$ using the FOIL method.

$$3x^2 + 17x + 14 = Ax^2 + 2Ax + 4A + Bx^2 - 2Bx + Cx - 2C$$

Step 4 Write both sides in descending powers, equate coefficients of like powers of x, and equate constant terms. The left side, $3x^2 + 17x + 14$, is in descending powers of x. We write the right side in descending powers of x

$$3x^2 + 17x + 14 = Ax^2 + Bx^2 + 2Ax - 2Bx + Cx + 4A - 2C$$

and express both sides in the same form.

$$3x^2 + 17x + 14 = (A + B)x^2 + (2A - 2B + C)x + (4A - 2C)$$

Equating coefficients of like powers of x and constant terms results in the following system of linear equations:

$$A + B = 3$$
$$2A - 2B + C = 17$$
$$4A - 2C = 14.$$

Step 5   Solve the resulting system for $A, B,$ and C.   Because the first equation involves A and B, we can obtain another equation in A and B by eliminating C from the second and third equations. Multiply the second equation by 2 and add equations. Solving in this manner, we obtain $A = 5$, $B = -2$, and $C = 3$.

Step 6   Substitute the values of A, B, and C and write the partial fraction decomposition.   With $A = 5$, $B = -2$, and $C = 3$, the required partial fraction decomposition is

$$\frac{3x^2 + 17x + 14}{(x - 2)(x^2 + 2x + 4)} = \frac{A}{x - 2} + \frac{Bx + C}{x^2 + 2x + 4} = \frac{5}{x - 2} + \frac{-2x + 3}{x^2 + 2x + 4}.$$

Check Point 3   Find the partial fraction decomposition of

$$\frac{8x^2 + 12x - 20}{(x + 3)(x^2 + x + 2)}.$$

Case 4:   The Partial Fraction Decomposition of a Rational Expression with a Prime, Repeated Quadratic Factor in the Denominator   Suppose that $(ax^2 + bx + c)^n$ is a factor of the denominator and that $ax^2 + bx + c$ cannot be factored further. This means that the quadratic factor $ax^2 + bx + c$ is repeated n times. When this occurs, the partial fraction decomposition will contain a linear numerator for each power of $ax^2 + bx + c$.

$$\frac{P(x)}{(ax^2 + bx + c)^n} = \frac{A_1 x + B_1}{ax^2 + bx + c} + \frac{A_2 x + B_2}{(ax^2 + bx + c)^2} + \frac{A_3 x + B_3}{(ax^2 + bx + c)^3} + \cdots + \frac{A_n x + B_n}{(ax^2 + bx + c)^n}$$

Include one fraction with a linear numerator for each power of $ax^2 + bx + c$.

Study Tip

When the denominator of a rational expression contains a **power of a linear factor,** set up the partial fraction decomposition with **constant numerators** (A, B, C, etc.). When the denominator of a rational expression contains a **power of a prime quadratic factor,** set up the partial fraction decomposition with **linear numerators** ($Ax + B$, $Cx + D$, etc.).

EXAMPLE 4   Partial Fraction Decomposition with a Repeated Quadratic Factor

Find the partial fraction decomposition of

$$\frac{5x^3 - 3x^2 + 7x - 3}{(x^2 + 1)^2}.$$

Solution

Step 1   Set up the partial fraction decomposition with the unknown constants.   Because the quadratic factor $x^2 + 1$ is repeated twice, we must include one fraction with a linear numerator for each power of $x^2 + 1$.

$$\frac{5x^3 - 3x^2 + 7x - 3}{(x^2 + 1)^2} = \frac{Ax + B}{x^2 + 1} + \frac{Cx + D}{(x^2 + 1)^2}$$

Step 2 **Multiply both sides of the resulting equation by the least common denominator.** We clear fractions, multiplying both sides by $(x^2 + 1)^2$, the least common denominator.

$$(x^2 + 1)^2 \left[\frac{5x^3 - 3x^2 + 7x - 3}{(x^2 + 1)^2} \right] = (x^2 + 1)^2 \left[\frac{Ax + B}{x^2 + 1} + \frac{Cx + D}{(x^2 + 1)^2} \right]$$

Now we multiply and simplify.

$$5x^3 - 3x^2 + 7x - 3 = (x^2 + 1)(Ax + B) + Cx + D$$

Step 3 **Simplify the right side of the equation.** We multiply $(x^2 + 1)(Ax + B)$ using the FOIL method.

$$5x^3 - 3x^2 + 7x - 3 = Ax^3 + Bx^2 + Ax + B + Cx + D$$

Step 4 **Write both sides in descending powers, equate coefficients of like powers of x, and equate constant terms.**

$$5x^3 - 3x^2 + 7x - 3 = Ax^3 + Bx^2 + Ax + Cx + B + D$$

$$5x^3 - 3x^2 + 7x - 3 = Ax^3 + Bx^2 + (A + C)x + (B + D)$$

Equating coefficients of like powers of x and constant terms results in the following system of linear equations:

$$A = 5$$
$$B = -3$$
$$A + C = 7 \qquad \text{With } A = 5, \text{ we immediately obtain } C = 2.$$
$$B + D = -3. \qquad \text{With } B = -3, \text{ we immediately obtain } D = 0.$$

Step 5 **Solve the resulting system for $A, B, C,$ and D.** Based on our observations in step 4, $A = 5, B = -3, C = 2,$ and $D = 0.$

Step 6 **Substitute the values of $A, B, C,$ and D and write the partial fraction decomposition.**

$$\frac{5x^3 - 3x^2 + 7x - 3}{(x^2 + 1)^2} = \frac{Ax + B}{x^2 + 1} + \frac{Cx + D}{(x^2 + 1)^2} = \frac{5x - 3}{x^2 + 1} + \frac{2x}{(x^2 + 1)^2}$$

> **Check Point 4** Find the partial fraction decomposition of $\dfrac{2x^3 + x + 3}{(x^2 + 1)^2}$.

Study Tip

When a rational expression contains a power of a factor in the denominator, be sure to set up the partial fraction decomposition to allow for every natural-number power of that factor less than or equal to the power.

Example:

$$\frac{2x + 1}{(x - 5)^2 x^3}$$

$$= \frac{A}{x - 5} + \frac{B}{(x - 5)^2} + \frac{C}{x} + \frac{D}{x^2} + \frac{E}{x^3}$$

Although $(x - 5)^2$ and x^2 are quadratic, they are expressed as powers of a linear factor, $x - 5$ and x. Thus, the numerator is constant.

EXERCISE SET 5.3

Practice Exercises

In Exercises 1–8, write the form of the partial fraction decomposition of the rational expression. It is not necessary to solve for the constants.

1. $\dfrac{11x - 10}{(x - 2)(x + 1)}$

2. $\dfrac{5x + 7}{(x - 1)(x + 3)}$

3. $\dfrac{6x^2 - 14x - 27}{(x + 2)(x - 3)^2}$

4. $\dfrac{3x + 16}{(x + 1)(x - 2)^2}$

5. $\dfrac{5x^2 - 6x + 7}{(x - 1)(x^2 + 1)}$

6. $\dfrac{5x^2 - 9x + 19}{(x - 4)(x^2 + 5)}$

7. $\dfrac{x^3 + x^2}{(x^2 + 4)^2}$

8. $\dfrac{7x^2 - 9x + 3}{(x^2 + 7)^2}$

In Exercises 9–42, write the partial fraction decomposition of each rational expression.

9. $\dfrac{x}{(x - 3)(x - 2)}$

10. $\dfrac{1}{x(x - 1)}$

11. $\dfrac{3x + 50}{(x - 9)(x + 2)}$

12. $\dfrac{5x - 1}{(x - 2)(x + 1)}$

13. $\dfrac{7x - 4}{x^2 - x - 12}$

14. $\dfrac{9x + 21}{x^2 + 2x - 15}$

15. $\dfrac{4}{2x^2 - 5x - 3}$

16. $\dfrac{x}{x^2 + 2x - 3}$

17. $\dfrac{4x^2 + 13x - 9}{x(x - 1)(x + 3)}$

18. $\dfrac{4x^2 - 5x - 15}{x(x + 1)(x - 5)}$

19. $\dfrac{4x^2 - 7x - 3}{x^3 - x}$

20. $\dfrac{2x^2 - 18x - 12}{x^3 - 4x}$

21. $\dfrac{6x - 11}{(x - 1)^2}$

22. $\dfrac{x}{(x + 1)^2}$

23. $\dfrac{x^2 - 6x + 3}{(x - 2)^3}$

24. $\dfrac{2x^2 + 8x + 3}{(x + 1)^3}$

25. $\dfrac{x^2 + 2x + 7}{x(x - 1)^2}$

26. $\dfrac{3x^2 + 49}{x(x + 7)^2}$

27. $\dfrac{x^2}{(x - 1)^2(x + 1)}$

28. $\dfrac{x^2}{(x - 1)^2(x + 1)^2}$

29. $\dfrac{5x^2 - 6x + 7}{(x - 1)(x^2 + 1)}$

30. $\dfrac{5x^2 - 9x + 19}{(x - 4)(x^2 + 5)}$

31. $\dfrac{5x^2 + 6x + 3}{(x + 1)(x^2 + 2x + 2)}$

32. $\dfrac{9x + 2}{(x - 2)(x^2 + 2x + 2)}$

33. $\dfrac{x + 4}{x^2(x^2 + 4)}$

34. $\dfrac{10x^2 + 2x}{(x - 1)^2(x^2 + 2)}$

35. $\dfrac{6x^2 - x + 1}{x^3 + x^2 + x + 1}$

36. $\dfrac{3x^2 - 2x + 8}{x^3 + 2x^2 + 4x + 8}$

37. $\dfrac{x^3 + x^2 + 2}{(x^2 + 2)^2}$

38. $\dfrac{x^2 + 2x + 3}{(x^2 + 4)^2}$

39. $\dfrac{x^3 - 4x^2 + 9x - 5}{(x^2 - 2x + 3)^2}$

40. $\dfrac{3x^3 - 6x^2 + 7x - 2}{(x^2 - 2x + 2)^2}$

41. $\dfrac{4x^2 + 3x + 14}{x^3 - 8}$

42. $\dfrac{3x - 5}{x^3 - 1}$

Application Exercises

43. Find the partial fraction decomposition for $\dfrac{1}{x(x + 1)}$ and use the result to find the following sum:

$$\frac{1}{1 \cdot 2} + \frac{1}{2 \cdot 3} + \frac{1}{3 \cdot 4} + \cdots + \frac{1}{99 \cdot 100}.$$

44. Find the partial fraction decomposition for $\dfrac{2}{x(x + 2)}$ and use the result to find the following sum:

$$\frac{2}{1 \cdot 3} + \frac{2}{3 \cdot 5} + \frac{2}{5 \cdot 7} + \cdots + \frac{2}{99 \cdot 101}.$$

Writing in Mathematics

45. Explain what is meant by the partial fraction decomposition of a rational expression.

46. Explain how to find the partial fraction decomposition of a rational expression with distinct linear factors in the denominator.

47. Explain how to find the partial fraction decomposition of a rational expression with a repeated linear factor in the denominator.

48. Explain how to find the partial fraction decomposition of a rational expression with a prime quadratic factor in the denominator.

49. Explain how to find the partial fraction decomposition of a rational expression with a repeated, prime quadratic factor in the denominator.

50. How can you verify your result for the partial fraction decomposition for a given rational expression without using a graphing utility?

Technology Exercises

51. A graphing utility can be used to check the partial fraction decomposition for a given rational expression. Graph $y_1 = $ *the given rational expression* and $y_2 = $ *its partial fraction decomposition* on the same screen. If the graphs are identical, the decomposition is correct. Use this method to verify any five of the decompositions that you obtained in Exercises 9–42.

52. As you worked Exercise 51, did you find that it took a while to determine the range setting that showed a graph for the rational function and its decomposition? Suggest another method for showing that $y_1 = y_2$ using your graphing utility. Use this method to check the results of the same five decompositions you worked with in Exercise 51.

Critical Thinking Exercises

53. Use an extension of the Study Tip on page 471 to describe how to set up the partial fraction decomposition of a rational expression that contains powers of a cubic factor in the denominator. Give an example of such a decomposition.

54. If a, b, and c are constants, find the partial fraction decomposition of

$$\frac{ax + b}{(x - c)^2}.$$

55. Find the partial fraction decomposition of

$$\frac{4x^2 + 5x - 9}{x^3 - 6x - 9}.$$

SECTION 5.4   *Systems of Nonlinear Equations in Two Variables*

Objectives

1. Recognize systems of nonlinear equations in two variables.
2. Solve nonlinear systems by substitution.
3. Solve nonlinear systems by addition.
4. Solve problems using systems of nonlinear equations.

Scientists debate the probability that a "doomsday rock" will collide with Earth. It has been estimated that an asteroid, a tiny planet that revolves around the sun, crashes into Earth about once every 250,000 years, and that such a collision would have disastrous results. In 1908 a small fragment struck Siberia, leveling thousands of acres of trees. One theory about the extinction of dinosaurs 65 million years ago involves Earth's collision with a large asteroid and the resulting drastic changes in Earth's climate.

Understanding the path of Earth and the path of a comet is essential to detecting threatening space debris. Orbits about the sun are not described by linear equations in the form $Ax + By = C$. The ability to solve systems that do not contain linear equations provides NASA scientists watching for troublesome asteroids with a way to locate possible collision points with Earth's orbit.

1 Recognize systems of nonlinear equations in two variables.

Systems of Nonlinear Equations and Their Solutions

A **system** of two **nonlinear equations** in two variables, also called a **nonlinear system,** contains at least one equation that cannot be expressed in the form $Ax + By = C$. Here are two examples:

$$x^2 = 2y + 10$$
$$3x - y = 9$$

> Not in the form $Ax + By = C$. The term x^2 is not linear.

$$y = x^2 + 3$$
$$x^2 + y^2 = 9.$$

> Neither equation is in the form $Ax + By = C$. The terms x^2 and y^2 are not linear.

A **solution** to a nonlinear system in two variables is an ordered pair of real numbers that satisfies all equations in the system. The **solution set** to the system is the set of all such ordered pairs. As with linear systems in two variables, the solution to a nonlinear system (if there is one) corresponds to the intersection point(s) of the graphs of the equations in the system. Unlike linear systems, the graphs can be circles, parabolas, or anything other than two lines. We will solve nonlinear systems using the substitution method and the addition method.

2 Solve nonlinear systems by substitution.

Eliminating a Variable Using the Substitution Method

The substitution method involves converting a nonlinear system to one equation in one variable by an appropriate substitution. The steps in the solution process are nearly the same as those used to solve a linear system by substitution. However, when you obtain an equation in one variable, this equation will not be linear. In our first example, this equation is quadratic.

EXAMPLE 1 **Solving a Nonlinear System by the Substitution Method**

Solve by the substitution method:

$$x^2 = 2y + 10 \quad \text{(The graph is a parabola.)}$$
$$3x - y = 9. \quad \text{(The graph is a line.)}$$

Solution

Step 1 **Solve one of the equations for one variable in terms of the other.** We begin by isolating one of the variables raised to the first power in either of the equations. By solving for y in the second equation, which has a coefficient of -1, we can avoid fractions.

$$3x - y = 9 \qquad \text{This is the second equation in the given system.}$$
$$3x = y + 9 \qquad \text{Add } y \text{ to both sides.}$$
$$3x - 9 = y \qquad \text{Subtract 9 from both sides.}$$

Step 2 **Substitute the expression from step 1 into the other equation.** We substitute $3x - 9$ for y in the first equation.

$$y = \boxed{3x - 9} \qquad x^2 = 2\,\boxed{y} + 10$$

This gives us an equation in one variable, namely

$$x^2 = 2(3x - 9) + 10.$$

The variable y has been eliminated.

Step 3 **Solve the resulting equation containing one variable.**

$$x^2 = 2(3x - 9) + 10 \qquad \text{This is the equation containing one variable.}$$
$$x^2 = 6x - 18 + 10 \qquad \text{Use the distributive property.}$$
$$x^2 = 6x - 8 \qquad \text{Combine numerical terms on the right.}$$
$$x^2 - 6x + 8 = 0 \qquad \text{Move all terms to one side and set the quadratic equation equal to 0.}$$
$$(x - 4)(x - 2) = 0 \qquad \text{Factor.}$$
$$x - 4 = 0 \quad \text{or} \quad x - 2 = 0 \qquad \text{Set each factor equal to 0.}$$
$$x = 4 \quad \text{or} \quad x = 2 \qquad \text{Solve for x.}$$

Step 4 **Back-substitute the obtained values into the equation from step 1.** Now that we have the x-coordinates of the solutions, we back-substitute 4 for x and 2 for x in the equation $y = 3x - 9$.

If x is 4, $y = 3(4) - 9 = 3$, so $(4, 3)$ is a solution.

If x is 2, $y = 3(2) - 9 = -3$, so $(2, -3)$ is a solution.

Step 5 **Check the proposed solutions in both of the system's given equations.** We begin by checking $(4, 3)$. Replace x with 4 and y with 3.

$$x^2 = 2y + 10 \qquad\qquad 3x - y = 9 \qquad \text{These are the given equations.}$$
$$4^2 \stackrel{?}{=} 2(3) + 10 \qquad 3(4) - 3 \stackrel{?}{=} 9 \qquad \text{Let x = 4 and y = 3.}$$
$$16 \stackrel{?}{=} 6 + 10 \qquad\quad 12 - 3 \stackrel{?}{=} 9 \qquad \text{Simplify.}$$
$$16 = 16 \;\checkmark \qquad\qquad\quad 9 = 9 \;\checkmark \qquad \text{True statements result.}$$

The ordered pair $(4, 3)$ satisfies both equations. Thus, $(4, 3)$ is a solution to the system.

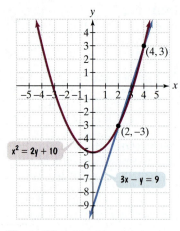

Figure 5.8 Points of intersection illustrate the nonlinear system's solutions.

Study Tip

Recall from Chapter 2 that $(x - h)^2 + (y - k)^2 = r^2$ describes a circle with center (h, k) and radius r.

Now let's check $(2, -3)$. Replace x with 2 and y with -3 in both given equations.

$x^2 = 2y + 10$	$3x - y = 9$	These are the given equations.
$2^2 \overset{?}{=} 2(-3) + 10$	$3(2) - (-3) \overset{?}{=} 9$	Let $x = 2$ and $y = -3$.
$4 \overset{?}{=} -6 + 10$	$6 + 3 \overset{?}{=} 9$	Simplify.
$4 = 4$ ✓	$9 = 9$ ✓	True statements result.

The ordered pair $(2, -3)$ also satisfies both equations and is a solution to the system. The solution set is $\{(4, 3), (2, -3)\}$. Figure 5.8 shows the graphs of the equations in the system and the solutions as intersection points.

Check Point 1 Solve by the substitution method:

$$x^2 = y - 1$$
$$4x - y = -1.$$

EXAMPLE 2 **Solving a Nonlinear System by the Substitution Method**

Solve by the substitution method:

$$x - y = 3 \qquad \text{(The graph is a line.)}$$
$$(x - 2)^2 + (y + 3)^2 = 4. \qquad \text{(The graph is a circle.)}$$

Solution Graphically, we are finding the intersection of a line and a circle with center $(2, -3)$ and radius 2.

Step 1 Solve one of the equations for one variable in terms of the other. We will solve for x in the linear equation — that is, the first equation. (We could also solve for y.)

$x - y = 3$	This is the first equation in the given system.
$x = y + 3$	Add y to both sides.

Step 2 Substitute the expression from step 1 into the other equation. We substitute $y + 3$ for x in the second equation.

$$x = \boxed{y + 3} \qquad (\boxed{x} - 2)^2 + (y + 3)^2 = 4$$

This gives an equation in one variable, namely

$$(y + 3 - 2)^2 + (y + 3)^2 = 4.$$

The variable x has been eliminated.

Step 3 Solve the resulting equation containing one variable.

$(y + 3 - 2)^2 + (y + 3)^2 = 4$	This is the equation containing one variable.
$(y + 1)^2 + (y + 3)^2 = 4$	Combine numerical terms in the first parentheses.
$y^2 + 2y + 1 + y^2 + 6y + 9 = 4$	Use the formula $(A + B)^2 = A^2 + 2AB + B^2$ to square $y + 1$ and $y + 3$.
$2y^2 + 8y + 10 = 4$	Combine like terms on the left.
$2y^2 + 8y + 6 = 0$	Subtract 4 from both sides and set the quadratic equation equal to 0.

$$y^2 + 4y + 3 = 0 \quad \text{Simplify by dividing both sides by 2.}$$
$$(y + 3)(y + 1) = 0 \quad \text{Factor.}$$
$$y + 3 = 0 \quad \text{or} \quad y + 1 = 0 \quad \text{Set each factor equal to 0.}$$
$$y = -3 \quad \text{or} \quad y = -1 \quad \text{Solve for y.}$$

Step 4   Back-substitute the obtained values into the equation from step 1.
Now that we have the y-coordinates of the solutions, we back-substitute -3 for y and -1 for y in the equation $x = y + 3$.

If $y = -3$: $\quad x = -3 + 3 = 0, \quad$ so $(0, -3)$ is a solution.
If $y = -1$: $\quad x = -1 + 3 = 2, \quad$ so $(2, -1)$ is a solution.

Step 5   Check the proposed solution in both of the system's given equations.
Take a moment to show that each ordered pair satisfies both equations. The solution set of the given system is $\{(0, -3), (2, -1)\}$.

Figure 5.9 shows the graphs of the equations in the system and the solutions as intersection points.

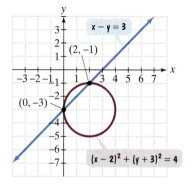

Figure 5.9 Points of intersection illustrate the nonlinear system's solutions.

Check Point 2   Solve by the substitution method:

$$x + 2y = 0$$
$$(x - 1)^2 + (y - 1)^2 = 5.$$

3 Solve nonlinear systems by addition.

Eliminating a Variable Using the Addition Method

In solving linear systems with two variables, we learned that the addition method works well when each equation is in the form $Ax + By = C$. For nonlinear systems, the addition method can be used when each equation is in the form $Ax^2 + By^2 = C$. If necessary, we will multiply either equation or both equations by appropriate numbers so that the coefficients of x^2 or y^2 will have a sum of 0. We then add equations. The sum will be an equation in one variable.

EXAMPLE 3   Solving a Nonlinear System by the Addition Method

Solve the system:

$$4x^2 + y^2 = 13 \quad \text{Equation 1}$$
$$x^2 + y^2 = 10. \quad \text{Equation 2}$$

Solution   We can use steps that are similar to those used to solve linear systems by the addition method.

Step 1   Write both equations in the form $Ax^2 + By^2 = C$.   Both equations are already in this form, so we can skip this step.

Step 2   If necessary, multiply either equation or both equations by appropriate numbers so that the sum of the x^2-coefficients or the sum of the y^2-coefficients is 0.   We can eliminate y^2 by multiplying Equation 2 by -1.

$$4x^2 + y^2 = 13 \xrightarrow{\text{No change}} 4x^2 + y^2 = 13$$
$$x^2 + y^2 = 10 \xrightarrow{\text{Multiply by } -1.} -x^2 - y^2 = -10$$

$4x^2 + y^2 = 13$ Equation 1

$x^2 + y^2 = 10$ Equation 2

The system we are solving, repeated

Steps 3 and 4   **Add equations and solve for the remaining variable.**

$$4x^2 + y^2 = 13$$
$$\underline{-x^2 - y^2 = -10}$$

Add: $3x^2 \qquad = 3$

$\qquad\qquad x^2 = 1$ Divide both sides by 3

$\qquad\qquad x = \pm 1$ Use the square root method: If $x^2 = c$, then $x = \pm\sqrt{c}$.

Step 5   **Back-substitute and find the values for the other variables.** We must back-substitute each value of x into either one of the original equations. Let's use $x^2 + y^2 = 10$, Equation 2. If $x = 1$,

$\qquad 1^2 + y^2 = 10$ Replace x with 1 in Equation 2.

$\qquad\qquad y^2 = 9$ Subtract 1 from both sides.

$\qquad\qquad y = \pm 3.$ Apply the square root method.

$(1, 3)$ and $(1, -3)$ are solutions. If $x = -1$,

$\qquad (-1)^2 + y^2 = 10$ Replace x with −1 in Equation 2.

$\qquad\qquad y^2 = 9$ The steps are the same as before.

$\qquad\qquad y = \pm 3.$

$(-1, 3)$ and $(-1, -3)$ are solutions.

Step 6   **Check.** Take a moment to show that each of the four ordered pairs satisfies Equation 1 and Equation 2. The solution set of the given system is $\{(1, 3), (1, -3), (-1, 3), (-1, -3)\}$.

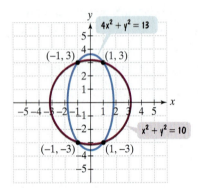

Figure 5.10 A system with four solutions

Figure 5.10 shows the graphs of the equations in the system and the solutions as intersection points.

Study Tip

When solving nonlinear systems, extra solutions may be introduced that do not satisfy both equations in the system. Therefore, you should get into the habit of checking all proposed pairs in each of the system's two equations.

Check Point 3   Solve the system:

$$3x^2 + 2y^2 = 35$$
$$4x^2 + 3y^2 = 48.$$

In solving nonlinear systems, we include only ordered pairs with real numbers in the solution set. We have seen that each of these ordered pairs corresponds to a point of intersection of the system's graphs.

EXAMPLE 4   **Solving a Nonlinear System by the Addition Method**

Solve the system:

$$y = x^2 + 3 \qquad \text{Equation 1} \quad \text{(The graph is a parabola.)}$$
$$x^2 + y^2 = 9. \qquad \text{Equation 2} \quad \text{(The graph is a circle.)}$$

Solution   We could use substitution because Equation 1 has y expressed in terms of x, but this would result in a fourth-degree equation. However, we can rewrite Equation 1 by subtracting x^2 from both sides and adding the equations to eliminate the x^2-terms.

Notice how like terms are arranged in columns.

$\qquad -x^2 + y \qquad\quad = 3$ Subtract x^2 from both sides of Equation 1.

$\qquad \underline{x^2 \qquad + y^2 = 9}$ This is Equation 2.

$\qquad\qquad\quad y + y^2 = 12$ Add the equations.

We now solve this quadratic equation.

$$y + y^2 = 12 \qquad \text{This is the equation containing one variable.}$$

$$y^2 + y - 12 = 0 \qquad \text{Subtract 12 from both sides and set the quadratic equation equal to 0.}$$

$$(y + 4)(y - 3) = 0 \qquad \text{Factor.}$$

$$y + 4 = 0 \quad \text{or} \quad y - 3 = 0 \qquad \text{Set each factor equal to 0.}$$

$$y = -4 \quad \text{or} \quad y = 3 \qquad \text{Solve for } y.$$

To complete the solution, we must back-substitute each value of y into either one of the original equations. We will use $y = x^2 + 3$, Equation 1. First, we substitute -4 for y.

$$-4 = x^2 + 3$$

$$-7 = x^2 \qquad \text{Subtract 3 from both sides.}$$

Because the square of a real number cannot be negative, the equation $x^2 = -7$ does not have real-number solutions. Thus, we move on to our other value for y, 3, and substitute this value into Equation 1.

$$y = x^2 + 3 \qquad \text{This is Equation 1.}$$

$$3 = x^2 + 3 \qquad \text{Back-substitute 3 for } y.$$

$$0 = x^2 \qquad \text{Subtract 3 from both sides.}$$

$$0 = x \qquad \text{Solve for } x.$$

We showed that if $y = 3$, then $x = 0$. Thus, $(0, 3)$ is the solution. Take a moment to show that $(0, 3)$ satisfies Equation 1 and Equation 2. The solution set of the given system is $\{(0, 3)\}$. Figure 5.11 shows the system's graphs and the solution as an intersection point.

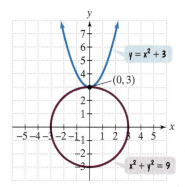

Figure 5.11 A system with one real solution

Check Point 4   Solve the system:

$$y = x^2 + 5$$

$$x^2 + y^2 = 25.$$

4   Solve problems using systems of nonlinear equations.

Applications

Many geometric problems can be modeled and solved by the use of systems of nonlinear equations. We will use our step-by-step strategy for solving problems using mathematical models that are created from verbal models.

EXAMPLE 5   An Application of a Nonlinear System

You have 36 yards of fencing to build the enclosure in Figure 5.12. Some of this fencing is to be used to build an internal divider. If you'd like to enclose 54 square yards, what are the dimensions of the enclosure?

Solution

Step 1   **Use variables to represent unknown quantities.** Let $x =$ the enclosure's length and $y =$ the enclosure's width. These variables are shown in Figure 5.12.

Step 2   **Write a system of equations describing the problem's conditions.** The first condition is that you have 36 yards of fencing.

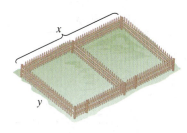

Figure 5.12 Building an enclosure

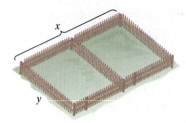

Figure 5.12, repeated

$$\boxed{\text{Fencing along both lengths}} \text{ plus } \boxed{\text{Fencing along both widths}} \text{ plus } \boxed{\text{Fencing for the internal divider}} \text{ equals } \boxed{\text{36 yards.}}$$

$$2x \quad + \quad 2y \quad + \quad y \quad = \quad 36$$

Adding like terms, we can express the equation that models the verbal conditions for the fencing as $2x + 3y = 36$.

The second condition is that you'd like to enclose 54 square yards. The rectangle's area, the product of its length and its width, must be 54 square yards.

$$\boxed{\text{Length}} \text{ times } \boxed{\text{width}} \text{ is } \boxed{\text{54 square yards.}}$$

$$x \quad \cdot \quad y \quad = \quad 54$$

Step 3 Solve the system and answer the problem's question. We must solve the system

$$2x + 3y = 36 \quad \text{Equation 1}$$
$$xy = 54. \quad \text{Equation 2}$$

We will use substitution. Because Equation 1 has no coefficients of 1 or -1, we will solve Equation 2 for y. Dividing both sides of $xy = 54$ by x, we obtain

$$y = \frac{54}{x}.$$

Now we substitute $\frac{54}{x}$ for y in Equation 1 and solve for x.

$2x + 3y = 36$	This is Equation 1.
$2x + 3 \cdot \dfrac{54}{x} = 36$	Substitute $\dfrac{54}{x}$ for y.
$2x + \dfrac{162}{x} = 36$	Multiply.
$x\left(2x + \dfrac{162}{x}\right) = 36 \cdot x$	Clear fractions by multiplying both sides by x.
$2x^2 + 162 = 36x$	Use the distributive property on the left side.
$2x^2 - 36x + 162 = 0$	Subtract $36x$ from both sides and set the quadratic equation equal to 0.
$x^2 - 18x + 81 = 0$	Simplify by dividing both sides by 2.
$(x - 9)^2 = 0$	Factor using $A^2 - 2AB + B^2 = (A - B)^2$.
$x - 9 = 0$	Set the repeated factor equal to zero.
$x = 9$	Solve for x.

We back-substitute this value of x into $y = \dfrac{54}{x}$.

$$\text{If } x = 9, \quad y = \frac{54}{9} = 6.$$

This means that the dimensions of the enclosure in Figure 5.12 are 9 yards by 6 yards.

Step 4 **Check the proposed solution in the original wording of the problem.** With a length of 9 yards and a width of 6 yards, take a moment to check that this results in 36 yards of fencing and an area of 54 square yards.

Check Point 5 Find the length and width of a rectangle whose perimeter is 20 feet and whose area is 21 square feet.

EXERCISE SET 5.4

Practice Exercises

In Exercises 1–18, solve each system by the substitution method.

1. $x + y = 2$
$y = x^2 - 4$

2. $x - y = -1$
$y = x^2 + 1$

3. $x + y = 2$
$y = x^2 - 4x + 4$

4. $2x + y = -5$
$y = x^2 + 6x + 7$

5. $y = x^2 - 4x - 10$
$y = -x^2 - 2x + 14$

6. $y = x^2 + 4x + 5$
$y = x^2 + 2x - 1$

7. $x^2 + y^2 = 25$
$x - y = 1$

8. $x^2 + y^2 = 5$
$3x - y = 5$

9. $xy = 6$
$2x - y = 1$

10. $xy = -12$
$x - 2y + 14 = 0$

11. $y^2 = x^2 - 9$
$2y = x - 3$

12. $x^2 + y = 4$
$2x + y = 1$

13. $xy = 3$
$x^2 + y^2 = 10$

14. $xy = 4$
$x^2 + y^2 = 8$

15. $x + y = 1$
$x^2 + xy - y^2 = -5$

16. $x + y = -3$
$x^2 + 2y^2 = 12y + 18$

17. $x + y = 1$
$(x - 1)^2 + (y + 2)^2 = 10$

18. $2x + y = 4$
$(x + 1)^2 + (y - 2)^2 = 4$

In Exercises 19–28, solve each system by the addition method.

19. $x^2 + y^2 = 13$
$x^2 - y^2 = 5$

20. $4x^2 - y^2 = 4$
$4x^2 + y^2 = 4$

21. $x^2 - 4y^2 = -7$
$3x^2 + y^2 = 31$

22. $3x^2 - 2y^2 = -5$
$2x^2 - y^2 = -2$

23. $3x^2 + 4y^2 - 16 = 0$
$2x^2 - 3y^2 - 5 = 0$

24. $16x^2 - 4y^2 - 72 = 0$
$x^2 - y^2 - 3 = 0$

25. $x^2 + y^2 = 25$
$(x - 8)^2 + y^2 = 41$

26. $x^2 + y^2 = 5$
$x^2 + (y - 8)^2 = 41$

27. $y^2 - x = 4$
$x^2 + y^2 = 4$

28. $x^2 - 2y = 8$
$x^2 + y^2 = 16$

In Exercises 29–42, solve each system by the method of your choice.

29. $3x^2 + 4y^2 = 16$
$2x^2 - 3y^2 = 5$

30. $x + y^2 = 4$
$x^2 + y^2 = 16$

31. $2x^2 + y^2 = 18$
$xy = 4$

32. $x^2 + 4y^2 = 20$
$xy = 4$

33. $x^2 + 4y^2 = 20$
$x + 2y = 6$

34. $3x^2 - 2y^2 = 1$
$4x - y = 3$

35. $x^3 + y = 0$
$x^2 - y = 0$

36. $x^3 + y = 0$
$2x^2 - y = 0$

37. $x^2 + (y - 2)^2 = 4$
$x^2 - 2y = 0$

38. $x^2 - y^2 - 4x + 6y - 4 = 0$
$x^2 + y^2 - 4x - 6y + 12 = 0$

39. $y = (x + 3)^2$
$x + 2y = -2$

40. $(x - 1)^2 + (y + 1)^2 = 5$
$2x - y = 3$

41. $x^2 + y^2 + 3y = 22$
$2x + y = -1$

42. $x - 3y = -5$
$x^2 + y^2 - 25 = 0$

In Exercises 43–46, let x represent one number and let y represent the other number. Use the given conditions to write a system of nonlinear equations. Solve the system and find the numbers.

43. The sum of two numbers is 10 and their product is 24. Find the numbers.

44. The sum of two numbers is 20 and their product is 96. Find the numbers.

45. The difference between the squares of two numbers is 3. Twice the square of the first number increased by the square of the second number is 9. Find the numbers.

46. The difference between the squares of two numbers is 5. Twice the square of the second number subtracted from three times the square of the first number is 19. Find the numbers.

Application Exercises

47. A planet's orbit follows a path described by $16x^2 + 4y^2 = 64$. A comet follows the parabolic path $y = x^2 - 4$. Where might the comet intersect the orbiting planet?

48. A system for tracking ships indicates that a ship lies on a path described by $2y^2 - x^2 = 1$. The process is repeated and the ship is found to lie on a path described by $2x^2 - y^2 = 1$. If it is known that the ship is located in the first quadrant of the coordinate system, determine its exact location.

49. Find the length and width of a rectangle whose perimeter is 36 feet and whose area is 77 square feet.

50. Find the length and width of a rectangle whose perimeter is 40 feet and whose area is 96 square feet.

Use the formula for the area of a rectangle and the Pythagorean Theorem to solve Exercises 51–52.

51. A small television has a picture with a diagonal measure of 10 inches and a viewing area of 48 square inches. Find the length and width of the screen.

52. The area of a rug is 108 square feet and the length of its diagonal is 15 feet. Find the length and width of the rug.

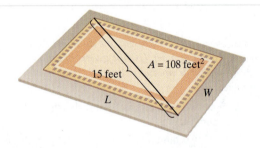

53. The figure at the top of the next column shows a square floor plan with a smaller square area that will accommodate a combination fountain and pool. The floor with the fountain-pool area removed has an area of 21 square meters and a perimeter of 24 meters. Find the dimensions of the floor and the dimensions of the square that will accommodate the pool.

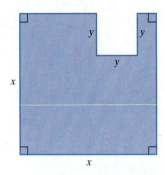

54. The area of the rectangular piece of cardboard shown below is 216 square inches. The cardboard is used to make an open box by cutting a 2-inch square from each corner and turning up the sides. If the box is to have a volume of 224 cubic inches, find the length and width of the cardboard that must be used.

Writing in Mathematics

55. What is a system of nonlinear equations? Provide an example with your description.

56. Explain how to solve a nonlinear system using the substitution method. Use $x^2 + y^2 = 9$ and $2x - y = 3$ to illustrate your explanation.

57. Explain how to solve a nonlinear system using the addition method. Use $x^2 - y^2 = 5$ and $3x^2 - 2y^2 = 19$ to illustrate your explanation.

58. The daily demand and supply models for a carrot cake supplied by a bakery to a convenience store are given by the demand model $N = 40 - 3p$ and the supply model $N = \dfrac{p^2}{10}$, in which p is the price of the cake and N is the number of cakes sold or supplied each day to the convenience store. Explain how to determine the price at which supply and demand are equal. Then describe how to find how many carrot cakes can be supplied and sold each day at this price.

Technology Exercises

59. Verify your solutions to any five exercises from Exercises 1–42 by using a graphing utility to graph the two equations in the system in the same viewing rectangle. Then use the trace or intersection feature to verify the solutions.

60. Write a system of equations, one equation whose graph is a line and the other whose graph is a parabola, that has no ordered pairs that are real numbers in its solution set. Graph the equations using a graphing utility and verify that you are correct.

Critical Thinking Exercises

61. Which one of the following is true?
   a. A system of two equations in two variables whose graphs represent a circle and a line can have four real solutions.
   b. A system of two equations in two variables whose graphs represent a parabola and a circle can have four real solutions.
   c. A system of two equations in two variables whose graphs represent two circles must have at least two real solutions.

   d. A system of two equations in two variables whose graphs represent a parabola and a circle cannot have only one real solution.

62. The points of intersection of the graphs of $xy = 20$ and $x^2 + y^2 = 41$ are joined to form a rectangle. Find the area of the rectangle.

63. Find a and b in this figure.

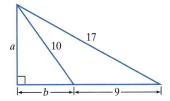

Solve the systems in Exercises 64–65.

64. $\log_y x = 3$
   $\log_y (4x) = 5$

65. $\log x^2 = y + 3$
   $\log x = y - 1$

SECTION 5.5 *Systems of Inequalities*

Objectives

1. Graph a linear inequality in two variables.

2. Graph a nonlinear inequality in two variables.

3. Graph a system of inequalities.

4. Solve applied problems involving systems of inequalities.

Had a good workout lately? If so, could you tell if you were overdoing it or not pushing yourself hard enough? In this section, we will use systems of inequalities in two variables to help you establish a target zone for your workouts.

Linear Inequalities in Two Variables and Their Solutions

We have seen that equations in the form $Ax + By = C$ are straight lines when graphed. If we change the $=$ sign to $>, <, \geq$, or $\leq$, we obtain a **linear inequality in two variables.** Some examples of linear inequalities in two variables are $x + y > 2$, $3x - 5y \leq 15$, and $2x - y < 4$.

   A **solution of an inequality in two variables,** x and y, is an ordered pair of real numbers with the following property: When the x-coordinate is substituted for x and the y-coordinate is substituted for y in the inequality, we obtain a true statement. For example, $(3, 2)$ is a solution of the inequality $x + y > 1$. When 3 is substituted for x and 2 is substituted for y, we obtain the true statement $3 + 2 > 1$, or $5 > 1$. Because there are infinitely many pairs of numbers that have a sum greater than 1, the inequality $x + y > 1$ has infinitely many solutions. Each ordered pair solution is said to **satisfy** the inequality. Thus, $(3, 2)$ satisfies the inequality $x + y > 1$.

1 Graph a linear inequality in two variables.

The Graph of a Linear Inequality in Two Variables

We know that the graph of an equation in two variables is the set of all points whose coordinates satisfy the equation. Similarly, the **graph of an inequality in two variables** is the set of all points whose coordinates satisfy the inequality.

Let's use Figure 5.13 to get an idea of what the graph of a linear inequality in two variables looks like. Part of the figure shows the graph of the linear equation $x + y = 2$. The line divides the points in the rectangular coordinate system into three sets. First, there is the set of points along the line, satisfying $x + y = 2$. Next, there is the set of points in the green region above the line. Points in the green region satisfy the linear inequality $x + y > 2$. Finally, there is the set of points in the pink region below the line. Points in the pink region satisfy the linear inequality $x + y < 2$.

A **half-plane** is the set of all the points on one side of a line. In Figure 5.13, the green region is a half-plane. The pink region is also a half-plane. A half-plane is the graph of a linear inequality that involves $>$ or $<$. The graph of an inequality that involves $\geq$ or $\leq$ is a half-plane and a line. A solid line is used to show that the line is part of the graph. A dashed line is used to show that a line is not part of a graph.

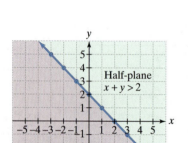

Figure 5.13

> ## Graphing a Linear Inequality in Two Variables
>
> 1. Replace the inequality symbol with an equal sign and graph the corresponding linear equation. Draw a solid line if the original inequality contains a $\leq$ or $\geq$ symbol. Draw a dashed line if the original inequality contains a $<$ or $>$ symbol.
>
> 2. Choose a test point in one of the half-planes that is not on the line. Substitute the coordinates of the test point into the inequality.
>
> 3. If a true statement results, shade the half-plane containing this test point. If a false statement results, shade the half-plane not containing this test point.

EXAMPLE 1 **Graphing a Linear Inequality in Two Variables**

Graph: $3x - 5y < 15$.

Solution

Step 1 **Replace the inequality symbol with $=$ and graph the linear equation.** We need to graph $3x - 5y = 15$. We can use intercepts to graph this line.

We set $y = 0$ to find the x-intercept:	We set $x = 0$ to find the y-intercept:
$3x - 5y = 15$	$3x - 5y = 15$
$3x - 5 \cdot 0 = 15$	$3 \cdot 0 - 5y = 15$
$3x = 15$	$-5y = 15$
$x = 5.$	$y = -3.$

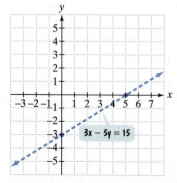

Figure 5.14 Preparing to graph $3x - 5y < 15$

The x-intercept is 5, so the line passes through $(5, 0)$. The y-intercept is -3, so the line passes through $(0, -3)$. The graph is indicated by a dashed line because the inequality $3x - 5y < 15$ contains a $<$ symbol, rather than $\leq$. The graph of the line is shown in Figure 5.14.

Step 2 **Choose a test point in one of the half-planes that is not on the line. Substitute its coordinates into the inequality.** The line $3x - 5y = 15$ divides the plane into three parts—the line itself and two half-planes. The points in one half-plane satisfy $3x - 5y > 15$. The points in the other half-plane satisfy $3x - 5y < 15$. We need to find which half-plane belong to the solution. To do so, we test a point from either half-plane. The origin, $(0, 0)$, is the easiest point to test.

$$3x - 5y < 15 \qquad \text{This is the given inequality.}$$
$$\text{Is} \quad 3 \cdot 0 - 5 \cdot 0 < 15? \qquad \text{Test } (0, 0) \text{ by substituting 0 for } x \text{ and 0 for } y.$$
$$0 - 0 < 15 \qquad \text{Multiply.}$$
$$0 < 15 \qquad \text{Subtract. This statement is true.}$$

Step 3 **If a true statement results, shade the half-plane containing the test point.** Because 0 is less than 15, the test point, $(0, 0)$, is part of the solution set. All the points on the same side of the line $3x - 5y = 15$ as the point $(0, 0)$ are members of the solution set. The solution set is the half-plane that contains the point $(0, 0)$, indicated by shading this half-plane. The graph is shown using green shading and a dashed blue line in Figure 5.15.

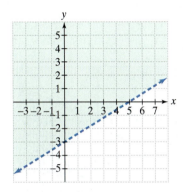

Figure 5.15 The graph of $3x - 5y < 15$

Check Point 1 Graph: $2x - 4y < 8$.

When graphing a linear inequality, test a point that lies in one of the half-planes and *not on the line dividing the half-planes*. The test point, $(0, 0)$, is convenient because it is easy to calculate when 0 is substituted for each variable. However, if $(0, 0)$ lies on the dividing line and not in a half-plane, a different test point must be selected.

EXAMPLE 2 **Graphing a Linear Inequality in Two Variables**

Graph: $y \leq \dfrac{2}{3}x$.

Solution

Step 1 **Replace the inequality symbol with = and graph the linear equation.** We need to graph $y = \frac{2}{3}x$. We can use the slope and the y-intercept to graph this line.

$$y = \frac{2}{3}x + 0$$

Slope $= \dfrac{2}{3} = \dfrac{\text{rise}}{\text{run}}$ $\qquad$ y-intercept $= 0$

The y-intercept is 0, so the line passes through $(0, 0)$. Using the y-intercept and the slope, the line is shown in Figure 5.16. A solid line is used because the inequality $y \leq \frac{2}{3}x$ contains a $\leq$ symbol, in which equality is included.

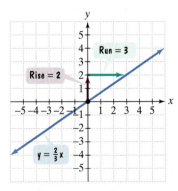

Figure 5.16 Preparing to graph $y \leq \dfrac{2}{3}x$

Step 2 **Choose a test point in one of the half-planes that is not on the line. Substitute its coordinates into the inequality.** We cannot use $(0, 0)$ as a test point because it lies on the line and not in a half-plane. Let's use $(1, 1)$, which lies in the half-plane above the line.

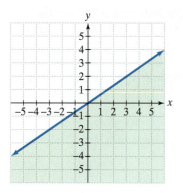

Figure 5.17 The graph of $y \leq \frac{2}{3}x$

$$y \leq \frac{2}{3}x \qquad \text{\color{blue}This is the given inequality.}$$

$$\text{Is} \quad 1 \leq \frac{2}{3} \cdot 1? \qquad \text{\color{blue}Test (1, 1) by substituting 1 for } x \text{ and 1 for } y.$$

$$1 \leq \frac{2}{3} \qquad \text{\color{blue}This statement is false.}$$

Step 3   If a false statement results, shade the half-plane not containing the test point.   Because 1 is not less than or equal to $\frac{2}{3}$, the test point $(1, 1)$ is not part of the solution set. Thus, the half-plane below the solid line $y = \frac{2}{3}x$ is part of the solution set. The solution set is the line and the half-plane that does not contain the point $(1, 1)$, indicated by shading this half-plane. The graph is shown using green shading and a blue line in Figure 5.17.

Technology

Most graphing utilities can graph inequalities in two variables with the $\boxed{\text{SHADE}}$ feature. The procedure varies by model, so consult your manual. For most graphing utilities, you must first solve for y if it is not already isolated. The figure shows the graph of $y \leq \frac{2}{3}x$. Most displays do not distinguish between dashed and solid boundary lines.

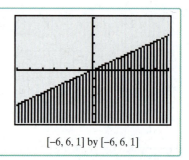

$[-6, 6, 1]$ by $[-6, 6, 1]$

Check Point 2   Graph: $y \geq \frac{1}{2}x$.

You can graph inequalities in the form $y > mx + b$ or $y < mx + b$ without using test points. The inequality symbol indicates which half-plane to shade.

- If $y > mx + b$, shade the half-plane above the line $y = mx + b$.
- If $y < mx + b$, shade the half-plane below the line $y = mx + b$.

In Chapter 1, we learned that $y = b$ graphs as a horizontal line, where b is the y-intercept. Similarly, the graph of $x = a$ is a vertical line, where a is the x-intercept. Half-planes can be separated by horizontal or vertical lines. For example, Figure 5.18 shows the graph of $y \leq 2$. Because $(0, 0)$ satisfies this inequality ($0 \leq 2$ is true), the graph consists of the half-plane below the line $y = 2$ and the line. Similarly, Figure 5.19 shows the graph of $x < 4$.

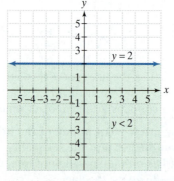

Figure 5.18 The graph of $y \leq 2$

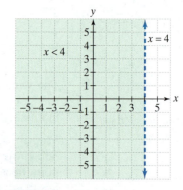

Figure 5.19 The graph of $x < 4$

2 Graph a nonlinear inequality in two variables.

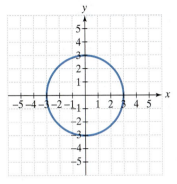

Figure 5.20 Preparing to graph $x^2 + y^2 \leq 9$

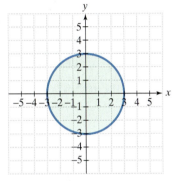

Figure 5.21 The graph of $x^2 + y^2 \leq 9$

3 Graph a system of inequalities.

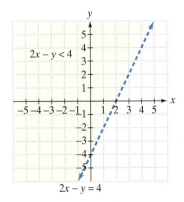

$2x - y < 4$

$2x - y = 4$

Figure 5.22 The graph of $2x - y < 4$

Graphing a Nonlinear Inequality in Two Variables

Example 3 illustrates that a nonlinear inequality in two variables is graphed in the same way that we graph a linear inequality.

EXAMPLE 3 Graphing a Nonlinear Inequality in Two Variables

Graph: $x^2 + y^2 \leq 9$.

Solution

Step 1 Replace the inequality symbol with = and graph the nonlinear equation. We need to graph $x^2 + y^2 = 9$. The graph is a circle of radius 3 with its center at the origin. The graph is shown in Figure 5.20 as a solid circle because equality is included in the $\leq$ symbol.

Step 2 Choose a test point in one of the regions that is not on the circle. Substitute its coordinates into the inequality. The circle divides the plane into three parts—the circle itself, the region inside the circle, and the region outside the circle. We need to determine whether the region inside or outside the circle is included in the solution. To do so, we will use the test point $(0, 0)$ from inside the circle.

$x^2 + y^2 \leq 9$	This is the given inequality.
Is $0^2 + 0^2 \leq 9$?	Test $(0, 0)$ by substituting 0 for x and 0 for y.
$0 + 0 \leq 9$	Square 0: $0^2 = 0$.
$0 \leq 9$	Add. This statement is true.

Step 3 If a true statement results, shade the region containing the test point. The true statement tells us that all the points inside the circle satisfy $x^2 + y^2 \leq 9$. The graph is shown using green shading and a solid blue circle in Figure 5.21.

Check Point 3 Graph: $x^2 + y^2 \geq 16$.

Systems of Inequalities in Two Variables

The **solution set of a system of inequalities** in two variables, x and y, is the set of all ordered pairs (x, y) that satisfy each inequality in the system. The **graph of a system of inequalities** in two variables is the graph of the system's solution set. Thus, to graph a system of inequalities in two variables, begin by graphing each individual inequality in the same rectangular coordinate system. Then find the region, if there is one, that is common to every graph in the system. This region of intersection gives a picture of the system's solution set.

EXAMPLE 4 Graphing a System of Linear Inequalities

Graph the solution set:

$$2x - y < 4$$
$$x + y \geq -1.$$

Solution We begin by graphing $2x - y < 4$. Because the inequality contains a $<$ symbol, rather than $\leq$, we graph $2x - y = 4$ as a dashed line. (If $x = 0$, then $y = -4$, and if $y = 0$, then $x = 2$. The x-intercept is 2 and the y-intercept is -4.) Because $(0, 0)$ makes the inequality $2x - y < 4$ true, we shade the half-plane containing $(0, 0)$, shown in yellow in Figure 5.22.

Now we graph $x + y \geq -1$ in the same rectangular coordinate system. Because the inequality contains a $\geq$ symbol, in which equality is included, we graph $x + y = -1$ as a solid line. (If $x = 0$, then $y = -1$, and if $y = 0$, then $x = -1$. The x-intercept and y-intercept are both -1.) Because $(0, 0)$ makes the inequality true, we shade the half-plane containing $(0, 0)$. This is shown in Figure 5.23 using green vertical shading.

The solution set of the system is shown graphically by the intersection (the overlap) of the two half-planes. This is shown in Figure 5.23 as the region in which the yellow shading and the green vertical shading overlap. The solution of the system is shown again in Figure 5.24.

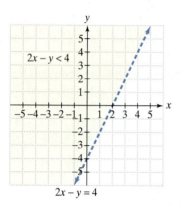

Figure 5.22, repeated
The graph of $2x - y < 4$

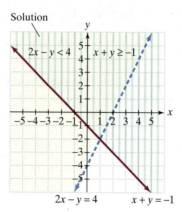

Figure 5.23 Adding the graph of $x + y \geq -1$

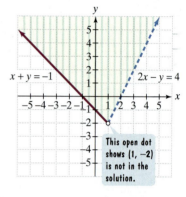

Figure 5.24 The graph of $2x - y < 4$ and $x + y \geq -1$

Check Point 4 Graph the solution set:

$$x + 2y > 4$$
$$2x - 3y \leq -6.$$

EXAMPLE 5 Graphing a System of Inequalities

Graph the solution set:

$$y \geq x^2 - 4$$
$$x - y \geq 2.$$

Solution We begin by graphing $y \geq x^2 - 4$. Because equality is included in $\geq$, we graph $y = x^2 - 4$ as a solid parabola. Because $(0, 0)$ makes the inequality $y \geq x^2 - 4$ true (we obtain $0 \geq -4$), we shade the interior portion of the parabola containing $(0, 0)$, shown in yellow in Figure 5.25.

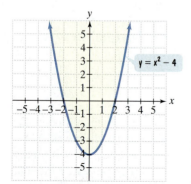

Figure 5.25 The graph of $y \geq x^2 - 4$

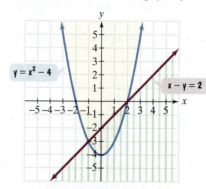

Figure 5.26 Adding the graph of $x - y \geq 2$

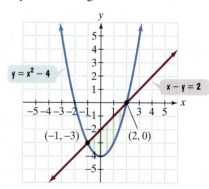

Figure 5.27 The graph of $y \geq x^2 - 4$ and $x - y \geq 2$

Now we graph $x - y \geq 2$ in the same rectangular coordinate system. First we graph the line $x - y = 2$ using its x-intercept, 2, and its y-intercept, -2. Because $(0, 0)$ makes the inequality $x - y \geq 2$ false (we obtain $0 \geq 2$), we shade the half-plane below the line. This is shown in Figure 5.26 using green vertical shading.

The solution of the system is shown in Figure 5.26 by the intersection (the overlap) of the solid yellow and green vertical shadings. The graph of the system's solution set consists of the region enclosed by the parabola and the line. To find the points of intersection of the parabola and the line, use the substitution method to solve the nonlinear system

$$y = x^2 - 4$$
$$x - y = 2.$$

Take a moment to show that the solutions are $(-1, -3)$ and $(2, 0)$, as shown in Figure 5.27.

Check Point 5 Graph the solution set:

$$y \geq x^2 - 4$$
$$x + y \leq 2.$$

A system of inequalities has no solution if there are no points in the rectangular coordinate system that simultaneously satisfy each inequality in the system. For example, the system

$$2x + 3y \geq 6$$
$$2x + 3y \leq 0$$

whose separate graphs are shown in Figure 5.28 has no overlapping region. Thus, the system has no solution. The solution set is Ø, the empty set.

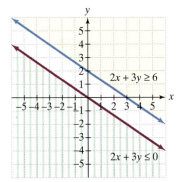

Figure 5.28 A system of inequalities with no solution

EXAMPLE 6   Graphing a System of Inequalities

Graph the solution set:

$$x - y < 2$$
$$-2 \leq x < 4$$
$$y < 3.$$

Solution We begin by graphing $x - y < 2$, the first given inequality. The line $x - y = 2$ has an x-intercept of 2 and a y-intercept of -2. The test point $(0, 0)$ makes the inequality $x - y < 2$ true, and its graph is shown in Figure 5.29.

Now, let's consider the second given inequality, $-2 \leq x < 4$. Replacing the inequality symbols by $=$, we obtain $x = -2$ and $x = 4$, graphed as vertical lines. The line of $x = 4$ is not included. Using $(0, 0)$ as a test point and substituting the x-coordinate, 0, into $-2 \leq x < 4$, we obtain the true statement $-2 \leq 0 < 4$. We therefore shade the region between the vertical lines. We've added this region to Figure 5.29, intersecting the region between the vertical lines with the yellow region in Figure 5.29. The resulting region is shown in yellow and green vertical shading in Figure 5.30.

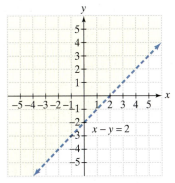

Figure 5.29 The graph of $x - y < 2$

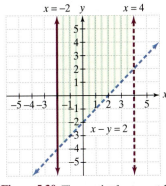

Figure 5.30 The graph of $x - y < 2$ and $-2 \leq x < 4$

Finally, let's consider the third given inequality, $y < 3$. Replacing the inequality symbol by =, we obtain $y = 3$, which graphs as a horizontal line. Because $(0, 0)$ satisfies $y < 3$ ($0 < 3$ is true), the graph consists of the half-plane below the line $y = 3$. We've added this half-plane to the region in Figure 5.30, intersecting the half-plane with this region. The resulting region is shown in yellow and green vertical shading in Figure 5.31. This region represents the graph of the solution set of the given system.

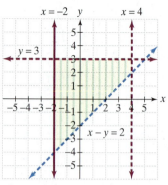

Figure 5.31 The graph of $x - y < 2$ and $-2 \le x < 4$ and $y < 3$

Check Point 6 Graph the solution set:

$$x + y < 2$$
$$-2 \le x < 1$$
$$y > -3.$$

4 Solve applied problems involving systems of inequalities.

Applications

Now we are ready to use a system of inequalities to establish a target zone for your workouts.

EXAMPLE 7 Inequalities and Aerobic Exercise

For people between ages 10 and 70, inclusive, the target zone for aerobic exercise is given by the following system of inequalities in which a represents one's age and p is one's pulse rate:

$$2a + 3p \ge 450$$
$$a + p \le 190.$$

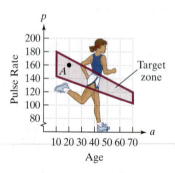

Figure 5.32

The graph of this target zone is shown in Figure 5.32 for $10 \le a \le 70$. Find your age. The line segments on the top and bottom of the shaded region indicate upper and lower limits for your pulse rate, in beats per minute, when engaging in aerobic exercise.

a. What are the coordinates of point A and what does this mean in terms of age and pulse rate?
b. Show that the coordinates of point A satisfy each inequality in the system.

Solution

a. Point A has coordinates $(20, 160)$. This means that a pulse rate of 160 beats per minute is within the target zone for a 20-year-old person engaged in aerobic exercise.

b. We can show that $(20, 160)$ satisfies each inequality by substituting 20 for a and 160 for p.

$2a + 3p \ge 450$	$a + p \le 190$
Is $\quad 2(20) + 3(160) \ge 450$?	Is $\quad 20 + 160 \le 190$?
$40 + 480 \ge 450$	$180 \le 190,$ true
$520 \ge 450,$ true	

The pair $(20, 160)$ makes each inequality true, so it satisfies each inequality in the system.

Check Point 7 Identify a point other than *A* in the target zone in Figure 5.32.

a. What are the coordinates of this point and what does this mean in terms of age and pulse rate?

b. Show that the coordinates of the point satisfy each inequality in the system in Example 7.

EXERCISE SET 5.5

Practice Exercises

In Exercises 1–22, graph each inequality.

1. $x + 2y \le 8$

2. $3x - 6y \le 12$

3. $x - 2y > 10$

4. $2x - y > 4$

5. $y \le \dfrac{1}{3}x$

6. $y \le \dfrac{1}{4}x$

7. $y > 2x - 1$

8. $y > 3x + 2$

9. $x \le 1$

10. $x \le -3$

11. $y > 1$

12. $y > -3$

13. $x^2 + y^2 \le 1$

14. $x^2 + y^2 \le 4$

15. $x^2 + y^2 > 25$

16. $x^2 + y^2 > 36$

17. $y < x^2 - 1$

18. $y < x^2 - 9$

19. $y \ge x^2 - 9$

20. $y \ge x^2 - 1$

21. $y > 2^x$

22. $y \le 3^x$

In Exercises 23–52, graph the solution set of each system of inequalities or indicate that the system has no solution.

23. $3x + 6y \le 6$
$2x + y \le 8$

24. $x - y \ge 4$
$x + y \le 6$

25. $2x - 5y \le 10$
$3x - 2y > 6$

26. $2x - y \le 4$
$3x + 2y > -6$

27. $y > 2x - 3$
$y < -x + 6$

28. $y < -2x + 4$
$y < x - 4$

29. $x + 2y \le 4$
$y \ge x - 3$

30. $x + y \le 4$
$y \ge 2x - 4$

31. $x \le 2$
$y \ge -1$

32. $x \le 3$
$y \le -1$

33. $-2 \le x < 5$

34. $-2 < y \le 5$

35. $x - y \le 1$
$x \ge 2$

36. $4x - 5y \ge -20$
$x \ge -3$

37. $x + y > 4$
$x + y < -1$

38. $x + y > 3$
$x + y < -2$

39. $x + y > 4$
$x + y > -1$

40. $x + y > 3$
$x + y > -2$

41. $y \ge x^2 - 1$
$x - y \ge -1$

42. $y \ge x^2 - 4$
$x - y \ge 2$

43. $x^2 + y^2 \le 16$
$x + y > 2$

44. $x^2 + y^2 \le 4$
$x + y > 1$

45. $x^2 + y^2 > 1$
$x^2 + y^2 < 4$

46. $x^2 + y^2 > 1$
$x^2 + y^2 < 9$

47. $x - y \le 2$
$x \ge -2$
$y \le 3$

48. $3x + y \le 6$
$x \ge -2$
$y \le 4$

49. $x \ge 0$
$y \ge 0$
$2x + 5y \le 10$
$3x + 4y \le 12$

50. $x \ge 0$
$y \ge 0$
$2x + y \le 4$
$2x - 3y \le 6$

51. $3x + y \le 6$
$2x - y \le -1$
$x \ge -2$
$y \le 4$

52. $2x + y \le 6$
$x + y \ge 2$
$1 \le x \le 2$
$y \le 3$

Application Exercises

The figure shows three kinds of regions—deserts, grasslands, and forests—that results from various ranges of temperature, T, and precipitation, P. Use the figure to solve Exercises 53–54.

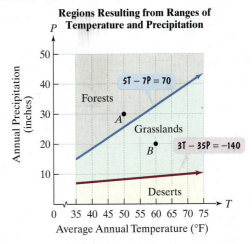

Regions Resulting from Ranges of Temperature and Precipitation

Source: A. Miller and J. Thompson, *Elements of Meteorology*

53. Use the figure on the previous page to write a system of inequalities that describe where forests occur. Then show that the coordinates of point A satisfy each inequality in the system.

54. Use the figure on the previous page to write a system of inequalities that describe where grasslands occur. Then show that the coordinates of point B satisfy each inequality in the system.

55. Many elevators have a capacity of 2000 pounds

a. If a child averages 50 pounds and an adult 150 pounts, write an inequality that describes when x children and y adults will cause the elevator to be overloaded.

b. Graph the inequality. Because x and y must be positive, limit the graph to quadrant I only.

c. Select an ordered pair satisfying the inequality. What are its coordinates and what do they represent in this situation?

56. A patient is not allowed to have more than 330 milligrams of cholesterol per day from a diet of eggs and meat. Each egg provides 165 milligrams of cholesterol. Each ounce of meat provides 110 milligrams.

a. Write an inequality that describes the patient's dietary restrictions for x eggs and y ounces of meat.

b. Graph the inequality. Because x and y must be positive, limit the graph to quadrant I only.

c. Select an ordered pair satisfying the inequality. What are its coordinates and what do they represent in this situation?

57. A person with no more than $15,000 to invest plans to place the money in two investments. One investment is high risk, high yield; the other is low risk, low yield. At least $2000 is to be placed in the high-risk investment. Furthermore, the amount invested at low risk should be at least three times the amount invested at high risk. Find and graph a system of inequalities that describes all possibilities for placing the money in the high- and low-risk investments.

58. Promoters of a rock concert must sell at least 25,000 tickets priced at $35 and $50 per ticket. Furthermore, the promoters must take in at least $1,025,000 in ticket sales. Find and graph a system of inequalities that describes all possibilities for selling the $35 tickets and the $50 tickets.

59. Use Figure 8.32 on page 750 to solve this exercise.

a. Find a pulse rate that lies within the target zone for a person your age engaged in aerobic exercise.

b. Express your answer in part (a) as an ordered pair. Show that the coordinates of this ordered pair satisfy each inequality.

The graph of an inequality in two variables is usually a region in the rectangular coordinate system. Regions in coordinate systems have numerous applications. For example, the regions in the two graphs at the top of the next column indicate whether a person is overweight, borderline overweight, or normal weight.

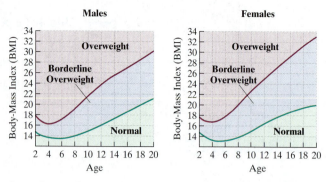

Source: Centers for Disease Control and Prevention

The horizontal axis shows a person's age. The vertical axis shows that person's body-mass index (BMI), computed using the following formula:

$$\text{BMI} = \frac{703W}{H^2}.$$

The variable W represents weight, in pounds. The variable H represents height, in inches. Use this information to solve Exercises 60–61.

60. A man is 20 years old, 72 inches (6 feet) tall, and weighs 200 pounds.

a. Compute the man's BMI. Round to the nearest tenth.

b. Use the man's age and his BMI to locate this information as a point in the coordinate system for males. Is this person overweight, borderline overweight, or normal weight?

61. A girl is 10 years old, 50 inches (4 feet, 2 inches) tall, and weighs 100 pounds.

a. Compute the girl's BMI. Round to the nearest tenth.

b. Use the girl's age and her BMI to locate this information as a point in the coordinate system for females. Is this person overweight, borderline overweight, or normal weight?

Writing in Mathematics

62. What is a half-plane?

63. What does a dashed line mean in the graph of an inequality?

64. Explain how to graph $2x - 3y < 6$.

65. Compare the graphs of $3x - 2y > 6$ and $3x - 2y \le 6$. Discuss similarities and differences between the graphs.

66. Describe how to solve a system of inequalities.

67. What does it mean if a system of linear inequalities has no solution?

Technology Exercises

Graphing utilities can be used to shade regions in the rectangular coordinate system, thereby graphing an inequality in two variables. Read the section of the user's manual for your graphing utility that describes how to shade a region. Then use your graphing utility to graph the inequalities in Exercises 68–73.

68. $y \leq 4x + 4$

69. $y \geq \dfrac{2}{3}x - 2$

70. $y \geq x^2 - 4$

71. $y \geq \dfrac{1}{2}x^2 - 2$

72. $2x + y \leq 6$

73. $3x - 2y \geq 6$

74. Does your graphing utility have any limitations in terms of graphing inequalities? If so, what are they?

75. Use a graphing utility with a $\boxed{\text{SHADE}}$ feature to verify any five of the graphs that you drew by hand in Exercises 1–22.

76. Use a graphing utility with a $\boxed{\text{SHADE}}$ feature to verify any five of the graphs that you drew by hand for the systems in Exercises 23–52.

Critical Thinking Exercises

77. Write a system of inequalities that has no solution.

78. Write a system of inequalities that describes the shaded region in the figure at the top of the next column.

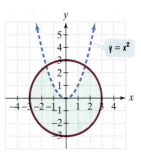

79. Sketch the graph of the solution set for the following system of inequalities:

$$y \geq nx + b \quad (n < 0, b > 0)$$
$$y \leq mx + b \quad (m > 0, b > 0).$$

80. Sketch the graph of the solution set for the following system of inequalities:

$$|x + y| \leq 3$$
$$|y| \leq 2.$$

SECTION 5.6 *Linear Programming*

Objectives

1. Write an objective function describing a quantity that must be maximized or minimized.

2. Use inequalities to describe limitations in a situation.

3. Use linear programming to solve problems.

West Berlin children at Tempelhof airport watch fleets of U.S. airplanes bringing in supplies to circumvent the Russian blockade. The airlift began June 28, 1948 and continued for 15 months.

The Berlin Airlift (1948–1949) was an operation by the United States and Great Britain in response to military action by the former Soviet Union: Soviet troops closed all roads and rail lines between West Germany and Berlin, cutting off supply routes to the city. The Allies used a mathematical technique developed during World War II to maximize the amount of supplies transported. During the 15-month airlift, 278,228 flights provided basic necessities to blockaded Berlin, saving one of the world's great cities.

In this section, we will look at an important application of systems of linear inequalities. Such systems arise in **linear programming,** a method for solving problems in which a particular quantity that must be maximized or minimized is limited by other factors. Linear programming is one of the most widely used tools in management science. It helps businesses allocate resources to manufacture products in a way that will maximize profit. Linear programming accounts for more than 50% and perhaps as much as 90% of all computing time used for management decisions in business. The Allies used linear programming to save Berlin.

1 Write an objective function describing a quantity that must be maximized or minimized.

Objective Functions in Linear Programming

Many problems involve quantities that must be maximized or minimized. Businesses are interested in maximizing profit. An operation in which bottled water and medical kits are shipped to earthquake victims needs to maximize the number of victims helped by this shipment. An **objective function** is an algebraic expression in two or more variables describing a quantity that must be maximized or minimized.

EXAMPLE 1   Writing an Objective Function

Bottled water and medical supplies are to be shipped to victims of an earthquake by plane. Each container of bottled water will serve 10 people and each medical kit will aid 6 people. Let x represent the number of bottles of water to be shipped and y the number of medical kits. Write the objective function that describes the number of people that can be helped.

Solution   Because each bottle of water serves 10 people and each medical kit aids 6 people, we have

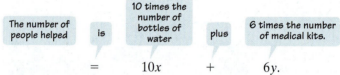

$$= \qquad 10x \qquad + \qquad 6y.$$

Using z to represent the objective function, we have

$$z = 10x + 6y.$$

Unlike the functions that we have seen so far, the objective function is an equation in three variables. For a value of x and a value of y, there is one and only one value of z. Thus, z is a function of x and y.

Check Point 1   A company manufactures bookshelves and desks for computers. Let x represent the number of bookshelves manufactured daily and y the number of desks manufactured daily. The company's profits are $25 per bookshelf and $55 per desk. Write the objective function that describes the company's total daily profit, z, from x bookshelves and y desks. (Check Points 2 through 4 are also related to this situation, so keep track of your answers.)

2 Use inequalities to describe limitations in a situation.

Constraints in Linear Programming

Ideally, the number of earthquake victims helped in Example 1 should increase without restriction so that every victim receives water and medical kits. However, the planes that ship these supplies are subject to weight and volume restrictions. In linear programming problems, such restrictions are called **constraints.** Each constraint is expressed as a linear inequality. The list of constraints forms a system of linear inequalities.

EXAMPLE 2 **Writing a Constraint**

Each plane can carry no more than 80,000 pounds. The bottled water weighs 20 pounds per container and each medical kit weighs 10 pounds. Let x represent the number of bottles of water to be shipped and y the number of medical kits. Write an inequality that describes this constraint.

Solution Because each plane can carry no more than 80,000 pounds, we have

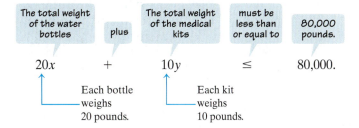

The plane's weight constraint is described by the inequality
$$20x + 10y \le 80,000.$$

Check Point 2 To maintain high quality, the company in Check Point 1 should not manufacture more than a combined total of 80 bookshelves and desks per day. Write an inequality that describes this constraint.

In addition to a weight constraint on its cargo, each plane has a limited amount of space in which to carry supplies. Example 3 demonstrates how to express this constraint.

EXAMPLE 3 **Writing a Constraint**

The total volume of supplies that a plane carries cannot exceed 6000 cubic feet. Each water bottle is 1 cubic foot and each medical kit also has a volume of 1 cubic foot. With x still representing the number of water bottles and y the number of medical kits, write an inequality that describes this second constraint.

Solution Because each plane can carry a volume of supplies that does not exceed 6000 cubic feet, we have

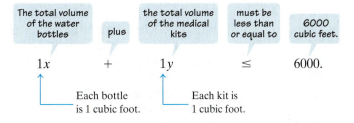

The plane's volume constraint is described by the inequality $x + y \le 6000$.

In summary, here's what we have described in this aid-to-earthquake-victims situation:

$$z = 10x + 6y$$

This is the objective function describing the number of people helped with x bottles of water and y medical kits.

$$20x + 10y \leq 80,000$$
$$x + y \leq 6000.$$

These are the constraints based on each plane's weight and volume limitations.

Check Point 3 To meet customer demand, the company in Check Point 1 must manufacture between 30 and 80 bookshelves per day, inclusive. Furthermore, the company must manufacture at least 10 and no more than 30 desks per day. Write an inequality that describes each of these sentences. Then summarize what you have described about this company by writing the objective function for its profits, and the three constraints.

3 Use linear programming to solve problems.

Solving Problems with Linear Programming

The goal in the earthquake situation described previously is to maximize the number of victims who can be helped, subject to the planes' weight and volume constraints. The process of solving this problem is called *linear programming*, based on a theorem that was proven during World War II.

> **Solving a Linear Programming Problem**
>
> Let $z = ax + by$ be an objective function that depends on x and y.
> Furthermore, z is subject to a number of constraints on x and y. If a maximum or minimum value of z exists, it can be determined as follows:
> 1. Graph the system of inequalities representing the constraints.
> 2. Find the value of the objective function at each corner, or **vertex**, of the graphed region. The maximum and minimum of the objective function occur at one or more of the corner points.

EXAMPLE 4   Solving a Linear Programming Problem

Determine how many bottles of water and how many medical kits should be sent on each plane to maximize the number of earthquake victims who can be helped.

Solution   We must maximize $z = 10x + 6y$ subject to the constraints:

$$20x + 10y \leq 80,000$$
$$x + y \leq 6000.$$

Step 1   Graph the system of inequalities representing the constraints. Because x (the number of bottles of water per plane) and y (the number of medical kits per plane) must be nonnegative, we need to graph the system of inequalities in quadrant I and its boundary only ($x \geq 0$ and $y \geq 0$). To graph the inequality $20x + 10y \leq 80,000$, we graph the equation $20x + 10y = 80,000$ as a solid blue line (Figure 5.33). Setting $y = 0$, the x-intercept is 4000 and setting $x = 0$, the y-intercept is 8000. Using $(0, 0)$ as a test point, the inequality is satisfied, so we shade below the blue line, as shown in yellow in Figure 5.33. Now we graph $x + y \leq 6000$ by first graphing $x + y = 6000$ as a solid red line. Setting $y = 0$, the x-intercept is 6000. Setting $x = 0$, the y-intercept is 6000. Using $(0, 0)$ as a test point, the inequality is satisfied, so we shade below the red line, as shown using green vertical shading in Figure 5.33.

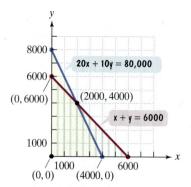

Figure 5.33 The region in quadrant I representing the constraints

$$20x + 10y \leq 80,000$$
$$x + y \leq 6000$$

We use the addition method to find the coordinates of the point where the lines $20x + 10y = 80{,}000$ and $x + y = 6000$ intersect.

$$\begin{aligned} 20x + 10y &= 80{,}000 \xrightarrow{\text{No change}} & 20x + 10y &= 80{,}000 \\ x + y &= 6000 \xrightarrow{\text{Multiply by } -10.} & -10x - 10y &= -60{,}000 \\ & \text{Add:} & 10x \quad &= 20{,}000 \\ & & x &= 2000 \end{aligned}$$

Figure 5.34

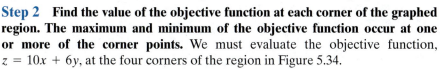

Back-substituting 2000 for x in $x + y = 6000$, we find $y = 4000$, so the intersection point is (2000, 4000).

The system of inequalities representing the constraints is shown by the region in which the yellow shading and the green vertical shading overlap in Figure 5.33. The graph of the system of inequalities is shown again in Figure 5.34. The red and blue line segments are included in the graph.

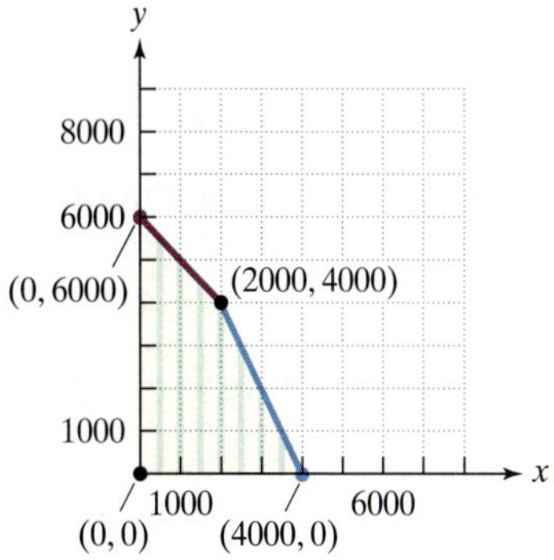

Step 2 Find the value of the objective function at each corner of the graphed region. The maximum and minimum of the objective function occur at one or more of the corner points. We must evaluate the objective function, $z = 10x + 6y$, at the four corners of the region in Figure 5.34.

Corner (x, y)	Objective Function $z = 10x + 6y$
(0, 0)	$z = 10(0) + 6(0) = 0$
(4000, 0)	$z = 10(4000) + 6(0) = 40{,}000$
(2000, 4000)	$z = 10(2000) + 6(4000) = 44{,}000$ ← maximum
(0, 6000)	$z = 10(0) + 6(6000) = 36{,}000$

Thus, the maximum value of z is 44,000 and this occurs when $x = 2000$ and $y = 4000$. In practical terms, this means that the maximum number of earthquake victims who can be helped with each plane shipment is 44,000. This can be accomplished by sending 2000 water bottles and 4000 medical kits per plane.

Check Point 4 For the company in Check Points 1–3, how many bookshelves and how many desks should be manufactured per day to obtain a maximum profit? What is the maximum daily profit?

EXAMPLE 5 Solving a Linear Programming Problem

Find the maximum value of the objective function

$$z = 2x + y$$

subject to the constraints:

$$\begin{aligned} x \ge 0,\ & y \ge 0 \\ x + 2&y \le 5 \\ x - \ &y \le 2. \end{aligned}$$

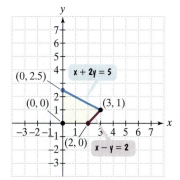

Figure 5.35 The graph of $x + 2y \le 5$ and $x - y \le 2$ in quadrant I

Solution We begin by graphing the region in quadrant I ($x \ge 0$, $y \ge 0$) formed by the constraints. The graph is shown by the closed yellow region in Figure 5.35.

Now we evaluate the objective function at the four vertices of this region.

Objective function: $z = 2x + y$

At $(0, 0)$: $\quad z = 2 \cdot 0 + 0 = 0$

At $(2, 0)$: $\quad z = 2 \cdot 2 + 0 = 4$

At $(3, 1)$: $\quad z = 2 \cdot 3 + 1 = 7$ ⟵ Maximum value of z

At $(0, 2.5)$: $\quad z = 2 \cdot 0 + 2.5 = 2.5$

Thus, the maximum value of z is 7, and this occurs when $x = 3$ and $y = 1$.

We can see why the objective function in Example 5 has a maximum value that occurs at a vertex by solving the equation for y.

$z = 2x + y$ — This is the objective function of Example 5.

$y = -2x + z$ — Solve for y. Recall that the slope-intercept form of a line is $y = mx + b$.

Slope = −2   y-intercept = z

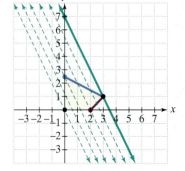

Figure 5.36 The line with slope −2 with the greatest y-intercept that intersects the shaded region passes through one of its vertices.

In this form, z represents the y-intercept of the objective function. The equation describes infinitely many parallel lines, each with a slope of −2. The process in linear programming involves finding the maximum z-value for all lines that intersect the region determined by the constraints. Of all the lines whose slope is −2, we're looking for the one with the greatest y-intercept that intersects the given region. As we see in Figure 5.36, such a line will pass through one (or possibly more) of the vertices of the region.

Check Point 5 Find the maximum value of the objective function $z = 3x + 5y$ subject to the constraints $x \geq 0, y \geq 0, x + y \geq 1, x + y \leq 6$.

Faster and Faster

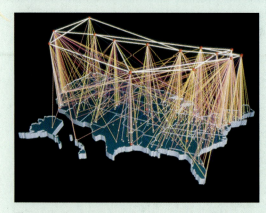

The network of computer linkages in the United States is growing exponentially.

The problems we solve nowadays have thousands of equations, sometimes a million variables. One of the things that still amazes me is to see a program run on the computer—and to see the answer come out. If we think of the number of combinations of different solutions that we're trying to choose the best of, it's akin to the stars in the heavens. Yet we solve them in a matter of moments. This, to me, is staggering. Not that we can solve them—but that we can solve them so rapidly and efficiently.

—George Dantzig
Inventor of the simplex method, a linear programming method

Problems in linear programming can involve objective functions with thousands of variables subject to thousands of constraints. Several nongeometric linear programming methods are available on software for solving such problems. And we continue to search for faster and faster linear programming methods. This area of applied mathematics has a direct impact on the efficiency and profitability of numerous industries, including telephone and computer communications, and the airlines.

EXERCISE SET 5.6

Practice Exercises

In Exercises 1–4, find the value of the objective function at each corner of the graphed region. What is the maximum value of the objective function? What is the minimum value of the objective function?

1. Objective Function $z = 5x + 6y$

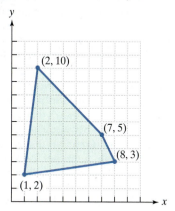

2. Objective Function $z = 3x + 2y$

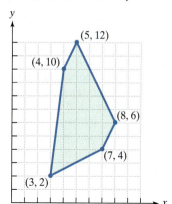

3. Objective Function $z = 40x + 50y$

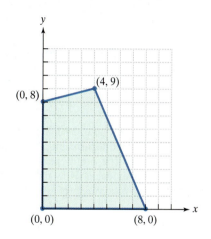

4. Objective Function $z = 30x + 45y$

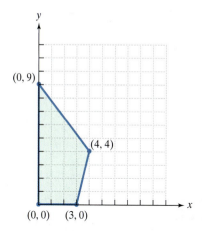

In Exercises 5–14, an objective function and a system of linear inequalities representing constraints are given.

a. *Graph the system of inequalities representing the constraints.*

b. *Find the value of the objective function at each corner of the graphed region.*

c. *Use the values in part (b) to determine the maximum value of the objective function and the values of x and y for which the maximum occurs.*

5. Objective Function $z = 3x + 2y$
   Constraints $x \geq 0, y \geq 0$
   $$2x + y \leq 8$$
   $$x + y \geq 4$$

6. Objective Function $z = 2x + 3y$
   Constraints $x \geq 0, y \geq 0$
   $$2x + y \leq 8$$
   $$2x + 3y \leq 12$$

7. Objective Function $z = 4x + y$
   Constraints $x \geq 0, y \geq 0$
   $$2x + 3y \leq 12$$
   $$x + y \geq 3$$

8. Objective Function $z = x + 6y$
   Constraints $x \geq 0, y \geq 0$
   $$2x + y \leq 10$$
   $$x - 2y \geq -10$$

9. Objective Function $z = 3x - 2y$
   Constraints $1 \leq x \leq 5$
   $$y \geq 2$$
   $$x - y \geq -3$$

10. Objective Function $z = 5x - 2y$
Constraints $0 \le x \le 5$
$0 \le y \le 3$
$x + y \ge 2$

11. Objective Function $z = 4x + 2y$
Constraints $x \ge 0, y \ge 0$
$2x + 3y \le 12$
$3x + 2y \le 12$
$x + y \ge 2$

12. Objective Function $z = 2x + 4y$
Constraints $x \ge 0, y \ge 0$
$x + 3y \ge 6$
$x + y \ge 3$
$x + y \le 9$

13. Objective Function $z = 10x + 12y$
Constraints $x \ge 0, y \ge 0$
$x + y \le 7$
$2x + y \le 10$
$2x + 3y \le 18$

14. Objective Function $z = 5x + 6y$
Constraints $x \ge 0, y \ge 0$
$2x + y \ge 10$
$x + 2y \ge 10$
$x + y \le 10$

Application Exercises

15. A television manufacturer makes console and wide-screen televisions. The profit per unit is $125 for the console televisions and $200 for the wide-screen televisions.

a. Let x = the number of consoles manufactured in a month and y = the number of wide-screens manufactured in a month. Write the objective function that describes the total monthly profit.

b. The manufacturer is bound by the following constraints:
- Equipment in the factory allows for making at most 450 console televisions in one month.
- Equipment in the factory allows for making at most 200 wide-screen televisions in one month.
- The cost to the manufacturer per unit is $600 for the console televisions and $900 for the wide-screen televisions. Total monthly costs cannot exceed $360,000.

Write a system of three inequalities that describes these constraints.

c. Graph the system of inequalities in part (b). Use only the first quadrant and its boundary, because x and y must both be nonnegative.

d. Evaluate the objective function for total monthly profit at each of the five vertices of the graphed region. [The vertices should occur at $(0, 0)$, $(0, 200)$, $(300, 200)$, $(450, 100)$, and $(450, 0)$.]

e. Complete the missing portions of this statement: The television manufacturer will make the greatest profit by manufacturing ___ console televisions each month and ___ wide-screen televisions each month. The maximum monthly profit is $ ___.

16. a. A student earns $10 per hour for tutoring and $7 per hour as a teacher's aid. Let x = the number of hours each week spent tutoring, and y = the number of hours each week spent as a teacher's aid. Write the objective function that describes total weekly earnings.

b. The student is bound by the following constraints:
- To have enough time for studies, the student can work no more than 20 hours per week.
- The tutoring center requires that each tutor spend at least three hours per week tutoring.
- The tutoring center requires that each tutor spend no more than eight hours per week tutoring.

Write a system of three inequalities that describes these constraints.

c. Graph the system of inequalities in part (b). Use only the first quadrant and its boundary, because x and y are nonnegative.

d. Evaluate the objective function for total weekly earnings at each of the four vertices of the graphed region. [The vertices should occur at $(3, 0)$, $(8, 0)$, $(3, 17)$, and $(8, 12)$.]

e. Complete the missing portions of this statement: The student can earn the maximum amount per week by tutoring for ___ hours per week and working as a teacher's aid for ___ hours per week. The maximum amount that the student can earn each week is $ ___.

Use the two steps for solving a linear programming problem, given in the box on page 496, to solve the problems in Exercises 17–23.

17. A manufacturer produces two models of mountain bicycles. The times (in hours) required for assembling and painting each model are given in the following table:

	Model A	Model B
Assembling	5	4
Painting	2	3

The maximum total weekly hours available in the assembly department and the paint department are 200 hours and 108 hours, respectively. The profits per unit are $25 for model A and $15 for model B. How many of each type should be produced to maximize profit?

18. A large institution is preparing lunch menus containing foods A and B. The specifications for the two foods are given in the following table:

Food	Units of Fat per Ounce	Units of Carbohydrates per Ounce	Units of Protein per Ounce
A	1	2	1
B	1	1	1

Each lunch must provide at least 6 units of fat per serving, no more than 7 units of protein, and at least 10 units of carbohydrates. The institution can purchase food A for $0.12 per ounce and food B for $0.08 per ounce. How many ounces of each food should a serving contain to meet the dietary requirements at the least cost?

19. Food and clothing are shipped to victims of a natural disaster. Each carton of food will feed 12 people, while each carton of clothing will help 5 people. Each 20-cubic-foot box of food weighs 50 pounds and each 10-cubic-foot box of clothing weighs 20 pounds. The commercial carriers transporting food and clothing are bound by the following constraints:

- The total weight per carrier cannot exceed 19,000 pounds.
- The total volume must be less than 8000 cubic feet.

How many cartons of food and clothing should be sent with each plane shipment to maximize the number of people who can be helped?

20. On June 24, 1948, the former Soviet Union blocked all land and water routes through East Germany to Berlin. A gigantic airlift was organized using American and British planes to supply food, clothing, and other supplies to the more than 2 million people in West Berlin. The cargo capacity was 30,000 cubic feet for an American plane and 20,000 cubic feet for a British plane. To break the Soviet blockade, the Western Allies had to maximize cargo capacity, but were subject to the following restrictions:

- No more than 44 planes could be used.
- The larger American planes required 16 personnel per flight, double that of the requirement for the British planes. The total number of personnel available could not exceed 512.
- The cost of an American flight was $9000 and the cost of a British flight was $5000. Total weekly costs could not exceed $300,000.

Find the number of American and British planes that were used to maximize cargo capacity.

21. A theater is presenting a program on drinking and driving for students and their parents. The proceeds will be donated to a local alcohol information center. Admission is $2.00 for parents and $1.00 for students. However, the situation has two constraints: The theater can hold no more than 150 people and every two parents must bring at least one student. How many parents and students should attend to raise the maximum amount of money?

22. You are about to take a test that contains computation problems worth 6 points each and word problems worth 10 points each. You can do a computation problem in 2 minutes and a word problem in 4 minutes. You have 40 minutes to take the test and may answer no more than 12 problems. Assuming you answer all the problems attempted correctly, how many of each type of problem must you do to maximize your score? What is the maximum score?

23. In 1978, a ruling by the Civil Aeronautics Board allowed Federal Express to purchase larger aircraft. Federal Express's options included 20 Boeing 727s that United Airlines was retiring and/or the French-built Dassault Fanjet Falcon 20. To aid in their decision, executives at Federal Express analyzed the following data:

	Boeing 727	Falcon 20
Direct Operating Cost	$1400 per hour	$500 per hour
Payload	42,000 pounds	6000 pounds

Federal Express was faced with the following constraints:

- Hourly operating cost was limited to $35,000.
- Total payload had to be at least 672,000 pounds.
- Only twenty 727s were available.

Given the constraints, how many of each kind of aircraft should Federal Express have purchased to maximize the number of aircraft?

Writing in Mathematics

24. What kinds of problems are solved using the linear programming method?

25. What is an objective function in a linear programming problem?

26. What is a constraint in a linear programming problem? How is a constraint represented?

27. In your own words, describe how to solve a linear programming problem.

28. Describe a situation in your life in which you would really like to maximize something, but you are limited by at least two constraints. Can linear programming be used in this situation? Explain your answer.

Technology Exercises

In Exercises 29–32, use a graphing utility to sketch the region determined by the constraints. Then determine the maximum value of the objective function subject to the contraints.

29. Objective Function $z = 6x + 8y$
     Constraints $x \geq 0, y \geq 0$
                     $x + 2y \leq 6$

30. Objective Function $z = 30x + 20y$
     Constraints $x \geq 0, y \geq 0$
                     $2x + y \leq 14$
                     $3x + y \leq 18$

31. Objective Function $z = 9x + 14y$
     Constraints $x \geq 0, y \geq 0$
                     $2x + y \leq 10$
                     $2x + 3y \leq 18$

32. Objective Function $z = 10x + 3y$
     Constraints $0 \leq x \leq 10, \quad y \geq 0$
                     $4x + 5y \leq 60$
                     $4x - 5y \geq -20$

Critical Thinking Exercises

33. Suppose that you inherit $10,000. The will states how you must invest the money. Some (or all) of the money must be invested in stocks and bonds. The requirements are that at least $3000 be invested in bonds, with expected returns of $0.08 per dollar, and at least $2000 be invested in stocks, with expected returns of $0.12 per dollar. Because the stocks are medium risk, the final stipulation requires that the investment in bonds should never be less than the investment in stocks. How should the money be invested so as to maximize your expected returns?

34. Consider the objective function $z = Ax + By$ ($A > 0$ and $B > 0$) subject to the following constraints: $2x + 3y \leq 9$, $x - y \leq 2$, $x \geq 0$, and $y \geq 0$. Prove that the objective function will have the same maximum value at the vertices $(3, 1)$ and $(0, 3)$ if $A = \frac{2}{3}B$.

Group Exercises

35. Group members should choose a particular field of interest. Research how linear programming is used to solve problems in that field. If possible, investigate the solution of a specific practical problem. Present a report on your findings, including the contributions of George Dantzig, Narendra Karmarkar, and L.G. Khachion to linear programming.

36. Members of the group should interview a business executive who is in charge of deciding the product mix for a business. How are production policy decisions made? Are other methods used in conjunction with linear programming? What are these methods? What sort of academic background, particularly in mathematics, does this executive have? Present a group report addressing these questions, emphasizing the role of linear programming for the business.

CHAPTER SUMMARY, REVIEW, AND TEST

Summary

DEFINITIONS AND CONCEPTS	EXAMPLES
5.1 Systems of Linear Equations in Two Variables	
a. Two equations in the form $Ax + By = C$ are called a system of linear equations. A solution to the system is an ordered pair that satisfies both equations in the system.	Ex. 1, p. 440
b. Systems of linear equations in two variables can be solved by eliminating a variable, using the substitution method (see the box on page 442) or the addition method (see the box on page 444).	Ex. 2, p. 441; Ex. 3, p. 442; Ex. 4, p. 444; Ex. 5, p. 445
c. Some linear systems have no solution and are called inconsistent systems; others have infinitely many solutions. The equations in a linear system with infinitely many solutions are called dependent. For details, see the box on page 446.	Ex. 6, p. 446; Ex. 7, p. 447
5.2 Systems of Linear Equations in Three Variables	
a. Three equations in the form $Ax + By + Cz = D$ are called a system of linear equations in three variables. A solution to the system is an ordered triple that satisfies all three equations in the system.	Ex. 1, p. 457
b. A system of linear equations in three variables can be solved by eliminating variables. Use the addition method to eliminate any variable, reducing the system to two equations in two variables.	Ex. 2, p. 458; Ex. 3, p. 459

DEFINITIONS AND CONCEPTS	EXAMPLES
Use substitution or the addition method to solve the resulting system in two variables. Details are found in the box on page 457.	

5.3 Partial Fraction Decomposition

a. Partial fraction decomposition is used on rational expressions in which the numerator and denominator have no common factors and the highest power in the numerator is less than the highest power in the denominator. The steps in partial fraction decomposition are given in the box on page 468.

b. Include one partial fraction with a constant numerator for each distinct linear factor in the denominator. Include one partial fraction with a constant numerator for each power of a repeated linear factor in the denominator.
Ex. 1, p. 467; Ex. 2, p. 468

c. Include one partial fraction with a linear numerator for each distinct prime quadratic factor in the denominator. Include one partial fraction with a linear numerator for each power of a prime, repeated quadratic factor in the denominator.
Ex. 3, p. 470; Ex. 4, p. 471

5.4 Systems of Nonlinear Equations in Two Variables

a. A system of two nonlinear equations in two variables contains at least one equation that cannot be expressed as $Ax + By = C$.

b. Systems of nonlinear equations in two variables can be solved algebraically by eliminating all occurrences of one of the variables by the substitution or addition methods.
Ex. 1, p. 475; Ex. 2, p. 476; Ex. 3, p. 477; Ex. 4, p. 478

5.5 Systems of Inequalities

a. A linear inequality in two variables can be written in the form $Ax + By > C$, $Ax + By \geq C$, $Ax + By < C$, or $Ax + By \leq C$.

b. The procedure for graphing a linear inequality in two variables is given in the box on page 484. A nonlinear inequality in two variables is graphed using the same procedure.
Ex. 1, p. 484; Ex. 2, p. 485; Ex. 3, p. 487

c. To graph the solution set to a system of inequalities, graph each inequality in the system in the same rectangular coordinate system. Then find the region, if there is one, that is common to every graph in the system.
Ex. 4, p. 487; Ex. 5, p. 488; Ex. 6, p. 489

5.6 Linear Programming

a. An objective function is an algebraic expression in three variables describing a quantity that must be maximized or minimized.
Ex. 1, p. 494

b. Constraints are restrictions, expressed as linear inequalities.
Ex. 2, p. 495; Ex. 3, p. 495

c. Steps for solving a linear programming problem are given in the box on page 496.
Ex. 4, p. 496; Ex. 5, p. 497

Review Exercises

5.1

In Exercises 1–5, solve by the method of your choice. Identify systems with no solution and systems with infinitely many solutions, using set notation to express their solution sets.

1. $y = 4x + 1$
$3x + 2y = 13$

2. $x + 4y = 14$
$2x - y = 1$

3. $5x + 3y = 1$
$3x + 4y = -6$

4. $2y - 6x = 7$
$3x - y = 9$

5. $4x - 8y = 16$
$3x - 6y = 12$

6. A company is planning to manufacture computer desks. The fixed cost will be $60,000 and it will cost $200 to produce each desk. Each desk will be sold for $450.
   a. Write the cost function, C, of producing x desks.
   b. Write the revenue function, R, from the sale of x desks.
   c. Determine the break-even point. Describe what this means.

7. The weekly demand and supply models for the video *Pearl Harbor* at a chain of stores that sells videos are given by the demand model $N = -60p + 1000$ and the supply model $N = 4p + 200$, in which p is the price of the video and N is the number of videos sold or supplied each week to the chain of stores. Find the price at which supply and demand are equal. At this price, how many copies of *Pearl Harbor* can be supplied and sold each week?

8. The graph makes Super Bowl Sunday look like a day of snack food binging in the United States. The number of pounds of guacamole consumed is ten times the difference between the number of pounds of potato and tortilla chips eaten on the same day. On Super Bowl Sunday, Americans also eat a total quantity of potato and tortilla chips that exceeds popcorn consumption by 7.3 million pounds. How many millions of pounds of potato chips and tortilla chips are consumed on Super Bowl Sunday?

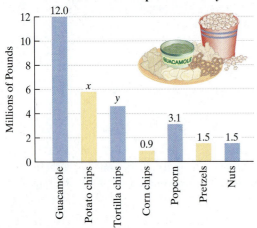

Millions of Pounds of Snack Food Consumed on Super Bowl Sunday

Source: Association of American Snack Foods

9. A travel agent offers two package vacation plans. The first plan costs $360 and includes 3 days at a hotel and a rental car for 2 days. The second plan costs $500 and includes 4 days at a hotel and a rental car for 3 days. The daily charge for the hotel is the same under each plan, as is the daily charge for the car. Find the cost per day for the hotel and for the car.

10. The calorie-nutrient information for an apple and an avocado is given in the table. How many of each should be eaten to get exactly 1000 calories and 100 grams of carbohydrates?

	One Apple	One Avocado
Calories	100	350
Carbohydrates (grams)	24	14

5.2

Solve each system in Exercises 11–12.

11. $2x - y + z = 1$
$3x - 3y + 4z = 5$
$4x - 2y + 3z = 4$

12. $x + 2y - z = 5$
$2x - y + 3z = 0$
$2y + z = 1$

13. Find the quadratic function $y = ax^2 + bx + c$ whose graph passes through the points $(1, 4)$, $(3, 20)$, and $(-2, 25)$.

14. The bar graph shows that the U.S. divorce rate increased between 1970 and 1985 and then decreased between 1985 and 1999.

U.S. Divorce Rates: Number of Divorces per 1000 People

Source: U.S. Census Bureau

a. Write the data for 1970, 1985, and 1999 as ordered pairs (x, y), where x is the number of years after 1970 and y is that year's divorce rate.

b. The three data points in part (a) can be modeled by the quadratic function $y = ax^2 + bx + c$. Write a system of linear equations in three variables that can be used to find values for a, b, and c. It is not necessary to solve the system.

15. The bar graph indicates countries in which ten or more languages have become extinct. The number of extinct languages in the United States, Colombia, and India combined is 50. The number of extinct languages in the United States exceeds the number in Colombia by 4 and is 2 more than twice that for India. How many languages have become extinct in the United States, Colombia, and India?

Countries Where Ten or More Languages Have Become Extinct (Number of Languages)

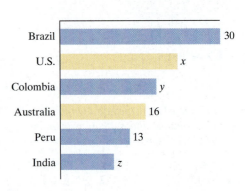

Source: Grimes

5.3

In Exercises 16–24, write the partial fraction decomposition of each rational expression.

16. $\dfrac{x}{(x-3)(x+2)}$

17. $\dfrac{11x-2}{x^2-x-12}$

18. $\dfrac{4x^2-3x-4}{x(x+2)(x-1)}$

19. $\dfrac{2x+1}{(x-2)^2}$

20. $\dfrac{2x-6}{(x-1)(x-2)^2}$

21. $\dfrac{3x}{(x-2)(x^2+1)}$

22. $\dfrac{7x^2-7x+23}{(x-3)(x^2+4)}$

23. $\dfrac{x^3}{(x^2+4)^2}$

24. $\dfrac{4x^3+5x^2+7x-1}{(x^2+x+1)^2}$

5.4

In Exercises 25–35, solve each system by the method of your choice.

25. $5y = x^2 - 1$
$x - y = 1$

26. $y = x^2 + 2x + 1$
$x + y = 1$

27. $x^2 + y^2 = 2$
$x + y = 0$

28. $2x^2 + y^2 = 24$
$x^2 + y^2 = 15$

29. $xy - 4 = 0$
$y - x = 0$

30. $y^2 = 4x$
$x - 2y + 3 = 0$

31. $x^2 + y^2 = 10$
$y = x + 2$

32. $xy = 1$
$y = 2x + 1$

33. $x + y + 1 = 0$
$x^2 + y^2 + 6y - x = -5$

34. $x^2 + y^2 = 13$
$x^2 - y = 7$

35. $2x^2 + 3y^2 = 21$
$3x^2 - 4y^2 = 23$

36. The perimeter of a rectangle is 26 meters, and its area is 40 square meters. Find its dimensions.

37. Find the coordinates of all points (x, y) that lie on the line whose equation is $2x + y = 8$, so that the area of the rectangle shown in the figure is 6 square units.

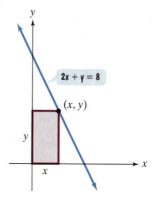

38. Two adjoining square fields with an area of 2900 square feet are to be enclosed with 240 feet of fencing. The situation is represented in the figure. Find the length of each side where a variable appears.

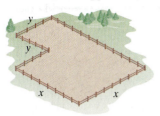

5.5

In Exercises 39–45, graph each inequality.

39. $3x - 4y > 12$

40. $y \le -\dfrac{1}{2}x + 2$

41. $x < -2$

42. $y \ge 3$

43. $x^2 + y^2 > 4$

44. $y \le x^2 - 1$

45. $y \le 2^x$

In Exercises 46–55, graph the solution set of each system of inequalities or indicate that the system has no solution.

46. $3x + 2y \ge 6$
$2x + y \ge 6$

47. $2x - y \ge 4$
$x + 2y < 2$

48. $y < x$
$y \le 2$

49. $x + y \le 6$
$y \ge 2x - 3$

50. $0 \le x \le 3$
$y > 2$

51. $2x + y < 4$
$2x + y > 6$

52. $x^2 + y^2 \le 16$
$x + y < 2$

53. $x^2 + y^2 \le 9$
$y < -3x + 1$

54. $y > x^2$
$x + y < 6$
$y < x + 6$

55. $y \ge 0$
$3x + 2y \ge 4$
$x - y \le 3$

5.6

56. Find the value of the objective function $z = 2x + 3y$ at each corner of the graphed region shown. What is the maximum value of the objective function? What is the minimum value of the objective function?

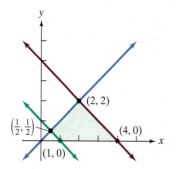

In Exercises 57–59, graph the region determined by the constraints. Then find the maximum value of the given objective function, subject to the constraints.

57. Objective Function $\quad z = 2x + 3y$
Constraints $\qquad\qquad x \geq 0, y \geq 0$
$\qquad\qquad\qquad\qquad x + y \leq 8$
$\qquad\qquad\qquad\qquad 3x + 2y \geq 6$

58. Objective Function $\quad z = x + 4y$
Contraints $\qquad\qquad 0 \leq x \leq 5, 0 \leq y \leq 7$
$\qquad\qquad\qquad\qquad x + y \geq 3$

59. Objective Function $\quad z = 5x + 6y$
Constraints $\qquad\qquad x \geq 0, y \geq 0$
$\qquad\qquad\qquad\qquad y \leq x$
$\qquad\qquad\qquad\qquad 2x + y \leq 12$
$\qquad\qquad\qquad\qquad 2x + 3y \geq 6$

60. A paper manufacturing company converts wood pulp to writing paper and newsprint. The profit on a unit of writing paper is $500 and the profit on a unit of newsprint is $350.

a. Let x represent the number of units of writing paper produced daily. Let y represent the number of units of newsprint produced daily. Write the objective function that models total daily profit.

b. The manufacturer is bound by the following constraints:
- Equipment in the factory allows for making at most 200 units of paper (writing paper and newsprint) in a day.
- Regular customers require at least 10 units of writing paper and at least 80 units of newsprint daily.

Write a system of inequalities that models these constraints.

c. Graph the inequalities in part (b). Use only the first quadrant, because x and y must both be positive. (*Suggestion:* Let each unit along the x- and y-axes represent 20.)

d. Evaluate the objective profit function at each of the three vertices of the graphed region.

e. Complete the missing portions of this statement: The company will make the greatest profit by producing ___ units of writing paper and ___ units of newsprint each day. The maximum daily profit is $ ___.

61. A manufacturer of lightweight tents makes two models whose specifications are given in the following table:

	Cutting Time per Tent	Assembly Time per Tent
Model A	0.9 hour	0.8 hour
Model B	1.8 hours	1.2 hours

On a monthly basis, the manufacturer has no more than 864 hours of labor available in the cutting department and at most 672 hours in the assembly division. The profits come to $25 per tent for model A and $40 per tent for model B. How many of each should be manufactured monthly to maximize the profit?

Chapter 5 Test

In Exercises 1–5, solve the system.

1. $x = y + 4$
$3x + 7y = -18$

2. $2x + 5y = -2$
$3x - 4y = 20$

3. $x + y + z = 6$
$3x + 4y - 7z = 1$
$2x - y + 3z = 5$

4. $x^2 + y^2 = 25$
$x + y = 1$

5. $2x^2 - 5y^2 = -2$
$3x^2 + 2y^2 = 35$

6. Find the partial fraction decomposition for
$$\frac{x}{(x + 1)(x^2 + 9)}.$$

In Exercises 7–10, graph the solution set of each inequality or system of inequalities.

7. $x - 2y < 8$

8. $x \geq 0, y \geq 0$
$3x + y \leq 9$
$2x + 3y \geq 6$

9. $x^2 + y^2 > 1$
$x^2 + y^2 < 4$

10. $y \leq 1 - x^2$
$x^2 + y^2 \leq 9$

11. Find the maximum value of the objective function $z = 3x + 5y$ subject to the following constraints: $x \geq 0$, $y \geq 0$, $x + y \leq 6$, $x \geq 2$.

12. Health experts agree that cholesterol intake should be limited to 300 mg or less each day. Three ounces of shrimp and 2 ounces of scallops contain 156 mg of cholesterol. Five ounces of shrimp and 3 ounces of scallops contain 45 mg of cholesterol less than the suggested maximum daily intake. Determine the cholesterol content in an ounce of each item.

13. A company is planning to produce and sell a new line of computers. The fixed cost will be $360,000, and it will cost $850 to produce each computer. Each computer will be sold for $1150.

a. Write the cost function, C, of producing x computers.

b. Write the revenue function, R, from the sale of x computers.

c. Determine the break-even point. Describe what this means.

14. Find the quadratic function whose graph passes through the points $(-1, -2)$, $(2, 1)$, and $(-2, 1)$.

15. The rectangular plot of land shown in the figure is to be fenced along three sides using 39 feet of fencing. No fencing is to be placed along the river's edge. The area of the plot is 180 square feet. What are its dimensions?

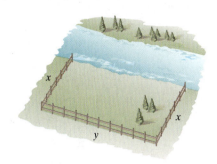

16. A manufacturer makes two types of jet skis, regular and deluxe. The profit on a regular jet ski is $200 and the profit on the deluxe model is $250. To meet customer demand, the company must manufacture at least 50 regular jet skis per week and at least 75 deluxe models. To maintain high quality, the total number of both models of jet skis manufactured by the company should not exceed 150 per week. How many jet skis of each type should be manufactured per week to obtain maximum profit? What is the maximum weekly profit?

Cumulative Review Exercises (Chapters 1–5)

Solve each equation or inequality in Exercises 1–8.

1. $\sqrt{x^2 - 3x} = 2x - 6$

2. $4x^2 = 8x - 7$

3. $\left| \dfrac{x}{3} + 2 \right| < 4$

4. $\dfrac{x + 5}{x - 1} > 2$

5. $2x^3 + x^2 - 13x + 6 = 0$

6. $6x - 3(5x + 2) = 4(1 - x)$

7. $\log (x + 3) + \log x = 1$

8. $3^{x+2} = 11$

In Exercises 9–12, graph each equation, function, or inequality in the rectangular coordinate system.

9. $f(x) = (x + 2)^2 - 4$

10. $2x - 3y \le 6$

11. $y = 3^{x-2}$

12. $f(x) = \dfrac{x^2 - x - 6}{x + 1}$

13. Expand and simplify: $\log_2 (8x^5)$.

14. What interest rate is required for an investment of $6000 subject to continuous compounding to grow to $18,000 in 10 years?

15. If $f(x) = 7x - 3$, find $f^{-1}(x)$.

16. If $f(x) = 7x - 3$ and $g(x) = 3x - 7$, find $g(f(x))$.

17. Explain why $x^2 + y^2 = 4$ does not represent y as a function of x.

18. Solve the system:

$$3x - y = -2$$
$$2x^2 - y = 0.$$

19. The length of a rectangle is 1 meter more than twice the width. If the rectangle's area is 36 square meters, find its dimensions.

20. The function $f(x) = 0.1x^2 - 3x + 22$ describes the distance, $f(x)$, in feet, needed for an airplane to land when its initial landing speed is x feet per second. Find and interpret $f(90)$. Will there be a problem if 550 feet of runway is available? Explain.

Appendix
Where Did That Come From? Selected Proofs

SECTION 4.3   *Properties of Logarithms*

The Product Rule

Let b, M, and N be positive real numbers with $b \neq 1$.

$$\log_b(MN) = \log_b M + \log_b N$$

Proof

We begin by letting $\log_b M = R$ and $\log_b N = S$.

Now we write each logarithm in exponential form.

$$\log_b M = R \quad \text{means} \quad b^R = M.$$
$$\log_b N = S \quad \text{means} \quad b^S = N.$$

By substituting and using a property of exponents, we see that

$$MN = b^R b^S = b^{R+S}.$$

Now we change $MN = b^{R+S}$ to logarithmic form.

$$MN = b^{R+S} \quad \text{means} \quad \log_b(MN) = R + S.$$

Finally, substituting $\log_b M$ for R and $\log_b N$ for S gives us

$$\log_b(MN) = \log_b M + \log_b N,$$

the property that we wanted to prove.

The quotient and power rules for logarithms are proved using similar procedures.

The Change-of-Base Property

For any logarithmic bases a and b, and any positive number M,

$$\log_b M = \frac{\log_a M}{\log_a b}.$$

Proof

To prove the change-of-base property, we let x equal the logarithm on the left side:

$$\log_b M = x.$$

Now we rewrite this logarithm in exponential form.

$$\log_b M = x \quad \text{means} \quad b^x = M.$$

Because b^x and M are equal, the logarithms with base a for each of these expressions must be equal. This means that

$$\log_a b^x = \log_a M$$

$$x \log_a b = \log_a M \qquad \text{Apply the power rule for logarithms on the left side.}$$

$$x = \frac{\log_a M}{\log_a b} \qquad \text{Solve for x by dividing both sides by } \log_a b.$$

In our first step we let x equal $\log_b M$. Replacing x on the left side by $\log_b M$ gives us

$$\log_b M = \frac{\log_a M}{\log_a b},$$

which is the change-of-base property.

Answers to Selected Exercises

CHAPTER P

Section P.1

Check Point Exercises

1. a. $\sqrt{2}-1$ **b.** $\pi-3$ **c.** 1 **2.** 9 **3.** 250; In 1990, the population of the United States was 250 million. **4.** $38x-19y$

Exercise Set P.1

1. a. $\sqrt{100}$ **b.** $0, \sqrt{100}$ **c.** $-9, 0, \sqrt{100}$ **d.** $-9, -\frac{4}{5}, 0, 0.25, 9.2, \sqrt{100}$ **e.** $\sqrt{3}$ **3. a.** $\sqrt{64}$ **b.** $0, \sqrt{64}$ **c.** $-11, 0, \sqrt{64}$

d. $-11, -\frac{5}{6}, 0, 0.75, \sqrt{64}$ **e.** $\sqrt{5}, \pi$ **5.** 0 **7.** Answers may vary. **9.** true **11.** true **13.** true **15.** 300 **17.** $12-\pi$

19. $5-\sqrt{2}$ **21.** -1 **23.** 4 **25.** 3 **27.** 7 **29.** -1 **31.** $|17-2|; 15$ **33.** $|5-(-2)|; 7$ **35.** $|-4-(-19)|; 15$
37. $|-1.4-(-3.6)|; 2.2$ **39.** 27 **41.** -19 **43.** 25 **45.** 10 **47.** -8 **49.** commutative property of addition
51. associative property of addition **53.** commutative property of addition **55.** distributive property of multiplication over addition
57. inverse property of multiplication **59.** $15x+16$ **61.** $27x-10$ **63.** $29y-29$ **65.** $8y-12$ **67.** $16y-25$
69. $14x$ **71.** $-2x+3y+6$ **73.** x **75.** yes **77.** Answers may vary. **79.** 21; In 2000, approximately 21% of American adults
smoked cigarettes. **81. a.** $132-0.6a$ **b.** 120 **89.** (c) is true. **91.** $<$ **93.** $>$

Section P.2

Check Point Exercises

1. -256 **2. a.** $\frac{1}{8}$ **b.** 36 **3. a.** 243 **b.** $\frac{1}{8}$ **c.** x^6 **4. a.** 729 **b.** y^{28} **c.** $\frac{1}{x^8}$ **5. a.** 9 **b.** $\frac{1}{x^7}$ **c.** y^9 **6.** $-64x^3$

7. a. $\frac{27}{64}$ **b.** $-\frac{32}{y^5}$ **8. a.** $16x^{12}y^{24}$ **b.** $-18x^3y^8$ **c.** $\frac{5y^6}{x^4}$ **d.** $\frac{y^8}{25x^2}$ **9. a.** 7,400,000,000 **b.** 0.000003017

10. a. 7.41×10^9 **b.** 9.2×10^{-8} **11.** \$12.86

Exercise Set P.2

1. 50 **3.** 64 **5.** -64 **7.** 1 **9.** -1 **11.** $\frac{1}{64}$ **13.** 32 **15.** 64 **17.** 16 **19.** $\frac{1}{9}$ **21.** $\frac{1}{16}$ **23.** $\frac{y}{x^2}$ **25.** y^5 **27.** x^{10}

29. x^5 **31.** x^{21} **33.** x^{-15} **35.** x^7 **37.** x^{21} **39.** $64x^6$ **41.** $-\frac{64}{x^3}$ **43.** $9x^4y^{10}$ **45.** $6x^{11}$ **47.** $18x^9y^5$ **49.** $4x^{16}$

51. $-5a^{11}b$ **53.** $\frac{2}{b^7}$ **55.** $\frac{1}{16x^6}$ **57.** $\frac{3y^{14}}{4x^4}$ **59.** $\frac{y^2}{25x^6}$ **61.** $-\frac{27\,b^{15}}{a^{18}}$ **63.** 1 **65.** 4700 **67.** 4,000,000 **69.** 0.000786

71. 0.00000318 **73.** 3.6×10^3 **75.** 2.2×10^8 **77.** 2.7×10^{-2} **79.** 7.63×10^{-4} **81.** 600,000 **83.** 0.123 **85.** 30,000

87. 0.021 **89.** $\frac{4.8\times10^{11}}{1.2\times10^{-4}}; 4\times10^{15}$ **91.** $\frac{(7.2\times10^{-4})(3\times10^{-3})}{2.4\times10^{-4}}; 9\times10^{-3}$ **93.** \$6800 **95.** $\$1.12\times10^{12}$ **97.** 1.06×10^{-18} gram
107. (b) is true. **109.** $A=C+D$

Section P.3

Check Point Exercises

1. a. 3 **b.** $5x\sqrt{2}$ **2. a.** $\frac{5}{4}$ **b.** $5x\sqrt{3}$ **3. a.** $17\sqrt{13}$ **b.** $-19\sqrt{17x}$ **4. a.** $17\sqrt{3}$ **b.** $10\sqrt{2x}$ **5. a.** $\frac{5\sqrt{3}}{3}$ **b.** $\sqrt{3}$

6. $\frac{32-8\sqrt{5}}{11}$ **7. a.** $2\sqrt[3]{5}$ **b.** $2\sqrt[5]{2}$ **c.** $\frac{5}{3}$ **8.** $5\sqrt[3]{3}$ **9. a.** 9 **b.** 3 **c.** $\frac{1}{2}$ **10. a.** 8 **b.** $\frac{1}{4}$ **11. a.** $10x^4$ **b.** $4x^{5/2}$ **12.** $\sqrt{x}$

Exercise Set P.3

1. 6 **3.** not a real number **5.** 13 **7.** $5\sqrt{2}$ **9.** $3|x|\sqrt{5}$ **11.** $2x\sqrt{3}$ **13.** $x\sqrt{x}$ **15.** $2x\sqrt{3x}$ **17.** $\dfrac{1}{9}$ **19.** $\dfrac{7}{4}$ **21.** $4x$

23. $5x\sqrt{2x}$ **25.** $2x^2\sqrt{5}$ **27.** $13\sqrt{3}$ **29.** $-2\sqrt{17x}$ **31.** $5\sqrt{2}$ **33.** $3\sqrt{2x}$ **35.** $34\sqrt{2}$ **37.** $20\sqrt{2} - 5\sqrt{3}$ **39.** $\dfrac{\sqrt{7}}{7}$

41. $\dfrac{\sqrt{10}}{5}$ **43.** $\dfrac{13(3-\sqrt{11})}{-2}$ **45.** $7(\sqrt{5}+2)$ **47.** $3(\sqrt{5}-\sqrt{3})$ **49.** 5 **51.** -2 **53.** not a real number **55.** 3 **57.** -3

59. $-\dfrac{1}{2}$ **61.** $2\sqrt[3]{4}$ **63.** $x\sqrt[3]{x}$ **65.** $3\sqrt[3]{2}$ **67.** $2x$ **69.** $7\sqrt[5]{2}$ **71.** $13\sqrt[3]{2}$ **73.** $-y\sqrt[3]{2x}$ **75.** $\sqrt{2}+2$ **77.** 6 **79.** 2

81. 25 **83.** $\dfrac{1}{16}$ **85.** $14x^{7/12}$ **87.** $4x^{1/4}$ **89.** x^2 **91.** $5x^2|y|^3$ **93.** $27y^{2/3}$ **95.** $\sqrt{5}$ **97.** x^2 **99.** $\sqrt[3]{x^2}$ **101.** $\sqrt[3]{x^2y}$

103. $20\sqrt{2}$ mph **105.** $\dfrac{\sqrt{5}+1}{2} \approx 1.62$ **107.** $\dfrac{7\sqrt{2\cdot2\cdot3}}{6} = \dfrac{7\sqrt{2^2\cdot3}}{6} = \dfrac{7\sqrt{2^2}\sqrt{3}}{6} = \dfrac{7\cdot2\sqrt{3}}{6} = \dfrac{7}{3}\sqrt{3}$

109. The duration of a storm whose diameter is 9 miles is 1.89 hours. **117.** 45.00, 23.76, 15.68, 11.33, 8.59, 6.70, 5.31, 4.25, 3.41, 2.73, 2.17, 1.70, 1.30, 0.95, 0.65, 0.38; The percentage of potential employees testing positive for illegal drugs is decreasing over time.
119. (d) is true. **121.** Let □ = 25 and □ = 14. **123. a.** > **b.** >

Section P.4

Check Point Exercises

1. a. $-x^3 + x^2 - 8x - 20$ **b.** $20x^3 - 11x^2 - 2x - 8$ **2.** $15x^3 - 31x^2 + 30x - 8$ **3.** $28x^2 - 41x + 15$
4. a. $49x^2 - 64$ **b.** $4y^6 - 25$ **5. a.** $x^2 + 20x + 100$ **b.** $25x^2 + 40x + 16$ **6. a.** $x^2 - 18x + 81$ **b.** $49x^2 - 42x + 9$
7. $2x^2y + 5xy^2 - 2y^3$ **8. a.** $21x^2 - 25xy + 6y^2$ **b.** $x^4 + 10x^2y + 25y^2$

Exercise Set P.4

1. yes; $3x^2 + 2x - 5$ **3.** no **5.** 2 **7.** 4 **9.** $11x^3 + 7x^2 - 12x - 4$; 3 **11.** $12x^3 + 4x^2 + 12x - 14$; 3 **13.** $6x^2 - 6x + 2$; 2
15. $x^3 + 1$ **17.** $2x^3 - 9x^2 + 19x - 15$ **19.** $x^2 + 10x + 21$ **21.** $x^2 - 2x - 15$ **23.** $6x^2 + 13x + 5$ **25.** $10x^2 - 9x - 9$
27. $15x^4 - 47x^2 + 28$ **29.** $8x^5 - 40x^3 + 3x^2 - 15$ **31.** $x^2 - 9$ **33.** $9x^2 - 4$ **35.** $25 - 49x^2$ **37.** $16x^4 - 25x^2$ **39.** $1 - y^{10}$
41. $x^2 + 4x + 4$ **43.** $4x^2 + 12x + 9$ **45.** $x^2 - 6x + 9$ **47.** $16x^4 - 8x^2 + 1$ **49.** $4x^2 - 28x + 49$ **51.** $x^3 + 3x^2 + 3x + 1$
53. $8x^3 + 36x^2 + 54x + 27$ **55.** $x^3 - 9x^2 + 27x - 27$ **57.** $27x^3 - 108x^2 + 144x - 64$ **59.** $7x^2y - 4xy$ is of degree 3
61. $2x^2y + 13xy + 13$ is of degree 3 **63.** $-5x^3 + 8xy - 9y^2$ is of degree 3 **65.** $x^4y^2 + 8x^3y + y - 6x$ is of degree 6
67. $7x^2 + 38xy + 15y^2$ **69.** $2x^2 + xy - 21y^2$ **71.** $15x^2y^2 + xy - 2$ **73.** $49x^2 + 70xy + 25y^2$ **75.** $x^4y^4 - 6x^2y^2 + 9$
77. $x^3 - y^3$ **79.** $9x^2 - 25y^2$ **81.** $49x^2y^4 - 100y^2$ **83.** 7.567; A person earning $40,000 feels underpaid by $7567.
85. 527.53; The number of violent crimes in the United States was 527.53 per 100,000 inhabitants in 2000. The calculated value is a good
approximation to the actual value, 524.7. **87.** $\dfrac{2}{3}t^3 - 2t^2 + 4t$ **89.** $6x + 22$ **99.** 61.2, 59.0, 56.8, 54.8, 52.8, 50.9, 49.3, 47.7, 46.4, 45.2, 44.3, 43.6, 43.1, 43.0, 43.1, 43.6, 44.4, 45.5, 47.0, 48.9, 51.2; The percentage of U.S. high school seniors who had ever used marijuana decreased from 1980, reached a low in 1993, then increased through 2000. **101.** $49x^2 + 70x + 25 - 16y^2$ **103.** $x^4 - y^4$

Section P.5

Check Point Exercises

1. a. $2x^2(5x - 2)$ **b.** $(x - 7)(2x + 3)$ **2.** $(x + 5)(x^2 - 2)$ **3. a.** $(x + 8)(x + 5)$ **b.** $(x - 7)(x + 2)$ **4.** $(3x - 1)(2x + 7)$
5. a. $(x + 9)(x - 9)$ **b.** $(6x + 5)(6x - 5)$ **6.** $(9x^2 + 4)(3x + 2)(3x - 2)$ **7. a.** $(x + 7)^2$ **b.** $(4x - 7)^2$

8. a. $(x + 1)(x^2 - x + 1)$ **b.** $(5x - 2)(25x^2 + 10x + 4)$ **9.** $3x(x - 5)^2$ **10.** $(x + 10 + 6a)(x + 10 - 6a)$ **11.** $\dfrac{2x - 1}{(x - 1)^{1/2}}$

Exercise Set P.5

1. $9(2x + 3)$ **3.** $3x(x + 2)$ **5.** $9x^2(x^2 - 2x + 3)$ **7.** $(x + 5)(x + 3)$ **9.** $(x - 3)(x^2 + 12)$ **11.** $(x^2 + 5)(x - 2)$
13. $(x - 1)(x^2 + 2)$ **15.** $(3x - 2)(x^2 - 2)$ **17.** $(x + 2)(x + 3)$ **19.** $(x - 5)(x + 3)$ **21.** $(x - 5)(x - 3)$
23. $(3x + 2)(x - 1)$ **25.** $(3x - 28)(x + 1)$ **27.** $(2x - 1)(3x - 4)$ **29.** $(2x + 3)(2x + 5)$ **31.** $(x + 10)(x - 10)$
33. $(6x + 7)(6x - 7)$ **35.** $(3x + 5y)(3x - 5y)$ **37.** $(x^2 + 4)(x + 2)(x - 2)$ **39.** $(4x^2 + 9)(2x + 3)(2x - 3)$ **41.** $(x + 1)^2$
43. $(x - 7)^2$ **45.** $(2x + 1)^2$ **47.** $(3x - 1)^2$ **49.** $(x + 3)(x^2 - 3x + 9)$ **51.** $(x - 4)(x^2 + 4x + 16)$
53. $(2x - 1)(4x^2 + 2x + 1)$ **55.** $(4x + 3)(16x^2 - 12x + 9)$ **57.** $3x(x + 1)(x - 1)$ **59.** $4(x + 2)(x - 3)$
61. $2(x^2 + 9)(x + 3)(x - 3)$ **63.** $(x - 3)(x + 3)(x + 2)$ **65.** $2(x - 8)(x + 7)$ **67.** $x(x - 2)(x + 2)$ **69.** prime
71. $(x - 2)(x + 2)^2$ **73.** $y(y^2 + 9)(y + 3)(y - 3)$ **75.** $5y^2(2y + 3)(2y - 3)$ **77.** $(x - 6 + 7y)(x - 6 - 7y)$

79. $(x + y)(3b + 4)(3b - 4)$ **81.** $(y - 2)(x + 4)(x - 4)$ **83.** $2x(x + 6 + 2a)(x + 6 - 2a)$ **85.** $x^{1/2}(x - 1)$ **87.** $\dfrac{4(1 + 2x)}{x^{2/3}}$

89. $-(x + 3)^{1/2}(x + 2)$ **91.** $\dfrac{x + 4}{(x + 5)^{3/2}}$ **93.** $\dfrac{4(4x - 1)^{1/2}(x - 1)}{3}$ **95. a.** $0.36\,x$ **b.** no; It is selling at 36% of the original price.

97. $16(4 + t)(4 - t)$ **99.** $(3x + 2)(3x - 2)$ **109.** $(x^n + 4)(x^n + 2)$ **111.** $(x - y)^3(x + y)$ **113.** $b = 8, -8, 16, -16$

Section P.6

Check Point Exercises

1. a. -5 **b.** $6, -6$ **2. a.** $x^2, x \neq -3$ **b.** $\dfrac{x - 1}{x + 1}, x \neq -1$ **3.** $\dfrac{x - 3}{(x - 2)(x + 3)}, x \neq 2, x \neq -2, x \neq -3$

4. $\dfrac{3(x - 1)}{x(x + 2)}, x \neq 1, x \neq 0, x \neq -2$ **5.** $-2, x \neq -1$ **6.** $\dfrac{2(4x + 1)}{(x + 1)(x - 1)}, x \neq 1, x \neq -1$ **7.** $(x - 3)(x - 3)(x + 3)$

8. $\dfrac{-x^2 + 11x - 20}{2(x - 5)^2}, x \neq 5$ **9.** $\dfrac{2(2 - 3x)}{4 + 3x}, x \neq 0, x \neq -\dfrac{4}{3}$

Exercise Set P.6

1. 3   **3.** $5, -5$ **5.** $-1, -10$ **7.** $\dfrac{3}{x - 3}, x \neq 3$ **9.** $\dfrac{x - 6}{4}, x \neq 6$ **11.** $\dfrac{y + 9}{y - 1}, y \neq 1, 2$ **13.** $\dfrac{x + 6}{x - 6}, x \neq 6, -6$ **15.** $\dfrac{1}{3}, x \neq 2, -3$

17. $\dfrac{(x - 3)(x + 3)}{x(x + 4)}, x \neq 0, -4, 3$ **19.** $\dfrac{x - 1}{x + 2}, x \neq -2, -1, 2, 3$ **21.** $\dfrac{x^2 + 2x + 4}{3x}, x \neq -2, 0, 2$ **23.** $\dfrac{7}{9}, x \neq -1$

25. $\dfrac{(x - 2)^2}{x}, x \neq 0, -2, 2$ **27.** $\dfrac{2(x + 3)}{3}, x \neq 3, -3$ **29.** $\dfrac{x - 5}{2}, x \neq 1, -5$ **31.** $\dfrac{(x + 2)(x + 4)}{x - 5}, x \neq -6, -3, -1, 3, 5$

33. $2, x \neq -\dfrac{5}{6}$ **35.** $\dfrac{2x - 1}{x + 3}, x \neq 0, -3$ **37.** $3, x \neq 2$ **39.** $\dfrac{3}{x - 3}, x \neq 3, -4$ **41.** $\dfrac{9x + 39}{(x + 4)(x + 5)}, x \neq -4, -5$

43. $-\dfrac{3}{x(x + 1)}, x \neq -1, 0$ **45.** $\dfrac{3x^2 + 4}{(x + 2)(x - 2)}, x \neq -2, 2$ **47.** $\dfrac{2x^2 + 50}{(x - 5)(x + 5)}, x \neq -5, 5$ **49.** $\dfrac{4x + 16}{(x + 3)^2}, x \neq -3$

51. $\dfrac{x^2 - x}{(x + 5)(x - 2)(x + 3)}, x \neq -5, 2, -3$ **53.** $\dfrac{x - 1}{x + 2}, x \neq -2, -1$ **55.** $\dfrac{1}{3}, x \neq 3$ **57.** $\dfrac{x + 1}{3x - 1}, x \neq 0, \dfrac{1}{3}$

59. $\dfrac{1}{xy}, x \neq 0, y \neq 0, x \neq -y$ **61.** $\dfrac{x}{x + 3}, x \neq -2, -3$ **63.** $\dfrac{x - 14}{7}, x \neq -2, 2$ **65. a.** 86.67,520, 1170; It costs \$86,670,000 to inoculate 40% of the population against this strain of flu, \$520,000,000 to inoculate 80% of the population, and \$1,170,000,000 to inoculate 90% of the population.   **b.** $x = 100$ **c.** increases rapidly; impossible to inoculate 100% of the population.

67. a. $\dfrac{100W}{L}$ **b.** round   **69.** $\dfrac{2r_1 r_2}{r_1 + r_2}$; 24 mph   **83. a.** $\dfrac{Pi(1 + i)^n}{(1 + i)^n - 1}$ **b.** \$527   **85.** $-4x - 1$ **87.** It cubes x.

Chapter P Review Exercises

1. a. $\sqrt{81}$ **b.** $0, \sqrt{81}$ **c.** $-17, 0, \sqrt{81}$ **d.** $-17, -\dfrac{9}{13}, 0, 0.75, \sqrt{81}$ **e.** $\sqrt{2}, \pi$ **2.** 103   **3.** $\sqrt{2} - 1$ **4.** $\sqrt{17} - 3$

5. $|4 - (-17)|; 21$ **6.** 20   **7.** 4   **8.** commutative property of addition   **9.** associative property of multiplication
10. distributive property of multiplication over addition   **11.** commutative property of multiplication   **12.** commutative property of multiplication   **13.** commutative property of addition   **14.** $23x - 23y - 2$ **15.** $2x$ **16.** -108 **17.** $\dfrac{5}{16}$ **18.** $\dfrac{1}{25}$ **19.** $\dfrac{1}{27}$

20. $-8x^{12}y^9$ **21.** $\dfrac{10}{x^8}$ **22.** $\dfrac{1}{16x^{12}}$ **23.** $\dfrac{y^8}{4x^{10}}$ **24.** 37,400   **25.** 0.0000745   **26.** 3.59×10^6 **27.** 7.25×10^{-3} **28.** 3.9×10^5

29. 2.3×10^{-2} **30.** 10^3 or 1000 yr   **31.** $\$4.2 \times 10^{10}$ **32.** $10\sqrt{3}$ **33.** $2|x|\sqrt{3}$ **34.** $2x\sqrt{5}$ **35.** $r\sqrt{r}$ **36.** $\dfrac{11}{2}$ **37.** $4x\sqrt{3}$

38. $20\sqrt{5}$ **39.** $16\sqrt{2}$ **40.** $24\sqrt{2} - 8\sqrt{3}$ **41.** $6\sqrt{5}$ **42.** $\dfrac{\sqrt{6}}{3}$ **43.** $\dfrac{5(6 - \sqrt{3})}{33}$ **44.** $7(\sqrt{7} + \sqrt{5})$ **45.** 5   **46.** -2

47. not a real number   **48.** 5   **49.** $3\sqrt[3]{3}$ **50.** $y\sqrt[3]{y^2}$ **51.** $2\sqrt[4]{5}$ **52.** $13\sqrt[3]{2}$ **53.** $x\sqrt[4]{2}$ **54.** 4

55. $\dfrac{1}{5}$ **56.** 5   **57.** $\dfrac{1}{3}$ **58.** 16   **59.** $\dfrac{1}{81}$ **60.** $20x^{11/12}$ **61.** $3x^{1/4}$ **62.** $25x^4$ **63.** $\sqrt{y}$ **64.** $8x^3 + 10x^2 - 20x - 4$; degree 3
65. $8x^4 - 5x^3 + 6$; degree 4   **66.** $12x^3 + x^2 - 21x + 10$ **67.** $6x^2 - 7x - 5$ **68.** $16x^2 - 25$ **69.** $4x^2 + 20x + 25$
70. $9x^2 - 24x + 16$ **71.** $8x^3 + 12x^2 + 6x + 1$ **72.** $125x^3 - 150x^2 + 60x - 8$ **73.** $-x^2 - 17xy - 3y^2$; degree 2
74. $24x^3y^2 + x^2y - 12x^2 + 4$; degree 5   **75.** $3x^2 + 16xy - 35y^2$ **76.** $9x^2 - 30xy + 25y^2$ **77.** $9x^4 + 12x^2y + 4y^2$

78. $49x^2 - 16y^2$ **79.** $a^3 - b^3$ **80.** $3x^2(5x + 1)$ **81.** $(x - 4)(x - 7)$ **82.** $(3x + 1)(5x - 2)$ **83.** $(8 - x)(8 + x)$ **84.** prime
85. $3x^2(x - 5)(x + 2)$ **86.** $4x^3(5x^4 - 9)$ **87.** $(x + 3)(x - 3)^2$ **88.** $(4x - 5)^2$ **89.** $(x^2 + 4)(x + 2)(x - 2)$
90. $(y - 2)(y^2 + 2y + 4)$ **91.** $(x + 4)(x^2 - 4x + 16)$ **92.** $3x^2(x - 2)(x + 2)$ **93.** $(3x - 5)(9x^2 + 15x + 25)$
94. $x(x - 1)(x + 1)(x^2 + 1)$ **95.** $(x^2 - 2)(x + 5)$ **96.** $(x + 9 + y)(x + 9 - y)$ **97.** $\dfrac{16(1 + 2x)}{x^{3/4}}$
98. $(x + 2)(x - 2)(x^2 + 3)^{1/2}(-x^4 + x^2 + 13)$ **99.** $\dfrac{6(2x + 1)}{x^{3/2}}$ **100.** $x^2, x \neq -2$ **101.** $\dfrac{x - 3}{x - 6}, x \neq -6, 6$ **102.** $\dfrac{x}{x + 2}, x \neq -2$
103. $\dfrac{(x + 3)^3}{(x - 2)^2(x + 2)}, x \neq 2, -2$ **104.** $\dfrac{2}{x(x + 1)}, x \neq 0, 1, -1, -\dfrac{1}{3}$ **105.** $\dfrac{x + 3}{x - 4}, x \neq -3, 4, 2, 8$ **106.** $\dfrac{1}{x - 3}, x \neq 3, -3$
107. $\dfrac{4x(x - 1)}{(x + 2)(x - 2)}, x \neq 2, -2$ **108.** $\dfrac{2x^2 - 3}{(x - 3)(x + 3)(x - 2)}, x \neq 3, -3, 2$ **109.** $\dfrac{11x^2 - x - 11}{(2x - 1)(x + 3)(3x + 2)}, x \neq \dfrac{1}{2}, -3, -\dfrac{2}{3}$
110. $\dfrac{3}{x}, x \neq 0, 2$ **111.** $\dfrac{3x}{x - 4}, x \neq 0, 4, -4$ **112.** $\dfrac{3x + 8}{3x + 10}, x \neq -3, -\dfrac{10}{3}$

Chapter P Test

1. $-7, -\dfrac{4}{5}, 0, 0.25, \sqrt{4}, \dfrac{22}{7}$ **2.** commutative property of addition   **3.** distributive property of multiplication over addition
4. 7.6×10^{-4} **5.** $85x + 2y - 15$ **6.** $\dfrac{5y^8}{x^6}$ **7.** $3r\sqrt{2}$ **8.** $11\sqrt{2}$ **9.** $\dfrac{3(5 - \sqrt{2})}{23}$ **10.** $2x\sqrt[3]{2x}$ **11.** $\dfrac{x + 3}{x - 2}, x \neq 2, 1$ **12.** $\dfrac{1}{243}$
13. $2x^3 - 13x^2 + 26x - 15$ **14.** $25x^2 + 30xy + 9y^2$ **15.** $(x - 3)(x - 6)$ **16.** $(x^2 + 3)(x + 2)$ **17.** $(5x - 3)(5x + 3)$
18. $(6x - 7)^2$ **19.** $(y - 5)(y^2 + 5y + 25)$ **20.** $(x + 5 + 3y)(x + 5 - 3y)$ **21.** $\dfrac{2x + 3}{(x + 3)^{3/5}}$ **22.** $\dfrac{2(x + 3)}{x + 1}, x \neq 3, -1, -4, -3$
23. $\dfrac{x^2 + 2x + 15}{(x + 3)(x - 3)}, x \neq 3, -3$ **24.** $\dfrac{5}{(x - 3)(x - 4)}, x \neq 3, 4$ **25.** $\dfrac{3 - x}{3}, x \neq 0$

CHAPTER 1

Section 1.1

Check Point Exercises

1.
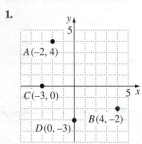

2.
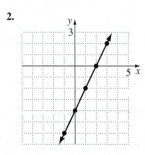

3. The minimum x-value is -100 and the maximum x-value is 100. The distance between consecutive tick marks is 50. The minimum y-value is -80 and the maximum y-value is 80. The distance between consecutive tick marks is 10.

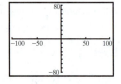

4. $21\dfrac{1}{2}$; 1900

Exercise Set 1.1

1.
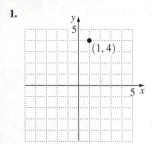

3.
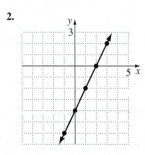

5.
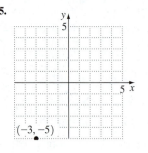

7.
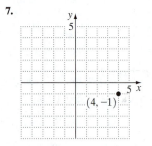

9.
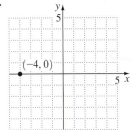

11.
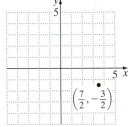

13.
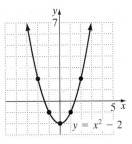

15.
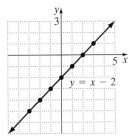

17.
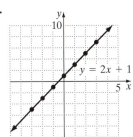

19.
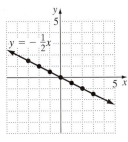

21.
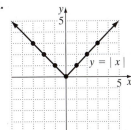

23.
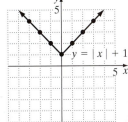

25.
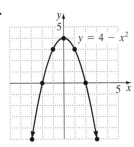

27.
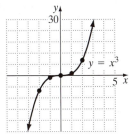

29. (c) **31.** (b) **33. a.** 2 **b.** −4 **35. a.** 1, −2 **b.** 2
37. a. −1 **b.** None **39.** A (2, 7); The football is 7 feet high
when it is 2 yards from the quarterback.

41. $C\left(6, 9\frac{1}{2}\right)$ **43.** 12 feet; 15 yards **45.** 0–4 years

47. year 4; 8% **57.** (c) gives a complete graph.
59. (c) gives a complete graph. **61.** (a) **63.** (b)

Section 1.2

Check Point Exercises

1. {16} **2.** {5} **3.** {−2} **4.** {3} **5.** ∅ **6.** identity

Exercise Set 1.2

1. {11} **3.** {7} **5.** {13} **7.** {2} **9.** {9} **11.** {−5} **13.** {6} **15.** {−2} **17.** {12} **19.** {24} **21.** {−15}

23. {5} **25.** $\left\{\dfrac{33}{2}\right\}$ **27.** {−12} **29.** $\left\{\dfrac{46}{5}\right\}$ **31. a.** 0 **b.** $\left\{\dfrac{1}{2}\right\}$ **33. a.** 0 **b.** {−2} **35. a.** 0 **b.** {2} **37. a.** 0

b. {4} **39. a.** 1 **b.** {3} **41. a.** −1 **b.** ∅ **43. a.** 1 **b.** {2} **45. a.** −2, 2 **b.** ∅ **47. a.** −1, 1 **b.** {−3}
49. a. −2, 4 **b.** ∅ **51.** identity **53.** inconsistent equation **55.** conditional equation **57.** inconsistent equation
59. {−7} **61.** not true for any real number, ∅ **63.** {−4} **65.** {8} **67.** {−1} **69.** not true for any real number, ∅

71. a. 250 mg/dl **b.** 375,000 annual deaths; 350,000 saved lives **73.** $409\frac{1}{5}$ ft **87.** inconsistent **89.** conditional; {−5}

91. $x = \dfrac{c - b}{a}$ **93.** Answers may vary. **95.** 20

Section 1.3

Check Point Exercises

1. 2008 **2.** *Saturday Night Fever* sold 11 million albums; *Jagged Little Pill* sold 16 million albums. **3.** 300 min

4. $15,000 at 9%; $10,000 at 12% **5.** width = 50 ft; length = 94 ft **6.** $m = \dfrac{y - b}{x}$ **7.** $C = \dfrac{P}{1 + M}$

Exercise Set 1.3

1. $x + 9$ **3.** $20 - x$ **5.** $8 - 5x$ **7.** $15 \div x$ **9.** $2x + 20$ **11.** $7x - 30$ **13.** $4(x + 12)$ **15.** $x + 40 = 450; \{410\}$
17. $5x - 7 = 123; \{26\}$ **19.** $9x = 3x + 30; \{5\}$ **21.** 40 years old; It is shown by the point $(40, 117)$ on the line for females.
23. approximately 41 years after 1960 in 2001 **25.** 196 lb **27.** $Waterworld = \$160$ million; $Titanic = \$200$ million
29. Miami $= 57$ hr; Los Angeles $= 82$ hr **31.** 800 mi **33.** 2005 **35. a.** total monthly cost with coupon book $= 21 + 0.50x$; total
monthly cost without coupon book $= 1.25x$ **b.** 28 times **37.** $\$600; \$580$ **39.** $\$31,250$ in noninsured bonds; $\$18,750$ in
government-insured certificates of deposit **41.** $\$6000$ at 12%; $\$2000$ at a 5% loss **43.** length $= 78$ ft; width $= 36$ ft

45. length $= 2$ ft; height $= 5$ ft **47.** 11 hr **49.** $\$31,000$ **51.** 7 oz **53.** $\$20,000$ **55.** 5 ft 7 in. **57.** $\omega = \dfrac{A}{l}$ **59.** $b = \dfrac{2A}{h}$

61. $p = \dfrac{I}{rt}$ **63.** $m = \dfrac{E}{c^2}$ **65.** $p = \dfrac{T - D}{m}$ **67.** $a = \dfrac{2A}{h} - b$ **69.** $r = \dfrac{S - P}{Pt}$ **71.** $S = \dfrac{F}{B} + V$ **73.** $I = \dfrac{E}{R + r}$

75. $f = \dfrac{pq}{p + q}$ **81.**

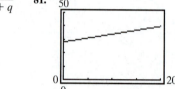

83. North campus had 600 students; South campus had 400 students.
85. Coburn $= 60$ years old; woman $= 20$ years old
87. $\$4000$ for the mother; $\$8000$ for the boy; $\$2000$ for the girl

The trace feature shows x to be about 15
when $y = 37$, so 2005. The trace feature
shows x to be 20 when $y = 40$, so 2010.

Section 1.4

Check Point Exercises

1. a. $8 + i$ **b.** $-10 + 10i$ **2. a.** $63 + 14i$ **b.** $58 - 11i$ **3.** $\dfrac{3}{5} + \dfrac{13}{10}i$ **4. a.** $7i\sqrt{3}$ **b.** $1 - 4i\sqrt{3}$ **c.** $-7 + i\sqrt{3}$

Exercise Set 1.4

1. $8 - 2i$ **3.** $-2 + 9i$ **5.** $24 + 7i$ **7.** $-14 + 17i$ **9.** $21 + 15i$ **11.** $-43 - 23i$ **13.** $-29 - 11i$ **15.** 34 **17.** 34
19. $-5 + 12i$ **21.** $\dfrac{3}{5} + \dfrac{1}{5}i$ **23.** $1 + i$ **25.** $-\dfrac{24}{25} + \dfrac{32}{25}i$ **27.** $\dfrac{7}{5} + \dfrac{4}{5}i$ **29.** $3i$ **31.** $47i$ **33.** $-8i$ **35.** $2 + 6i\sqrt{7}$
37. $-\dfrac{1}{3} + \dfrac{\sqrt{2}}{6}i$ **39.** $-\dfrac{1}{8} - \dfrac{\sqrt{3}}{24}i$ **41.** $-2\sqrt{6} - 2i\sqrt{10}$ **43.** $24\sqrt{15}$ **53.** (d) is true. **55.** $\dfrac{14}{25} - \dfrac{2}{25}i$ **57.** 0

Section 1.5

Check Point Exercises

1. a. $\{0, 3\}$ **b.** $\left\{-1, \dfrac{1}{2}\right\}$ **2. a.** $\{-\sqrt{7}, \sqrt{7}\}$ **b.** $\{-5 + \sqrt{11}, -5 - \sqrt{11}\}$ **3.** $49; (x - 7)^2$ **4.** $\{1 + \sqrt{3}, 1 - \sqrt{3}\}$
5. $\left\{\dfrac{-1 + \sqrt{3}}{2}, \dfrac{-1 - \sqrt{3}}{2}\right\}$ **6.** $\{1 + i, 1 - i\}$ **7.** -56; two complex imaginary solutions **8.** 1998; a good approximation **9.** 12 in.

Exercise Set 1.5

1. $\{-2, 5\}$ **3.** $\{3, 5\}$ **5.** $\left\{-\dfrac{5}{2}, \dfrac{2}{3}\right\}$ **7.** $\left\{-\dfrac{4}{3}, 2\right\}$ **9.** $\{-4, 0\}$ **11.** $\left\{0, \dfrac{1}{3}\right\}$ **13.** $\{-3, 1\}$ **15.** $\{-3, 3\}$
17. $\{-\sqrt{10}, \sqrt{10}\}$ **19.** $\{-7, 3\}$ **21.** $\left\{-\dfrac{5}{3}, \dfrac{1}{3}\right\}$ **23.** $\left\{\dfrac{1 - \sqrt{7}}{5}, \dfrac{1 + \sqrt{7}}{5}\right\}$ **25.** $\left\{\dfrac{4 - 2\sqrt{2}}{3}, \dfrac{4 + 2\sqrt{2}}{3}\right\}$
27. $36; x^2 + 12x + 36 = (x + 6)^2$ **29.** $25; x^2 - 10x + 25 = (x - 5)^2$ **31.** $\dfrac{9}{4}; x^2 + 3x + \dfrac{9}{4} = \left(x + \dfrac{3}{2}\right)^2$
33. $\dfrac{49}{4}; x^2 - 7x + \dfrac{49}{4} = \left(x - \dfrac{7}{2}\right)^2$ **35.** $\dfrac{1}{9}; x^2 - \dfrac{2}{3}x + \dfrac{1}{9} = \left(x - \dfrac{1}{3}\right)^2$ **37.** $\dfrac{1}{36}; x^2 - \dfrac{1}{3}x + \dfrac{1}{36} = \left(x - \dfrac{1}{6}\right)^2$ **39.** $\{-7, 1\}$
41. $\{1 + \sqrt{3}, 1 - \sqrt{3}\}$ **43.** $\{3 + 2\sqrt{5}, 3 - 2\sqrt{5}\}$ **45.** $\{-2 + \sqrt{3}, -2 - \sqrt{3}\}$ **47.** $\left\{\dfrac{-3 + \sqrt{13}}{2}, \dfrac{-3 - \sqrt{13}}{2}\right\}$

49. $\left\{\frac{1}{2}, 3\right\}$ **51.** $\left\{\frac{1 + \sqrt{2}}{2}, \frac{1 - \sqrt{2}}{2}\right\}$ **53.** $\left\{\frac{1 + \sqrt{7}}{3}, \frac{1 - \sqrt{7}}{3}\right\}$ **55.** $\{-5, -3\}$ **57.** $\left\{\frac{-5 + \sqrt{13}}{2}, \frac{-5 - \sqrt{13}}{2}\right\}$

59. $\left\{\frac{3 + \sqrt{57}}{6}, \frac{3 - \sqrt{57}}{6}\right\}$ **61.** $\left\{\frac{1 + \sqrt{29}}{4}, \frac{1 - \sqrt{29}}{4}\right\}$ **63.** $\{3 + i, 3 - i\}$ **65.** 36; 2 unequal real solutions

67. 97; 2 unequal real solutions   **69.** 0; 1 real solution   **71.** 37; 2 unequal real solutions   **73.** $\left\{-\frac{1}{2}, 1\right\}$ **75.** $\left\{\frac{1}{5}, 2\right\}$

77. $\{-2\sqrt{5}, 2\sqrt{5}\}$ **79.** $\{1 + \sqrt{2}, 1 - \sqrt{2}\}$ **81.** $\left\{\frac{-11 + \sqrt{33}}{4}, \frac{-11 - \sqrt{33}}{4}\right\}$ **83.** $\left\{0, \frac{8}{3}\right\}$ **85.** $\{2\}$ **87.** $\{-2, 2\}$

89. $\{3 + 2i, 3 - 2i\}$ **91.** $\{2 + i\sqrt{3}, 2 - i\sqrt{3}\}$ **93.** $\left\{0, \frac{7}{2}\right\}$ **95.** $\{2 + \sqrt{10}, 2 - \sqrt{10}\}$ **97.** $\{-5, -1\}$

99. 19 year olds and 72 year olds; fairly well   **101.** 1994   **103.** $(4, 27)$; This is the graph's highest point; During this time period, the greatest number of recipients was 27 million in 1994.   **105.** 1990; $(10, 740)$ **107.** 127.28 ft   **109.** 34 ft   **111.** width = 15 ft; length = 20 ft   **113.** 10 in.   **115.** 9.3 in. and 0.7 in.   **117.** 2 in.   **129.** (c) is true.   **131.** $x^2 - 2x - 15 = 0$ **133.** 2.4 m; Yes

Section 1.6

Check Point Exercises

1. $\{-\sqrt{3}, 0, \sqrt{3}\}$ **2.** $\left\{-2, -\frac{3}{2}, 2\right\}$ **3.** $\{-1, 3\}$ **4.** $\{4\}$ **5. a.** $\{\sqrt[3]{25}\}$ or $\{5^{2/3}\}$ **b.** $\{-8, 8\}$ **6.** $\{-\sqrt{3}, -\sqrt{2}, \sqrt{2}, \sqrt{3}\}$

7. $\left\{-\frac{1}{27}, 64\right\}$ **8.** $\{-2, 3\}$

Exercise Set 1.6

1. $\{-4, 0, 4\}$ **3.** $\left\{-2, -\frac{2}{3}, 2\right\}$ **5.** $\left\{-\frac{1}{2}, \frac{1}{2}, \frac{3}{2}\right\}$ **7.** $\left\{-2, -\frac{1}{2}, \frac{1}{2}\right\}$ **9.** $\{0, 2, -1 + i\sqrt{3}, -1 - i\sqrt{3}\}$ **11.** $\{6\}$ **13.** $\{6\}$

15. $\{-6\}$ **17.** $\{10\}$ **19.** $\{12\}$ **21.** $\{8\}$ **23.** $\varnothing$ **25.** $\varnothing$ **27.** $\left\{\frac{13 + \sqrt{105}}{6}\right\}$ **29.** $\{4\}$ **31.** $\{13\}$ **33.** $\{\sqrt[5]{4}\}$

35. $\{-60, 68\}$ **37.** $\{-4, 5\}$ **39.** $\{-2, -1, 1, 2\}$ **41.** $\left\{-\frac{4}{3}, -1, 1, \frac{4}{3}\right\}$ **43.** $\{25, 64\}$ **45.** $\left\{-\frac{1}{4}, \frac{1}{5}\right\}$ **47.** $\{-8, 27\}$ **49.** $\{1\}$

51. $\left\{\frac{1}{4}, 1\right\}$ **53.** $\{2, 12\}$ **55.** $\{-3, -1, 2, 4\}$ **57.** $\{-8, -2, 1, 4\}$ **59.** $\{-8, 8\}$ **61.** $\{-5, 9\}$ **63.** $\{-2, 3\}$ **65.** $\left\{-\frac{5}{3}, 3\right\}$

67. $\left\{-\frac{2}{5}, \frac{2}{5}\right\}$ **69.** $\varnothing$ **71.** $\left\{\frac{1}{2}\right\}$ **73.** $\{-1, 3\}$ **75.** $\{1\}$ **77.** $\{0\}$ **79.** $\left\{\frac{5}{2}\right\}$ **81.** $\{-8, -6, 4, 6\}$ **83.** $\{-1, 1, 2\}$

85. 2018   **87.** 36 years old; $(36, 40{,}000)$ **89.** 149 million km   **91.** either 1.2 feet or 7.5 feet from the base of the 6 foot pole

101. $\{-3, -1, 1\}$ **103.** $\{-2\}$ **105.** (d) is true.   **107.** $\left\{\frac{2}{5}, \frac{1}{2}\right\}$ **109.** $\{0, 1\}$

Section 1.7

Check Point Exercises

1. a.  **b.** **c.**

2. a. $\{x|-2 \le x < 5\}$ **b.** $\{x|1 \le x \le 3.5\}$ **c.** $\{x|x < -1\}$ **3.** $[-1, \infty)$ or $\{x|x \ge -1\}$

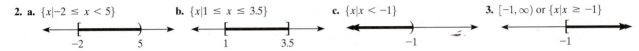

4. $[1, \infty)$ or $\{x|x \ge 1\}$ **5.** $[-1, 4)$ or $\{x|-1 \le x < 4\}$ **6.** $(-3, 7)$ or $\{x|-3 < x < 7\}$ **7.** $(-\infty, 1]$ or $[4, \infty)$ or $\{x|x \le 1 \text{ or } x \ge 4\}$

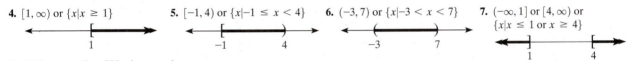

8. driving more than 720 mi per week

Exercise Set 1.7

1.
     6

3.   −4

5.   −3

7.   4

9.   −2   5

11.   −1   4

13. $1 < x \le 6$

15. $-5 \le x < 2$

17. $-3 \le x \le 1$

19. $x > 2$

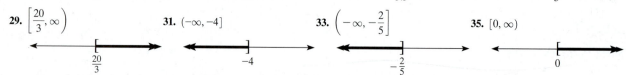

21. $x \ge -3$

23. $x < 3$

25. $x < 5.5$

27. $(-\infty, 3)$

29. $\left[\dfrac{20}{3}, \infty\right)$

31. $(-\infty, -4]$

33. $\left(-\infty, -\dfrac{2}{5}\right]$

35. $[0, \infty)$

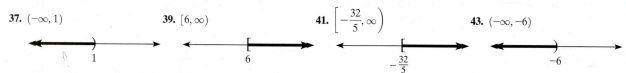

37. $(-\infty, 1)$

39. $[6, \infty)$

41. $\left[-\dfrac{32}{5}, \infty\right)$

43. $(-\infty, -6)$

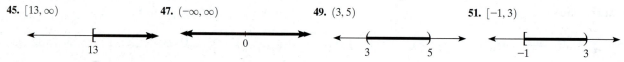

45. $[13, \infty)$

47. $(-\infty, \infty)$

49. $(3, 5)$

51. $[-1, 3)$

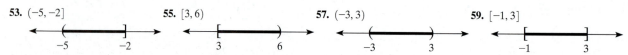

53. $(-5, -2]$

55. $[3, 6)$

57. $(-3, 3)$

59. $[-1, 3]$

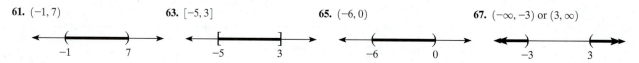

61. $(-1, 7)$

63. $[-5, 3]$

65. $(-6, 0)$

67. $(-\infty, -3)$ or $(3, \infty)$

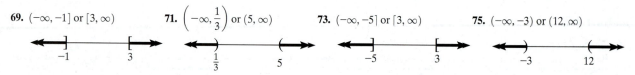

69. $(-\infty, -1]$ or $[3, \infty)$

71. $\left(-\infty, \dfrac{1}{3}\right)$ or $(5, \infty)$

73. $(-\infty, -5]$ or $[3, \infty)$

75. $(-\infty, -3)$ or $(12, \infty)$

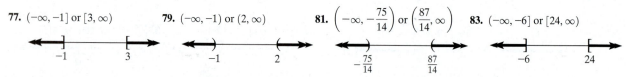

77. $(-\infty, -1]$ or $[3, \infty)$

79. $(-\infty, -1)$ or $(2, \infty)$

81. $\left(-\infty, -\dfrac{75}{14}\right)$ or $\left(\dfrac{87}{14}, \infty\right)$

83. $(-\infty, -6]$ or $[24, \infty)$

85. sports events and playing sports   **87.** amusement parks, gardening, movies, and exercise   **89.** gardening and movies
91. home improvement, amusement parks, and gardening   **93.** $x > 20$; all years after 2008   **95.** between 59°F and 95°F inclusive
97. $58.6 \le x \le 61.8$; Between 58.6% and 61.8% of U.S. households watched the "M*A*S*H" episode.   **99.** $h \le 41$ or $h \ge 59$
101. $50 + 0.20x < 20 + 0.50x$; more than 100 mi   **103.** $1800 + 0.03x < 200 + 0.08x$; greater than \$32,000
105. $2x > 10,000 + 0.40x$; more than 6250 tapes   **107.** $265 + 65x \le 2800$; at most 39 bags
109. a. $\dfrac{86 + 88 + x}{3} \ge 90$; at least a 96   **b.** $\dfrac{86 + 88 + x}{3} < 80$; a grade less than 66

121.

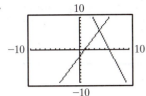

$x < 4$

123.

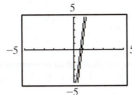

The graph of the left side of the inequality is never above the graph of the right side, therefore there is no solution; You get a statement that is always false.

125. a. $C = 4 + 0.10x; C = 2 + 0.15x$
b.

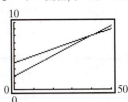

c. 41 or more checks
d. $x > 40$

127. Because $x > y, y - x$ represents a negative number, so when both sides are multiplied by $(y - x)$, the inequality must be reversed.
129. at least $500, but no more than $2500

Section 1.8

Check Point Exercises

1. $(-3, 1)$

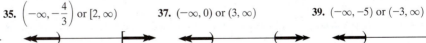

2. $(-\infty, -4]$ or $[5, \infty)$

3. $(-\infty, -2)$ or $(5, \infty)$

4. $(-1, 1]$

5. between 1 and 4 sec

Exercise Set 1.8

1. $(-\infty, -2)$ or $(4, \infty)$

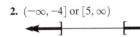

3. $[-3, 7]$

5. $(-\infty, 1)$ or $(4, \infty)$

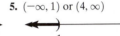

7. $(-\infty, -4)$ or $(-1, \infty)$

9. $\varnothing$

11. $[2, 4]$

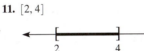

13. $\left[-4, \frac{2}{3}\right]$

15. $\left(-3, \frac{5}{2}\right)$

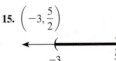

17. $\left(-1, -\frac{3}{4}\right)$

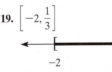

19. $\left[-2, \frac{1}{3}\right]$

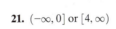

21. $(-\infty, 0]$ or $[4, \infty)$

23. $\left(-\infty, -\frac{3}{2}\right)$ or $(0, \infty)$

25. $[0, 1]$

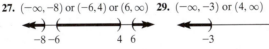

27. $(-\infty, -8)$ or $(-6, 4)$ or $(6, \infty)$

29. $(-\infty, -3)$ or $(4, \infty)$

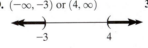

31. $(-4, -3)$

33. $[2, 4)$

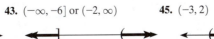

35. $\left(-\infty, -\frac{4}{3}\right)$ or $[2, \infty)$

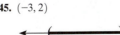

37. $(-\infty, 0)$ or $(3, \infty)$

39. $(-\infty, -5)$ or $(-3, \infty)$

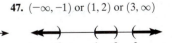

41. $\left(-\infty, \frac{1}{2}\right)$ or $\left[\frac{7}{5}, \infty\right)$

43. $(-\infty, -6]$ or $(-2, \infty)$

45. $(-3, 2)$

47. $(-\infty, -1)$ or $(1, 2)$ or $(3, \infty)$

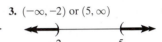

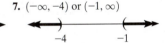

49. between 2 and 3 sec   **51.** 3.46 sec   **53. a.** 200 beats/min   **b.** [0, 4) or (12,   ); (12,   ); [0, 4); heart rate will plateau when it reaches its normal level.   **55.** from 2008 on   **57.** They must produce at least 20,000 wheelchairs; the x-values of all points on the graph which lie below $y = 425$ are solutions to the inequality.   **63.** $\left[-3, \dfrac{1}{2}\right]$ **65.** $(1, 4]$ **67.** Answers may vary.

69. Because the square of any number other than zero is positive, the solution includes all real numbers except 2.

71. Because the square of any number is positive, the solution is $\varnothing$.   **73.** $(-2, -1)$ or $(2, \infty)$ **75. a.** $(-\infty, \infty)$ **b.** $\varnothing$

Chapter 1 Review Exercises

1.

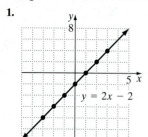

2.

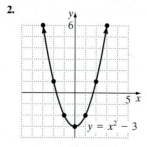

3.

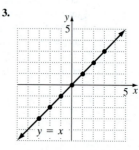

4.

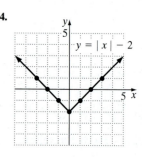

5.

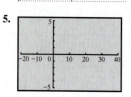

6. x-intercept: -2; y-intercept: 2   **7.** x-intercepts: 2, -2; y-intercept: -4

8. x-intercept: 5; y-intercept: none   **9.** 20%   **10.** 85 years

11. The percentage of Americans with Alzheimer's disease increases with age.

12. $\{6\}$ **13.** $\{-10\}$ **14.** $\{5\}$ **15.** $\{-13\}$ **16.** $\{-3\}$ **17.** $\{-1\}$ **18.** $\{2\}$ **19.** $\{2\}$

20. $\left\{\dfrac{72}{11}\right\}$ **21.** $\{-12\}$ **22.** $\left\{\dfrac{77}{15}\right\}$ **23. a.** 0   **b.** $\{2\}$ **24. a.** 5   **b.** $\varnothing$ **25. a.** $-1, 1$ **b.** all real numbers except 1 and -1

26. a. $-2, 4$ **b.** $\{7\}$ **27.** inconsistent equation   **28.** identity   **29.** conditional equation   **30.** 2005   **31.** 2000

32. low = \$174 thousand, middle = \$237 thousand, high = \$345 thousand   **33.** 9 years; 2009   **34.** 500 min

35. \$6250 at 8%; \$3750 at 12%   **36.** length = 120 m; width = 53 m   **37.** 20 times   **38.** \$10,000   **39.** 95 concerts   **40.** $h = \dfrac{3V}{B}$

41. $M = \dfrac{f - F}{f}$ **42.** $g = \dfrac{T}{r + vt}$ **43.** $-9 + 4i$ **44.** $-12 - 8i$ **45.** $29 + 11i$ **46.** $-7 - 24i$ **47.** 113   **48.** $\dfrac{15}{13} - \dfrac{3}{13}i$

49. $\dfrac{1}{5} + \dfrac{11}{10}i$ **50.** $i\sqrt{2}$ **51.** $-96 - 40i$ **52.** $2 + i\sqrt{2}$ **53.** $\left\{-8, \dfrac{1}{2}\right\}$ **54.** $\{-4, 0\}$ **55.** $\{-8, 8\}$ **56.** $\left\{\dfrac{4 + 3\sqrt{2}}{3}, \dfrac{4 - 3\sqrt{2}}{3}\right\}$

57. $100; (x + 10)^2$ **58.** $\dfrac{9}{4}; \left(x - \dfrac{3}{2}\right)^2$ **59.** $\{3, 9\}$ **60.** $\left\{2 + \dfrac{\sqrt{3}}{3}, 2 - \dfrac{\sqrt{3}}{3}\right\}$ **61.** $\{1 + \sqrt{5}, 1 - \sqrt{5}\}$

62. $\{1 + 3i\sqrt{2}, 1 - 3i\sqrt{2}\}$ **63.** $\left\{\dfrac{-2 + \sqrt{10}}{2}, \dfrac{-2 - \sqrt{10}}{2}\right\}$ **64.** -36; 2 complex imaginary solutions   **65.** 81; 2 unequal real solutions

66. $\left\{\dfrac{1}{2}, 5\right\}$ **67.** $\left\{-2, \dfrac{10}{3}\right\}$ **68.** $\left\{\dfrac{7 + \sqrt{37}}{6}, \dfrac{7 - \sqrt{37}}{6}\right\}$ **69.** $\{-3, 3\}$ **70.** $\{-2, 8\}$ **71.** $\left\{\dfrac{1}{6} + i\dfrac{\sqrt{23}}{6}, \dfrac{1}{6} - i\dfrac{\sqrt{23}}{6}\right\}$

72. 20 weeks   **73.** 10 years   **74.** $(10, 7250)$ **75.** length = 5 yd; width = 3 yd   **76.** approximately 134 m   **77.** $\{-5, 0, 5\}$

78. $\left\{-3, \dfrac{1}{2}, 3\right\}$ **79.** $\{2\}$ **80.** $\{8\}$ **81.** $\{16\}$ **82.** $\{132\}$ **83.** $\{-2, -1, 1, 2\}$ **84.** $\{16\}$ **85.** $\{-4, 3\}$ **86.** $\{-5, 11\}$

87. $\left\{-1, -\dfrac{2\sqrt{6}}{9}, \dfrac{2\sqrt{6}}{9}, 1\right\}$ **88.** $\{2\}$ **89.** $\{1, 4\}$ **90.** $\{-3, -2, 3\}$ **91.** 1250 ft

92.   **93.**   **94.**   **95.** $-2 < x \le 3$

96. $-1.5 \le x \le 2$ **97.** $x > -1$ **98.** $[-2, \infty)$ **99.** $\left[\dfrac{3}{5}, \infty\right)$

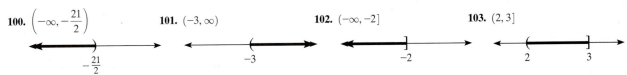

100. $\left(-\infty, -\dfrac{21}{2}\right)$ **101.** $(-3, \infty)$ **102.** $(-\infty, -2]$ **103.** $(2, 3]$

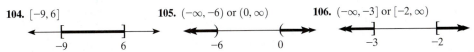

104. $[-9, 6]$ **105.** $(-\infty, -6)$ or $(0, \infty)$ **106.** $(-\infty, -3]$ or $[-2, \infty)$

107. Most people sleep between 5.5 and 7.5 hours.   **108.** between 50°F and 77°F inclusively   **109.** more than 50 checks

110. at least 93

111. $\left[-4, \dfrac{1}{2}\right]$ **112.** $\left(-\infty, \dfrac{3 - \sqrt{3}}{2}\right)$ or $\left(\dfrac{3 + \sqrt{3}}{2}, \infty\right)$ **113.** $(-\infty, -2)$ or $(6, \infty)$

114. $(-\infty, 4)$ or $\left[\dfrac{23}{4}, \infty\right)$ **115.** from 1 to 2 sec

Chapter 1 Test

1

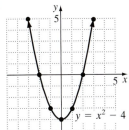

2. x-intercept: 2; y-intercept: 3   **3.** 1992; 7.8%

4. $\{-1\}$ **5** $\{-6\}$ **6.** $\{5\}$

7. $\left\{-\dfrac{1}{2}, 2\right\}$ **8.** $\left\{\dfrac{1 - 5\sqrt{3}}{3}, \dfrac{1 + 5\sqrt{3}}{3}\right\}$ **9.** $\{1 - \sqrt{5}, 1 + \sqrt{5}\}$

10. $\left\{1 + \dfrac{1}{2}i, 1 - \dfrac{1}{2}i\right\}$ **11.** $\{-1, 1, 4\}$ **12.** $\{7\}$ **13.** $\{5\}$

14. $\{\sqrt[3]{4}\}$ **15.** $\{1, 512\}$ **16.** $\{6, 12\}$

17. $(-\infty, 12]$ **18.** $\left[\dfrac{21}{8}, \infty\right)$ **19.** $\left[-7, \dfrac{13}{2}\right)$

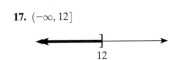 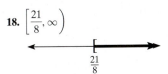 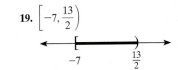

20. $\left(-\infty, -\dfrac{5}{3}\right]$ or $c\dfrac{1}{3}, \infty\right)$ **21.** $(-3, 4)$ **22.** $(3, 10)$

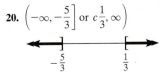 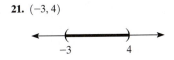 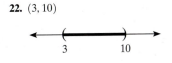

23. $47 + 16i$ **24.** $2 + i$ **25.** $38i$ **26.** $h = \dfrac{3V}{lw}$ **27.** $x = x_1 + \dfrac{y - y_1}{m}$ **28.** 2004; very well   **29.** 2007; very well

30. New York City: 55 days; Los Angeles: 213 days   **31.** 26 yr; \$33,600   **32.** \$3000 at 8%; \$7000 at 10%

33. length = 12 ft; width = 4 ft   **34.** 10 ft   **35.** \$47,500

CHAPTER 2

Section 2.1

Check Point Exercises

1. a. 6   **b.** $-\dfrac{7}{5}$ **2.** $y + 5 = 6(x - 2); y = 6x - 17$ **3.** $y + 1 = -5(x + 2); y = -5x - 11$

4.
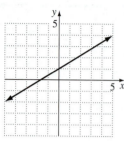

5.
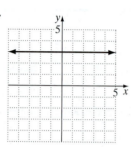

6.
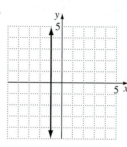

7. slope: $-\dfrac{1}{2}$; y-intercept: 2
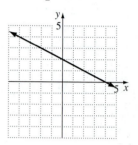

Exercise Set 2.1

1. $\dfrac{3}{4}$; rises   **3.** $\dfrac{1}{4}$; rises   **5.** 0; horizontal   **7.** -5; falls   **9.** undefined; vertical   **11.** $y - 5 = 2(x - 3); y = 2x - 1$

13. $y - 5 = 6(x + 2); y = 6x + 17$ **15.** $y + 3 = -3(x + 2); y = -3x - 9$ **17.** $y - 0 = -4(x + 4); y = -4x - 16$

19. $y + 2 = -1\left(x + \dfrac{1}{2}\right); y = -x - \dfrac{5}{2}$ **21.** $y - 0 = \dfrac{1}{2}(x - 0); y = \dfrac{1}{2}x$ **23.** $y + 2 = -\dfrac{2}{3}(x - 6); y = -\dfrac{2}{3}x + 2$

25. using $(1, 2), y - 2 = 2(x - 1); y = 2x$ **27.** using $(-3, 0), y - 0 = 1(x + 3); y = x + 3$

29. using $(-3, -1), y + 1 = 1(x + 3); y = x + 2$ **31.** using $(-3, -2), y + 2 = \dfrac{4}{3}(x + 3); y = \dfrac{4}{3}x + 2$

33. using $(-3, -1), y + 1 = 0(x + 3); y = -1$ **35.** using $(2, 4), y - 4 = 1(x - 2); y = x + 2$

37. using $(0, 4), y - 4 = 8(x - 0); y = 8x + 4$

39. $m = 2; b = 1$
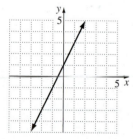

41. $m = -2; b = 1$
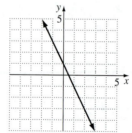

43. $m = \dfrac{3}{4}; b = -2$
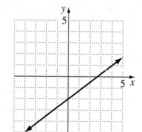

45. $m = -\dfrac{3}{5}; b = 7$
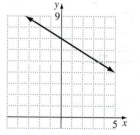

47.
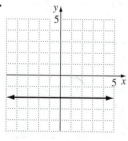

49.
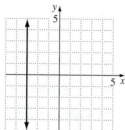

51.
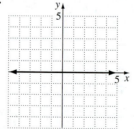

53. a. $y = -3x + 5$
b. $m = -3; b = 5$
c.
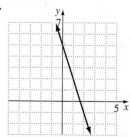

55. a. $y = -\dfrac{2}{3}x + 6$

b. $m = -\dfrac{2}{3}; b = 6$

c.

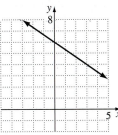

57. a. $y = 2x - 3$

b. $m = 2; b = -3$

c.

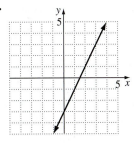

59. a. $x = 3$

b. m is undefined; no y-intercept

c.

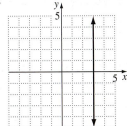

61. $y + 10 = -4(x + 8); y = -4x - 42$ **63.** $y + 3 = -5(x - 2); y = -5x + 7$ **65.** $y - 2 = \dfrac{2}{3}(x + 2); y = \dfrac{2}{3}x + \dfrac{10}{3}$

67. $y + 7 = -2(x - 4); y = -2x + 1$ **69.** $y = 15$ **71.** 111; The federal budget surplus is increasing $111 billion each year.

73. a. 16, In 1950, there were 16 workers per Social Security beneficiary.   **b.** −0.24; The number of workers per Social Security beneficiary is decreasing by 0.24 workers each year.   **c.** $y = -0.24x + 16$ **d.** 1.6; 5   **75. a.** $y - 30 = 4(x - 2)$ **b.** $y = 4x + 22$

c. 74   **77.** $y = -2.3x + 255$, where x is the percentage of adult females who are literate and y is under-five mortality per thousand; For each percent increase is adult female literacy, under-five mortality decreases by 2.3 per thousand.   **79.** $y = -0.7x + 60$; y represents the percentage of U.S. adults who read a newspaper x years after 1995.   **81.** $y = -500x + 29,500$; 4500 shirts

93. $m = -3$

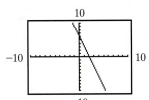

95. $m = \dfrac{3}{4}$

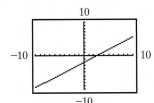

97. (c) is true.

99. a. m_1, m_3, m_2, m_4
     b. b_2, b_1, b_4, b_3

Section 2.2

Check Point Exercises

1. 5   **2.** $\left(4, -\dfrac{1}{2}\right)$ **3.** $x^2 + y^2 = 16$ **5.** center: $(-3, 1)$; radius: 2   **6.** $(x + 2)^2 + (y - 2)^2 = 9$

4. $(x - 5)^2 + (y + 6)^2 = 100$

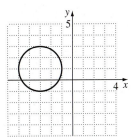

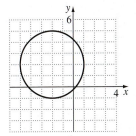

Exercise Set 2.2

1. 13   **3.** $2\sqrt{2} \approx 2.83$ **5.** 5   **7.** $\sqrt{29} \approx 5.39$ **9.** $4\sqrt{2} \approx 5.66$ **11.** $2\sqrt{5} \approx 4.47$ **13.** $2\sqrt{2} \approx 2.83$ **15.** $\sqrt{93} \approx 9.64$

17. $\sqrt{5} \approx 2.24$ **19.** $(4, 6)$ **21.** $(-4, -5)$ **23.** $\left(\dfrac{3}{2}, -6\right)$ **25.** $(-3, -2)$ **27.** $(1, 5\sqrt{5})$ **29.** $(2\sqrt{2}, 0)$ **31.** $x^2 + y^2 = 49$

33. $(x - 3)^2 + (y - 2)^2 = 25$ **35.** $(x + 1)^2 + (y - 4)^2 = 4$ **37.** $(x + 3)^2 + (y + 1)^2 = 3$ **39.** $(x + 4)^2 + (y - 0)^2 = 100$

41. center: $(0, 0)$
radius: 4

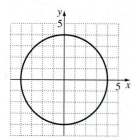

43. center: $(3, 1)$
radius: 6

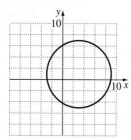

45. center: $(-3, 2)$
radius: 2

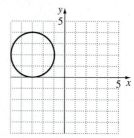

47. center: $(-2, -2)$
radius: 2

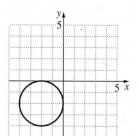

49. $(x + 3)^2 + (y + 1)^2 = 4$
center: $(-3, -1)$
radius: 2

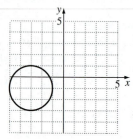

51. $(x - 5)^2 + (y - 3)^2 = 64$
center: $(5, 3)$
radius: 8

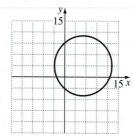

53. $(x + 4)^2 + (y - 1)^2 = 25$
center: $(-4, 1)$
radius: 5

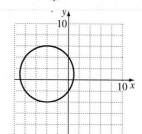

55. $(x - 1)^2 + (y - 0)^2 = 16$
center: $(1, 0)$
radius: 4

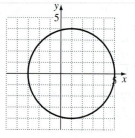

57. 0.5hr; 30 min   **59.** $x^2 + (y - 82)^2 = 4624$ **69.**

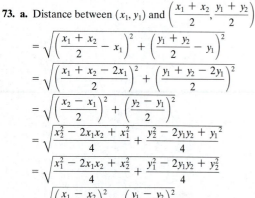

71. (d) is true.

73. a. Distance between (x_1, y_1) and $\left(\dfrac{x_1 + x_2}{2}, \dfrac{y_1 + y_2}{2} \right)$

$$= \sqrt{\left(\frac{x_1 + x_2}{2} - x_1 \right)^2 + \left(\frac{y_1 + y_2}{2} - y_1 \right)^2}$$

$$= \sqrt{\left(\frac{x_1 + x_2 - 2x_1}{2} \right)^2 + \left(\frac{y_1 + y_2 - 2y_1}{2} \right)^2}$$

$$= \sqrt{\left(\frac{x_2 - x_1}{2} \right)^2 + \left(\frac{y_2 - y_1}{2} \right)^2}$$

$$= \sqrt{\frac{x_2^2 - 2x_1x_2 + x_1^2}{4} + \frac{y_2^2 - 2y_1y_2 + y_1^2}{4}}$$

$$= \sqrt{\frac{x_1^2 - 2x_1x_2 + x_2^2}{4} + \frac{y_1^2 - 2y_1y_2 + y_2^2}{4}}$$

$$= \sqrt{\left(\frac{x_1 - x_2}{2} \right)^2 + \left(\frac{y_1 - y_2}{2} \right)^2}$$

$$= \sqrt{\left(\frac{x_1 + x_2 - 2x_2}{2} \right)^2 + \left(\frac{y_1 + y_2 - 2y_2}{2} \right)^2}$$

$$= \sqrt{\left(\frac{x_1 + x_2}{2} - x_2 \right)^2 + \left(\frac{y_1 + y_2}{2} - y_2 \right)^2}$$

$= $ Distance between (x_2, y_2) and $\left(\dfrac{x_1 + x_2}{2}, \dfrac{y_1 + y_2}{2} \right)$

b. $\sqrt{\left(\dfrac{x_2 - x_1}{2} \right)^2 + \left(\dfrac{y_2 - y_1}{2} \right)^2} + \sqrt{\left(\dfrac{x_2 - x_1}{2} \right)^2 + \left(\dfrac{y_2 - y_1}{2} \right)^2}$

$$= 2\sqrt{\left(\frac{x_2 - x_1}{2} \right)^2 + \left(\frac{y_2 - y_1}{2} \right)^2}$$

$$= 2\sqrt{\frac{(x_2 - x_1)^2 + (y_2 - y_1)^2}{4}}$$

$$= \sqrt{(x_2 - x_1)^2 + (y_2 - y_1)^2}$$

$= $ Distance from (x_1, y_1) to (x_2, y_2)

75. $(x + 3)^2 + (y - 2)^2 = 16$; $x^2 + y^2 + 6x - 4y - 3 = 0$ **77.** $y + 4 = \dfrac{3}{4}(x - 3)$

Section 2.3

Check Point Exercises

1. domain: $\{5, 10, 15, 20, 25\}$; range: $\{12.8, 16.2, 18.9, 20.7, 21.8\}$ **2. a.** not a function   **b.** function   **3. a.** $y = 6 - 2x$; function
b. $y = \pm\sqrt{1 - x^2}$, not a function   **4. a.** 42   **b.** $x^2 + 6x + 15$ **c.** $x^2 + 2x + 7$ **5. a.** $x^2 + 2hx + h^2 - 7x - 7h + 3$
b. $2x + h - 7$ **6. a.** 28   **b.** 33   **7. a.** $(-\infty, \infty)$ **b.** $\{x|x \neq -7, x \neq 7\}$ **c.** $[3, \infty)$.

Exercise Set 2.3

1. function; $\{1, 3, 5\}$; $\{2, 4, 5\}$ **3.** not a function; $\{3, 4\}$; $\{4, 5\}$ **5.** function; $\{-3, -2, -1, 0\}$; $\{-3, -2, -1, 0\}$
7. not a function; $\{1\}$; $\{4, 5, 6\}$ **9.** y is a function of x.   **11.** y is a function of x.   **13.** y is not a function of x.
15. y is not a function of x.   **17.** y is a function of x.   **19.** y is a function of x.   **21. a.** 29   **b.** $4x + 9$ **c.** $-4x + 5$
23. a. 2   **b.** $x^2 + 12x + 38$ **c.** $x^2 - 2x + 3$ **25. a.** 13   **b.** 1   **c.** $x^4 - x^2 + 1$ **d.** $81a^4 - 9a^2 + 1$ **27. a.** 3   **b.** 7
c. $\sqrt{x} + 3$ **29. a.** $\dfrac{15}{4}$ **b.** $\dfrac{15}{4}$ **c.** $\dfrac{4x^2 - 1}{x^2}$ **31. a.** 1   **b.** -1 **c.** 1   **33.** $4, h \neq 0$ **35.** $3, h \neq 0$ **37.** $2x + h, h \neq 0$

39. $2x + h - 4, h \neq 0$ **41.** $0, h \neq 0$ **43.** $-\dfrac{1}{x(x + h)}, h \neq 0$ **45. a.** -1 **b.** 7   **c.** 19   **47. a.** 3   **b.** 3   **c.** 0

49. a. 8   **b.** 3   **c.** 6   **51.** $(-\infty, \infty)$ **53.** $(-\infty, 4)$ or $(4, \infty)$ **55.** $(-\infty, -4)$ or $(-4, 4)$ or $(4, \infty)$
57. $(-\infty, -3)$ or $(-3, 7)$ or $(7, \infty)$ **59.** $(-\infty, -8)$ or $(-8, -3)$ or $(-3, \infty)$ **61.** $(-\infty, \infty)$ **63.** $[3, \infty)$ **65.** $(3, \infty)$
67. $[-7, \infty)$ **69.** $(-\infty, 12]$ **71.** $(-\infty, -2]$ or $[7, \infty)$ **73.** $[2, 5)$ or $(5, \infty)$
75. $\{(1, 31), (2, 53), (3, 70), (4, 86), (5, 86)|\}$; $\{1, 2, 3, 4, 5\}$; $\{31, 53, 70, 86\}$; Yes; Each member of the domain corresponds to exactly one
member of the range.   **77.** No; There is a member of the domain that corresponds to more than one member of the range.
79. 1713; There were 1713 gray wolves in the U.S. in 1990; Very well.   **81.** 19; Very well.   **83.** 5; Okay.
85. $f(0) = 200$; There were 200 thousand lawyers in the United States in 1951.   **87.** $f(50) = 1058$; There were 1058 thousand or
1,058,000 lawyers in the United States in the year 2001.   **89.** 8873; A person earning $40,000 owed $8873 in taxes.
91. $C = 100,000 + 100x$, where x is the number of bicycles produced; $C(90) = 109,000$; It cost $109,000 to produce 90 bicycles.
93. $T = \dfrac{40}{x} + \dfrac{40}{x + 30}$, where x is the rate on the outgoing trip; $T(30) = 2$; It takes 2 hours, traveling 30 mph outgoing and 60 mph returning.
103. $[1, \infty)$ **105.** $(-\infty, 5]$ **107.** Answers may vary.   **109.** $f(r_1) = 0$; r_1 is a solution to the equation $ax^2 + bx + c = 0$.

Section 2.4

Check Point Exercises

1. $(-3, 7), (-2, 2), (-1, -1),$
$(0, -2), (1, -1), (2, 2), (3, 7)$

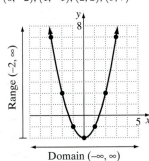

2. $f(4) = 1$; domain: $[0, 6)$; range: $(-2, 2]$
3. a. function
   b. function
   c. not a function
4. a. $f(10) \approx 16$ **b.** $x \approx 8$
5. increasing on $(-\infty, -1)$, decreasing on $(-1, 1)$, increasing on $(1, \infty)$
6. a. 1
   b. 7
   c. 4
7. a. even
   b. odd
   c. neither

Exercise Set 2.4

1. $(-3, 11), (-2, 6), (-1, 3),$
$(0, 2), (1, 3), (2, 6), (3, 11)$

3. $(0, -1), (1, 0), (4, 1), (9, 2)$

5. $(1, 0), (2, 1), (5, 2), (10, 3)$

7. $(-3, 2), (-2, 1), (-1, 0),$
$(0, -1), (1, 0), (2, 1), (3, 2)$

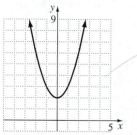

Domain: $(-\infty, \infty)$
Range: $[2, \infty)$

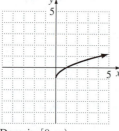

Domain: $[0, \infty)$
Range: $[-1, \infty)$

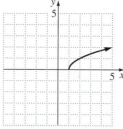

Domain: $[1, \infty)$
Range: $[0, \infty)$

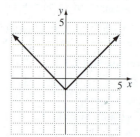

Domain: $(-\infty, \infty)$
Range: $[-1, \infty)$

9. $(-3, 4), (-2, 3), (-1, 2),$
$(0, 1), (1, 0), (2, 1), (3, 2)$

11. $(-3, 5), (-2, 5), (-1, 5),$
$(0, 5), (1, 5), (2, 5), (3, 5)$

13. $(-2, -10), (-1, -3),$
$(0, -2), (1, -1), (2, 6)$

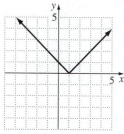

Domain: $(-\infty, \infty)$
Range: $[0, \infty)$

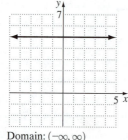

Domain: $(-\infty, \infty)$
Range: $\{5\}$

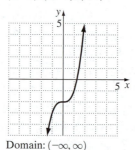

Domain: $(-\infty, \infty)$
Range: $(-\infty, \infty)$

15. a. $(-\infty, \infty)$ **b.** $[-4, \infty)$ **c.** -3 and 1 **d.** -3 **17. a.** $(-\infty, \infty)$ **b.** $[1, \infty)$ **c.** none **d.** 1
e. $f(-1) = 2$ and $f(3) = 4$ **19. a.** $[0, 5)$ **b.** $[-1, 5)$ **c.** 2 **d.** -1 **e.** $f(3) = 1$ **21. a.** $[0, \infty)$ **b.** $[1, \infty)$ **c.** none
d. 1 **e.** $f(4) = 3$ **23. a.** $[-2, 6]$ **b.** $[-2, 6]$ **c.** 4 **d.** 4 **e.** $f(-1) = 5$ **25. a.** $(-\infty, \infty)$ **b.** $(-\infty, -2]$ **c.** none
d. -2 **e.** $f(-4) = -5$ and $f(4) = -2$ **27. a.** $(-\infty, \infty)$ **b.** $(0, \infty)$ **c.** none **d.** 1 **29. a.** $\{-5, -2, 0, 1, 3\}$ **b.** $\{2\}$
c. none **d.** 2 **31.** function **33.** function **35.** not a function **37.** function **39. a.** increasing: $(-1, \infty)$
b. decreasing: $(-\infty, -1)$ **c.** constant: none **41. a.** increasing: $(0, \infty)$ **b.** decreasing: none **c.** constant: none
43. a. increasing: none **b.** decreasing: $(-2, 6)$ **c.** constant: none **45. a.** increasing: $(-\infty, -1)$ **b.** decreasing: none
c. constant: $(-1, \infty)$ **47. a.** increasing: $(-\infty, 0)$ or $(1.5, 3)$ **b.** decreasing: $(0, 1.5)$ or $(3, \infty)$ **c.** constant: none
49. a. increasing: $(-2, 4)$ **b.** decreasing: none **c.** constant: $(-\infty, -2)$ or $(4, \infty)$ **51. a.** $0; f(0) = 4$

b. $-3, 3; f(-3) = f(3) = 0$ **53. a.** $-2; f(-2) = 21$ **b.** $1; f(1) = -6$ **55.** 3 **57.** 10 **59.** $\dfrac{1}{5}$ **61.** odd **63.** neither

65. even **67.** even **69.** even **71.** odd **73.** even **75.** odd

77. $f(1.06) = 1$ **79.** $f\left(\dfrac{1}{3}\right) = 0$ **81.** $f(-2.3) = -3$ **83.** $f(60) \approx 3.1$; In 1960, Jewish Americans made up about 3.1% of the U.S.

population. **85.** $x \approx 19$ and $x \approx 64$; In 1919 and 1964, Jewish Americans made up about 3% of the U.S. population. **87.** 1940; 3.7%
89. Each year corresponds to only one percentage. **91.** increasing: $(45, 74)$; decreasing: $(16, 45)$; The number of accidents occurring per
50,000 miles driven increases with age starting at age 45, while it decreases with age starting at age 16. **93.** Answers will vary; an example
is 16 and 74 years old. For those ages, the number of accidents is 526.4 per 50 million miles.
95.

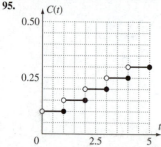

97. Answers may vary.
109. a.

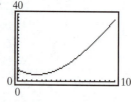

The number of doctor visits decreases during childhood
and then increases as you get older. The minimum is
$(20.29, 3.99)$, which means that the minimum number of
annual doctor visits, about 4, occurs at around age 20.

111.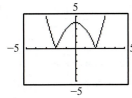

Increasing: $(-2, 0)$ or $(2, \infty)$
Decreasing: $(-\infty, -2)$ or $(0, 2)$

113.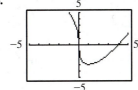

Increasing: $(1, \infty)$
Decreasing: $(-\infty, 1)$

115.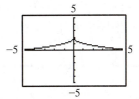

Increasing: $(-\infty, 0)$
Decreasing: $(0, \infty)$

117. (c) is true.
119. Answers may vary.

121.

Weight at least	Cost
0 oz.	$0.37
1	0.60
2	0.83
3	1.06
4	1.29

Section 2.5

Check Point Exercises

1.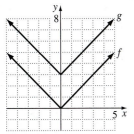

2.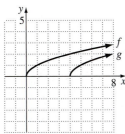

3.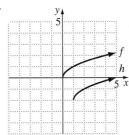

4.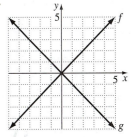

5.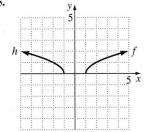

6.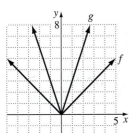

7.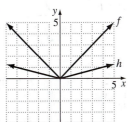

8.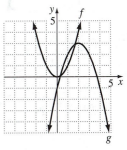

Exercise Set 2.5

1.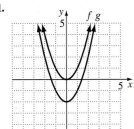

3.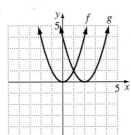

5.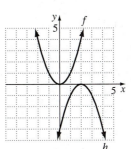

7.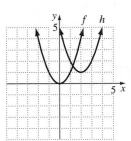

9.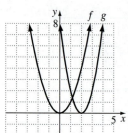

11.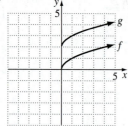

13.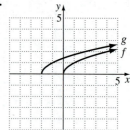

15.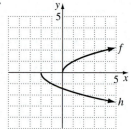

17.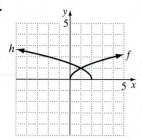

19.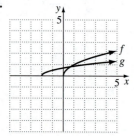

21.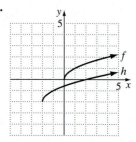

23.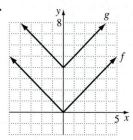

25.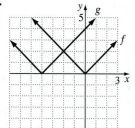

27.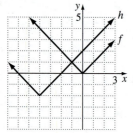

29.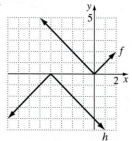

31.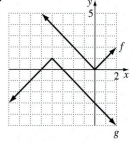

33.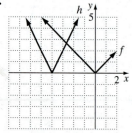

35.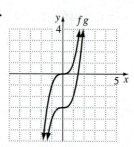

37.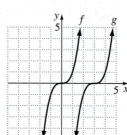

39.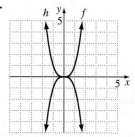

41.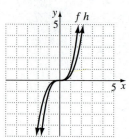

43.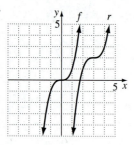

45.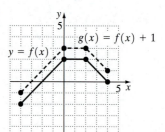

$g(x) = f(x) + 1$

$y = f(x)$

47.

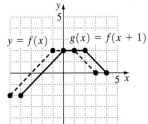

$y = f(x)$ $g(x) = f(x + 1)$

49.

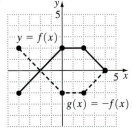

$y = f(x)$

$g(x) = -f(x)$

51.

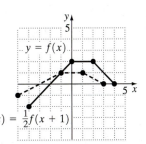

$y = f(x)$

$g(x) = \frac{1}{2}f(x + 1)$

53. $y = \sqrt{x - 2}$
55. $y = (x + 1)^2 - 4$

57. a. First, vertically stretch the graph of $f(x) = \sqrt{x}$ by the factor 2.9; then, vertically shift the result up 20.1 units.
b. 40.2 in.; Very well.   **c.** 0.9 in. per month   **d.** 0.2 in. per month; This is a much smaller rate of change; The graph is not as steep between 50 and 60 as it is between 0 and 10.

65. a.

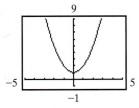

b.

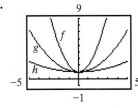

c. Answers may vary.
d. Answers may vary.
e. Answers may vary.

67. $g(x) = -(x + 4)^2$
69. $g(x) = -\sqrt{x - 2} + 2$
71. $(-a, b)$
73. $(a + 3, b)$

Section 2.6

Check Point Exercises

1. a. $(f + g)(x) = 3x^2 + 6x + 6$ **b.** $(f + g)(4) = 78$ **2. a.** $(f + g)(x) = \sqrt{x - 3} + \sqrt{x + 1}$ **b.** $[3, \infty)$

3. a. $(f - g)(x) = -x^2 + x - 4$ **b.** $(fg)(x) = x^3 - 5x^2 - x + 5$ \ **5. a.** $(f \circ g)(x) = \dfrac{4x}{1 + 2x}$ **b.** $\left\{x \middle| x \neq \text{ and } x \neq -\dfrac{1}{2}\right\}$

6. $f(x) = \sqrt{x}$ and $g(x) = x^2 + 5$, then $h(x) = (f \circ g)(x)$

Exercise Set 2.6

1. a. $(f + g)(x) = 2x^2 + 3x + 2$ **b.** $(f + g)(4) = 46$ **3. a.** $(f + g)(x) = \sqrt{x - 6} + \sqrt{x + 2}$ **b.** Domain: $[6, \infty)$
5. $(f + g)(x) = 3x + 2$; Domain: $(-\infty, \infty)$; $(f - g)(x) = x + 4$; Domain: $(-\infty, \infty)$; $(fg)(x) = 2x^2 + x - 3$;

Domain: $(-\infty, \infty)$; $\left(\dfrac{f}{g}\right)(x) = \dfrac{2x + 3}{x - 1}$; Domain: $(-\infty, 1)$ or $(1, \infty)$ **7.** $(f + g)(x) = 3x^2 + x - 5$;

Domain: $(-\infty, \infty)$; $(f - g)(x) = -3x^2 + x - 5$; Domain: $(-\infty, \infty)$; $(fg)(x) = 3x^3 - 15x^2$; Domain: $(-\infty, \infty)$; $\left(\dfrac{f}{g}\right)(x) = \dfrac{x - 5}{3x^2}$;

Domain: $(-\infty, 0)$ or $(0, \infty)$ **9.** $(f + g)(x) = 2x^2 - 2$; Domain: $(-\infty, \infty)$; $(f - g)(x) = 2x^2 - 2x - 4$;

Domain: $(-\infty, \infty)$; $(fg)(x) = 2x^3 + x^2 - 4x - 3$; Domain: $(-\infty, \infty)$; $\left(\dfrac{f}{g}\right)(x) = 2x - 3$; Domain: $(-\infty, -1)$ or $(-1, \infty)$

11. $(f + g)(x) = \sqrt{x} + x - 4$; Domain: $[0, \infty)$; $(f - g)(x) = \sqrt{x} - x + 4$; Domain: $[0, \infty)$; $(fg)(x) = \sqrt{x}(x - 4)$;

Domain: $[0, \infty)$; $\left(\dfrac{f}{g}\right)(x) = \dfrac{\sqrt{x}}{x - 4}$; Domain: $[0, 4)$ or $(4, \infty)$ **13.** $(f + g)(x) = \dfrac{2x + 2}{x}$; Domain: $(-\infty, 0)$ or $(0, \infty)$; $(f - g)(x) = 2$;

Domain: $(-\infty, 0)$ or $(0, \infty)$; $(fg)(x) = \dfrac{2x + 1}{x^2}$; Domain: $(-\infty, 0)$ or $(0, \infty)$; $\left(\dfrac{f}{g}\right)(x) = 2x + 1$; Domain: $(-\infty, 0)$ or $(0, \infty)$

15. $(f + g)(x) = \sqrt{x + 4} + \sqrt{x - 1}$; Domain: $[1, \infty)$; $(f - g)(x) = \sqrt{x + 4} - \sqrt{x - 1}$; Domain: $[1, \infty)$; $(fg)(x) = \sqrt{x^2 + 3x - 4}$;

Domain: $[1, \infty)$; $\left(\dfrac{f}{g}\right)(x) = \dfrac{\sqrt{x + 4}}{\sqrt{x - 1}}$; Domain: $(1, \infty)$ **17. a.** $(f \circ g)(x) = 2x + 14$ **b.** $(g \circ f)(x) = 2x + 7$ **c.** $(f \circ g)(2) = 18$

19. a. $(f \circ g)(x) = 2x + 5$ **b.** $(g \circ f)(x) = 2x + 9$ **c.** $(f \circ g)(2) = 9$ **21. a.** $(f \circ g)(x) = 20x^2 - 11$
b. $(g \circ f)(x) = 80x^2 - 120x + 43$ **c.** $(f \circ g)(2) = 69$ **23. a.** $(f \circ g)(x) = x^4 - 4x^2 + 6$ **b.** $(g \circ f)(x) = x^4 + 4x^2 + 2$
c. $(f \circ g)(2) = 6$ **25. a.** $(f \circ g)(x) = \sqrt{x - 1}$ **b.** $(g \circ f)(x) = \sqrt{x} - 1$ **c.** $(f \circ g)(2) = 1$ **27. a.** $(f \circ g)(x) = x$
b. $(g \circ f)(x) = x$ **c.** $(f \circ g)(2) = 2$ **29. a.** $(f \circ g)(x) = \dfrac{2x}{1 + 3x}$ **b.** $\left\{x \middle| x \neq 0 \text{ and } x \neq -\dfrac{1}{3}\right\}$ **31. a.** $(f \circ g)(x) = \dfrac{4}{4 + x}$
b. $\{x | x \neq 0 \text{ and } x \neq -4\}$ **33. a.** $(f \circ g)(x) = \sqrt{x + 3}$ **b.** $\{x | x \geq 0\}$ **35. a.** $(f \circ g)(x) = 5 - x$ **b.** $\{x | x \leq 1\}$

37. a. $(f \circ g)(x) = 8 - x^2$ **b.** $\{x | x < -2 \text{ or } x > 2\}$ **39.** $f(x) = x^4, g(x) = 3x - 1$ **41.** $f(x) = \sqrt[3]{x}, g(x) = x^2 - 9$

43. $f(x) = |x|, g(x) = 2x - 5$ **45.** $f(x) = \dfrac{1}{x}, g(x) = 2x - 3$ **47.** 0   **49.** 10   **51.** 2   **53.** 0   **55.** 4   **57.** −6   **59.** 20; In

2000, veterinary costs in the U.S. for dogs and cats were about $20 billion.   **61.** Domain of $D + C = \{1983, 1987, 1991, 1996, 2000\}$
63. $f + g$ represents the total world population in year x.   **65.** $(f + g)(2000) \approx 6$ billion people   **67.** $(R - C)(20,000) = -200,000$;
The company lost $200,000 since costs exceeded revenues; $(R - C)(30,000) = 0$; The company broke even since revenues equaled cost;
$(R - C)(40,000) = 200,000$; The company made a profit of $200,000.   **69. a.** f gives the price of the computer after a $400 discount. g
gives the price of the computer after a 25% discount.   **b.** $(f \circ g)(x) = 0.75x - 400$. This models the price of a computer after first a 25%
discount and then a $400 discount.   **c.** $(g \circ f)(x) = 0.75(x - 400)$. This models the price of a computer after first a $400 discount and
then a 25% discount.   **d.** The function $f \circ g$ models the greater discount, since the 25% discount is taken on the regular price first.

77.

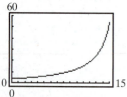

The per capita costs are
increasing over time.

79.

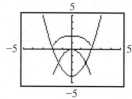

Domain of $f \circ g$ is $[-2, 2]$.

81. Assume f and g are even; then $f(-x) = f(x)$ and
$g(-x) = g(x)$. $(fg)(-x) = f(-x) \cdot g(-x) = f(x) \cdot g(x)$
$= (fg)(x)$, so fg is even.

83.

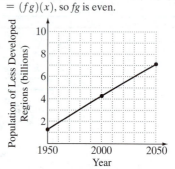

Section 2.7

Check Point Exercises

1. $f(g(x)) = x; g(f(x)) = x; f$ and g are inverses.
2. $f(g(x)) = x; g(f(x)) = x; f$ and g are inverses.

3. $f^{-1}(x) = \dfrac{x - 7}{2}$

4. $f^{-1}(x) \stackrel{?}{=} \sqrt{\dfrac{x + 1}{4}}$

5. (b) and (c) have inverse functions.

6.

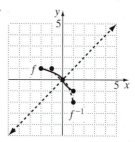

Exercise Set 2.7

1. $f(g(x)) = x; g(f(x)) = x; f$ and g are inverses.   **3.** $f(g(x)) = x; g(f(x)) = x; f$ and g are inverses.

5. $f(g(x)) = \dfrac{5x - 56}{9}; g(f(x)) = \dfrac{5x - 4}{9}; f$ and g are not inverses.   **7.** $f(g(x)) = x; g(f(x)) = x; f$ and g are inverses.

9. $f(g(x)) = x; g(f(x)) = x; f$ and g are inverses.   **11.** $f^{-1}(x) = x - 3$ **13.** $f^{-1}(x) = \dfrac{x}{2}$ **15.** $f^{-1}(x) = \dfrac{x - 3}{2}$

17. $f^{-1}(x) = \sqrt[3]{x - 2}$ **19.** $f^{-1}(x) = \sqrt[3]{x} - 2$ **21.** $f^{-1}(x) = \dfrac{1}{x}$ **23.** $f^{-1}(x) = x^2, x \ge 0$ **25.** $f^{-1}(x) = \sqrt{x - 1}; x \ge 1$

27. $f^{-1}(x) = \dfrac{3x + 1}{x - 2}; x \ne 2$ **29.** $f^{-1}(x) = (x - 3)^3 + 4$

31. The function is not one-to-one, so it does not have an inverse function.

33. The function is not one-to-one, so it does not have an inverse function.

35. The function is one-to-one, so it does have an inverse function.

37.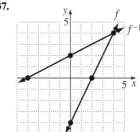

39.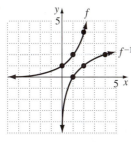

41. a. $f = \{($Zambia, $-7.3), ($Colombia, $-4.5), ($Poland, $-2.8),$
($\,$Italy, $-2.8), ($United States, $-1.8)\}$
b. $f^{-1} = \{(-7.3,$ Zambia$), (-4.5,$ Colombia$), (-2.8,$ Poland$),$
$(-2.8,$ Italy$), (-1.8,$ United States$)\}$; No; One member of the
domain, -2.8, corresponds to more than one member of the
range, Poland and Italy.

43. a. f is a one-to-one function. **b.** $f^{-1}(0.25)$ is the number of people in a room for a 25% probability of two people sharing a birthday.
$f^{-1}(0.5)$ is the number of people in a room for a 50% probability of two people sharing a birthday. $f^{-1}(0.7)$ is the number of people in a
room for a 70% probability of two people sharing a birthday.

45. $f(g(x)) = \dfrac{9}{5}\left[\dfrac{5}{9}(x - 32)\right] + 32 = x$ and $g(f(x)) = \dfrac{5}{9}\left[\left(\dfrac{9}{5}x + 32\right) - 32\right] = x$

53.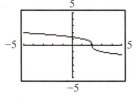

one-to-one

55.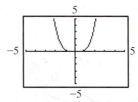

not one-to-one

57.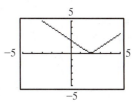

not one-to-one

59.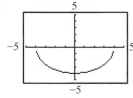

not one-to-one

61.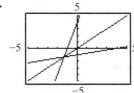

f and g are inverses.

63.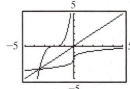

f and g are inverses.

65. $(f \circ g)^{-1}(x) = \dfrac{x - 15}{3}; (g^{-1} \circ f^{-1})(x) = \dfrac{x}{3} - 5 = \dfrac{x - 15}{3}$

67. No; The space craft was at the same height, $s(t)$, for two different values of t-once during the ascent and once again during the descent.

Chapter 2 Review Exercises

1. $m = -\dfrac{1}{2}$; falls **2.** $m = 1$; rises **3.** $m = 0$; horizontal **4.** $m =$ undefined; vertical **5.** $y - 2 = -6(x + 3); y = -6x - 16$

6. using $(1, 6), y - 6 = 2(x - 1); y = 2x + 4$

7. Slope: $\dfrac{2}{5}$; y-intercept: -1 **8.** Slope: -4; y-intercept: 5 **9.** Slope: $-\dfrac{2}{3}$; y-intercept: -2 **10.** Slope: 0; y-intercept: 4

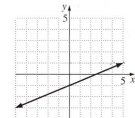

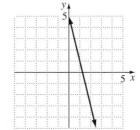

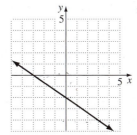

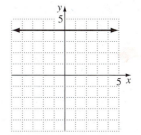

11. a. $y - 480 = 40(x - 2)$ **b.** $y = 40x + 400$ **c.** $1200 billion. **12. a.** Answers may vary. **b.** Answers may vary.
c. Answers may vary. **13.** $y + 7 = -3(x - 4); y = -3x + 5$ **14.** $y - 6 = -3(x + 3); y = -3x - 3$ **15.** 13

16. $2\sqrt{2} \approx 2.83$ **17.** $(-5, 5)$ **18.** $\left(-\dfrac{11}{2}, -2\right)$ **19.** $x^2 + y^2 = 9$ **20.** $(x + 2)^2 + (y - 4)^2 = 36$

21. Center: $(0, 0)$; radius: 1

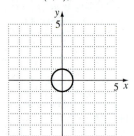

22. Center: $(-2, 3)$; radius: 3

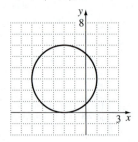

23. Center: $(2, -1)$; radius: 3

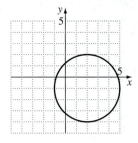

24. Function; Domain: $\{2, 3, 5\}$; Range: $\{7\}$ **25.** Function; Domain: $\{1, 2, 13\}$; Range: $\{10, 500, \pi\}$

26. Not a function; Domain: $\{12, 14\}$; Range: $\{13, 15, 19\}$ **27.** y is a function of x.   **28.** y is a function of x.

29. y is not a function of x.   **30. a.** $f(4) = -23$ **b.** $f(x + 3) = -7x - 16$ **c.** $f(-x) = 5 + 7x$

31. a. $g(0) = 2$ **b.** $g(-2) = 24$ **c.** $g(x - 1) = 3x^2 - 11x + 10$ **d.** $g(-x) = 3x^2 + 5x + 2$

32. a. $f(a) = 4a - 3$ **b.** $f(a + h) = 4a + 4h - 3$ **c.** $\dfrac{f(a + h) - f(a)}{h} = 4$ **d.** $f(a) + f(h) = 4a + 4h - 6$

33. a. $g(13) = 3$ **b.** $g(0) = 4$ **c.** $g(-3) = 7$ **34.** 8   **35.** $2x + h - 13$ **c.** $f(2) = 3$ **36.** $(-\infty, \infty)$

37. $(-\infty, 7)$ or $(7, \infty)$ **38.** $(-\infty, 4]$ **39.** $(-\infty, -1)$ or $(-1, 1)$ or $(1, \infty)$ **40.** $[2, 5)$ or $(5, \infty)$

41. Ordered pairs: $(-1, 9)$, $(0, 4)$, $(1, 1)$, $(2, 0)$, $(3, 1)$, $(4, 4)$.

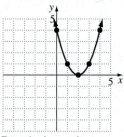

Domain: $(-\infty, \infty)$
Range: $[0, \infty)$

42. Ordered pairs: $(-1, 3)$, $(0, 2)$, $(1, 1)$, $(2, 0)$, $(3, 1)$, $(4, 2)$.

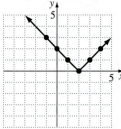

Domain: $(-\infty, \infty)$
Range: $[0, \infty)$

43. a. Domain: $[-3, 5)$
b. Range: $[-5, 0]$
c. x-intercept: -3
d. y-intercept: -2
e. increasing: $(-2, 0)$ or $(3, 5)$
   decreasing: $(-3, -2)$ or $(0, 3)$
f. $f(-2) = -3$ and $f(3) = -5$

44. a. Domain: $(-\infty, \infty)$
b. Range: $(-\infty, \infty)$
c. x-intercepts: -2 and 3
d. y-intercept: 3
e. increasing: $(-5, 0)$;
   decreasing: $(-\infty, -5)$ or $(0, \infty)$
f. $f(-2) = 0$ and $f(6) = -3$

45. a. Domain: $(-\infty, \infty)$ **b.** Range: $[-2, 2]$ **c.** x-intercept: 0   **d.** y-intercept: 0   **e.** increasing: $(-2, 2)$; constant: $(-\infty, -2)$ or $(2, \infty)$ **f.** $f(-9) = -2$ and $f(14) = 2$ **46. a.** 0; $f(0) = -2$ **b.** $-2, 3$; $f(-2) = -3$, $f(3) = -5$

47. a. 0; $f(0) = 3$ **b.** -5; $f(-5) = -6$ **48.** not a function   **49.** function   **50.** function   **51.** not a function   **52.** 10

53. about $1167   **54.** odd; symmetric with respect to the origin   **55.** even; symmetric with respect to the y-axis

56. odd; symmetric with respect to the origin   **57. a.** yes; The graph passes the vertical line test.   **b.** Decreasing: $(3, 12)$;
The vulture descended.   **c.** Constant: $(0, 3)$ and $(12, 17)$; The vulture's height held steady during the first 3 seconds and the vulture
was on the ground for 5 seconds.   **d.** Increasing: $(17, 30)$; The vulture was ascending.

58.

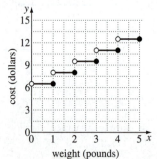

59.

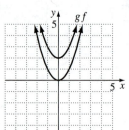

60.

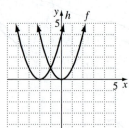

61.

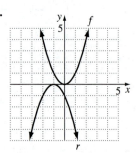

62.
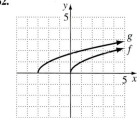

63.
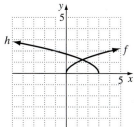

64.
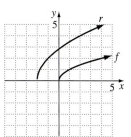

65.
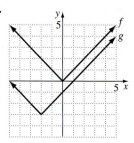

66.
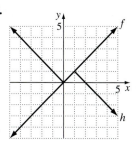

67.
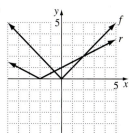

68.
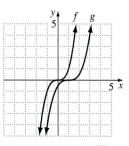

69.
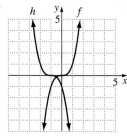

70.
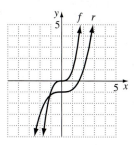

71.
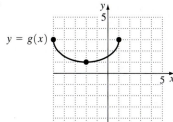

72.
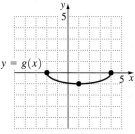

73.
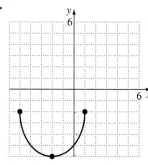

74. $(f + g)(x) = 4x - 6$; Domain: $(-\infty, \infty)$; $(f - g)(x) = 2x + 4$; Domain: $(-\infty, \infty)$;

$(fg)(x) = 3x^2 - 16x + 5$; Domain: $(-\infty, \infty)$; $\left(\dfrac{f}{g}\right)(x) = \dfrac{3x - 1}{x - 5}$; Domain: $(-\infty, 5)$ or $(5, \infty)$

75. $(f + g)(x) = 2x^2 + x$; Domain: $(-\infty, \infty)$; $(f - g)(x) = x + 2$; Domain: $(-\infty, \infty)$;

$(fg)(x) = x^4 + x^3 - x - 1$; Domain: $(-\infty, \infty)$; $\left(\dfrac{f}{g}\right)(x) = \dfrac{x^2 + x + 1}{x^2 - 1}$;

Domain: $(-\infty, -1)$ or $(-1, 1)$ or $(1, \infty)$

76. $(f + g)(x) = \sqrt{x + 7} + \sqrt{x - 2}$; Domain: $[2, \infty)$; $(f - g)(x) = \sqrt{x + 7} - \sqrt{x - 2}$;

Domain: $[2, \infty)$; $(fg)(x) = \sqrt{x^2 + 5x - 14}$; Domain: $[2, \infty)$; $\left(\dfrac{f}{g}\right)(x) = \dfrac{\sqrt{x + 7}}{\sqrt{x - 2}}$;

Domain: $(2, \infty)$

77. a. $(f \circ g)(x) = 16x^2 - 8x + 4$ **b.** $(g \circ f)(x) = 4x^2 + 11$ **c.** $(f \circ g)(3) = 124$ **78. a.** $(f \circ g)(x) = \sqrt{x + 1}$

b. $(g \circ f)(x) = \sqrt{x} + 1$ **c.** $(f \circ g)(3) = 2$ **79. a.** $\dfrac{1 + x}{1 - 2x}$ **b.** $\{x | x \neq 0 \text{ and } x \neq \frac{1}{2}\}$ **80. a.** $\sqrt{x + 2}$ **b.** $\{x | x \geq -2\}$

81. $f(x) = x^4, g(x) = x^2 + 2x - 1$ **82.** $f(x) = \sqrt[3]{x}, g(x) = 7x + 4$ **83.** $f(g(x)) = x - \dfrac{7}{10}; g(f(x)) = x - \dfrac{7}{6}; f$ and g are not

inverses of each other.   **84.** $f(g(x)) = x; g(f(x)) = x; f$ and g are inverses of each other.   **85.** $f^{-1}(x) = \dfrac{x + 3}{4}$

86. $f^{-1}(x) = x^2 - 2$ for $x \geq 0$ **87.** $f^{-1}(x) = \sqrt[3]{\dfrac{x - 1}{8}}$

88. Inverse function exists.
89. Inverse function does not exist.
90. Inverse function exists.
91. Inverse function does not exist.

92.

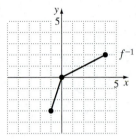

Chapter 2 Test

1. using $(2, 1)$, $y - 1 = 3(x - 2)$; $y = 3x - 5$ **2.** $y - 6 = 4(x + 4)$; $y = 4x + 22$ **3. a.** $y = 42x + 320$ **b.** 866
4. Center: $(-2, 3)$; radius: 4   **5.** b, c, d   **6.** $f(x - 1) = x^2 - 4x + 8$ **7.** $g(-1) = 4$; $g(7) = 2$ **8.** Domain: $(-\infty, 4]$
9. $2x + h + 11$ **10. a.** $f(4) - f(-3) = 5$ **b.** Domain: $(-5, 6]$ **c.** Range: $[-4, 5]$ **d.** Increasing: $(-1, 2)$
e. Decreasing: $(-5, -1)$ or $(2, 6)$ **f.** 2; $f(2) = 5$ **g.** -1; $f(-1) = -4$ **h.** -4, 1, and 5   **i.** -3 **11.** 48   **12.** $f(x)$ is even and is
symmetric with respect to the y-axis. The graph in the figure is symmetric with respect to the origin.   **13.** The graph of f is shifted 3 to
the right to obtain the graph of g. Then the graph of g is stretched by a factor of 2 and reflected about the x-axis to obtain the graph of h.

14.

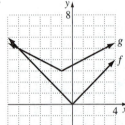

15. $(f - g)(x) = x^2 - 2x - 2$.   **16.** $\left(\dfrac{f}{g}\right)(x) = \dfrac{x^2 + 3x - 4}{5x - 2}$; Domain: $\left(-\infty, \dfrac{2}{5}\right)$ or $\left(\dfrac{2}{5}, \infty\right)$

17. $(f \circ g)(x) = 25x^2 - 5x - 6$ **18.** $(g \circ f)(x) = 5x^2 + 15x - 22$.

19. $f(g(2)) = 84$.   **20.** $\dfrac{7x}{2 - 4x}$; $\{x | x \neq 0 \text{ or } x \neq \frac{1}{2}\}$.   **21.** $f(x) = x^7$, $g(x) = 2x + 13$

22. $f^{-1}(x) = x^2 + 2$ for $x \geq 0$

23. a. The graph of f passes the Horizontal Line Test.
  b. $f(80) = 2000$
  c. $f^{-1}(2000)$ is the income, in thousands of dollars, for those who give $2000 to charity.

24.

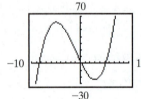

  a. not one-to-one
  b. neither; f is not symmetric about the origin or the y-axis
  c. $(-\infty, \infty)$
  d. Increasing: $(-\infty, -5)$ or $(3, \infty)$
  e. Decreasing: $(-5, 3)$
  f. -5; $f(-5) = \dfrac{184}{3}$
  g. 3; $f(3) = -24$

Cumulative Review Exercises (Chapters P–2)

1. $\dfrac{2y^4}{x^3}$ **2.** $\dfrac{5\sqrt{2}}{8}$ **3.** $(x - 4)(x^2 + 2)$ **4.** $\dfrac{2x^2 - x + 6}{(x + 4)(x - 2)}$ **5.** $\dfrac{2x + 1}{2x - 1}$ **6.** $x = -4$ or $x = 5$ **7.** $x = \dfrac{25}{18}$ **8.** $x = 4$
9. $x = -8$ or $x = 27$ **10.** $x \leq 20$; $(-\infty, 20]$ **11.** $(-\infty, 2)$ or $[7, \infty)$ **12.** $y - 5 = 4(x + 2)$; $y = 4x + 13$

13.

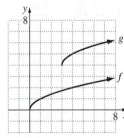

14. $f^{-1}(x) = (x - 2)^2 + 3$
15. $\dfrac{3 - (x + h)^2 - (3 - x^2)}{h} = -2x - h$
16. $c = \dfrac{Ad}{d - A}$
17. $1500 at 7\%, $4500 at 9\%
18. $2000
19. 3 ft by 8 ft
20. You must make an 85 on the final exam to have an average score of 80.

CHAPTER 3

Section 3.1

Check Point Exercises

1.
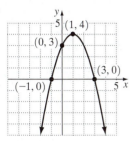

2.
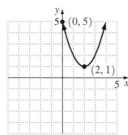

3.
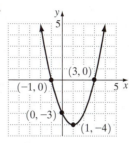

4. $45; 190$

Exercise Set 3.1

1. $h(x) = (x - 1)^2 + 1$ **3.** $j(x) = (x - 1)^2 - 1$ **5.** $h(x) = x^2 - 1$ **7.** $g(x) = x^2 - 2x + 1$ **9.** $(3, 1)$ **11.** $(-1, 5)$
13. $(2, -5)$ **15.** $(-1, 9)$

17. Domain: $(-\infty, \infty)$
Range: $[-1, \infty)$
axis of symmetry: $x = 4$

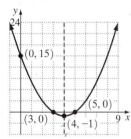

19. Domain: $(-\infty, \infty)$
Range: $[2, \infty)$
axis of symmetry: $x = 1$

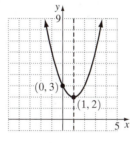

21. Domain: $(-\infty, \infty)$
Range: $[1, \infty)$
axis of symmetry: $x = 3$

23. Domain: $(-\infty, \infty)$
Range: $[-1, \infty)$
axis of symmetry: $x = -2$

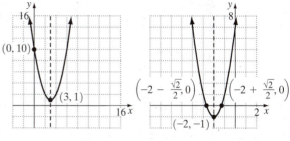

25. Domain: $(-\infty, \infty)$

Range: $(-\infty, 4]$

axis of symmetry: $x = 1$

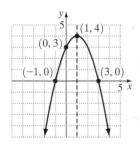

27. Domain: $(-\infty, \infty)$

Range: $[-4, \infty)$

axis of symmetry: $x = 1$

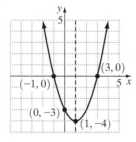

29. Domain: $(-\infty, \infty)$

Range: $\left[-\dfrac{49}{4}, \infty\right)$

axis of symmetry: $x = -\dfrac{3}{2}$

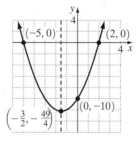

31. Domain: $(-\infty, \infty)$

Range: $(-\infty, 4]$

axis of symmetry: $x = 1$

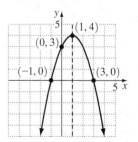

33. Domain: $(-\infty, \infty)$

Range: $(-\infty, -1]$

axis of symmetry: $x = 1$

35. minimum; $(2, -13)$

37. maximum; $(1, 1)$

39. minimum; $\left(\dfrac{1}{2}, -\dfrac{5}{4}\right)$

41. 1968; 4238 cigarettes per person; Yes

43. 6.25 s; 629 ft

45. The graph has the shape of a parabola.

47. 30 ft; 60 ft; 1800 ft^2

49. 5 in.

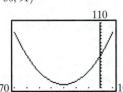

59. a.

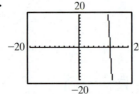

You can only see a little of the parabola.

b. $(20.5, -120.5)$
c. Ymax $= 750$
d. You can choose Xmin and Xmax so the x-value of the vertex is in the center of the graph. Choose Ymin to include the y-value of the vertex.

61. $(2.5, 185)$

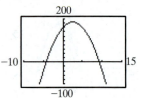

63. $(-30, 91)$

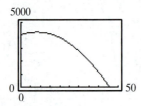

65. a.

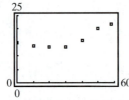

The data decrease and then increase.

b. $y = 0.01x^2 - 0.22x + 15.10$
c. $x \approx 18$; $1940 + 18 = 1958$;
worst year: 1958;
fuel efficiency: about 13.1 mpg

d.

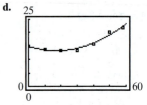

67. Answers may vary.

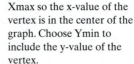

69. $x = 3$; $(0, 11)$

Section 3.2

Check Point Exercises

1. The graph rises to the left and to the right.
2. Since n is odd and the leading coefficient is negative, the function falls to the right. Since the ratio cannot be negative, the model won't be appropriate.
3. No; the graph should fall to the left, but doesn't appear to.
4. $\{-2, 2\}$
5. $\{-2, 0, 2\}$

6.

Exercise Set 3.2

1. polynomial function; degree: 3   **3.** polynomial function; degree: 5   **5.** not a polynomial function   **7.** not a polynomial function
9. not a polynomial function   **11.** polynomial function   **13.** not a polynomial function   **15.** (c)   **17.** (b)   **19.** (a)
21. falls to the left and rises to the right   **23.** rises to the left and to the right   **25.** falls to the left and to the right
27. $x = 5$ has multiplicity 1; The graph crosses the x-axis; $x = -4$ has multiplicity 2; The graph touches the x-axis and turns around.
29. $x = 3$ has multiplicity 1; The graph crosses the x-axis; $x = -6$ has multiplicity 3; The graph crosses the x-axis.
31. $x = 0$ has multiplicity 1; The graph crosses the x-axis; $x = 1$ has multiplicity 2; The graph touches the x-axis and turns around.
33. $x = 2$, $x = -2$ and $x = -7$ have multiplicity 1; The graph crosses the x-axis.

35. a. $f(x)$ rises to the right and falls to the left.
  b. $x = -2$, $x = 1$, $x = -1$;
   $f(x)$ crosses the x-axis at each.
  c. The y-intercept is -2..
  d. neither
  e.

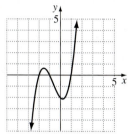

37. a. $f(x)$ rises to the left and the right.
  b. $x = 0$, $x = 3$, $x = -3$;
   $f(x)$ crosses the x-axis at -3 and 3;
   $f(x)$ touches the x-axis at 0.
  c. The y-intercept is 0.
  d. y-axis symmetry
  e.

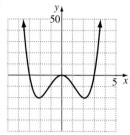

39. a. $f(x)$ falls to the left and the right.
  b. $x = 0$, $x = 4$, $x = -4$;
   $f(x)$ crosses the x-axis at -4 and 4;
   $f(x)$ touches the x-axis at 0.
  c. The y-intercept is 0.
  d. y-axis is symmetry
  e.

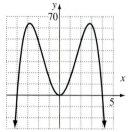

41. a. $f(x)$ rises to the left and the right.
  b. $x = 0$, $x = 1$;
   $f(x)$ touches the x-axis at 0 and 1.
  c. The y-intercept is 0.
  d. neither
  e.

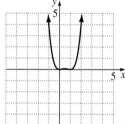

43. a. $f(x)$ falls to the left and the right.
  b. $x = 0$, $x = 2$;
   $f(x)$ crosses the x-axis at 0 and 2.
  c. The y-intercept is 0.
  d. neither
  e.

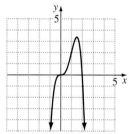

45. a. $f(x)$ rises to the left and falls to the right.
  b. $x = 0$, $x = \pm\sqrt{3}$;
   $f(x)$ crosses the x-axis at $(0, 0)$;
   $f(x)$ touches the x-axis at $\sqrt{3}$ and $-\sqrt{3}$.
  c. The y-intercept is 0.
  d. origin symmetry
  e.

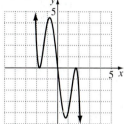

47. a. $f(x)$ rises to the left and falls to the right.
  b. $x = 0$, $x = 3$;
   $f(x)$ crosses the x-axis at 3;
   $f(x)$ touches the x-axis at 0.
  c. The y-intercept is 0.
  d. neither
  e.

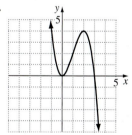

49. a. $f(x)$ falls to the left and the right.
  b. $x = 1$, $x = -2$, $x = 2$;
   $f(x)$ crosses the x-axis at -2 and 2;
   $f(x)$ touches the x-axis at 1.
  c. The y-intercept is 12.
  d. neither
  e.

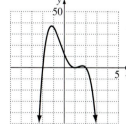

51. a. Leading coefficient test suggests the elk population will decline and eventually will die off.

b.

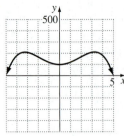

c.

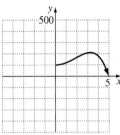

The population reaches extinction at the end of 5 years.

53. No; eventually the function would predict a negative number of larceny thefts, which is impossible.

55. degree 4; positive; the graph rises to the left and to the right

71. Answers may vary.

73. Answers may vary.

75.

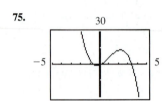

77.

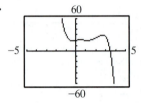

79.

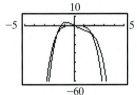

81. $f(x) = x^3 + x^2 - 12x$

Section 3.3

Check Point Exercises

1. $x + 5$ **2.** $2x^2 + 3x - 2 + \dfrac{1}{x - 3}$ **3.** $2x^2 + 7x + 14 + \dfrac{21x - 10}{x^2 - 2x}$ **4.** $x^2 - 2x - 3$ **5.** -105 **6.** $\left\{-1, -\dfrac{1}{3}, \dfrac{2}{5}\right\}$

Exercise Set 3.3

1. $x + 3$ **3.** $x^2 + 3x + 1$ **5.** $2x^2 + 3x + 5$ **7.** $4x + 3 + \dfrac{2}{3x - 2}$ **9.** $2x^2 + x + 6 - \dfrac{38}{x + 3}$

11. $4x^3 + 16x^2 + 60x + 246 + \dfrac{984}{x - 4}$ **13.** $2x + 5$ **15.** $6x^2 + 3x - 1 - \dfrac{3x - 1}{3x^2 + 1}$ **17.** $2x + 5$ **19.** $3x - 8 + \dfrac{20}{x + 5}$

21. $4x^2 + x + 4 + \dfrac{3}{x - 1}$ **23.** $6x^4 + 12x^3 + 22x^2 + 48x + 93 + \dfrac{187}{x - 2}$ **25.** $x^3 - 10x^2 + 51x - 260 + \dfrac{1300}{x + 5}$

27. $x^4 + x^3 + 2x^2 + 2x + 2$ **29.** $x^3 + 4x^2 + 16x + 64$ **31.** $2x^4 - 7x^3 + 15x^2 - 31x + 64 - \dfrac{129}{x + 2}$ **33.** -25 **35.** 4729

37. $x^2 - 5x + 6$; $x = -1, x = 2, x = 3$ **39.** $\left\{-\dfrac{1}{2}, 1, 2\right\}$ **41.** $\left\{-\dfrac{3}{2}, -\dfrac{1}{3}, \dfrac{1}{2}\right\}$ **43.** $x^3 + 5x^2 - 9x - 45$ **45. a.** 70

b. $80 + \dfrac{800}{x - 110}$; $f(30) = 70$; yes   **c.** No, f is a rational function because it is a quotient of two polynomials.

57.

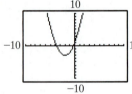

The division is correct.

59.

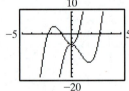

The division is not correct.
The right side should be 3x‹-8x -5.

61. $k = -12$

63. $x^{2n} - x^n + 1$

Section 3.4

Check Point Exercises

1. $\pm 1, \pm 2, \pm 3, \pm 6$ **2.** $\pm 1, \pm 3, \pm \frac{1}{2}, \pm \frac{1}{4}, \pm \frac{3}{2}, \pm \frac{3}{4}$ **3.** $\{-5, -4, 1\}$ **4.** $\{1, 2 - 3i, 2 + 3i\}$
5. 4, 2, or 0 positive zeros, no possible negative zeros

Exercise Set 3.4

1. $\pm 1, \pm 2, \pm 4$ **3.** $\pm 1, \pm 2, \pm 3, \pm 6, \pm \frac{1}{3}, \pm \frac{2}{3}$ **5.** $\pm 1, \pm 2, \pm 3, \pm 6, \pm \frac{1}{2}, \pm \frac{1}{4}, \pm \frac{3}{2}, \pm \frac{3}{4}$ **7.** $\pm 1, \pm 2, \pm 3, \pm 4, \pm 6, \pm 12$ **9. a.** $\pm 1, \pm 2, \pm 4$

b. 2 is a zero **c.** $\{2, -2, -1\}$ **11. a.** $\pm 1, \pm 2, \pm 3, \pm 6, \pm \frac{1}{2}, \pm \frac{3}{2}$ **b.** 3 is a zero **c.** $\left\{3, \frac{1}{2}, -2\right\}$

13. a. $\pm 1, \pm 2, \pm 4, \pm 8, \pm \frac{1}{3}, \pm \frac{2}{3}, \pm \frac{4}{3}, \pm \frac{8}{3}$ **b.** 2 is a zero **c.** $\left\{2, -\frac{1}{3}, -4\right\}$ **15. a.** $\pm 1, \pm 2, \pm 3, \pm 4, \pm 6, \pm 12$ **b.** 4 is a root

c. $\{-3, 1, 4\}$ **17. a.** $\pm 1, \pm 2, \pm 3, \pm 4, \pm 6, \pm 12$ **b.** -2 is a root **c.** $\{-2, 1 + \sqrt{7}, 1 - \sqrt{7}\}$ **19. a.** $\pm 1, \pm 5, \pm \frac{1}{2}, \pm \frac{5}{2}, \pm \frac{1}{3}, \pm \frac{5}{3}, \pm \frac{1}{6}, \pm \frac{5}{6}$

b. -5 is a root **c.** $\left\{-5, \frac{1}{2}, \frac{1}{3}\right\}$ **21. a.** $\pm 1, \pm 2, \pm 4$ **b.** 2 is a root **c.** $\{-2, 2, 1 + \sqrt{2}, 1 - \sqrt{2}\}$

23. no positive real roots; 3 or 1 negative real roots **25.** 3 or 1 positive real roots; no negative real roots

27. 2 or 0 positive real roots; 2 or 0 negative real roots **29.** $x = -2, x = 5, x = 1$ **31.** $\left\{-\frac{1}{2}, \frac{1 + \sqrt{17}}{2}, \frac{1 - \sqrt{17}}{2}\right\}$

33. $\{-1, -2 + 2i, -2 - 2i\}$ **35.** $\{-1, -2, 3 + \sqrt{13}, 3 - \sqrt{13}\}$ **37.** $x = -1, x = 2, x = -\frac{1}{3}, x = 3$ **39.** $\left\{1, -\frac{3}{4}, i\sqrt{2}, -i\sqrt{2}\right\}$

41. $\left\{-2, \frac{1}{2}, \sqrt{2}, -\sqrt{2}\right\}$ **43. a.** $x = 40$; at age 40, about 27% of art productivity occurs **b.** degree 2; leading coefficient: negative

45. $W = 3$ mm **47.** 2 in. by 9 in. by 4 in. **57.** $\frac{1}{2}, \frac{2}{3}, 2$ **59.** $\pm \frac{1}{2}$ **61.** 5, 3, or 1 positive real roots exist **63.** (d) is true. **65.** 3 in.

Section 3.5

Check Point Exercises

1.

2	2	11	-7	-6
		4	30	46
	2	15	23	40

All the numbers are nonnegative.

-7	2	11	-7	-6
		-14	21	-98
	2	-3	14	-104

The signs alternate.

2. $f(-3) = -42; f(-2) = 5$
3. $\{-3, 7, 2 + i, 2 - i\}$
4. a. $(x^2 - 5)(x^2 + 1)$
　　b. $(x + \sqrt{5})(x - \sqrt{5})(x^2 + 1)$
　　c. $(x + \sqrt{5})(x - \sqrt{5})(x + i)(x - i)$
5. $f(x) = x^3 + 3x^2 + x + 3$

Exercise Set 3.5

1.

-4	1	-5	11	33	-18
		-4	36	-188	620
	1	-9	47	-155	602

Since signs alternate, -4 is a lower bound.

7	1	-5	11	33	-18
		7	14	175	1456
	1	2	25	208	1438

Since no sign is negative, 7 is an upper bound.

3.

-4	2	5	-8	-7
		-8	12	-16
	2	-3	4	-23

Since signs alternate, -4 is a lower bound.

2	2	5	-8	-7
		4	18	20
	2	9	10	13

Since no sign is negative, 2 is an upper bound.

5. a. $\pm 1, \pm 2, \pm 3, \pm 4, \pm 6, \pm 12$ **b.** 1 is not a root. 1 is an upper bound. **c.** Eliminate all positive possible rational roots.
d. -3 is not a root. -3 is a lower bound. **e.** Eliminate $-3, -4, -6$ and -12. **7.** $f(1) = -1; f(2) = 5; 1.3$
9. $f(-1) = -1; f(0) = 1; -0.5$ **11.** $f(-3) = -11; f(-2) = 1; -2.1$ **13.** $f(-3) = -42; f(-2) = 5; -2.2$ **15.** $\{-2i, 2i, 2\}$

17. $\left\{1 - i, 1 + i, \dfrac{1}{3}\right\}$ **19.** $\{2 - i, 2 + i, -2 + i, -2 - i\}$ **21.** $\{2 - i, 2 + i, -3, 7\}$ **23. a.** $(x^2 - 5)(x^2 + 4)$

b. $(x + \sqrt{5})(x - \sqrt{5})(x^2 + 4)$ **c.** $(x + \sqrt{5})(x - \sqrt{5})(x + 2i)(x - 2i)$ **25. a.** $(x^2 - 2)(x^2 + 3)$

b. $(x + \sqrt{2})(x - \sqrt{2})(x^2 + 3)$ **c.** $(x + \sqrt{2})(x - \sqrt{2})(x + i\sqrt{3})(x - i\sqrt{3})$ **27. a.** $(x - 3)(x + 1)(x^2 + 4)$

b. $(x - 3)(x + 1)(x^2 + 4)$ **c.** $(x - 3)(x + 1)(x + 2i)(x - 2i)$ **29.** $f(x) = 2x^3 - 2x^2 + 50x - 50$

31. $f(x) = x^3 - 3x^2 - 15x + 125$ **33.** $f(x) = x^4 + 10x^2 + 9$ **35.** $f(x) = x^4 - 9x^3 + 21x^2 + 21x - 130$

37. $x = 1; x = \pm 5i; f(x) = (x - 1)(x - 5i)(x + 5i)$ **39.** $x = 2; x = 3 \pm 2i; f(x) = (x - 2)(x - 3 + 2i)(x - 3 - 2i)$

41. $x = \pm 6i; x = \pm i; f(x) = (x - 6i)(x + 6i)(x - i)(x + i)$

43. $x = -2; x = \dfrac{3}{4}; x = -\dfrac{1}{2} \pm i; f(x) = (x + 2)(4x - 3)(2x + 1 - 2i)(2x + 1 + 2i)$ **45.** ≈ 3 yr   **47.** Answers may vary.

49. Answers may vary.

55.

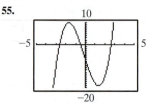

−3 is a lower bound;
3 is an upper bound

57. a. As x, a person's age,
increases, y, the number of
visits, increases.

b. 60

c.

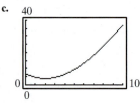

59.

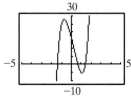

3 real zeros, 2 nonreal
complex zeros

61.

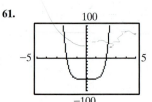

2 real zeros, 4 nonreal
complex zeros

63. 3   **65.** 5   **67.** Answers may vary.

Section 3.6

Check Point Exercises

1. a. $\{x | x \neq 5\}$ **b.** $\{x | x \neq -5, x \neq 5\}$ **c.** all real numbers   **2. a.** $x = 1, x = -1$ **b.** $x = -1$ **c.** none

3. a. $y = 3$ **b.** $y = 0$ **c.** none

4.

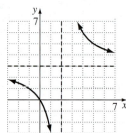

5.

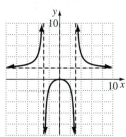

6.

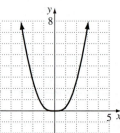

7. $y = 2x - 1$

8. a. $C(x) = 500x + 600,000$ **b.** $\overline{C}(x) = \dfrac{500x + 600,000}{x}$

c. $\overline{C}(1000) = 1100$, when 1000 new systems are produced, it costs \$1100 to produce each system; $\overline{C}(10,000) = 560$, when 10,000 new systems are produced, it costs \$560 to produce each system, $\overline{C}(100,000) = 506$, when 100,000 new systems are produced, it costs \$506 to produce each system.

d. $y = 500$; The cost per system approaches \$500 as more systems are produced.

Exercise Set 3.6

1. $\{x | x \neq 4\}$ **3.** $\{x | x \neq 5, x \neq -4\}$ **5.** $\{x | x \neq 7, x \neq -7\}$ **7.** All real numbers **9.** $-\infty$ **11.** $-\infty$ **13.** 0 **15.** $+\infty$

17. $-\infty$ **19.** 1 **21.** $x = -4$ **23.** $x = 0, x = -4$ **25.** $x = -4$ **27.** no vertical asymptotes **29.** $y = 0$ **31.** $y = 4$

33. no horizontal asymptote **35.** $y = -\dfrac{2}{3}$

37. 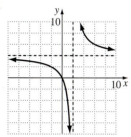 **39.** 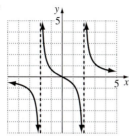 **41.** 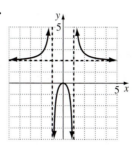 **43.**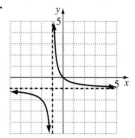

45. 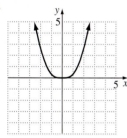 **47.** 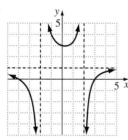 **49.** 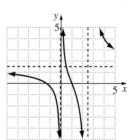 **51.**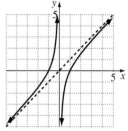

53. 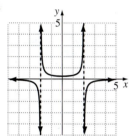 **55.** 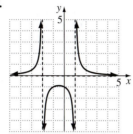 **57.** 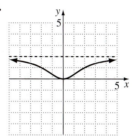 **59. a.** Slant asymptote: $y = x$ **b.**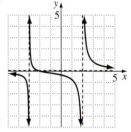

61. a. Slant asymptote: $y = x$ **b.**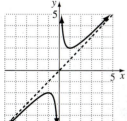

63. a. Slant asymptote: $y = x + 4$ **b.**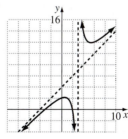

65. a. Slant asymptote: $y = x - 2$ **b.**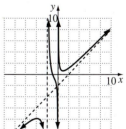

67. a. $C(x) = 100x + 100{,}000$ **b.** $\overline{C}(x) = \dfrac{100x + 100{,}000}{x}$

  c. $\overline{C}(500) = 300$, when 500 bicycles are produced, it costs \$300 to produce each bicycle; $\overline{C}(1000) = 200$, when 1000 bicycles are produced, it costs \$200 to produce each bicycle; $\overline{C}(2000) = 150$, when 2000 bicycles are produced, it costs \$150 to produce each bicycle; $\overline{C}(4000) = 125$, when 4000 bicycles are produced, it costs \$125 to produce each bicycle.

  d. $y = 100$; The cost per bicycle approaches \$100 as more bicycles are produced.

69. a. $M(x) = \dfrac{190.9x + 2413.99}{0.234x + 12.54}$ **b.** 355.65; $M(19) \approx 355.65$ on the graph

c. $y = \dfrac{190.9}{0.234} \approx 816$; The cost of textbooks per college student approaches \$816 as the years progress.

71. 90; An incidence ratio of 10 means 90% of the deaths are smoking related.

73. $y = 100$; The percentage of deaths cannot exceed 100% as the incidence ratios increase.

75. a. After 1 day: 35 words; after 5 days: about 12 words; after 15 days: about 7 words
  b. $N(1) = 35$ words; This is the same as the estimate from the graph.; $N(5) = 11$ words; This is a little less than the estimate from the graph.; $N(15) = 7$ words; This is the same as the estimate from the graph.
  c. The graph indicates the students will remember 5 words over a long period of time.
  d. $y = 5$; The horizontal asymptote indicates the students will remember 5 words over a long period of time.

87.

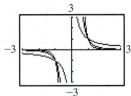

The graph approaches the horizontal asymptote faster and the vertical asymptote slower as n increases.

89.

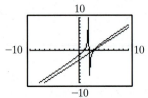

$g(x)$ is the graph of a line whereas $f(x)$ is the graph of a rational function with a slant asymptote; In $g(x)$, $x - 2$ is a factor of $x^2 - 5x + 6$.

91. (d) is true.   **93.** Answers may vary.   **95.** Answers may vary.

Section 3.7

Check Point Exercises

1. a. $L = kN$ **b.** $L = 4N$ **c.** 68 in.   **2. a.** $W = kL$ **b.** $k = \dfrac{75}{6}$ **c.** $W = \dfrac{75L}{6}$ **d.** 200 lb   **3.** 137.5 lb/in^2

4. about 556 ft   **5.** \$26 per barrel   **6.** 24 min   **7.** 96π ft^3

Exercise Set 3.7

1. $g = kh$ **3.** $a = kb^2$ **5.** $r = \dfrac{k}{t}$ **7.** $a = \dfrac{k}{b^3}$ **9.** $r = \dfrac{ks}{v}$ **11.** $s = kgt^2$ **13.** $k = 25$ **15.** $k = 5$ **17.** $k = 5000$

19. $k = 3$ **21.** $k = 2$ **23.** 84   **25.** 25   **27.** $\dfrac{5}{6}$ **29.** 240   **31. a.** $G = kW$ **b.** $G = 0.02W$ **c.** 1.04 in.   **33.** \$60

35. 2442 mph   **37.** 607 lb   **39.** 0.5 hr   **41.** 6.4 lb   **43.** 31.78; index: about 32; not in the desirable range

45. 11.11 foot-candles   **47.** 72 erg   **49.** The average number of phone calls is about 126.

59. The destructive power is four times as much.   **61.** Reduce the resistance by a factor of $\dfrac{1}{3}$.

Chapter 3 Review Exercises

1.

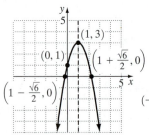

axis of symmetry: $x = 1$

2.

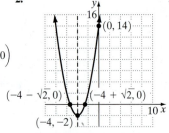

axis of symmetry: $x = -4$

3.

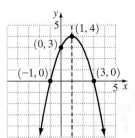

axis of symmetry: $x = 1$

4.

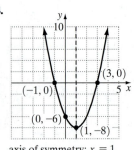

axis of symmetry: $x = 1$

5. after 2 seconds, the ball reaches a maximum height of 144 feet   **6.** (20, 5.4); In 1980, the divorce rate reached a maximum of 5.4%.
7. 250 yd by 500 yd; maximum area is 125,000 yd^2 **8. c**   **9. b**   **10. a**   **11. d**   **12.** Because the degree is odd and the leading coefficient is negative, the graph falls to the right. Therefore, the model indicates that the percentage of families below the poverty level will

eventually be negative, which is impossible.   **13.** Since the degree is even and the leading coefficient is negative, the graph falls to the right. Therefore, the model indicates a patient will eventually have a negative number of viral bodies, which is impossible.

14. $x = 1$, multiplicity 1, crosses; $x = -2$, multiplicity 2, touches; $x = -5$, multiplicity 3, crosses

15. $x = -5$, multiplicity 1, crosses; $x = 5$, multiplicity 2, touches

16. a. The graph falls to the
left and rises to the right.
b. no symmetry
c.

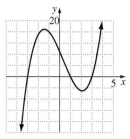

17. a. The graph rises to the
left and falls to the right.
b. origin symmetry
c.

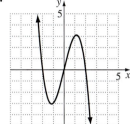

18. a. The graph falls to the
left and rises to the right.
b. no symmetry
c.

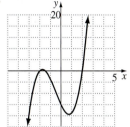

19. a. The graph falls to the
left and to the right.
b. y-axis symmetry
c.

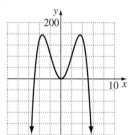

20. a. The graph falls to the
left and to the right.
b. no symmetry
c.

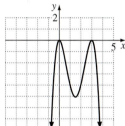

21. a. The graph rises to the
left and to the right.
b. no symmetry
c.

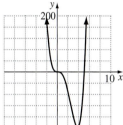

22. $4x^2 - 7x + 5 - \dfrac{4}{x + 1}$ **23.** $2x^2 - 4x + 1 - \dfrac{10}{5x - 3}$ **24.** $2x^2 + 3x - 1$ **25.** $3x^3 - 4x^2 + 7$

26. $3x^3 + 6x^2 + 10x + 10 + \dfrac{20}{x - 2}$ **27.** -5697 **28.** $2, \dfrac{1}{2}, -3$ **29.** $\{4, -2 \pm \sqrt{5}\}$ **30.** $\pm 1, \pm 5$

31. $\pm 1, \pm 2, \pm 4, \pm 8, \pm \dfrac{8}{3}, \pm \dfrac{4}{3}, \pm \dfrac{2}{3}, \pm \dfrac{1}{3}$ **32.** 2 or 0 positive solutions; no negative solutions

33. 3 or 1 positive real roots; 2 or 0 negative solutions   **34.** No sign variations exist for either $f(x)$ or $f(-x)$, so no real roots exist.

35. a. $\pm 1, \pm 2, \pm 4$ **b.** 1 positive real zero; 2 or no negative real zeros   **c.** 1 is a zero   **d.** $\{1, -2\}$

36. a. $\pm 1, \pm \dfrac{1}{2}, \pm \dfrac{1}{3}, \pm \dfrac{1}{6}$ **b.** 2 or 0 positive real zeros; 1 negative real zero   **c.** -1 is a zero   **d.** $\left\{-1, \dfrac{1}{3}, \dfrac{1}{2}\right\}$

37. a. $\pm 1, \pm 3, \pm 5, \pm 15, \pm \dfrac{1}{2}, \pm \dfrac{1}{4}, \pm \dfrac{1}{8}, \pm \dfrac{3}{2}, \pm \dfrac{3}{4}, \pm \dfrac{3}{8}, \pm \dfrac{5}{2}, \pm \dfrac{5}{4}, \pm \dfrac{5}{8}, \pm \dfrac{15}{2}, \pm \dfrac{15}{4}, \pm \dfrac{15}{8}$

b. 3 or 1 positive real solutions; no negative real solutions   **c.** $\dfrac{1}{2}$ is a zero   **d.** $\left\{\dfrac{1}{2}, \dfrac{3}{2}, \dfrac{5}{2}\right\}$

38. a. $\pm 1, \pm 2, \pm 3, \pm 6$ **b.** 2 or zero positive real solutions; 2 or zero negative real solutions   **c.** -2 is a zero   **d.** $\{-2, -1, 1, 3\}$

39. a. $\pm 1, \pm 2, \pm \dfrac{1}{2}, \pm \dfrac{1}{4}$ **b.** 1 positive real root; 1 negative real root   **c.** $\dfrac{1}{2}$ is a zero   **d.** $\left\{\dfrac{1}{2}, -\dfrac{1}{2}, i\sqrt{2}, -i\sqrt{2}\right\}$

40. a. $\pm 1, \pm 2, \pm 4, \pm \dfrac{1}{2}$ **b.** 2 or no positive zeros; 2 or no negative zeros   **c.** $x = 2$ is a zero   **d.** $\left\{2, -2, \dfrac{1}{2}, -1\right\}$

41.

-2	2	-7	-5	28	-12
		-4	22	-34	12
	2	-11	17	-6	0

-2 is a root and a lower bound.

6	2	-7	-5	28	-12
		12	30	150	1068
	2	5	25	178	1056

6 is an upper bound, but not a zero.

$\pm 1, \pm 2, 3, 4, \pm \dfrac{1}{2}, \pm \dfrac{3}{2}$

42. a. $\pm 1, \pm 2, \pm 3, \pm 4, \pm 6, \pm 12, \pm\dfrac{1}{2}, \pm\dfrac{3}{2}$ **b.** 2 is not a root but is an upper bound.   **c.** -2 is not a root but is a lower bound.

d. Possible roots are $\pm 1, \pm\dfrac{1}{2},$ and $\pm\dfrac{3}{2}$.   **43.** $f(1) = -2; f(2) = 3; x \approx 1.6$ **44.** $f(-3) = -32; f(-2) = 7; x \approx -2.3$

45. $\left\{-\dfrac{1}{4}, 6 \pm 5i\right\}$ **46.** $\{1 \pm 3i, 1 \pm i\}$ **47.** $\left\{-\dfrac{1}{2}, 1, 4 \pm 7i\right\}$ **48.** $f(x) = x^3 - 6x^2 + 21x - 26$

49. $f(x) = 2x^4 + 12x^3 + 20x^2 + 12x + 18$ **50.** $f(x) = x^4 - 3x^3 + 6x^2 + 2x - 60$

51. $-2, \dfrac{1}{2}, \pm i; f(x) = (x - i)(x + i)(x + 2)\left(x - \dfrac{1}{2}\right)$ **52.** $-1, 4; g(x) = (x + 1)^2(x - 4)^2$

53. 4 real zeros, one with multiplicity two   **54.** 3 real zeros; 2 nonreal complex zeros

55. 2 real zeros, one with multiplicity two; 2 nonreal complex zeros   **56.** 1 real zero; 4 nonreal complex zeros

57. Vertical asymptote: $x = 3$ and $x = -3$ horizontal asymptote: $y = 0$

58. Vertical asymptote: $x = -3$ horizontal asymptote: $y = 2$

59. Vertical asymptotes: $x = 3, -2$ horizontal asymptote: $y = 1$

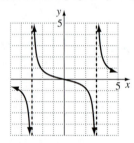

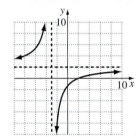

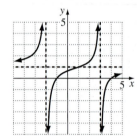

60. Vertical asymptote: $x = -2$ horizontal asymptote: $y = 1$

61. Vertical asymptote: $x = -1$ no horizontal asymptote slant asymptote: $y = x - 1$

62. Vertical asymptote: $x = 3$ no horizontal asymptote slant asymptote: $y = x + 5$

63. No vertical asymptote no horizontal asymptote slant asymptote: $y = -2x$

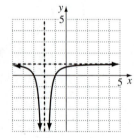

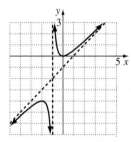

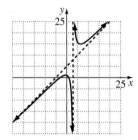

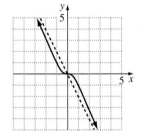

64. Vertical asymptote: $x = \dfrac{3}{2}$

no horizontal asymptote

slant asymptote: $y = 2x - 5$

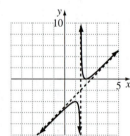

65. a. $C(x) = 25x + 50{,}000$ **b.** $\overline{C}(x) = \dfrac{25x + 50{,}000}{x}$

c. $\overline{C}(50) = 1025$, when 50 calculators are manufactured, it costs $1025 to manufacture each; $\overline{C}(100) = 525$, when 100 calculators are manufactured, it costs $525 to manufacture each; $\overline{C}(1000) = 75$, when 1000 calculators are manufactured, it costs $75 to manufacture each; $\overline{C}(100{,}000) = 25.5$, when 100,000 calculators are manufactured, it costs $25.50 to manufacture each.

d. $y = 25$; Minimum costs will approach $25.

66. a. 1600; The difference in cost of removing 90% versus 50% of the contaminants is 16 million dollars.

b. $x = 100$; No amount of money can remove 100% of the contaminants, since $C(x)$ increases without bound as x approaches 100.

67. $y = 3000$; The number of fish in the pond approaches 3000.

68. $y = 0$; As the number of years of education increases the percentage rate of unemployment approaches zero.

69. a. $f(x) = \dfrac{1.96x + 3.14}{3.04x + 21.79}$ **b.** $y = \dfrac{49}{76}$; As the years increase, the fraction of nonviolent prisoners approaches $\dfrac{49}{76}$.

c. Answers may vary.   **70.** $154   **71.** 1600 ft   **72.** 5 hr   **73.** 112 decibels   **74.** 16 hr   **75.** 800 ft^3

Chapter 3 Test

1.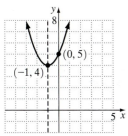

axis of symmetry: $x = -1$

2.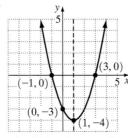

axis of symmetry: $x = 1$

3. maximum; $(3, 2)$

4. 23 computers; maximum daily profit $= \$16{,}900$

5. a. $5, 2, -2$
b.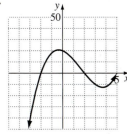

6. Since the degree of the polynomial is odd and the leading coefficient is positive, the graph of f should fall to the left and rise to the right. The x-intercepts should be -1 and 1.

7. a. 2 **b.** $\frac{1}{2}, \frac{2}{3}$

8. $\pm 1, \pm 2, \pm 3, \pm 6, \pm\frac{1}{2}, \pm\frac{3}{2}$

9. 3 or 1 positive real zeros; no negative real zeros.

10. $\{-5, -3, 2\}$

11. a. $\pm 1, \pm 3, \pm 5, \pm 15, \pm\frac{1}{2}, \pm\frac{3}{2}, \pm\frac{5}{2}, \pm\frac{15}{2}$ **b.** $\left\{-1, \frac{3}{2}, \pm\sqrt{5}\right\}$

12.

-3	3	4	-7	-2	-3
		-9	15	-24	78
	3	-5	8	-26	75

-3 is a lower bound.

2	3	4	-7	-2	-3
		6	20	26	48
	3	10	13	24	45

2 is an upper bound.

13. $\{2, 3, 1 + i, 1 - i\}$ **14.** $(x - 1)(x + 2)^2$

15. domain: $\{x \mid x \neq 4, x \neq -4\}$ **16.** domain: $\{x \mid x \neq 2\}$ **17.** domain: $\{x \mid x \neq -3, x \neq 1\}$ **18.** domain: all real numbers

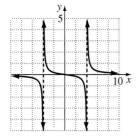 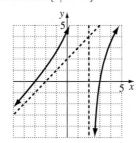 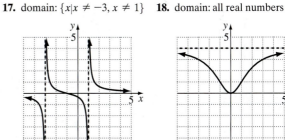 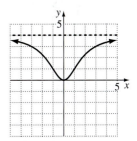

19. a. 0.9 **b.** 11 **c.** $y = 1$; as the number of learning trials increases, the proportion of correct responses approaches 1.
20. 45 foot-candles

Cumulative Review Exercises (Chapters P–3)

1. $2 + \sqrt{3}$ **2.** $-3x^2 - 11x + 11$ **3.** $15\sqrt{2}$ **4.** $x^5(x - 1)(x + 1)$ **5.** $\{2, -1\}$ **6.** $\left\{\frac{5 + \sqrt{13}}{6}, \frac{5 - \sqrt{13}}{6}\right\}$ **7.** $\left\{\frac{1}{3}, -\frac{2}{3}\right\}$

8. $\{-3, -1, 2\}$ **9.** $(-\infty, 1)$ or $(4, \infty)$ **10.** $(-\infty, -1)$ or $\left(\frac{5}{3}, \infty\right)$

11. Center: $(1, -2)$; radius: 3

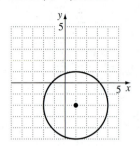

12. $t = 1 - \dfrac{V}{C}$

13. $(-\infty, 5]$

14. $x^2 - 2x - 4$

15. $16x^2 - 6$

16. -9

17. a. $\{-1, 1, 4\}$

b.

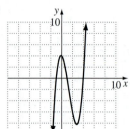

18.

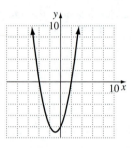

19.

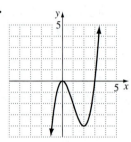

20.

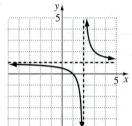

CHAPTER 4

Section 4.1

Check Point Exercises

1. 1 O-ring

2.

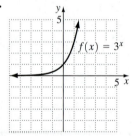

3.

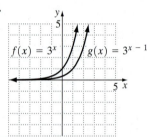

4.

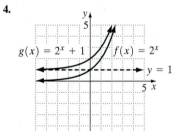

5. 11.49 billion

6. a. $14,859.47

   b. $14,918.25

Exercise Set 4.1

1. 10.556   **3.** 11.665   **5.** 0.125   **7.** 9.974   **9.** 0.387

11.

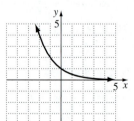

13.

15.

17.

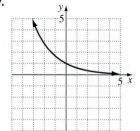

19. $H(x) = -3^{-x}$ **21.** $F(x) = -3^x$ **23.** $h(x) = 3^x - 1$

25.
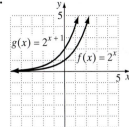

27.
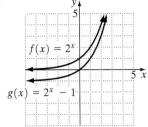

29.
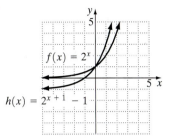

31.
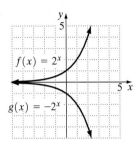

33.
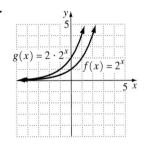

35.
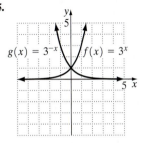

37.
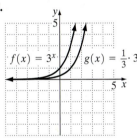

39.
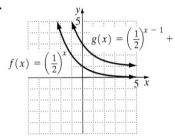

41. a. \$13,116.51
b. \$13,140.67
c. \$13,157.04
d. \$13,165.31

43. 7% compounded monthly

45. a. 67.38 million
b. about 134.74 million
c. about 269.46 million
d. 538.85 million
e. appears to double every 27 yr

47. $f(10) \approx 48$; 10 minutes after 8:00, 48 people have heard the rumor.   **49.** \$116,405.10

51. 3.249009585; 3.317278183; 3.321880096; 3.321995226; 3.321997068; $2^{\sqrt{3}} \approx 3.321997085$; The closer the exponent is to $\sqrt{3}$, the closer the value is to $2^{\sqrt{3}}$.   **53.** 175.6   **55. a.** 100%   **b.** $\approx 68.5\%$ **c.** $\approx 30.8\%$ **d.** 20%

57. a. 1429   **b.** 24,546   **c.** Growth is limited by the population; The entire population will eventually become ill.

65.
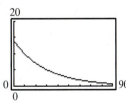

no; Nearly 4 O-rings are expected to fail.

67. a.
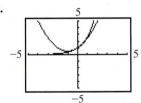

b.
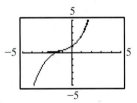

c.
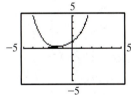

d. Answers may vary.

69. $y = 3^x$ is (d); $y = 5^x$ is (c); $y = \left(\dfrac{1}{3}\right)^x$ is (a); $y = \left(\dfrac{1}{5}\right)^x$ is (b).

71.
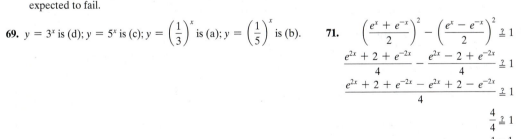

$$\left(\frac{e^x + e^{-x}}{2}\right)^2 - \left(\frac{e^x - e^{-x}}{2}\right)^2 \stackrel{?}{=} 1$$

$$\frac{e^{2x} + 2 + e^{-2x}}{4} - \frac{e^{2x} - 2 + e^{-2x}}{4} \stackrel{?}{=} 1$$

$$\frac{e^{2x} + 2 + e^{-2x} - e^{2x} + 2 - e^{-2x}}{4} \stackrel{?}{=} 1$$

$$\frac{4}{4} \stackrel{?}{=} 1$$

$$1 = 1$$

Section 4.2

Check Point Exercises

1. a. $7^3 = x$ **b.** $b^2 = 25$ **c.** $4^y = 26$ **2. a.** $5 = \log_2 x$ **b.** $3 = \log_b 27$ **c.** $y = \log_e 33$ **3. a.** 2   **b.** 1   **c.** $\dfrac{1}{2}$

4. a. 1   **b.** 0   **5. a.** 8   **b.** 17   **6.**

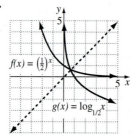

7. $(5, \infty)$
8. 80%
9. 4.0
10. a. $(-\infty, 4)$
    b. $(-\infty, 0)$ or $(0, \infty)$
11. a. $25x$
    b. $\sqrt{x}$
12. 4.6 ft per sec

Exercise Set 4.2

1. $2^4 = 16$ **3.** $3^2 = x$ **5.** $b^5 = 32$ **7.** $6^y = 216$ **9.** $\log_2 8 = 3$ **11.** $\log_2 \dfrac{1}{16} = -4$ **13.** $\log_8 2 = \dfrac{1}{3}$

15. $\log_{13} x = 2$ **17.** $\log_b 1000 = 3$ **19.** $\log_7 200 = y$ **21.** 2   **23.** 6   **25.** $\dfrac{1}{2}$ **27.** -3 **29.** $\dfrac{1}{2}$ **31.** 1

33. 0   **35.** 7   **37.** 19

39.

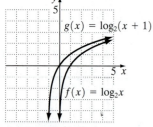

41.

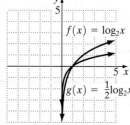

43. $H(x) = 1 - \log_3 x$
45. $h(x) = \log_3 x - 1$
47. $g(x) = \log_3(x - 1)$

49.

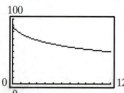

x-intercept: $(0, 0)$
vertical asymptote: $x = -1$

51.

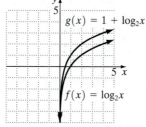

x-intercept: $(0.5, 0)$
vertical asymptote: $x = 0$

53.

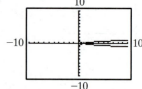

x-intercept: $(1, 0)$
vertical asymptote: $x = 0$

55. $(-4, \infty)$ **57.** $(-\infty, 2)$ **59.** $(-\infty, 2)$ or $(2, \infty)$ **61.** 2   **63.** 7   **65.** 33   **67.** 0   **69.** 6   **71.** -6 **73.** 125

75. $9x$ **77.** $5x^2$ **79.** $\sqrt{x}$ **81.** 95.4%   **83.** \$5.65 billion   **85.** $\approx$ 188 db; yes

87. a. 88
   b. 71.5; 63.9; 58.8; 55; 52; 49.5
   c.

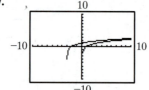

Material retention decreases
as time passes.

97.

g(x) is f(x) shifted left 3 units left.

99.

g(x) is f(x) reflected about the x-axis.

101.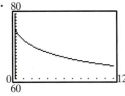

The score falls below 65 after 9 months.

103. $y = \ln x$, $y = \sqrt{x}$, $y = x$, $y = x^2$, $y = e^x$, $y = x^x$

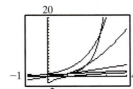

105. $\dfrac{4}{5}$

107. $\log_3 40 > \log_4 60$

Section 4.3

Check Point Exercises

1. a. $\log_6 7 + \log_6 11$ **b.** $2 + \log x$ **2. a.** $\log_8 23 - \log_8 x$ **b.** $5 - \ln 11$ **3. a.** $9 \log_6 3$ **b.** $\dfrac{1}{3} \ln x$

4. a. $4 \log_b x + \dfrac{1}{3} \log_b y$ **b.** $\dfrac{1}{2} \log_5 x - 2 - 3 \log_5 y$ **5. a.** 2 **b.** $\log \dfrac{7x + 6}{x}$

6. a. $\ln x^2 \sqrt[3]{x + 5}$ **b.** $\log \dfrac{(x - 3)^2}{x}$ **c.** $\log_b \dfrac{\sqrt[4]{x}\, y^{10}}{25}$ **7.** 4.02 **8.** 4.02

Exercise Set 4.3

1. $\log_5 7 + \log_5 3$ **3.** $1 + \log_7 x$ **5.** $3 + \log x$ **7.** $1 - \log_7 x$ **9.** $\log x - 2$ **11.** $3 - \log_4 y$ **13.** $2 - \ln 5$

15. $3 \log_b x$ **17.** $-6 \log N$ **19.** $\dfrac{1}{5} \ln x$ **21.** $2 \log_b x + \log_b y$ **23.** $\dfrac{1}{2} \log_4 x - 3$ **25.** $2 - \dfrac{1}{2} \log_6(x + 1)$

27. $2 \log_b x + \log_b y - 2 \log_b z$ **29.** $1 + \dfrac{1}{2} \log x$ **31.** $\dfrac{1}{3} \log x - \dfrac{1}{3} \log y$ **33.** $\dfrac{1}{2} \log_b x + 3 \log_b y - 3 \log_b z$

35. $\dfrac{2}{3} \log_5 x + \dfrac{1}{3} \log_5 y - \dfrac{2}{3}$ **37.** $3 \ln x + \dfrac{1}{2} \ln(x^2 + 1) - 4 \ln(x + 1)$ **39.** $\left(1 + 2 \log x + \dfrac{1}{3} \log(1 - x1)\right) - (\log 7 + 2 \log(x + 1))$

41. 1 **43.** $\ln(7x)$ **45.** 5 **47.** $\log\left(\dfrac{2x + 5}{x}\right)$ **49.** $\log(xy^3)$ **51.** $\ln(x^{1/2}y)$ or $\ln(y\sqrt{x})$ **53.** $\log_b(x^2 y^3)$ **55.** $\ln\left(\dfrac{x^5}{y^2}\right)$

57. $\ln\left(\dfrac{x^3}{y^{1/3}}\right)$ or $\ln\left(\dfrac{x^3}{\sqrt[3]{y}}\right)$ **59.** $\ln\dfrac{(x + 6)^4}{x^3}$ **61.** $\ln\left(\dfrac{x^3 y^5}{z^6}\right)$ **63.** $\log\sqrt{xy}$ **65.** $\log_5\left(\dfrac{\sqrt{xy}}{(x + 1)^2}\right)$ **67.** $\ln\sqrt[3]{\dfrac{(x + 5)^2}{x(x^2 - 4)}}$

69. $\log\left(\dfrac{7x(x^2 - 1)}{x + 1}\right)$ or $\log(7x(x - 1))$ **71.** 1.5937 **73.** 1.6944 **75.** -1.2304 **77.** 3.6193

79.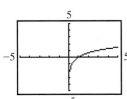

81.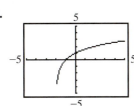

83. a. $D = 10 \log \dfrac{I}{I_0}$ **b.** 20 decibels louder

93. a.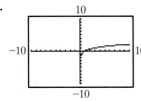

b.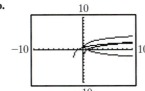

$y = 2 + \log_3 x$ shifts the graph of $y = \log_3 x$ two units upward; $y = \log_3(x + 2)$ shifts the graph of $y = \log_3 x$ two units left; $y = -\log_3 x$ reflects the graph of $y = \log_3 x$ about the x-axis.

95.

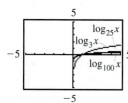

a. top graph: $y = \log_{100} x$; bottom graph: $y = \log_3 x$
b. top graph: $y = \log_3 x$; bottom graph: $y = \log_{100} x$
c. The graph of the equation with the largest b will be on the top in the interval $(0, 1)$ and on the bottom in the interval $(1, \infty)$.

101. (d) is true.   **103.** $\dfrac{2A}{B}$ **105.** Answers may vary.

Section 4.4

Check Point Exercises

1. $\left\{ \dfrac{\ln 134}{\ln 5} \right\}$; ≈ 3.04 **2.** $\left\{ \dfrac{\ln 9}{2} \right\}$; ≈ 1.10 **3.** $\left\{ \dfrac{\ln 2088 + 4\ln 6}{3\ln 6} \right\}$; ≈ 2.76 **4.** $\{0, \ln 7\}$; $\ln 7 \approx 1.95$ **5.** $\{12\}$ **6.** $\{5\}$

7. $\left\{ \dfrac{e^2}{3} \right\}$ **8.** 0.01   **9.** 16.2 yr   **10.** 2149

Exercise Set 4.4

1. $\left\{ \dfrac{\ln 3.91}{\ln 10} \right\}$; ≈ 0.59 **3.** $\{\ln 5.7\}$; ≈ 1.74 **5.** $\left\{ \dfrac{\ln 17}{\ln 5} \right\}$; ≈ 1.76 **7.** $\left\{ \ln \dfrac{23}{5} \right\}$; ≈ 1.53 **9.** $\left\{ \dfrac{\ln 659}{5} \right\}$; ≈ 1.30

11. $\left\{ \dfrac{\ln 793 - 1}{-5} \right\}$; ≈ -1.14 **13.** $\left\{ \dfrac{\ln 10{,}478 + 3}{5} \right\}$; ≈ 2.45 **15.** $\left\{ \dfrac{\ln 410}{\ln 7} - 2 \right\}$; ≈ 1.09 **17.** $\left\{ \dfrac{\ln 813}{0.3 \ln 7} \right\}$; ≈ 11.48

19. $\left\{ \dfrac{3\ln 5 + \ln 3}{\ln 3 - 2\ln 5} \right\}$; ≈ -2.80 **21.** $\{0, \ln 2\}$; $\ln 2 \approx 0.69$ **23.** $\left\{ \dfrac{\ln 3}{2} \right\}$; ≈ 0.55 **25.** $\{0\}$ **27.** $\{81\}$ **29.** $\{59\}$ **31.** $\left\{ \dfrac{109}{27} \right\}$

33. $\left\{ \dfrac{62}{3} \right\}$ **35.** $\left\{ \dfrac{5}{4} \right\}$ **37.** $\{6\}$ **39.** $\{6\}$ **41.** $\{5\}$ **43.** $\{2, 12\}$ **45.** $\{e^2\}$; ≈ 7.39 **47.** $\left\{ \dfrac{e^4}{2} \right\}$; ≈ 27.30 **49.** $\{e^{-1/2}\}$; ≈ 0.61

51. $\{e^2 - 3\}$; ≈ 4.39 **53.** about 0.11   **55. a.** 18.9 million   **b.** ≈ 2006 **57.** 8.2 yr   **59.** 16.8%   **61.** 8.7 yr   **63.** 15.7%

65. 1995   **67.** 2.8 days; Yes, the point (2.8, 50) appears to lie on the graph of P.   **69.** $10^{-2.4}$; 0.004 moles per liter

75. $\{2\}$ **77.** $\{4\}$ **79.** $\{2\}$ **81.** $\{-1.391606, 1.6855579\}$

83.

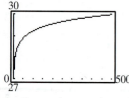

As distance from eye increases, barometric air pressure increases, leveling off at about 30 inches of mercury.

85.

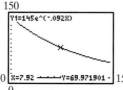

about 7.9 min

87. (c) is true.   **89.** $\{1, e^2\}$, $e^2 \approx 7.389$ **91.** $\{e\}$, $e \approx 2.718$

Section 4.5

Check Point Exercises

1. a. $A = 643\,e^{0.023t}$ **b.** 2039   **2. a.** $A = A_0 e^{-0.0248t}$ **b.** about 72 yr   **3. a.** 0.4 correct responses   **b.** 0.7 correct responses
c. 0.8 correct responses   **4.** $y = 4e^{(\ln 7.8)x}$; $y = 4e^{2.054x}$

Exercise Set 4.5

1. 203 million   **3.** 2005   **5.** 2.6%   **7.** 2014   **9. a.** $A = 158{,}700e^{0.053t}$ **b.** 2007   **11.** $A = 6.04e^{0.01t}$ **13.** 8.01 g

15. 8 g; 4 g; 2 g; 1 g; 0.5 g   **17.** 15,679 years old   **19. a.** $\dfrac{A_0}{2} = A_0 e^{k(1.31)}$; $\dfrac{1}{2} = e^{1.31k}$; $\ln \dfrac{1}{2} = \ln e^{1.31k}$; $\ln \dfrac{1}{2} = 1.31k$; $k = \dfrac{\ln \frac{1}{2}}{1.31} \approx -0.52912$

b. 107 million years   **21.** $2A_0 = A_0 e^{kt}$; $2 = e^{kt}$; $\ln 2 = \ln e^{kt}$; $\ln 2 = kt$; $t = \dfrac{\ln 2}{k}$ **23.** 63 yr   **25. a.** about 20 people

b. about 1080 people   **c.** 100,000 people   **27.** about 3.7%   **29.** about 48 years old   **31.** $y = 100e^{(\ln 4.6)x}$; $y = 100e^{1.526x}$

33. $y = 2.5e^{(\ln 0.7)x}$; $y = 2.5e^{-0.357x}$ **45.** $y = 1.740(1.037)^x$; $r \approx 0.971$, a very good fit

47. $y = 0.112x + 1.547$; $r = 0.989$; a very good fit   **49.** The model of best fit is the linear model; 2022

51. The logarithmic model, $y = -905,231.353 + 119,204.060 \ln x$, best fits the data. Answers for prediction may vary.

53. Answers may vary.

Chapter 4 Review Exercises

1. $g(x) = 4^{-x}$ **2.** $h(x) = -4^{-x}$ **3.** $r(x) = -4^{-x} + 3$ **4.** $f(x) = 4^x$

5.
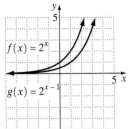

6.
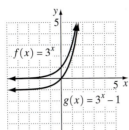

7.
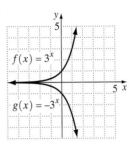

8.
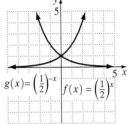

9. 5.5% compounded semiannually   **10.** 7% compounded monthly

11. a. $200°$ **b.** $120°$; $119°$ **c.** $70°$; The temperature in the room is $70°$.   **12.** $49^{1/2} = 7$ **13.** $4^3 = x$ **14.** $3^y = 81$

15. $\log_6 216 = 3$ **16.** $\log_b 625 = 4$ **17.** $\log_{13} 874 = y$ **18.** 3   **19.** -2 **20.** $\varnothing$; $\log_b x$ is defined only for $x > 0$.   **21.** $\dfrac{1}{2}$

22. 1   **23.** 8   **24.** 5   **25.** 0   **26.**
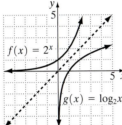
   27.
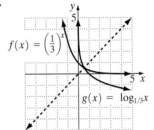
   28. $g(x) = \log(-x)$

29. $r(x) = 1 + \log(2 - x)$ **30.** $h(x) = \log(2 - x)$ **31.** $f(x) = \log x$

32.
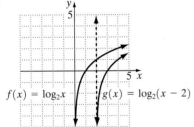

x-intercept: $(3, 0)$
vertical asymptote: $x = 2$

33.
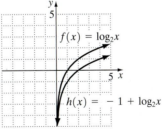

x-intercept: $(2, 0)$
vertical asymptote: $x = 0$

34.
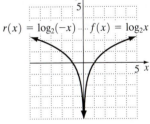

x-intercept: $(-1, 0)$
vertical asymptote: $x = 0$

35. $(-5, \infty)$ **36.** $(-\infty, 3)$ **37.** $(-\infty, 1) \cup (1, \infty)$ **38.** $6x$ **39.** $\sqrt{x}$ **40.** $4x^2$ **41.** 3.0

42. a. 76
b. $\approx 67, \approx 63, \approx 61, \approx 59, \approx 56$

c.

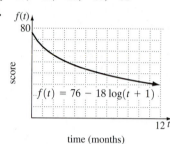

time (months)

Retention decreases as time passes.

43. about 9 weeks   **44.** $2 + 3 \log_6 x$ **45.** $\dfrac{1}{2} \log_4 x - 3$

46. $\log_2 x + 2 \log_2 y - 6$ **47.** $\dfrac{1}{3} \ln x - \dfrac{1}{3}$ **48.** $\log_b 21$ **49.** $\log \dfrac{3}{x^3}$

50. $\ln(x^3 y^4)$ **51.** $\ln \dfrac{\sqrt{x}}{y}$ **52.** 6.2448   **53.** -0.1063

54. $\left\{ \dfrac{\ln 12{,}143}{\ln 8} \right\}; \approx 4.523$ **55.** $\left\{ \dfrac{1}{5} \ln 141 \right\}; \approx 0.990$

56. $\left\{ \dfrac{12 - \ln 130}{5} \right\}; \approx 1.426$ **57.** $\left\{ \dfrac{\ln 37{,}500 - 2 \ln 5}{4 \ln 5} \right\}; \approx 1.136$

58. $\{\ln 3\}; \approx 1.099$ **59.** $\{23\}$ **60.** $\{5\}$ **61.** $\varnothing$ **62.** $\left\{ \dfrac{1}{e} \right\}$

63. $\left\{ \dfrac{e^3}{2} \right\}$ **64.** 2042   **65.** 2086   **66.** 2005   **67.** 7.3 yr   **68.** 14.6 yr   **69.** about 21.97%

70. a. 0.045   **b.** 55.6 million   **c.** 2012   **71.** about 15,679 years old   **72. a.** 200 people   **b.** about 45,411 people   **c.** 500,000 people
73. $y = 73e^{(\ln 2.6)x}; y = 73e^{0.956x}$ **74.** $y = 6.5e^{(\ln 0.43)x}; y = 6.5e^{-0.844x}$
75. high: exponential; medium: linear; low: quadratic; Explanations will vary; negative; The parabola opens downward.
76. The exponential model, $y = (3.460)(1.024)^x$, is the best fit; about 116 million

Chapter 4 Test

1.

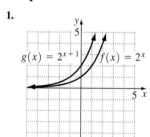

2.

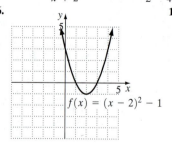

3. $5^3 = 125$ **4.** $\log_{36} 6 = \dfrac{1}{2}$ **5.** $(-\infty, 3)$

6. $3 + 5 \log_4 x$ **7.** $\dfrac{1}{3} \log_3 x - 4$ **8.** $\log(x^6 y^2)$

9. $\ln \dfrac{7}{x^3}$ **10.** 1.5741   **11.** $\left\{ \dfrac{\ln 1.4}{\ln 5} \right\}$

12. $\left\{ \dfrac{\ln 4}{0.005} \right\}$ **13.** $\{0, \ln 5\}$ **14.** $\{54.25\}$

15. $\{5\}$ **16.** $\left\{ \dfrac{e^4}{3} \right\}$

17. 6.5% compounded semiannually; $221.15 more   **18.** 120 db   **19. a.** about 89%   **b.** decreasing; $k = -0.004 < 0$ **c.** 1995
20. $A = 509e^{0.036t}$ **21.** about 24,758 years ago   **22. a.** 14 elk   **b.** about 51 elk   **c.** 140 elk

Cumulative Review Exercises (Chapters 1–4)

1. $\left\{ \dfrac{2}{3}, 2 \right\}$ **2.** $\{3, 7\}$ **3.** $\{-2, -1, 1\}$ **4.** $\left\{ \dfrac{\ln 128}{5} \right\}$ **5.** $\{3\}$ **6.** $(-\infty, 4]$ **7.** $[1, 3]$
8. using $(1, 3), y - 3 = -3(x - 1); y = -3x + 6$ **9.** $(f \circ g)(x) = (x + 2)^2; (g \circ f)(x) = x^2 + 2$
10. $f^{-1}(x) = \dfrac{1}{2}x + \dfrac{7}{2}$ **11.** $x^2 + 3x - 3 + \dfrac{-4}{x + 2}$ **12.** $\pm 1, \pm \dfrac{1}{2}, \pm \dfrac{1}{4}, \pm 3, \pm \dfrac{3}{2}, \pm \dfrac{3}{4}$ **13.** 300   **14.** $\{1 + i, 1 - i, 2\}$
15.

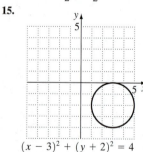

$(x - 3)^2 + (y + 2)^2 = 4$

16.

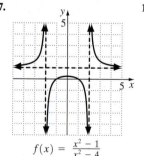

$f(x) = (x - 2)^2 - 1$

17.

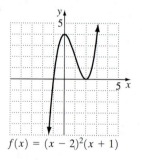

$f(x) = \dfrac{x^2 - 1}{x^2 - 4}$

18.

$f(x) = (x - 2)^2(x + 1)$

19. $12 per hr   **20.** $\dfrac{0.5}{\ln 4} \approx 0.361$; about $\dfrac{361}{1000}$ of the people

CHAPTER 5

Section 5.1

Check Point Exercises

1. solution   **2.** $\{(3,2)\}$ **3.** $\{(1,-2)\}$ **4.** $\{(2,-1)\}$ **5.** $\left\{\left(\frac{23}{16},\frac{3}{8}\right)\right\}$ **6.** $\varnothing$ **7.** $\{(x,y)|y=4x-4\}$ **8. a.** $C(x)=300,000+30x$
b. $R(x)=80x$ **c.** $(6000,480,000)$; The company will breakeven if it produces and sells 6000 pairs of shoes.   **9.** \$30; 400 units

Exercise Set 5.1

1. solution   **3.** not a solution   **5.** $\{(1,3)\}$ **7.** $\{(5,1)\}$ **9.** $\{(-22,-5)\}$ **11.** $\{(0,0)\}$ **13.** $\{(3,-2)\}$ **15.** $\{(5,4)\}$

17. $\{(7,3)\}$ **19.** $\{(2,-1)\}$ **21.** $\{(3,0)\}$ **23.** $\{(-4,3)\}$ **25.** $\{(3,1)\}$ **27.** $\{(1,-2)\}$ **29.** $\left\{\left(\frac{7}{25},\frac{1}{25}\right)\right\}$ **31.** $\varnothing$

33. $\{(x,y)|y=3x-5\}$ **35.** $\{(1,4)\}$ **37.** $\{(x,y)|x+3y=2\}$ **39.** $\{(-5,-1)\}$ **41.** $\left\{\left(\frac{29}{22},-\frac{5}{11}\right)\right\}$

43. $x+y=7$; $x-y=-1$; 3 and 4   **45.** $3x-y=1$; $x+2y=12$; 2 and 5   **47. a.** $C(x)=18,000+20x$ **b.** $R(x)=80x$
c. $(300,24,000)$; This means the company will break even if it produces and sells 300 canoes.   **49. a.** $C(x)=30,000+2500x$
b. $R(x)=3125$ **c.** $(48,150,000)$; The play will break even if 48 sold-out performances are produced.   **51. a.** 6500 tickets can be sold.
6200 tickets can be supplied.   **b.** \$50; 6250 tickets   **53.** 2004; $12\frac{8}{33}$; The lines intersect at $\left(39\frac{13}{33},12\frac{8}{33}\right)$.   **55. a.** $E(x)=508+25x$
b. $E(x)=345+9x$ **c.** 26; 2011; \$1158 for college graduates, \$579 for high school graduates   **57.** Pan pizza: 1120 calories; beef
burrito: 430 calories   **59.** Scrambled eggs: 366 mg cholesterol; Double Beef Whopper: 175 mg cholesterol   **61.** 50 rooms with kitchen
facilities, 150 rooms without kitchen facilities   **63.** 100 ft long by 80 ft wide   **65.** Rate rowing in still water: 6 mph; rate of the current:
2 mph   **67.** $x=55$, $y=35$ **81.** Answers may vary.   **83.** the twin who always lies

Section 5.2

Check Point Exercises

1. $(-1)-2(-4)+3(5)=22$; $2(-1)-3(-4)-5=5$; $3(-1)+(-4)-5(5)=-32$ **2.** $\{(1,4,-3)\}$ **3.** $\{(4,5,3)\}$
4. $y=3x^2-12x+13$

Exercise Set 5.2

1. solution   **3.** solution   **5.** $\{(2,3,3)\}$ **7.** $\{(2,-1,1)\}$ **9.** $\left\{\left(\frac{1}{3},-\frac{2}{5},\frac{1}{2}\right)\right\}$ **11.** $\{(3,1,5)\}$ **13.** $\{(1,0,-3)\}$

15. $\{(1,-5,-6)\}$ **17.** $\left\{\left(\frac{1}{2},\frac{1}{3},-1\right)\right\}$ **19.** 7, 4 and 5   **21.** $y=2x^2-x+3$ **23.** $y=2x^2+x-5$

25. a. $(0,1180)$, $(1,1070)$, $(2,1230)$ **b.** $c=1180$ $a+b+c=1070$ $4a+2b+c=1230$ **c.** $y=135x^2-245x+1180$
27. a. $y=-16x^2+40x+200$ **b.** $y=0$ when $x=5$; The ball hit the ground after 5 seconds   **29.** Carnegie: \$100 billion; Vanderbilt:
\$96 billion; Gates: \$60 billion   **31.** 200 \$8 tickets; 150 \$10 tickets; 50 \$12 tickets   **33.** \$1200 at 8%, \$2000 at 10%, and \$3500 at 12%
35. $x=60$, $y=55$, $z=65$ **43.** Answers may vary.

Section 5.3

Check Point Exercises

1. $\frac{2}{x-3}+\frac{3}{x+4}$ **2.** $\frac{2}{x}-\frac{2}{x-1}+\frac{3}{(x-1)^2}$ **3.** $\frac{2}{x+3}+\frac{6x-8}{x^2+x+2}$ **4.** $\frac{2x}{x^2+1}+\frac{-x+3}{(x^2+1)^2}$

Exercise Set 5.3

1. $\frac{A}{x-2}+\frac{B}{x+1}$ **3.** $\frac{A}{x+2}+\frac{B}{x-3}+\frac{C}{(x-3)^2}$ **5.** $\frac{A}{x-1}+\frac{Bx+C}{x^2+1}$ **7.** $\frac{Ax+B}{x^2+4}+\frac{Cx+D}{(x^2+4)^2}$ **9.** $\frac{3}{x-3}-\frac{2}{x-2}$

11. $\frac{7}{x-9}-\frac{4}{x+2}$ **13.** $\frac{24}{7(x-4)}+\frac{25}{7(x+3)}$ **15.** $\frac{4}{7(x-3)}-\frac{8}{7(2x+1)}$ **17.** $\frac{3}{x}+\frac{2}{x-1}-\frac{1}{x+3}$ **19.** $\frac{3}{x}+\frac{4}{x+1}-\frac{3}{x-1}$

21. $\frac{6}{x-1}-\frac{5}{(x-1)^2}$ **23.** $\frac{1}{x-2}-\frac{2}{(x-2)^2}-\frac{5}{(x-2)^3}$ **25.** $\frac{7}{x}-\frac{6}{x-1}+\frac{10}{(x-1)^2}$ **27.** $\frac{1}{4(x+1)}+\frac{3}{4(x-1)}+\frac{1}{2(x-1)^2}$

29. $\frac{3}{x-1}+\frac{2x-4}{x^2+1}$ **31.** $\frac{2}{x+1}+\frac{3x-1}{x^2+2x+2}$ **33.** $\frac{1}{4x}+\frac{1}{x^2}+\frac{x+4}{4(x^2+4)}$ **35.** $\frac{4}{x+1}+\frac{2x-3}{x^2+1}$ **37.** $\frac{x+1}{x^2+2}-\frac{2x}{(x^2+2)^2}$

39. $\dfrac{x-2}{x^2-2x+3}+\dfrac{2x+1}{(x^2-2x+3)^2}$ **41.** $\dfrac{3}{x-2}+\dfrac{x-1}{x^2+2x+4}$ **43.** $\dfrac{1}{x}-\dfrac{1}{x+1};\dfrac{99}{100}$

53. When the denominator of a rational expression contains a power of a cubic factor, set up a partial fraction decomposition with quadratic numerators. $(Ax^2+Bx+C,Dx^2+Ex+F,\text{etc.})$. For example:

$\dfrac{x^3+1}{(x^3+2)^2}=\dfrac{Ax^2+bx+C}{x^3+2}+\dfrac{Dx^2+Ex+F}{(x^3+2)^2}=\dfrac{1}{x^3+2}+\dfrac{-1}{(x^3+2)^2}.$ **55.** $\dfrac{2}{x-3}+\dfrac{2x+5}{x^2+3x+3}$

Section 5.4

Check Point Exercises

1. $\{(0,1),(4,17)\}$ **2.** $\left\{\left(-\dfrac{6}{5},\dfrac{3}{5}\right),(2,-1)\right\}$ **3.** $\{(3,2),(3,-2),(-3,2),(-3,-2)\}$ **4.** $\{(0,5)\}$

5. length: 7 ft; width: 3 ft or length: 3 ft; width: 7 ft

Exercise Set 5.4

1. $\{(-3,5),(2,0)\}$ **3.** $\{(1,1),(2,0)\}$ **5.** $\{(4,-10),(-3,11)\}$ **7.** $\{(4,3),(-3,-4)\}$ **9.** $\left\{\left(-\dfrac{3}{2},-4\right),(2,3)\right\}$

11. $\{(-5,-4),(3,0)\}$ **13.** $\{(3,1),(-3,-1),(1,3),(-1,-3)\}$ **15.** $\{(4,-3),(-1,2)\}$ **17.** $\{(0,1),(4,-3)\}$

19. $\{(3,2),(3,-2),(-3,2),(-3,-2)\}$ **21.** $\{(3,2),(3,-2),(-3,2),(-3,-2)\}$ **23.** $\{(2,1),(2,-1),(-2,1),(-2,-1)\}$

25. $\{(3,4),(3,-4)\}$ **27.** $\{(0,2),(0,-2),(-1,\sqrt{3}),(-1,-\sqrt{3})\}$ **29.** $\{(2,1),(2,-1),(-2,1),(-2,-1)\}$

31. $\{(-2\sqrt{2},-\sqrt{2}),(-1,-4),(1,4),(2\sqrt{2},\sqrt{2})\}$ **33.** $\{(2,2),(4,1)\}$ **35.** $\{(0,0),(-1,1)\}$ **37.** $\{(0,0),(-2,2),(2,2)\}$

39. $\left\{(-4,1),\left(-\dfrac{5}{2},\dfrac{1}{4}\right)\right\}$ **41.** $\left\{\left(\dfrac{12}{5},-\dfrac{29}{5}\right),(-2,3)\right\}$ **43.** 4 and 6   **45.** 2 and 1, 2 and -1, -2 and 1, or -2 and -1

47. $(0,-4),(-2,0),(2,0)$ **49.** 11 ft and 7 ft   **51.** width: 6 in.; length: 8 in.   **53.** $x=5$ m, $y=2$ m   **61.** (b) is true.

63. $b=6,a=8$ **65.** $\{(10{,}000,5)\}$

Section 5.5

Check Point Exercises

1.

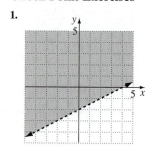

2.

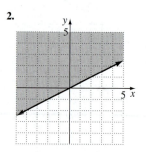

3.

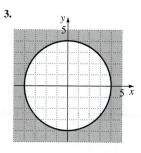

4.

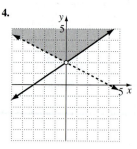

5.

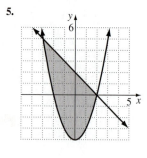

6.

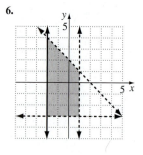

Exercise Set 5.5

1.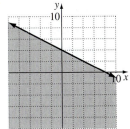

3.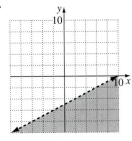

5.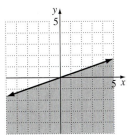

7.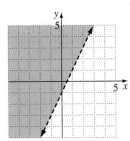

9.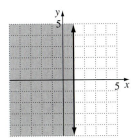

11.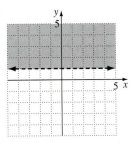

13.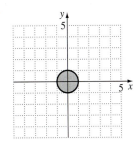

15.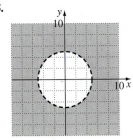

17.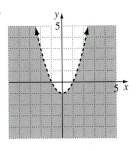

19.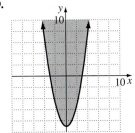

21.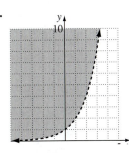

23.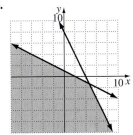

25.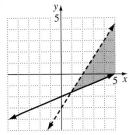

27.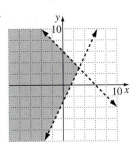

29.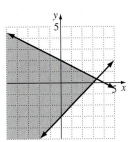

31.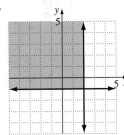

33.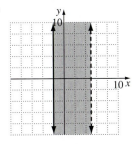

35.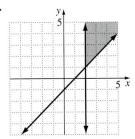

37. no solution

39.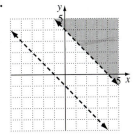

41.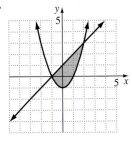

43.

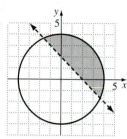

45.

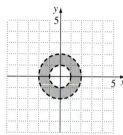

47.

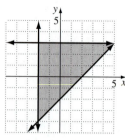

49.

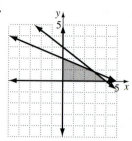

51.

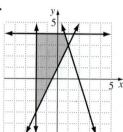

53. $5T - 7P \le 70, T \ge 0, P \ge 0; A(50, 30): 5(50) - 7(30) = 40 \le 70$

55. a. $50x + 150y > 2000$

b.

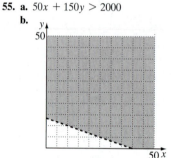

c. $(20, 20)$: 20 children and 20 adults will cause the elevator to be overloaded.

57. $x + y \le 15{,}000$
$x \ge 2{,}000$
$y \ge 3x$
$x \ge 0$
$y \ge 0$

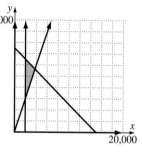

59. Answers may vary   **61. a.** 28.1   **b.** overweight

69.

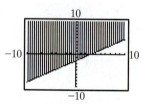

71.

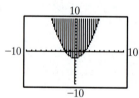

73.

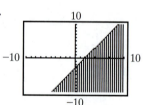

75. Answers may vary.

77. Answers may vary.

79.

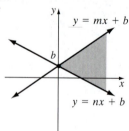

Section 5.6

Check Point Exercises

1. $z = 25x + 55y$ **2.** $x + y \le 80$ **3.** $30 \le x \le 80; 10 \le y \le 30$; objective function: $z = 25x + 55y$;

constraints: $x + y \le 80; 30 \le x \le 80; 10 \le y \le 30$ **4.** 50 bookshelves and 30 desks; $2900   **5.** 30

Exercise Set 5.6

1. $(1, 2): 17; (2, 10): 70; (7, 5): 65; (8, 3): 58$; maximum: $z = 70$; minimum: $z = 17$

3. $(0, 0): 0; (0, 8): 400; (4, 9): 610; (8, 0): 320$; maximum: $z = 610$; minimum: $z = 0$

5. a.

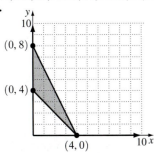

b. $(0, 8): 16; (0, 4): 8; (4, 0): 12$
c. maximum value: 16 at $x = 0$
and $y = 8$

7. a.

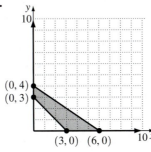

b. $(0, 4): 4; (0, 3): 3; (3, 0): 12; (6, 0): 24$
c. maximum value: 24 at $x = 6$ and
$y = 0$

9. a.

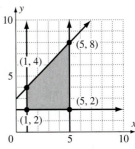

b. $(1, 2): -1; (1, 4): -5; (5, 8): -1; (5, 2): 11$
maximum value: 11 at $x = 5$ and $y = 2$

11. a.

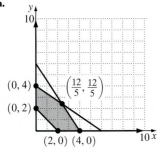

b. $(0, 4): 8; (0, 2): 4; (2, 0): 8; (4, 0): 16;$

$\left(\dfrac{12}{5}, \dfrac{12}{5}\right): \dfrac{72}{5}$

c. maximum value: 16 at $x = 4$ and $y = 0$

13. a.

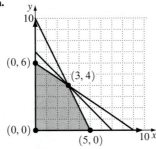

b. $(0, 6): 72, (0, 0): 0; (5, 0): 50; (3, 4): 78$

c. maximum value: 78 at $x = 3$ and $y = 4$

15. a. $z = 125x + 200y$
b. $x \le 450; y \le 200; 600x + 900y \le 360{,}000$
c.

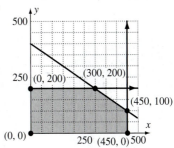

d. $(0, 0): 0; (0, 200): 40{,}000; (300, 200): 77{,}500;$
$(450, 100): 76{,}250; (450, 0): 56{,}250$
e. 300; 200; $77,500

17. 40 model A bicycles and no model B bicycles   **19.** 300 cartons of food and 200 cartons of clothing   **21.** 50 students and 100 parents

23. 10 Boeing 727s and 42 Falcon 20s

29.

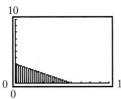

31.

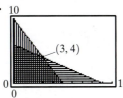

33. $5000 in stocks and $5000 in bonds

Chapter 5 Review Exercises

1. $\{(1, 5)\}$ **2.** $\{(2, 3)\}$ **3.** $\{(2, -3)\}$ **4.** $\varnothing$ **5.** $\{(x, y)|3x - 6y = 12\}$ **6. a.** $C(x) = 60{,}000 + 200x$ **b.** $R(x) = 450x$
c. $(240, 108{,}000)$; This means the company will break even if it produces and sells 240 desks.   **7.** 250 copies can be supplied and sold for
\$12.50 each.   **8.** 5.8 million pounds of potato chips, 4.6 million pounds of tortilla chips   **9.** \$80 per day for the room, \$60 per day for the car
10. 3 apples and 2 avocados   **11.** $\{(0, 1, 2)\}$ **12.** $\{(2, 1, -1)\}$ **13.** $y = 3x^2 - 4x + 5$ **14. a.** $(0, 3.5), (15, 5.0), (29, 4.1)$
b. $c = 3.5$ $225a + 15b + c = 5.0$ $841a + 29b + c = 4.1$ **15.** United States: 22; Colombia: 18; India: 10

16. $\dfrac{3}{5(x - 3)} + \dfrac{2}{5(x + 2)}$ **17.** $\dfrac{6}{x - 4} + \dfrac{5}{x + 3}$ **18.** $\dfrac{2}{x} + \dfrac{3}{x + 2} - \dfrac{1}{x - 1}$ **19.** $\dfrac{2}{x - 2} + \dfrac{5}{(x - 2)^2}$

20. $\dfrac{4}{x - 1} + \dfrac{4}{x - 2} - \dfrac{2}{(x - 2)^2}$ **21.** $\dfrac{6}{5(x - 2)} + \dfrac{-6x + 3}{5(x^2 + 1)}$ **22.** $\dfrac{5}{x - 3} + \dfrac{2x - 1}{x^2 + 4}$ **23.** $\dfrac{x}{x^2 + 4} - \dfrac{4x}{(x^2 + 4)^2}$

24. $\dfrac{4x + 1}{x^2 + x + 1} + \dfrac{2x - 2}{(x^2 + x + 1)^2}$ **25.** $\{(4, 3), (1, 0)\}$ **26.** $\{(0, 1), (-3, 4)\}$ **27.** $\{(1, -1), (-1, 1)\}$

28. $\{(3, \sqrt{6}), (3, -\sqrt{6}), (-3, \sqrt{6}), (-3, -\sqrt{6})\}$ **29.** $\{(2, 2), (-2, -2)\}$ **30.** $\{(9, 6), (1, 2)\}$ **31.** $\{(-3, -1), (1, 3)\}$

32. $\left\{\left(\dfrac{1}{2}, 2\right), (-1, -1)\right\}$ **33.** $\left\{\left(\dfrac{5}{2}, -\dfrac{7}{2}\right), (0, -1)\right\}$ **34.** $\{(2, -3), (-2, -3), (3, 2), (-3, 2)\}$

35. $\{(3, 1), (3, -1), (-3, 1), (-3, -1)\}$ **36.** 8 m and 5 m   **37.** $(1, 6), (3, 2)$ **38.** $x = 46$ and $y = 28$ or $x = 50$ and $y = 20$

39. 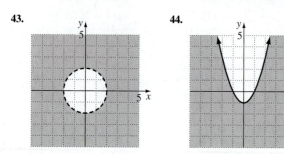   **40.** 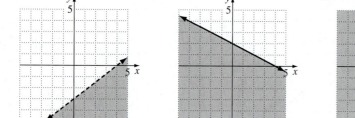   **41.** 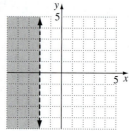   **42.**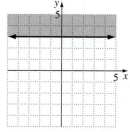

43. 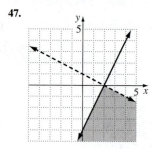   **44.** 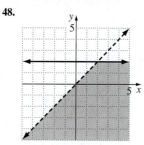   **45.** 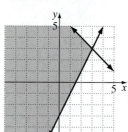   **46.**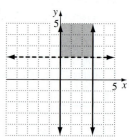

47. 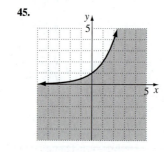   **48.**   **49.** 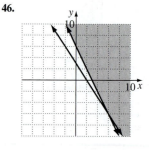   **50.**

51. no solution

52.
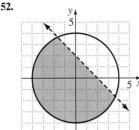

53.
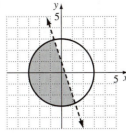

54.
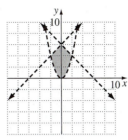

55.
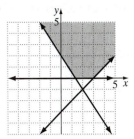

56. $(2, 2)$: 10; $(4, 0)$: 8; $\left(\frac{1}{2}, \frac{1}{2}\right)$: $\frac{5}{2}$; $(1, 0)$: 2; maximum value: 10; minimum value: 2

57.
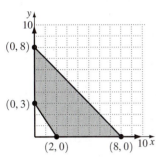

58.
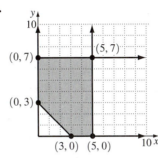

59.
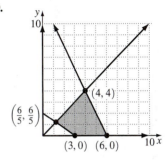

60. a. $z = 500x + 350y$

b. $x + y \leq 200$; $x \geq 10$; $y \geq 80$

c.
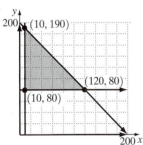

d. $(10, 80)$: 33,000; $(10, 190)$: 71,500; $(120, 80)$: 88,000

e. 120; 80; 88,000

61. 480 of model A and 240 of model B

Chapter 5 Test

1. $\{(1, -3)\}$ **2.** $\{(4, -2)\}$ **3.** $\{(1, 3, 2)\}$ **4.** $\{(4, -3), (-3, 4)\}$ **5.** $\{(3, 2), (3, -2), (-3, 2), (-3, -2)\}$

6. $\dfrac{-1}{10(x + 1)} + \dfrac{x + 9}{10(x^2 + 9)}$

7.
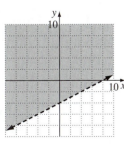

8.
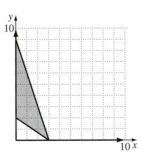

9.
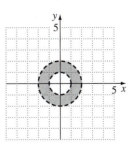

10.
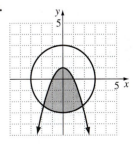

11. 26   **12.** Shrimp: 42 mg; scallops: 15mg   **13. a.** $C(x) = 360,000 + 850x$ **b.** $R(x) = 1150x$ **c.** $(1200, 1,380,000)$; The company will break even if it produces and sells 1200 computers.   **14.** $y = x^2 - 3$ **15.** $x = 7.5$ ft and $y = 24$ ft or $x = 12$ ft and $y = 15$ ft

16. 50 regular and 100 deluxe jet skis; $35,000

Cumulative Review Exercises (Chapters 1–5)

1. $\{3, 4\}$ **2.** $\left\{\dfrac{2 + i\sqrt{3}}{2}, \dfrac{2 - i\sqrt{3}}{2}\right\}$ **3.** $(-18, 6)$ **4.** $(1, 7)$ **5.** $\left\{-3, \dfrac{1}{2}, 2\right\}$ **6.** $\{-2\}$ **7.** $\{2\}$ **8.** $\{-2 + \log_3 11\}$

9.

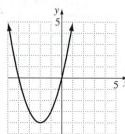

10.

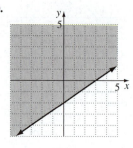

11.

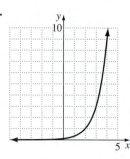

12.

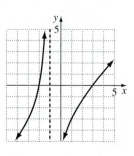

13. $3 + 5 \log_2 x$ **14.** 10.99% **15.** $f^{-1}(x) = \dfrac{1}{7}x + \dfrac{3}{7}$ **16.** $g(f(x)) = 21x - 16$ **17.** Answers may vary.

18. $\left\{\left(-\dfrac{1}{2}, \dfrac{1}{2}\right), (2, 8)\right\}$ **19.** 4 m by 9 m **20.** A plane with an initial landing speed of 90 ft per second needs 562 ft to land. There is a problem since 550 ft is not enough.

Subject Index

Photo Credits

2. Quadratic Function: $f(x) = ax^2 + bx + c, a \neq 0$

Graph is a parabola with vertex at $x = -\dfrac{b}{2a}$.

Quadratic Function: $f(x) = a(x - h)^2 + k$
In this form, the parabola's vertex is (h, k).

3. nth-Degree Polynomial Function: $f(x) = a_n x^n + a_{n-1} x^{n-1} + a_{n-2} x^{n-2} + \cdots + a_1 x + a_0, a_n \neq 0$
For n odd and $a_n > 0$, graph falls to the left and rises to the right.
For n odd and $a_n < 0$, graph rises to the left and falls to the right.
For n even and $a_n > 0$, graph rises to the left and to the right.
For n even and $a_n < 0$, graph falls to the left and to the right.

4. Rational Function: $f(x) = \dfrac{p(x)}{q(x)}$, $p(x)$ and $q(x)$ are polynomials, $q(x) \neq 0$

5. Exponential Function: $f(x) = b^x, b > 0, b \neq 1$
Graphs:

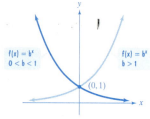

6. Logarithmic Function: $f(x) = \log_b x, b > 0, b \neq 1$
$y = \log_b x$ is equivalent to $x = b^y$.
Graph:

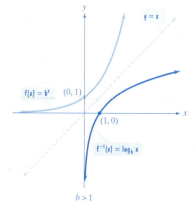

$b > 1$

PROPERTIES OF LOGARITHMS

1. $\log_b(MN) = \log_b M + \log_b N$

2. $\log_b\left(\dfrac{M}{N}\right) = \log_b M - \log_b N$

3. $\log_b M^p = p \log_b M$

4. $\log_b M = \dfrac{\log_a M}{\log_a b} = \dfrac{\ln M}{\ln b} = \dfrac{\log M}{\log b}$

5. $\log_b b^x = x; \quad \ln e^x = x$

6. $b^{\log_b x} = x; \quad e^{\ln x} = x$

INVERSE OF A 2 × 2 MATRIX

If $A = \begin{bmatrix} a & b \\ c & d \end{bmatrix}$, then $A^{-1} = \dfrac{1}{ad - bc}\begin{bmatrix} d & -b \\ -c & a \end{bmatrix}$, where $ad - bc \neq 0$.

CRAMER'S RULE

If

$$a_{11}x_1 + a_{12}x_2 + a_{13}x_3 + \cdots + a_{1n}x_n = b_1$$
$$a_{21}x_1 + a_{22}x_2 + a_{23}x_3 + \cdots + a_{2n}x_n = b_2$$
$$a_{31}x_1 + a_{32}x_2 + a_{33}x_3 + \cdots + a_{3n}x_n = b_3$$
$$\vdots$$
$$a_{n1}x_1 + a_{n2}x_2 + a_{n3}x_3 + \cdots + a_{nn}x_n = b_n$$

then $x_i = \dfrac{D_i}{D}, D \neq 0.$

D: determinant of the system's coefficients

D_i: determinant in which coefficients of x_i are replaced by $b_1, b_2, b_3, \ldots, b_n$.

CONIC SECTIONS

Circle

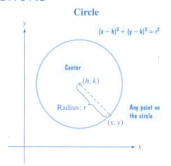

Ellipse

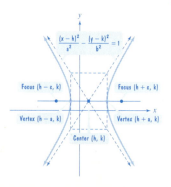

Hyperbola

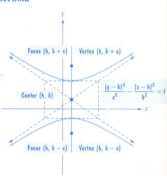